KU-198-317

PROCEEDINGS of the INTERNATIONAL NUCLEAR PHYSICS CONFERENCE

HARROGATE U.K.
25-30 August 1986

Volume 2
Invited Papers

Edited by J L Durell, J M Irvine and G C Morrison

Institute of Physics Conference Series Number 86
Institute of Physics, Bristol

CODEN IPHSAC 86 1 – 609 (1987)

British Library Cataloguing in Publication Data

International Nuclear Physics Conference (*1986: Harrogate*)
Proceedings of the international nuclear physics conference,
Harrogate U.K., 25/30 August 1986.—(Institute of Physics
conference series, ISSN 0305–2346; v. 86).
Vol. 2: Invited papers
1. Nuclear Physics
I. Title II. Series
539.7 QC776

ISBN 0-85498-174-8

Editors
J L Durell, J M Irvine, G C Morrison

Volume 1 of the Conference Proceedings contains abstracts of all contributions accepted by the Scientific Programme Committee up to 30 June 1986. Volume 2 contains the text of the Opening Session plus all the invited talks and a list of the titles and authors of the oral and poster contributions. Additional copies of Volume 1 can be purchased from the Meetings Officer, Institute of Physics, 47 Belgrave Square, London SW1X 8QX, UK and Volume 2 from Stonebridge Distribution Ltd, 7 Great Western Way, Bristol BS1 6HE, UK.

Published under the Institute of Physics imprint by IOP Publishing Ltd
Techno House, Redcliffe Way, Bristol BS1 6NX, England

Printed in Great Britain by J W Arrowsmith Ltd, Bristol

INTERNATIONAL NUCLEAR PHYSICS CONFERENCE

Organized by: The Institute of Physics

Programme Administration: SERC Daresbury Laboratory

Sponsored by: The International Union of Pure and Applied Physics

Supported by:

The British Council

The Institute of Physics

The International Atomic Energy Agency

The Royal Society

The Science and Engineering Research Council

British Caledonian Airways

John Caunt Scientific Ltd

Institute of Physics Publishing Division

E.G. and G. Instruments Ltd

Rank Xerox (U.K.) Ltd

POLICY COMMITTEE

L L Green (Daresbury, Chairman)

K W Allen (Oxford)
G Brown (Bradford)
J L Durell (Manchester)
J P Elliott (Sussex)
J M Irvine (Manchester)
G C Morrison (Birmingham)
R O Owens (Glasgow)
W R Phillips (Manchester)
P J Twin (Daresbury)
S A Lowndes (Daresbury, Secretary)
D G Mayston (Institute of Physics)

INTERNATIONAL ADVISORY COMMITTEE

A Arima (Toyko)
D R Bes (Buenos Aires)
J P Blaser (SIN)
D A Bromley (Yale)
N Cindro (Zagreb)
M El Nadi (Cairo)
T E O Ericson (CERN)
A Faessler (Tübingen)
H Feshbach (MIT)
G N Flerov (Dubna)
G T Garvey (Los Alamos)
W Greiner (Frankfurt)
W Haeberli (Wisconsin)
E W Hamburger (Sao Paolo)
O Hansen (Brookhaven)
J C Hardy (Chalk River)
E Hayward (NBS)
P Kienle (Munich)
M Lefort (Orsay)
T Mayer-Kuckuk (Bonn)
M K Mehta (IAEA, Vienna)
B R Mottelson (Copenhagen)
S Nagamiya (Tokyo)
J Nemeth (Budapest)
J O Newton (Canberra)
B Povh (Heidelberg)
R A Ricci (Padua)
J P Schiffer (Argonne)
Li Shounan (Peking)
R H Siemssen (Groningen)
V G Soloviev (JINR, Dubna)
F S Stephens (Berkeley)
I Talmi (Weizmann)
G Van Middelkoop (Amsterdam)
E Vogt (TRIUMF)
D Walecka (Stanford)
H Weidenmuller (Heidelberg)
D H Wilkinson (Sussex)
Yang Chang-Zhong (Lanzhou)

SCIENTIFIC PROGRAMME COMMITTEE

G C Morrison (Birmingham, Chairman)

R R Betts (Oxford)
P Blasi (Florence)
D Branford (Edinburgh)
D M Brink (Oxford)
H Doubre (GANIL)
J L Durell (Manchester)
J P Elliott (Sussex)
J M Irvine (Manchester)
R C Johnson (Surrey)
J Lowe (Birmingham)
J E Lynn (Harwell)
M H MacFarlane (Indiana)
C Mahaux (Liège)
M K Mehta (IAEA, Vienna)
V Metag (Giessen)
M A Nagarajan (Daresbury)
R O Owens (Glasgow)
W R Phillips (Manchester)
J F Sharpey-Schafer (Liverpool)
I Sick (Basle)
U P Smilansky (Weizmann)
H-J Specht (Heidelberg)
P J Twin (Daresbury)
P M Walker (Daresbury)
D L Watson (Bradford)
A Winther (Copenhagen)
S A Lowndes (Daresbury, Secretary)

ORGANISING COMMITTEE

J M Irvine (Manchester, Chairman)

G Brown (Bradford)
M D Cohler (Bradford)
J L Durell (Manchester)
P J Twin (Daresbury)
D G Mayston (Institute of Physics, Secretary)
S Lippman (Institute of Physics)

Preface

The International Nuclear Physics Conference held in Harrogate, UK, 25–30 August 1986 was the latest in the triennial series of conferences sponsored by the International Union of Pure and Applied Physics, following those held in Florence (1983) and Berkeley (1980). Twenty-five years had elapsed since the UK last hosted a conference in this series. It was a timely return marking, as it did, the 75th anniversary of the discovery of the nucleus by Ernest Rutherford in Manchester.

The Scientific Programme Committee was extremely conscious of the many exciting developments currently taking place in nuclear physics. The traditional areas are yielding impressive new results with the use of novel detector systems and improved experimental techniques. At the same time our community is poised to exploit the potential of new facilities such as kaon factories, high-duty factor electron accelerators and relativistic heavy ions. There is growing interest in the role of sub-nucleon degrees of freedom in nuclear physics. These developments are reflected in the programme of this conference. We are indebted to our International Advisory Committee for their initial suggestions and to our colleagues on the Scientific Programme Committee for their assistance in composing the programme recorded in these proceedings.

The success of this programme was due, in no small way, to the excellence of the Invited Talks and we should like to express our thanks to the speakers, who reflected our intentions admirably. This conference introduced the concept of 'short' invited talks to provide a focus for discussion in the parallel sessions. This too worked well and our thanks are also due to the speakers and contributors who responded to this new initiative. Although not included in this volume, an evening talk by David Schramm on the interaction between nuclear physics and cosmology was one of the highlights. In all, the week at Harrogate proved lively and interesting and it is our hope that both volumes 1 and 2 of these proceedings will provide a substantial contribution to the literature.

A conference of this size requires assistance from many people including session chairmen, scientific secretaries and student assistants. Although too numerous to name individually here, we append a list of all those involved to whom we owe our sincere gratitude.

The conference was organised by The Institute of Physics and we acknowledge the efforts of their staff, especially Delia Mayston and Sue Lippman. The Secretariat for the Scientific Programme Committee was provided by SERC Daresbury Laboratory. It is a pleasure to acknowledge the assistance of Shirley Lowndes and her colleagues at all stages in the programme of the conference up to and including the preparation of this volume.

Finally, the Organisers would like to gratefully acknowledge the support generously provided by The British Council, The Institute of Physics, The International Atomic Energy Agency, The Royal Society, The Science and Engineering Research Council, British Caledonian Airways, John Caunt Scientific Ltd, The Institute of Physics Publishing Division, EG and G Instruments Ltd, LeCroy Ltd and Rank Xerox (UK) Ltd.

G C Morrison (Chairman, Scientific Programme Committee)
J L Durell
J M Irvine

SESSION CHAIRMEN

K Amos (Melbourne)
N Austern (Pittsburgh)
P Blasi (Florence)
F E Close (RAL)
H Doubre (GANIL)
J P Elliott (Sussex)
T E O Ericson (CERN)
H Feshbach (MIT)
G N Flerov (JINR, Dubna)
J D Garrett (Niels Bohr)
G T Garvey (Los Alamos)
G Goldring (Weizmann)
A Klein (Pennsylvania)
E L Lomon (MIT)
V Metag (Giessen)
M A Nagarajan (Daresbury)
M K Pal (SAHA, Calcutta)
B A Robson (Canberra)
S P Rosen (Los Alamos)
H Ryde (Lund)
F D Santos (Lisbon)
G R Satchler (ORNL)
A P Stamp (Auckland)
Sun Zuxun (Beijing)
I Talmi (Weizmann)
L Wilets (Washington)
D H Wilkinson (Sussex)
B Zeidman (Argonne)

SCIENTIFIC SECRETARIES, PARALLEL SESSION SUPERVISORS

S C Allcock (Oxford)
J R M Annand (Glasgow)
M D Cohler (Bradford)
J L Durell (Manchester)
B R Fulton (Birmingham)
W Gelletly (Manchester)
G D Jones (Liverpool)
F Khazaie (Manchester)
A Kirwan (Liverpool)
D J G Love (Daresbury)
W D M Rae (Oxford)
J Simpson (Daresbury)
R Wadsworth (Bradford)
P M Walker (Daresbury)
P J Woods (Birmingham)

STUDENT ASSISTANTS

S Abu-Kamar (Surrey)
S C Allcock (Oxford)
J Copnell (Sussex)
H Cranmer-Gordon (Liverpool)
M J Fawcett (Manchester)
N P Hessey (Birmingham)
D Hook (Manchester)
J A McGovern (Manchester)
R Moscrop (Manchester)
J M O'Donnell (Bradford)
C Ogilvie (Birmingham)
C N Pass (Oxford)
S Sergiwa (Manchester)
T L Shaw (Oxford)
A M Street (Oxford)
P Wallace (Glasgow)
K L Ying (Liverpool)
J Yorkston (Edinburgh)

INSTITUTE OF PHYSICS

D Mayston
S Lippman
C Jones
S Greenwood
I Zawiewski

SERC DARESBURY LABORATORY

S A Lowndes
B Jackson
G R Berry
J Johnson
V A Ireland

Contents

Note: Underline indicates author presenting this paper.

OPENING SESSION

Message of Welcome

Kenneth Baker, Secretary of State for Education and Science

'I am indeed sorry that circumstances prevent me from being with you today, but I would like to take this opportunity to welcome you all. Our country has a long tradition in the advancement of science and the need to understand the world we live in plays an essential role in man's life. Physics is at the heart of the process of unravelling the laws and rules which determine nature's behaviour. Thus I can appreciate the intriguing nature and complexity of the nucleus and its powerful influence on other sciences. Your science is truly international in allowing for the free exchange of ideas and providing an atmosphere in which both collaboration and competition stimulate progress. This week many of you will hear reports of research based on such international collaborations sharing not just ideas but also experimental facilities in many countries. I am delighted to hear of the highly successful programme at our Nuclear Structure Facility and proud that we in the UK have a world class facility contributing so positively to scientific advancement.

The sharing of research facilities, and the long tradition of sharing research results through open publication, makes sound economic, as well as scientific sense. Those concerned with the management of research appreciate the need for more sophisticated, albeit expensive, instruments and welcome international partnerships when the scale is right for them.

Competition, too, is important in research and is an essential indicator of worth at the highest international level. The funding of research depends very much on an assessment of such elements in evaluating the best way to allocate scarce resources. In this country we place considerable responsibility, through the Research Councils, on the scientists themselves - in universities, research institutes and industry - in deciding on research to be supported. You, the scientists, will naturally be looking within a framework of pure scientific merit but you must also be aware that the Government, representing society as a whole, must add a further dimension - a 'return' on investment. When applied to fundamental research, timescales are extended and the strands connecting basic research to the market place are long, each development progressing from basic, through strategic and applied stages in the research process. Nevertheless there are some returns which are immediately apparent in both education and information.

I recognise that one element of the return from research is the excitement of discovery, a pride in the intellectual achievements of our scientists. The public want to share in your work and scientists need to convey, through different media, the excitement and meaning of it. The laymen can then participate in a more informed way in the debate about technical changes which are such an important force in our society. This is especially true where nuclear science and technology are concerned. The public must be brought into this and we all have a responsibility here to help to remove the fear of technology and the mystique surrounding it.

Without basic research there can be no applied research, no development and nothing to sell in the market place. Your endeavours are therefore essential to both modern life and progress. I know you work with utmost enthusiasm and dedication in your field and I have sincere respect for your achievements. May I wish you a very fruitful Conference.'

Inst. Phys. Conf. Ser. No. 86
Paper presented at Int. Nucl. Phys. Conf., Harrogate, UK, 1986

Opening Remarks

D.A. Bromley, President of the International Union of Pure and Applied Physics and Henry Ford II Professor of Physics, Yale University, New Haven, Ct.

It is both a privilege and a pleasure to add my welcome to this the 10th International Conference on Nuclear Physics to be held under the auspices of the International Union of Pure and Applied Physics (IUPAP). As most of you know, the Union is the world co-ordinating body for physics and was founded in 1922, with 16 members, under the Presidency of Sir William Henry Bragg. In the intervening years the size and membership have increased until now there are over 40 organisations included within the Union. Its work is carried out through some 19 International Commissions including, of course, the Commission on Nuclear Physics, whose current Chairman, Professor Feshbach and the Secretary, Dr. Bartholomew, are here today. Other Commissions extend from particle physics and astrophysics to magnetism and acoustics - spanning the entire range of physics. In addition there are Commissions on Education in Physics and on Physics and its Role in Development. IUPAP's major activity, as you know, is the organisation and coordination of conferences of various sizes reflecting the fact that the Union is a truly international organisation. However, there are a number of other activities, for example, under the Commission on Elementary Particle Physics we have an International Commission on Future Accelerators, which brings together people on a regular basis from around the world to consider the state of affairs at the very highest energy frontier. There is a move at the moment to create a similar Commission under the aegis of the Nuclear Physics Commission to consider the complementary high intensity frontier: the question of kaon factories. To quote just another example we have recently empanelled an *ad hoc* Commission, chaired by Professor Sir Denys Wilkinson, which is charged with developing the criteria to be used by both physics and chemistry in naming the transfermium elements.

As a member of the International Council of Scientific Unions (ICSU), IUPAP is formally committed to the free circulation of scientists and we are called upon regularly each year to take action regarding what is happily a dwindling number of violations of these free circulation principles. Thus far I can report to you that we have been remarkably successful. The very real threat of removal of international conferences from any particular region until the violation has been adequately explained, and until mechanisms have been put in place to prevent its recurrence, is taken very seriously indeed. Since, 1922 IUPAP has insisted on keeping scientific questions separate from political questions and only by focusing on the science has it been possible for IUPAP to maintain international co-operation and communication through both hot wars and cold. I believe very deeply that it is essential to maintain this tradition in IUPAP, so it is with the greatest pleasure that I welcome you on behalf of IUPAP to this 10th International Conference on our science.

Let me now turn to nuclear physics itself, 75 years after Rutherford. I think that many of you here will agree that this is a particularly exciting year in nuclear physics; perhaps the most remarkable or exciting one since 1932. That was an astonishing year when within 12 months we learned in the literature about the discovery of the neutron; the deuteron; the positron; the development of the Cockcroft-Walton, the Van de Graaff and the cyclotron accelerators; the idea of isospin; and the experimental proof of the validity of general relativity - not bad for one year. In a very real sense, nuclear physics as we know it now dates back to 1932 and, not surprisingly, in the great discoveries of that year we saw illustrated the truly international character of nuclear physics. We also saw the United Kingdom - building upon Rutherford's work with the remarkable group that he established at the Cavendish Laboratory - well represented by the pioneering work of Chadwick on the neutron and of Cockcroft and Walton on accelerated particles and their use in the production of artificial radioactivity.

Many of us came away from the 9th of these Conferences, three years ago in Florence, with a sense of new vigour, new enthusiasm and new direction in our science. I think you will agree that the past three years have borne out this optimism in terms of new discoveries, new instrumentation and new facilities, now either coming into operation or in the final stages design. In nuclear structure itself one of the most remarkable recent experimental studies has undoubtedly been the work of Peter Twin and his associates at Daresbury. As the first reward for their intensive activities in developing new ways to look at discrete gamma-rays they have made a major breakthrough, almost doubling the angular momenta that can be unambiguously determined. From this work they have found at least 20 states in the superdeformed band in dysprosium-152. Later at this meeting we will hear of an even more recent development, that of Dr. Lister and his associates: the discovery of zirconium-80. It is clear from such work that in this vital area of nuclear structure one of the truly world-class facilities is here in the United Kingdom at Daresbury.

Paralleling the experimental breakthroughs, the theoretical understanding of symmetries - and possible super-symmetries - that underlie the details of nuclear structure and dynamics has deepened and become more fundamental with, for example, the recognition by Ginnochio and others that the fermion 0(7) symmetry underlies those so successful in the interacting boson models. However, let me also emphasise that how the nuclear many-body system responds to heating, to increased angular momentum, to extreme conditions, is still very much an open question, and many of the problems at the frontiers of this field are open to relatively low-energy but high-precision facilities in laboratories around the world.

Major progress has also been made in utilising electromagnetic probes in understanding deep nuclear structure. It is gratifying to see that work with electron accelerators shows energy resolution which even a few years ago was considered to be the exclusive domain of the Van de Graaffs.

These electron accelerators also provide the freedom to vary energy loss and momentum transfer independently, and through its participation in, and support of, the excellent programme on the Mainz accelerator the United

Kingdom is active on the frontiers of this area of our science as well. Beautiful experiments on the electro-disintegration of the deuteron from Saclay and their subsequent analyses have shown, to the surprise of many, that these results can be explained in completely classical nuclear physics terms without requiring quarks. This points to the fact that perhaps we have been too enthusiastic to leap for a quark when we see something a little surprising, and I take the French work as an important cautionary note in avoiding literature pollution. Nevertheless, the path to quark matter is one of great interest in our subject and although the interpretation of the data is still very much open to question, the European Muon Collaboration and work at the Stanford Linear Accelerator Centre on the inelastic scattering of electrons and muons have provided tantalising evidence that at higher energies there is at least a partial deconfinement of the quarks within the nuclei, and evidence for the possible appearance of new entities containing six or more quarks - we simply do not know more at this point. The transition from hadronic to quark matter, however, is going to provide us with an exceedingly complex and interesting area of study in the next few years. New facilities such as the Continuous Electron Beam Accelerator Facility, now under construction in Virginia, will be one of those providing progress in this area.

Again, in parallel with the experimental work, the theoretical work at least holds the promise that quantum chromodynamics (QCD) can provide a basis for understanding the phenomena of nuclear physics comparable to that achieved by quantum electrodynamics (QED) in understanding atomic and electromagnetic phenomena. Incidentally, current QED measurements have reached the stage of precision which corresponds to measuring the distance from New York to Los Angeles to within one hundredth of the diameter of a hydrogen atom. More surprisingly, QED can quite readily reproduce measurements made at that level of precision. The complete transition from hadronic to quark matter to an entirely new state of matter, the 5th one, can occur only at very high energies, very high temperatures, or very high pressures. These are the temperatures and pressures which existed instants after the creation of our universe and the Big Bang, and are conditions which exist in the final death throes of giant stars as they are crushed inexorably in gravitational collapse. We tend to forget just how remarkable it is that we can recreate these conditions, if only for the fleeting times required to characterise them. Current QCD lattice gauge calculations suggest that there should be a phase transition from hadronic to quark matter at about 200 MeV, that is 2.5 gigadegrees. Someone mentioned to me recently that they couldn't remember whether that was degrees Celsius or Kelvin. Obviously as the density is increased this critical transition temperature will decrease and, here again, since we do not know the equation of state of nuclear matter very well, at somewhere between 5 and 10 times normal nuclear density it should be possible to achieve this transition in an experimentally accessible regime. This, of course, brings us to high energy, heavy-ion collisions.

One of the very important results during the past three years has been that of Poskanzer and his collaborators at Berkeley who showed that in niobium-on-niobium and in gold-on-gold collisions, where one separates out the central collisions from the much more numerous peripheral ones, one sees quite convincing evidence for hydrodynamic behaviour and therefore for higher densities in the shock waves - higher densities which, at least in principle, can take us through this important transition to quark matter. Unfortunately the Bevalac energies and those of the Synchrophasotron at Dubna are simply not high enough to give us access to this transition. However, later this year, at Brookhaven with the AGS, we expect to have beams at 15.2 GeV per nucleon, and at CERN at 225 GeV per nucleon, both of oxygen and of sulphur. Even so, not too much should be expected of these preliminary measurements in either case. We must remember that even the CERN energy is marginal to show the effects for which we are looking and sulphur is, after all, a very small, light nucleus. For that reason there is great interest in the relativistic heavy-ion collider proposed for Brookhaven, which will give collisions between 100 GeV/nucleon beams up to gold. This facility is now in a very advanced stage of design, but has not yet been funded. Even though there is great interest in going to these very high energies there is still an enormous amount of work to be done at the energies represented by the Bevalac, by the Synchrophasotron and by what is, in effect, a very modern Bevalac, the SIS project now under construction at GSI.

There is also growing interest (in Europe, Japan, the U.S. and Canada) in going not to the high energy frontier but to the high intensity one. These would be the kaon factories: enormously productive machines, not only in allowing definitive searches for the rare decay modes of the kaon, which could provide critical tests for the now standard model and tell us something about the generation problem, time reversal invariance; but also, and more immediately connected to nuclear physics, they could inject strangeness into the nuclear many-body system. The pioneering work of Bogdan Povh at CERN and of Peter Barnes and his collaborators at Brookhaven have shown that in this way it is possible to probe the deep shell-model structure. It is also possible to find out the strong interaction corrections to the weak interaction in a totally unique way. The last point I would make is that it has recently been understood that the ambiguities in the interpretation of the muon and electron inelastic scattering can be removed if one does inelastic scattering at very high energies with either protons or kaons and compares those results with the lepton scattering. In this way one can become sensitive to the antiquark component in the nucleus and sensitive also, independently, to the different quark flavours.

All of this relates to accelerator physics, of course, and I would be very much remiss if I did not at least mention the fact that, as demonstrated by Rutherford, there is a tremendous amount of good physics still to be done without any accelerator whatsoever. I should mention the studies directed towards determining the mass of the neutrino, and the double beta-decay experiments of the modern genre, both of which will throw light on the standard model; the possibility of neutron-antineutron oscillations; the search for a finite lifetime of the proton. I could go on at length. Nuclear physics, however, does have a fundamental unity and all the areas mentioned - from those of relatively low energies to those of very high energies and to those that involve major accelerators and those that involve no accelerator at all - are complementary. They are all important and they all deserve our support and continuing interest.

It has been traditional at meetings in this series to focus on our fundamental understanding of nuclear physics as such, and I am sure that this will predominate again at our Conference this week. However, it is important to remember that our science, nuclear physics, has had a very vital impact on the societies that support it. I need not emphasise the importance of nuclear energy, nor the importance of nuclear energy in military matters. Much is made, and rightly so, of the terrors of nuclear war, but at the same time it does bear emphasis that the 40 years since the end of World War II represents the longest time in the recorded history of our race in which we have not been embroiled in a conflict that has caught up a very large part of the civilised world. Nuclear medicine continues to be one of the most rapidly growing areas in that same science - where the techniques, the instrumentation, the isotopes of nuclear physics, are brought to bear on diagnostic and clinical problems in human medicine. It bears remembering that Phillip Handler, a recent President of our US National Academy and a distinguished biophysicist in his own right, said a few years ago that 'knowledge of the life sciences has moved forward 50 years in the past five, purely through the application of physics techniques and physics instrumentation'. This use of nuclear techniques and instrumentation far transcends medical use however, and now pervades technology and even the arts. For example, neutron and ion beams are increasingly the method chosen by art historians to take us back to the time when the Old Masters were at work, enabling us literally to follow them day by day by unfolding layer after layer of the structure of their paintings and showing how their ideas and techniques changed during the flowering of their creativity and genius.

Just as nuclear physics is a seamless fabric, so also is physics - and all of science. Nuclear physics has much to contribute to its sister sciences and the largely artificial boundaries that have separated us from particle physics, which were never as high in Europe as in the US, are beginning to be eroded, to everyone's advantage. Nuclear physics has always been intimately related to astronomy. Ion beams from nuclear facilities have opened up entirely new domains in the study of solids and of surfaces, as in the case of medicine. This goes far beyond diagnosis. Ion beams are now being used to create new alloys through ion beam mixing, among them A15 alloys which are simply impossible to create chemically. Ion beams are also being used to create three-dimensional architectures in the most advanced, very large-scale integrated circuits.

I have talked about the instrumentation, the techniques, the concepts. The impact of our science on our societies has been enormous, but even so the most important part of that impact has, in fact, been the <u>people</u> of nuclear physics. I recall a small bit of history which may be new to many of you. Just at the turn of the century when Rutherford was still at McGill University, Montreal, he was offered a professorship at Yale, my University. He accepted the offer and after several trips to New Haven he finally bought a home there. However, on his way back to Montreal through the Vermont mountains, Rutherford changed his mind and instead of coming to Yale he decided to accept Schuster's chair at Manchester, with results familiar to us all. Many years later his biographer, Eve, asked him 'Ernest, why did you do that? What made you change your mind about going to Yale?' Rutherford's answer was 'Oh, it was easy. I got the distinct impression that in New Haven they were running the damn place for the students'. We would like to think that we still do - of course, no one was more interested in his students than Rutherford - and I want to take a moment to talk about this question of nuclear physics as a training ground in science.

In one sense it is an accident of size and of history but nonetheless nuclear physics is unique in that on the one hand, problems which are on the very frontier of human understanding are still within the intellectual grasp of a single good graduate student. On the other hand, if that student is going to make progress he has to take a systems approach: to develop personal expertise in a wide variety of techniques - electronics, thin films, vacuum technology and the like - and to organise the time, talents and energies of engineers, technicians, fellow students and Faculty, to progress. This approach, this training, and the fact that the students can work on the frontiers of knowledge and yet feel that they are totally in control of the investigation, gives them a sense of confidence - the confidence that they can handle unstructured and new problems and this serves them very well in a great many fields. Increasingly, industry around the world is recognising that this training makes such people extremely well suited as partners in the search for the broad solutions to unstructured societal problems. Few of us here today are in nuclear physics because of our interest in its applications, but I think it is critically important that we maintain a sensitivity to them - and to continue training our students with that sensitivity. Indeed, I feel that if all our students were now to be nuclear physicists it would be a signal that we had failed miserably; but it so happens that nuclear physics is a great place to educate creative physicists. It is for that reason that in my own country and in a number of others, I have been arguing vigorously for the importance of maintaining strong university-based programmes in nuclear physics in parallel with frontier national facilities. Only by having these strong university-based programmes will we see the continued flow of the bright young people we need in our field.

Let me end by emphasising the fact that thus far in our studies of nuclear physics we have been very much limited; we have, in a major sense, only scratched the surface. We have looked at cold nuclei near their ground states; low-spin nuclei where centrifugal effects were negligible; to nuclear species near stability; to the surfaces of nuclei, and particle and collective degrees of freedom involving primarily those surfaces; and to the consideration of nuclei as complexes of neutrons and protons only. But all of this is changing, and changing extremely rapidly. With new instrumentation, new understanding and new enthusiasm the next few years between this conference and the next meeting in this series will, I think, be tremendously exciting ones in nuclear physics. Nuclear physics remains a frontier field; it remains a field with unique interaction with our society, and it remains an excellent training field.

It is, therefore, entirely appropriate that this 10th International Conference on Nuclear Physics should be held here in Harrogate, so close to Manchester where nuclear physics was born. Nuclear physics has continued to benefit from its birth here in the UK and from a continued stream of critical discoveries and developments from this country. Recent work by UK scientists at Mainz, CERN, and at Daresbury with their collaborators drawn from around the world, shows us conclusively that we can continue to look to the UK for seminal contributions to our science. As the first of the general conferences in nuclear physics to be held in the UK since the establishment of the Nuclear Structure Facility at Daresbury, this Conference is also a celebration of the completion of that world-class facility; its formal welcome into the international nuclear physics community. I know that all of you here will join with me in wishing our colleagues at Daresbury many long, exciting and productive years with this marvellous facility, and in thanking Professor Leslie Green and his colleagues in the UK nuclear physics community for their efforts in organising this Conference in so efficient and pleasant a fashion and in such a beautiful town. We look forward to a stimulating and memorable week.

PLENARY SESSIONS

Inst. Phys. Conf. Ser. No. 86
Paper presented at Int. Nucl. Phys. Conf., Harrogate, UK, 1986

Stages on the way to Harrogate

Ben R. Mottelson

NORDITA, Blegdamsvej 17, 2100 Copenhagen Ø, Denmark

We are all, of course, eager to get on with the main business of this meeting, our triennial exchange of experience and opinions concerning the most recent achievements of our field. However, as a preliminary, the Organizers have invented a new category of contributions, asking Erich Vogt and me to attempt a broad assessment of nuclear physics as a whole which might help to provide a framework or at least a starting point for our coming discussions. I must confess that I feel that it is probably impossible to live up to the challenge of this charge; there are simply too many different dimensions to the nuclear physics experience. The differences of perspective on the many different vigorous frontiers are too varied and too incommensurate to admit of any neat summary that should be both useful, and comprehensive, and at the same time fit within three quarters of an hour.

I shall attempt to tackle this impossible problem by presenting an historically centred survey; perhaps we can derive useful inspiration by reminding ourselves of some of the significant 'Stages on the way to Harrogate', remembering some of the major themes that have dominated the history of studies of the atomic nucleus and how these themes have again and again been transformed under the impact of beautiful discoveries and inventions.

In trying to view the whole of the history of nuclear physics I find it useful to discern three major periods distinguished by the rather different character of the central issues being studied:

1. The discovery of the nucleus and its constituents (1896-1935). The achievement of this period was the identification of the 'nuclear problem' as involving a composite non-relativistic quantal system built out of neutrons and protons, held together by a new force of nature - the 'strong' interaction.

2. Defining the nuclear paradigm (1932-1952). The developments of this period led to the recognition of nuclear structure as based on independent-particle motion capable of supporting a rich variety of collective dynamics.

3. Discovering the feel of the nuclear stuff (1948-present).

Some of you may have seen this schema before, since I gave several lectures last year employing this threefold division as a framework for discussing the history of the compound nucleus idea. Unfortunately, it turned out every single time that I became so involved in the content of periods 1 and 2 that, not even once, was there time to discuss the lessons of the most recent period. This was indeed unfortunate since in a significant sense, the third period is by far the most challenging and illuminating. After all, the nuclei have to be made out of something, and so it is a rather simple question to ask, 'what are the elementary constituents of nuclei?' Similarly, having found the building blocks, there are, in fact, only relatively few general plans that are possible (as far as we know) for such finite many-body systems in their ground states. But the question of how 20 or 200 particles in interaction can give rise to an emergent microcosm of qualities and phenomena; there is a question to challenge you for the rest of your life and a question in which nuclear physics meets almost all the other branches of contemporary quantal physics, indeed of most of natural science. Thus I have resolved that these questions of the third period shall receive a major focus in this address, but, of course, this means that my treatment of ancient history will be little more than a birds-eye-view. Let us, nonetheless start at the beginning.

1. The Discovery of the Nucleus and its Constituents

The first hint of the existence of atomic nuclei was provided by Becquerel's discovery of those still faintly glowing ashes, remnants of the ancient cosmic fireworks that created the chemical elements of which our solid earth and our living bodies are made. Fifteen years later Rutherford was able to exploit the simple orbital motion of alpha particles from natural radioactivity as a marvellous microscope able to resolve the atoms into their open planetary electronic structure surrounding the small, dense, enigmatic atomic nucleus.

To begin with the nuclei appeared to be 'elementary particles', indivisible and immutable. But gradually, as is the way with 'elementary particles' the nuclei lost this holy status as closer acquaintance revealed the unmistakable ordering of nuclei with the integer quantum numbers of charge and mass. It became possible to produce nuclear reactions transforming one nucleus into another and it was clear that nuclei are composite systems built out of more elementary constituents.

This period is rich with profound and inspiring discoveries and inventions (see Table 1), of which I shall have time to refer in the very briefest terms to only a single episode (the discovery of the neutron) illustrating the style and methods of this broader development.

Table 1. Stages in the discovery of the atomic nucleus and its constituents (1896-1935).

1.	Discovery of radioactivity:	Becquerel (1896)
2.	Resolution of nucleus and atom:	Rutherford (1911)
3.	Discovery of the neutron:	Chadwick (1932)
4.	Theory of beta decay:	Fermi (1934)
5.	Meson theory:	Yukawa (1935)

We may remember that at the end of the 20's and the beginning of the 30's all efforts to understand the available knowledge about nuclei as composite systems built out of the known particles, electrons and protons, met with uniform, profound, and far reaching disaster. The spin and statistics came out wrong, and the attempt to confine electrons within the nuclear volume led to relativistic disasters. These paradoxes led Bohr to suggest that the question of nuclear stability might very well require a new revolution in the foundations of physics as far reaching as had been the consequences of the quantum of action. It was this crisis that was solved by the discovery of the neutron. The main steps in that remarkable discovery, compressed into a mere telegraphic report are:

1. Rutherford (1920) predicted the existence of the neutron (in his Bakerian lecture) by arguing that if heavy nuclei could form tightly bound states with electrons, as revealed in the difference between the atomic number A and the positive charge number Z, one could very well expect a single proton to unite with a single electron to produce a neutral and very unusual nuclear system. He felt, also, the need for such neutral nuclear systems in order to account for the building up of the heavy elements. This vision by Rutherford appears to be the first successful prediction of an elementary particle.

2. Chadwick joins Rutherford (1920-32) in a wide ranging research programme aimed at producing and exhibiting the expected neutron.

3. Bethe and Becker (1930) observe a penetrating radiation produced in Be + He reactions and interpret this as a high energy γ-ray. Joliot and Curie (1932) observe that the new radiation produces energetic recoils when passed through paraffin, but continue to interpret the radiation as a high energy γ-ray.

4. Chadwick (1932) compares the recoils in H, He, and in N to determine the mass of the new radiation and finds $M_{rad} \approx M_{prot}$ and thus the neutron is discovered at last!

After the discovery of the neutron, and the final banishment of electrons from the nuclear interior achieved with Fermi's theory of beta-decay, the 'nuclear problem' could be recognized as a non-relativistic many-body system built out of neutrons and protons held together by a new force of nature - the strong interaction (Heisenberg 1932), and we thus enter stage 2 on the way to Harrogate. However, before going on to stage 2, I would like to pause briefly to notice that already at this turning point, Yukawa's (1935) interpretation of the strong interactions as mediated by massive quanta, provided a measure of the limitation that must be encountered in a description involving only the degrees of freedom of neutrons and protons in nuclei. This limitation was thus known, was sometimes discussed, but in practice played a rather minor role in the development of the subject for the next 40 years (despite an occasional heretical voice); indeed it would appear that most of the fruitful developments of this period hinged explicitly on the possibility of ignoring these 'irrelevant' degrees of freedom. However, as we all know, the study of sub-nucleonic degrees of freedom and the associated relativistic effects has become a major item in the current agenda (for example, about half of the contributions to the present meeting are concerned with just these questions). This evolution of research interest in the field of nuclear physics, provides, I believe, a highly illuminating chapter for those interested in the history and real life practice of scientific method. What are the roles of the subject's internal criteria, of individual leaders, of fad, of technology, of the spirit of the age in this shift of interest? The answers to these questions can only be assessed after the new interests have been more fully developed and explored, but, in any case, the tension between these old and new interests is a very palpable element in the present stage of nuclear physics giving richness and excitement to our meeting.

2. Defining the Nuclear Paradigm

The first task that confronts any rational science trying to understand the emergent properties of a newly-discovered many-body system is to find the grand design that defines the relation of the individual components to the whole: are the particles in the nucleus delocalized and almost independent (as in a gas) or are they strongly interacting and correlated, perhaps localized (as in a crystalline solid), or are there, perhaps, still other fundamental architectures possible for quantal systems near their groundstate? It was, in fact, the unsettled status of this set of questions that lay behind most of the problems and paradoxes of the next 20 years of nuclear physics. In order to keep my promise of a serious concern with the third historical stage, I must drastically compress my discussions of the fascinating shifts in viewpoint encountered during the second stage. The main landmarks are listed in Table 2, and I shall only very briefly mention some of the most important arguments that were employed in connection with the developments of these ideas.

Table 2. Some major events contributing to the nuclear paradigm (1932-52)

1.	Study of neutron reactions	(1934)
	Analysis on basis of potential model	(1935)
2.	Neutron resonances	(1935)
	$\sigma_{capt} >> \sigma_{scat}$	(1935)
3.	Compound nucleus (short mean free path)	(1936)
	Liquid drop model	(1937)
	Statistical concepts; temperature etc.	(1937)
4.	Nuclear fission	(1939)
5.	'Magic' numbers	(1947)
6.	Shell model	(1948)
7.	Long mean free path in reactions	(1952)
	Optical model	(1953)

At the beginning an independent-particle model was employed by Heisenberg and was much further developed (especially by Bethe) in order to interpret the neutron cross sections. This static potential model was inspired, apparently, largely by the possibility of an analogy with the more familar problem of calculating electron scattering on atoms, but there was no particular reason to believe that this description would be appropriate for nuclei. In any case it had the enormous advantage that it made definite predictions and the ongoing experiments immediately disproved these predictions. Bohr looking at this conflict between theory and experiment recognized that the great density of narrow resonances being observed required the participation of many degrees of freedom. This led him to consider a picture in which the mean free path of the incoming particle is small compared to the nuclear radius; the neutron will then quickly share its energy with all the particles of the target creating the many-body state with energy statistically distributed, called the compound nucleus.

This picture led directly to the fruitful application of statistical concepts such as temperature and evaporation, and to the analogy with a liquid drop with its collective modes of capillary and compressive oscillations. The dramatic, and fateful, discovery of the fission reaction fitted brilliantly into this picture and just this set of concepts received their detailed and undeniable verification.

The unbroken successes of the compound nucleus over a period of more than a decade led to an almost universal acceptance of the idea of a short mean free path for nucleons in the nucleus. So much more, then, was the shock when Goepert-Mayer (1948) pointed out the inescapable evidence for shell structure contained in the systematics of nuclear binding energies. The sense of paradox in this discovery is preserved in the expression 'magic numbers' for the positions of the closed shells. It is very difficult for most of us brought up in a later age to fully appreciate the scope of this shock and the resulting disorientation since there has not been a comparable overthrow of established orthodoxy in any major field of quantal physics for almost 30 years.

The belief in a short mean free path could not be eliminated at once and thus there developed a number of popular half-way positions: for example, arguing that the independent particle motion might apply to the nuclear ground states, but for the excited states encountered in nuclear reactions, the mean free path was truly short.

Such arguments were not swept away until the early 1950s when Barschall (1952) summarized the experimental cross sections for scattering of neutrons of a few MeV incident on heavy nuclei and pointed out that these cross sections exhibited marked structure that would be impossible if the incident wave were strongly absorbed in the nuclear interior. This structure was then interpreted by Feshbach, Porter, and Weisskopf (1953) in terms of an interference between the incident wave and a wave transmitted through the nucleus; the intensity of the transmitted wave permitted a determination of the mean-free-path of the neutron in the nucleus and it was found to be much greater than the nuclear radius!

There still remained the problem of reconciling the long mean free path with the success of the compound nucleus. Indeed, it seems to me that this point is often not too clearly presented even in modern textbooks and so I would like to take a minute to remind you of the important relationships (see Table 3). The condition for shell structure is a mean free path long compared to the nuclear radius, or equivalently a collision time long compared to the nuclear traversal time ($\tau_{col} > R/v$). The

Table 3. Time scales in nuclear reactions

1.	Traversal time	$\tau_0 = \frac{2R_0}{v_{in}} \sim 10^{-22}$ s
2.	Collision time	$t_{col} = \frac{\hbar}{W} \sim 6 \times 10^{-22}$ s W = absorption potential $\sim$ 1 MeV
3.	Residence time for slow neutron reaction:	$t_{in} = \tau_0/T$ T = transmission coefficient of nuclear surface $\sim \frac{v_{out}}{v_{in}}$ $\sim 10^{-4}$ (slow neutron)
	thus	$t_{in} \sim 10^{-18}$ s
	Physical pictures:	black nucleus $t_{col} < \tau_0$ shell structure $t_{col} > \tau_0$ compound nucleus $t_{col} < t_{in}$

condition for the formation of the compound nucleus is the collision time short compared with the residence time of the incident neutron in the nucleus. The residence time is longer than the traversal time because of the strong reflection of a slow neutron in passing through the nuclear surface (the transmitted flux is reduced compared with the incident flux by a factor of order v_{out}/v_{in} which is of order 10^{-4} for slow neutrons). Thus the presence of quantal reflection makes possible the simultaneous applicability of the compound nucleus ideas and the long mean free path of the independent particle model.

With this discussion of the relation of mean free path to the compound nucleus we have, in fact, entered the universe of ideas that has been most characteristic of the third stage on the way to Harrogate. It is only a slight exaggeration to say that most of the nuclear physics efforts for the past 30 years have been engaged in the fascinating adventure of finding out how collective phenomena can arise in nuclei in the presence of independent particle motion, how the shell structure limits and forms these collective effects, and how the collective effects in turn react on the individual particles, profoundly modify their motion and give subtlety and depth to the nuclear independent-particle dynamics.

3. Discovering the Feel of the Nuclear Stuff (1948 to the present day)

During the past 30 years nuclei have been subjected to a wide ranging array of different probes, violent as well as delicate and precise; nuclei have been shocked and vaporised as well as tickled by electrons and other gentle inelastic exciters; single particles or pairs have been surgically added or taken away; and these systems have been set in rotation with velocities all the way up to the maximum set by the centrifugal instabilities; new nuclear species have been produced at an unabated pace of about 40 new isotopes each year until today we know of about 2400 different nuclear species (this is still only about 40% of the theoretically estimated number of particle-stable species).

What is the image of the nucleus that is revealed by all this very ingenious manipulation and synthesising of the nuclear stuff? I find it difficult to give a direct answer to that simple question because the nucleus turns out to be rich, too subtle, too many-faceted a universe to be simply described in terms of any single picture: it is both an independent-particle system, and at the same time a strongly coupled system exhibiting profound collective phenomena; it is both a non-relativistic many-body system composed of neutrons and protons and at the same time a complicated medium reflecting the transient presence of mesons, excited baryons, and perhaps quarks hopping between the different nucleon bags; it is both a highly ordered state of matter with a wealth of conserved quantum numbers and also a system exhibiting the most exquisite features of quantal chaos and statistical behaviour. In trying to describe the image suggested by such contrasting experiences I am reminded of a beautiful vision described by David Bohm in another context*. Thus I would like to suggest that we compare the nucleus with 'an object having a very large number of sides, having facets within facets, facets reflecting facets, facets consisting of mosaics of facets etc. To know what the object is, we must have a large number of different kinds of views and cross sections. Each view or cross section then contributes to our understanding of many aspects of the object. The relationship between the views are, however, equally important for they serve to correct the errors which arise as a result of regarding one or a limited number of views as a complete representation of the whole object; and they also indicate qualitatively new properties not apparent in the separate views ....'.

As a first reconnaissance of this fascinating object, I thought that it might be useful to draw up a catalogue of the discoveries which have so far revealed significant facets of the nucleus. It was not hard to identify some 10 - 15 discoveries that represented major turning points in the stages on the way to Harrogate. I found this catalogue to be both an inspiration, but also a cause for intense frustration, since even with the inclusion of some 15 items it represented only a grotesquely truncated caricature of the total development, and in any case with so many items it was quite excluded that I could in the few remaining minutes even begin to describe the fascinating content of each of these significant steps. Thus I have decided not to present you with the concrete physical representation of my catalogue, but rather let it be like the omitted figure of the Buddha in early pictures and sculptures. In these pictures the Buddha is present but is not explicitly represented; rather an empty chair, or an open book reminds us of his presence and each of us must imagine in his own mind the figure that is central to our thoughts - so let it be with the catalogue of discoveries. It is, in fact, much easier for me to discuss the general features of this catalogue, without being bound to a particular subjective representation of it. A number of features of the catalogue are indeed striking and worthy of comment.

i) Notice that the discoveries are chronologically distributed fairly evenly over the whole 40 year period; the catalogue contains some 15 items and indeed each decade has provided its 3 or 4 significant contributions. This lack of saturation, it seems to me, reflects the almost inexhaustible richness of the quantal many-body problems; the more we know about this

---

*D. Bohm, Causality and Chance in Modern Physics, p.31, University of Pennsylvania Press, Philadelphia (1971)

remarkable substance, the more deeply we probe into its structure, the more new facets have been revealed.

ii) A second striking feature of the catalogue is the decisive role of <u>specific</u> nuclear reactions in the discovery and exploration of the basic degrees of freedom of the nuclear stuff (see Table 4). From the first of the items, the giant dipole resonance with its inseparable connections to the nuclear photo-effect, to the last item, the isovector monopole mode, whose discovery depended essentially on the ingenious use of the π-meson charge-exchange reaction, there appeared each time a very specific reaction that could excite the mode and measure the amplitude of the collective motion. Without these specific reactions our understanding of the nuclear stuff would be enormously impoverished, yet before the discoveries there was no promise that the necessary specificity at all existed; its existence seems to me a marvellous and inexplicable gift of nature. Of course, it is a gift that has been most diligently, ingeniously, and effectively developed and exploited by the experimental nuclear physicists. In daring to explore some of the mysteries of nature there is always an act of faith that nature will respond with sufficient specificity; we can only observe that so far the nuclear adventure has been fortunate beyond expectation in this respect. It might also appear that this generosity has been closely connected with the possibility of studying the elementary modes of the nuclear stuff, in terms of sharply defined (quantal) states of the nuclear spectrum.

Table 4. Specificity and nuclear elementary modes

Elementary mode	Specific reaction
Giant dipole resonance	(γ,n)
Single particle config.	one particle transfer
Shape oscillation	(x,x')
Pair correlations	two particle transfer
Compound nucleus	neutron resonances
Rotation: low I	Coulomb excitation
high I	(heavy ion, $Xn\gamma$)
Isobaric analogue	(p,n)
	(p,p') resonance
GT mode	(p,n)
Compression modes	(α,α') (T=0 monopole)
	$(\pi^{\pm},\pi^0)$ (T=1 monopole)

iii) My last general comment on the catalogue concerns the recurrent, almost omnipresent role of spectroscopy in these cardinal developments. The construction of a nuclear term diagram goes back to the very earliest days of nuclear physics and played a central role in Bohr's discussion of the compound nucleus. This remarkably naive, yet sophisticated, way of looking at the nucleus appears as a phoenix in the development of the subject. Again and again, the spectra have appeared to have outlived their importance and the subject has been burned to ashes; yet as often again and again, the bird has been re-born and risen from the ashes to provide our field yet again with profound inspiration and insight into the many-faceted object we are trying to understand (see Table 5). This profound rôle of spectroscopy in the elucidation of the atomic nucleus

Table 5. The many reincarnations of nuclear spectroscopy

1.	Basic information on nuclear shell structure (spherical) isomers coupling schemes closed shell nuclei
2.	States in deformed nuclei (Nilsson)
3.	Neutron resonance spectroscopy (quantum chaos)
4.	Fission isomers and spectroscopy in the 'second' well
5.	States in rapidly rotating nuclei
6.	States in 'superdeformed' rotating nuclei

seems to me to reflect something very deep and characteristic about these systems - many body, yes, but conspicuously finite and emitting exquisitely sharp line spectra that can be efficiently and precisely measured; unravelling the coded message in these radiations continues to be a central challenge.

I cannot resist, as a last point, to return from these rather general considerations to a more down to earth, though brief, discussion of the latest rebirth of the spectroscopic phoenix brought about by the remarkable discovery of the so-called superdeformed band in ^{152}Dy (see Table 6). The underlying background for this phenomenon is the fact that the centrifugal distortion of a rapidly rotating body can in some circumstances favour an elongated shape (rather than an oblate flattening) as an equilibrium form on the way to the fission instability. This classical effect is strongly modulated by the quantal shell structure that selects very specific numbers of particles and shapes (in particular prolate shapes with ratio of axes 2:1) as energetically favourable. The

Table 6. Opening a new frontier in nuclear spectroscopy

Macroscopic effect:	rapid rotation centrifugal instability (fission) triaxial equilibrium shape of liquid drop
Shell structure:	quantal significance of 2:1 shape
Experimental innovation:	TESSA 3 50 element BGO ball 12 BGO suppressed Ge detectors ^{108}Pd (^{48}Ca, 4n) ^{152}Dy
Perspective:	single particle motion under new and extreme conditions collective modes nuclear matter on the verge of centrifugal instability

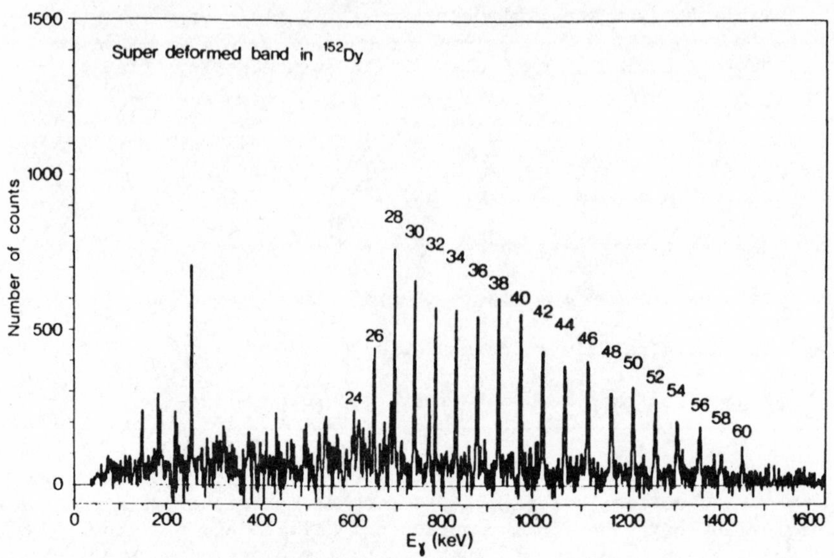

Fig. 1. Gamma-ray spectra in ^{152}Dy obtained by summation of gates set on most members of the superdeformed band.

recent discovery represents the intersection of these theoretical ideas with the marvellous development of experimental techniques associated with multiple-array BGO suppressed Ge detectors. Using such techniques to detect γ-rays from the ^{108}Pd(^{48}Ca,4n)^{152}Dy reaction, the Daresbury group have produced the spectrum shown in Fig. 1, representing the resolution of a rotational band extending from $I = 24\hbar$ to $I = 60\hbar$. The detailed study of this structure can be expected to extend significantly our understanding of centrifugal effects acting on nuclear matter at the edge of fission instability. This discovery will be discussed in more detail in presentations at this conference by Nolan and by Twin.

Inst. Phys. Conf. Ser. No. 86
Paper presented at Int. Nucl. Phys. Conf., Harrogate, UK, 1986

Paths from Harrogate: Towards new physics with new tools

Erich Vogt

TRIUMF, 4004 Wesbrook Mall, Vancouver, B.C., Canada V6T 2A3

Abstract. The impact of the standard model for quarks and leptons and for unified forces on the future development of nuclear physics is described. The past experience in nuclear physics with the very successful models based on the nucleon picture gives some indication how progress might be made with new experimental tools. Now nuclear physics has expanded to include all of strong interaction physics and especially the effect of nucleon substructure and quark confinement on the dynamics of the nucleus. The important new questions will be addressed with new cw electron accelerators, with relativistic heavy-ion colliders and with kaon factories.

1. Introduction

In the preceding paper Mottelson (these Proceedings) has described the intellectual basis of our field and the stages of development of nuclear physics up to the present Harrogate conference. The programme of this conference goes much beyond those of the previous international conferences in the same series and attempts to describe the very great stimulus which nuclear physics has received from the advent of the standard model for quarks and leptons and for unified forces. In this paper an attempt is made to provide a perspective for that programme: to describe the ideas of the standard model, to outline the urgent new questions for nuclear physics which result from it and to assess the new tools intended to address those questions. In this perspective the various papers which follow then point toward an exciting future for nuclear physics: the alternative paths from Harrogate.

To guide our approach and normalize our expectations to the standard model and its new questions, it is useful to reflect on what has been learned from other simple models which have emerged in the eventful history of nuclear physics. In each of the cases to which we shall refer there were important theoretical impediments to the emergence of the model and new experimental tools which brought about its discovery.

The nuclear shell model - or its concomitant for nuclear reactions, the optical model - triumphed over the ideas which underlay the compound nucleus picture (Mottelson, these Proceedings). The nucleon-nucleon force was known to be very strong and very short range. Rough theoretical estimates suggested that the forces were too strong for any shell model states to survive long enough for the model to be useful. Yet new data with neutrons from reactors and with protons and neutrons from the new post-war accelerators (Barschall et al., 1952, 1953) overwhelmed the impediments.

Somewhat later the experience was repeated for the model pertaining to isobaric analogue states. Extrapolating from the known value of the imaginary part of the nucleon-nucleus optical potential and the known ratio of Coulomb interactions to nucleon interactions in a heavy nucleus one could predict that the spreading width of the isobaric analogue states would be too large for them to be seen. The estimate was wrong, the states turned out to be much sharper and the simple model emerged with the help of high-resolution beams from the new generation of tandems.

A further example is the molecular states observed in low-energy heavy-ion reactions about which Betts (these Proceedings) will inform us in one of the plenary papers of this conference. When first seen twenty-six years ago by Bromley and his colleagues (Almqvist et al., 1960) in the total cross sections of $^{12}C + ^{12}C$ reactions these states were startling because of their narrowness. Any estimates based on the residual interactions of two heavy ions suggested that possible molecular states would be washed out. Nothing should disturb the smoothly varying cross section behaviour of two such liquid drops. The experimental evidence for the simple molecular states emerged, again, because of the advent of high-resolution heavy-ion beams from the new tandem accelerators combined with the new solid-state detectors.

These examples drawn from the history of nuclear physics tell us that:

- simple phenomenological models for nuclear dynamics tend to work much better than predicted;
- nuclear physics is an experimental science;
- new tools in nuclear physics often provide new science.

These lessons may be useful in coping with the new ideas and questions of the standard model.

2. The Standard Model and its Impact on Nuclear Physics

The standard model constitutes a remarkable leap forward in the understanding of the fundamental particles and forces of our world. It consolidates the knowledge gained from several decades of experiments in particle physics and it presents both particle physics and nuclear physics with urgent new questions. These are heroic times for both fields because extraordinarily important questions are to be answered with extraordinarily difficult new experiments. To discuss the impact of these new ideas on nuclear physics we first give the elements of the standard model and describe how the standard model is viewed by particle physicists.

2.1 The standard model

The elementary particles of the standard model are three pairs (generations) of quarks and leptons:

$$\text{quarks:} \quad \begin{pmatrix} u \\ d \end{pmatrix}, \begin{pmatrix} c \\ s \end{pmatrix}, \begin{pmatrix} t \\ b \end{pmatrix}$$

$$\text{leptons:} \quad \begin{pmatrix} e \\ \nu_e \end{pmatrix}, \begin{pmatrix} \mu \\ \nu_\mu \end{pmatrix}, \begin{pmatrix} \tau \\ \nu_\tau \end{pmatrix}$$

The quarks participate in all of the fundamental interactions while the leptons do not have strong interactions. Each of the fundamental particles has an antiparticle. The masses of the quarks and charged leptons are known (except for t) as are many of their quantum numbers, for example, charge, isospin, baryon number, etc. The six quarks each occur in three different colours. All of the known hadrons - the baryons and mesons - can be viewed as composites (colour singlets) of quarks (and/or antiquarks).

The standard model also unifies three of the fundamental interactions. Using Maxwell's unification of electricity and magnetism into electromagnetism as a template, the weak and the strong forces of nuclear physics have been unified in a common framework of renormalizable local gauge theories. The quanta of the theory are gauge bosons of which the massless photon is an example. In each case there is a local choice of gauge - a powerful symmetry which drives the renormalizability of the theory. In electromagnetism the gauge pertains to the local choice of potential. The combination of Maxwell's classical electromagnetism with quantum theory and the proof that gauge symmetry guarantees its renormalizability produced, four decades ago, the very elegant and beautiful theory of quantum electrodynamics (QED).

Bringing the weak force into the same framework required great theoretical ingenuity. In this case the choice of gauge pertains to isospin and the weak force is very short range. Massive field quanta are needed and the gauge symmetry is hidden. An auxiliary field, the Higgs field with its own quanta (Higgs bosons), hides the symmetry, aids in the renormalizability and provides mass to the field quanta. These ideas, all encompassed in the concepts associated with spontaneous symmetry breaking, eventually succeeded in unifying the weak interaction with electromagnetism into a single renormalizable local gauge theory, the electroweak theory. In this unified theory the photon was joined by three massive vector bosons, the $W^{\pm}$ and Z^0 particles, whose masses were predicted. The 1983 discovery of the $W^{\pm}$ and Z^0 at CERN by the UA1 and UA2 experiments constituted an experimental tour de force for modern physics and firmly established the standard model.

The unification of the strong force with the electroweak interaction into a single grand unified theory (GUT) is still under way. Quarks have been seen inside hadrons through deep inelastic electron scattering. There is no doubt that the quarks are the carriers of the strong interaction, that the "charge" of the interaction is the quantity called colour, that the quanta of the strong field are massless gluons and that the gluons themselves carry colour. All of this is embodied in the modern theory (Isgur, these Proceedings) of the strong interaction called quantum chromodynamics (QCD) in analogy to QED. However, the fact that the gluons carry colour, while the photons do not carry electric charge, makes QCD have essential differences from QED. At least eight colour-carrying gluons are required for QCD and the choice of colour is the gauge of the theory.

The candidates for GUT bring in additional symmetry and predict that the four vector bosons (γ, $W^{\pm}$ and Z^0) of the electroweak theory are joined by at least twelve new vector bosons called X particles. The X are predicted to be very massive - about 10^{15} times more massive than the nucleon. Some of the successes and problems of GUT are discussed below.

The standard model, with its quarks, leptons, vector bosons, Higgs, electroweak interaction and QCD, fits all of the present data. Yet it can be viewed in alternative ways.

2.2 The particle physics viewpoint of the standard model.

It is essential to understand how the view of the standard model and its very success by particle physics differs from that which is appropriate for nuclear physics. In particle physics one questions the basis, boundaries and extensions of the standard model while in nuclear physics (Sec. 2.3) one seeks to employ a more fundamental understanding of the basic building blocks and the strong force to describe the varied properties of systems of strongly interacting particles.

In spite of the fact that the standard model fits all the present data in particle physics one begins by noting that it is not complete and it is too arbitrary. One asks the following general questions:

- How can the standard model be improved?
- What are its boundaries?
- How does one go beyond it?

In turn, these general questions lead to a number of specific questions:

- Why are there so many quarks and leptons? (6 leptons plus 6×3 = 18 quarks).
- How many generations of quarks and leptons exist? This is a question for which astrophysics (Schramm, these Proceedings) is beginning to provide some answers complementing the information from colliders.
- What is the origin of the mass of the quarks and leptons? There are fundamental arguments which suggest that particles responsible for the mass should exist near 1 TeV and this is one of the main driving forces for the supercolliders now being proposed.
- Are there better ways to hide gauge symmetry? The two mechanisms - Higgs and technicolour - which hide the gauge symmetry of the electroweak theory both provide mass to the $W^{\pm}$ and Z. Both mechanisms predict the existence of particles not yet found, especially technicolour which predicts a rich spectroscopy of states near 1 TeV. Both have serious problems especially if extended to give mass to the quarks and leptons. There appear to be missing principles or ideas for these two mechanisms or perhaps one needs some entirely new mechanism with new phenomenological parameters.
- Is it possible to achieve greater unification among the interactions? First of all the unification of the strong interaction with the electroweak by the grand unified theories (GUT) needs improvement. GUT successfully predicts the Weinberg angle ($\sin^2\theta_W \sim 0.21$) and it also predicts proton decay, which has not yet been observed. (GUT does not necessarily require the proton lifetime to lie anywhere close to the level attainable by present experiments.) Several crucial problems for GUT remain, especially the hierarchy problem which refers to the fact that the very massive X particles are allowed to mix with the $W^{\pm}$ and Z. When mixing is possible how can such very different mass scales coexist. A possible answer lies in supersymmetry (SUSY), a higher symmetry which could remove the offending matrix elements for mixing. The second unification issue - and the most important specific question now confronting the standard model - pertains to including gravity among the unified forces. Efforts to achieve such unification within the framework of renormalized local gauge theories have been plagued by anomalies and infinities. Here SUSY, the ultimate symmetry and with its natural relationship to geometry, may help. A different approach,

superstrings, is currently exciting the interest of the world community of particle theorists. Instead of adding new symmetries or new phenomenology for new physics this approach looks at mathematical frameworks which in themselves are beautiful and trouble free and enquires whether or not such new frameworks might not describe everything - all particles plus all interactions - in some physical world, hopefully ours. Superstrings considers the quantum mechanics of extended objects (strings) in many dimensions (currently 10). There are mathematical miracles in that the quantum inconsistencies disappear and massless spin-2 modes (gravitons?) drop out of the theory but it is still very far from making predictions about experiments in particle physics or, indeed, masses in the vicinity of 1 TeV. These are still very early days for the quantum mechanics of extended objects and superstring theory promises to engage the attention of theorists for a long time.

2.3 The new nuclear physics - The physics of systems of strongly interacting particles

The application of the standard model to the physics of strongly interacting systems of particles is a major new opportunity in physics and constitutes the new nuclear physics. It deals with the structure of all hadrons: baryons, mesons, nuclei, neutron stars, etc., and with the use of hadrons as tools in exploring fundamental symmetries.

This important new opportunity is being entirely bypassed by particle physics and falls naturally into the traditional interest of our own field, nuclear physics. This does not mean at all that nuclear physics should feast on the neglected leftovers of particle physics. In the understanding of our beautiful physical world there is much of major interest beyond fundamental interactions and building blocks. Long after the discovery of the Coulomb force we are still left with the need to know how those Coulomb forces lead to the description of water, a magnificent substance with, under normal conditions, two phase transitions separated by only a hundred degrees. (This understanding is especially important in Harrogate which is so dominated by water and so precariously perched in temperature near the lower of these phase transitions.) When presented with new opportunities to understand our diverse world, physics must seek to seize them.

Much is now to be added to the long-standing nuclear structure interests of nuclear physics. Through the use of many low-energy accelerators a great deal has been achieved in understanding atomic nuclei as a system of nucleons interacting through meson exchange. The pursuit of fundamental symmetries in nuclei described in such terms has also been very fruitful. To such studies one now seeks to add:

- hadronic physics - the effect of nucleon substructure and of quark confinement on nuclear dynamics, pursued with intermediate energy electron and proton accelerators;
- multiquark spectroscopy - the structure of mesons, baryons and other simple quark clusters, pursued with electron-positron colliders and kaon factories;
- quark-gluon matter - the properties of new phases of matter at higher temperatures and densities, to be pursued at relativistic heavy-ion colliders, especially RHIC but with initial work at the Bevalac, GSI, the Brookhaven AGS and the CERN SPS.

This program encompasses all of the physics of (and in) the composites of quarks. This expansion of nuclear physics will attract many particle physicists and form an interesting new cultural mosaic. Perhaps we should follow Isgur's suggestion (these Proceedings) and rename this expanded field simply "strong interaction physics". It is interesting to note that Harrogate is the first in this series of nuclear physics conference to have a programme which extends reasonably across the whole domain of strong interaction physics. Therefore in my viewpoint of our field I am, I hope, serving to set the stage for what follows.

The application of the standard model to systems of quarks is far from trivial. We know that nucleons have quarks as subconstituents and that QCD - the exchange of massless coloured gluons between massive coloured quarks - is the strong force. But at quark separations (0.5 - 1.5 fm) appropriate for nucleons or nuclei QCD is nonperturbative. In this situation the only proper calculations - lattice gauge calculations - are very cumbersome and very rough. There is then a dilemma about how one should proceed. Should we use purely phenomenological models, such as bag models, to describe quark confinement or should one dismiss such models as foundationless and attempt descriptions, for example, baryons as topolical solitons of meson fields, which are somewhat more closely tied to the pathological properties of QCD? It is here that our past history might teach us something. In complicated many-body systems intuitive simple models have often worked much better than they had any reason to do so initially. At the other extreme one does not wish to encourage wild models taking no cognizance of our past successes. The dilemma remains.

There are many questions now confronting the new strong interaction physics:

1) What is the origin and best description of the nucleon-nucleon force within QCD?
2) What is the basis for (and limits) of the nucleonic description of low-energy phenomena?
3) What new nuclear properties arising from quark degrees of freedom should we expect to find at intermediate energies?
   a) How does one describe quark confinement?
   b) What symmetries (chiral, etc.) are crucial for the physics?
   c) Do the quarks wander beyond their confinement volumes?
   d) What nuclear properties are likely to be especially sensitive to the new physics - e.g. the Gamow-Teller piece of the nuclear response function?
4) What are the properties (phase transitions, etc.) of nuclear matter at high density and temperature?
5) What is the validity of simple models - for example, nonrelativistic quark models - for meson and baryon spectroscopy?

These, and many similar such questions, will be important throughout this conference.

Because of such questions and the new strong-interaction interests, the coming years promise to be a heroic age for nuclear physics. This does not mean that our traditional interests are passé, the oscillations, vibrations, rotations and single-nucleon behaviour of systems of neutrons and protons at low energy. Rather, some aspects of these studies have achieved new importance because one needs to know the limits of the nucleon picture. It is the continued pursuit of the old interests in disregard of the above questions which has lost its central position and it is the direct search for answers to these questions which has moved to centre stage and provided opportunities for heroism.

3. Some Open Doorways

The major paths from Harrogate involve a number of large new accelerators now planned or under construction. Before describing these paths we select here some current examples which can be viewed as doorways to the new nuclear physics. Many other examples of such doorways could have served equally well.

3.1 Electron scattering from ^{2}H and ^{3}H

At intermediate energy electron scattering, especially from few-nucleon systems, is the most clinical tool we have for probing the limits of the nucleonic description of nuclei. Two recent experiments, on electrons scattered from ^{2}H and from ^{3}H, serve to illustrate what one can learn from present single-arm experiments and perhaps hint at how much more the new generation of cw electron accelerators at intermediate energy (Sec. 4.1) is likely to provide.

During the past year Arnold et al. (unpublished) have used the NPAS facility at SLAC to measure the form factors of the deuteron at high momentum transfer. Largely with this experiment in mind SLAC built a new injector which can provide high-intensity, low duty-factor electron beams up to 5 GeV. The ^{2}H magnetic form factor has now been measured past the diffraction minimum. These results will put strong constraints on models for meson-exchange currents (MEC) and for 6-quark components in the deuteron wave function.

At Saclay (Juster et al., 1985; see also Sick, these Proceedings) and also, more recently, at Bates (unpublished, 1986) electron scattering from ^{3}H and ^{3}He has provided the three-nucleon form factors. The three-nucleon system has many advantages as a laboratory for the new physics. It is a simple system: the nonrelativistic three-body wave function can be solved with arbitrary precision, including the effect of three-nucleon potentials. These systems have isospin symmetry: the isoscalar and isovector components of the nuclear current can be determined unambiguously. By using longitudinal and transverse separations the charge and magnetization densities can be separately determined.

Figure 1 shows the charge and magnetic form factors for electron scattering from ^{3}H. Much of the data is from the Saclay experiments which will be reviewed by Sick (these Proceedings). The figure shows several different calculations for MEC effects. The impulse approximation (IA) alone does not agree with the data. The magnetic form factor is especially sensitive to different treatment of MEC effects, in this case to changing the nucleon form factor in the meson current. These are examples of the new generation of electron scattering experiments and of the courageous use of large tritium targets.

3.2 Pion propagation and charge exchange

Central to all of the issues of the new nuclear physics is the question of how the pion propagates in the nuclear medium. Does the pion change? Is the nuclear vacuum polarized by the pion? Does it join the nucleon to propagate as a delta? There is some hope that new generations of detectors

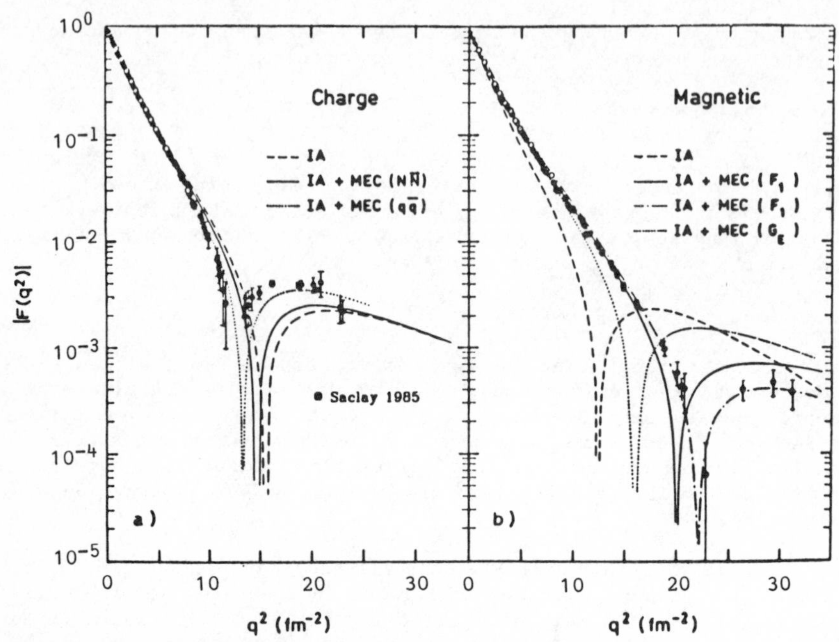

Fig. 1. The charge and magnetic form factor data for ^3H (Juster 1985; Sick, these Proceedings; Bates Laboratory 1986).

coupled with present pion facilities can make some progress on these matters.

Much interesting work has been done on pion elastic and inelastic scattering and only little on reactions induced by pions. Even at the meson factories the fluxes of pions are not large enough to make pion reactions to definite final states the kind of facile probe that one has with electron and proton beams. In pion spectrometers some price in resolution must be paid to achieve sufficient acceptance.

Illustrating both the promise and the problems of pions as a probe, Fig. 2 shows data for recent experiments on pion double charge exchange (DCX) on ^{14}C and ^{18}O, taken both at TRIUMF (Altman et al., 1985; Navon et al., 1984) and at LAMPF (Cooper et al., 1984). The data are compared to models for six-quark components in the target wave function (Miller, 1984) and to more conventional models (Karapiperis and Kobayashi, 1985). These low-energy DCX cross sections attracted considerable attention because they were so much higher than anticipated from earlier work at energies near the delta resonance. They may yet provide important data for understanding either pion propagation or the quark-nature of the nucleus. The point here is that such experiments are just now beginning to emerge. The first of these measurements (on ^{14}C) was carried out with the TPC at TRIUMF, a modern detector which had been built at TRIUMF for a muon conversion experiment. The TRIUMF TPC has very much larger acceptance than any spectrometer but, unfortunately, its energy resolution is only about 4 MeV and therefore, in DCX experiments, it is limited to ^{14}C as a target! The later experiments were forced to use pion spectrometers. I have a dual prejudice with regard to pion-nucleus experiments. First I believe that much will be learned if proper tools can be devised to study individual

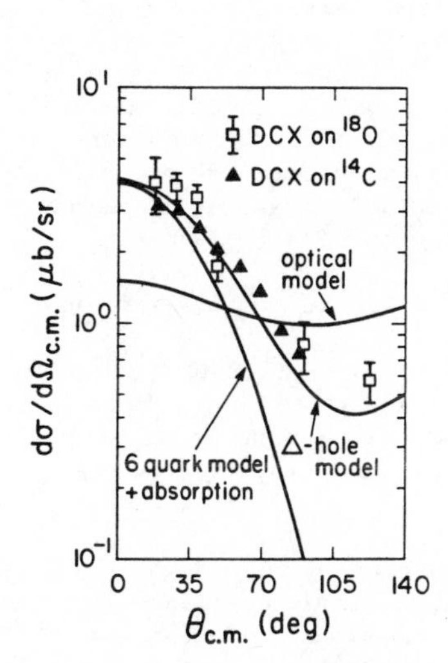

Fig. 2. The angular distribution for $^{18}O(\pi^+,\pi^-)^{18}Ne$ and $^{14}C(\pi^+,\pi^-)$ data (Altman et al., 1985; Navon et al., 1984; Cooper et al., 1984). Also shown are the theoretical calculations with a six-quark bag (Miller, 1984) and for the Δ-hole model (Karapiperis and Kobayashi, 1985).

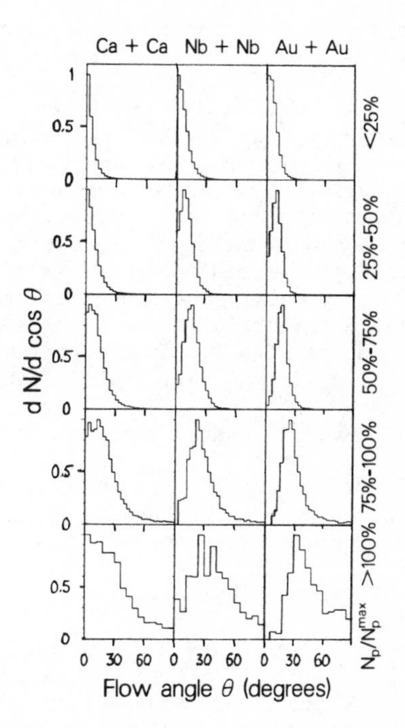

Fig. 3. Frequency distributions of the flow angle (θ) for Bevalac data (Gustafsson et al., 1984) at 400 MeV/nucleon. Different beam-target combination and multiplicity bins are shown.

pion reaction channels. Secondly I believe (as the above TPC experiment suggests) that we have much to learn about applying the new particle physics detectors to nuclear physics problems. Perhaps the time has come to replace pion spectrometers by multi-track detectors specifically designed for pion experiments.

3.3 Collective flow in relativistic heavy-ion reactions

One recent compelling picture, Fig. 3, from the pioneering work (Gustafsson et al., 1984) of the Berkeley Bevalac, provides evidence for important collective effects in the central collisions of relativistic heavy ions. As indicated by the data, the phenomena of collective flow, at these energies, does not occur in medium weight nuclei (Ca + Ca) but does occur in collisions of very heavy nuclei (Nb + Nb, Au + Au). For each pair of nuclei the events are subdivided according to their total charged-fragment multiplicity; the highest multiplicity (lowest row of figures) corresponds to the most central collisions. The number of events is plotted versus the flow angle. The flow angle gives the direction of the effective jet axis, the direction relative to which the transverse kinetic energy of the observed participant fragments is minimized. Clearly the most central collisions attain nonzero flow angles. Comparison of the data with models suggests that some form of compressional (potential) energy must be involved here.

The collective flow data is an important precursor for the relativistic heavy-ion work which lies ahead (Sec. 4.1). We should also remember that it points to the need for large multitrack detectors in modern strong interaction physics. For more than a decade the Bevalac has been unique on the world scene in its provision of relativistic heavy-ion beams. However, it was not until the recent commissioning and use of the large crystal ball plus crystal wall detector that it was possible to detect collective flow. Probably the impact of the Bevalac would have been even greater and come even earlier if such large detectors had been provided earlier. Again we can benefit from the experience of our particle physics colleagues when we plan for the detectors of our next generation of accelerators.

3.4 Nucleon charge exchange and the Gamow-Teller response function

My final example of a doorway is drawn from our own TRIUMF programme of nucleon charge exchange. I choose this example not only because I am familiar with it but because it is important new physics and I am not here poaching seriously on territory of the speakers who follow me in the Harrogate conference.

Several years ago the Indiana work of Goodman et al. (1984) on (p,n) reactions at 200 MeV draw a great deal of attention because it clearly brought out the Gamow-Teller (GT) nuclear response function, at 0°, and gave its strength distribution. The data indicated quenching of the GT giant resonance by a factor approaching 50%. The magnitude and origin of this quenching and of the spin-flip isospin-flip (GT) response function as a possible "chiral filter", especially sensitive to hadronic or quark effects, remains a topic of considerable theoretical interest. Careful (p,p') inelastic scattering (Häusser, 1986) with polarized beams and focal plane polarimeters has obtained similar GT quenching. Now the new TRIUMF data on (n,p) charge exchange has considerably expanded the experimental knowledge.

The full GT sum rule requires knowledge of both the (p,n) and (n,p) reactions. The sum rule is

$$S = \Sigma B(GT^-) - \Sigma B(GT^+) = 3(N-Z) \qquad (1)$$

with $\Sigma B(GT^-)$ obtained from (p,n) and $\Sigma B(GT^+)$ from (n,p).

At TRIUMF the proton spectrometer (MRS) is used as a recoil spectrometer for (n,p) reactions (Alford et al., 1986). In the (n,p) mode, shown on Fig. 4, a ^{7}Li target generates the neutrons which then charge-exchange in the segmented secondary target, only 0.9 m downstream, at the spectrometer pivot. The emerging protons are then analyzed by the MRS. The use of the segmented target, in which wire chambers pinpoint the segment of origin of the protons, brings three enormous benefits: it multiplies the rate by making the total target thicker; it maintains high resolution (1 MeV or less) and it normalizes the incident neutron intensity by having CH_2 as the first segment and using the known H(n,p) reaction. The same facility can be used to study (p,n) reactions. For this the ^{7}Li target is removed, the primary target is placed at the MRS pivot and the segmented target is replaced by a scintillator which serves as an n→p converter and whose light output is used to correct the proton momentum as determined by the MRS. Again a resolution of less than 1 MeV is achieved.

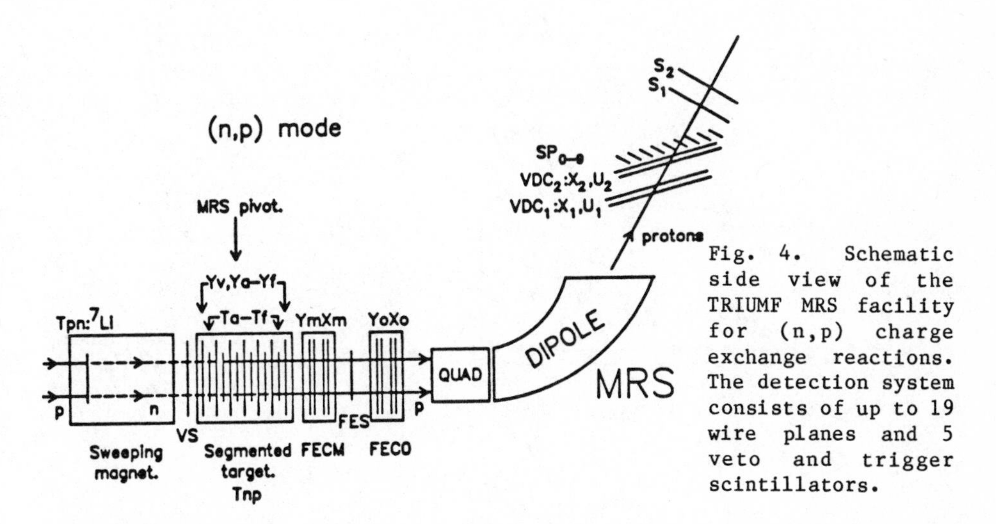

Fig. 4. Schematic side view of the TRIUMF MRS facility for (n,p) charge exchange reactions. The detection system consists of up to 19 wire planes and 5 veto and trigger scintillators.

Figure 5 gives the new TRIUMF results (Alford, 1986) for the ratio of the isovector interaction strengths at intermediate energies for the $^{14}C(p,n)^{14}N$ reactions. The cross sections to the 2.31 MeV (0^+, T=1, Fermi) and to the 3.95 MeV (1^+, T=0, GT) excited states are measured. The interaction strengths are derived from the 0° cross-section ratios using the factorized DWIA. The new TRIUMF results are in clear disagreement with the Love-Franey interaction (1955) (solid line of Fig. 5).

Figure 6 gives the new TRIUMF data (Häusser, 1986) for the (n,p) reactions on ^{12}C, Fig. 7 that for ^{26}Mg and Fig. 8 that for ^{54}Fe. For ^{12}C the $B(GT^+)$ strength is known from B^- decay and serves to normalize the cross sections for other targets. Figure 6 also shows the typical quality of the raw spectra. The ^{26}Mg and ^{54}Fe data are both compared to shell model calculations (Franey and Love, 1955; Bloom and Fuller, 1985) for which a

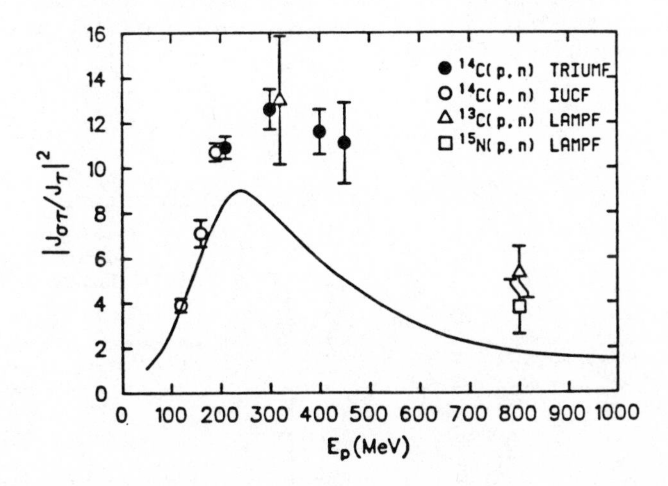

Fig. 5. Ratio of effective isovector interaction strengths derived from (p,n) reactions (Alford, 1986) at various incident energies.

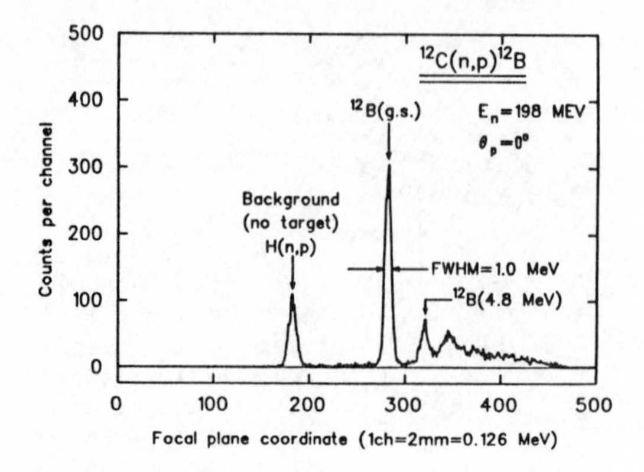

Fig. 6. Spectrum (Häusser, 1986) from the $^{12}C(n,p)^{12}B$ reaction at 200 MeV observed at 0°.

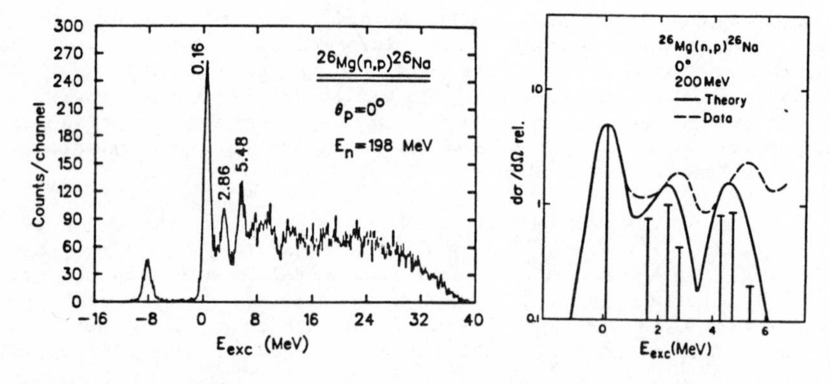

Fig. 7. Spectrum (Häusser, 1986) from the $^{24}Mg(n,p)$ reaction at 0° and E_p = 200 MeV (left). The data are compared to a theoretical calculation of the GT component only (right).

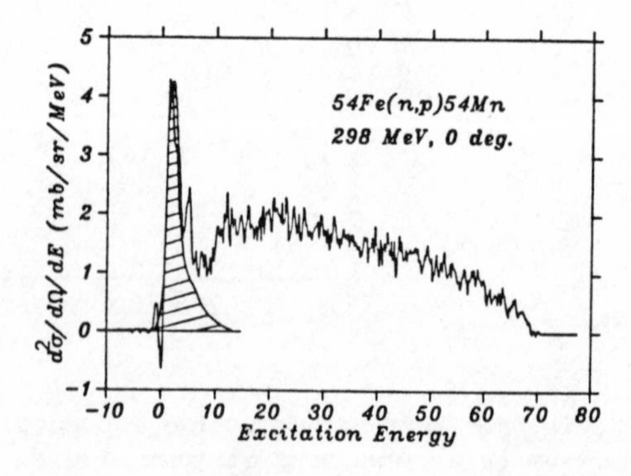

Fig. 8. Spectrum (Häusser, 1986) for the $^{54}Fe(n,p)^{54}Mn$ reaction at 0° and E_p = 300 MeV. The cross-hatched area is the theoretical GT distribution (Bloom and Fuller, 1985) multiplied by a quenching factor of 0.6.

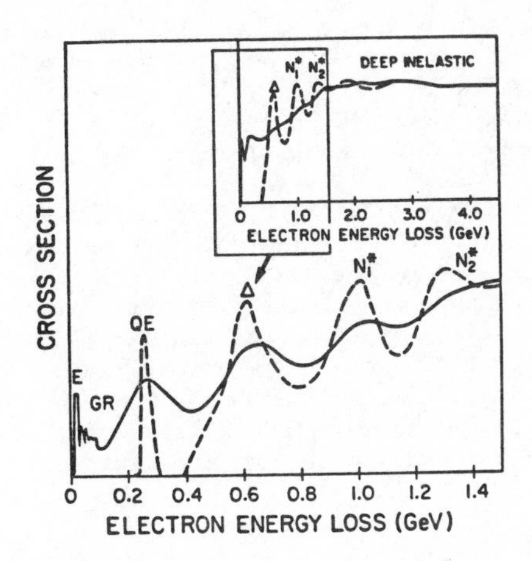

Fig. 9. Idealized inclusive (e,e') spectrum. Some nucleon resonances and the quasielastic (QE) and giant resonance (GR) peaks are indicated.

quenching factor of 0.6 is required to give the indicated agreement with the data. For ^{54}Fe a test of the GT sum-rule gives the following results:

$B(GT^-) = 7.8 \pm 1.9$ from previous (p,n) data (Rapaport et al., 1984)
$B(GT^+) = 5.1 \pm 0.6$ from the (n,p) experiment (Häusser, 1986)
S = difference $= 2.7 \pm 2.0$ compared to $3(N-Z) = 6$.

Therefore one can conclude that a sizeable fraction of the GT strength is shifted out of the low-energy shell-model region.

This new (n,p) work is the first data from a facility which should be very productive in the next few years. Los Alamos is planning a similar facility.

4. The New Accelerators

Much of the action of the new strong interaction physics will take place at a new generation of large accelerator facilities. In discussing the various projects and plans, below, I will not be encyclopaedic but rather give examples of each type with some bias toward the projects with which I am most familiar.

4.1 The new cw electron accelerators

The advent of appropriate superconducting technology has made possible the construction of many new cw electron accelerators covering a wide range of energies. With such a big leap forward in duty factor many new kinds of experiments will soon be possible.

Figure 9 gives an idealized (e,e') spectrum indicating the different physics in different regimes of energy loss. The dashed line represents the scattering from a free proton and the solid line the scattering from a nucleus. One can distinguish the following regimes:

- Confinement physics, with momentum transfers greater than 1 GeV/c. Here one will study nucleon structure, the modification of the nucleon in the nuclear medium and the effect of the nuclear medium on quark confinement. The main facility for this regime will be that of CEBAF, in Newport News, Virginia, which is now firmly scheduled, by the U.S. Department of Energy, to begin construction. It uses the new superconducting technology to achieve high-intensity cw electron beams up to 5-6 GeV. CEBAF has for some time been the first priority for new construction in the U.S. Long Range Plan for Nuclear Physics.

- Hadronic physics, with momentum transfers of a few hundred MeV/c up to a few GeV/c. Here the interest lies in the role of mesons and excited nucleons in the nuclear response. Several new facilities are likely to be constructed. Mainz is under way with a microtron providing cw electrons up to 800 MeV. Bates would upgrade its present accelerator and use a pulse stretcher ring (PSR) to achieve cw electrons up to 1 GeV. Bates is not yet funded but it is so cost effective and provides so much new science that it seems clearly destined for early approval. Plans at NIKHEF, Saclay and Japan for machines in this regime are also well developed.

- Nucleonic physics, with momentum transfers up to a few hundred MeV/c. The new accelerators in this regime will study single particle and collective motions of an assembly of nucleons. Saskatoon (300 MeV) is rapidly completing construction of a PSR to achieve a facility in this regime. There will likely be many other cw electron machines world wide in this regime, including Illinois (450 GeV) which is awaiting final approval.

In each of the regimes the cw electron accelerators will explore new territory, especially reactions involving coincidence measurements. The sensitivity of the recent form factor measurements (Sec. 3.2) to hadronic effects gives some hints of the physics likely to emerge. Special mention should be made of the possibility, at Bates, of using polarized internal targets in the PSR with polarized electron beams. This combination promises to be a veritable "multipole meter" sorting out the different pieces of the nuclear response.

4.2 Relativistic heavy-ion accelerators

The driving force for heavy-ion colliders is the search for the quark-gluon plasma which will be reviewed by Baym (these Proceedings). The nature of the search can be characterized by the now familiar phase diagram, Fig. 10, for the properties of hadronic matter at various temperatures and densities. In the early universe and in stellar collapse one attains the likely conditions for such a phase change.

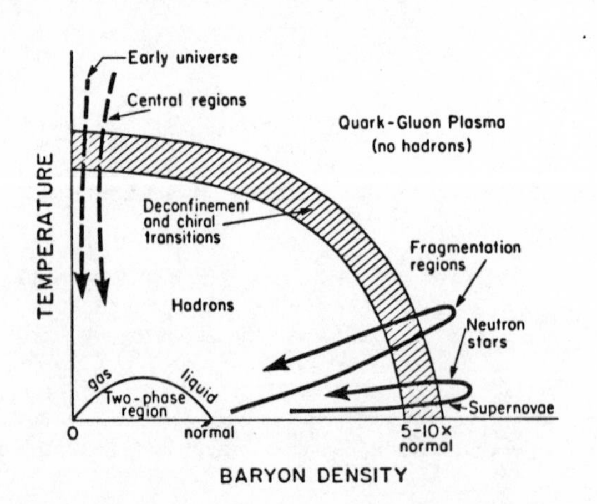

Fig. 10. Temperature-baryon density phase diagram of hadronic matter.

Can it be replicated in the laboratory? What are its properties? What signals will identify it? What behaviour occurs before the advent of the phase change? These and many other similar questions are now being confronted. The plans include initial skirmishes, some major battles and a frontal assault.

The nature of the relevant phase transitions is beginning to emerge from lattice gauge calculations (Baym). There are two phase transitions in question. In one the quarks become asymptotically free or deconfined, in the other they become undressed as chiral-invariance is restored. The two phase transitions seem to occur at the same place in the phase diagram. Is it a coincidence, or perhaps an amusing parable of the human condition that the quarks become naked as soon as they become free or are we seeing repetition of the wonderful story of water and its two-phase transitions. The QCD calculations appear to say that the transitions are second order.

Using recent data on high-energy proton-proton and proton-nucleus collisions some confidence is being gained about predicting the energy deposited in relativistic heavy-ion collisions and, consequently, in the regions of the phase diagram likely to be reached by accelerator experiments. The energies intended for RHIC (100 GeV/nucleon) appear to be sufficient to attain quark-gluon matter. In central collisions it will yield a baryon-free regime at very high temperature. RHIC appears to have been recommended for funding by DOE.

Much progress has been made on defining the signals pertaining to the phase transition toward a quark-gluon plasma and on designing the requisite detectors. The signals include photon and dilepton production, kaon production, etc. The detectors will need to measure everything - similar to the combination of drift chambers and calorimeters recently developed in particle physics. The experiments to begin in late 1986 at the Brookhaven AGS and the CERN SPS with relativistic heavy ions of moderate weight bombarding fixed targets should provide very valuable experience for the later work at RHIC and also should achieve temperatures and densities significantly greater than normal.

Figure 11 shows how a number of different heavy-ion accelerators will provide beams spanning the periodic table and covering the entire range of energies from the present low-energy machines for nucleonic physics up to the phase-transition studies of RHIC. The advent of the new facility at GSI (Darmstadt) - and possibly of an upgraded Bevalac - will provide intermediate-energy beams of much higher intensity than those used to achieve the collective flow data of Fig. 3. Such intermediate-energy machines not only fill a gap and act as a base camp for subsequent further forays into the plane of the phase diagram (Fig. 10) but they have an importance of their own. At such energies the collision dynamics reflect the energy needed to achieve compression, that is, to the equation of state of nuclear matter, even though the NN collisions (which dominate at higher energies) play a significant role. The theoretical analyses permit treatment of mean field effects and NN collisions on an equal footing. Further, it is found that the observed hydrodynamical flow of nuclear matter is very sensitive to the equation of state employed in the theoretical analyses.

Clearly in heavy-ion physics great opportunities for new discovery lie just ahead.

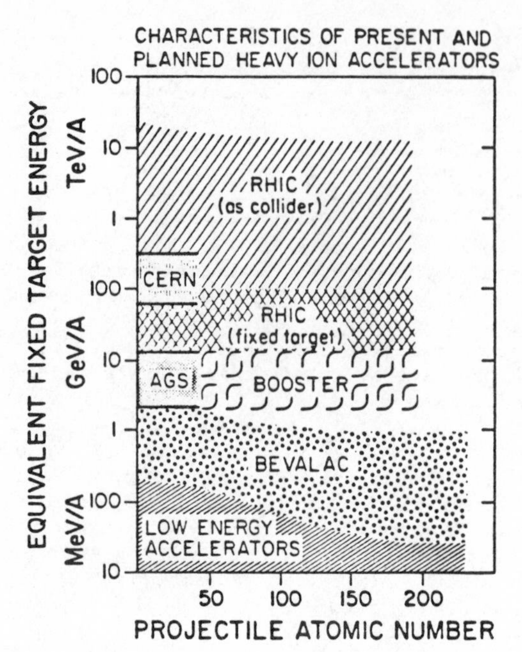

Fig. 11. Characteristics of present and planned heavy-ion accelerators.

4.3 The kaon factories

The widest range of new strong interaction physics studies will belong to the kaon factories. The new cw electron machines provide a clinical tool for proceeding beyond the present nucleonic picture of the nucleus, especially in few-nucleon systems. The relativistic heavy-ion collider (RHIC) is an imaginative gamble about very specific new physics to be found at extreme temperatures and densities. The electron and relativistic heavy-ion facilities will be complemented by the kaon factory (or factories) which will provide a whole arsenal of new tools, each important for the new interests of nuclear physics.

Plans for kaon factories abound worldwide and some proposals are at an advanced stage. In each case one seeks to provide a primary beam of very intense protons (cw ~100 μA) at energies of 30 GeV or more. The primary beam, in turn, produces intense secondary beams of kaons, antinucleons, other hadrons (π, Λ, etc.) and neutrinos.

To achieve such a facility one needs a pulsed injector adapted from those of the present meson factories (500 to 800 MeV) followed by synchrotrons boosting the energy in two steps. The complex of injector plus multiple synchrotron rings will look something like the complexes of stone circles which occurred all over the northern hemisphere three thousand years ago and for which Britain provided the most outstanding examples.

A detailed account of the accelerators of the various planned kaon factories and of their physics can be found elsewhere (Los Alamos, TRIUMF, unpublished; Vogt, 1985). To indicate the range of their physics I shall discuss only several of the important experiments of the kaon factories.

The kaons themselves are such a wonderful laboratory for physics. For example, the many rare-decay modes of charged kaons include some flavour-changing decays (e.g., $K^+ \to \pi^+\nu\nu$) which, the standard model tells us, would be brought within reach by the intense kaon beams of a kaon factory. Although in the past the study of kaon decays has been viewed as part of particle physics it is an excellent example of the new nuclear physics - strong interaction physics - in which the kaon serves as a laboratory for the study of fundamental interactions and symmetries. It is a logical extension of our present interest (Kündig, these Proceedings) in Harrogate in tritium decay for the study of neutrino masses. The kaon is not an isolated quark but a bound quark-antiquark pair.

The violation of CP invariance (and hence, possibly time-reversal non-variance) in the decay of neutral kaons will be a major research area of the kaon factories. At present several very high priority experiments at Fermilab and at CERN are looking at the ratio of $K_L^0 \to \pi^0\pi^0$ to $K_L \to \pi^+\pi^-$ to determine accurately the parameters of neutral kaon decay. Such experiments could help to distinguish among the several current models which describe this noninvariance. With their higher intensity and with dedicated neutral kaon beam lines the kaon factories can improve significantly on the present generation of neutral kaon experiments. CP noninvariance remains one of the most puzzling features of the standard model.

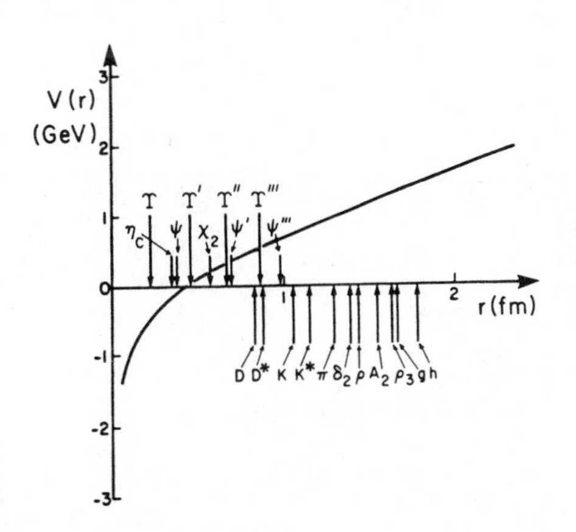

Fig. 12. The rms radii (arrows) of various selected mesons resulting from the nonrelativistic QCD quark model calculation of Godfrey and Isgur (1985).

For meson and baryon spectroscopy the kaon factories will help greatly. Fig. 12 shows the radii of various mesons predicted by the nonrelativistic quark model of Godfrey and Isgur (1985). The range of radii spans the regime of importance for the study of quarks confinement. Belief in the specific model of Godfrey and Isgur is not necessary to firmly associate quark masses with confinement issues: any model of mesons and baryons is bound to produce systems of the size indicated on Fig. 12.

For hypernuclear physics the kaon factories' hundredfold increase in intensity over the present facilities of Brookhaven and KEK would bring this science into full flourish, addressing questions such as the quark structure of kaons, the structure of Λ and Σ hypernuclei, the properties of double hypernuclei, etc.

The use of hadrons from kaon factories as nuclear probes will be addressed by Garvey (these Proceedings) at this conference. Of special interest is the Drell-Yan process for looking at sea-quarks.

Among the plans for kaon factories, the European community (the EHF project) has achieved an accelerator design and is preparing a proposal. In Japan there is great interest in kaon factory physics and the present interest focusses on an upgrade which might be viewed as the first stage of a future kaon factory, with kaon intensities of the first stage only modestly above those available at present. At Los Alamos a full proposal is ready but has not yet been submitted. In Canada (TRIUMF) a kaon factory proposal has been submitted to the government, has received strong backing from a technical and scientific panel which reviewed it for the government and active pursuit of funding is under way.

5. Conclusions

This conference at Harrogate should signal the emergence of a vigorous new period for nuclear physics. It has greatly expanded new interests covering the whole domain of strong interaction physics. There are heroic new issues to be addressed as a result of the advent of the standard model. Important new accelerator facilities are emerging and careful attention will need to be given to the provision of appropriate new detectors. Here we stand to gain from the recent technical advances in detectors for particle physics. Clearly this conference has been carefully organized to inform us about the exciting paths which will lead from Harrogate.

Acknowledgements

I am grateful to many colleagues who provided me with unpublished material but especially to Otto Häusser who provided much of the material for Sec. 3.4.

References

Alford W P et al. 1986 Phys. Rev. Lett. (in press)
Almqvist E et al. 1960 Phys. Rev. Lett. **4** 515
Altman A et al. 1985 Phys. Rev. Lett. **55** 1273
Arnold R G et al. unpublished NPAS data
Barschall H et al. 1952 Phys. Rev. **86** 431; 1953 Phys. Rev. **87** 775
Bates Laboratory 1986 unpublished report
Baym G Plenary paper proceedings of this Harrogate conference
Betts R R Plenary paper proceedings of this Harrogate conference
Bloom S D and Fuller G M 1985 Nucl. Phys. **A440** 511
Cooper M D et al. 1984 Phys. Rev. Lett. **52** 1100
Franey M A and Love W G 1955 Phys. Rev. C **31** 448
Garvey G Contributed paper proceedings of this Harrogate conference
Godfrey S and Isgur N 1985 Phys. Rev. D **32** 187
Goodman C G 1984 Spin Excitations in Nuclei (New York:Plenum) p 143
Gustafsson H A et al. 1984 Phys. Rev. Lett. **52** 1590
Häusser O 1986 Proc. Int. Symp. on Electromagnetic and Weak Interactions, Heidelberg
Isgur N Plenary paper proceedings of this Harrogate conference
Juster F D et al. 1985 Phys. Rev. Lett. **55** 2261
Karapiperis T and Kobayashi M 1985 Phys. Rev. Lett. **54** 1230
Kündig W Plenary paper proceedings of this Harrogate conference
Los Alamos report, The Physics and a Plan for a 45 GeV Facility that Extends the High-Intensity Capability in Nuclear and Particle Physics (unpublished)
Miller G A 1984 Phys. Rev. Lett. **53** 2008
Mottelson B R Keynote paper proceedings of this Harrogate conference
Navon I et al. 1984 Phys. Rev. Lett. **52** 105
Rapaport J et al. 1984 Nucl. Phys. **A427** 332
Schramm D N Special invited paper proceedings of this Harrogate conference
Sick I Plenary paper proceedings of this Harrogate conference
TRIUMF Kaon Factory Proposal (unpublished)
Vogt E W 1985 Proc. Brookhaven Hypernuclei Conf.

Inst. Phys. Conf. Ser. No. 86
Paper presented at Int. Nucl. Phys. Conf., Harrogate, UK, 1986

Review of proton scattering and reactions

J. M. Moss

Los Alamos National Laboratory, Los Alamos, New Mexico 87545, USA

1. Introduction

In the past few years proton-induced reactions (elastic scattering, inelastic scattering and charge-exchange reactions) have given us a remarkable amount of new physics. The discovery of the Gamow-Teller (GT) and related spin excitations, and the demonstration of the "need" for relativity in nuclear physics are prime examples. Looking back at the evolution of proton-nucleus physics since the International Nuclear Physics Conferences at Berkeley (1980)[1] and Florence(1983)[2] one can see two major trends:

1) Discrete Final States (including giant resonances) → Continuum States

A few years ago experiments focused on various elementary nuclear excitations, e.g., giant resonances and high-spin states. Recently, considerably more attention has been paid to the continuum region from the threshold of nucleon emission to well above that for pion production.

2) Cross Sections → Polarization Observables

At the major proton spectrometer facilities at LAMPF, TRIUMF, IUCF, and RCNP (Osaka) focal-plane polarimeters[3] have made measurements of polarization transfer (PT) observables nearly routine. Neutron polarimeters are similarly in common use at IUCF and RCNP and will be employed for much more extensive polarization measurements in the near future at LAMPF.

On the theoretical side, particularly significant work has been done in the following areas:

1) The Dirac equation and its relation to nonrelativistic models.[4] The data most relevant to this area, at present, is elastic polarization.

2) The connections between the operators appearing in hadronic reactions and those encountered in electroweak experiments.[5,6]

3) The description of continuum inelastic scattering and charge exchange reactions.[7-9]

The first of these will be dealt with in another plenary talk and will not be discussed here, except in its relation to existing or planned investigations of inelastic scattering and charge exchange reactions.

I will begin the main body of this review with a brief summary of simple proton-nucleus theory. Following this, in Section 3, I will display my prejudice for continuum experiments and discuss, in detail, four such measurements. Finally, in Section 4, I will mention some possible future developments.

2. Simple Theory and Terminology

In the most general distorted-waves formulation the expressions for the various polarization transfer observables are very complex. Fortunately in recent years a great deal of attention has been focused on deriving approximate expressions where the physics content of the spin observables is much more transparent.[10-13] This has been done in the context of both the Schroedinger and Dirac equations. A notable by-product of these analyses has been a much more obvious connection between the effective one-body operators of nucleon scattering and those appearing in electromagnetic and semi-leptonic weak interaction transitions.

In this talk I will use the simplest of the approximate forms. In this model the transition amplitude for N-nucleus scattering is

$$M_{J\mu} = \langle J\mu | M(q) e^{-i\vec{q}\cdot\vec{r}} | 0 \rangle \tag{1a}$$

where $M(q)$ is the NN scattering amplitude, and μ is the projection of the total angular momentum transfer J along the q axis. Specifically,

$$M(q) = A + B\sigma_{1n}\sigma_{2n} + C(\sigma_{1n} + \sigma_{2n}) + E\sigma_{1q}\sigma_{2q} + F\sigma_{1p}\sigma_{2p} \tag{1b}$$

with $\hat{n} = \vec{k} \times \vec{k}'$, $\hat{q} = \vec{k}' - \vec{k}$, and $\hat{p} = \hat{q} \times \hat{n}$; $\vec{k}(\vec{k}')$ is the incident (outgoing) nucleon momentum. Equation (1a) is the plane-wave Born approximation if one calculates cross sections. For spin observables, however, it implies only local plane waves since an assumed nuclear attenuation factor (spin-independent) would factor out; thus Eq. (1a) is more accurately described as the eikonal form. Using standard methods to evaluate the spin observables one finds for

(a) unnatural parity states

$$\sigma_o D_{nn} = \Sigma_T^2(C^2 + B^2 + F^2) - \Sigma_L^2 E^2 \quad , \tag{2a}$$

$$\sigma_o D_{qq} = \Sigma_T^2(C^2 - B^2 - F^2) + \Sigma_L^2 E^2 \quad , \tag{2b}$$

$$\sigma_o D_{pp} = \Sigma_T^2(C^2 - B^2 + F^2) - \Sigma_L^2 E^2 \quad , \tag{2c}$$

$$\sigma_o = \Sigma_T^2(C^2 + B^2 + F^2) + \Sigma_L^2 E^2 \quad , \tag{2d}$$

and for

(b) natural parity states

$$\sigma_o D_{nn} = \frac{1}{2} \Sigma_T^2 (C^2 + B^2 - F^2) + \rho^2 (A^2 + C^2) \quad , \tag{3a}$$

$$\sigma_o D_{qq} = \frac{1}{2} \Sigma_T^2 (C^2 - B^2 - F^2) + \rho^2 (A^2 - C^2) \quad , \tag{3b}$$

$$\sigma_o D_{pp} = \frac{1}{2} \Sigma_T^2 (C^2 - B^2 + F^2) + \rho^2 (A^2 - C^2) \quad , \tag{3c}$$

$$\sigma_o = \frac{1}{2} \Sigma_T^2 (C^2 + B^2 + F^2) + \rho^2 (A^2 + C^2) \quad . \tag{3d}$$

In the approximations we have used there are no spin-current couplings of the two nucleons, with the result that there are only three nuclear matrix elements. They are

$$\text{spin transverse, } \Sigma_T = \langle \mu | \vec{\sigma} x \vec{q} e^{-i\vec{q}\cdot\vec{r}} | 0 \rangle \quad , \tag{4a}$$

$$\text{spin longitudinal, } \Sigma_L = \langle \mu | \vec{\sigma}\cdot\vec{q} e^{-i\vec{q}\cdot\vec{r}} | 0 \rangle \quad , \tag{4b}$$

$$\text{and scalar, } \Sigma = \langle \mu | e^{i\vec{q}\cdot\vec{r}} | 0 \rangle \quad . \tag{4c}$$

In spite of the simple assumptions made, these equations provide a reasonably quantitative view of much of the existing polarization transfer data. An example is given in Fig. 1.

The relation between the PT observables of Eqs. (2) and (3) in the $\vec{n}$, $\vec{p}$, $\vec{q}$ (center-of-mass) and the usual laboratory frame is straightforward.[10] In the lab.frame the quantities L, N($N=\vec{n}$), and S refer to the longitudinal (beam direction), normal, and sideways ($\hat{N} \times \hat{L}$) directions. P(θ) is the polarization function and A is the analyzing power of the reaction. D_{NN}, D_{SS}, D_{SL}, D_{LS}, and D_{LL} are identical to the Woltenstein parameters D, R, R′, A, and A′.

For small angle scattering one has the approximate relations, $\hat{q} \to \hat{S}$, $\hat{p} \to \hat{L}$ which allow use of Eqs. (2) and (3); of course $D_{NN} \equiv D_{nn}$.

3. Continuum Experiments

3.1 Missing GT Strength in the (p,n) Continuum

The GT resonance observed in the (p,n) reaction exhausts only about 60% of the GT sum rule. Although this fact has been known experimentally for some years, its theoretical understanding is far from being settled. If the

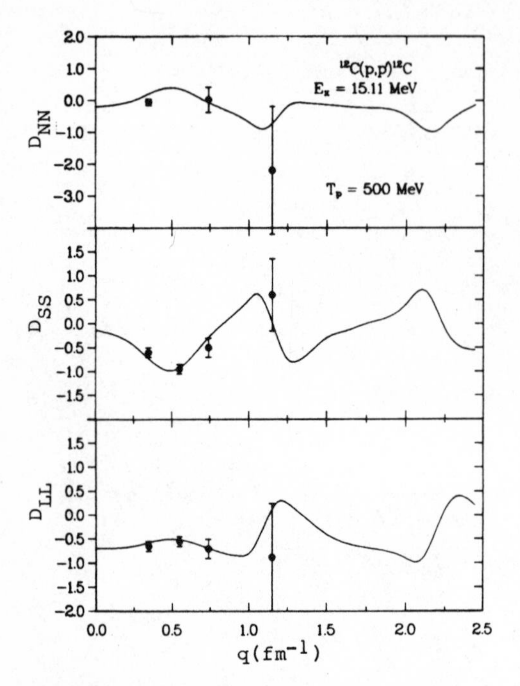

Fig. 1. Comparison of the simple equations of section 2 to polarization transfer data at E_p = 500 MeV.

quenching is due to admixtures of delta-isobar-hole configurations,[14] then the missing GT strength may lie at very high excitation energy and be, consequently, nearly unobservable experimentally. An opposing school, however, identifies the predominant mechanism as mixing of the 1p-1h GT states with a background of 2p-2h excitations.[15,16] In this view, the missing strength is likely to be found in the structureless continuum just above the GT resonance. The newly developed technique of polarization transfer in the (p,n) reaction at intermediate energies has considerable potential in addressing this problem.

3.1.1 Neutron Polarimetry

Neutron polarimeters employing liquid helium analyzers have been in common use for energies below 50 MeV for many years. For neutrons in the E > 100 MeV range, however, a more practical solution to the analyzer problem is to use the $^1H(n,n)^1H$ reaction with either solid or liquid scintillator providing the hydrogen.[17] Figure 2 shows a schematic representation of a polarimeter based on liquid scintillator being constructed at Los Alamos. Proton recoil tracks in two detector planes with no intermediate charged track define an $^1H(n,n)^1H$ event. Position sensitivity in one plane (the horizontal in Fig. 2) is achieved by timing between the ends of the rectangular cell. In the vertical direction position and consequently scattering angle are defined by segmentation. In this manner polarization

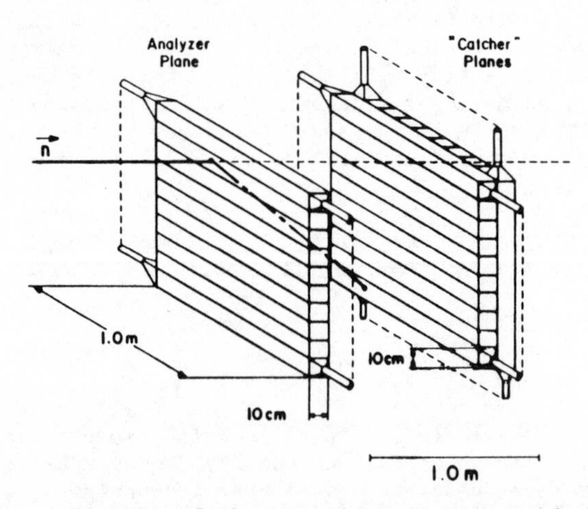

Fig. 2. Schematic of the Los Alamos neutron polarimeter.

analysis may be simultaneously achieved for planes perpendicular to the neutron's momentum. Double-scattering efficiencies in the range of 10^{-3} with effective analyzing powers of ~0.3 will be attained with the new Los Alamos polarimeter.[18]

3.1.2 Measurements of D_{NN} at IUCF

During the past two years there have been numerous measurements of D_{NN} mostly at 0^{0} for the (p,n) reaction[19,20] in the 100 to 200 MeV range at the Indiana University Cyclotron Facility (IUCF). For GT transitions at 0°, dominance of $\ell = 0$ transfer, $[\Sigma_L \alpha \Sigma_T$ in Eqs. (2)-(4)] leads to a simplification of Eq. (2a)

$$D_{NN} = \frac{-1}{1 + 2B^2/F^2} \quad .$$

The further assumption of a central interaction (no spin orbit or tensor terms) gives B = F and hence $D_{NN} = -1/3$. As is seen in Fig. 3 the experimental data correspond to this simplified case very well. Angular distributions when available are likewise in good agreement with the results of Eq. (2a). More complex distorted-wave Born approximation (DWBA) calculations verify the general validity of the simple model of Section 2.

Going beyond pure GT transitions the model of Section 2 predicts the general results for an interaction which is purely central

$$D_{NN}(0^{o}) = \begin{cases} 1 & \Delta S = 0 \quad \text{natural parity} \\ 0 & \Delta J = \Delta L,\ \Delta S = 1 \quad \text{natural parity} \\ < -1/3 & \Delta J = \Delta L \pm 1 \quad \text{unnatural parity} \end{cases}$$

These values can now be used to examine the distribution of spin-flip strength in the continuum. Results of the first such study[19] are shown in Figs. 4. Going from right to left in Fig. 4 one sees the isobaric

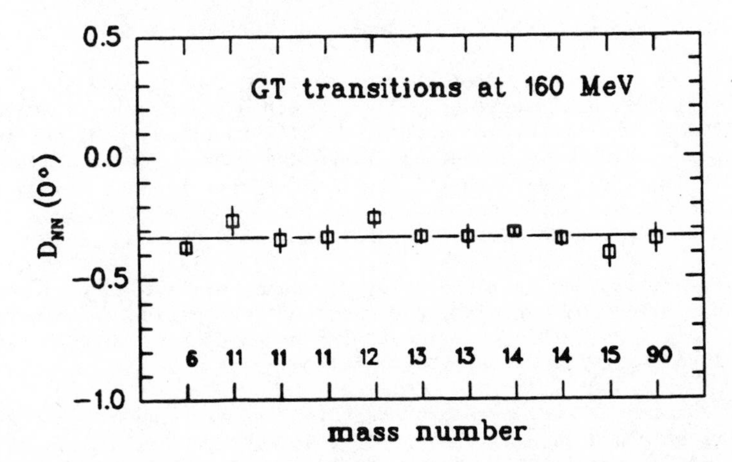

Fig. 3. Measurements of D_{NN} (0°) for the (p,n) reaction populating GT states.

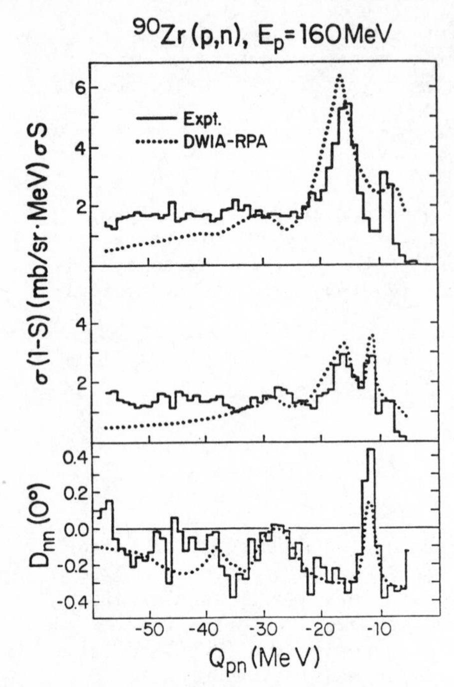

Fig. 4. Comparison of the RPA calculations of Ref. 9 with experimental data for D_{NN} (0°).

analogue state (IAS) with $D_{NN} = 1$, the GT resonance with $D_{NN} = -1/3$ and the region of spin-flip dipole strength near $E_x = 20$ MeV. The latter is dominated by $\Delta J = 1$, $\Delta S = 1$, and hence, $D_{NN} = 0$.

The dashed line of Fig. 4 is a distorted-wave impulse approximation calculation using a random-phase approximation response function.[21] The IAS, GT and spin dipole resonances described above are well accounted for by the RPA. What is most significant, however, is that the RPA contains all of the strength allowed by the GT sum rule, i.e., there is no missing strength in the theory, and theory agrees with the experiment! The same calculations, when compared to the spin-flip cross section (Fig. 4a) does indicate an overestimate of the size of the GT resonances. However, a slight redistribution obtained by arbitrarily taking ~20% of the GT strength from the $E_x < 20$ MeV region and spreading it uniformly into the region $E_x = 22$–40 MeV results in much better overall agreement with the cross section without "spoiling the previous agreement between theoretical and observed polarization transfer coefficients."[21]

Although it cannot be argued at this point that the continuum D_{NN} measurements have definitely solved the missing GT strength issue, it is clear that these experiments and calculations have brought strong new evidences to bear on the problem.

3.2 Spin-flip Strength in the (p,p′) Reactions at Small Momentum Transfer

Several years ago, somewhat after the discovery of the GT resonance, a series of beautiful small angle (p,p′) experiments at Orsay[22] revealed systematically a giant resonance with properties favoring an M1 assignment. Over the years considerable debate has raged over the apparent absence of this strength in (e,e′) reactions.[23] It is now agreed by most that a combination of several very different aspects of these two reactions can at least qualitatively account for these differences.[23,24]

An important verification that the Orsay resonance was actually spin-flip strength was provided by a LAMPF measurement of the transverse spin-flip probability (Fig. 5).[25] This experiment also revealed considerable additional continuum spin-flip strength at small q.

In spite of the recent successes many questions remain unanswered regarding the spin-flip strength seen in the (p,p′) reaction. Unlike the (p,n) reaction D_{NN} itself in the (p,p′) continuum does not give much information. The cross section is always dominated by the scalar-isoscalar cross

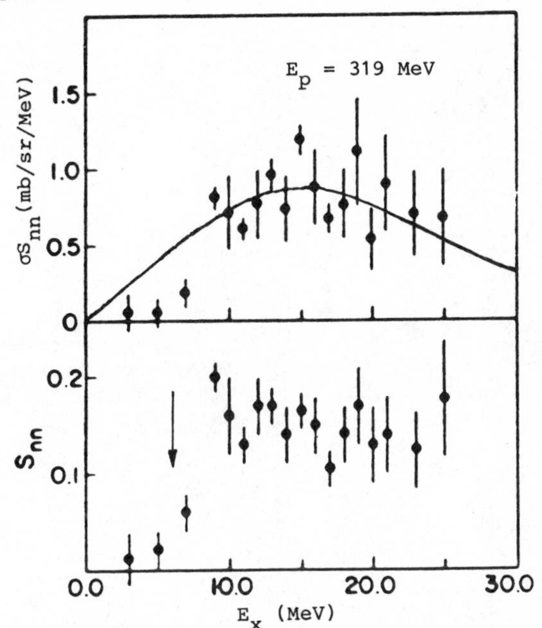

Fig. 5. Spin-flip probability and cross section for the $^{90}Zr(p,p')$ reaction from Ref. 25. The curve at the top is a calculation from Ref. 7. The arrow at the bottom indicates the beginning of the M1 resonance peak seen in the cross section data.

strength (for which $S_{NN} = 0$) so the total spin-flip probability is always small. More definitive information on multipolarity could be obtained by angular distribution measurements. For the M1 component this has been limited by the inability to make measurements of D_{NN} at 0° where the cross section peaks.

Figure 6 from a recent LAMPF experiment[26] shows that this situation is well on its way to being eliminated. In the near future one may anticipate sensitive searches for M1 strength in nuclei up to perhaps the ^{90}Zr region.

Measurements of the spin-flip cross section at larger q will also be invaluable in unravelling the physics of the continuum. Figure 7 shows recent results from a LAMPF experiment. Although very preliminary in terms of comparison with theory,[8] Fig. 8 suggests that new, higher multipolarity giant resonances may await discovery via such measurements.

3.3 Search for Excess Pions -- Connection to the EMC Effect

The search for collectivity associated with the nuclear pion field has had a long and somewhat checkered history.[28] It is now very clear that nuclei in their ground states are not in grave danger of being tickled across the threshold to the hypothetical pion condensate. Precursors, signaling the proximity to such a critical point, were sought several years with only negative results.[29] Nevertheless, the possibility of some weaker form of pionic collectivity could not be ruled out based on experimental evidence available before 1984.

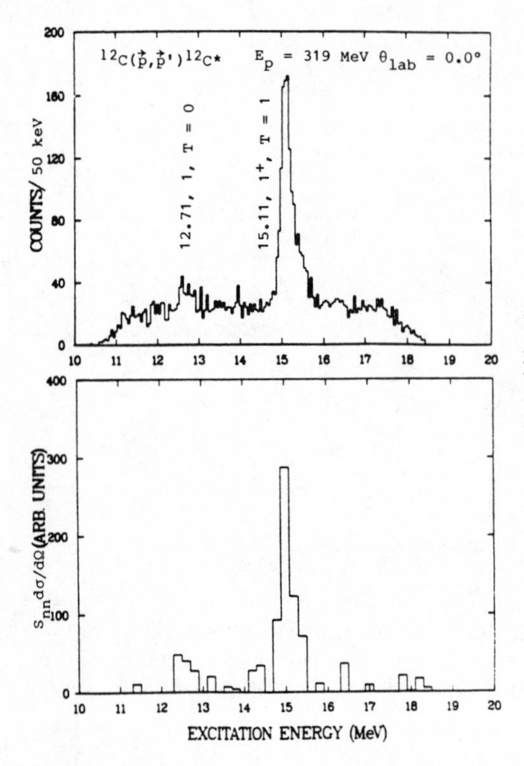

Fig. 6. Zero-degree cross section and spin-flip cross section for the $^{12}C(p,p')$ reaction taken at LAMPF (Ref. 26).

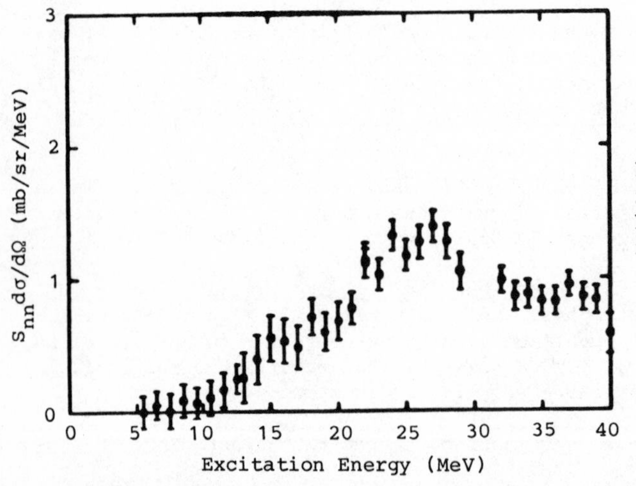

Fig. 7. $^{90}Zr(p,p)$ spin flip cross section at E_p = 319 MeV, $\theta_{lab} = 7^{o}$ from Ref. 27.

The notion of a small collective enhancement in the pion field was revived by Llewellyn-Smith[30] and Ericson and Thomas[31] about three years ago in connection with the EMC (European Muon Collaboration) effect.[32] The basis for this idea is illustrated in Fig. 9. High-energy deep-inelastic lepton scattering may occur from either valence quarks (Fig. 9a) or from the quark-antiquark sea (Fig. 9b). One component of the sea is due to the nuclear pion-field, which if enhanced, could certainly provide excess

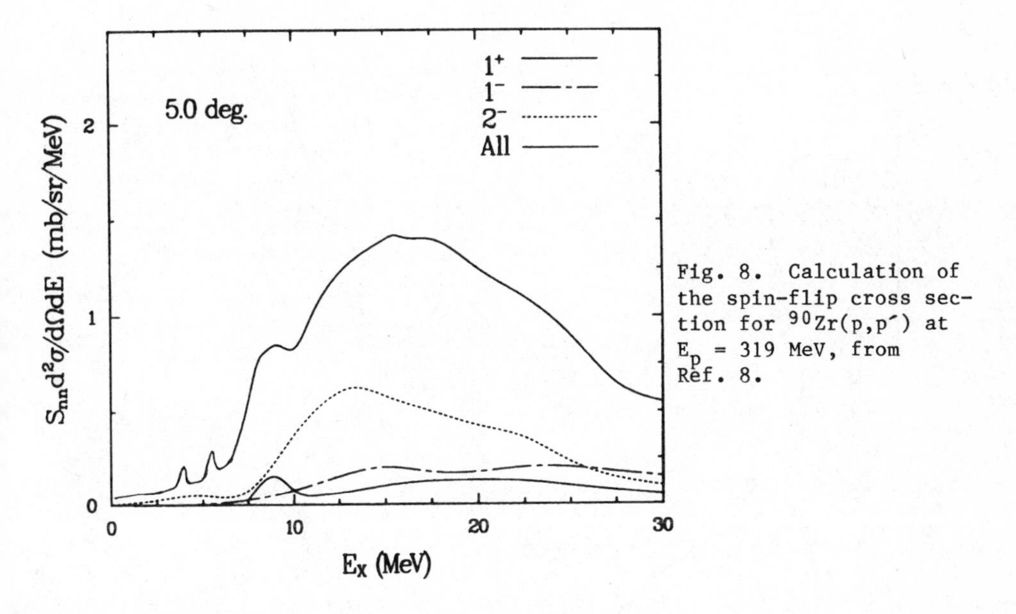

Fig. 8. Calculation of the spin-flip cross section for $^{90}Zr(p,p')$ at E_p = 319 MeV, from Ref. 8.

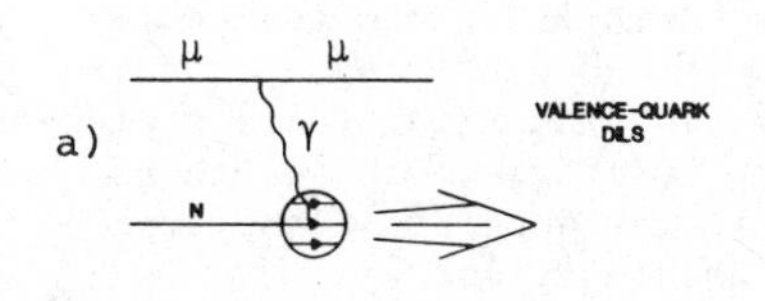

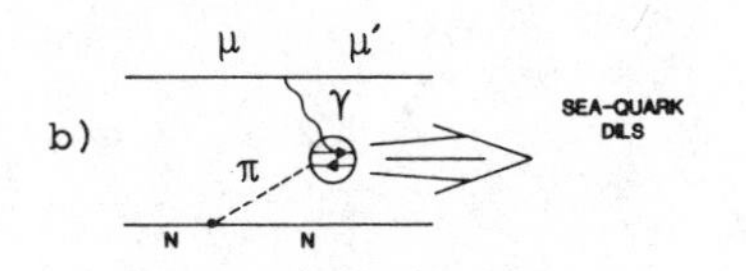

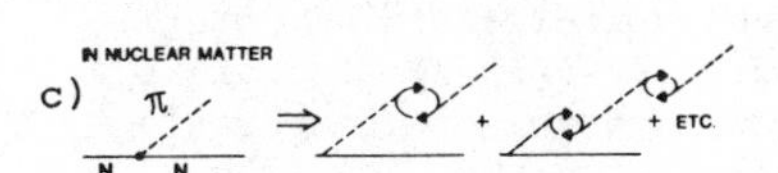

Fig. 9. Diagrams showing contributions of valence and sea-quarks to deep inelastic lepton scattering. The hypothetical many-body enhancement of the pion field is shown in c.

quarks and antiquark which might serve to "explain" the small x region of the EMC effect (Fig. 10). The original EMC effect implied a pion excess of something like 15%.

3.3.1 Inclusive Quasi-Free Scattering and the Spin-Isospin Response Function at Large Momentum Transfer

The physics associated with hypothetical pionic collectivity in nuclei is expected to be manifest in the domain of much larger momentum transfer than was dealt with in Sections 3.1 and 3.2. This is illustrated schematically

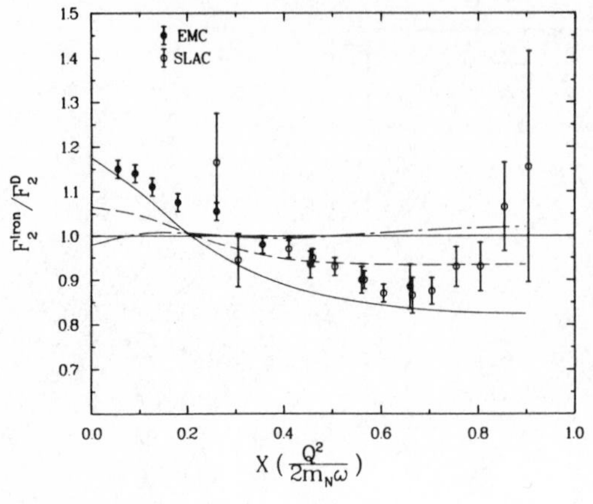

Fig. 10. The EMC effect as given by the data of Refs. 32. The calculations are based on the pionic enhancement model. The three curves correspond to values of g′= 0.55 (solid), 0.7 (dash), and 0.9 (long dash-short dash).

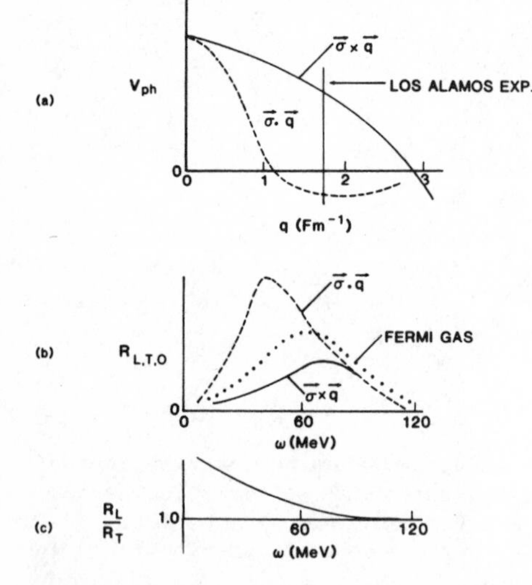

Fig. 11. The particle-hole forces in the spin-isospin channels (a). The corresponding response functions (b), including the noninteracting Fermi gas. The ratio of spin-longitudinal to transverse response functions (c).

in Fig. 11 where the transverse and longitudinal spin-isospin NN interactions are shown (11a) along with the corresponding response functions (11b). The attractive (hypothetical) pionic interaction (longitudinal) near $q \cong 2\ fm^{-1}$ softens the corresponding response function, pushing strength to smaller excitation energy with respect to a non-interacting Fermi gas. In contrast the transverse response, characterized by a repulsive NN interaction is hardened. The ratio of the two responses near $q = 2\ fm^{-1}$ would have the ω dependence shown in Fig. 11c.

The experiment done by the Los Alamos group[33] to investigate possible evidence for this form of pionic collectivity differs significantly from previous "precursor" experiments. First, as in the EMC experiment, the point of reference is deuterium. One compares the spin-dependent response functions for a heavy target (Pb or Ca) and ^2H using identical experimental techniques. If the predicted many-body effects are present in the heavy targets even at a very small level, they should be detectable in a precise ratio experiment.

Second, the technique of complete polarization transfer[10] is used to separate the spin-longitudinal ($\vec{\sigma}\cdot\vec{q}$) and spin-transverse ($\vec{\sigma}\times\vec{q}$) response in the continuum as a function of ω (across the entire quasielastic peak, ω = 20 to 100 MeV). The responses are measured at a momentum transfer $q = 1.75\ fm^{-1}$ which corresponds to the maximum predicted enhancement of $R_L(q,\omega)$ in most models.

The experiment consists of precise determinations of the polarization transfer coefficients D_{LL}, D_{SS}, and D_{NN} for 500 MeV protons inelastically scattered from Pb and Ca, and D at $q = 1.75\ fm^{-1}$. The experiment utilized longitudinal (L), sideways (S), and normal (N) polarized beams from LAMPF in conjunction with final polarization analysis from the focal-plane polarimeter of the high-resolution spectrometer. The quantities constructed from the above data are the longitudinal and transverse spin-flip probabilities defined by

$$S_L = \frac{1}{4}(1 - D_{NN} + (D_{SS} - D_{LL})\sec\theta_{lab}) \quad ,$$

$$S_T = \frac{1}{4}(1 - D_{NN} - (D_{SS} - D_{LL})\sec\theta_{lab}) \quad . \tag{5}$$

With the NN amplitude of Eq. (1b) and the above definitions one can form the spin longitudinal and transverse cross sections for N-nucleus scattering in terms of the analogous quantities for NN scattering

$$\begin{aligned} \sigma_o S_L &= \sigma_o^{NN} S_L^{NN} R_L(q,\omega) N_e \\ \sigma_o S_T &= \sigma_o^{NN} S_T^{NN} R_T(q,\omega) N_e \\ \sigma_o &= \sigma_o^{NN} R(q,\omega) N_e \end{aligned} \tag{6}$$

with the spin-longitudinal, transverse, and total response functions defined as

$$\begin{aligned} R_L(q,\omega) &= \left|\langle q,\omega\middle|\vec{\sigma}\cdot\vec{q}\ e^{-i\vec{q}\cdot\vec{r}}\middle|0\rangle\right|^2 , \\ R_T(q,\omega) &= \left|\langle q,\omega\middle|\vec{\sigma}\times\vec{q}\ e^{-i\vec{q}\cdot\vec{r}}\middle|0\rangle\right|^2 , \\ R(q,\omega) &= \frac{C^2 + B^2 + F^2}{I^{NN}} R_T + \frac{E^2}{I^{NN}} R_L + \frac{A^2 + C^2}{I^{NN}} R_0 \quad , \end{aligned} \tag{7}$$

where

$$R_0 = |\langle q,\omega|\ e^{-i\vec{q}\cdot\vec{r}}|0\rangle|^2 \quad .$$

N_e is the effective number of participating nucleons. The approximations implied in Eqs. (6) and (7) are well satisfied for forward-angle scattering of 500 MeV protons.[7,33]

With the assumption that scattering from deuterium represents free pp plus pn scattering, we have $S^D = S^{NN}$, and from Eqs. (7) one finds

$$S_L^{Pb}/S_L^D = R_L(q,\omega)/R(q,\omega) \quad , \tag{8}$$

$$S_T^{Pb}/S_T^D = R_T(q,\omega)/R(q,\omega) \quad , \tag{9}$$

and

$$\frac{S_L^{Pb}/S_L^D}{S_T^{Pb}/S_T^D} = R_L(q,\omega)/R_T(q,\omega) \quad . \tag{10}$$

Thus the simple ratios [Eqs. (8) and (9)] depend only on ratios of response functions for Pb or Ca. The super ratio of Eq. (10) can be used to contrast the two spin-dependent response functions of the heavy targets.

3.3.2 The Spin-Longitudinal Response Function and Excess Pions

In order to understand the connection between the longitudinal response function and the pionic enhancement model of the EMC effect we give the equations relating them. We use the version of the model given by Stump, Bertsch, and Pumlin.[34] The difference between the F_2 structure functions of iron and deuterium is

$$\delta F_2^{Fe}(x) - \delta F_2^D(x) = \int_x^1 dz\ \delta f_\pi(z)\ F_2^\pi(x/z)$$
$$+ \int_0^{1-x} dz\ \delta f_\pi(z)\ F_2^N(x/(1-z)) - \delta n_\pi\ F_2^N(x) \tag{11}$$

where

$$\delta f_\pi(z) = f_\pi^{Fe}(z) - f_\pi^{D}(z) \qquad (12)$$

is the pion excess in Fe, and

$$\delta n_\pi = \int_0^1 dz\, \delta f_\pi(z) \qquad (13)$$

is the integrated pion excess. In all calculations presented below, the simple assumption that $F_2^\pi(x) = F_2^N(x)$ has been made. The calculation of the pion excess distribution in a heavy target closely resembles that for a free nucleon given many years ago by Sullivan.[35] Specifically

$$f_\pi(z) = \frac{3g^2_{\pi NN}}{16\pi^2} z \int_0^\infty dq_\perp^2 \int_0^\infty d\omega \frac{|\Gamma(t)|^2 t}{(t + m^2_\pi)^2} R_L(q,\omega) \qquad (14)$$

with $q_z = zm_N + \omega$, $q = (q_\perp^2 + q_z^2)^{1/2}$, and $t = q^2 - \omega^2$; where $g_{\pi NN}$ and $\Gamma(t)$ are the coupling constant and form factor at the πNN vertex, and $R_L(q,\omega)$ is the spin-longitudinal response function--the latter of course being the object of measurement of the Los Alamos experiment. When $R_L(q,\omega)$ is assumed to be enhanced due to collectivity in the pion-exchange interaction, δf_π is large accordingly. Typically the maximum contribution to δf_π comes from the momentum transfer range of 300 to 400 MeV/c, the region in which pionic collectivity is expected to occur. This then is the range toward which the Los Alamos experiment is directed.

It is important to point out that the EMC data come from muon scattering which probes the entire nuclear volume. The proton scattering data are more localized to the nuclear surface. The response functions associated with the two types of experiments are, therefore, not identical. These differences can be accounted for very precisely, however, using the techniques described in Ref. 33.

3.3.3 Comparison of the Pionic Model with Experiment

Consistent calculations of R_L with comparison to both the EMC and Los Alamos Experiments are shown in Figs. 10 and 12. The details may be found in Ref. 33. In the $\pi + \rho$ exchange $+g'$ model for the nuclear response, the delta function range strength g' determines the degree of pionic collectivity. For smaller values, e.g., $g' = 0.55$ there is a large pion excess (~17%) and the ratio of R_L/R_T becomes large at small ω. Such small values of g' are clearly at variance with the Los Alamos experiment, leading to the conclusion[33] that the degree of pionic collectivity implied by the pionic model low-x EMC effect is not consistent with the Los Alamos data.

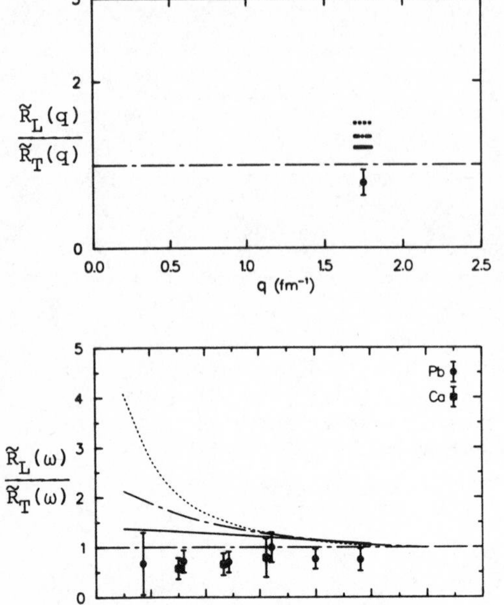

Fig. 12. The results for $\tilde{R}_L/\tilde{R}_T$ from the Los Alamos experiment (Ref. 34). The tildes indicate the appropriate isospin averaged quantities. The curves correspond to the calculations of Fig. 10.

3.3.4 New Experiments and the Low-x EMC Effect

Very recently the EMC[36] and BCDMS[37] collaboration have presented new data which cast significant doubt on the normalization of the original EMC experiment. Low-x enhancements of 5 to 10% now seem more likely.

Also in recent publications[38,39] new theoretical calculations using various versions of the pionic excess model would seem to imply rather little excess deep-inelastic scattering from heavy targets at low x. Clearly new experiments are required to bring order into this field where the ratio of experimental data to theoretical ideas is about 1/100. A very promising approach appears to be a study of the A dependence of the Drell-Yan process induced by protons. Since this falls in the general category of proton-nucleus physics, I will illustrate the potential of this type of experiment to distinguish between two models of the EMC effect which give equivalent agreement with experiment (Fig. 13). The key here is that while deep inelastic scattering is sensitive to both quarks and antiquarks, the Drell-Yan process can be made sensitive to <u>only</u> the antiquark component. Hence, the two types of experiments are complementary.

3.4 Medium Effects on Continuum Polarization Observables

Inclusive quasi-free scattering of medium-energy protons holds promise for revealing effects of the nuclear medium on the fundamental nucleon-nucleon interaction. At large momentum transfer, Pauli-blocking is of minor importance and the quasi-elastic peak well developed. Here it is possible to compare a dilute system, such as ^{2}H, with a heavy target, possessing high-density nuclear matter, and thereby accurately and directly assess the

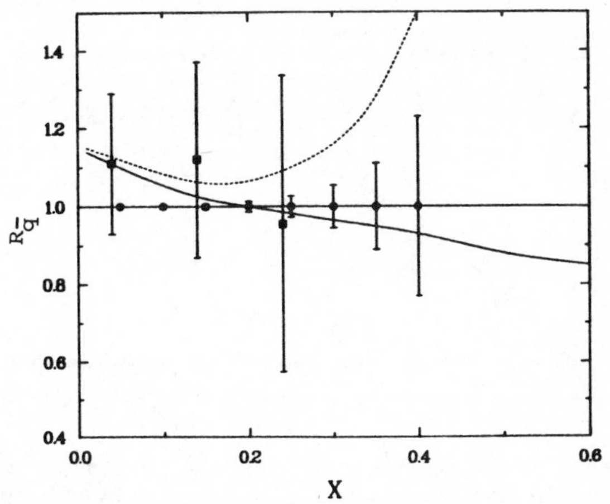

Fig. 13. Comparison of the pionic enhancement model (dashed line) and the rescaling model (solid line) with charged-current neutrino scattering data for the A-dependence of the antiquark distribution ($R = q^{Fe}(x)/q^{N}_{(x)}$) from Ref. 41. Also shown are estimates at the error bars obtainable in Fermilab experiment E772 (points along the line R=1), ref. 42

effect of the nuclear medium on the quasi-elastic scattering. Polarization transfer (PT) observables are able to provide even finer detail regarding the effect of the target medium on specific parts of the N-N interaction. This, of course, is precisely the type of experiment presented in Section 3.3 where no medium effect (pionic collectivity) was found. There is, however, a very significant effect of the nuclear medium on one particular continuum polarization observable--the analyzing power. This may be seen in Fig. 14 where the data[43] (square points) are observed to be significantly lower than the prediction of free p-p and p-n scattering (dashed line).

To our knowledge this observation is not reproduced by any multiple scattering calculation[44,45] except the one described below.

Horowitz and Iqbal (HI)[46] have developed a model of inclusive quasi-free scattering with a close connection to relativistic mean-field theory. It also includes an important optical effect, due to spin-orbit absorption, which is absent in early quasi-elastic models.[44]

A simplified picture of a quasi-free event in this model is that of a proton scattering from a nucleon in a nucleus which has an effective mass

$$\tilde{M} = M + S$$

where M is the free nucleon mass and S is the Dirac mean-field scalar potential. At central nuclear density $\tilde{M} = 0.56$ M. The effective mass is more typically 0.86 - 0.9 M for medium-energy proton nucleus scattering where the reaction is surface peaked.

The HI model accounts extremely well for the observed reduction in A_N for the data of Ref. 43 (Fig. 14). The crucial feature appears to be the $\tilde{M}$ effect.

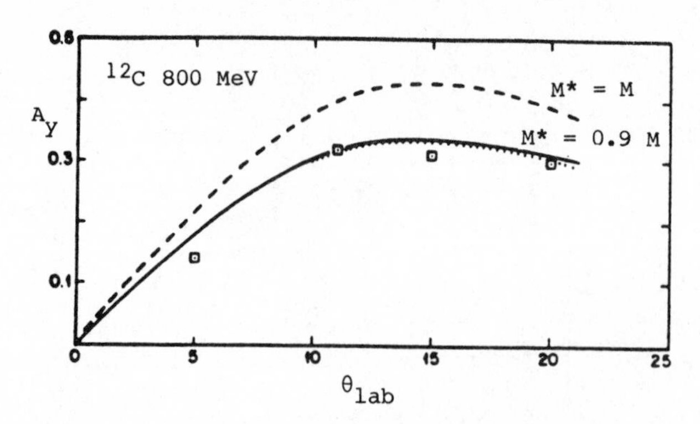

Fig. 14. Analyzing power measurements for the quasielastic peak for $^{12}C(p,p')$ at E_p = 800 MeV (Ref. 43). The calculations are from Ref. 46.

The HI model can also be compared to experiment for the other PT observables from the 500 MeV experiment described in Section 3.3. This is shown in Table 1. Here the notation HI1 refers to a calculation in which $\tilde{M} = M$, and HI2 to the complete ($\tilde{M} \neq M$, full distortion) calculation. The most significant target-mass effect observed is in the analyzing power, where the values for Ca and Pb are significantly lower than either ^{2}H or the phase shift solutions. Comparison to HI1 shows that little of this effect can be ascribed simply to spin-orbit distortion. This is confirmed by more elaborate calculations by Smith and Wallace[45] for 800 MeV proton scattering from ^{12}C. They likewise found little effect when the spin-dependence of optical distortion was included. Their model was similarly unable to account for the small analyzing power of the ^{12}C data measured by McGill et al.[43] Here, the HI2 model gives an impressive account of the reduction in A_N for Ca and Pb. At present it is the only model we are aware of which yields this effect.

The HI2 model also predicts a large medium effect for D_{LS}. In sharp contrast to A_N, the data show no significant differences between ^{2}H and the heavy targets.

Comparison of the diagonal D's, which have generally larger error bars, does not clarify the medium-effect issue. All three parameters for Ca are somewhat higher than for ^{2}H. However, the differences between ^{2}H and Pb are not significant. This is at odds with the HI2 predictions, particularly in the case of D_{SS}. Note for all three diagonal D's the N/Z ratio effect is small.

The HI model in the present formulation treats only single-step scattering. Models which explicitly evaluate double and triple scattering[44,45] contributions typically find that 5 to 10% of the cross section at the QE peak is due to processes higher than first order. This may lead to changes in polarization observables as large as 0.05 (see Ref. 45 for detailed predictions at 800 MeV). However, it is unlikely that the observed analyzing power effect or the predicted but unobserved change in D_{LS}, both of which are in the range >0.1, can be understood in terms of high order processes.

One may summarize by the following observations:

1) There is a large reduction in the analyzing power for the heavy targets. This effect is not due simply to spin-dependent distortion. The Dirac equation model of Horowitz and Iqbal,[46] however, can reproduce the magnitude of the effect.

2) There is no medium effect observed for D_{LS}. Here, in contrast to 1), the large medium effect predicted by the Horowitz-Iqbal model is in significant disagreement with the data.

3) The data for D_{NN}, D_{SS}, and D_{LL} show no strong medium effect, but at the present level of statistical precision, do not provide much additional model selectively.

It is clear that more extensive data of this type are needed if the effect of the medium on the NN interaction is to be understood. Inclusive QE scattering has many features which complement elastic and inelastic scattering studies which aim at probing the implications of relativistic models of nuclear structure and reactions.

4. Future Studies

I have purposely limited this review to investigations of continuum reactions since this area has seen extensive and exciting new results in recent years. In the near future it is easy to predict that the charge exchange programs being undertaken at TRIUMF and LAMPF will produce exciting results in the areas of (n,p) cross sections and PT observables in the $(\vec{p},\vec{n})$ reaction. Extremely challenging zero-degree measurements in (p,p´) reactions are also just around the corner. In both inelastic scattering and charge exchange reactions, difficult but feasible studies of polarization observables such as P-A and $D_{LS} + D_{SL}$ will probe the nonstatic parts of the NN interaction.[5] Here, the contrast between relativistic and nonrelativistic models is likely to be great.

Finally, high-energy p-nucleus physics in the form of high-mass dimuon production (the Drell-Yan process) will probe the quark structure and hopefully enlighten us on the effect of the nuclear medium on quark confinement.

Table 1

Comparison of ^{2}H, Ca, and Pb polarization observables at the quasielastic peak with the average free pp and pn phase shift solutions and with calculations described in the text. The Pb data have been adjusted to an equivalent N=Z nucleus. To recover the original data one divides the Pb values given by the last column of the table.

	^{2}H	Ca	Pb	Phase shift	HI1	HI2	Pb corr. factor
A_N	0.400±0.007	0.296±0.012	0.287±0.007	0.43	0.39	0.26	1.042
D_{LS}	0.384±0.012	0.367±0.018	0.406±0.012	0.41	0.45	0.47	0.869
D_{NN}	0.773±0.024	0.804±0.027	0.720±0.027	0.80	0.81	0.81	0.975
D_{SS}	0.476±0.016	0.517±0.026	0.465±0.016	0.46	0.43	0.58	0.996
D_{LL}	0.417±0.023	0.486±0.030	0.400±0.022	0.41	0.36	0.47	1.011

Acknowledgements

I would like to acknowledge many useful discussions about the theoretical issues discussed here with W. Alberico, E. L. Berger, G. F. Bertsch, G. E. Brown, F. Coester, H. Esbensen, C. Horowitz, M. B. Johnson, A. Molinari, E. Oset, E. Rost, J. R. Shepard, J. Speth, A. W. Thomas, H. Toki, and R. B. Wiringa. I also want to acknowledge my experimental colleagues T. A. Carey, K. W. Jones, J. B. McClelland, L. B. Rees, N. Tanaka, and A. D. Bacher without whom the experimental work in which I was involved would not have been accomplished.

References

1. Proc. of the International Nuclear Physics Conf., Berkeley, CA, 1980, Eds. R. M. Diamond and J. O. Rasmussen, (North Holland Publishing, Amsterdam).

2. Proc. of the International Nuclear Physics Conf., Florence, Italy, 1983, Eds. P. Blasi and R. A. Ricci, (Tipografia Compositori - Bologna).

3. Proc. of the 6th Int. Conf. on Polarization Phenomena in Nuclear Physics, Osaka, Japan, 1985, Ed. M. Kondo, et al., (Physical Society of Japan) Tokyo.

4. Proc. at the LAMPF Workshop on Dirac Approaches to Nuclear Physics, Eds. J. R. Shepard, C. Y. Cheung, and R. L. Boudrie, (Los Alamos Report LA-10438-C).

5. J. R. Shepard, Ref. 4, p. 300; E. Rost, p. 182.

6. W. G. Love and A. Klein, Ref. 4, p. 220.

7. H. Esbensen and G. F. Bertsch, Ann. Phys. 157, 255 (1984).

8. F. Osterfeld, D. Cha, and J. Speth, Phys. Rev. C 31, 372 (1985); F. Osterfeld et al., preprint.

9. A. Klein and W. G. Love, Phys. Rev. C 33, 1920 (1986).

10. J. M. Moss, "International Conference on Spin Excitations in Nuclei, eds. F. Petrovich et al.,(Plenum Press, New York, 1984) p355 J. M. Moss, Phys. Rev. C26:2063 (1982).

11. E. Bleszynski et al., Phys. Rev. C26:2063 (1982).

12. J. R. Shepard in Proc. of the International Conf. Antinucleon on Nucleon-Nucleus Interactions, Telluride, CO, 1985, ed. G. Walker et al., (Plenum Press, New York), p. 97.

13. J. A. McNeil, Ref. 4, p. 160.

14. M. Rho in Proc. of the Int. Conf. on Spin Excitations in Nuclei, eds. F. Petrovich et al., (Plenum Press, New York, 1984) p. 111.

15. A. Arima, Ref. 14, p. 7.

16. G. F. Bertsch and I. Hammamoto, Phys. Rev. C 26, 1326 (1982).

17. T. Taddeucci et al., Phys. Rev. Lett. 52, 1960 (84).

18. LAMPF proposal E881, T. A. Carey, C. D. Goodman, and J. B. McClelland, Spokesmen.

19. T. N. Taddeucci et al., Phys. Rev. C 33, 746 (1986).

20. J. W. Watson et al., Ref. 3, p. 940.

21. W. G. Love and A. Klein, Ref. 3, p. 78.

22. G. M. Crawley et al., Phys. Rev. C 26, 87 (1982).

23. A. Richter, Ref. 2, p. 189.

24. C. Glashausser, Ref. 3, p. 293.

25. S. Nanda et al., Phys. Rev. Lett. 51, 1526 (1983).

26. S. Nanda and J. B. McClelland, private communication.

27. S. Nanda and C. Glashausser, private communication.

28. J. Meyer-ter-Vehn, Phys. Reports 74, 281 (1981).

29. J. M. Moss, Nucl. Phys. A374, 229c (1982).

30. C. H. Llewellyn-Smith, Phys. Lett. 128B, 107 (1983).

31. M. Ericson and A. W. Thomas, Phys. Lett. 128B, 112 (1983).

32. J. J. Aubert et al., Phys. Lett. 123B, 275 (1983); A. Bodek et al., Phys. Rev. Lett. 50, 1431 (1983).

33. T. A. Carey et al., Phys. Rev. Lett. 53, 144 (1984); L. B. Rees et al., Phys. Rev. C (in Press).

34. D. Stump, G. F. Bertsch, and J. Pumplin in "Hadron Substructure in Nuclear Physics," (American Institute of Physics, New York, 1984) p. 339.

35. J. D. Sullivan, Phys. Rev. D5, 1732 (1972).

36. G. Taylor, International Conference on the Inter-Sections between Particle and Nuclear Physics, Lake Louise, Alberta, May 1986 (proceedings to be published).

37. K. Freudenreich, private communication.

38. R. P. Bickerstaff et al., Phys. Rev. D 33, 3228 (1986).

39. M. B. Johnson and J. Speth, to be published.

40. R. P. Bickerstaff et al., Phys. Rev. Lett. 53, 2532 (1984).

41. E. Berger, Nucl. Phys. B267, 231 (1986).

42. Fermilab proposal P772, unpublished.

43. J. A. McGill et al., Phys. Lett. 134B, 157 (1984).

44. Y. Tzeng and T. Tamura, Phys. Lett. 129B, 379 (1983).

45. R. O. Smith and S. J. Wallace, Phys. Rev. C 32, 1654 (1985).

46. C. J. Horowitz and M. J. Iqbal, Phys. Rev. C 33, 2059 (1986).

Inst. Phys. Conf. Ser. No. 86
Paper presented at Int. Nucl. Phys. Conf., Harrogate, UK, 1986

Probing single-particle structure with the (e,e′p) reaction

P.K.A. de Witt Huberts
NIKHEF-K, P.O. Box 41882, 1009 DB Amsterdam
and
Fysisch Laboratorium, Rijksuniversiteit Utrecht
P.O. Box 80.000, 3508 TA Utrecht

Abstract

Recent developments in experimentation and in the analysis of the quasi-free coincidence proton knockout reaction (e,e'p) are discussed. The DWIA analysis of high resolution (e,e'p) data yields spectroscopic factors for orbitals near the Fermi surface in a series of nuclides ranging from ^{12}C to ^{208}Pb that are appreciably smaller than the shell-model sumrule value. A similar reduction of spectroscopic strength is obtained if the radial wave function of the bound state deduced from (e,e'p) is used in the analysis of (d,^{3}He) data. Various sources of uncertainty and model dependence in the (e,e'p) data analysis are discussed. In particular the anomalous Coulomb Response function in the quasi-elastic regime is considered. Our results constitute evidence for partial occupancy of shell model states near the Fermi-surface.

Introduction

The empirical observation of the existence of single-particle properties of nuclei implies that basically the shell model concept of nucleons moving independently in an average potential is a good first-order approximation of the nuclear many-body problem. The single-particle spectroscopic information provided by the various reactions in which one nucleon is added to or removed from the target nucleus consists essentially of the spectroscopic factor $C^2S^f_\alpha$, i.e. the overlap of the many-body wavefunction ψ_A with a basis state $|\alpha>$ of the spherical shell model characterized by total and orbital angular momentum, j and l respectively. Upon adding a coupling of single-particle to other degrees of freedom (e.g. surface vibrations), dynamics is included in the shell model. Empirically such dynamics is manifest in the spreading width of single-particle states. For states in the vicinity of the Fermi-surface, it is usually found that the observed single-particle spectroscopic strength is spread over a fairly small energy range. In neutron stripping experiments on ^{208}Pb, in contrast, some single-particle states in ^{209}Pb appear to be almost pure, i.e. they are not spread out in energy. This interpretation, however, should be regarded with care; the states may appear pure because they are also coupled to configurations far away in energy and thus single-particle strength is removed into a region where it cannot easily be detected. With this observation I have come to the main theme, the question of absolute spectroscopic factors (De Witt Huberts 1985 , Wagner 1986). Absolute spectroscopic factors reflect the amount of overlap of the meanfield single-particle state with the real wavefunction and therefore the question lies at the heart of the shell model approximation.

The quest for absolute spectroscopic factors is an old one but has not yet been realized in practice due to both experimental limitations and uncertainties in the calibration of the scale of spectroscopic strength. Recently it has been argued (Papanicolas 1986) that elastic and inelastic electron scattering form factors of high multipolarity provide a growing body of evidence for the notion that partial occupancy of valence shells occurs systematically in nature. The recent series (Lapikás 1986 , Van der Steenhoven et al. 1985 , Den Herder et al. 1985) of high resolution coincidence studies of the quasi-free proton knockout reaction (e,e'p) also point in the direction of a partial depletion of shell-model orbits. I believe however that, regarded upon in isolation, neither of the above mentioned pieces of evidence has yet sufficient persuasiveness. I will therefore review the various sources of experimental information in an attempt to obtain a coherent picture including a realistic estimate of errors and model dependence. In particular it is considered of prime importance to attempt and achieve a synthesis of spectroscopic results obtained with the electromagnetic probe and the results of the traditional spectroscopic tool par excellence, the one-nucleon transfer reaction (d,^{3}He).

The present paper is organized as follows. In section 2 the basics of the (e,e'p) reaction is reviewed and the common and complementary features of the (e,e'p) and (d,^{3}He) reaction are illustrated. Recent experimental data of the (e,e'p) reaction on a series of nuclei is discussed in section 3. In addition the consequence of the use of the radial information on the bound state wave function, deduced from the electron scattering data, in the analysis of the (d,^{3}He) data is investigated for two representative cases. In section 4 the anomalous Coulomb coupling observed in the inclusive quasi-elastic reaction is considered as input to a rescaling procedure of the (e,e'p) spectroscopic factors. Subsequently recent (e,e'p) data pertaining to the longitudinal and transverse coupling of virtual photons to bound nucleons is discussed. In the last section I summarize the present results and discuss briefly some open problems.

2. Description of the (e,e'p) and (d,^{3}He) reactions

In the impulse approximation the cross section of the quasifree (e,e'p) reaction $A + \mathbf{e} \rightarrow B + \mathbf{e}' + \mathbf{p}$ in which 3-momentum $\mathbf{q} = \mathbf{e} - \mathbf{e}'$ and energy $\omega = e - e'$ is transferred to a proton, can be written (Frullani and Mougey 1984) as follows:

$$\frac{d\sigma}{dp_o d\Omega_p de' d\Omega_e} = K \sigma_{ep} S(E_m p_m). \tag{1}$$

Here K is a phase space factor, σ_{ep} is the electron-proton cross section and $S(E_m, p_m)$ is the spectral function, defined as the joint probability to have a proton with binding energy E_m and momentum p_m in the nucleus. Now,

$$\sqrt{S(E_m p_m)} \propto {}_0\int^{\infty} \chi^*_-(\mathbf{r}_p) \exp(i\, \mathbf{q}.\mathbf{r}_p) \langle\psi_B|\psi_A\rangle\, d\mathbf{r}_p \tag{2}$$

where $\chi_-(\mathbf{r}_p)$ is the distorted wave function of the proton, knocked out with an (asymptotic) momentum $\mathbf{p}$. The overlap function $\langle\psi_B|\psi_A\rangle$ for the knockout of a bound proton characterized by a set of quantum numbers $\alpha = \{nlj\}$ with orbital (total) angular momentum l(j) may be written

$$\langle\psi_B|\psi_A\rangle = \sum_n \sqrt{S_{nlj}}\, R_{nlj}(\mathbf{r}_p)$$

with the spectroscopic factor S_{nlj} and the radial wave function $R_{nlj}(\mathbf{r})$ of a given single-particle basis.This expression embodies unfortunately a rather unpractical proposition because shell-model calculations containing more than one orbit with a given (lj) are at present impossibly large. In practice therefore one projects on a 0 $\hbar\omega$ basis, i.e. a single n value is selected. Since in the Plane Wave Impulse Approximation (PWIA) $\chi^*_-(r) = \exp(-i\, \mathbf{p}.\mathbf{r}_p)$, one obtains the familiar result that $\sqrt{S(E_m \mathbf{p}_m)} = \sqrt{S_{nlj}} \int \exp(i\, \mathbf{p}_m.\mathbf{r}_p)$ $R_{nlj}(\mathbf{r}_p) d\, r_p = \sqrt{S_{nlj}}\, \phi(p_m)$, i.e. the (e,e'p) scattering amplitude is proportional to the Fourier transform of the bswf, where $\mathbf{p}_m = \mathbf{q} - \mathbf{p}$ is the missing momentum. The spectroscopic factor S_α is obtained by integrating the spectral function over final states at $E^f{}_m(\alpha)$, which yields the momentum distribution $\rho_\alpha(\mathbf{p}_m)$, and then performing the operation $S_\alpha = 4\pi\int\rho_\alpha(p_m)p^2{}_m dp_m$. In the experiments to be discussed presently the experimental conditions typically involve the following values of variables: q = 350 - 500 MeV/c, p_m ranges from 0 to 320 MeV/c and protons with kinetic energy T_p of either 70 or 100 MeV are detected.

Let us now recall the basic ingredients of the DWBA treatment of the single nucleon transfer reaction, in particular the (d,^{3}He) reaction $d + A \rightarrow B + {}^3He(\tau)$. The DWBA transition amplitude is written as

$$T_{AB} = \iint \chi_\tau^-(\mathbf{k}_\tau \mathbf{r}_\tau) \langle B,\tau|V_{eff}|A,d\rangle \chi_d^+(\mathbf{k}_d \mathbf{r}_d)\, d^3\tau\, d^3 r_d \qquad (3)$$

The effective transition potential driving the reaction can be written $V_{eff} = V_{dp}$. The distorted waves χ^+ and χ^-, occurring in the above expression, are generated by fitting elastic d- and τ-nucleus scattering data at the appropriate energy in the ingoing and outgoing channels. The effective matrix element separates into a product of two form factors $\langle B,\tau|V_{dp}|A,d\rangle = \langle\psi_B|\psi_A\rangle\langle\tau|V_{dp}|d\rangle$. The projectile formfactor $\langle\tau|V_{dp}|d\rangle$ is usually evaluated with non-locality and finite range corrections included. The formfactor of the transferred nucleon is an overlap integral where the integration is over the internal coordinates of the core-nucleus B, $\phi_{BA}(\mathbf{r}_p) = \langle\psi_B|\psi_A\rangle$. As before the radial part of the formfactor $\phi_{BA}(\mathbf{r}_p)$ is written as $\phi_{BA}(\mathbf{r}_p) = \sqrt{S_{nlj}}\, R_{nlj}(\mathbf{r}_p)$.

The structure of the expression for the amplitude of the (e,e'p) and (d,^{3}He) reaction is seen to be quite analogous in principle. In practice however there is a great difference due to the distinct coordinate-space sampling functions occurring in the integral expression (2) and (3). For the knockout reaction, the Fourier sampling function $\exp(i\mathbf{p}_m \cdot \mathbf{r}_p)$ essentially applies since the distortion effects in the outgoing-proton channel are moderate and do not basically modify the sampling function. Thus, by mapping out the spectral function $S(E_m p_m)$ for a range of p_m values, the bswf $R_{nlj}(r_p)$ is sampled in regions of coordinate space where its amplitude is large.

Due to the strong interaction acting on the projectile and ejectile the sampling function of the (d,^{3}He) reaction peaks in the tail of the nuclear density in such a way that, essentially, the asymptotic normalisation of R_{nlj} is sampled. The r-space sensitivity of a given reaction can be illustrated by evaluating the reaction amplitude as a function of the lower limit of integration r_c, i.e.

$$\sqrt{S_\alpha(r_c)} \propto \int_{r_c}^{\infty} f_s(r_p) R_\alpha(r_p) dr_p.$$

Examples are shown in fig. 1 and fig. 2, for l = 3 pickup in the ^{51}V(d,^{3}He)^{50}Ti reaction at T_d = 52 MeV and 2p-knockout from ^{90}Zr. It is seen that the main contribution to the (d,^{3}He) cross section in the first maximum of the angular distribution comes from the region (r = 6 - 8) fm, i.e. from the tail of $R_{13}(r)$. The double-humped 2p-wave function $R_{21}(r)$ is mainly sampled at the nuclear surface at p_m = 50 MeV/c and in addition also in the nuclear interior at a value p_m = 150 MeV/c in the (e,e'p) reaction (see fig. 2).

The main consequences of these observations are the following:

i) Spectroscopic factors from the (d,^{3}He) reaction are highly sensitive to the rms radius $\langle r_\alpha^2\rangle^{1/2}$ assumed for the bound state wave function; e.g. for l = 3 pickup from ^{51}V one obtains $\Delta S/S$ = 10 % for a one percent change of rms. This acute sensitivity is the

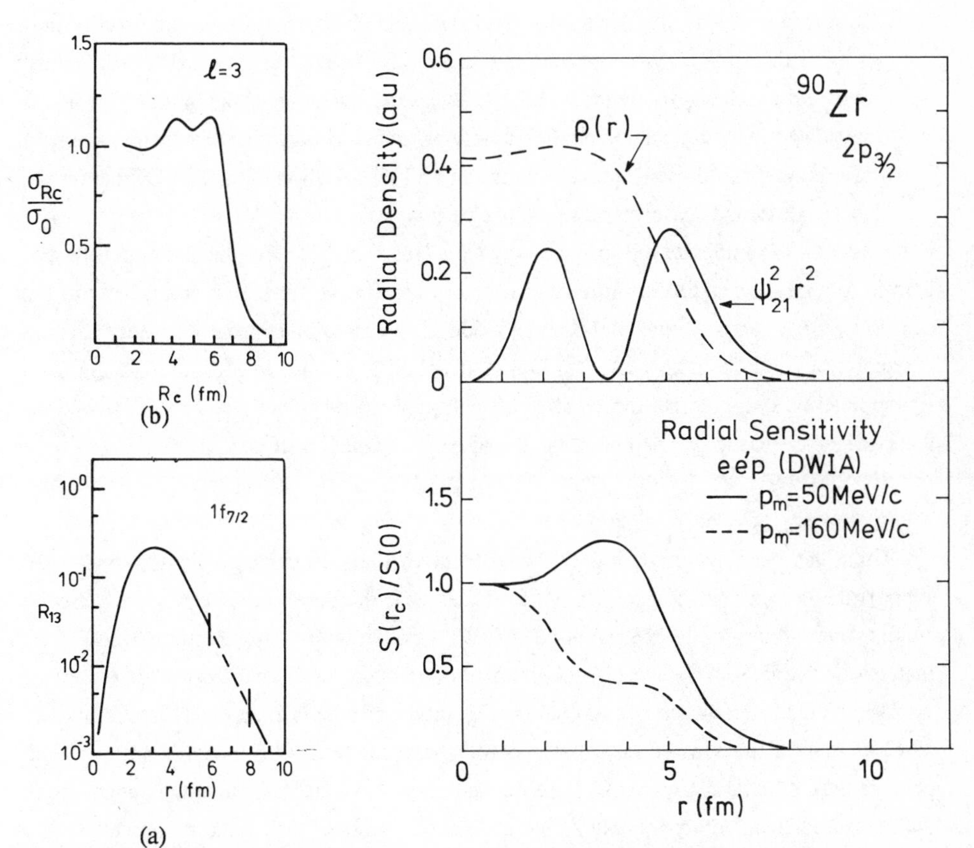

Fig. 1 Wave function of the 1f orbit in ^{51}V (a) and (b); the radial sensitivity of the (d,^{3}He) reaction illustrated with the radial cut-off method.

Fig. 2 Radial probability distribution of the 2p orbit and the total density of ^{90}Zr (not to scale) (upper panel). The r-space sampling function of the (e,e'p) cross section illustrated by the radial cut-off function (see text).

main impediment to obtain accurate absolute spectroscopic factors with the (d,^{3}He) reaction.

ii) Since the bswf is sampled in r-space the spectroscopic factor deduced from (e,e'p) is quite insensitive to the Ansatz for the bswf. Radial information on the bswf ensues from fitting $\rho(p_m)$ with the radial scale of the bound state (e.g. Woods-Saxon) potential as a free parameter. In this procedure the distortion of the outgoing proton is calculated in an optical potential(OP) with parameters constrained by the requirement of a good fit to proton-nucleus scattering and polarisation data. The OP parameter space is extensively searched in the (e,e'p) analysis.

In the attempt to synthesize the spectroscopy from (e,e'p) and (d,^{3}He) the strategy followed here is to apply the radial information on bound state wave functions obtained with the electromagnetic probe in the analysis of (d,^{3}He) data and to investigate whether the same spectroscopy, both in an absolute and in a relative sense, ensues. In addition this approach will enable us to appreciate the validity of the approximations involved in the (reaction) theory of either reaction, in particular the impulse approximation Ansatz.

3 .Occupancy of valence orbitals

There has been accumulating evidence from various electromagnetic observables to support the notion that the occupancy of orbitals near the Fermi surface is systematically smaller than is implied by the mean-field (MF) approximation. By selecting formfactors, both elastic and inelastic, of the largest multipolarity allowed in a given major shell (stretched transitions) the interference due to $0\hbar\omega$-configuration mixing is excluded (Platchkov et al. 1982). A quite persuasive case has recently been made on the basis of several such observables in the Pb-region by C. Papanicolas (1986). Thus the systematic quenching of the measured formfactors relative to the shell-model (meanfield) value is indicative of a reduced occupation probability of the orbital involved. The question arises then whether similar indications ensue from one nucleon removal reactions. We proceed to first present spectroscopic information from recent high-resolution (e,e'p) data . In the discussion a synthesis of spectroscopic information from (e,e'p) and (d,^{3}He) is attempted.

Spectroscopic factors from the (e,e'p) reaction

The following examples of (e,e'p) data drawn from a larger database will be discussed: ^{3}He(e,e'p)^{2}H, ^{51}V(e,e'p)^{50}Ti, ^{90}Zr(e,e'p)^{89}Y, Pb(e,e'p)Tl and ^{205}Tl(e,e'p)^{204}Hg.

Two-body breakup of ^{3}He

Recent coincidence studies performed at Saclay (Jans et al. 1982) and NIKHEF

(Keizer 1986) of the electro-disintegration of ^{3}He with good missing-mass resolution (ΔE_m $\approx$ 0.5 MeV) have reached a level of accuracy for the absolute cross sections of better than ten percent in the range of recoil momentum p_m = (0 - 300) MeV/c. An accurate test of the spectral function calculated with the Faddeev technique starting from a fundamental nucleon-nucleon interaction may thus be performed. Interestingly the trinucleon systems encompass already various aspects of the many-body dynamics of heavier systems. For instance the two-body breakup ^{3}He $\rightarrow$ p + d may be viewed as an analog of the removal of a valence nucleon in heavier systems, whereas the three-body breakup continuum ^{3}He $\rightarrow$ p + n + p constitutes a manifestation of correlations that spread out spectroscopic strength to high excitation energy in the residual system. In Fig. 3 is shown the spectral function $S_2(p_m)$ for two-body breakup measured at Saclay and NIKHEF. The two datasets show excellent consistency. The three-momentum transfer q involved is typically 450 MeV/c and the energy transfer ω ranges from 60 to 100 MeV leading to a (p-d) relative energy in the recoil center of mass system of either 65 MeV (p_m = 0 - 150 MeV/c) or 106 MeV (p_m = 150 - 300 MeV/c). In the figure two theoretical spectral functions are shown, calculated with the Paris nucleon-nucleon potential (Meier-Hajduk 1983)) and with the Tjon-potential (Tjon 1986). In the latter approach the correct binding energy E_b of the trinucleon system is achieved and consequently an improved description of the (e,e'p) data is obtained at low p_m ($\leq$ 50 MeV/c), where $S_2(p_m)$ is acutely sensitive to E_b. One notes that both calculations overshoot the data appreciably at p_m-values larger than 50 MeV/c. Most likely the prime mechanism responsible for the major part of the discrepancy is the final state interaction (FSI). Indeed, recent developments in the treatment of secondary reaction processes beyond the plane wave impulse approximation indicate non-negligible effects of FSI in the present kinematical domain. The processes illustrated with the diagrams shown in the inset of fig. 4 have been calculated by J.M. Laget (1985). The rescattering process (diagram d,e) dominant in the (0 - 300) MeV/c range of p_m leads to a reduction of the plane wave cross section σ_{pw} and thus of the spectral function. Casting the calculations in a partial sumrule representation $\Sigma(p_m) = 4\pi \int_0^{p_m} S_2(p)\, p^2 dp$ and treating the data likewise the results shown in fig. 4 are obtained.

Two observations ensue: i) The FSI reduces the integrated spectral function by $\approx$ 20 %, ii) the data fall (15 $\pm$ 10) % below the distorted spectral function calculation. Given the freedom of $\approx$ 5% in the various brands of Faddeev calculations for $S_2(p_m)$, in the prescription for the electron-proton off-shell cross section σ^*_{ep} ($\approx$ 4 %) (De Forest, 1983) and upon consideration of the approximations involved in the diagrammatic expansion of the three-body reaction mechanism, the conclusion is justified that the calculated (e,e'p) cross section is in good agreement with the data. In consequence there is no signature of a

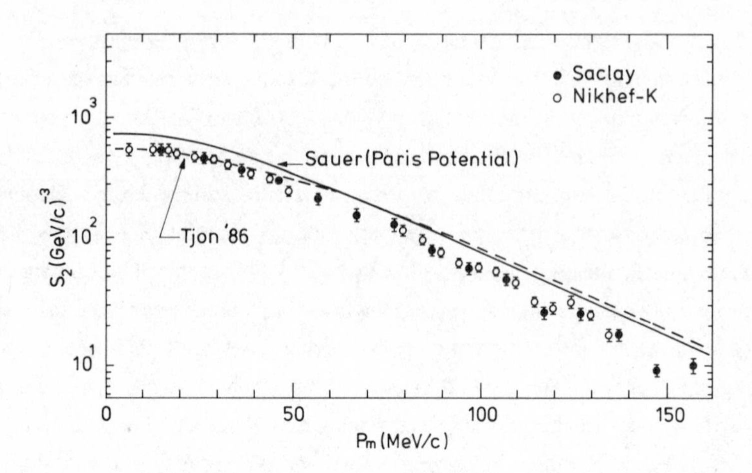

Fig. 3 Experimental and theoretical spectral functions for two-body breakup of ^{3}He.

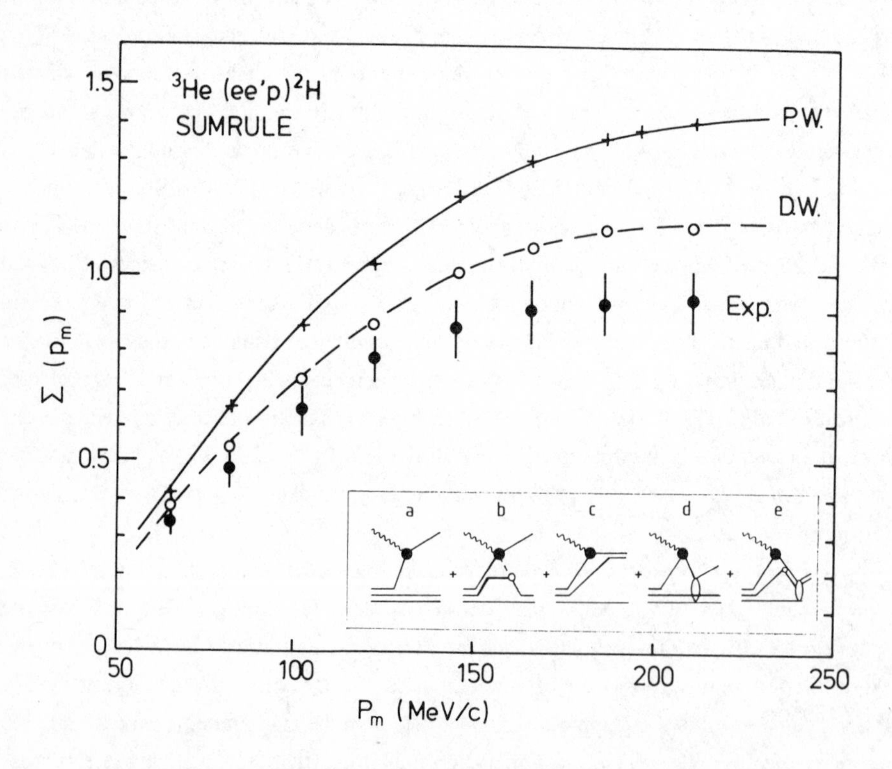

Fig. 4 Two-body breakup sumrule as a function of p_m. Results of Faddeev calculations in plane wave (PW) and with FSI (DW) are shown.

breakdown of the distorted wave impulse description of the (e,e'p) reaction on ^{3}He with its large central density (≈ 80 % of nuclear matter density).

The reaction $^{51}V(e,e'p)^{50}Ti$ ($J^\pi = 0^+, 2^+, 4^+, 6^+$)

The reaction in which a proton is removed from the main $(1f_{7/2})^3$ configuration in ^{51}V leading to the 0^+, 2^+, 4^+, 6^+ quadruplet in ^{50}Ti constitutes a classic in support of the seniority shell model (Hinterberger et al. 1967). In view of this and of the availability of radial information for the $1f_{7/2}$-orbit from elastic magnetic M7 scattering (Platchkov et al., 1982) the proton knockout reaction on ^{51}V has been investigated with 120 keV resolution (Den Herder et al. 1986). Concomitantly the ^{51}V(d,^{3}He)^{50}Ti reaction has been remeasured recently with the spectrometer setup of the KVI (Groningen) (Kramer et al. 1986) with an excellent energy resolution of 35 keV in order to provide the other ingredient for a synthesis of spectroscopic information from (e,e'p) and (d,^{3}He). A sample excitation spectrum of ^{50}Ti and the momentum distribution of the $(1f_{7/2})^2$ quadruplet 0^+ (gs), 2^+ (1.55 MeV), 4^+ (2.67 MeV), 6^+ (3.20 MeV) in ^{50}Ti are shown in figs. 5 and 6. In order to gauge the energy

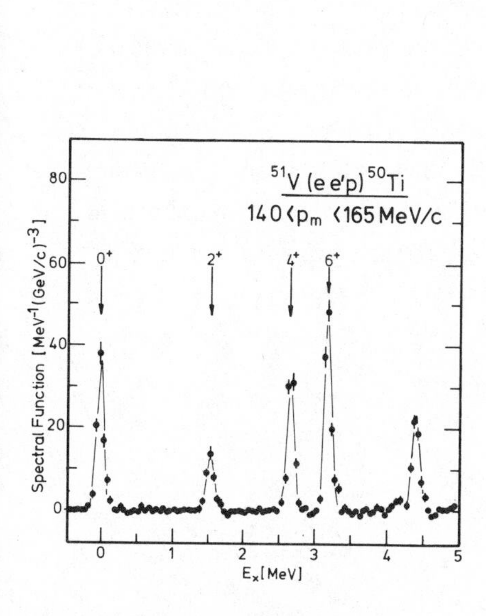

Fig. 5 Excitation energy spectrum of ^{50}Ti resulting from the ^{51}V(e,e'p) reaction.

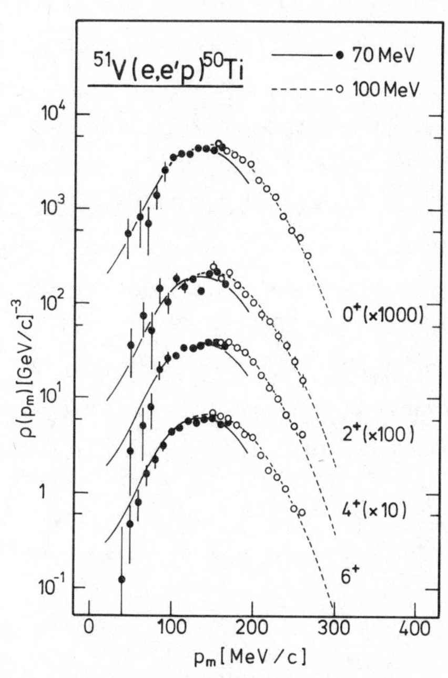

Fig. 6 Momentum distributions for l=3 knockout leading to 0^+, 2^+, 4^+, 6^+ states in ^{50}Ti.

dependence of the FSI two kinetic energy values for the knocked out proton were used, T_p = 70 MeV (for p_m < 170 MeV/c) and T_p = 100 MeV. The slight difference in shape of the two data sets reflects the differential effect of distortion. Following the procedure discussed in section 2, the two data sets were fitted simultaneously with the spectroscopic factor and the radius r_0 of the bound-state potential (either local or non-local) as free parameters. The results shown in Table I indicate that the $1f_{7/2}$ rms radius obtained is constant for the four transitions within the one percent error. This comes about as a result of r_0 ranging from r_0 = 1.286 fm to 1.331 fm, in going from the 0^+-state to the 6^+-state, and the compensating effect of increased binding ($E_A - E^f_{A-1}$). Note that this result does not support the separation-energy recipe usually employed in the analysis of (d,^{3}He) data, that would yield a rms radius for the 6^+ transition 3 % smaller than that of the 0^+-transition.

Evaluated in the laboratory-fixed potential frame the average value of the rms radius of the $1f_{7/2}$ orbit $<r^2>^{1/2}_{1f7/2}$ = 4.20 ± 0.14 fm is obtained. The error now reflects the cumulative effect of all sources of uncertainty in the analysis, the main contribution being due to the 2 % effect of the different possible FSI prescriptions. The quoted value is in agreement with various brands of Hartree-Fock calculations: 4.24 fm (DME) (Negele and Vautherin 1972), 4.14 fm (DDHF) and 4.11 fm (DDHFB) (Platchkov et al. 1982)). The rms ($1f_{7/2}$) is slightly larger than the result 4.06(5) fm of the elastic magnetic M7 scattering experiment still being in fair agreement though. The relative spectroscopic factors normalised to the summed strength, as extracted from the (e,e'p) measurement (see Table I) are 0.27(2), 0.12(1), 0.24(2) and 0.37(3), in good agreement with the seniority shell model (0.26, 0.12, 0.25, 0.37). Remarkably the summed absolute strength amounts to Σ = 1.11(14), i.e. 37 % of the single particle sumrule value of three. Adding to this the 10 % strength fractionation observed in a 7 MeV range of excitation energy in the (d,^{3}He) experiment one obtains Σ = 1.22(14), i.e. 40(5) % of the sumrule value of three protons, postulated by the static shell model. Let us for the moment defer the discussion of this surprisingly small number, a near-heresy presumably in the mind of the true believers of the shell model, and compare first the relative spectroscopic factors with the results from the ^{51}V(d,^{3}He) reaction.

Table I Absolute and relative spectroscopic factors for l = 3 states deduced from the (e,e'p) and (d,^{3}He) reaction on ^{51}V. Errors shown are statistical, except for the entry sum strength Σ.

^{50}Ti (J^π)	E_x (MeV)	$\langle r^2\rangle^{1/2}_{1f}$ (fm)	(e,e'p) S(1f) Abs.	(e,e'p) S(1f) Norm.	(d,^{3}He) S(1f) (I)	(d,^{3}He) S(1f) (II)	(d,^{3}He) S(1f) Norm.(II)	Shell Model Norm.
0^+	0	4.19(6)	0.30(2)	0.27(2)	0.41	0.41	0.27	0.26
2^+	1.55	4.20(10)	0.13(1)	0.12(1)	0.22	0.20	0.13	0.12
4^+	2.67	4.20(6)	0.27(2)	0.24(2)	0.41	0.35	0.23	0.25
6^+	3.20	4.22(5)	0.41(3)	0.37(3)	0.65	0.53	0.36	0.37
Sum strength Σ		-	1.11(14)	-	1.69	1.49	-	-

Comparison of l = 3 spectroscopic factors in ^{51}V from (e,e'p) and (d,^{3}He)

High-precision data of the ^{51}V(d,^{3}He) reaction with 52.9 MeV deuterons from the KVI cyclotron in Groningen were recently obtained with the magnetic spectrograph (Kramer et al. 1986), yielding the excellent energy resolution of 25 keV. Angular distributions in the range $\theta_{lab} = 4^o$ - 40° were analysed with the full sophistication of non-local finite range DWBA calculations. It was found that the freedom in the choice of optical-model parameters to describe the elastic scattering of projectile and ejectile leads to a 20 % uncertainty in the calculated cross section scale. Usually in the (d,^{3}He) analysis the separation energy (SE) recipe is applied to account for the effect of increased binding in going from the 0^+ (0 MeV) to the 6^+ (3.20 MeV) final-state. The SE method implies keeping the geometry constant of the Woods-Saxon well in which the transferred nucleon is bound, resulting in a shrinking rms radius of the bound-state wave function with increased binding. By requiring consistency of (e,e'p) and (d,^{3}He) results for the relative l = 3 spectroscopic factors of the (0^+, 2^+, 4^+, 6^+) quadruplet the validity of the SE-method, for which no theoretical justification has been forwarded yet, may be gauged. As is shown in Table I the rms radius extracted from (e,e'p) is constant within one percent for the four transitions. The (d,^{3}He) data were analysed with either the SE method (I) or the rms = constant method (II). In the latter the rms ($1f_{7/2}$) value of 4.06 fm deduced from the elastic magnetic M7 formfactor (Platchkov et al. 1982) was used. Results are shown in Table I. The relative spectroscopic strength was normalised to a summed strength of unity. Clearly the relative spectroscopic factors of method II agree better with both the (e,e'p) results and the (1f,2p) space shell-model calculation. The integrated $l = 3$ strength in the quadruplet deduced from (d,^{3}He) is $\Sigma = 1.49$ and $\Sigma = 1.69$ for method II and I, respectively. Due to the excellent energy resolution additional $l = 3$ strength has been found up to 7 MeV excitation energy in the (d,^{3}He) spectra, that amounts to $\approx$ 10 % of the quadruplet strength. This leads to the interesting observation that the total strength in a one $\hbar\omega$ energy range, as found from a DWBA analysis of (d,^{3}He) in which the radial information on the bswf from electron scattering is used exhaust 63 % (54 % for method II) of the shell model sumrule strength of three 1f-protons in ^{51}V.

Spectroscopy of the ^{90}Zr(e,e'p) reaction

Near the Z = 38 shell closure the 1f and 2p single-particle orbitals are quite close energetically. Knockout of protons from either of these quantum states in ^{90}Zr leads to discrete low-lying states in ^{89}Y, $1/2^-$ (g.s.), $3/2^-$ (1.51 MeV), $5/2^-$ (1.74 MeV). By virtue of the r-space sampling function of the (e,e'p) process (see fig. 2) the double-humped 2p wave function can be investigated both at the surface and in the nuclear interior. Results for 1f proton knockout have been discussed in (Den Herder 1985 , Lapikàs 1986). The data for knockout of $2p_{1/2}$ and $2p_{3/2}$ protons, shown in fig. 7 were obtained at outgoing proton energies, $T_p = 70$ and 100 MeV. The notable feature concerns the ratio of strength in the first ($p_m = 50$ MeV/c) over that in the second ($p_m = 160$ MeV/c) maximum of the momentum distribution that appears to be larger for the $T_p = 100$ MeV data. The $\rho(p_m)$, calculated with scaled meanfield single-particle wavefunctions, shown in fig. 7 fail to reproduce this feature in spite of an extensive search in a large space of parameters for the optical potential used to calculate the distortion of the knocked out proton. However, in the search the constraint of a simple Woods-Saxon radial shape was imposed (inset of fig. 7). Upon allowing for a non-standard (double) Woods-Saxon shape shown in the inset of fig. 8 a good fit to the data is achieved. Effectively an appreciably larger and more slender imaginary potential in the nuclear interior is required at $T_p = 100$ MeV than at $T_p = 70$ MeV. It should be noted that in the parameter search procedure always the constraint was imposed that elastic proton-nucleus scattering and polarisation data at the appropriate energies remained well accounted for. The achieved effect can be simply understood from the observation that at $p_m \approx 160$ MeV/c the proton of 100 MeV emerges with a large probability from the nuclear interior and has to pierce a stronger imaginary potential on its way out. Having achieved a good fit to the shape of the momentum distributions the spectroscopic information shown in Table II was extracted. The values for the rms radius of the $2p_{1/2}$- and $2p_{3/2}$-orbit are in fair agreement

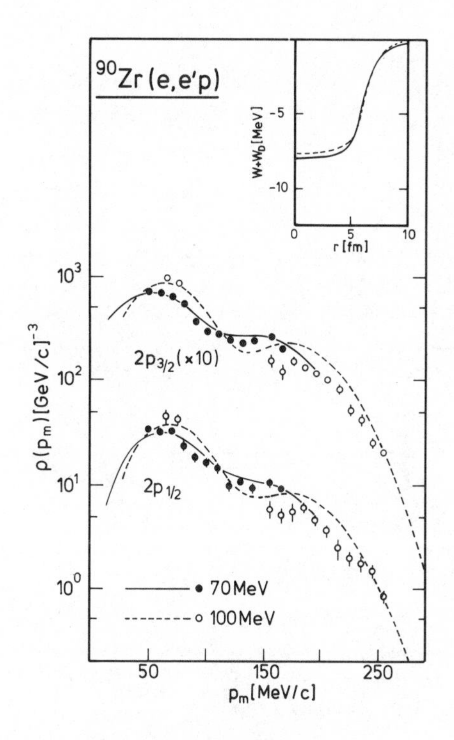

Fig. 7 Momentum distributions for $2p_{1/2}$ and $2p_{3/2}$ knockout. Fits with a standard imaginary optical potential (see inset) are shown.

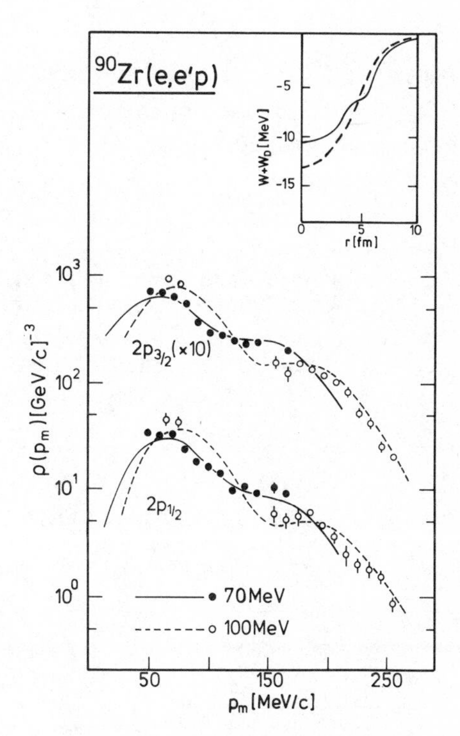

Fig. 8 Idem with fits using the non-standard optical potential shape shown in the inset.

with various brands of Hartree-Fock calculations with preference for the DME calculation (Negele and Vautherin 1972). The summed 2p spectroscopic strength amounts to $\Sigma(2p)$ = 2.33(23) where the error now includes the systematic error in the absolute cross section scale and the effect of different FSI prescriptions.

Adding 2p-strength that has been found from a multipole decomposition of the (e,e'p) spectrum up to E_x = 20 MeV in ^{89}Y, one obtains $\Sigma(2p)$ = 2.93(30), i.e. 50(5) % of the sumrule limit. Thus, like in the case of ^{51}V an appreciable reduction of spectroscopic strength is observed. Given the appreciable loss of strength observed in the (e,e'p) spectroscopic factors some reservation concerning the extracted rms radius values is called for. On the other hand the error, typically one percent statistical and (1 - 2) % systematic, is such that consistency is observed with the magnetic orbital-radius values (Platchkov et al. 1982) which I believe are the most accurate available at present. Using the Woods-Saxon geometry found for the $1g_{9/2}$ proton orbit in ^{93}Nb and correcting for the A-dependence, we obtain $\langle r^2\rangle^{1/2}_{2p}$ = 4.49 fm. By employing this value in a non-local finite range DWBA analysis of $^{90}Zr(d,^3He)$ data (Stuirbrink et al. 1980) we obtain $S(2p_{1/2}) = 0.8$ and $S(2p_{3/2}) = 1.3$. Therefore also the $(d,^3He)$ reaction, when analysed with a proper radial form factor, yields appreciably reduced spectroscopic factors that are fairly consistent with the (e,e'p) results.

Table II Spectroscopic factor and rms radius for $2p_{1/2}$ and $2p_{3/2}$ knockout from ^{90}Zr are compared with results from a standard analysis and from a reanalysis of the $^{90}Zr(d,^3He)$ reaction and with Hartree-Fock results. Statistical errors are shown.

J^π	rms (fm)	rms (fm) DME 1)	DDHFB 2)	S(e,e'p)	$S(d,^3He)$ 3)	$S(d,^3He)$ 4)
$1/2^-$	4.72(8)	4.63	4.46	0.64(3)	1.8	0.83
$3/2^-$	4.63(5)	4.57	4.40	1.69(5)	3.9	1.33
				$\Sigma S_{2p} = 2.93(30)$ 5)		

1) Ref. Negele (1972)
2) Ref. Platchkov (1982)
3) Ref. Stuirbrink et al. (1980)
4) Data reanalyzed with rms(2p) = 4.49 fm.
5) Summed 2p strength in interval E_x = (0 -20) MeV, (error includes systematic error).

Occupancy of the $3s_{1/2}$ proton orbit in the Pb-region

Common wisdom suggests that the mean field approximation should be optimal for the doubly magic nucleus ^{208}Pb. The part of the vast database for this mass region that involves stripping and pickup reactions is suggestive of nearly pure shell-model orbitals near the Fermi-edge. In contrast, several observables measured with elastic and inelastic electron scattering are incompatible with Hartree-Fock meanfield calculations in the sense that systematically, a reduction (quenching) of (transition) strength is observed. The key observables involve the density of charge in the centre of ^{206}Pb and ^{208}Pb (Frois, 1983), that is ≈ 30 % smaller than the Hartree-Fock prediction, the missing strength in the elastic M1-formfactor of both ^{207}Pb and ^{205}Tl and the quenching of high-multipole magnetic M12 and M14 transitions in ^{208}Pb (Papanicolas, 1986). Such quenchings may be understood in a natural way if partial occupancy of single-particle states in assumed. In order to obtain a measure of the occupation probability of the $3s_{1/2}$-orbit in the lead region a joint venture was recently initiated by groups from Tübingen, Saclay, Indiana and NIKHEF-K. The basic idea is to measure relative spectroscopic factors for the removal of a $3s_{1/2}$-proton in the pairs of nuclides ^{208}Pb-^{206}Pb, ^{206}Pb-^{205}Tl, with both the (e,e'p) and the $(d,^3He)$ reaction, in order to avoid the uncertainties inherent in the determination of absolute spectroscopic factors. The absolute scale is assessed by the application of a sumrule (Grabmayr, 1985)

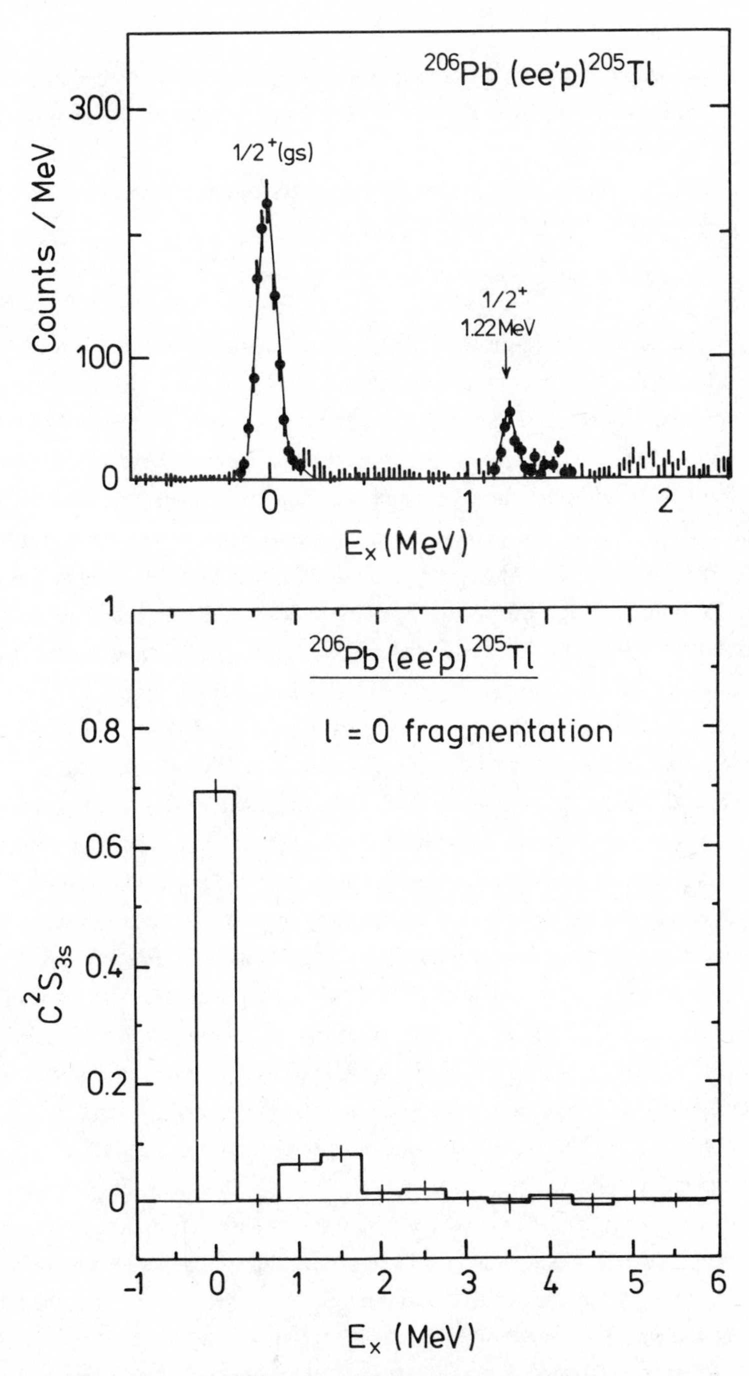

Fig. 9 Spectrum of the ^{206}Pb(e,e'p)^{205}Tl reaction at $p_m = 15$ MeV/c (upper panel). Fragmentation of 3s-knockout strength deduced from a multipole decomposition of momentum distributions (lower panel).

that relates spectroscopic factors to the contribution of the $3s_{1/2}$ orbit to the charge density difference of ^{206}Pb and ^{205}Tl, $\Delta\rho = \rho(206) - \rho(205)$, as deduced from the ratio of elastic charge scattering formfactors.

The data from this important experiment (Frois, 1983) are compatible with z = 0.7 3s-protons contributing to $\Delta\rho(r)$. The sumrule can be written

$$z = \sum_f (S^f(206) - S^f(205)) = n(206) - n(205)$$

where n(A) is the number of 3s-protons in the ground state of nuclide A, $S^f(A)$ is the spectroscopic factor for 3s knockout leading to final state (f) in (A-1). The need to assess absolute spectroscopic factor is circumvented by putting $z/n(206) = 1 - (\Sigma S^f(205)/\Sigma S^f(206)) = (1 - R(205/206))$. The latter ratio can be determined in a way that is largely independent of the reaction model. In addition, the reasonable assumption is made that the fraction of spectroscopic strength fragmented beyond the experimental window of excitation energy is equal for the neighbour nuclides. Thus the quantity n(206) may be determined and by using the measured (Quint et al., 1986) ratio $\Sigma S(208)/\Sigma S(206)$ also n(208). In the (e,e'p) experiments the pronounced $l = 0$ selectivity at $p_m \cong 15$ MeV/c has been exploited to locate 3s-strength up to 5.5 MeV excitation energy in the residual system. In figs. 9 and 10 missing mass (E_m) spectra at low p_m are shown with a typical resolution of 120 keV. By performing an angular momentum decomposition of momentum distributions $\rho(p_m)$ obtained in 500 keV wide E_m-slices in the range $p_m = (15 - 160)$ MeV/c the distribution of $l = 0$ strength up to $E_x = 5.5$ MeV shown in figs. 9 and 10 was obtained. Clearly, the amount of fragmentation of strength to excited states in the residual system is appreciable and appears stronger in ^{204}Hg, a trend that reflects increasing deformation further away from shell closure. The direct result obtained from the (e,e'p) analysis is R(205/206) = 0.49(2) and R(206/208) = 0.83(5). The systematic error of these ratios amounts to 2 %. Upon application of the sumrule and adopting the value z = 0.7 (Frois, 1983), the following $3s_{1/2}$ occupation probabilities are deduced n(208) = 1.65(12), n(206) = 1.37(7), n(205) = 0.67(6). Adopting an estimate of a 10 % error on z one finds a depletion of the $3s_{1/2}$ proton orbit in doubly-magic ^{208}Pb of 20 ± 10 % and 30 ± 8 % in ^{206}Pb. We note that the present results are compatible with those of a similar experiment with the (d,^{3}He) reaction on ^{205}Tl and ^{208}Pb (Clement, 1986).

The partial occupancy implied by these results when imposed on an occupation number constrained meanfield calculation would lead to a natural and consistent description of such diverse observables as the elastic M1 formfactors of ^{205}Tl and ^{207}Pb, the charge distribution at the nuclear center of ^{206}Pb and ^{208}Pb and M12 and M14 transition formfactors in the lead region (Papanicolas, 1986). Recent theoretical estimates (Pandharipande, 1985 ; Jaminon, 1985) for the effect of nuclear matter and RPA

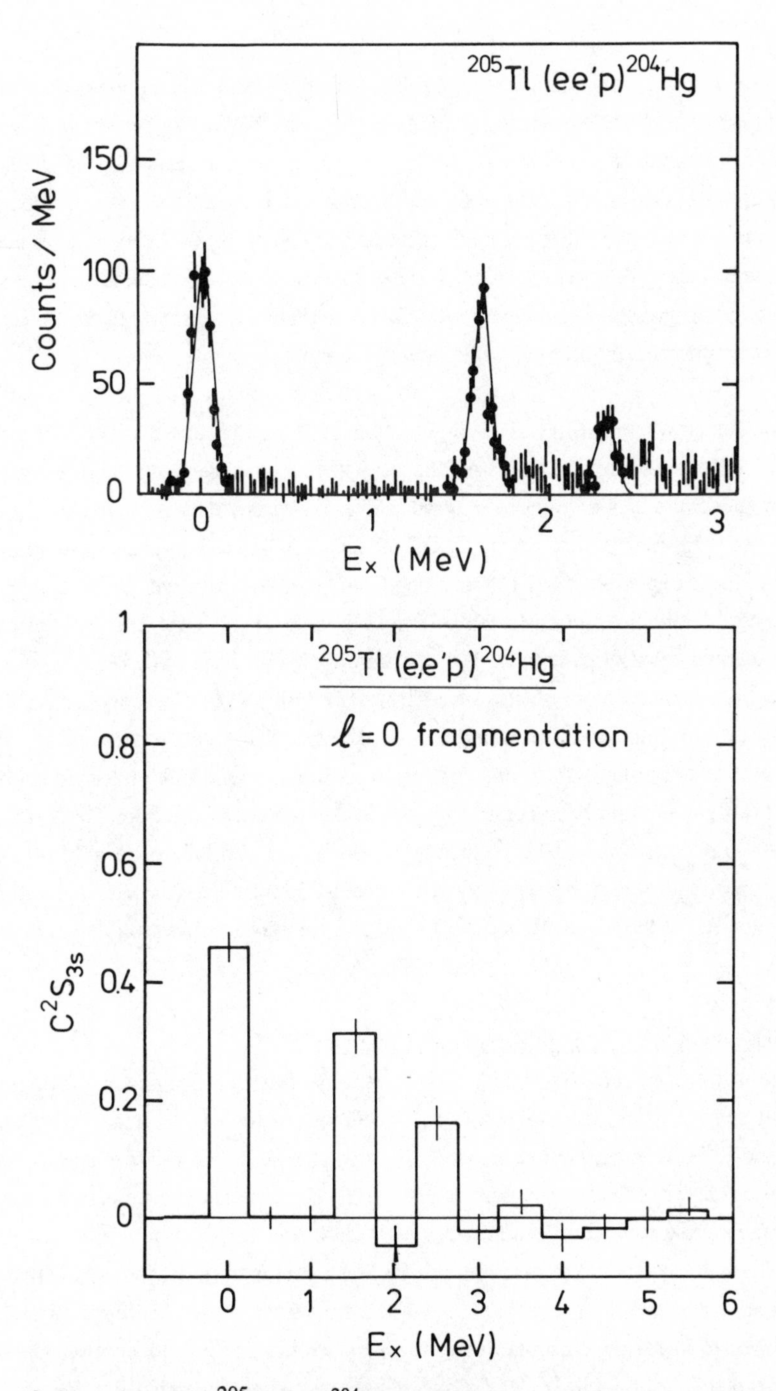

Fig. 10 Spectrum of the ^{205}Tl(e,e'p)^{204}Hg reaction at p_m = 15 MeV/c (upper panel). Fragmentation of 3s-knockout strength deduced from a multipole decomposition of momentum distributions (lower panel).

correlations on occupation numbers η of orbits near the Fermi surface yield $\eta = 0.7 - 0.8$, quite compatible with the present results. In conclusion we believe that the quenching that is manifest in so many observables is due at least in part to partial occupancy of shell-model states. Strictly speaking this comes as no surprise since partial occupancy induced by correlations is a basic ingredient of nuclear meanfield theory, as elaborated e.g. in the work of Brueckner and collaborators. What the experimental results discussed above should convey is that progress has been made towards a quantitative assessment in the Pb-region of the limitations inherent in the closed-shell meanfield approximation.

From the above examination of a selection of (e,e'p) results a picture of partially occupied shell-model orbits has emerged that may further our understanding of the generally observed quenching of 'single-particle' observables. Whereas the 3s-depletion of (20 - 30%) deduced in the Pb-region via a nearly model-independent sumrule appears quite reasonable theoretically, the depletion of ≈ 50 % resulting from the DWIA analysis for 1f- and 2p-orbits in ^{51}V and ^{90}Zr seems uncomfortably large. However, it is intriguing to note that the DWBA analysis of (d,^{3}He) data leads to a similar large value provided that a boundstate wavefunction deduced from electron scattering is employed in the analysis. Among the elements of the impulse approximation treatment of the (e,e'p) reaction - virtual photon-proton coupling (σ_{ep}), and distortion of the knocked-out proton in an optical potential (OP) - the latter has been scrutinized in great detail (see section 3). It was found that, with the constraint to reproduce proton-nucleus scattering data imposed, the spectroscopic factors vary by typically ten percent. Therefore I will briefly examine an experimental test of the adequacy of the off-shell electron proton cross section σ_{ep} in the next section.

4. Calibration of electron proton coupling in (e,e'p)

Data of the (e,e') inclusive quasi-elastic response function, separated in longitudinal (Coulomb) ($R_L(q)$) and transverse ($R_T(q)$) components for ^{12}C, ^{40}Ca,^{48}Ca (Meziani 1985) and ^{56}Fe (Deady 1983) provide a testing ground of the quasi-elastic process in which the influence of the final state interaction (FSI) is expected to be suppressed. Indeed, a detailed theoretical treatment (Horikawa, 1980) indicates a reduction due to FSI of (e,e') cross sections by typically ten percent. The R_L data, cast in a model independent sumrule representation $C_L(q) = \int R_L(q,\omega)[Z^*G_E(q]^{-1}d\omega$ are shown in fig. 11. Surprisingly, since one believes to know the Coulomb coupling very well, a general quenching is observed ranging from a factor 0.75 in ^{12}C to 0.6 in ^{48}Ca. Now assume the redistribution of strength in excitation energy (ω) due to correlations to remain contained within the experimental

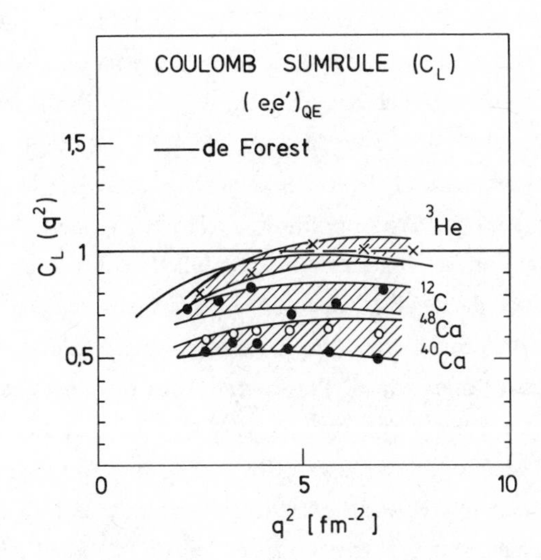

Fig. 11 Coulomb sumrule from the inclusive quasielastic (e,e') reaction. The De Forest calculation is for ^{12}C (De Forest, Jr, 1984)

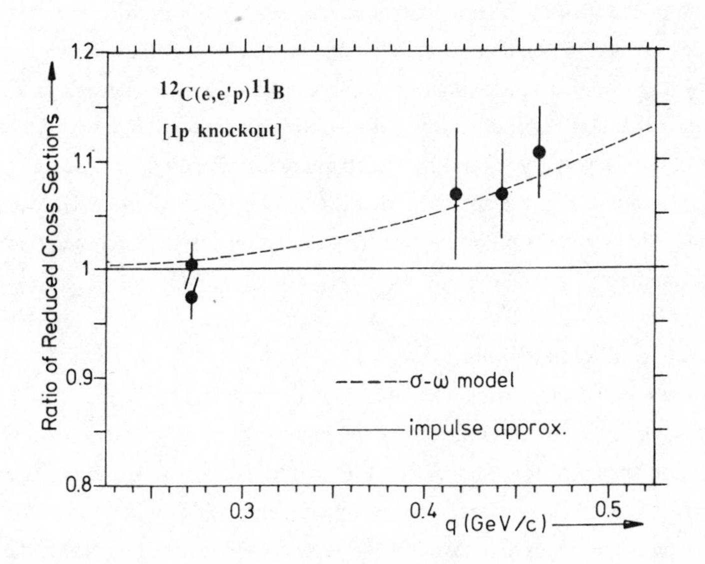

Fig. 12 Ratio of reduced cross sections for the reaction $^{12}C(e,e'p)^{11}B$(g.s.) as a function of 3-momentum transfer.

window $\Delta\omega \approx 150$ MeV of the inclusive spectra. Then the observed reduction of C_L would imply that the quasi-elastic Coulomb coupling needs to be renormalized. Since in the (e,e'p) experiments discussed here the Coulomb coupling is dominant (80 % of σ_{ep}) the spectroscopic factors deduced should in consequence be renormalized. Upon application of a 10 % FSI correction and using the data shown in fig. 11 the fraction of the sumrule value of 1f spectroscopic strength for $^{51}V(e,e'p)^{50}Ti$ is 0.65 ± 0.10, a quite reasonable number in view of the theoretical estimate of depletion of surface orbitals. However, before firm conclusions may be drawn the above preliminary inferences need be corroborated by additional inclusive data on heavier nuclides, combined with a longitudinal-transverse separation method in which the Coulomb distortion effects are treated accurately.

As a final topic I will discuss a recent experiment (Van der Steenhoven, 1986) in which a longitudinal transverse separation of the 1p-knockout process was obtained for the $^{12}C(e,e'p)^{11}B$ reaction. Precise(e,e'p) cross sections were measured in a range of q values ranging from 300 - 480 MeV/c. At each q-value two kinematics were used, one at backward scattering angle thus enhancing the transverse electron-proton coupling, and one at forward angle. Typically the values of σ_{ep} at the two kinematics differ by a factor five to ten. By dividing the coincidence cross section by σ_{ep} a reduced cross section σ_R is obtained. As is shown in fig. 12 the ratio of backward over forward angle σ_R's follows to better than ten percent the impulse approximation prescription. While referring for a more detailed interpretation of the data to (Van der Steenhoven, 1986), the implication relevant in the present context is that the change of the elementary σ_{ep} by an order of magnitude is well described by the impulse approximation. Whereas this does not necessarily preclude a possible renormalization of the coupling, it does strongly support that basically the kinematic dependence of the elementary photon-proton coupling is correctly described by σ_{ep}. In order to resolve the Coulomb sumrule anomaly what is clearly needed is a more detailed theoretical investigation of the adequacy of the impulse approximation description of the (e,e'p) reaction in quasifree kinematics.

Summary and conclusions

With a few selected examples I have illustrated the potential of the high resolution proton knockout reaction as a spectroscopic tool. One major goal of these (e,e'p) experiments is to deduce absolute spectroscopic factors, i.e. the degree of overlap of the real wavefunction with a basis state of the shell model. Since an exclusive reaction is considered with one observed hadron in the final state inevitably the aspect of the reaction mechanism enters in the analysis. Within the confines of the Distorted Wave Impulse Approximation the final state interaction (FSI) calculated with a proton-nucleus optical potential was shown to be well under control. The spectroscopic factors for states near the Fermi edge deduced from the (e,e'p) cross sections indicate an uncomfortably small partial occupation probability of ≈ 50%. It is intriguing though that an analysis of (d,^{3}He) data in which a boundstate wavefunction deduced from electron scattering data is used yields similarly small numbers.

In the lead region the application of a nearly model-independent sumrule yields an occupancy of 80 ± 10 % of the 3s proton orbit in ^{208}Pb. Such partial occupancy allows to understand in a natural way the quenching of a great many 'single-particle' observables in the lead region. In spite of the solid progress made in experimental technique and in the interpretation of (e,e'p) data the following main points remain to be investigated. Concerning the FSI a consistent treatment of the bound-state wavefunction and the scattering state distortion in one, possibly microscopic, framework is needed. The applicability of the impulse approximation in the present kinematics must be scrutinized in view of the seemingly anomalous Coulomb coupling of the virtual photon with the proton. We note finally that future progress in the understanding of nuclear structure beyond the meanfield approximation will benefit greatly from the enlarged kinematic window and increased duty factor of the new generation of electron machines.

The work reported here is the result of a concerted effort of the coincidence group members of NIKHEF-K. Fruitful collaboration with H. Nann (Indiana), B. Frois (Saclay), G. Wagner (Tübingen) and his group is acknowledged. I would like to thank H.P. Blok (Free University, Amsterdam) for his essential contribution, in particular to the merging of spectroscopic information from (e,e'p) and (d,^{3}He). Penetrating discussions with B. Frois helped in the preparation of the manuscript. Mrs. Marijke Oskam-Tamboezer showed diligence and patience in typing the manuscript. The graphic-art support of R. Arink and H. Beumer is gratefully acknowledged.

This work is part of the research program of the National Institute for Nuclear Physics and High-Energy Physics (NIKHEF, section K), made possible by financial support from the Foundation for Fundamental Research on Matter (FOM) and the Netherlands Organization for the Advancement of Pure Research (ZWO).

References

Clement, H. et al., 1986 , preprint University of Tübingen
De Forest, T., Jr., 1983 , Nucl. Phys. **A392**, 232
De Forest, T., Jr., 1984 , Phys. Rev. Lett. **53**, 895
De Witt Huberts, P.K.A., 1985 , Nucl. Phys. **A446**, 301c
Deady, M. et al., 1983 , Phys. Rev. **C28**, 631
Den Herder, J.W. et al., 1985 , Phys. Lett. **161B**, 65
Den Herder, J.W. et al., 1986 , to be published in Phys. Rev. Lett.
Frois, B, 1983 , Nucl. Phys. **A396**, 409c
Frullani, S. and Mougey, J., 1984 , Advances in Nuclear Physics **14**, 12
Grabmayr , P. et al., 1985 , Phys. Lett. **164B**, 15
Hinterberger, F. et al., 1967, Z. Physik **202**, 236
Horikawa, Y. et al., 1980 , Phys. Rev. **C22**, 1680
Jaminon, M. et al., 1985 , Nucl. Phys. **A440**, 228
Jans, E. et al., 1982 , Phys. Rev. Lett. **49**, 974
Keizer, P.H.M., 1986 , Ph.D. Thesis, University of Amsterdam (unpublished)
Kramer, G.J. et al., 1986 , Preprint Free University Amsterdam (to be published in Nucl. Phys. B).
Laget, J.M., 1985 , Phys. Lett. **151B**, 325
Lapikàs, L., 1986 , Proc. Fourth Miniconference 'Nuclear Structure in the 1p shell', NIKHEF-K, EMIN 86-04
Meier-Hadjuk, H. et al., 1983 , Nucl. Phys. **A395**, 332
Meziani, Z.E., 1985 , Nucl. Phys. **A446**, 113c
Negele, J.W. and Vautherin, D., 1972, Phys. Rev. **C5**, 1472
Pandharipande, V.R., Papanicolas, C.N. and Wambach, J., 1985 , Nucl. Phys. **A446**, 189c
Papanicolas, C.N., 1986 , Nuclear Structure at high spin excitation and momentum transfer, AIP Conf. Proc. 142, Ed. H. Nann
Platchkov, S. et al., 1982 , Phys. Rev. **C**, vol. 25(2), 2318
Quint, E.N.M. et al., 1986 , Phys. Rev. Lett. **57**, 186
Stuirbrink, A. et al., 1980 , Z. Physik **297**, 307
Tjon, J.A., 1986 , private communication
Van der Steenhoven, G. et al., 1985 , Phys. Lett. **156B**, 146
Van der Steenhoven, G. et al., 1986 , Phys. Rev. Lett. **57**, 182
Wagner, G.J., 1986 , Proc. AIP Conference ' Nuclear Structure at high spin, excitation and momentum transfer; Editor: H. Nann

Inst. Phys. Conf. Ser. No. 86
Paper presented at Int. Nucl. Phys. Conf., Harrogate, UK, 1986

The pion as a probe of nuclear structure, what we have and have not learned

J. David Bowman

Los Alamos National Laboratory, Los Alamos, NM 87545, U.S.A.

I will discuss some experiments that have used the pion as a probe of nuclear structure. My talk will not be a comprehensive review. I have chosen examples of experiments that use the qualitative features of the π-nucleon interaction to study selectively certain aspects of nuclear structure. The talk is divided into five parts: (1) properties of the pion and the pion-nucleon interaction, (2) studies of the isospin mixing and neutron-to-proton matrix element ratios in inelastic scattering, (3) studies of isovector giant resonances in pion single-charge-exchange reactions, (4) the search for six-quark structures in the nucleus using pion double-charge-exchange reactions, and (5) concluding remarks.

A theme that runs through all work that seeks to use strongly interacting probes to observe the structure of the nucleus is the interplay of the incompletely known interaction mechanism and structure of the nuclear transition being studied. The cross sections that we measure involve both structure and interaction mechanism. The separation of these two aspects of the problem must begin with the qualitative features of the probe. Based on these qualitative features the experimenter can focus on those types of nuclear transitions that are best suited to study using a given probe. He may then design his experimental observation so as to minimize interaction-mechanism uncertainties.

Properties of the Pion and the Pion-Nucleon Interaction

The pion has isospin 1, hence it exists in three charge states, π^+, π^-, and π^0. These particles have a mass about 1/7, that of the nucleon mass. Pions having a kinetic energy comparable to their mass interact strongly with the nucleon forming a resonance (the (3,3) resonance) having isospin 3/2 and angular momentum 3/2. The pion-nucleon cross sections are shown as functions of momentum in Figure 1. The (3,3) resonance dominates the π-nucleon interaction at intermediate energies. At the peak of the (3,3) resonance the semiclassical estimate of the pion mean free path λ in nucleon matter is $\lambda = \frac{1}{\rho\sigma} = 0.7$ fm. Here ρ is the density of nucleons per cubic femtometer and σ is the isospin-averaged pion-nucleon total cross section. On the basis of this short mean free path it is to be expected that resonance energy pions will interact primarily in the nuclear surface. At energies above and below resonance energies the pion mean free path exceeds that of the nucleon and contributions to the scattering amplitude from the nuclear interior are to be expected.

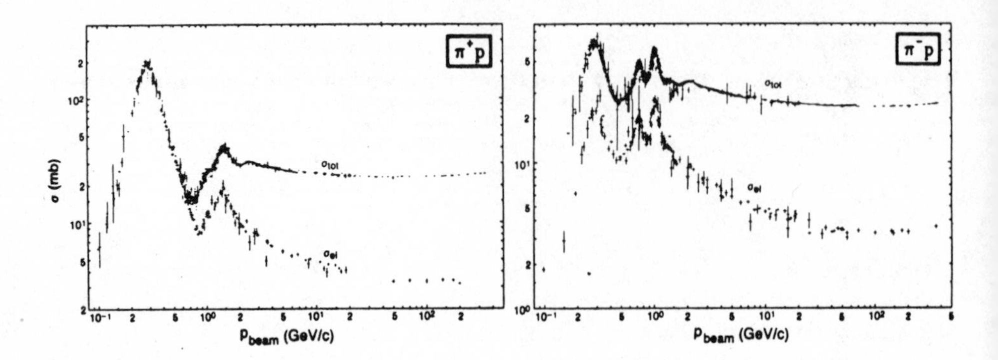

Figure 1. Pion nucleon total and elastic cross sections as functions of pion momentum.

Since the pion exists in three charge states a rich variety of reactions is possible. Both positive and negative pions may be used for elastic and inelastic scattering $\left((\pi^+, \pi^{+\prime})\right.$ and $\left.(\pi^-, \pi^{-\prime})\right)$. For nuclear targets both isospin-raising and -lowering charge-exchange reactions are possible $\left((\pi^-, \pi^0) \text{ and } (\pi^+, \pi^0)\right)$. Two types of reactions are possible for pions that are not possible for the nucleon. The first is double charge exchange (π^+, π^-) and (π^-, π^+) and the second is pion absorption.

The pion-nucleon scattering amplitude may be written as

$$f = f_s + g_s \sigma \cdot \frac{k \times k'}{|k|^2}$$

$$f_v T \cdot \tau + g_v T \cdot \tau \sigma \cdot \frac{k \times k'}{|k|^2} \quad .$$

Here k and k' are the initial and final pion momenta, σ is the nucleon spin, T and τ are the pion and nucleon isospin and f_s, g_s, f_v, and g_v are coupling constants. At energies near the (3,3) resonance the coupling constants have the ratios

$$f_s : g_s : f_v : g_v = 4 : 2 : 2 : 1$$

and pion-nucleon cross sections have the ratios

$$\sigma(\pi^+ p \to \pi^+ p) : \sigma(\pi^- p \to \pi^0 n) : \sigma(\pi^- p \to \pi^- p) = 9 : 3 : 1 \ .$$

Since the pion has spin 0 the nucleon spin couples to the pion orbital angular momentum. At forward scattering angles only non-spin-flip nuclear states are excited. At angles near 90° both spin-flip and non-spin-flip states are excited.

The experimental results that I will discuss were guided by the above considerations. In the inelastic scattering of positive pions the projectile couples nine times more strongly to the protons in the nucleus than to the neutrons. The opposite situation pertains for negative pions scattered by nucleons. In pion charge-exchange reactions only the $T \cdot \tau$ terms in the scattering amplitude are active, hence only isovector states are excited. In double-charge-exchange reactions at least two nucleons must be involved since only one unit of charge can be transferred in a pion nucleon collision. Therefore two-step processes play a leading role in double-charge-exchange reactions whereas two-step processes are higher order corrections in most other nuclear reactions.

The Study of Isospin Mixing and Neutron-to-Proton Matrix-Element Ratios in Pion Inelastic Scattering

In this section I will summarize work carried out by groups from the University of Minnesota, Los Alamos National Laboratory, the University of Texas, and the University of Pennsylvania. The studies were motivated by the 9:1 ratios of π^+p to π^+n and π^-n to π^-p scattering cross sections at (3,3) resonance energies. Both non-spin-flip and spin-flip transitions were studied.

In order to verify that the above 9:1 ratio pertains in the nucleus as well as for the nucleon, inelastic cross sections for both positive and negative pions were measured for well known $T = 0$ states in the $T = 0$ nuclei ^{12}C and ^{40}Ca.[1] Some of the results are shown in Figure 2. Since neutrons and protons contribute equally and symmetrically to the wave functions of $T = 0$ states in $T = 0$ nuclei one would predict that the differential cross sections for positive and negative pions would be the same. The experimental results verify this prediction.

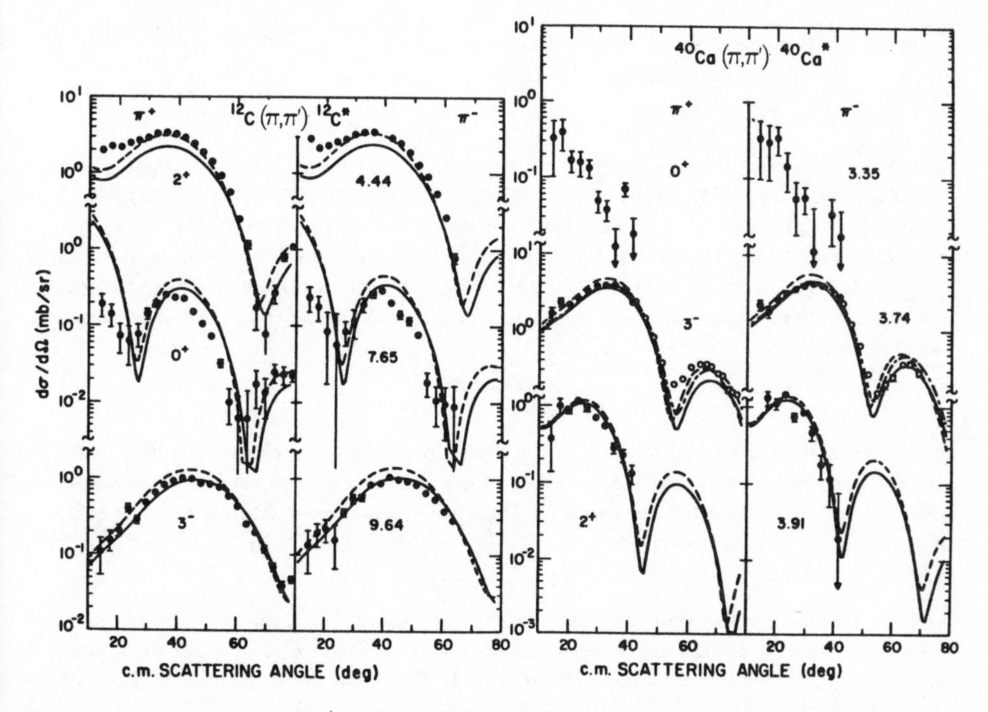

Figure 2. Comparison of π^+ and π^- differential cross sections for $T = 0$ states in ^{12}C and ^{40}Ca.

Two additional experiments give confidence that the ratio of 9:1 applies in the nuclear medium. The $9/2^+$ at 9.5 MeV in ^{13}C is formed by promoting a $1p3/2$ neutron to the $1d5/2$ shell. Detailed shell-model calculations[(2)] indicate that this state is a pure neutron excitation. Therefore the ratio of π^+ to π^- cross section is expected to be 1:9. Figure 3 shows the experimental results. The ratio is indeed 1:9.

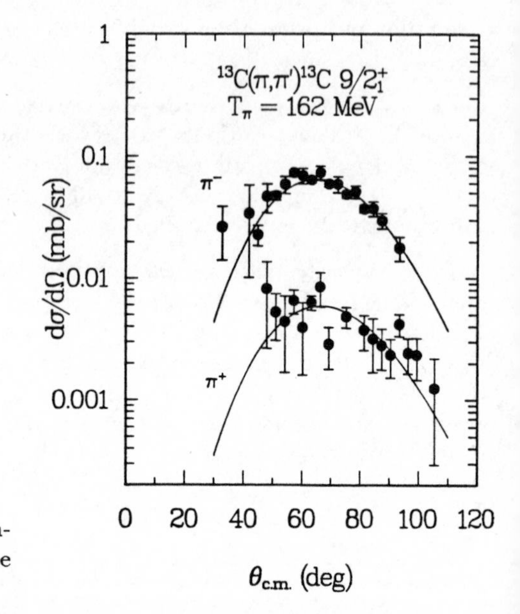

Figure 3. Comparison of π^- and π^+ inelastic differential cross sections to the $9/2^+$ state in ^{13}C at 9.5 MeV.

A quantitative test of the ability of $\pi^+ : \pi^-$ matrix elements to yield isospin mixing ratios has been provided by the $J^\pi = 1^+$ states in ^{12}C at 12.71 and 15.11 MeV. The lower state is primarily $T = 0$ and the upper state is $T = 1$. These states are known to be isospin mixed and the amount of mixing has been measured by Flanz et al.[(3)] using 180° electron scattering. The ratio of 180° electron cross sections for these two states in the absence of isospin mixing would be the ratio of isoscalar to isovector magnetic moments squared,

$$\frac{\sigma_0}{\sigma_1} = \left|\frac{\mu_n + \mu_p}{\mu_n - \mu_p}\right|^2 = \frac{1}{28} \quad .$$

From the measured ratio Flanz et al. deduce a mixing matrix element of $H_{10} = 140 \pm 35$ keV. The π^+ to π^- cross-section ratio for these two states has been measured by Morris et al.[(4)] From the deviation of the ratios from unity Morris et al. deduce a mixing matrix element of $H_{10} = 148 \pm 28$ keV in agreement with the value from 180° electron scattering.

As indicated above the pion inelastic scattering near 90° is a good probe of isoscalar spin-flip strength. A class of states in p-shell nuclei having $\Delta J^\pi = 4^-$, $\Delta L = 3$, $\Delta S = 1$ has been discovered in pion inelastic scattering. These states are formed by transitions of the type $[1d3/2^-, 1d5/2]_{4^-}$. An example is the 4^- doublet in ^{12}C at 19.25 and 19.65 MeV. The π^+ to π^- cross-section ratios for these states are shown in Figure 4. Remarkably the π^+ to π^- ratio for the 19.25 MeV state is close to 9:1. The π^+ to π^- ratio for the 19.65 MeV state is close to 1:9. Therefore these states are nearly pure proton and neutron excitations due to strong isospin mixing.

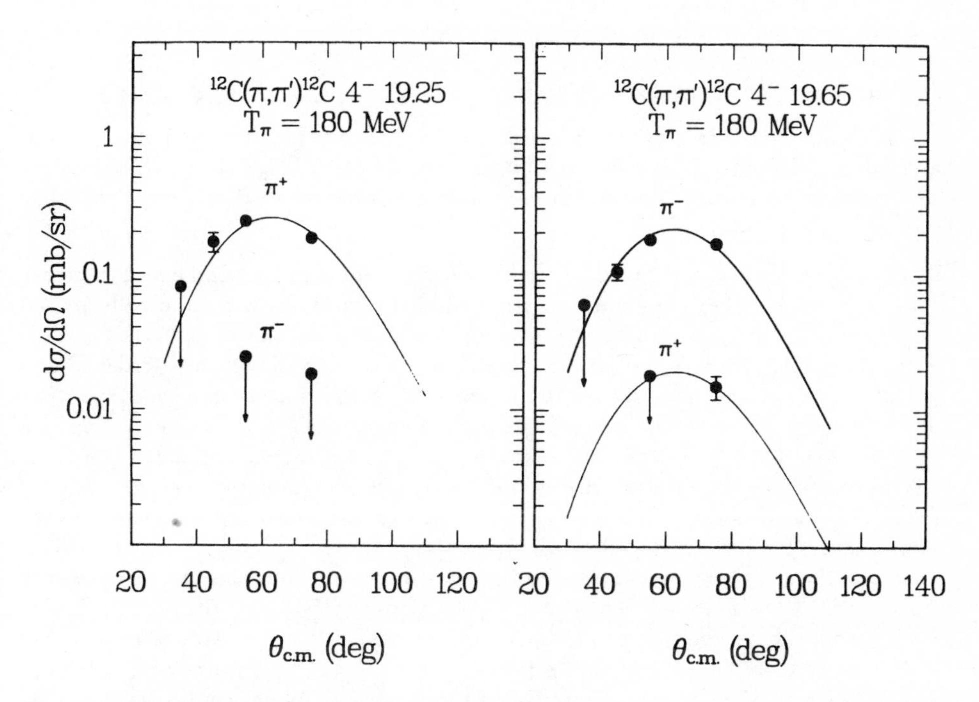

Figure 4. Comparison of π^+ and π^- cross differential cross sections for the isospin mixed doublet in ^{12}C.

Efforts to study the isospin structure of isoscalar quadrupole giant resonance in heavy nuclei have yielded interesting results.(5,6) The interpretation of $\pi^+ : \pi^-$ cross-section ratios for nuclei with a large neutron excess will be difficult for two reasons. First, the isovector distortion of the incident pion waves is not well known. For nuclei with a large neutron excess the π^- waves will be absorbed more than the π^+ waves and increase the $\pi^+ : \pi^-$ cross-section ratios. Second, the Coulomb force will cause the proton part of the transition density to extend to larger radii than the neutron part of the transition density. This effect will also increase the π^+ to π^- cross-section ratio since the resonance-energy pions interact primarily in the surface. The observed $\pi^- : \pi^+$ cross-section ratio for the collective 3^- first excited state of ^{208}Pb in which neutrons and protons participate in the ratio $N : Z$ has been measured to be 1.8 ± 0.1.(6) The measured $\pi^- : \pi^+$ ratio for the isoscalar quadrupole giant resonance peak in ^{208}Pb has a much larger value of 2.7 ± 0.5.(6) This result would indicate a substantial amount of isovector admixture in the isoscalar quadrupole resonance, but one must bear in mind the above reservations. A better characterized isovector optical potential is necessary for the quantitative interpretation of π^- to π^+ cross-section ratios in nuclei having a large neutron excess.

In summary the measurement of $\pi^+ : \pi^-$ cross sections in inelastic scattering has been used to study isospin mixing in $T = 0$ nuclei and to measure neutron-to-proton matrix-element ratios. A class of isospin-mixed $J^\pi = 4^-$ states in *p*-shell nuclei has been discovered and studied. The extension of this technique to nuclei having large neutron excess has yielded intriguing results but the quantitative interpretation of measured ratios is made ambiguous by interaction-mechanism uncertainties.

The Study of Isovector Resonances in Pion Single-Charge-Exchange Reactions

In this section I will discuss the results of a series of experiments carried out by a collaboration of Tel-Aviv University and Los Alamos National Laboratory, which studied the excitation of $L = 0$, 1, and 2 isovector giant resonances in pion charge-exchange reactions $(\pi^{\pm}, \pi^{0})$.[7]

Giant resonances are excitations of the nucleus in which large numbers of nucleons move collectively. They are simple modes of nuclear excitation that can be interpreted microscopically or macroscopically. Their experimental observation and study as well as their theoretical interpretation are important for models of nuclear excitation and the knowledge of the nucleon-nucleon interaction in the nuclear environment. The experimental properties that characterize a giant resonance are the concentration of a large fraction of the total available transition strength with specific quantum numbers in a narrow region of excitation energy, the occurrence of resonances in a wide range of nuclei, and the smooth variation of excitation energy and width of the resonance with nuclear mass A. In contrast to the isoscalar electric modes, which have been extensively studied in the scattering of hadronic probes, the $L = 1$, $T = 1$ or giant dipole resonance (GDR), which has been studied with electromagnetic probes, and the $L = 0$, $T = 1$, $S = 1$ Gamow-Teller resonance, which has been studied in the (p, n) reaction, the $L = 0$ $T = 1$ isovector monopole resonance (IVM) and $L = 2$, $T = 1$ isovector quadrupole resonance (IVQ) were poorly characterized before pion charge-exchange studies. The study of the IVM was of particular importance. Its existence had been predicted by both macroscopic[8] and microscopic[9] theories, but it had not been observed. The IVM plays a central role in Coulomb effects such as isospin mixing in nuclear ground states, Coulomb displacement energies and widths of analog states.

The quantum numbers and dynamical properties of resonance-energy pions make the pion charge-exchange reactions $(\pi^{\pm}, \pi^{0})$ ideal for the study of electric isovector resonance especially the IVM. First, the use of a charge-exchange reaction that excites only isovector states eliminates the excitation of isoscalar states that dominate the excitation spectra of inelastic scattering processes. Second, at forward angles, where the angular distribution of the IVM peaks, pion charge exchange excites primarily electric, or non-spin-flip transitions (in contrast to the (p, n) reaction), reducing spin-flip backgrounds. Third, the strong absorption of the pion is essential for the excitation of a monopole state for which the volume integral of the transition density is zero. The angular distributions produced by the surface-related diffractive pion scattering process [similar to $((\alpha, \alpha')$] oscillate sharply and characteristically with angle.[10]

$$\frac{d\sigma}{d\Omega} \sim J_0^2(kR\theta) \quad \text{for} \quad L = 0$$

$$\frac{d\sigma}{d\Omega} \sim J_1^2(kR\theta) \quad \text{for} \quad L = 1$$

$$\frac{d\sigma}{d\Omega} \sim J_0^2(kR\theta) + 3J_2^2(kR\theta) \quad \text{for} \quad L = 2 \quad ,$$

where k is the π momentum, θ is the scattering angle and R is the strong absorption radius. This rapidly oscillating behavior serves to identify the multipolarity of the transition and to distinguish the giant resonance signals from the nonresonant background. In Figure 5 representative angular distributions for the ^{60}Ni (π^-, π^0) reaction at 230 MeV are shown. The qualitative patterns of the angular distributions do not depend on the details of the reaction model used so long as the pion waves are strongly absorbed.

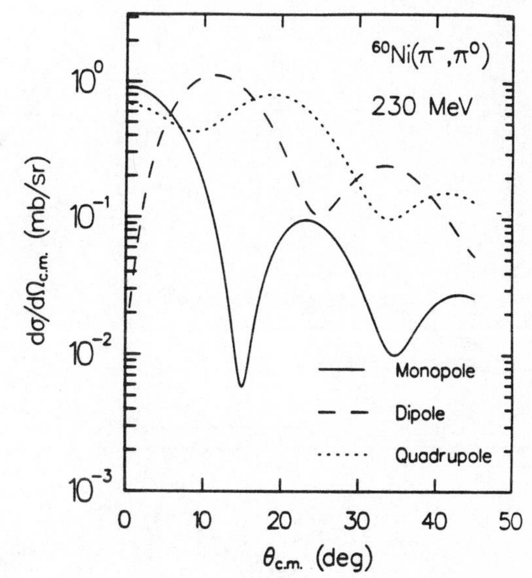

Figure 5. Angular distribution for $L =$ 0, 1, and 2 collective transitions for 230 MeV (π^-, π^0) reactions on ^{60}Ni.

Fourth, the Coulomb energy shift for states populated by the (π^-, π^0) reaction is advantageous. This point is illustrated in Figure 6 where the analog state relations are shown for an isovector resonance built on a target ground state of isospin T. For nuclei with $T \gg 1$, transitions to the state of lowest total isospin are strongly favored by isospin coupling coefficients. The state of total isospin $T+1$ in the (π^-, π^0) daughter is shifted down by the Coulomb displacement energy relative to its analog in the parent nucleus. Thus the $T+1$ state has a relatively low excitation energy in the (π^-, π^0) daughter and occurs at an excitation energy where the density of states of the same isospin is small. The opposite situation pertains for the (π^+, π^0) reaction. The $T+1$ component populated in the (π^-, π^0) reaction is expected to be narrow and to occur at a low excitation energy while the $T-1$ component populated in the (π^+, π^0) reaction is expected to be wide and to occur at a high excitation energy.

In the experiments well-understood, spherical, even-even nuclei were studied in the $(\pi^{\mp}, \pi^0)$ reactions at 120, 165, and 230 MeV. Double differential cross sections up to excitation energies of 60 MeV were measured out to angles extending well past the second maximum of the IVQ angular distribution. Data for the ^{120}Sn target with a 165 MeV π^- beam are shown in Figures 7a and 7b. At the most forward angle, 4.5°, the IVM cross section is expected to be the largest. The second angle, 11°, is chosen to be near the first minimum of the monopole angular distribution. The GDR cross section is small at the forward angle and has a maximum near 11°. Figure 7c shows the results of subtracting the 11° spectrum from the 4.5° spectrum. This subtraction suppresses the approximately isotropic non-resonant background. The IVM signal is the positive-going hump and the small GDR signal is the negative-going hump.

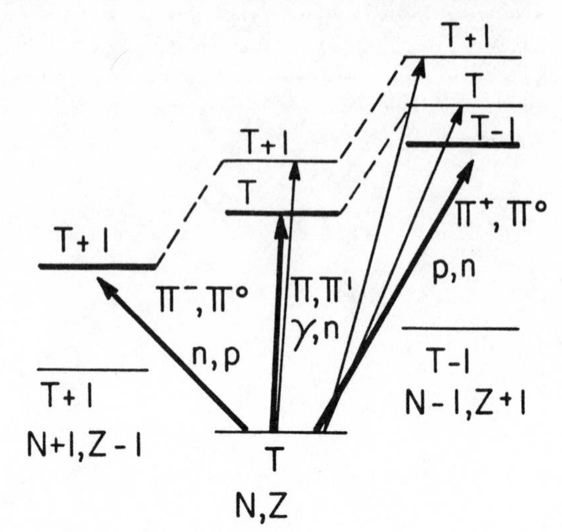

Figure 6. Analog relationships for isovector resonances built on a $T \geq 1$ target nucleus.

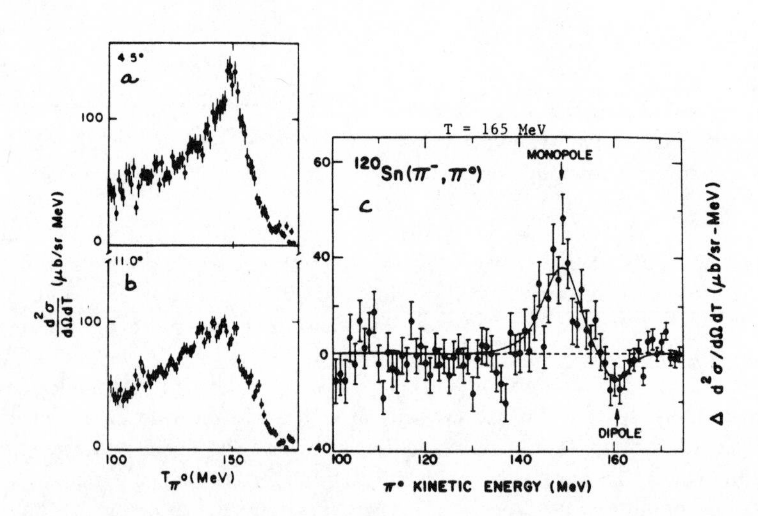

Figure 7. Figures 7a and b show the double differential cross section as functions of T_{π^0} In the ^{120}Sn(π^-, π^0) reaction at a forward angle where the IVM cross section is large (a) and at a larger angle where the IVM cross section is small (b). Figure 7c shows the difference of the 4.5° and 11° spectra.

The above analysis shows the existence of peaks above an approximately isotropic background in the $d^2\sigma/d\Omega dE$ versus pion kinetic energy spectra. The IVM peak is visible in a comparison of the 4.5° and 11° spectra but the GDR peak is not. The subtraction procedure makes the weak GDR peak visible. To investigate the degree of isotropy of the non-resonant background it is necessary to look at the dependence of different regions of excitation energy on scattering angle or momentum transfer

q. Figure 8 shows the forward-angle data for ^{60}Ni (π^-, π^0) taken at 230 MeV. As before, the IVM is large in the most forward angle where the GDR is small. The IVM is small at the second angle where the GDR is large and both are small at the largest angle. In general cross sections depend on energy loss ν and momentum transfer q^2. Figure 9 shows $d\sigma/d\Omega$ obtained by integrating over the three regions indicated in Figure 8. Region one emphasizes the IVM, region two emphasizes the GDR and region three contains mostly nonresonant background. Each plot can be represented as a superposition of a background linear in q^2 and a component having the q^2 dependence expected for $d\sigma/d\Omega$ for a $L = 0$ or 1 resonance. Although the background is not strictly isotropic its dependence on q^2 is much less rapid than that of the giant resonances.

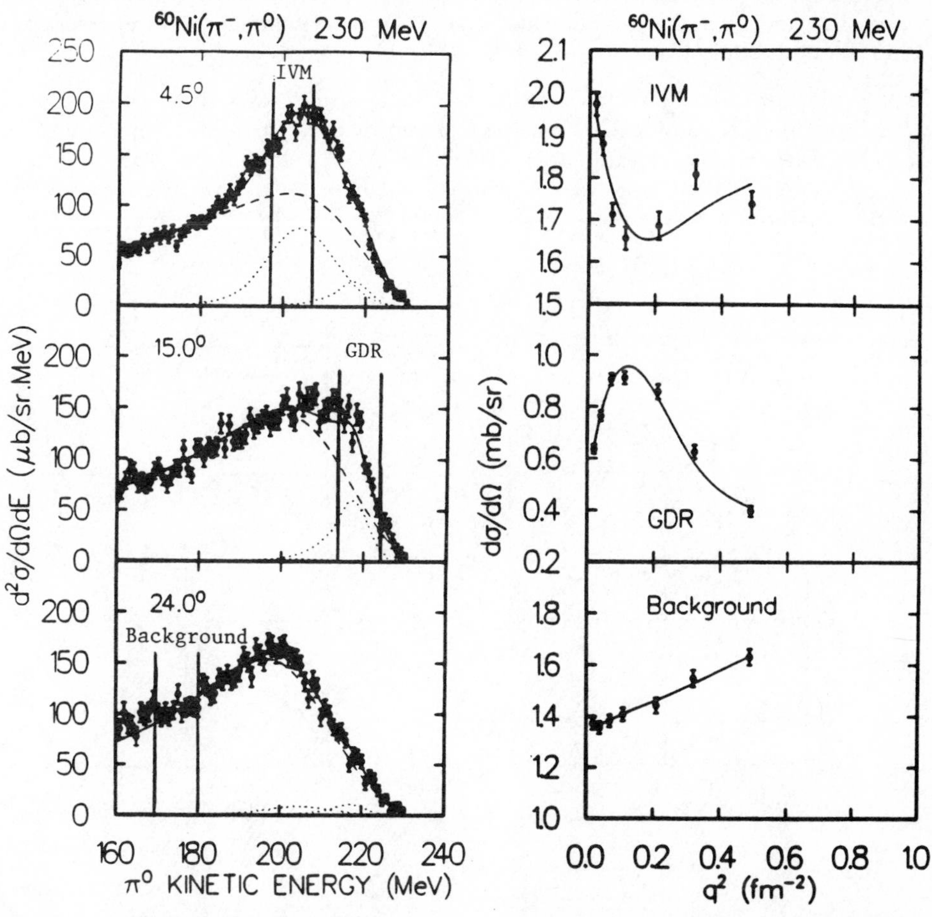

Figure 8. Doubly differential cross sections as functions of T_{π^0} for scattering angles of 4.5°, 15.0°, and 24.0° for the ^{60}Ni(π^-, π^0) reaction at 230 MeV.

Figure 9. Singly differential cross sections obtained by integrating over the T_{π^0} regions shown in Figure 8, as functions of momentum transfer squared. The solid lines are fits to a linear background added to a $L = 0$ or $L = 1$ angular distribution for the IVM and GDR respectively.

In order to extract quantitatively excitation energies, widths, and maximum cross section, a least-squares fitting procedure was followed. The double differential cross section as a function of q^2 and ν was written as a sum of two Gaussian peaks at an angle-independent excitation energy. The q^2 variation of the peaks was taken to be that of distorted wave impulse approximation calculations using random phase approximation (RPA)[11] transition densities. The sizes of the maximum cross sections were varied. The background was written as a function having a smooth ν-dependence and a quadratic q^2-dependence. Resonance and background parameters were varied to fit the data for each target. For targets where data were taken at different bombarding energies, the extracted resonance energies and widths were consistent although the background shapes were different. The background function and the resonance components shown in Figure 8 were obtained in this way.

Figure 10 shows the extracted maximum cross sections, excitation energies, and widths for the IVM and GDR resonances with results of random-phase-approximation distorted-wave-impulse-approximation (RPA-DWIA) calculations.[11] The A dependence of the 1 $\hbar\omega$ GDR cross sections can be understood as follows. In the (π^-,π^0) reaction a proton is turned into a neutron and is promoted by one major shell. For

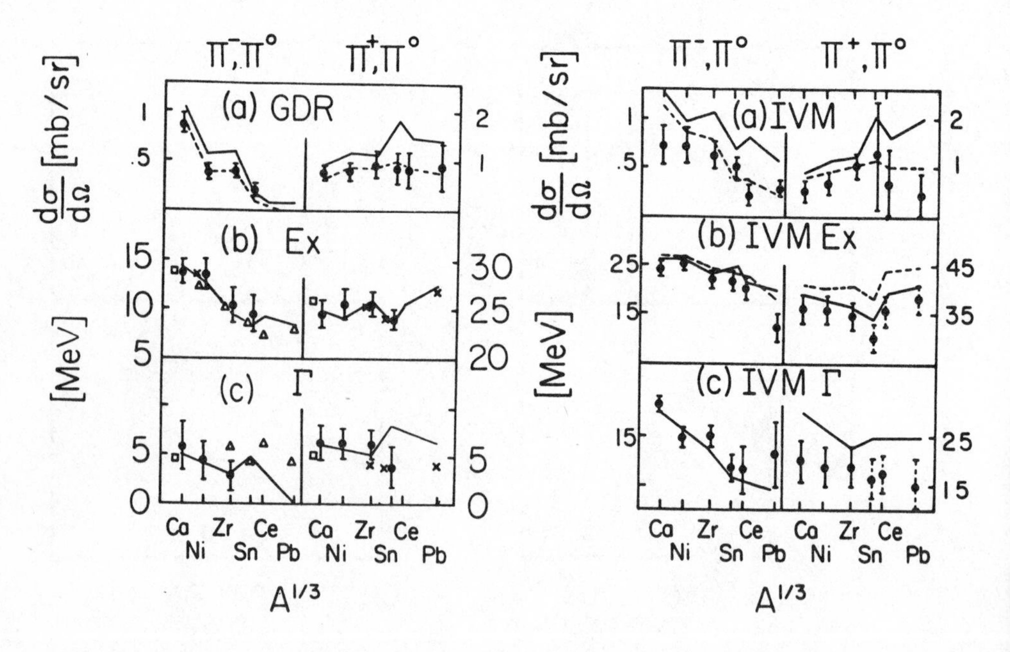

Figure 10. Extracted maximum cross sections, excitation energies, and widths for the GDR and IVM resonances. The lines are the results of random phase approximation[11] calculations done before the data were available.

the $T = 0$ nucleus, ^{40}Ca, there is no blocking and the π^+ and π^- cross sections are comparable. For ^{208}Pb the neutron shell is fully occupied and the (π^-,π^0) cross section for the GDR is zero. The same effect is seen to a lesser extent for the A dependence of the 2 $\hbar\omega$ IVM maximum cross sections. Here the (π^-,π^0) IVM cross sections decrease by about a factor of two from ^{40}Ca to ^{208}Pb while the (π^+, π^0) IVM cross sections are approximately constant. As expected, the widths and excitation energies of $T+1$ states are larger than those of the corresponding $T-1$ states. Where data are available from other reactions, they are also shown. The solid curves give the theoretical results for multiple strength weighting and the dashed curves give the results for cross section weighting. The cross-section-weighted RPA theory using the Skyrme III residual interaction[11] gives a reasonable description of the data.

Remarkably, no IVQ peak was necessary to fit the experimental double differential cross section data. In order to quantify the amount of IVQ cross section present in the data, a third Gaussian peak was added to the fitting function. The excitation energy, width, and q^2-dependence of the peak were taken from RPA-DWIA calculations. The data were refitted and 90% confidence level upper limits were deduced for the presence of an IVQ component. These upper limits were 0.18, 0.30, and 0.15 of the RPA-DWIA estimate of the peak IVQ cross section for ^{40}Ca, ^{60}Ni, ^{90}Zr (π^-, π^0) respectively. For the IVM and GDR the observed cross sections were typically 0.7 of the RPA-DWIA calculations. If a much larger width was assumed the upper limit for the IVQ cross section was increased and became consistent with the RPA-DWIA estimate.

An interesting explanation of the absence of isovector quadrupole strength in the pion charge-exchange reaction has been proposed by Leonardi et al.[12] They argue in a sum rule framework that the inclusion of non-local terms in the residual interaction would have little effect on the properties of $L = 0$, 1, and 2 isovector giant resonances in $\Delta T_z = 0$ channels, and on the properties of the $L = 0$ and $L = 1$ isovector resonances in charge-exchange channels. Thus RPA calculations with local interaction could correctly describe the properties of these isovector giant resonances even if the true residual interaction were non-local. However, Leonardi et al. argue that the inclusion of non-local residual interaction could radically broaden and weaken the isovector quadrupole resonance in the charge-exchange channels. The absence of isovector quadrupole strength in the pion charge-exchange reactions has been a problem and Leonardi's explanation is intriguing.

To summarize this section, the isovector monopole resonance has been observed in the pion charge-exchange reaction. The properties of the charge-exchange components of the isovector monopole and dipole resonances have been studied and found to be in good agreement with RPA predictions. No charge-exchange isovector quadrupole strength was observed. An explanation in terms of non-local residual interaction has been suggested.

Search for Six-Quark States in Nuclei Using Double Charge Exchange

The search for six-quark clusters in pion double charge exchange was motivated by the weakness of the pion-nucleon interaction at 50 MeV, which allows the pion to probe the nuclear interior at this energy, and by the existence of a cancellation between the s and p wave amplitudes for pion-nucleon single charge exchange at zero degrees. Both the s and p wave amplitudes are nearly real at low energies. The s-wave amplitude increases as k while the p-wave amplitude increases as k^3. Their signs are opposite and a cancellation occurs at 50 MeV. This cancellation also dominates the energy dependence of the π-nucleus charge charge-exchange reaction leading to the isobaric analog state. Figure 11 compares the bombarding energy dependence of nucleon charge exchange cross sections at zero degrees and the reaction leading to the isobaric analog state $^{14}C(\pi^+,\pi^0)^{14}O$ (IAS) at zero degrees.[13] This interference minimum has been studied for nuclei from ^7Li to ^{120}Sn.[14] Remarkably the minimum persists even in heavy nuclei. Due to the *s-p* interference at 50 MeV, the angular distribution for pion charge exchange on the nucleon is backward peaked. Angular distributions of single charge exchange reactions leading to the IAS have been measured for the nuclei ^7Li, ^{14}C, and ^{15}N[15] at 50 MeV. These angular distributions were also found to be backward peaked, reflecting the behavior of the elementary process. If the isobaric analog state were the dominant intermediate state for double charge exchange leading to the double-isobaric-analog state one would expect backward peaked angular distribution for this process as well.

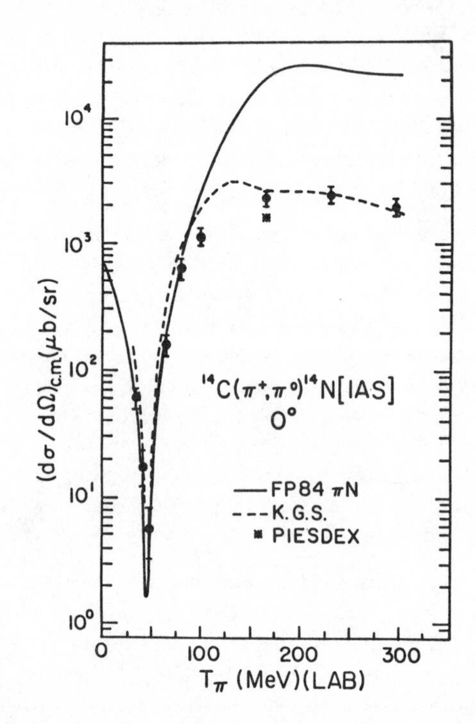

Figure 11. Zero-degree cross section for the reaction $\pi^- p \to \pi^0 n$ (solid line) compared to the zero-degree cross section for the reaction $^{14}C(\pi^-,\pi^0)^{14}N$ (IAS)[13] as functions of the π^- kinetic energy.

Measurements of the double-isobaric-analog state transition by Navon et al.[16] on ^{14}C, $^{14}C(\pi^+,\pi^-)^{14}O$ showed an unexpectedly large cross section. Furthermore the differential cross section was increasing as the scattering angle decreased. The most forward angle of this measurement was 50°. Distorted wave impulse approximation[16,17] calculations based on the idea that the isobaric analog state was the dominant intermediate state predicted a zero-degree cross section of about 0.2 μb/sr at zero degrees, which increased to a few hundred μb/sr at backward angles.

Miller[18] proposed a new interaction mechanism that could lead to large forward-peaked double-isobaric-analog state cross sections while keeping the single-charge-exchange isobaric-analog-state cross section small. He argued that when the two valence neutrons in ^{14}C were closer than about 1 fm (about 6% of the time) they would exist in a six-quark state. In his calculations for the structure of this state a π^+ could flip the isospin of two down quarks to up and emerge in the forward direction as a π^-. The zero-degree cross section was calculated to be 11 μb/sr. An experiment was carried out at Los Alamos that measured the zero-degree cross section to be 4 μb/sr. The results of this experiment as well as various theoretical calculations are shown in Figure 12. Clearly Miller's approach better explained the experimental results than did the sequential isobaric-analog-state hypothesis. The inclusion of a realistic amount of absorption brought Miller's original calculation of the zero-degree cross section down to 4 μb/sr in agreement with the experiment.

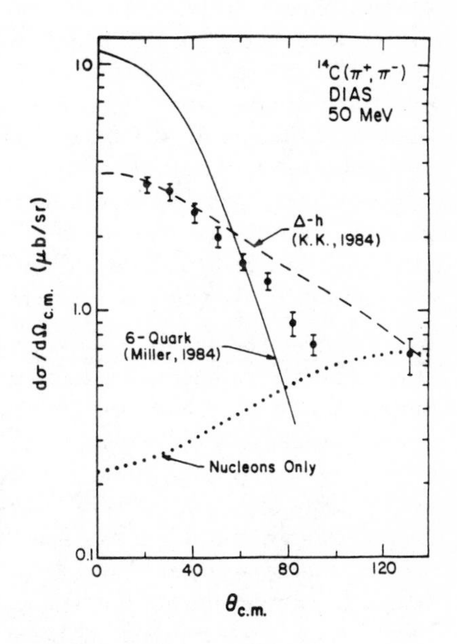

Figure 12. Measured $^{14}C(\pi^+, \pi^-)^{14}O$ DIAS cross sections compared with three theoretical calculations, see text.

Karapiperis and Kahagashi[19] proposed an explanation for the large zero-degree double-isobaric-analog-state cross section, which does not require non-nucleonic degrees of freedom. These authors calculated the double-isobaric-analog-state cross section in a delta-hole approach. They included non-analog intermediate states in the closure approximation. Their calculation is shown in Figure 12 and compares well to the experimental data. Gibbs[20] also obtained large forward-peaked cross sections by including non-analog intermediate states. Both groups point out the importance of sequential single-charge-exchange scattering processes at large angles. Although the zero-degree single-charge-exchange cross section is small, the ninety-degree cross section is not. Two sequential scatterings at ninety degrees can lead to double charge exchange at zero degrees. An interesting feature that emerges from the work of Gibbs[20] is the important role of short-range nucleon-nucleon correlations. About half the scattering amplitude comes from double scatterings where the two struck nucleons are separated by less than 1 fm.

To summarize this section one must say that double-charge-exchange studies have not led to an unambiguous identification of six-quark structures in the nucleus. Attempts to understand the large forward-peaked double-isobaric-analog-state cross sections have brought to our attention the need for a better understanding of the role of non-analog intermediate states in the double-charge-exchange reaction. The expectation that double charge exchange is a good probe of nucleon pairs at short distances retains its validity. However, in order to use the double-charge-exchange reaction as a quantitative probe of short-range correlations, a better understanding of the pion nucleus reaction mechanism than we have at present is required.

Conclusion

The isospin structure of the pion makes it a specific probe of interesting aspects of the structure of nuclei. Successful investigations of separate neutron and proton components in nuclear transitions have been carried out in pion inelastic scattering. The isovector-monopole giant resonance has been discovered in pion charge-exchange scattering and the isospin structure of the isovector-monopole and isovector-dipole giant resonances have been studied. In these investigations as well as in studies of short-range nucleon-nucleon correlations in double charge exchange, reaction-mechanism uncertainties have been a serious problem in the quantitative interpretation of data, although experiments have been designed to exploit qualitative features of the pion-nucleon-scattering process, which minimize reaction-mechanism uncertainties.

In my opinion the field of pion-nucleus physics would benefit from experiments aimed at constraining reaction-mechanism uncertainties. Excellent elastic and inelastic pion-nucleus cross-section data over a wide range of scattering angles and bombarding energies are available. Double-charge-exchange experiments are more difficult, but a body of data exists for analog and non-analog transitions for a range of nuclei, energies above, near, and below the (3,3) resonance. For pion charge exchange the experimental data are much less complete. Only forward-angle cross sections leading to the isobaric analog state have been measured due to the limitations of existing instrumentation. Measurements of isobaric-analog-state cross sections to large angles are needed. Double-charge-exchange studies have indicated the importance of non-analog intermediate states. To study these effects it is important to study inelastic, single charge exchange and double charge exchange to final inelastic nuclear states, which are themselves related by isospin symmetry. I believe that the availability of such data would not only lead to a better phenomenological characterization of the isospin dependence of the pion-nucleus optical potential, but also to an understanding of the interesting underlying pion dynamics in the nuclear medium.

References

1. C. L. Morris et al., Phys. Rev. C24 (1981) 231.
2. D. J. Millener and D. Kurath, Nucl. Phys. A255 (1975) 315.
3. J. B. Flanz et al., Phys. Rev. Lett. 43 (1979) 1922.
4. C. L. Morris et al., Phys. Lett. (1981) 387.
5. J. L. Ullman et al., Phys. Rev. C31 (1985) 177.
6. S. J. Seestrom-Morris et al., Phys. Rev. C33 (1986) 1847; S. J. Seestrom-Morris, private communication; S. J. Seestrom-Morris, D. B. Holtcamp, and W. B. Cottingame, "Spin Excitations in Nuclei," Petrovich, Brown, Garvey, Goodman, Lindgren, and Love, eds., Plenum, 1984, p. 291.

7. A. Erell et al., Phys. Rev. C34 (1986) 1822.
8. A. Bohr and B. R. Mottelson, Nuclear Structure (Benjamin, New York, 1975).
9. G. F. Bertsch and S. F. Tsai, Phys. Rev. C18 (1975) 125; K. F. Liu and G. E. Brown, Nucl. Phys. A265 (1976) 385; N. Auerbach, Nucl. Phys. A182 (1972) 247.
10. J. D. Bowman et al., Phys. Rev. Lett. 46 (1981) 1614; A Gal, Phys. Rev. C25 (1982) 2680.
11. N. Auerbach and A. Klein, Phys. Rev. C28 (1983) 2075; N. Auerbach and A. Klein, Nucl. Phys. A395 (1983) 77; A. Klein, Ph.D. Thesis, Tel-Aviv University, 1984.
12. Leonardi et al., preprint (submitted to Phys. Lett.).
13. F. Irom et al., Phys. Rev. C28 (1983) 2565.
14. F. Irom et al., Phys. Rev. Lett. 55 (1985) 1862.
15. M. D. Cooper et al., Phys. Rev. Lett. 52 (1984) 1100; F. Irom et al., Phys. Rev. C28 (1984) 2565, and Los Alamos National Laboratory Report No. LA-UR-84-2451 (unpublished); J. L. Ullman et al., private communication.
16. I. Navon et al., Phys. Rev. Lett. 52 (1984) 105.
17. M. B. Johnson and E. R. Siciliano, Phys. Rev. C27 (1983) 1647.
18. G. A. Miller, Phys. Rev. Lett. 53 (1984) 2008.
19. T. Karapiperis and M. Kobayashi, Phys. Rev. Lett. 54 (1985) 1230.
20. Proceedings of Pion Double Exchange Workshop, Los Alamos, New Mexico, 1985 (unpublished).

Inst. Phys. Conf. Ser. No. 86
Paper presented at Int. Nucl. Phys. Conf., Harrogate, UK, 1986

Strange probes of the nucleus

Carl B. Dover

Brookhaven National Laboratory, Upton, NY, USA 11973

Abstract. Recent experimental and theoretical advances in hypernuclear physics are reviewed. An appraisal is given of various suggestions for using strange probes to test partial quark deconfinement in nuclei and meson exchange vs. quark-gluon exchange descriptions of baryon-baryon interactions.

1. Introduction

In the past two to three years, there has been slow but steady progress in the study of the spectroscopy and decays of strangeness S = -1 hypernuclei. The $(K^-,\pi^-\gamma)$ reaction has been used to obtain energy level splittings of p-shell Λ hypernuclei, enabling one to derive constraints on the spin dependence of the Λ-nucleon effective interaction. The (π^+,K^+) reaction has proven to be a promising alternative to the usual (K^-,π^-) process for producing selected hypernuclear states, particularly those of higher spin: the spectrum of ${}^{12}_{\Lambda}C$ has already been measured, and an approved proposal exists for the study of ${}^{28}_{\Lambda}Si$ and ${}^{40}_{\Lambda}Ca$. The spectroscopy of Σ hypernuclei has been explored via the $(K^-,\pi^{\pm})$ reaction on a variety of targets (most recently ${}^7Li$ and ${}^{12}C$). The interpretation of possible narrow Σ structures seen above a sizable quasifree background is a topic of active discussion, and remains controversial.

The weak decays of hypernuclei provide access to the rates for the non-mesonic process $\Lambda N \rightarrow NN$. Recently, lifetimes for the weak decay of ${}^{11}_{\Lambda}B$ and ${}^{12}_{\Lambda}C$ have become available, and the analysis of results for ${}_{\Lambda}He$ is underway. The ratios of the rates for $\Lambda n \rightarrow nn$ and $\Lambda p \rightarrow np$, as well as that for mesonic $(\Lambda \rightarrow N\pi)$ to non-mesonic processes present a strong constraint on the theoretical treatment. After reviewing the newer data and their interpretation, we proceed to a discussion of some future prospects for hypernuclear research, for instance S = -2 systems.

The conventional treatment of the interaction of strange particles with nucleons and nuclei is phrased in terms of meson exchanges. In terms of the underlying theory of strong interactions, namely quantum chromodynamics (QCD), one would rather like to express the low and medium energy hadron-hadron amplitudes which enter our problem in terms of quark-gluon degrees of freedom. So far, perturbative (one gluon exchange) calculations have been most numerous. For several specific examples, namely K^+N, ΣN, and ΛN, we compare meson and quark-gluon exchange models, particularly with respect to the spin-flavor dependence of the amplitudes and the spin-orbit components. The explicit multiquark description of the S = -1 and -2 baryon-baryon interaction leads to predictions of long-lived dibaryon bound states or resonances. Experimental searches for such objects via $(K^-,\pi^{\pm})$ or

(K^-,K^+) reactions on deuterium and ^3He targets are reviewed.

One can look for signatures of quark-gluon effects in K^+-nucleus scattering and hyperon-nucleus binding energies and spin-orbit potentials as well. Recent suggestions of a sort of "EMC Effect" for K^+ scattering are evaluated, as well as the idea that the strange quark in a hypernucleus may be partially deconfined from the hyperon.

2. The $(K^-,\pi^-\gamma)$ Reaction and the Spin Dependence of the ΛN Interaction

Emission of γ rays following Λ hypernuclear formation via the (K^-,π^-) reaction has been studied at the Brookhaven Alternating Gradient Synchrotron (AGS). The results of experiments with ^7Li and ^9Be targets were reported by May et al (1983), and preliminary data for ^{10}Be and ^{16}O were given by May (1986).

The utility of the energy level splitting deduced from the observed γ transitions lies in providing constraints on the spin dependence of the Λ-nucleon (ΛN) effective interaction. For 1s Λ states in the nuclear 1p shell, there are three ΛN matrix elements, Δ, S_Λ, and T, corresponding to spin-spin, spin-orbit, and tensor interactions, respectively (Dalitz and Gal 1978). Theoretical estimates based on meson exchange models were given by Millener et al (1985) and found to be consistent with the data. For instance, a 3.1 MeV γ line was observed by May et al (1983) in $^9_\Lambda Be$. this is interpreted as the transition from a $3/2^+$, $5/2^+$ doublet $[s_\Lambda \times {}^8Be(2^+)]$ to the ground state of $^9_\Lambda Be$. An upper limit of 100 keV was placed on the energy splitting of the doublet, which strongly constrains the size of the $\underset{\sim}{s}_\Lambda \cdot \underset{\sim}{\ell}_N$ spin-orbit effective interaction, namely $|S_\Lambda| \leq 0.04$ MeV. This in turn provides the best limit on the one-body Λ-nucleus spin-orbit splitting:

$$\left|\varepsilon(p^\Lambda_{1/2}) - \varepsilon(p^\Lambda_{3/2})\right| \lesssim 0.25 \text{ MeV} . \tag{1}$$

More data are required to more fully constrain Δ and T, particularly γ transitions between members of $1s^\Lambda_{1/2}\ 1p^{-1}_{3/2}$ and $1s^\Lambda_{1/2}\ 1p^{-1}_{1/2}$ ground state doublets. Good cases are $^7_\Lambda Li$, where the $1/2^+ - 3/2^+$ splitting is largely determined by the spin-spin coupling Δ and the $0^- - 1^-$ splitting in $^{16}_\Lambda O$, which is sensitive to T.

3. Production of Hypernuclei via the (π^+,K^+) Reaction

The high momentum transfer ($q > p_F$) (π,K) reaction was suggested by Dover et al (1980) as an effective way of producing high spin Λ and Σ hypernuclear states. Experimental data on the reaction $^{12}C(\pi^+,K^+)^{12}_\Lambda C$ have recently been obtained at Brookhaven (Milner et al 1985). The relevant natural parity hypernuclear configurations are the $(1s^\Lambda_{1/2}\ 1p^{-1}_{3/2})$ ground state and the $(1p^\Lambda_{1/2}\ 1p^{-1}_{3/2})2^+$, $(1p^\Lambda_{3/2}\ 1p^{-1}_{3/2})0^+,2^+$ excited states at about 11 MeV. The measured angular distributions are shown in Fig. 1. The solid curves represent DWBA results, using a Fermi-averaged free space $\pi^+n \to K^+\Lambda$ cross section as input, and optical potentials for the π^+ and K^+ distorted waves which were fitted to elastic scattering data (Marlow et al 1982, Marlow et al 1984) on ^{12}C at 800 MeV/c. The agreement with the data, both absolute magnitude and angular shape, is good. The (K^-,π^-) and (π^+,K^+) reactions excite the same class of natural parity hypernuclear states, but are distinguished by their momentum transfer q, the (π^+,K^+) reaction having

$q > p_F$ (p_F = Fermi momentum) for all θ, while the (K^-,π^-) reaction corresponds to small q for small θ. Consequently, the (K^-,π^-) reaction at 800 MeV/c (Chrien et al 1979) predominantly excites the 0^+ member of the 11 MeV cluster of $(p^\Lambda\ p^{-1})$ states, with the 2^+ states contributing a shoulder at larger θ (Auerbach et al 1983), while the (π^+,K^+) cross section is dominated by the 2^+ components at all θ, as shown in Fig. 1.

The virtue of the (π^+,K^+) reaction lies in its sensitivity to high spin states. The work of Milner et al (1985) will be extended in a run at the AGS with ^{16}O, ^{28}Si, and ^{40}Ca targets (proposal of Bart et al 1984). For instance in $^{28}_{\Lambda}Si$ one expects $(d^{\Lambda}_{5/2,3/2}\ d^{-1}_{5/2})_{4^+}$ states to be most strongly excited.

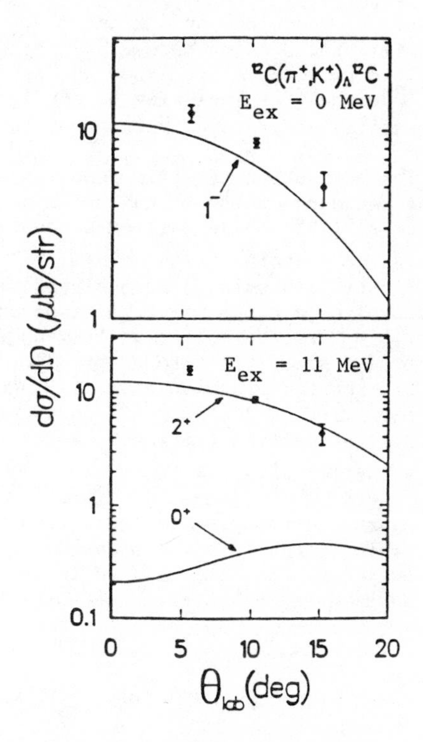

Fig. 1. Differential cross sections for the excitation of the ground state and 11 MeV cluster of states in $^{12}_{\Lambda}C$ in the (π^+,K^+) reaction at 1.05 GeV/c, from Milner et al (1985). The solid curves are DWBA results.

4. Σ Hypernuclear States

Since the discovery of narrow Σ structures in the $^9Be(K^-,\pi^-)^9_\Sigma Be$ reaction by the CERN group (Bertini et al 1980), the study of Σ hypernuclei has been actively pursued. I will concentrate on recent developments, since several reviews are available (Hungerford 1986, Yamazaki 1986, Gal 1986).

The $^{12}C(K^-,\pi^+)^{12}_\Sigma Be$ reaction has received the greatest amount of attention. Yamazaki et al (1985) reported three narrow peaks ($\Gamma \leq 4$ MeV), with formation rates of order 1% per stopped kaon. This experiment is being repeated by a Heidelberg-KEK group, but results are not yet available. The same process at 715 MeV/c has been measured at Brookhaven, and preliminary data were shown by Hungerford (1986). The "in flight" data are consistent with the existence of the three peaks seen by Yamazaki et al (1985), but the statistics are not sufficiently good for a firm conclusion.

One of the main focuses of interest in Σ spectroscopy has been the attempt to extract the value of the Σ-nucleus spin-orbit splitting $\varepsilon_p^\Sigma = \varepsilon(p_{1/2}) - \varepsilon(p_{3/2})$ in the p-shell. Contradictory values $\varepsilon_p^\Sigma \sim 5$ MeV and 12 MeV were suggested by Yamazaki et al (1985) and Bertini et al (1985), respectively, the latter on the basis of a combined analysis of results for (K^-,π) reactions on ^{12}C and ^{16}O. These analyses assume the jj coupling limit for Σ eigenstates. and neglect the ΣN residual interaction $V^{\Sigma N}$. In this picture, there is no natural explanation for a strong third peak in $^{12}_{\Sigma}Be$, but the inclusion of configuration mixing and all relevant core excited states does not appear to cure this problem.

The extraction of ε_p^{Σ} is complicated by the expected strong spin-isospin dependence of $V^{\Sigma N}$ if the nuclear core is not spin-isospin saturated. Dover et al (1986) have suggested several more favorable examples. For instance, in the reaction $^7Li(K^-,\pi^+)^7_{\Sigma}H, 1/2^-$ and $3/2^-$ states corresponding to $p^{\Sigma}_{1/2,3/2}$ + $^6He(0^+)$ should be strongly excited. The $3/2^-$ strength is peaked at 0° (orbital angular momentum transfer $\Delta L = 0$) while the $1/2^-$ peaks at 10-15° ($\Delta L = 2$). The splitting between these two peaks would be a direct measure of ε_p^{Σ}, since $V^{\Sigma N}$ here exerts little influence. Preliminary data from Brookhaven were discussed by Hungerford (1986). Unfortunately, the data for $^7Li(K^-,\pi^+)$ at $\theta = 12°$ do not appear to show a peak structure which is statistically significant, so it has not yet been possible to deduce a value for ε_p^{Σ}.

Another difficulty in extracting relative intensities and energies of possible Σ states is the presence of a sizable quasi-free background which is peaked in the same mass region as the candidates for narrow structures. An example, taken from the work of Chrien et al (1986), is shown in Fig. 2. This calculation goes beyond the standard Fermi gas model, and incorporates the full momentum dependence of the $K^-p \to \pi^+\Sigma^-$ amplitude, the empirical distribution of hole strengths in the recoiling nucleus, and the experimental nucleon momentum distribution as taken from (e,e'p) studies. As seen in Fig. 2, the quasi-free continuum accounts for the centroid and the width of the structure seen in the $^{16}O(K^-,\pi^+)$ reaction. It is then a moot point whether one can claim two separate narrow peaks at the energies indicated in Fig. 2, which was crucial in deducing the very large value of ε_p^{Σ} favored by Bertini et al (1985). In my view, the question of the Σ spin-orbit splitting is still an open one. In Section 8, the theoretical expectations for ε_p^{Σ} in the context of quark-gluon and meson exchange models will be compared.

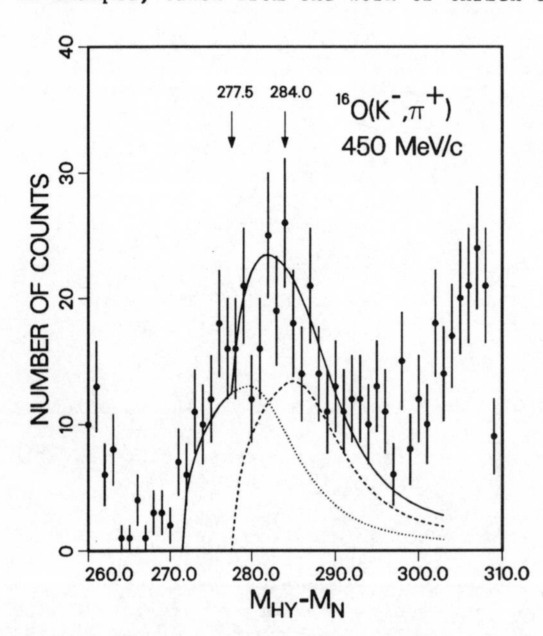

Fig. 2. The calculated quasi-free Σ^- spectrum, from Chrien et al (1986), is shown as a solid line. The dotted and dashed curves correspond to the contributions of the $p^{-1}_{1/2}$ and $p^{-1}_{3/2}$ core states. The curves are arbitrarily normalized to the data, taken from Bertini et al (1985).

For the cases of $^{6}Li(K^{-},\pi^{+})$ and $^{12}C(K^{-},\pi^{+})$, the quasi-free spectrum displays peaks in the correct mass region (with the expected q dependence), but the widths are too large to account fully for the data, which appear to display some narrower structures. The origin of the Σ width suppression and the dynamics of continuum Σ hypernuclear production remain mysterious, in spite of the efforts of many authors (Chrien 1986, Proceedings of Brookhaven conference).

5. Weak Decays of Hypernuclei

Grace et al (1985) have recently measured the weak decay lifetime of several light hypernuclei at Brookhaven. In units of the free space Λ decay width Γ_{Λ}, they report total weak decay widths Γ of

$$\frac{\Gamma}{\Gamma_{\Lambda}} = \begin{cases} 1.25 \pm 0.18 & (^{12}_{\Lambda}C) \\ 1.37 \pm 0.16 & (^{11}_{\Lambda}B) \end{cases} \tag{2}$$

An experiment on $^{4}_{\Lambda}He$ and $^{5}_{\Lambda}He$ weak decays has also been completed at Brookhaven (Barnes et al 1983), but results are not yet available. Heavy hypernuclei have been produced by the annihilation of antiprotons in ^{238}U. A lifetime of about 10^{-10} sec has been determined by observing the fission mode (Bocquet et al 1986).

The widths Γ for all but the lightest hypernuclei are dominated by the non-mesonic decay $\Lambda N \to NN$. Since the free decay $\Lambda \to N\pi$ involves a low momentum recoil nucleon, the Pauli principle suppresses this mode of decay in hypernuclei. The study of weak hypernuclear lifetimes is thus of considerable intrinsic interest, since it provides a unique access to the four fermion $\Lambda N \to NN$ process, which cannot be studied in free space.

The width Γ may be written as

$$\Gamma = \Gamma_{nm} + \Gamma_{\pi} \tag{3}$$

where the non-mesonic part Γ_{nm} is a sum of contributions Γ_n and Γ_p from the reactions Λ+n $\to$ n+n and Λ+p $\to$ n+p, respectively, and $\Gamma_{\pi} = \Gamma_{\pi^-} + \Gamma_{\pi^\circ}$ ($\Lambda \to p\pi^-$, $n\pi^\circ$). Assuming $\Gamma_{\pi^\circ}/\Gamma_{\pi^-} = 7/6$, Barnes (1986) gives the following ratios for $^{12}_{\Lambda}C$:

$$n = \frac{\Gamma_n}{\Gamma_n + \Gamma_p} \approx 0.57 {}^{+0.14}_{-0.23}$$

$$Q^- = \frac{\Gamma_{nm}}{\Gamma_{\pi^-}} \approx 22 {}^{+43}_{-12} \tag{4}$$

corresponding to the values

$$\Gamma_p = \left(0.49 {}^{+0.3}_{-0.2}\right) \Gamma_{\Lambda}$$

$$\Gamma_n = \left(0.65 {}^{+0.2}_{-0.3}\right) \Gamma_{\Lambda} \tag{5}$$

$$\Gamma_{\pi^-} = \left(0.05 {}^{+0.06}_{-0.03}\right) \Gamma_{\Lambda}$$

In heavy hypernuclei, there is some evidence that $\Gamma_n \gg \Gamma_p$, and hence η approaches unity [see Dover and Walker (1982) for a list of references].

There has been a flurry of theoretical papers on hypernuclear weak decays in the past year or two, stimulated by the above experimental results (Bando and Takaki 1984,1985; Cheung, Heddle and Kisslinger 1983; Heddle and Kisslinger 1986; Dubach 1986; McKellar and Gibson 1984; Oset and Salcedo 1985). Most of these calculations give reasonable agreement with total rates for the non-mesonic process, i.e., $\Gamma_{nm} \sim \Gamma_\Lambda$. This order of magnitude arises from one pion exchange (OPE). The models differ in how they treat the short range correlations. In one boson exchange (OBE) models, one adds contributions of heavier mesons, namely the ρ (McKellar and Gibson 1984) or also η, ω, K, K*, σ (Dubach 1986). In the hybrid quark model (Kisslinger and collaborators 1983,1986) one supplements OPE for $r < r_0 \sim 1$ fm by an explicit six quark contribution. Both the OBE and quark models can be made consistent with the experimental result $\Gamma_{nm} = (1.14 \pm 0.2)\Gamma_\Lambda$ (Barnes 1986), so total rates alone do not enable us to choose between models.

Although more or less consistent with Γ, existing models do not account for the ratio η of Eq. 4. Indeed, the most naive OPE model would predict $\eta \sim 1/10$, while short range meson exchanges only raise this to $\eta \sim 1/4$ (Dubach 1986), still short of the experimental value (4). The use of a six quark prescription at short distances does not appear to cure this problem. The difficulty is that the tensor part of the long range OPE potential strongly favors the $\Lambda p(^3S_1) \to np(^3D_1)$ transition [note that $\Lambda n(^3S_1) \to nn(^3D_1)$ is forbidden by the Pauli principle] and hence yields $\Gamma_{\Lambda p \to np} \gg \Gamma_{\Lambda n \to nn}$, while the data in $^{12}_{\Lambda}C$ indicate $\Gamma_{\Lambda p \to np} \approx \Gamma_{\Lambda n \to nn}$. Inclusion of the $\Lambda n(^3P_1) \to nn(^3P_1)$ transition may alleviate the problem; published calculations include only s-wave ΛN initial states.

The pionic decay mode of hypernuclei has been emphasized by Bando and Takaki (1984,1985) and by Oset and Salcedo (1985). In the most naive Fermi gas model, where nucleon single-particle orbits are either fully occupied or empty, one would get $\Gamma_\pi = 0$ because of Pauli blocking. In a real finite nucleus, the observation of pionic decays thus probes nucleon occupation probabilities in the hypernuclear ground state wave function. Neglecting short range correlation effects, Bando and Takaki (1984) find a suppression factor

$$S_\pi = 1 - 1/2 \sum_{n\ell j} N_{n\ell j}\, \eta_{n\ell j}$$

$$\eta_{n\ell j} = \left|\langle \phi_\Lambda(r) \left| j_\ell(qr) \right| \phi_{n\ell j}(r)\rangle\right|^2 \tag{6}$$

where $N_{n\ell j}$ denotes the occupation number for nucleons in orbit $\{n\ell j\}$ and q is the momentum release in the decay. For A~100, Eq. 6 gives $\Gamma_{\pi^-}/\Gamma_\Lambda \sim 10^{-3}$ whereas the data of Lagnaux et al (1964) suggest $\Gamma_{\pi^-}/\Gamma_\Lambda \sim (1-3)\times10^{-2}$, an order of magnitude larger. Bando and Takaki (1985) propose to cure this problem by including short range correlations (via the usual "wound integral" κ) and pion final state interactions via an absorption parameter a. Thus, Eq. 6 is replaced by

$$S_\pi = a - \left(\frac{1-\kappa}{2}\right) \sum_{n\ell j} N_{n\ell j}\, \tilde{\eta}_{n\ell j} \tag{7}$$

where $\tilde{n}_{n\ell j}$ is calculated with pion distorted waves rather than $j_\ell(qr)$. For heavy hypernuclei (A $\geq$ 100), Bando and Takaki (1985) find a saturation effect of $\Gamma_{\pi^-}/\Gamma_\Lambda$; they obtain $S_\pi \simeq a\kappa$ and $\Gamma_{\pi^-}/\Gamma_\Lambda \sim 10^{-2}$, in agreement with the old data (Lagnaux et al 1964). For ^{13}C, they get $\Gamma_{\pi^-}/\Gamma_\Lambda = 0.069(\kappa=0)$ or $0.11(\kappa=0.1)$, more or less consistent with Eq. 5.

Oset and Salcedo (1985) have also found that $\Gamma_{\pi^-}/\Gamma_\Lambda$ is sensitive to the strong interaction of the π^- with the nuclear medium. They obtain an <u>order of magnitude</u> enhancement of $\Gamma_{\pi^-}/\Gamma_\Lambda$ due to this effect, considerably larger than Bando and Takaki (1985), and inconsistent with the very small experimental value of Eq. 5.

In summary, there does not seem to be a theoretical model at present which is consistent with the experimental data on weak hypernuclear decays. The ratios n and Q^- of Eq. 4 pose a particular problem. Precise data on very light systems ($^4_\Lambda He$, $^5_\Lambda He$) will be very useful to unravel the spin dependence of the $\Lambda N \to NN$ process, while more data on heavy systems are needed to map out the A dependence of n and $\Gamma_{\pi^-}/\Gamma_\Lambda$.

6. Strange Dibaryons

In the bag model, an ample spectrum of six quark dibaryon states is predicted, [see Dover (1986) for a review] but only a few candidates with strangeness S = -1 or -2 are possibly long-lived with respect to strong decay. Recently, a Brookhaven experiment was mounted to search for S = -1 dibaryons X in the $d(K^-,\pi^-)X$ reaction at 870 MeV/c. Preliminary results were discussed by Piekarz (1986). In addition to the well known cusp effect at the ΣN threshold (2129 MeV/c^2 for the missing mass X), a second maximum was seen at about 2140 MeV/c^2. Unlike the cusp cross section, which falls smoothly as θ_π increases, this structure peaks at finite angle, characteristic of a p-wave hyperon-nucleon system. If confirmed as a narrow peak clearly separated in mass from the cusp, this object might be interpreted as the spin one dibaryon D_t of quark cluster structure $Q^4 \times Q^2$, which is predicted in this mass region (Dover 1986). Note that ordinary p-wave Λp final state interactions would also produce a kinematical broadening of the cusp behavior as the momentum transfer increases, and for some potential models (although not those currently available) one might also find narrow $^3P_1$ ΛN-ΣN resonant states. Thus, it will be difficult to disentangle the quark structure from conventional final state interactions. A key point is the prediction of a second $Q^4 \times Q^2$ dibaryon D_s in bag models, a spin singlet state ($^1P_1$) lying some 30 MeV/c^2 below D_t. This mass splitting, resulting from the spin dependence of one gluon exchange, is presumably a more reliable quantity than the masses themselves. The production of D_s will be negligible in the $d(K^-,\pi^-)$ reaction, since spin flip is required. Instead, the $^3He(K^-,\pi^+)nD_s$ reaction has been proposed (Piekarz et al 1985), and data are expected within the next year.

An experiment to study the reaction $pp \to K^+X$ is underway at the Saturne facility at Saclay (Frascaria 1986). Preliminary analysis indicates tantalizing peaks in the mass spectrum of X, one of which is around 2140 MeV/c^2, where a peak is seen in $d(K^-,\pi^-)X$. Note that the $pp \to K^+X$ process could lead to production of both D_s and D_t.

The best candidate for a stable dibaryon is the famous H proposed by Jaffe (1977). The H is a 0^+ dibaryon, S = -2, of ssuudd quark structure, and an SU(3) flavor singlet. This configuration benefits maximally from the color magnetic attraction generated by single gluon exchange between quarks. Franklin (1986) has recently reviewed the experimental aspects of H searches. Two experiments have been approved for the Brookhaven AGS (Franklin et al 1985,1986). One involves tagged Ξ^- production via the $K^-p \rightarrow K^+\Xi^-$ reaction, followed by slowing down the Ξ^- in a moderator and capturing it in a Ξ^-d atomic state. The idea is to search for a mono-energetic neutron from the $\Xi^-d \rightarrow Hn$ reaction. The second involves production of the H in the $^3He(K^-,K^+)nH$ reaction. These experiments require a K^- beam in the momentum range 1.5 - 2 GeV/c, in order to produce Ξ^-'s. The search for the H provides a very strong motivation for the construction of such a beam line. As a side benefit, one could explore other aspects of S = -2 physics, for instance Ξ hypernuclear spectroscopy. In analogy to Σ's, certain Ξ states could be long-lived, in spite of strong $\Xi N \rightarrow \Lambda\Lambda$ conversion (Dover and Gal 1983).

7. Quark/gluon vs Meson Exchange Description of Spin-orbit Potentials for Strange Particles

For some time, the spin-orbit splitting for Σ hypernuclei has been discussed as a means of distinguishing between quark-gluon and meson exchange models of the YN two-body interaction. This hope was based on the paper by Pirner (1979), which yielded for the ratio of baryon-nucleus spin-orbit well depths V^B_{LS} the result

$$V^N_{LS} : V^\Sigma_{LS} : V^\Lambda_{LS} : V^\Xi_{LS} = 1 : 4/3 : 0 : -1/3 \ . \qquad (8)$$

Although meson exchange models fitted to YN data (Dover and Gal 1984) also give a very small value of V^Λ_{LS}, they lead to $V^\Sigma_{LS}/V^N_{LS} < 1/2$, in contrast to Eq. 8. The analyses of Bertini et al (1985) and Yamazaki et al (1985) appeared to favor a large value of V^Σ_{LS}, and hence the quark picture.

The result (8) follows from an evaluation of graph (a) of Fig. 3, treating the strange quark as a spectator. This one gluon exchange (OGE) approximation to the two-body YN spin-orbit potential, when summed over the density of nucleons, yields V^Y_{LS}.

It was pointed out by Dover and Gal (1984) that Eq. 8 is also obtained in an (unrealistic) meson exchange model, when only ω exchange is included, with SU(6) coupling constants. The close relation between vector meson exchange (VME) and OGE has recently been clarified by Wang and Wong (1985) and He et al (1986). They point out that graph (c) of Fig. 3, which represents meson exchange between quarks, has the spin-flavor structure

$$(\underset{\sim}{\sigma}_3 + \underset{\sim}{\sigma}_4)\ (V^\omega_0 + V^\rho_0\ \underset{\sim}{\tau}_3 \cdot \underset{\sim}{\tau}_4) \qquad (9)$$

for $\{\rho,\omega\}$ exchange (the quarks are numbered 1 to 6 from left to right). If we assume SU(6) symmetry, then $V^\omega_0 = V^\rho_0$, and Eq. 9 becomes

$$2V^\omega_0(\underset{\sim}{\sigma}_3 + \underset{\sim}{\sigma}_4)\ P^\sigma_{34} P^\tau_{34} \ . \qquad (10)$$

However, this is just the factor which enters in graph 3a. The same argument applies for K* exchange. Thus VME and OGE models (graphs c and a) have the same spin-flavor structure. Hence a model with $\rho + \omega + K^*$ exchange and SU(6) symmetry for coupling constants cannot be distinguished from quark-gluon exchange (Fig. 3a) at either the baryon-baryon (BB) or baryon-nucleus levels, except possibly through the radial or energy dependence of the potentials (we return to this point).

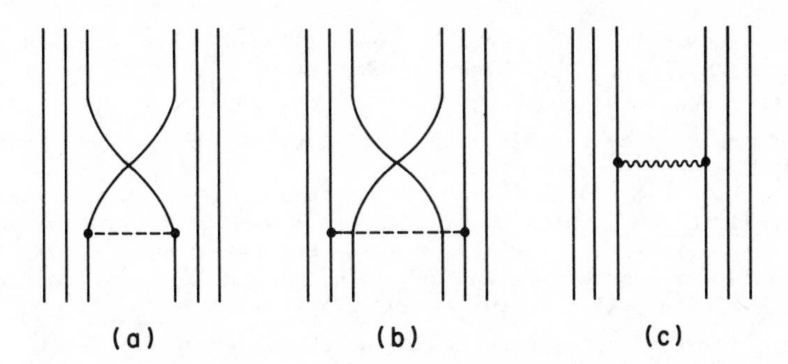

Fig. 3. Gluon exchange (dashed lines) or meson exchange (wavy line) diagrams which contribute to the baryon-baryon spin-orbit potential.

The spin-orbit part t_{BN} of the baryon-nucleon (BN) scattering amplitude receives contributions from both the symmetric and antisymmetric parts of the OGE spin-orbit potential $V_{ij}(r)$ between quarks. The symmetric part V_{ij}^{LS} has the form

$$V_{ij}^{LS}(r) = \frac{-\alpha_c \lambda_i \cdot \lambda_j}{8r^3} \left(\underset{\sim}{\sigma}_i + \underset{\sim}{\sigma}_j\right) \cdot \left(\underset{\sim}{r} \times \underset{\sim}{\xi}_{rel}\right)$$
$$\underset{\sim}{\xi}_{rel} \equiv \underset{\sim}{p}_i/m_i^2 - \underset{\sim}{p}_j/m_j^2 + 2\left(\underset{\sim}{p}_i - \underset{\sim}{p}_j\right)/m_i m_j \tag{11}$$

where α_c is the strong coupling constant (note some authors use $\alpha_s = 4\alpha_c$) and $\{m_i, m_j\}$ are the quark masses. The symmetric spin-orbit amplitude t_{BN}^{LS} which arises from V_{ij}^{LS} in Born approximation is of the form (He et al 1986)

$$t_{BN}^{LS}(E,q) = \frac{2\pi xy}{R^2}\{A_{BN}e^{-x^2/9}H(y/3) + \frac{9B_{BN}}{8\sqrt{3}}\, e^{-z^2/9}H\left(\frac{x}{2\sqrt{3}}\right) + C_{BN}e^{-4y^2/9}H(x/6) + \frac{9D_{BN}}{8\sqrt{3}}\, e^{-z^2/9}\, H(y/\sqrt{3})\} \tag{12}$$

where $x = qR$, $y = KR$, $z = kR$, $q = |\underset{\sim}{k} - \underset{\sim}{k}'|$ is the momentum transfer, $k^2 = k'^2 = m_N E/2$ (for NN system), E = lab kinetic energy, $K = (k^2 - q^2/4)^{1/2}$, R is the bag radius for oscillator quark wave functions, and

$$H(t) \equiv 3/2t^2 \left(1 - \frac{1}{t}\, e^{-t^2} \int_0^t e^{y^2} dy\right) . \tag{13}$$

Note that t_{BN}^{LS} is energy dependent, and goes to zero for large E. The factors A and C in Eq. 12 come from graph 3a, while B and D arise from 3b. Other independent quark-gluon exchange graphs occur for the central potential, but only 3a and 3b contribute to the spin-orbit part. For the VME model, graph 3c, the result is

$$t_{BN}^{LS}(E,q) = 24\pi xy V_0^{\omega}\{A_{BN}e^{-x^2/9}(x^2 + \tilde{m}_V^2)^{-1} + C_{BN}e^{-4y^2/9}(4y^2 + \tilde{m}_{K*}^2)^{-1}\} \tag{14}$$

in the SU(6) limit, where $\tilde{m} \equiv mR$. The constants A and C are the same as in Eq. 12.

Note that a symmetric quark-quark LS potential also gives rise to an antisymmetric LS amplitude t^{ALS}_{YN}. For the NN system, t^{ALS}_{NN} vanishes. For YN, t^{ALS}_{YN} has the same form (He et al 1986) as Eqs. 12 and 14, upon replacement of the constants $\{A,B\}$ by $\{E,F\}$ and setting C=D=0 (double quark exchange).

As shown by He et al (1986) for the NN system, diagram 3a, which generates A and C, dominates 3b, which corresponds to B and D. Diagram 3b does not have the same spin-flavor factors as meson exchange (3c). A comparison of Eq. 12 and 14 reveals that 3a and 3c yield the same form of amplitude, with the replacements $H(y/3) \to (x^2+\tilde{m}_V^2)^{-1}$ and $H(x/6) \to (4y^2+\tilde{m}_{K*}^2)^{-1}$. The VME model has a term from ω exchange which does not vanish for large E. Thus, in principle one could distinguish the OGE and VME models by determining the energy dependence of BN spin-orbit potentials. However, for small E and q, $H \approx 1$ and $\{x,y\} \ll m_V R$, so the OGE and VME models exhibit essentially the same form proportional to $A_{BN}e^{-x^2/9} + C_{BN}e^{-4y^2/9}$.

If we take $k \to o$, appropriate to Λ and Σ nucleus states near zero binding, we find

$$R^{LS}_{BNi} = \begin{cases} 1 & (NN\omega) \\ 5/9 & (NN\rho) \\ 1/3 & (\Lambda N\omega) \\ 1 & (\Sigma N\omega) \\ 7/9 & (\Sigma N\rho) \\ m/m_s & (\Lambda NK^*) \\ -m/9m_s & (\Sigma NK^*) \end{cases} \qquad R^{ALS}_{BNi} = \begin{cases} 1/3 & (\Lambda N\omega) \\ -1/3 & (\Sigma N\omega) \\ 1/3 & (\Sigma N\rho) \end{cases} \tag{15}$$

for the ratios $R^{LS}_{BNi} = t^{LS}_{BNi}/t^{LS}_{NN\omega}$ and $R^{ALS}_{BNi} = t^{ALS}_{BNi}/t^{LS}_{NN\omega}$ of symmetric and antisymmetric spin-orbit Born amplitudes for mesons $\{i\}$, where we have assumed $m_{K*} = m_\rho = m_\omega$, and $m/m_s \approx 0.6$ is the ratio of non-strange to strange constituent quark masses.

To obtain B-nucleus potentials from Eq. 15, we spin-isospin average over the nucleons, and the ρ contribution drops out. For the ω, we have

$$V^B_{LS}/V^N_{LS} = R^{LS}_{BN\omega} - R^{ALS}_{BN\omega} \tag{16}$$

and we recover Eq. 8 for Λ and Σ. Note, however, that the K* contribution (equivalent to OGE with strange quark exchange), neglected by Pirner (1979), would give a non-negligible contribution to V^Λ_{LS}. See also the work of Morimatsu et al (1984) for an improved quark model estimate of V^B_{LS}. They obtain

$$V^N_{LS} : V^\Sigma_{LS} : V^\Lambda_{LS} \approx 1 : 0.55 : 0.21 \,, \tag{17}$$

not very different from the results of realistic meson exchange models. The contribution of an antisymmetric quark-quark spin-orbit potential V^{ALS}_{ij} has been considered by He et al (1986) and Suzuki and Hecht (1984). They use

$$V_{ij}^{ALS}(r) = \frac{\alpha_c}{8r^3} \lambda_i \cdot \lambda_j \, (\underset{\sim}{\sigma}_i - \underset{\sim}{\sigma}_j) \cdot (\underset{\sim}{r} \times \underset{\sim}{\xi}_{tot})$$
$$\underset{\sim}{\xi}_{tot} \equiv 2(\underset{\sim}{p}_i + \underset{\sim}{p}_j)/m_i m_j - \underset{\sim}{p}_i/m_i^2 - \underset{\sim}{p}_j/m_j^2 \tag{18}$$

In the NN system, V_{ij}^{LS} and V_{ij}^{ALS} make contributions of the <u>same sign</u> to t_{NN}^{LS}. Thus, if α_c is readjusted to fit the phenomenological strength of t_{NN}^{LS} in the Paris model (Lacombe et al 1980), then α_c is reduced by about 30% (He et al 1986). Note that α_c thus determined is still a factor of two larger than the value required by a fit to p-wave meson mass splittings (He et al 1986). This presumably reflects the fact that isoscalar s-wave ππ exchange (the σ) generates a significant spin-orbit potential in meson exchange models (and also supplies the needed medium range attraction in the central part), so clearly OGE is not the whole story.

The Born amplitudes t_{BN}^{LS} and t_{BN}^{ALS} calculated by He et al (1986) are displayed in Fig. 4 as a function of q for fixed E = 300 MeV. The effect of the antisymmetric quark-quark spin-orbit potential V_{ij}^{ALS} (not shown) is not large, typically of order 1/3 of V_{ij}^{LS}. If α_c is readjusted to fit t_{NN}^{LS} for I = 1, as in Fig. 4, most results for $t_{BN}^{LS,ALS}$ remain largely unchanged. An exception is $t_{\Sigma N}^{LS}(I = 3/2)$, which is decreased by about 40% by the inclusion

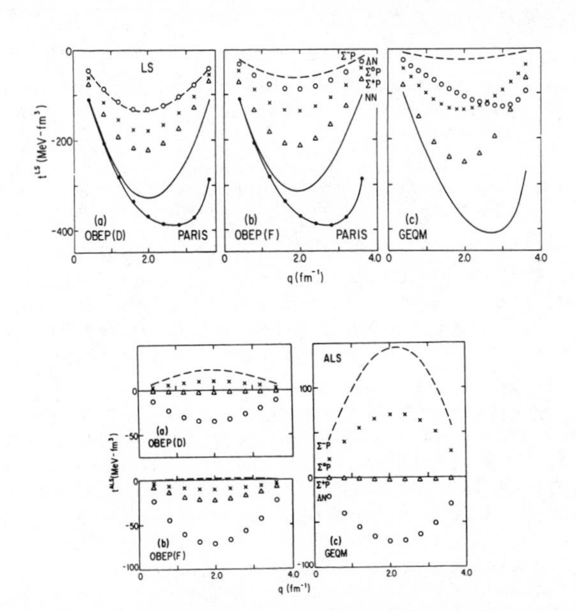

Fig. 4. Baryon-nucleon symmetric and antisymmetric spin-orbit amplitudes t^{LS} and t^{ALS} in Born approximation, from He et al (1986). Results are shown for the Nijmegen meson exchange model F (Nagels et al 1979) and Paris model (Lacombe et al 1980). Results labelled GEQM refer to the gluon exchange quark model (graphs 3a, b), including only the symmetric quark-quark spin orbit potential of (11). Dashed lines refer to the Σ^-p channel, circles to Λp, crosses to $\Sigma^\circ p$, triangles to Σ^+p, solid lines to NN (isospin one).

of V_{ij}^{ALS}. This leads to a significant decrease in V_{LS}, which is proportional to an integral of the combination (using $t_{\Sigma^+p} = t_{\Sigma^-n}$)

$$V_{LS}^{\Sigma} \sim t_{\Sigma^+p}^{LS} + t_{\Sigma^-p}^{LS} - t_{\Sigma^+p}^{ALS} - t_{\Sigma^-p}^{ALS} \qquad (19)$$

Each of these terms contributes with the same sign (if V_{ij}^{LS} and V_{ij}^{ALS} are included), while for the Λ we have

$$V_{LS}^{\Lambda} \sim t_{\Lambda N}^{LS} - t_{\Lambda N}^{ALS} \qquad (20)$$

and the two contributions tend to cancel. In the limit where only graph 3a and V_{ij}^{LS} are included, this cancellation is complete, and one recovers Eq. 8. In a more realistic quark model calculation, however, a non-zero value for V_{LS}^{Λ} remains (Morimatsu et al 1984, He et al 1986). If we take the maxima of the t's of Fig. 4 as a guide, we estimate roughly

$$\frac{V_{LS}^{\Lambda}}{V_{LS}^{\Sigma}} \sim \begin{cases} 1/4 \ (V_{ij}^{LS} + V_{ij}^{ALS}) \\ 1/7 \ (V_{ij}^{LS} \text{ only}) \end{cases} \qquad (21)$$

If the full quark model calculation is to respect the experimental constraint $V_{LS}^{\Lambda}/V_{LS}^{N} \lesssim \frac{1}{20}$ (see Eq. 1), then it will also predict V_{LS}^{Σ} to be considerably smaller than V_{LS}^{N}, in contrast to Eq. 8. Thus, if the large values of V_{LS}^{Σ} claimed by Yamazaki et al (1985) and Bertini et al (1985) are taken seriously, it would be a major surprise both in the context of meson exchange and quark-gluon models.

Since the spin-flavor structure of the dominant OGE graph 3a and the VME graph 3c are in fact the same, it will clearly be difficult to distinguish these two models on the basis of $V_{LS}^{\Sigma}/V_{LS}^{N}$, as originally thought. The experimental values of $V_{LS}^{\Lambda,\Sigma}$ would provide important constraints, however, on the contribution of isoscalar two pion exchange to the spin-orbit part of the YN potential. In Nijmegen model D (Nagels et al 1977), for instance, the σ is treated as an SU(3) singlet, so except for differences in recoil factors $(M_\sigma/M_B)^2$, it makes much the same contribution to V_{LS}^{N}, V_{LS}^{Λ}, and V_{LS}^{Σ}. Eq. 1 may already rule out such a treatment of the σ. Recently, Büttgen et al (1985,1986) have presented an improved relativistic meson exchange treatment of KN and YN scattering, in which the σ is replaced by box diagrams involving various intermediate states. It would be interesting to calculate the Λ and Σ nucleus spin-orbit potentials in this model.

Unfortunately, the existing two-body ΛN and ΣN data tell us very little about spin-orbit forces. Although various models provide good fits to YN differential and total cross section data, their predictions for spin observables are untested. In fitting YN data, as for the NN case, medium range central attraction is required. This is not generated by the OGE graphs. In an interesting recent paper, Fujiwara and Hecht (1985) attempt to obtain such attraction in the NN potential by building in $Q^3(Q\bar{Q})$ components in the nucleon wavefunction. They enter at the 20% level (amplitude) in one gluon approximation. Some attraction results, but not as much as required by NN phenomenology. This points to the non-perturbative aspects of the baryon-baryon interaction at distances of order 0.5 – 0.8 fm.

Graphs such as 3a and 3b can only be dominant for the very short distance behavior of the interaction.

In the above discussion, we have omitted a possible contribution to the BN spin-orbit potentials due to confinement. The sign of such a term depends on whether confinement is treated as a scalar potential or as a requirement on the mass of a Dirac particle. For a quadratic scalar quark-quark confinement potential of the form

$$V_{ij}^{conf} = C\lambda_i \cdot \lambda_j r_{ij}^2 \tag{22}$$

one can show (Suzuki and Hecht 1984, He et al 1986) that there is no net contribution to BN spin-orbit potentials. This is unlike the situation for a single baryon, where a scalar confinement potential tends to cancel the spin-orbit effects of OGE (Isgur and Karl 1978).

A problem with the V_{ij}^{conf} of Eq. 22 is that it leads to unwanted long-range Van der Waals forces. In a "flip-flop" model of confinement (Lenz et al 1986, Koike et al 1986), this problem is cured. In this model, the confinement potential gives a non-vanishing contribution to BB spin-orbit potentials, with a sign opposite to that of OGE.

The question of meson exchange vs OGE descriptions has also been discussed in detail for K^+N scattering. Realistic meson exchange models have been constructed by Davis et al (1976) and Büttgen et al (1985, 1986). In the latter, σ exchange was replaced by higher order contributions involving Δ and K* intermediate states. This ameliorated a problem with the $I = 0$ $s_{1/2}$ phase shift (which was too attractive) in the work of Davis et al (1976). The quark calculations of K^+N scattering were pursued by Pirner and Povh (1982), Bender et al (1984) and Mukhopadhyay and Pirner (1985). They are also qualitatively successful in explaining the data, particularly the isospin dependence of the phase shifts.

8. Strangeness as a Probe of Partial Deconfinement of Quarks

Strange probes have recently been discussed in connection with various tests of quark confinement in nuclei. We critically examine several of these here, and examine the necessity of invoking quark degrees of freedom to explain some "anomalies" in strange particle scattering and hypernuclear binding.

8.1 K^+ nucleus scattering

The two-body K^+N cross section is rather small (~10 mb) at low momentum ($\lesssim$ 500 MeV/c), and hence the K^+ has a long mean free path in nuclear matter. The K^+ should be a sensitive probe of neutron/proton densities ρ_n and ρ_p, since the first order approximation $V \sim \rho_n t_{K^+n} + \rho_p t_{K^+p}$ should be very good, with small and calculable higher order corrections. The K^+, as a "weakly" interacting hadronic probe, thus occupies a unique position (Dover and Moffa 1977).

The first significant tests of these ideas came after measurements of K^+ elastic scattering from ^{12}C and ^{40}Ca at 800 MeV/c (Marlow et al 1982). Although the first order optical potential provided an acceptable fit to the Ca^{40} data, the theoretical predictions for $K^+ + ^{12}C$ elastic scattering, using the free space K^+N amplitudes of Martin (1975), fell short of the data, as shown in Fig. 5. The same difficulty appears in the total cross section for $K^+ + ^{12}C$.

Although first order multiple scattering theory agrees very well with the data (Bugg et al 1968) above 1.2 GeV/c, where the K^+N amplitudes involve several partial waves (s, p, d, f) and the K^+N cross section is of normal hadronic size, the theory falls consistently below the data at lower momentum. Equivalently, the theory predicts substantial "shadowing" in the 0.7 to 1 GeV/c region, whereas the total cross section (σ^T) data for $K^+ + {}^{12}C$ lie close to $6 \times \sigma^T_{K^+d}$, i.e., no shadowing.

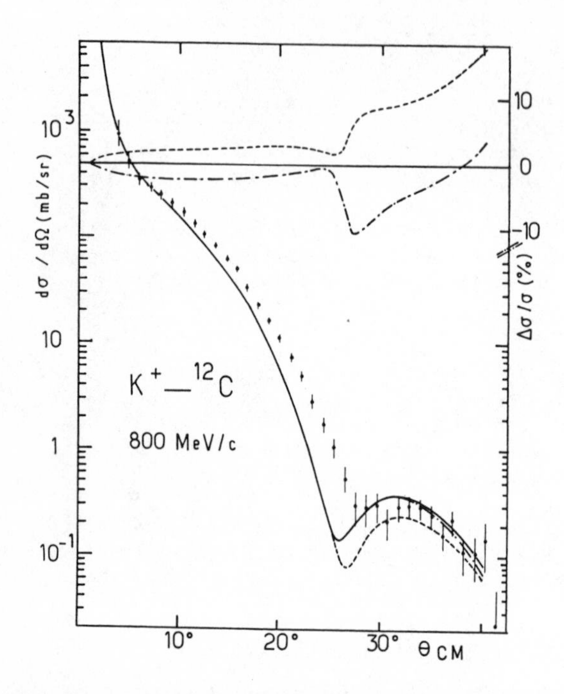

Fig. 5. Differential cross sections for K^+ elastic scattering on ${}^{12}C$ at 800 MeV. The data are from Marlow et al (1982), and the theoretical calculations from Abgrall and Labarsouque (1984). The lower dashed line represents scattering from uncorrelated nucleons, while the solid line includes c.m. and Pauli corrections and coupling to collective states. In the upper part (scale on right) are shown fractional changes $\Delta\sigma/\sigma$ due to Pauli correlations (dashed curve) and coupling to collective states (dash-dot curve).

Siegel et al (1984, 1985) have tried to understand the discrepancy between the first order theory and experiment for $K^+ + {}^{12}C$ elastic and total cross sections, by including a variety of higher order corrections. The summed effect of these corrections is rather small, and represents the theoretical error band in Fig. 6. Similar results were found by Abgrall and Labarsouque (1984).

Another strategy is to regard K^+ nucleus scattering as a test of the correctness of one's model for the K^+N amplitude, and thus parametrize the K^+N t-matrix so as to simultaneously fit K^+N and K^+ nucleus scattering. Here one simply assumes that the effective K^+N amplitude in the nuclear medium is identical to the free space amplitude. This has been done in a KMT framework by Chaumeaux and Lemaire (1983). Reasonable fits to the elastic $K^+ + {}^{12}C$ elastic data are obtained. It was found that the oscillatory behavior in the angular distribution is sensitive to the (unknown) relative phase of the K^+p and K^+n amplitudes. The ratio of real to imaginary parts of the K^+n amplitude, which is also essentially unknown experimentally, exerts an important influence on the elastic $K^+ + {}^{12}C$ cross section. Thus, one might take refuge in an enlarged theoretical error bar in Fig. 6, due to ambiguities in the elementary free space amplitudes, particularly that for K^+n.

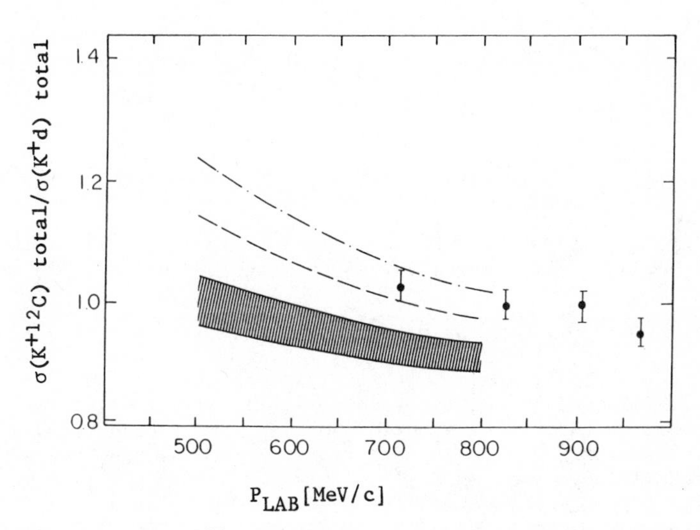

Fig. 6. The ratio of $K^+ + {}^{12}C$ to K^+d total cross sections per nucleon $R_T/6$ as a function of lab momentum. The data are from Bugg et al (1968), the shaded area represents the theoretical error bars associated with a conventional multiple scattering calculation (Siegel et al 1984), and the dashed curves are obtained by increasing the S_{11} K^+N phase shift in the $K^+ + {}^{12}C$ calculation (Siegel et al 1985). The figure is taken from the AGS proposal of Alster et al (1986).

An alternative way to account for the discrepancy between multiple scattering theory and data for $K^+ + {}^{12}C$ was suggested by Siegel et al (1985). They argue, in analogy to one interpretation of the EMC effect (Aubert et al 1983, Bodek et al 1983), that a nucleon in a nucleus is "swollen", and hence the S_{11} K^+ nucleon phase shift δ, which in free space is well parametrized by

$$\delta(S_{11}) = -kR \ , \tag{23}$$

where $R \approx 0.32$ fm is an effective hard sphere radius (Dover and Walker 1982), is modified to $\delta(S_{11}) = -k\tilde{R}$, where $\tilde{R} > R$, in the nucleus. They find that an increase of $\delta(S_{11})$ by 15% ($\tilde{R}/R = 1.15$), consistent with an increase of confinement scale of 10-30%, produces agreement with the $K^+ + {}^{12}C$ data (dashed band in Fig. 6). Note that below 500 MeV/c, the S_{11} phase shift dominates the K^+N interaction. At higher momentum, the P-wave cross section grows rapidly (perhaps reflecting exotic Z* resonances) and the K^+ interaction becomes more absorptive.

In order to understand the results of Siegel et al (1985) in a simple way, we recall the Glauber approximation for the total cross section

$$\sigma^T = 4\pi \int_0^\infty bdb \left\{ 1 - e^{-2\mathrm{Im}\chi(b)} \cos\left(2\mathrm{Re}\ \chi(b)\right)\right\} \tag{24}$$

where

$$\chi(b) = \langle\sigma\rangle\ (i + \beta)\ T(b)/4$$

$$T(b) = \int_{-\infty}^{+\infty} dz\ \rho(r)$$

$$\beta = \mathrm{Re}\ f_{K^+N}/\mathrm{Im}\ f_{K^+N} \text{ at } \Theta = 0° \tag{25}$$

$$\langle\sigma\rangle = (\sigma^T_{K^+p} + \sigma^T_{K^+n})/2$$

Here $\rho(r)$ is the nuclear density and $f_{K^+N} = (f_{K^+n} + f_{K^+p})/2$. Since $\langle\sigma\rangle$ is relatively small, we may expand Eq. 24 to second order in $\chi(b)$, obtaining

$$\sigma^T = A\langle\sigma\rangle + \frac{3}{8\pi}\left(\frac{\langle\sigma\rangle}{r_0}\right)^2 A^{4/3}(\beta^2 - 1) \tag{26}$$

where we have assumed a Gaussian form for $\rho(r)$, with $\langle r^2\rangle_A^{1/2} = r_0 A^{1/3}$. If β is small, which is true at high energies, the second term is negative, and we get shadowing, i.e. $\sigma^T/A\langle\sigma\rangle < 1$. For low momenta, the K^+N amplitude is predominantly real, i.e.,

$$\beta \approx \cot\delta \approx -1/kR > 1 \tag{27}$$

Thus, for small k, the double scattering term is positive, and we get "anti-shadowing", i.e. $\sigma^T/A\langle\sigma\rangle > 1$. This explains the behavior of the multiple scattering result in Fig. 6, since if $\langle\sigma\rangle$ is taken to be the same for K^+d and K^+A, we get

$$R_T = \sigma^T(K^+ + A)/\sigma^T(K^+ + d) \approx \frac{A}{2}\left[1 + \frac{3}{8\pi}A^{1/3}\frac{\langle\sigma\rangle}{r_0^2}(\beta^2 - 1)\right] \tag{28}$$

Thus, for $K^+ + {}^{12}C$, we pass from $R_T > 6$ for small k to $R_T < 6$ for large k. For $\beta = 1$, the double scattering term vanishes, and $R_T = 6$. In our simple model, this occurs when $\delta = -\pi/4$ or $k = \pi/4R \approx 500$ MeV/c for R = 0.32 fm.

The suggestion of Siegel et al (1985) is that $\langle\sigma\rangle$ is replaced by an effective cross section $\langle\tilde{\sigma}\rangle$ for ^{12}C, but remains the same for deuterium. Thus,

$$R_T \approx \frac{A}{2}\frac{\langle\tilde{\sigma}\rangle}{\langle\sigma\rangle}\left[1 + \frac{3}{8\pi}A^{1/3}\langle\tilde{\sigma}\rangle(\tilde{\beta}^2 - 1)/r_0^2\right] \tag{29}$$

where

$$\frac{\langle\tilde{\sigma}\rangle}{\langle\sigma\rangle} \approx \frac{\sin^2(k\tilde{R})}{\sin^2(kR)} \tag{30}$$

in the hard sphere model. If we now choose $\tilde{R} = 0.35$ fm, R = 0.32 fm, we find that $\langle\tilde{\sigma}\rangle/\langle\sigma\rangle$ varies from 1.15 at 500 MeV/c to 0.9 at 1 GeV/c. This explains the behavior of the dashed curves in Fig. 6. Note that the double scattering term has the opposite effect, i.e., $\tilde{\beta} = -1$ occurs at a <u>lower</u> momentum than $\beta = -1$. This would push the crossover point from shadowing to antishadowing to lower momentum, contrary to the desired effect.

Note that the double scattering term in Eq. 29 is small (of order 10%) for light nuclei, since $\langle\tilde{\sigma}\rangle/r_0^2 \approx 1$. The main effect in Fig. 6 thus arises from the modification of the single scattering term $(\langle\sigma\rangle \to \langle\tilde{\sigma}\rangle)$.

To eliminate the dependence on the somewhat uncertain ratio $\tilde{\beta}$, one can consider the ratio R_r of total reaction cross sections rather than R_T. From Glauber theory, we obtain, to second order in $\langle\tilde{\sigma}\rangle$, the following expressions for the total elastic and reaction cross sections σ^E and σ^R:

$$\sigma^E \approx \frac{3A^{4/3}}{8\pi}\left(\frac{\langle\tilde{\sigma}\rangle\tilde{\beta}}{r_0}\right)^2$$
$$\sigma^R \approx A\langle\tilde{\sigma}\rangle - \frac{3A^{4/3}}{8\pi}\left(\frac{\langle\tilde{\sigma}\rangle}{r_0}\right)^2 \tag{31}$$

thus

$$R_r = \frac{\sigma^R(K^+ + A)}{\sigma^R(K^+ + d)} \approx \frac{A}{2}\frac{\langle\tilde{\sigma}\rangle}{\langle\sigma\rangle}\left[1 - \frac{3A^{1/3}}{8\pi}\frac{\langle\tilde{\sigma}\rangle}{r_0^2}\right] \tag{32}$$

independent of $\tilde{\beta}$. Here, if $\langle\tilde{\sigma}\rangle = \langle\sigma\rangle$, we always get shadowing ($R_r < A/2$). In this case, anti-shadowing can only be produced by having $\tilde{\sigma}/\sigma > 1$, not by a rapid momentum dependence of $\tilde{\beta}$, as in Eq. 29.

For $\langle\tilde{\sigma}\rangle \to o$ and fixed $\tilde{\beta}$, we expect that

$$\frac{\sigma^E}{\sigma^R} \approx \frac{3A^{1/3}}{8\pi}\frac{\langle\tilde{\sigma}\rangle\tilde{\beta}^2}{r_0^2} \tag{33}$$

is <u>small</u>, since σ^E does not contain a term linear in $\langle\tilde{\sigma}\rangle$. Thus, if we regard $\langle\tilde{\sigma}\rangle$ as determined from R_r, then the ratio σ^E/σ^R should also be measured and used to determine $\tilde{\beta}$.

Marlow et al (1982) provide $K^+-{}^{12}C$ elastic data only for $\theta_{c.m.} < 41.1°$, but this already comprises most of the cross section. To obtain a rough estimate of σ^E, we fit the data with a standard exponential form

$$\left(\frac{d\sigma}{d\Omega}\right) \approx Ce^{bt} \tag{34}$$

where $t = -2k^2(1-\cos\theta_{c.m.})$. The choice $C \approx 486$ mb/sr, $b \approx 1.96$ fm^{-2}, is consistent with the rate of falloff of $d\sigma/d\Omega$ for $\theta_{c.m.} > 6°$, beyond the Coulomb-nuclear interference region. We then find $\sigma^E \approx 13$ mb for $K^+ + {}^{12}C$ at 800 MeV/c, and hence

$$\sigma^E/\sigma^R \approx 0.08 \ , \tag{35}$$

using $\sigma_T \approx 177$ mb from Bugg et al (1968). This is consistent with Eq. 33.

The idea of testing partial deconfinement by K^+ nucleus scattering is attractive, but of course there may also be a more conventional explanation of the discrepancies seen in Figs. 5 and 6. The advantage of the K^+ over other hadronic probes is that $\langle\sigma\rangle$ is relatively small (long mean free path) and hence the multiple scattering expansion converges rapidly, and theoretical uncertainties due to conventional medium modifications are small.

8.2 Hypernuclear Binding Energies and Decays

Several features of Λ hypernuclear binding energies and decay modes have been suggested as possible tests of partial deconfinement of the strange quarks. So far these discussions have remained on a very qualitative level.

One problem concerns the binding energy of $^5_\Lambda$He, which is less than that calculated using central two-body ΛN potentials fitted to scattering lengths. Some repulsive mechanism is needed to correct this difficulty. Hungerford and Biedenharn (1984) suggested that this may be due to Pauli repulsion at the quark level. That is, the u and d quarks in the $\Lambda = (ud)_{I=0} \times s$ are Pauli blocked in $^5_\Lambda$He from occupying the 1s state in a quark shell model of the hypernucleus. Quantitative calculations of this effect are not available, but the challenge for the quark model is to repair the $^5_\Lambda$He problem without losing the fit to the binding energies of A = 3,4 hypernuclei, which are well fitted by conventional potential models (Gibson 1985). One conventional solution to the $^5_\Lambda$He problem has been discussed by Bodmer and Usmani (1986). They introduce a repulsive three-body ΛNN potential and an attractive two-body ΛN potential, both consistent with meson exchange models, and obtain a fit to Λp scattering, all s-shell hypernuclear binding energies (including $^5_\Lambda$He), $^9_\Lambda$Be, and the Λ well depth. They require both ππ three-body potentials and repulsive dispersive ΛNN interactions. Another solution to the $^5_\Lambda$He overbinding problem has been suggested by Shinmura et al (1984), through the effect of tensor forces.

Goldman (1984) has suggested that the s quark in a hypernucleus may be at least partially deconfined, i.e., no longer localized in a cluster with a $(ud)_{I=0}$ diquark. The consequences of this idea have not been worked out quantitatively, but clearly there are numerous experimental constraints on such a quark delocalization which should be considered. For instance, the shell spacing $\hbar\omega_\Lambda$ varies essentially as $A^{-1/3}$ in hypernuclei, suggesting a Λ single-particle wave function resembling that of a nucleon. A delocalized Λ wave function, spread over the whole nucleus, would produce a different A dependence of $\hbar\omega_\Lambda$ ($A^{-2/3}$?). Another aspect is isospin. If the s quark is delocalized, the ground state of a hypernucleus need not have good I , i.e., the s u d three-quark system outside a nuclear core would in general be a mixture of Λ and Σ°. This should have consequences for weak decay rates, for example $^4_\Lambda H \rightarrow \pi^- + {}^4He$. These rates give bounds on isospin mixing in the initial state. The ratio of mesonic to non-mesonic weak decays can change by an order of magnitude depending on whether one uses localized or delocalized Λ wave funtions. The binding energies of heavy hypernuclei are important to measure, in order to extract the Λ well depth. Gal (1986) has pointed out that nucleon emission widths of hypernuclear states of $s_\Lambda s_N^{-1}$ structure would be reduced in the quark shell model, since the non-strange quarks of the Λ can partially fill the nucleon hole.

It seems clear that the hypernucleus, like an ordinary nucleus, is closer to a system of color singlet three-quark clusters than a quark gas. Nevertheless, one should pursue the search for experimental signatures, as suggested above, which provide constraints on the degree of strange quark delocalization.

9. Conclusions and Prospects

The field of strange particle nuclear physics is a rich one. On the one hand, one can use strange particles (for example, the K^+) to probe aspects of conventional nuclear structure (neutron densities, etc.). On the other hand, one or more strange quarks can be embedded in a many nucleon system, giving rise to a spectroscopy with distinctive features. Strangeness may offer special advantages for the study of quark phenomena, for instance partial deconfinement. However, the signatures of such phenomena are likely to be subtle, and require careful analysis.

Only a few topics have been selected for discussion here. Many more are treated in the proposals for the LAMPF II and TRIUMF kaon factories. For instance, the use of $K^{\pm}$ beams to study Drell-Yan phenomena is an exciting prospect for probing the virtual sea quark distributions in nuclei. Thus far, the pace of experimentation in this field has been frustratingly slow. A dedicated medium energy kaon beam line, such as the 1-2 GeV/c facility proposed for Brookhaven, would already represent a significant step forward. In the long term, however, the exploitation of this interesting area of physics requires the construction of a high intensity kaon factory.

Acknowledgement

This work was supported by contract DE-AC02-76CH00016 with the U.S. Department of Energy.

References

Abgrall Y and Labarsouque J 1984 Nucl. Phys. A426 431
Alster J et al 1986 Brookhaven AGS Research Proposal 835
Aubert J J et al 1983 Phys. Lett. 123B 275
Auerbach E H et al 1983 Ann. Phys. (NY) 148 381
Bando H and Takaki H 1984 Prog. Theor. Phys. 72 106
Bando H and Takaki H 1985 Phys. Lett. 150B 409
Barnes P et al 1983 Brookhaven AGS Research Proposal 788
Barnes P 1986 Nucl. Phys. A450 43c
Bart S et al 1984 Brookhaven AGS Research Proposal 798
Bender I, Dosch H G, Pirner H J and Kruse H G 1984 Nucl. Phys. A414 359
Bertini R et al 1980 Phys. Lett. 90B 375
Bertini R et al 1985 Phys. Lett. 158B 19
Bocquet J P et al 1986 CERN-EP/86-95, submitted to Phys. Lett. B.
Bodek A et al 1983 Phys. Rev. Lett. 51 534
Bodmer A R and Usmani Q N 1986 Nucl. Phys. A450 257c
Bugg D et al 1968 Phys. Rev. 168 1466
Büttgen R, Holinde K and Speth J 1985 Phys. Lett. 163B 305
Büttgen R, Holinde K, Holzenkamp B and Speth J 1986 Nucl. Phys. A450 403c
Chaumeaux A and Lemaire M-C 1983 Phys. Rev. C28 772
Cheung C-Y, Heddle D P and Kisslinger L S 1983 Phys. Rev. C27, 335
Chrien R E et al 1979 Phys. Lett. 89B 31
Chrien R E, Hungerford E V and Kishimoto T 1986 Brookhaven preprint
Chrien R E (Editor) 1986 Proc. of the International Symposium on Hypernuclear and Kaon Physics, Brookhaven National Laboratory, Nucl. Phys. A450 1c-586c
Dalitz R H and Gal A 1978 Ann. Phys. (NY) 116 167
Davis A C, Cottingham W N and Alcock J W 1976 Nucl. Phys. B111 233
Dover C B and Moffa P J 1977 Phys. Rev. C16 1087
Dover C B, Ludeking L and Walker G E 1980 Phys. Rev. C22 2073
Dover C B and Walker G E 1982 Phys. Reports 89 1
Dover C B and Gal A 1983 Ann. Phys. (NY) 146 309
Dover C B and Gal A 1984 Prog. Part. and Nucl. Phys. 12 171
Dover C B, Gal A, Klieb L and Millener D J 1986 Phys. Rev. Lett. 56 119
Dover C B 1986 Nucl. Phys. A450 95c
Dubach J F 1986 Nucl. Phys. A450 71c
Franklin G B et al 1985 Brookhaven AGS Research Proposal 813
Franklin G B 1986 Nucl. Phys. A450 117c
Frascaria R 1986 private communication
Fujiwara Y and Hecht K T 1985 Nucl. Phys. A444 541
Gal A 1986 Proc. Conf. on the Intersection between Particle and Nuclear Physics, Lake Louise, Canada (in press)

References, continued

Gibson B F 1985 Proc. Conf. on Hadronic Probes and Nuclear Interactions eds J R Comfort et al (New York: American Inst. of Phys.) pp 390-1
Goldman T 1984 Proc. 1st Conf. on Intersections Between Particle and Nuclear Physics ed R E Mischke (New York: American Inst. of Phys.) pp 799-805
Grace R et al 1985 Phys. Rev. Lett. 55 1055
He Y, Wang F and Wong C W 1986 Nucl. Phys. A448 652, A451 653, A454 541
Heddle D P and Kisslinger L S 1986 Phys. Rev. C33 608
Hungerford E V and Biedenharn L C 1984 Phys. Lett. 142B 232
Hungerford E V 1986 Nucl. Phys. A450 157c
Isgur N and Karl G 1978 Phys. Rev. D18 4187
Jaffe R L 1977 Phys. Rev. Lett. 38 195 and 1617(E)
Koike Y 1986 Nucl. Phys. A454 509
Koike Y, Morimatsu O and Yazaki K 1986 Nucl. Phys. A449 635
Lacombe M et al 1980 Phys. Rev. C21 861
Lagnaux J P et al 1964 Nucl. Phys. 60 97
Lenz F et al 1986 Ann. Phys. (NY) 170 65
Marlow D et al 1982 Phys. Rev. C25 2619
Marlow D et al 1984 Phys. Rev. C30 1662
Martin B R 1975 Nucl. Phys. B94 413
May M et al 1983 Phys. Rev. Lett. 51 2085
May M 1986 Nucl. Phys. A450 179c
McKellar B H J and Gibson B F 1984 Phys. Rev. C30 322
Millener D J, Gal A, Dover C B and Dalitz R H 1985 Phys. Rev. C31 499
Milner C et al 1985 Phys. Rev. Lett. 54 1237
Morimatsu O, Ohta S, Shimizu K and Yazaki K 1984 Nucl. Phys. A420 573
Mukhopadhyay D and Pirner H J 1985 Nucl. Phys. A442 605
Nagels M M, Rijken T A and de Swart J J 1977 Phys. Rev. D15 2547
Nagels M M, Rijken T A and de Swart J J 1979 Phys. Rev. D20 1633
Oset E and Salcedo L L 1985 Nucl. Phys. A443 704
Piekarz H et al 1985 Brookhaven AGS Research Proposal 820
Piekarz H 1986 Nucl. Phys. A450 85c
Pirner H J 1979 Phys. Lett. 85B 190
Pirner H J and Povh B 1982 Phys. Lett. 114B 308
Shinmura S, Akaishi Y and Tanaka H 1984 Prog. Theor. Phys. 71 546
Siegel P B, Kaufmann W B and Gibbs W R 1984 Phys. Rev. C30 1256
Siegel P B, Kaufmann W B and Gibbs W R 1985 Phys. Rev. C31 2184
Suzuki Y and Hecht K T 1984 Nucl. Phys. A420 525 and A446 749(E)
Wang F and Wong C W 1985 Nucl. Phys. A438 620
Yamazaki T et al 1985 Phys. Rev. Lett. 54 102
Yamazaki T 1986 Nucl. Phys. A450 1c

Inst. Phys. Conf. Ser. No. 86
Paper presented at Int. Nucl. Phys. Conf., Harrogate, UK, 1986

Achievements and goals of recent shell-model calculations

B. Alex Brown

National Superconducting Cyclotron Laboratory, Michigan State University, East Lansing, Michigan, 48824

Abstract. Recent work on large-basis multi-shell wave functions for the 0s-1d shell is discussed. Binding energies and excitation energies for nuclei with A=16-40 are used to provide constraints on the residual interaction. The derived wave functions are used to calculate essentially all measured electromagnetic and beta-decay observables. Emphasis is placed on information concerning non-sd shell components obtained from a comparison of experimental and calculated Gamow-Teller and M1 matrix elements. E2 and E4 matrix elements, (e,e') form factors, and nuclei far from stability are also discussed. Prospects for improvements and extensions are summarized.

1. Introduction

Experimental data on spins, parities, isospins, binding energies and excitation energies have usually been the primary consideration for the success or failure of nuclear structure models. The first considerations of these quantities led to the single-particle shell-model of Mayer and Jensen (1955). Since then, much effort has gone towards explicitly including as many configurations as possible into multi-particle wave functions. The remaining configurations must be taken into account, either microscopically or phenomenologically, in the context of the effective operators and interactions. The simplest tests for these wave functions concern the observables related to the creation or destruction of one or two nucleons. Comparison with such experimental observables provides important feedback for the correctness of the wave functions and the assumed forms for the effective operators. Exotic nuclei, e.g. those far from the valley of stability, also provide important tests and constraints on the model. Having passed these traditional tests, one can have some confidence in using the wave functions to calculate more complex processes and in the application to "special interest" aspects of nuclear physics such as isospin mixing, parity mixing and double beta decay.

For most cases of interest there are large numbers of nearly degenerate basis states. Improvements in large-basis multi-shell calculations rely on several factors: (1) better methods and models to predict which of the basis states are most important, (2) improved mathematical and computational techniques for angular momentum projection and setting up and diagonalizing large matrices and (3) the use of less restrictive and more realistic assumptions concerning the residual interactions. These improvements usually have been complemented by the use of faster computers with larger memory. Advances on many of these fronts have been made by a number of groups. In this talk I will concentrate on one particular set of calculations for which a great deal of progress has recently been made namely, those for the 0s-1d shell nuclei - those with neutron and proton numbers between 8 and 20. The problems associated with the sd-shell model space and Hamiltonian are discussed in section 2.

In section 3, I discuss the comparison between experimental and theoretical Gamow-Teller and M1 observables. These one-body operators provide one of the most direct tests of the sd-shell wave functions and provide important information on the non-sd shell components. Observables for more complicated one-body operators are discussed in subsequent sections: E2 matrix elements in section 4, and electron scattering form factors in section 5. Prospects and problems for the future are considered in section 6.

2. The sd-shell Hamiltonian

Even though there are only 24 active m-states in the $1s_{1/2}$-$0d_{5/2}$-$0d_{3/2}$ model space, the full basis dimensions are quite large; in the m scheme up to 93,710 for the 12-particle M=0, T_z=0 configuration of ^{28}Si, and in the J-T scheme up to 6706 for the for the 12-particle J=3, T=1 configuration. Although this placed a limitation on the early sd-shell calculations, current computational techniques and computer facilities can accommodate the full basis. For such large matrix dimensions it is essential to use the Lanczos algorithm to obtain the eigenfunctions (Whitehead et al 1977). Some of the first full basis calculations for the sd shell were carried out with the Glasgow m-scheme code (Whitehead et al 1977). Most of the results presented here are based on J-T coupled wave functions obtained with the MSU version of the Rochester Oak-Ridge code (French et al 1969).

All of the results presented here were obtained with wave functions with good isospin. We consider binding energies relative to the ^{16}O closed-shell configuration with the Coulomb energy contribution subtracted in an approximate manner (Chung 1976). Recently (Ormand 1986), the charge-dependent and charge-asymmetric interactions have been added to the isospin-conserving interaction discussed below in order to study analogue displacement energies, isospin mixing corrections to Fermi decays (Ormand and Brown 1985) and isospin forbidden proton and neutron decays (Ormand and Brown 1986) for the sd-shell nuclei. I will not have time to go into the interesting results from these calculations in this talk.

In a spherical basis the spectra for nuclei in the middle of the sd shell converge slowly. This point is illustrated in Fig. 1 which shows the binding energies of the ^{28}Si spectrum starting with a $(d_{5/2})^{12}$ closed-shell configuration (order 0) and allowing for an increasing number (order) of excitations to the $s_{1/2}$ and $d_{3/2}$ orbitals (order 1 to 12). The same interaction (the W interaction discussed below) has been used for all calculations. The ground state m-scheme dimension (i.e. the total number of states in each spectrum) is also given, and the low-lying (T=0) states are labeled by the J-value. Note in particular the slow excitation energy convergence for the J=2 and the excited J=0 states.

Given the complete basis, the primary problem has been to obtain an appropriate residual interaction. The microscopic effective interaction approach based upon a G matrix plus core-polarization corrections for the

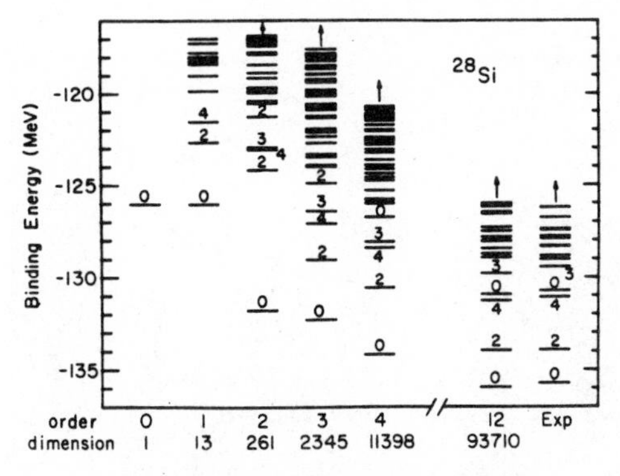

Fig. 1 Binding energies of levels in ^{28}Si as a function of the number of excitations from the $d_{5/2}$ orbital

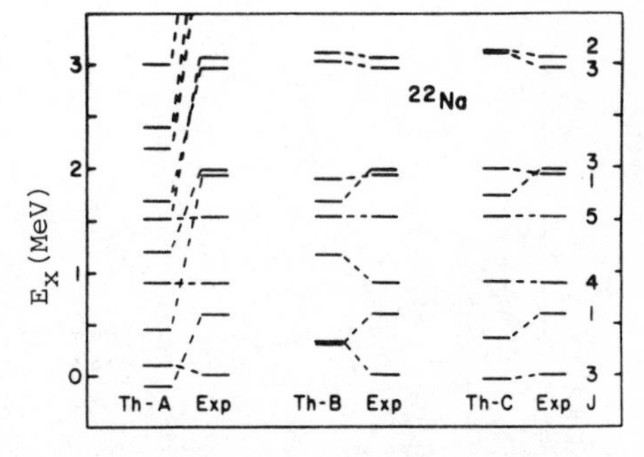

Fig. 2 Low-lying excited states of ^{22}Na compared with three theoretical calculations discussed in the text

non-sd shell components has been qualitatively successful in describing the experimental spectra, especially at the beginning of the shell (Kuo and Brown 1966, Shurpin et al 1983). However, there are shortcomings to this microscopic "no-parameter" approach (McGrory and Wildenthal 1980). As a "worst-case" example, I show in Fig. 2 the spectrum of ^{22}Na based upon the G-matrix obtained from the Paris potential and a folded-diagram approach for the core-polarization corrections (Th-A) (Shurpin et al 1983).

From a more phenomenological approach, it turns out to be possible to obtain much improved spectra for ^{22}Na (Th-C in Fig. 2) as well for other nuclei. The essence of this approach is to treat the two-body matrix elements and single-particle energies as parameters in a least-squares fit to experimental binding energies and excitation energies. Since there are 63 such two-body matrix element parameters, this is a formidable task. However, the work of Wildenthal and his collaborators has culminated in the remarkable achievement of fitting the binding and excitation energies of about 440 sd-shell states (typically with JT-basis dimensions of about 1000 or less) to within an rms deviation of 150 keV (Wildenthal 1984). (States with larger dimensions were not included in the fit due to computational time limitations. However, subsequent calculations for these cases with the final fitted interaction gave comparable good agreement with experiment - see Fig. 3.) The parameters of this fit were actually the 47 best determined linear combinations of two-body matrix elements. The remaining poorly-determined linear combinations were kept fixed to Kuo (1967) values. I will refer to the two-body matrix elements obtained from this fit as the "W-interaction".

Over the mass region A=16-40 one should expect some mass dependence in the values of the two-body matrix elements. For example, with a delta-function interaction and harmonic-oscillator radial wave functions with $\hbar\omega=41\ A^{-1/3}$ MeV, one would obtain a mass dependence proportional to $A^{-1/2}$ for the two-body matrix elements. The W interaction includes a mass dependence of the form $\langle V\rangle(A) = \langle V\rangle(A{=}18)\ (A/18)^{-0.3}$ where the power -0.3 was an informed guess. This was one of the key ingredients in the improvement over the previous work of Chung and Wildenthal (Chung 1976) which required two

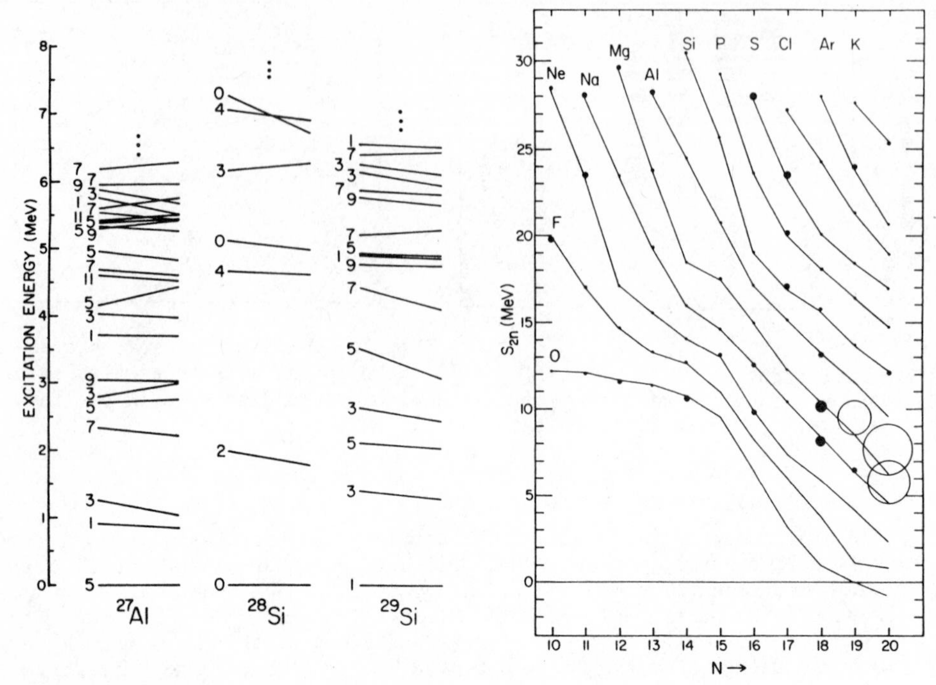

Fig. 3 Experimental (right) vs theoretical (left) energy levels calculated with the W interaction

Fig. 4 Two-neutron separation energies as a function of N

different sets of mass-independent two-body matrix elements, one for the lower and another for the upper part of the sd shell. The <V>(A=18) matrix elements of the W interaction are given in Table 1 of Wildenthal (1984).

Spectra for ^{27}Al, ^{28}Si and ^{29}Si are shown in Fig. 3. For each nucleus the experimental excitation energies are plotted on the right-hand side with a line joining to the theoretical energies obtained with the W interaction plotted on the left-hand side. A flat line thus indicates perfect agreement between experiment and theory. The slopes of the lines are typical of the 150 keV rms deviation obtained in the least-squares fit described above. (In fact, most of the levels shown here were not included in the least-squares fit because of their large dimensions.)

Two-neutron separation energies are shown in Fig. 4. The values obtained with the W interaction are given by the thin solid line connecting the chains of isotopes. The deviation between experiment and theory is indicated by the size of the circles. Most of the circles are small and consistent with the 150 keV rms. A few quite large deviations appear for the most neutron rich Na and Mg isotopes which may indicate the disappearance of the N=20 shell closure.

Much remains to be studied concerning the effective interactions. For the sd shell, why does the W interaction work as well as it does? Can one find an even better phenomenological interaction? How can the differences between the W interaction and the microscopic interactions be understood?

Do explicit many-body interactions and/or density-dependent interactions need to be introduced? What should be done for other model spaces? Can an sd-shell interaction be extrapolated to other model spaces?

Work which may help to answer some of these questions has recently been initiated (Brown et al 1986a). It is useful to start with a spin-tensor decomposition of the 63 j-j coupled two-body matrix elements (Brown et al 1985). Then, for each channel (central, spin orbit, tensor and antisymmetric spin orbit), we consider one-boson exchange potentials (OBEP) with strengths adjusted to fit the data. The results can be compared to analogous fits to the G matrix based, for example, on the Paris potential (Hosaka et al 1985). When isospin and parity are conserved, the antisymmetric spin-orbit terms vanish. Indeed, the 18 matrix elements of this type for the sd shell are small (Brown et al 1985), and if these terms are constrained to be zero, the rms deviation in the fit discussed above increases only by about 15% (Brown et al 1986a). To the OBEP we add the possibility of density dependence and multipole-multipole terms to take into account the model-space truncation and nuclear medium modification effects. From these considerations we have arrived at a 14 parameter density-dependent OBEP potential which can reproduce the 440 sd-shell data considered above with an rms deviation of about 250 keV. The spectrum of ^{22}Na calculated with this interaction is shown in Fig. 2 (Th-B). Applications of this potential to the fp shell and other mass regions are being considered (Brown et al 1986a).

It would also be important to unify the interactions used for Hartree-Fock calculations of closed-shell nuclei with those used for the valence spectra. One of the most successful and commonly used phenomenological interactions for Hartree-Fock calculations is the zero-range density-dependent Skyrme-type interaction. Many of the Skyrme parameterizations do not give the correct pairing property when applied to valence spectra, but this problem is corrected with the SGII parameterization (Sagawa et al 1985). Much work remains to be done concerning improvement and unification of effective interactions in nuclei.

3. Gamow-Teller and M1 Matrix Elements

Magnetic moments, M1 gamma decay and Gamow-Teller (GT) beta decay provide clear and direct information about nuclear structure because of the simple forms of the associated operators. We have compared essentially all such data for the sd shell with calculations based on the W interaction discussed in section 2. In the following I will discuss comparisons to experimental data whose matrix elements have a 10% or less uncertainty. In the figures discussed below, the experimental data plotted on the y axis is compared with two different theoretical calculations plotted on the x axis: one based on the "free-nucleon" operator (left-hand side) and the other based upon an "effective" operator (right-hand side). Agreement between experiment and theory would be indicated by the points on the plot clustering around the 45° line.

The free-nucleon theory is based upon using the standard one-body operators: GT = $|g_A/g_V|\ \underline{s}$ and M1 = $g_s\ \underline{s} + g_\ell\ \underline{\ell}$, where g_A/g_V = 1.251 and g_s and g_ℓ are the spin and orbital g factors for the free nucleon. The effective operator is meant to take into account the non-sd shell components of the wave functions. For the one-particle (A=17) and one-hole (A=39) cases, the effective operator can be expressed in terms of a correction involving the three general types of rank-one operators: $\underline{x} \rightarrow \underline{x} + g_x(\delta_s \underline{s} + \delta_\ell \underline{\ell} + \delta_p \underline{p})$ where $\underline{p} = (8\pi)^{1/2}[Y^{(2)} \otimes \underline{s}]^{(1)}$ and where $g_x = |g_A/g_V|$ for x = GT and $g_x = g_s$ for x = M1. The calculated δ parameters are typically 30-40% larger for A=39 compared to A=17 (Arima et al 1986, Towner and Khanna 1983). We will assume that this effective operator can be interpolated between A=17 and A=39 by making the δ parameters mass dependent in the form $\delta(A) = \delta(A=28)\ (A/28)^{0.35}$ (Brown and Wildenthal 1985). (In principle, one should also introduce effective many-body operators for A=18-38; however, empirically there appears to be little need for them beyond that incorporated implicitly into the mass dependence.)

The final results for the effective operator are based upon a least-squares fit of the data to the three parameters δ_s, δ_ℓ and δ_p. Three sets of δ parameters are considered. One set for the Gamow-Teller matrix elements (Brown and Wildenthal 1985), one set for the isovector M1 matrix elements as determined from a fit to M1 transition and magnetic moment data (Brown et al 1986b), and one for the isoscalar M1 matrix elements as determined from a fit to isoscalar moments (Brown and Wildenthal 1983).

First, I will discuss the comparison of the data with the free-nucleon and effective operators. Then I will discuss the significance of the extracted values for the δ parameters.

Experimental and theoretical GT matrix elements R(GT) are compared in Fig. 5. The experimental data are from a recent compilation (Brown and Wildenthal 1985). The quantity R(GT) is given by M(GT)/W with $W = |g_A/g_V|[(2J_i+1)3(N_i-Z_i)]^{1/2}$ for $N_i \neq Z_i$ and $W = |g_A/g_V|[(2J_f+1)3(N_f-Z_f)]^{1/2}$ for $N_i = Z_i$. The ft values are given by 6170/[B(F)+B(GT)] with $B(GT) = M(GT)^2/(2J_i+1)$. The quantity W takes into account the 3(N-Z) sum-rule unit.

The experimental GT matrix elements are systematically smaller by a factor of about 0.77 (indicated by the lower line on the left-hand side of Fig. 5) than those calculated with the free-nucleon operator. This indicates that this quenching is dominated by the δ_s parameter. The effective operator

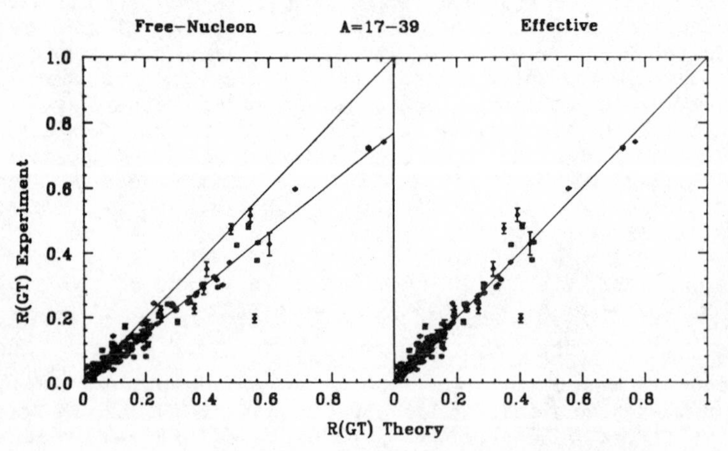

Fig. 5 Theoretical vs Experimental R(GT) matrix elements

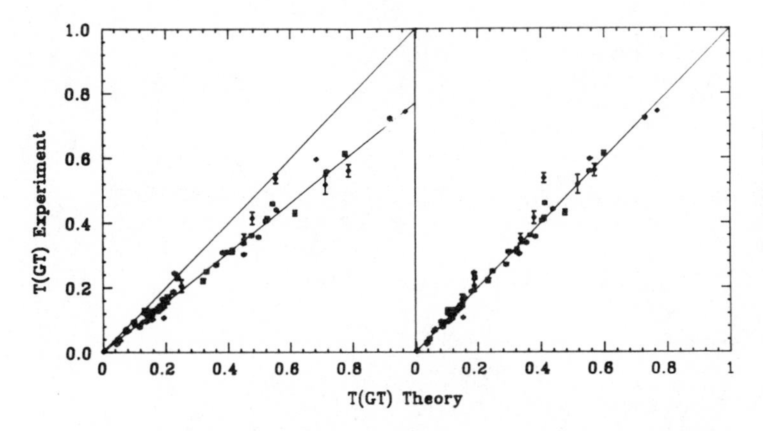

Fig. 6
Theoretical vs Experimental T(GT) matrix elements

accounts extremely well for most of the data. Some of the remaining deviations can be accounted for by the mixing of close-lying final states. This can be removed by summing the strength over final states as shown by the comparison of T(GT) in Fig. 6, where $T(GT) = [\Sigma_f R(GT)^2]^{1/2}$. The size of T(GT) is determined mainly by how much of the 3(N-Z) sum rule lies inside the beta decay Q-value window. For two cases, $^{18}Ne \rightarrow {}^{18}F$ and $^{19}Ne \rightarrow {}^{19}F$, the sum rule is nearly exhausted by a single final state, while in others only a very small fraction is found.

Our beta-decay calculations have been extended to the very neutron rich sd-shell (Wildenthal et al 1983) and p-shell (Curtin et al 1986) nuclei. For the sd-shell nuclei, most experimental and calculated half lives are in good agreement. However, there are some outstanding discrepancies such as for ^{31}Mg (see Fig. 1 in Wildenthal et al 1983), which again may indicate the disappearance of the N=20 shell closure. The beta decay of very proton-rich and neutron-rich nuclei such as ^{32}Ar (Bjornstad et al 1985) and ^{29}Na (Baumann et al 1986) recently studied at the ISOLDE facility as well as those currently being studied at the new NSCL (Curtin et al 1986) and GANIL (Langevin et al 1986) facilities will provide many important new tests of the nuclear structure models. In particular, detailed decay studies are needed in order to elucidate the levels and spins of these exotic nuclei.

Recent (p,n) experiments have provided valuable information on the GT strength by extending the "Q-value window" up to much higher excitation energies. A nice example is provided by a recent $^{26}Mg(p,n)$ experiment. The extracted experimental GT strength distribution (Madey et al 1986) is compared with the calculated distribution in Fig. 7. Both theory and experiment have been averaged over 2 MeV in order to bring out the gross structure in the comparison. Only the peak around 1 MeV in excitation lies within the Q-value window of the GT beta decay of the analogue nucleus ^{26}Si. The calculated free-nucleon distribution multiplied by $(0.77)^2 = 0.6$ is in good qualitative agreement with experiment. The "missing" GT strength is not to be found, at least up to an excitation energy of about 20 MeV, for the sd-shell nuclei. The GT strength extracted from (p,n) experiments on nuclei up to the Pb region is systematically "quenched" relative to 3(N-Z) by this factor of about 0.6 (Goodman and Bloom 1984). New (n,p) experiments such as those being started at TRIUMF (Alford and Hausser 1986) will provide valuable information on the final state isospin dependence of the GT strength.

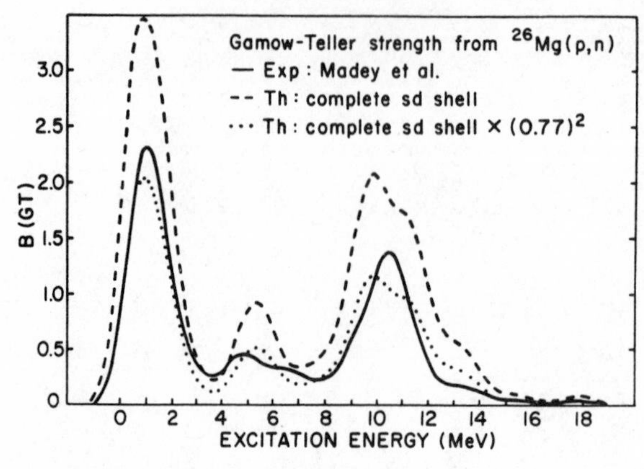

Fig. 7 Theoretical vs experimental Gamow-Teller strength functions for ^{26}Mg → ^{26}Al

The experimental M1 matrix elements derived from gamma decay transition data (Wildenthal and Keinonen 1986) are compared with theory in Fig. 8; $B(M1) = M(M1)^2/(2J_i+1)$. There appears to be no need for quenching relative to the free-nucleon calculation. However, the scatter of the points is reduced significantly with the three-parameter effective operator. Likewise, the magnetic moments data (Lederer and Shirley 1978) shown in Fig. 9

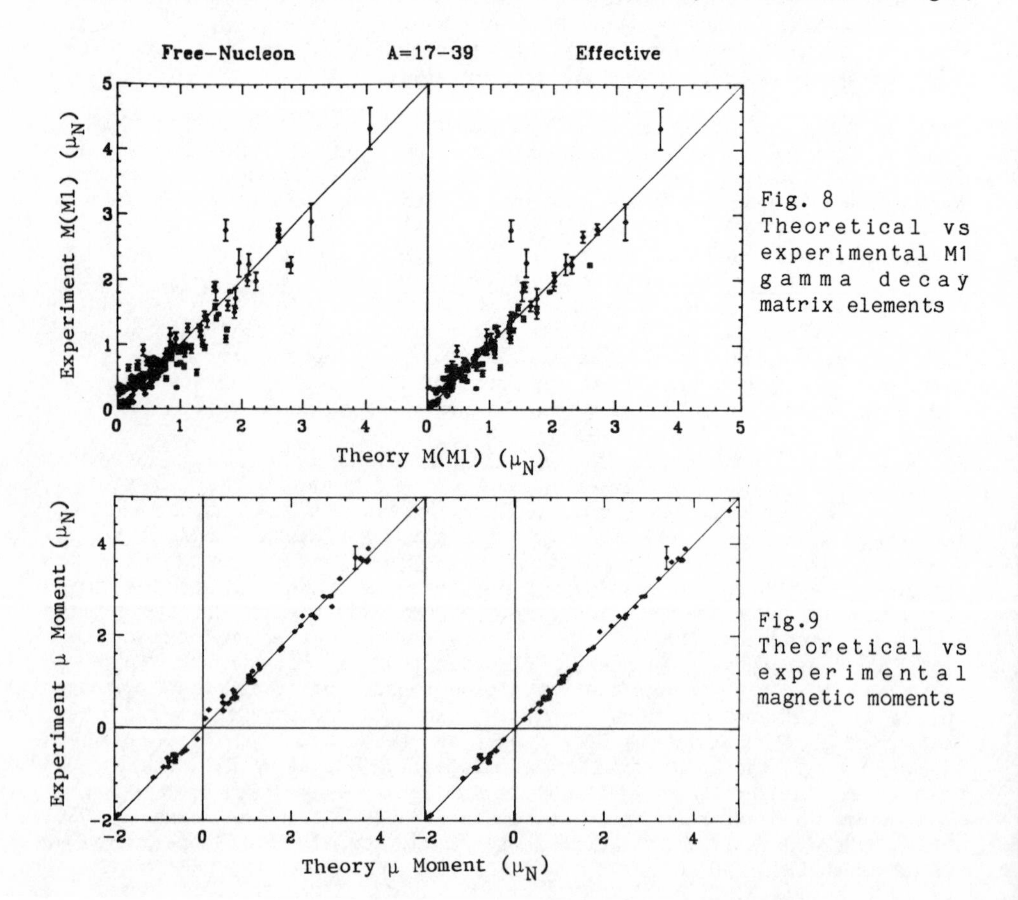

Fig. 8 Theoretical vs experimental M1 gamma decay matrix elements

Fig.9 Theoretical vs experimental magnetic moments

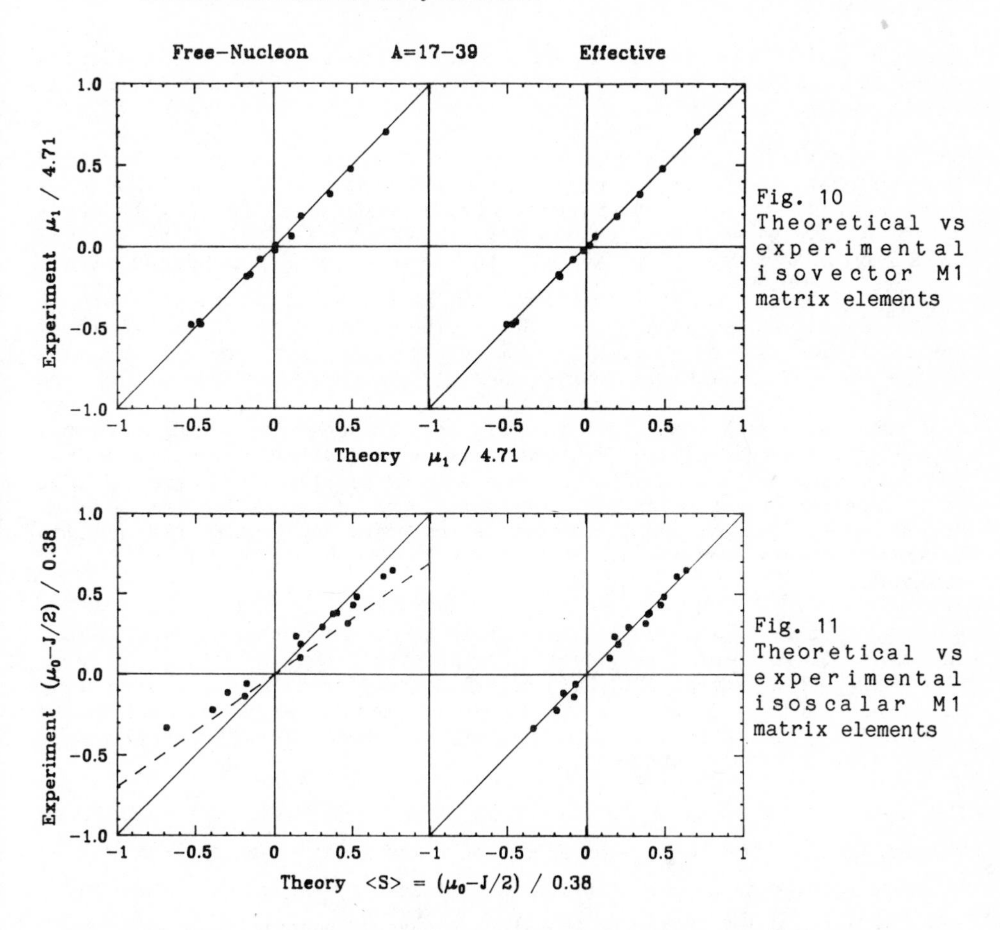

Fig. 10 Theoretical vs experimental isovector M1 matrix elements

Fig. 11 Theoretical vs experimental isoscalar M1 matrix elements

are qualitatively explained by the free-nucleon theory and the quantitative agreement with the effective-operator theory is impressive. In view of the fact that the M1 operator is dominated by the isovector spin term (since the isovector spin g factor is large), it is surprising to find that the the free-nucleon comparisons for the GT and M1 matrix elements give such different results.

The isovector and isoscalar components of the M1 operator can be isolated by looking at the difference and sum, respectively, of the matrix elements for mirror moments and transitions. These data for the moments are shown in Fig. 10 for the isovector case, $\mu_1 = [\mu(T_z{=}T)-\mu(T_z{=}-T)]/2$, and in Fig. 11 for the isoscalar case, $\mu_0 = [\mu(T_z{=}T)+\mu(T_z{=}-T)]/2$. In addition to the data covered by Brown and Wildenthal (1983), I include results based on the recent measurement of the ^{33}Cl magnetic moment (Rogers et al 1986). [μ_0 has been converted into an equivalent spin matrix element for Fig. 11 and for the extracted δ parameters, see Wildenthal and Brown (1983).] The isovector comparison is as expected. However, the isoscalar moments show a considerable deviation from the free-nucleon theory. The deviations cannot be simply explained by an overall reduction factor which would correspond to the rotated dashed line on the left-hand side of Fig. 11. Both δ_s and δ_ℓ components of the effective isoscalar operator are important. The effective operator derived from isoscalar moments is consistent with the quenching found for isoscalar M1 transitions (Anantaraman et al 1984 and Brown et al 1986b).

Extensive calculations for the non-sd shell corrections to the GT and M1 operators have been carried out by Towner and Khanna (1983) and Arima et al (1986). In Fig. 12 I compare our empirical values for δ with these calculations. The empirical values derived from the fits to the A=17-39 mass region are scaled to A=39 by the assumed mass dependence $\delta(A) = \delta(A{=}28)\ (A/28)^{0.35}$ and are compared to the calculations based on one hole in a ^{40}Ca closed-shell configuration. The error bars on the empirical results reflect the rms deviation between experiment and theory in addition to the error in the experimental data. The calculated corrections are broken down into three components: (1) those arising from high-order configuration mixing (CM), (2) those arising from mesonic exchange currents involving the Δ isobar, and (3) those arising from other mesonic exchange currents such as the one-pion exchange and the pionic pair diagrams (MEC). It is important to remember that the first-order configuration mixing which dominates the non-sd shell corrections for operators like E2 (see section 4) are not present for the simple M1 and GT operators which have no radial dependence.

A number of interesting conclusions can be drawn from these comparisons. The empirical value of $\delta_s(\mu_0)$ derived from isoscalar moments is quite large and is in good agreement with both calculations. The calculated quenching comes mostly from CM with a little enhancement from MEC - the Δ isobar contributions vanish in this case because the particle-hole vertex must change isospin by one unit.

The empirical value of $\delta_s(GT)$ is about 50% larger than the calculated values. Again, the calculated quenching comes mostly from CM. For both $\delta_s(\mu_0)$ and $\delta_s(GT)$ the CM contribution is dominated by the tensor interaction, and the ratio $\delta_s(GT,CM)/\delta_s(\mu_0,CM) \approx 0.42$ is very insensitive to details of the CM calculation (Arima et al 1986). Thus, with CM alone it is difficult to explain the additional empirical GT quenching without destroying the agreement for the isoscalar moments.

The importance of the Δ contribution to $\delta_s(GT)$, however, turns out to extremely controversial and many calculations have been made. For example, Oset and Rho (1979) obtain $\delta_s(GT,\Delta) = -0.15$, compared to -0.038 from Towner and Khanna and 0.004 from Arima et al. Differences arise from the assumptions made about short-range correlations, exchange terms and crossing terms. Agreement with our empirical value would be obtained with $\delta_s(GT) \approx -0.10$, somewhere intermediate to the calculated extremes.

The calculated δ_s values differ between the GT and the isovector M1 (μ_1) operators only in the MEC component. The CM and Δ contributions are equal essentially because the operator is just $\sigma\tau$ in both cases. The MEC contribution on the other hand depends upon the underlying relativistic structure, i.e. axial-vector for GT and vector for M1. It turns out that the MEC corrections for the GT operator, arising mainly from the ρ-π diagram, are small. The empirical difference $\delta_s(\mu_1)-\delta_s(GT) \approx 0.12$ is in good agreement with the MEC calculations.

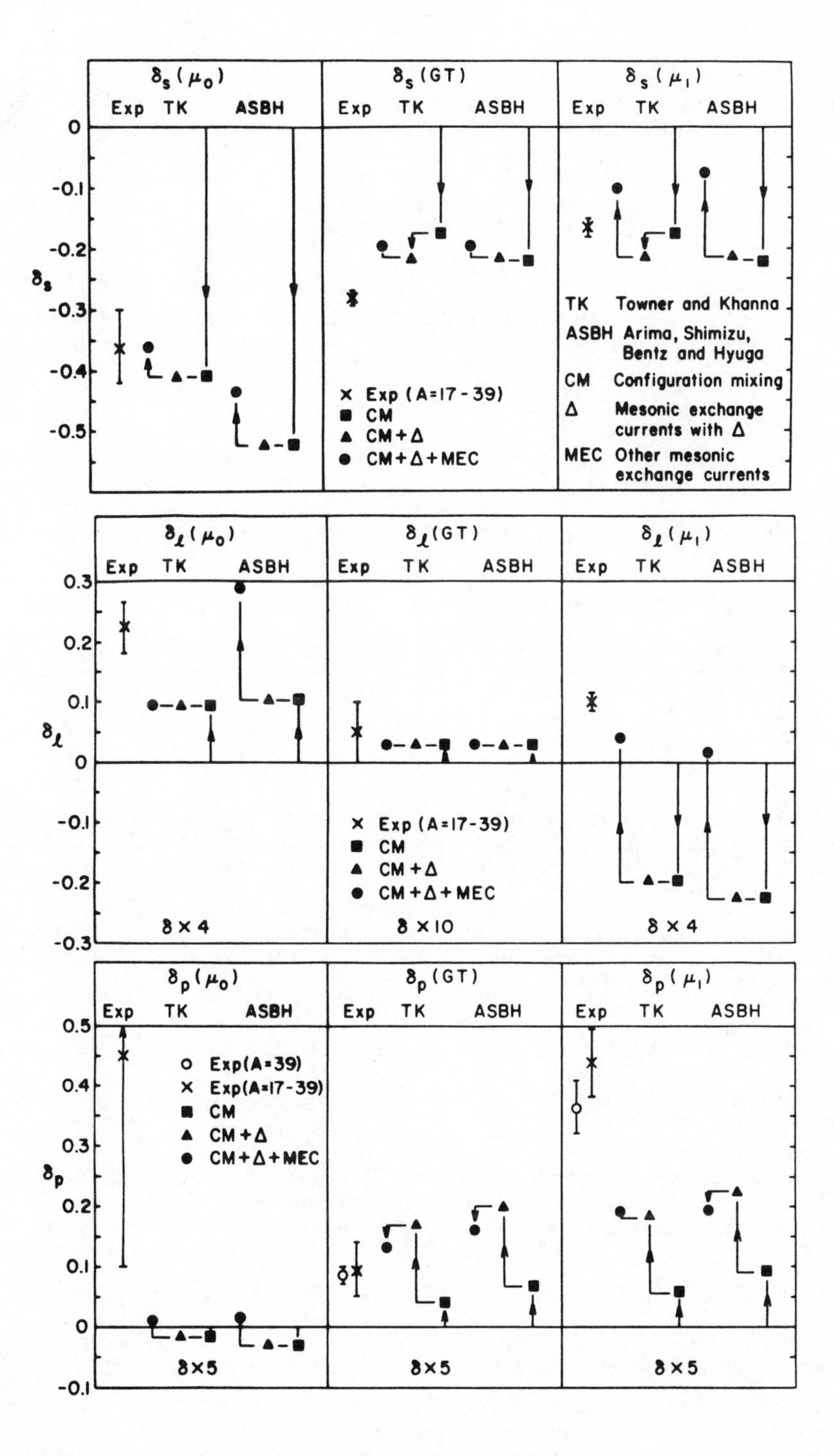

Fig. 12 Empirical δ parameters (scaled to A = 39) compared with two recent calculations

The empirical δ_ℓ values are relatively small and in fair agreement with the calculations. At first it appeared that the relativistic σ-ω model (Serot and Walecka 1986) would give extremely large corrections, related to the M*/M factor, to the orbital operator (Bouyssy et al 1984). However, now it appears that there are vertex corrections, "backflow" diagrams, which essentially bring the result back close to the nonrelativistic model, at least for the isoscalar term (McNeil et al 1986 and Arima et al 1986).

The empirical value for $\delta_\ell(\mu_0)$ is about twice as large as the CM calculations. The addition of new MEC calculations by Arima et al somewhat overshoots the empirical value. For $\delta_\ell(\mu_1)$ the agreement between experiment and theory seems good given the very large cancellation between the CM and MEC contributions.

The δ_p comparison in Fig. 12 includes values derived from the "single-particle" $3/2^+$ to $1/2^+$ transitions in A=39 (Adelberger et al 1984) which are consistent with those obtained in our global (A=17-39) analysis. The large error in the empirical $\delta_p(\mu_0)$ value precludes any meaningful comparison with the small theoretical corrections. The calculated δ_p corrections for GT and μ_1 are similar in value and dominated by the Δ isobar correction. The agreement for GT is good, but the empirical correction for the isovector M1 transitions is nearly twice as large as that calculated.

4. E2 matrix elements

Experimental E2 matrix elements derived from gamma decay transition data (Wildenthal and Keinonen 1986) are compared with theory in Fig. 13; $B(E2) = M(E2)^2/(2J_i+1)$. Only matrix elements which have an experimental uncertainty of 10% or less are considered in the comparison. The comparison (Carchidi et al 1986) for the quadrupole moment data is shown in Fig. 14. These data include the atomic measurements (Lederer and Shirley 1978) (crosses), as well as more recent data for the 2^+ states from Coulomb excitation experiments (Spear 1981) (circles with error bars) and from muonic atoms (Weber et al 1982 and Jeckelmann et al 1983) (squares with error bars). Error bars are not given for the atomic data because of the unknown uncertainties associated with the Sternheimer corrections (Sternheimer and Peierls 1971).

The E2 operator is more complicated than M1 or GT because some assumptions must be made about the radial behavior of the wave functions. We have considered many parameterizations for the radial wave functions (Brown et al 1977, Brown et al 1982, Carchidi et al 1986). However, for simplicity here we use the harmonic-oscillator form with $\hbar\omega = 45\ A^{-1/3} - 25\ A^{-2/3}$.

On the left-hand sides of Figs. 13 and 14 experiment is compared with the theoretical values obtained with the free-nucleon charges e_p=1e and e_n=0. The experimental data are uniformly larger than those calculated by a factor of about 1.8 (represented by the dashed line in Fig. 13 and the line with an increased slope in Fig. 14). A least-squares fit gives values for the "effective charges" of $e_p+\delta e_p+\delta e_n$=1.78(3)e for the isoscalar and

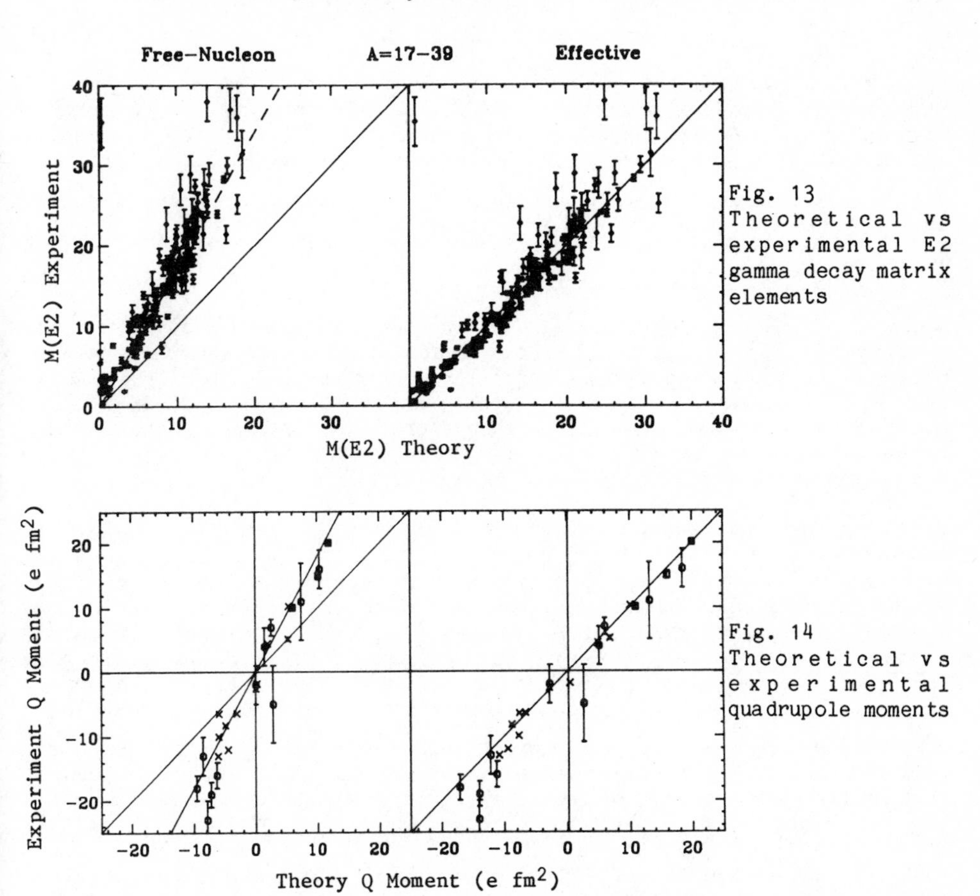

Fig. 13 Theoretical vs experimental E2 gamma decay matrix elements

Fig. 14 Theoretical vs experimental quadrupole moments

$e_p+\delta e_p-\delta e_n$=0.8(1)e for the isovector. These values have been used for the effective-operator calculation in the comparison on the right-hand side of the figures. With a few exceptions out of the 143 cases, the agreement between experiment and theory is excellent. (The exceptional point in the upper corner corresponds to the transition from the third 4^+ to the second 2^+ state in ^{28}Si.) The matrix elements between low-lying states are dominated by the isoscalar component and for this reason the isoscalar effective charge is well-determined. The isovector effective charge is poorly determined because it "lies in the noise" of the experimental and theoretical errors in the isoscalar component.

The isovector term can be isolated by looking at the differences between transitions in mirror nuclei (Brown et al 1982). In this analysis, the isovector effective charge turned out to be quite sensitive to the radial wave functions because many of the strong isovector transitions occur between loosely bound states in the lower part of the shell. The results ranged from $e_p+\delta e_p-\delta e_n \approx 1.0$e with harmonic-oscillator radial wave functions to $e_p+\delta e_p-\delta e_n \approx 0.65$e with the more realistic Woods-Saxon radial wave functions. The use of the Woods-Saxon wave functions reduces the isoscalar effective charge in these analyses to about 1.60e.

It is well known that the first-order coupling of the valence states with the $2\hbar\omega$ giant quadrupole states is the primary source for the charge renormalization. Recent calculations based on the SGII interaction (Sagawa and Brown 1984) gave average values of $e_p+\delta e_p+\delta e_n=1.68e$ and $e_p+\delta e_p-\delta e_n \approx 0.78e$, in good agreement with the ranges for the empirical values quoted above.

5. Electron scattering form factors

Electron scattering provides an extension to the M1 and E2 gamma decay data in both multipolarity and momentum transfer. For the sd shell we can consider the C0, C2 and C4 longitudinal form factors and the M1, E2, M3, E4 and M5 transverse form factors. The elastic C0 data (and related rms radii data) are used primarily to parameterize the phenomenological radial potentials (Brown et al 1983). Inelastic C0 transitions appear to be dominated by non-sd shell components (Blok et al 1984).

The inelastic longitudinal (e,e') spectra for the sd-shell nuclei are dominated by the the C2 and C4 multipoles. At the first maximum the experimental C2 form factors are typically a factor of three larger than those calculated with the free-nucleon charges. However, the C2 form factors can be well understood by combining the contribution from the valence orbits with a Tassie model form factor for the collective contribution normalized to reproduce the E2 effective charges parameters δe_p and δe_n discussed in section 4 (Brown et al 1983). The experimental C4 form factors are typically a factor of four larger than the free-nucleon calculation. Quantitative agreement in nearly all cases can be obtained by the addition of a collective term normalized to the effective charges of $e_p+\delta e_p+\delta e_n=2.0e$ and $e_p+\delta e_p-\delta e_n=1.0e$ (Brown et al 1983 and Wildenthal et al 1985). A recent example (Wildenthal et al 1985) which shows the level of agreement which can be obtained for the shape and magnitude of relatively strong C2 and C4 form factors is illustrated by the $^{32}S(e,e')$ example shown in Fig. 15. The individual multipoles are indicated by the crosses (C2) and dashed line (C4). The form factor for this doublet is dominated by the C4 term and is in excellent agreement with experiment.

It may seem surprising that the isoscalar E4 effective charge (2.0e) is larger than the E2 effective charge (1.6-1.8e). However, this is consistent with a recent calculation for the coupling of the sd-shell valence orbits to the E4 giant resonance (Sagawa and Brown 1985). It would be interesting to continue such comparisons for the higher-multipole ($L>4$) longitudinal form factors in heavier nuclei.

Given these state and mass independent effective charge ingredients, the C2 and C4 strength distributions are usually in excellent agreement with experiment up to excitation energies of about 6-8 MeV. As an illustration, I show in Fig. 16 the comparison of the calculated (Radhi et al 1983) and measured (Ryan et al 1983) $^{27}Al(e,e')$ spectrum at 90° for E_e = 177 MeV ($q \approx 1.3 fm^{-1}$). At this angle the spectrum is dominated by the longitudinal C2 and C4 terms. The states in the experimental spectrum (bottom) labeled with arrows are known to be positive parity states. The theoretical spectrum (top) has been plotted with a resolution to match that of the experiment.

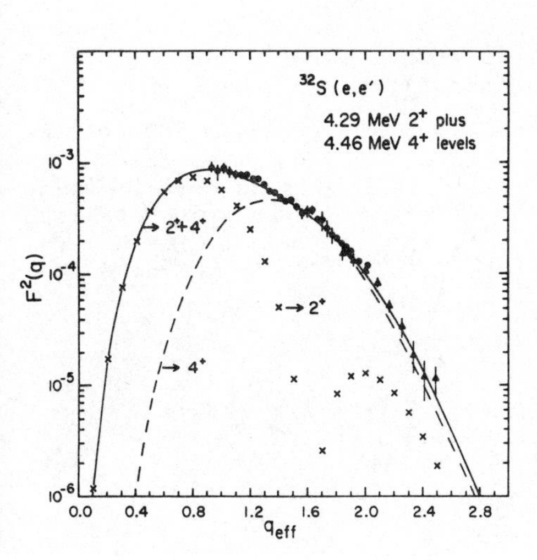

Fig. 15 Longitudinal (e,e') form factor for the unresolved 4.29 - 4.46 MeV doublet in ^{32}S

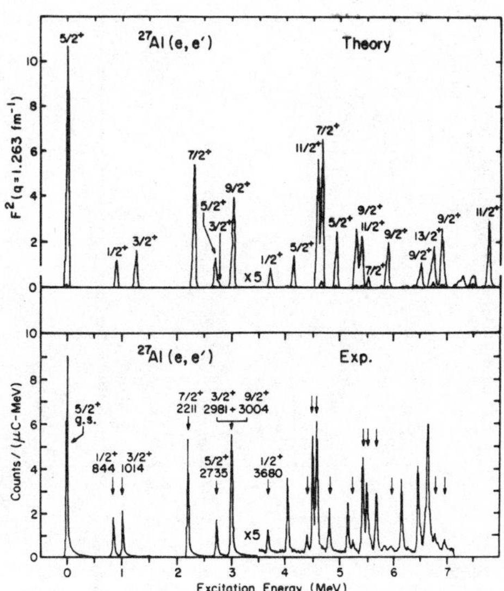

Fig. 16 Theoretical vs experimental spectrum for ^{27}Al(e,e') at 90° and E_e = 177 MeV

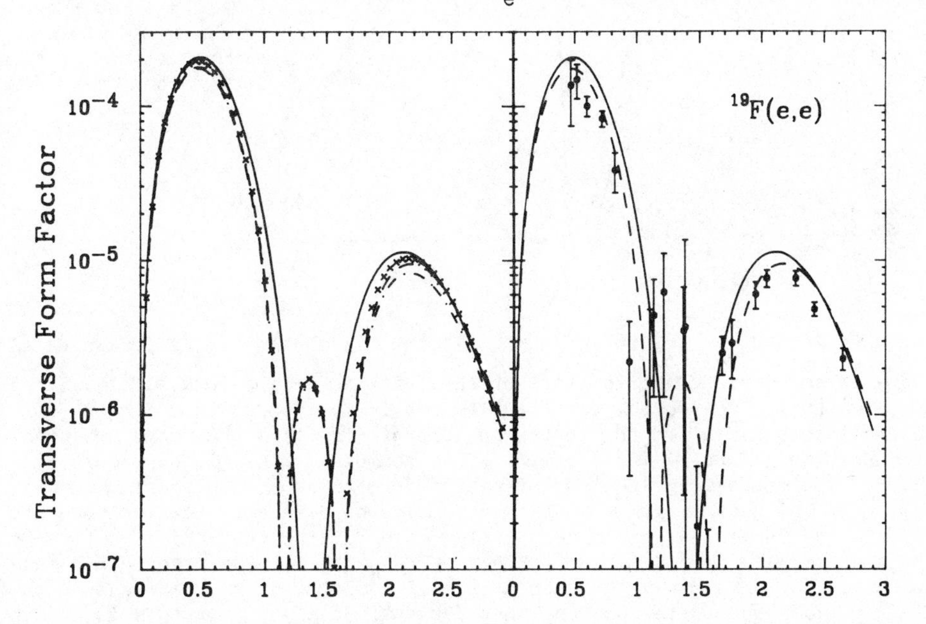

Fig. 17 Theoretical vs experimental transverse elastic form factor for ^{19}F.

Transverse M1 form factors provide an important extension of the M1 moment and gamma-decay data discussed in section 3. Such data has recently been obtained from the transverse elastic scattering from the stable $1/2^+$ ground states of ^{19}F (Donne et al 1986), ^{29}Si and ^{31}P (Miessen et al 1984). The ^{19}F data is compared to our calculations in Fig. 17. The free-nucleon shell-model calculation using harmonic-oscillator radial wave functions

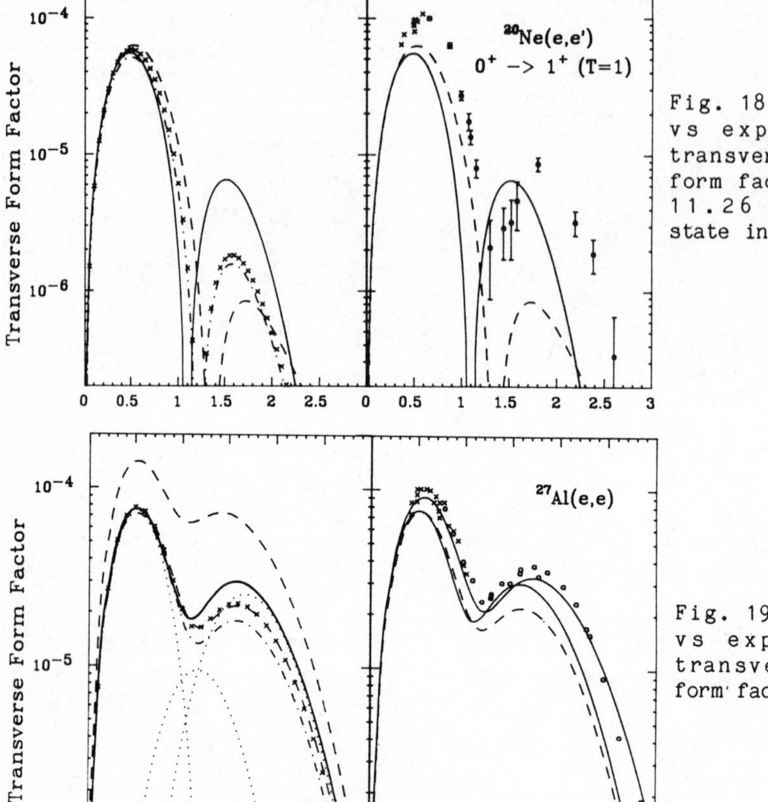

Fig. 18 Theoretical vs experimental transverse inelastic form factor for the 11.26 MeV 1^+ T=1 state in ^{20}Ne

Fig. 19 Theoretical vs experimental transverse elastic form factor for ^{27}Al

(HORWF) with b_{rms}=1.83 fm chosen to fit the rms charge radius (Brown et al 1980) is shown by the solid lines on the left- and right-hand sides of the figure. It is, of course, important to put in the corrections for the non-sd shell components. On the left-hand side of Fig. 17 I show some of these corrections (Blunden 1986) based on the formulation of Blunden and Castel (1985). The crosses include the effect of first-order (1p1h) configuration mixing (FCM) (which goes to zero at q=0), the dash-dot line includes the effects of FCM plus higher-order configuration mixing (FCM+CM), and finally the dashed line includes the effect of FCM plus CM plus mesonic exchange currents (FCM+CM+δ+MEC). The dashed line is compared with experiment on the right-hand side. It is apparent that FCM dominates and that this is enough to qualitatively explain the double minimum suggested by the data.

Similar comparisons are made in Fig. 18 for the ^{20}Ne inelastic M1 form factor for the transition to the T=1 1^+ state at 11.26 MeV with data from Darmstadt (crosses) and Mainz (circles) (Bergstrom 1983). We have used HORWF with b_{rms}=1.87 fm. In this case MEC are relatively more important. The non-sd shell components improve the agreement with experiment; however, there is still a big disagreement between experiment and theory in this case. Clearly something is missing in the calculation. The calculation predicts that this state has a large orbital component at small q (Knupfer and Metsch 1983 and Willis et al 1986), and this has been confirmed in a recent (p,p') experiment (Willis et al 1986).

The elastic ^{27}Al transverse form factor shown in Fig. 19 illustrates an example involving all three terms M1, M3 and M5 (the individual contributions are shown by the dotted lines on the left-hand side). The data are taken from the review of Donnelly and Sick (1984). The calculations are as described above using HORWF with b_{rms}=1.80 fm. The upper dashed line on the left-hand side is the form factor for the pure single-particle $d_{5/2}$ configuration. In this case the non-sd shell components worsen the agreement with experiment. It is interesting to note that the agreement with experiment can be made nearly perfect if we simply modify the free-nucleon sd-shell calculation by reducing the oscillator length by 10% (the upper solid line on the right-hand side). This change, while not large, is difficult to justify however. It is reminiscent of the famous Nolen-Schiffer anomaly problem for displacement energies (Nolen and Schiffer 1969).

These examples are typical of the problems faced in understanding transverse form factors. More data is needed, especially for the pure E2, M3 and E4 multipoles which can be isolated in the transverse scattering from $J=0^+$ targets.

6. Prospects for the future

The success of the comparisons presented above puts the extended shell-model approach on very solid ground, at least empirically. There is probably still room for improvement in the sd-shell calculations, but it is not clear from what direction they will come. Perhaps some further adjustments of the two-body matrix elements will help. Feedback from the properties of exotic nuclei as well as the electromagnetic and beta decay data should help to tie down some of the poorly determined two-body matrix elements. For other data such as C0 form factors and some of the transverse magnetic form factors, it appears that there must be important non-sd shell contributions that are as yet not understood.

The sd-shell results should provide a touchstone for other extended shell-model calculations. The fp shell is the next logical challenge. Although much progress has been made in the fp shell, away from the A=40, 48 and 56 $f_{7/2}$ closed shells most of the calculations are still carried out at the level equivalent to "order 1" in Fig. 1 and only a few at the level of "order 2." Hopefully a way can be found to unify the effective interactions from one shell to another. Just beyond the fp shell the situation is further complicated by the presence of the "intruder" high-spin orbitals from the next higher major shell.

Cross-shell excitations in light nuclei present another challenge. The $1\hbar\omega$ spectrum for the p-sd region (Millener and Kurath 1975 and van Hees and Glaudemans 1983) seems fairly well under control and similar work is being started for the sd-pf region (Warburton et al 1986). However for $2\hbar\omega$ and higher, there is again the problem of "order" convergence (Glaudemans 1986), along with the problems of removing spurious center of mass motion and allowing for the Hartree-Fock condition concerning the mixture of 1p1h into the closed-shell ground state.

To make further progress, better techniques for picking out the important basis states are important, such as those incorporated in the MONSTER-VAMPIRE approach (Schmid et al 1986) or those implicit in the variety of collective models. Also, it would help to have one more order of magnitude increase in the size of the conventional spherical basis; up to about 50,000 in the J-T scheme or about 1,000,000 in the m-scheme. This may be possible with continued improvements in methods used in the modern codes such as OXBASH (Etchegoyen et al 1985) or RITSSCHIL (Zwarts 1985). Or, perhaps, completely new methods will be found.

Acknowledgment

This research was supported by the U.S. National Science Foundation under grant no. PHY 83-12245.

References

Adelberger E G, Osborne J L, Swanson H E and Brown B A 1984 Nucl. Phys. A417 269
Alford W P and Hausser O 1986 private communication
Anantaraman N, Brown B A, Crawley G M, Galonsky A, Djalali C, Marty N, Morley M, Willis A, Jourdain J C and Wildenthal B H 1984 Phys. Rev. Lett. 52 1409
Arima A, Shimizu K, Bentz W and Hyuga H 1986 Adv. Nucl. Phys. to be published
Bauman P et al 1986 Int. Symp. Weak and Electromagnetic Interactions in Nuclei Heidelberg, Abst. Invited and Contributed Papers p 187
Bergstrom J C 1983 private communication
Bjornstad T et al 1985 Nucl. Phys. A443 283
Blok H et al 1984 Phys. Lett. B149 441
Blunden P G and Castel B 1985 Nucl. Phys. A445 742
Blunden P G 1986 private communication
Bouyssy A, Marcos S and Mathiot J F 1984 Nucl. Phys. A415 497
Brown B A, Arima A and McGrory J B 1977 Nucl. Phys. A277 77
Brown B A, Richter W A and Godwin N S 1980 Phys. Rev. Lett. 45 1681
Brown B A et al 1982 Phys. Rev. C26 2247
Brown B A and Wildenthal B H 1983 Phys. Rev. C28 2397
Brown B A, Radhi R and Wildenthal B H 1983 Phys. Rep. 101 313
Brown B A, Richter W A and Wildenthal B H 1985 J. Phys. G11 1191 (Note that the scale for the matrix elements shown in Fig. 2 of this paper should be reduced by a factor of three and scale for the matrix elements shown in Figs. 3 and 4 should be reduced by a factor of four)
Brown B A and Wildenthal B H 1985 At. Data Nucl. Data Tables 33 347
Brown B A, Richter W A, Van der Merwe and Julies R E 1986a contribution to this Conference, Vol. 1 B50
Brown B A, Etchegoyen M C, Etchegoyen A and Wildenthal B H 1986b unpublished
Carchidi M, Wildenthal B H and Brown B A 1986 preprint
Chung W 1976 PhD thesis Michigan State Univ. East Lansing
Curtin M S et al 1986 Phys. Rev. Lett. 56 34
Donne A J H et al 1986 Nucl. Phys. A455 453
Donnelly T W and Sick I 1984 Rev. Mod. Phys. 56 461
Etchegoyen A, Rae W D M, Godwin N S, Richter W A, Zimmerman C H, Brown B A, Ormand W E and Winfield J S 1985 MSU-NSCL report # 524

French J B, Halbert E C, McGrory J B and Wong S S M 1969 Adv. Nucl. Phys. 3 193
Glaudemanns P W M 1986 Nuclear Structure at High Spin, Excitation, and Momentum Transfer ed. H. Nann, AIP Conference Proc. 142, p 316
Goodman C D and Bloom S D 1984 Spin Excitations in Nuclei, edited by F. Petrovich, G.E. Brown, G.T. Garvey, C.D. Goodman, R.A. Lindgren and W.G. Love (New York: Plenum) p 143
van Hees A G M and Glaudemanns P W M 1983 Z. Phys. A314 323, 315 223
Hosaka A, Kubo K I and Toki H 1985 Nucl. Phys. A444 76
Jeckelmann B et al 1983 Nucl. Phys. A408 495
Knupfer W and Metsch B C 1983 Phys. Rev. C27 2487
Kuo T T S and Brown G E 1966 Nucl. Phys. 85 40
Kuo T T S 1967 Nucl. Phys. A103 71
Langevin et al 1986 Nucl. Phys. A455 149
Lederer C M and Shirley V S 1978 Table of Isotopes (New York: Wiley) Appendix VII
Madey R, Flanders B S, Anderson B D, Baldin A R, Austin S M, Galonsky A, Wildenthal B H and Foster C C 1986 Phys. Rev. C to be published
Mayer M G and Jensen J H D 1955 Elementary theory of nuclear structure (New York: John Wiley)
McGrory J B and Wildenthal B H 1980 Ann. Rev. Nucl. Sci. 30 383
McNeil J A et al 1986 Phys. Rev. C34 746
Miessen H et al 1984 Nucl. Phys. A430 189
Millener D J and Kurath D 1975 Nucl. Phys. A255 315
Nolen J A and Schiffer J P 1969 Ann. Rev. Nucl. Sci. 19 471
Ormand W E 1986 PhD thesis Michigan State Univ. East Lansing
Ormand W E and Brown B A 1985 Nucl. Phys. A440 274
Ormand W E and Brown B A 1986 Phys. Lett. B174 128
Oset E and Rho M 1979 Phys. Rev. Lett. 42 47
Radhi R, Brown B A and Wildenthal B H 1983 unpublished; and Radhi R 1983 PhD thesis Michigan State Univ. East Lansing
Rogers W F, Clark D L, Dutta S B and Martin A G 1986 Bull. Am. Phys. Soc. 31 771
Ryan P J et al 1983 Phys. Rev. C27 2515
Sagawa H and Brown B A 1984 Nucl. Phys. A430 84
Sagawa H and Brown B A 1985 Phys. Lett. B150 247
Sagawa H, Brown B A and Scholten O 1985 Phys. Lett. B159 228
Schmid K W, Hammaren E and Grummer F 1986 Nuclear Structure at High Spin, Excitation, and Momentum Transfer ed. H. Nann, AIP Conference Proc. 142, p 327
Serot B and Walecka J D 1986 Adv. in Nucl. Phys. 16 1
Shurpin J, Kuo T T S and Strottman D 1983 Nucl. Phys. A408 310
Spear R H 1981 Phys. Rep. 73 369
Sternheimer R M and Peierls R F 1971 Phys. Rev. A4 1722
Towner I S and Khanna F C 1983 Nucl. Phys. A399 334
Warburton E K et al 1986 Phys. Rev. C to be published
Weber R et al 1982 Nucl. Phys. A377 361
Whitehead R R, Watt A, Cole B J and Morrison I 1977 Adv. Nucl. Phys. 9 123
Wildenthal B H, Curtin M S and Brown B A 1983 Phys. Rev. C28 1343
Wildenthal B H 1984 Progress in Particle and Nuclear Physics 11 edited by D H Wilkinson (Oxford: Pergamon) p 5
Wildenthal B H, Brown B A and Sick I 1985 Phys. Rev. C32 2185
Wildenthal B H and Keinonen J 1986 unpublished
Willis A et al 1986 Nucl. Phys. A to be published
Zwarts D 1985 Comp. Phys. Comm. 38 365

Inst. Phys. Conf. Ser. No. 86
Paper presented at Int. Nucl. Phys. Conf., Harrogate, UK, 1986

Achievements, limitations and future of the interacting boson model

A.E.L. Dieperink
Kernfysisch Versneller Instituut, 9747 AA Groningen, The Netherlands

1. Introduction

One can distinguish a great variety of approaches to describe nuclear collective motion. In the fully microscopic methods one attempts to take into account all nucleonic degrees of freedom and then applies a series of systematic approximations (such as the mean field approximation). In the Interacting Boson Approximation (IBA) model a rather pragmatic approach is taken in the sense that a priori a few degrees of freedom are selected which are supposed to dominate the nuclear collective properties and moreover are treated as decoupled from the other microscopic degrees of freedom. This has the advantage that the few selected building blocks obey a simple algebraic structure which greatly facilitates phenomenological applications. On the other hand justification a postiori of the microscopic validity is less straightforward.

In discussing the microscopic foundation of the IBA model [1-3] three basic steps can be distinguished. First it is assumed that in good approximation the collective properties of low-lying states in even-even nuclei are determined by the valence nucleons (outside a closed spherical core) only; the second assumption is that the collective properties are dominated by correlated (Cooper) pairs in this shell model space, coupled to L=0 and L=2. Finally this fermion many-body problem is mapped onto a bosonic one, whereby in practice the drastic assumption is made that the boson image of the hamiltonian consists of one- and two-body terms only. The justification of these three steps is by no means obvious, and appreciable efforts have been made to examine the microscopic validity of the IBA model. Since I believe that this is a topic of considerable interest I will begin my talk with a few comments on the present situation.

For a detailed description of the IBA model, its group structure, microscopic foundation and applications I refer to several review articles [1-3]. In this talk I will be mainly concerned with the IBA-2 model for even-even nuclei, in which one distinguishes the neutron and proton degrees of freedom. The building blocks are the s and d_μ ($\mu=\pm2,\pm1,0$) bosons for neutrons and protons. The 36+36 operators $d^\dagger_\mu s$, $d_\mu s^\dagger$, $d^\dagger_\mu d_{\mu'}$ and $s^\dagger s$ are the generators of the symmetry group $U^{(\nu)}(6) \times U^{(\pi)}(6)$. The basis states belong to the irreps $[N_\pi] \times [N_\nu]$ where N_π, N_ν are the numbers of proton and neutron bosons.

A popular semi-realistic schematic hamiltonian has the form [1-3]

$$H = \varepsilon(d^\dagger_\pi \cdot d_\pi + d^\dagger_\nu \cdot d_\nu) + \kappa\, Q^{(2)}_\pi \cdot Q^{(2)}_\nu + \lambda \hat{M} \quad , \tag{1.1}$$

where the first term represents the energy difference between s and d bosons, the second term the neutron-proton quadrupole-quadrupole interaction; $\hat{M}$, usually referred to as a Majorana interaction, is related to the quadratic Casimir invariant of the subgroup $U^{(\pi+\nu)}(6)$ and can be regarded as a neutron-proton symmetry energy. A generalization of the IBA model by including an L=4 pair (g-boson) is straightforward; the corresponding symmetry group is U(15) (or $U^{(\pi)}(15) \times U^{(\nu)}(15)$ in the neutron-proton case).

2. On the limitations of the IBA model

2.1 SD dominance

The SD dominance that forms a basic assumption of the IBA model has been questioned at many occasions. It is difficult to give a quantitative discussion of this problem since one seems never in the happy situation of a complete decoupling of the SD degrees of freedom and the less collective ones. Moreover the validity of SD dominance appears to vary from vibrational to rotational nuclei and also from ground state to excited states.

Let me briefly discuss the role of the L=4 (G) pair in the vibrational regime. If the SD dominance point of view is correct a low-lying L=4 state would have predominantly the two-broken pair configuration $|S^{N-2}D^2[\alpha L]\rangle$ and a one-broken pair configuration of the form $|GS^{N-1}L{=}4\rangle$ would be regarded as an intruder. It seems of interest to investigate whether these two types of states have experimental counterparts.

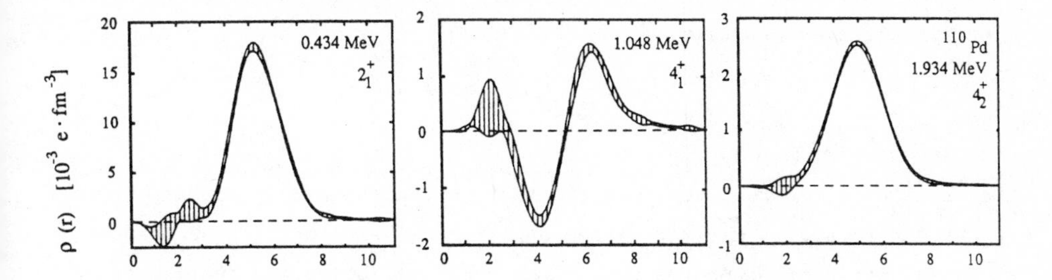

Fig. 1 Transition charge densities for low-lying collective states in ^{110}Pd obtained from a Bessel-Fourier fit to electron scattering cross sections [4].

A recent electron scattering experiment [4] on the Pd isotopes (which can be described as perturbed vibrational U(5) nuclei) in which the form factors for the lowest $0^+ \rightarrow 4_i^+$ (i=1,2,3) transitions could be measured with high accuracy shows that transition densities for the 4_1^+ states have the form of a second derivative of the elastic charge distribution, whereas the densities for the 4_2^+ states are characterized by a first derivative (see fig. 1). On general grounds one expects that the transition densities associated with the S → G and D → D matrix elements of the E4 operator behave like a first and second derivative, respectively. Therefore one may conclude [5] that there is strong indication that the lowest L=4 state in vibrational nuclei has predominantly a two-phonon character, and that a rather pure G-pair state is located at higher excitation energy.

In deformed nuclei the validity of SD dominance is less clear. Several authors have analyzed the structure of the intrinsic ground-state band of axially symmetric deformed nuclei [6-9]. To this end the intrinsic ground-state wave function is written as a number projected HFB wave function

$$\Psi_0 \sim (\Gamma_\nu^+)^{N_\nu} (\Gamma_\pi^+)^{N_\pi} |0\rangle, \tag{2.1}$$

and the pair creation operators Γ_ρ^+ ($\rho=\pi,\nu$) of deformed Cooper pairs are expanded in terms of collective pairs with good angular momentum

$$\Gamma^+ = \sum_L \lambda_L \Gamma_L^+ \tag{2.2}$$

where the multipole creation operators Γ_L^+ can be expressed as pairs of spherical fermion orbitals:

$$\Gamma_L^+ = \sum_L \sum_{jj'} \alpha_{jj'}^L \; [a_j^+ a_{j'}^+]_0^{(L)} \; . \tag{2.3}$$

The fermion pair structure coefficients $\alpha_{jj'}^L$ and the multipole parameters λ_L are determined by minimizing the shell-model hamiltonian with respect to (2.1). Depending upon the details of the hamiltonian and the number of fermions it has been shown [6-9] that a large percentage of the Cooper pair (80-90%) is already given by the S(L=0) and D(L=2) pairs. The contribution of the higher angular momentum pairs to observables like the moment of inertia is found to be much larger (up to 50%) [9]. It should be noted that to a large extent the effect of the L=4 pair in the ground-state band (in the boson space) can be absorbed by a renormalization [10] of the parameters in the SD space. This explains that in phenomenological applications of the IBA-2 model little need for the explicit inclusion of g-bosons has been found.

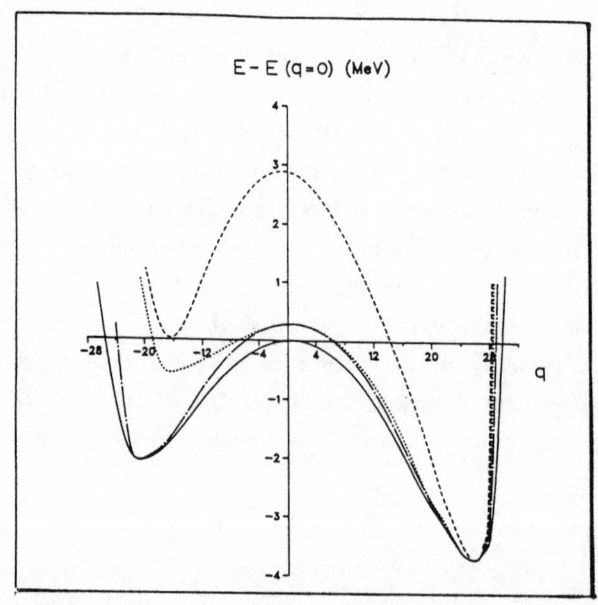

Fig. 2. The energy surface E-E(q=0) as a function of $q = \langle\psi|Q_0^{(2)}|\psi\rangle$ for ^{156}Gd. The full curve is the result of minimization of H with respect to a full variation of λ_L and α^L in (2.3); the dashed curve is obtained with fixed α^L, and the dashed-dotted curve gives the result if an additional S' (orthogonal to S) pair is included.

A more subtle question that has been studied recently [11,12] concerns the structure of excited bands in deformed nuclei. To examine the structure of K=0 bands the collective energy surface is constructed as a function of the quadrupole mass deformation $q = \langle\Psi|Q_0^{(2)}|\Psi\rangle$ by minimizing $E(q) = \langle\Psi|H|\Psi\rangle - \frac{1}{2}(q-q_0)^2$ as a function of q_0. It is found [12] that the multipole parameters λ_L vary strongly with q_0 while the coefficients $\alpha^L_{jj'}$ show a much weaker variation. However, it is seen from fig. 2 where the resulting energy surface for fixed $\alpha^L_{jj'}$ (i.e. fixed structure of S, D and G pairs) is compared with those of a free variation, that not only the oblate minimum ($q<0$) is off by 2 MeV but also the width of the prolate minimum is not correctly given.

To show the effect on observables the orthogonalized excited K=0 bands have been calculated using the generator coordinate method by solving the Hill-Wheeler equation $\langle\Psi(q)|H|\Phi\rangle = E\langle\Psi(q)|\Phi\rangle$ with $|\Phi\rangle = \int dq\; f(q)|\Psi(q)\rangle$. From the resulting excitation energies and Q-moments, given in table 1, one sees that the properties of the excited bands depend strongly on whether or not the structure of the S, D and G pairs is kept fixed.

Table 1. Properties of K=0 bands

	λ_L, α_L free		α_L fixed, λ_L free	
band	Energy (MeV)	Q/b^2	Energy (MeV)	Q/b^2
g.s.	0	25.7	0	25.8
β band	1.90	21.5	2.35	23.8
oblate band	1.60	-20.1	3.45	-17.8

Another point which remains controversal is the nature of the higher excited bands in deformed nuclei, which in the IBA model and also in other phenomenological approaches are interpreted as double phonon excitations: ββ, βγ and γγ. On the other hand Soloviev et al. [13] have pointed out using the framework of the quasi-particle phonon model (in which the particle-hole rather than particle-particle correlations are emphasized) that two-phonon states in deformed nuclei are not well developed. In fact these authors suggest that because of the Pauli principle most low-lying bands have predominantly a two quasi-particle character. The experimental data on transfer reactions [14] on ^{168}Er indicate that some of the higher excited K=0 and K=2 bands have appreciable two quasi-particle components (~20%).

2.2 Number of valence pairs N_π, N_ν

A characteristic feature of the IBA approach, which distinguishes it from most other collective models, is the emphasis on the number of pairs of valence nucleons that take part in the collective motion. At the phenomenological level it has recently been demonstrated by Casten [15] that collective quantities like the energy of the 2_1^+ state and the ratio $E(4_1^+)/E(2_1^+)$ can be parametrized successfully in terms of the product $N_\pi \cdot N_\nu$ of the number of valence proton and neutron pairs counted to the nearest closed shell. Deviations from this rule have been found near pronounced subshell closures; for example, from an analysis of magnetic moments [16] it seems that for neutron number N<88 the valence protons must be taken with respect to Z=64 rather than Z=50. That N_π and N_ν are meaningful quantities in the spherical region does not seem to be surprising. However, a glance at the Nilsson model does not lend much support for conservation of the shell structure in strongly deformed nuclei. Therefore it is worthwhile to discuss the validity of the concept of N_π, N_ν a bit more. For this purpose an appropriate observable appears to be the matrix elements of the M1 operator since it is basically a $0\hbar\omega$ operator. For the following discussion it is convenient to consider the most general one-body M1 operator in the neutron-proton sdg-space can be written as [17]

$$T(M1) = \sqrt{\frac{3}{4\pi}} \sum_{\rho=\pi,\nu} \left(g_{d,\rho} L_{d,\rho}^{(1)} + g_{g,\rho} L_{g,\rho}^{(1)}\right) \tag{2.4}$$

where the g_d and g_g are the pair g-factors, and $L_d^{(1)} = \sqrt{10}(d^\dagger\tilde{d})^{(1)}$, $L_g^{(1)} = \sqrt{60}(g^\dagger\tilde{g})^{(1)}$. For the physics it is more convenient to introduce total and relative angular momenta:

$$L_\rho^{(1)} = L_{d,\rho}^{(1)} + L_{g,\rho}^{(1)}, \quad \Lambda_\rho^{(1)} = \frac{1}{7}\left(4\, L_{d,\rho}^{(1)} - 3\, L_{g,\rho}^{(1)}\right)$$

$$L^{(1)} = L_\pi^{(1)} + L_\nu^{(1)}, \quad L_a^{(1)} = \frac{1}{N}\left(N_\nu L_\pi^{(1)} - N_\pi L_\nu^{(1)}\right), \quad \Lambda^{(1)} = \Lambda_\pi^{(1)} + \Lambda_\nu^{(1)} \tag{2.5}$$

in terms of which one can rewrite eq. (2.4) as

$$T(M1) = \sqrt{\frac{3}{4\pi}}\,[gL^{(1)} + (g_\pi - g_\nu)\,L_a^{(1)} + \frac{N_\pi h_\pi + N_\nu h_\nu}{N}\,\Lambda^{(1)} + \ldots] \tag{2.6}$$

$$\text{with } g_\rho = \frac{1}{7}(3\,g_{d,\rho} + 4\,g_{g,\rho}),\; h_\rho = (g_{d,\rho} - g_{g,\rho}),\; g = (N_\pi g_\pi + N_\nu g_\nu)/N \tag{2.7}$$

The second and third terms at the r.h.s. of (2.6) can be associated with the neutron-proton degree of freedom and the d-g degree of freedom, respectively, and will be discussed below (section 3). The first term at the r.h.s. of eq. (2.6) is purely diagonal and thus contributes only to magnetic moments. Assuming that the ground state band is totally symmetric one obtains from (2.6) for the g-factors [18]

$$g = g_R = (g_\pi N_\pi + g_\nu N_\nu)/N. \tag{2.8}$$

It was shown by Wolf et al. [19] that eq. (2.8) describes the N_π, N_ν dependence of g-factors of 2_1^+ states in the rare-earth region quite well with fixed g_π, g_ν (see fig. 3). However, the values of the best fit boson g-factors ($g_\pi \simeq 0.63$, $g_\nu \simeq -0.05\ \mu_N$) cannot easily be explained from a

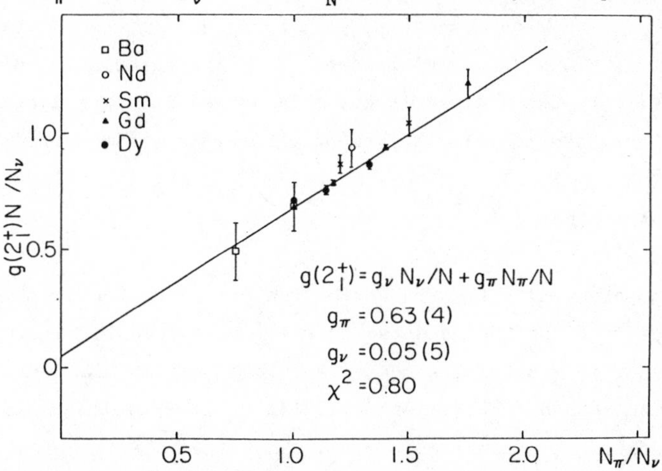

Fig. 3 Plot of $g(2_1) \cdot N/N_\nu$ against N_π/N_ν (see eq. (2.8)); from ref. [19].

microscopic point of view, in which the boson g-factors are considered as the image of fermionic g-factors for a collective D pair. In fact microscopic calculations show that the net spin contribution in the fermionic M1 matrix element is very small (<5%) and thus one expects $g_\pi \sim 1.0$ and $g_\nu \sim 0$ [μ_N]. This suggests that the naive interpretation of N_π, N_ν is not fully correct and that there is a non-negligible contribution from the core nucleons which is rather constant over a large mass region.

To illustrate this point it is instructive to consider the well known approximate expression for the g_R-factor of axially symmetric deformed nuclei in terms of an angular momentum projected intrinsic wave function [20]

$$g_R^{PHF} \sim \langle L_\perp^2 \rangle_\pi \; / \; \langle L_\perp^2 \rangle_{\pi+\nu} + \Delta_{spin}^{PHF}, \tag{2.9}$$

where the net spin contribution Δ_{spin} has been shown to be small. Where only the valence nucleons contribute to $\langle L_\perp^2 \rangle$ one has $g_R \sim N_\pi/N$; on the other hand if there are no pairing effects and all nucleons contribute $g_R \sim Z/A$. Microscopic calculations suggest that in reality one has an in-between situation [20].

Here I note that the success of the parametrization (2.8) suggests that to a large extent the core contribution can be absorbed in effective g-factors. For example, if one makes the plausible assumptions that $\langle L_\perp^2 \rangle_{\pi,val}/\langle L_\perp^2 \rangle_{\nu,val} = N_\pi/N_\nu$ and that the core contribution to $\langle L_\perp^2 \rangle = \langle L_\perp^2 \rangle_{val} + \langle L_\perp^2 \rangle_{\nu,core}$ is proportional to the number of valence pairs, $\langle L_\perp^2 \rangle_{\nu,core} \sim \alpha N$, one can rewrite (2.8) as $g_R = g_\pi^{eff} N_\pi/N$ with $g_\pi^{eff} = g_\pi/(1 + \alpha)$. In a recent microscopic cranking calculation of the rare-earth nuclei by the Tübingen group [21] indeed an appreciable contribution of the core neutrons to the moment of inertia was reported.

2.3 Fermion-boson mapping

Unlike the problem of the truncation the procedure for the fermion - boson mapping is well defined and can formally be solved in an exact fashion. However, in general the equivalent boson hamiltonian will contain higher-order many-boson interactions and will be non-hermitian. Therefore in practice additional approximations are made, the validity of which is not easy to judge. For a more detailed discussion of this subject I refer to the literature. Here I restrict myself to one remark. In the IBA approach one has successfully emphasized the mapping of states rather than the mapping of operators [1,2]. In the vibrational (or U(5)) limit of IBA this leads to a correspondence between the generalized seniority quantum number v and the number of d-bosons, $n_d = \frac{1}{2}v$. On the other hand, in the rotational (SU(3)) region seniority is not a good quantum number, and it has been suggested that the mapping of intrinsic states be considered. In that case one could attempt to label the bosonic states with the well known (λ,μ) labels of the Elliott model. Then the problem arises that with increasing particle number in the fermion sector the Pauli principle does not allow ground state representations with $\mu = 0$; for example, in the 2s1d shell the ground-state configuration for ^{24}Mg has $(\lambda_F,\mu_F) = (8,4)$, i.e. is triaxial in nature. However, in the boson sector there exists no simple two-body hamiltonian which has a lowest eigenstate with $(\lambda,\mu \neq 0)$. Although for larger shell degeneracies the Pauli principle becomes less effective it is not straightforward to find a satisfactory solution to this problem.

3. M1 transitions

3.1 The isovector M1 excitation

An interesting recent development has been the observation [22-24] of collective M1 strength around $E_x \simeq 3$ MeV in deformed nuclei in the rare-earth and transactinide regions in (e,e') and (γ,γ') experiments. A detailed analysis of the experimental information strongly suggests that this M1 strength $(B(M1,\uparrow) \sim 2\ \mu_N^2)$ to the K=1 bands has an isovector orbital (rather than spinflip) character, and is concentrated in a narrow energy interval of about 0.5 MeV. In the rotational [SU(3)] limit of the IBA model [25] and also the two-rotor model [26] a single orbital $L^\pi=1^+$ state had been predicted (the socalled "scissors mode") corresponding to the response of a deformed nucleus to the isovector M1 operator, $T(M1) \sim \frac{1}{N}\left(N_\nu L_\pi^{(1)} - N_\pi L_\nu^{(1)}\right)$ (note that the isoscalar M1 operator produces a spurious state). In the IBA-2 model the total strength is given by [25]

$$B(M1) \sim \frac{4\pi}{3}\,\frac{4N_\pi N_\nu}{N}\,(g_\pi - g_\nu)^2 \ . \tag{3.1}$$

More recently several groups have calculated the M1 response function using a microscopic Nilsson approach [27-30]. In general the M1 strength distribution is found to depend very sensitively on the details of the residual ph interaction used; these calculations have in common that they predict a larger strength with more fragmentation than observed experimentally and moreover appreciable spin-flip strength which does not seem consistent with the results of high resolution (p,p') experiments [31]. One may conclude that on the one hand the IBA model tends to oversimplify the picture by neglecting the coupling with 2 quasi-particle states; on the other hand in the particle-hole approach in which the unperturbed M1 strength is strongly fragmented due to the large Nilsson splitting it appears difficult to reproduce the experimentally observed concentration of orbital M1 strength.

The discovery of the large orbital M1 strength around 3 MeV in a series of deformed nuclei is interesting also for other reasons. It is well known that in general giant resonances, through the coupling with less collective states, do affect properties of these states. Therefore one may speculate whether M1 transitions between low-lying excited bands (e.g. β, γ and g-state bands) may originate from a mixing with the isovector M1 state. In fact a similar idea has been discussed long ago by Greiner [38] who suggested that collective M1 transitions originate from a difference in deformation between neutrons and protons. At present, with the position and the M1 strength of the K = 1 band known, this hypothesis can be investigated more quantitatively (see below).

3.2 M1 transitions between low-lying states

Although the experimental information on absolute B(M1) values and magnetic moments in γ-bands in deformed nuclei is rather limited there is strong evidence for their collective nature. For example, they seem to follow the leading order intensity rule, which in the geometric model can be expressed as [32]

$$\langle \gamma L' \| T(M1) \| \gamma L \rangle = \sqrt{\frac{3}{4\pi}}\, [g_R \sqrt{L(L+1)(2L+1)}\ \delta_{LL'} + (g_K - g_R)\ \sqrt{(2L+1)(2L'+1)}\ \begin{pmatrix} L' & 1 & L \\ K & 0 & K \end{pmatrix} K] \qquad (3.2)$$

The question of interest is what the physical origin of the intrinsic $(g_K - g_R)$ contribution is. The microscopy underlying the IBA-2 model suggests two possible mechanisms for M1 transitions between low-lying states.

(i) F-spin mixing

For IBA-2 hamiltonians in which the neutron-neutron, proton-proton and neutron-proton interactions have equal strengths the eigenstates can be classified according to their neutron-proton symmetry, in practice referred to as F-spin [34]. In more realistic hamiltonians the neutron-proton quadrupole-quadrupole interaction is stronger than that between the like bosons and the low-lying states will not have a pure F-spin; in that case the second term in (2.6) can effectively contribute to M1 transitions. In perturbation theory one can derive closed expressions for the M1 matrix elements. For example, in a perturbed $U^{(p+n)}(6) \supset SU(3) \supset O(3)$ scheme with a hamiltonian $H = H_0 + \Delta H$, where the F-spin symmetric hamiltonian $H_0=\kappa(Q_\pi^{(2)}+Q_\nu^{(2)}) \cdot (Q_\pi^{(2)}+Q_\nu^{(2)}) + \lambda\hat{M}$ has U(6) and SU(3) symmetry, and $\Delta H=\Delta\kappa\, Q_\pi^{(2)} \cdot Q_\nu^{(2)}$ breaks U(6) (F-spin) symmetry one finds for $\gamma\to\gamma$ M1 matrix elements (for large N_π, N_ν) an expression with the same structure as (3.2) with $g_K-g_R = 6\Delta\kappa/\Delta E\ (g_\pi-g_\nu)\ (N_\pi-N_\nu)\ N_\pi N_\nu/N^2$, where ΔE is the energy difference between the symmetric γ-band and the mixed-symmetric one. An analogous expression can be derived for $\gamma\to g$ M1 transitions which occur through $\Delta K=1$ mixing of the collective K=1 band discussed above in the g.s. and γ bands. An interesting result is that the ratio of $B(M1,\gamma\to g)/B(M1,\gamma\to\gamma) \sim 1/N$, which agrees qualitatively with the experimental trend.

(ii) Effect from L=4 pair.

If the G-pair is included in IBA another source of M1 transitions can be taken into account, namely the possibility that the g-factors from the D and G pairs are different which leads to non-vanishing contributions of the third term at the r.h.s. of (2.6). Its effect can be evaluated in the SU(3) limit of IBA using expressions for the intrinsic states for U(15) [35]. Again the matrix elements for $\gamma-\gamma$ M1 transitions take on the form (3.2) with $g_K-g_R = (h_\pi N_\pi + h_\nu N_\nu)/N$. The value of $h = g_d-g_g$ can be estimated by a simple mapping procedure based upon the equality of matrix elements with one L=2 and one L=4 pair in boson and fermion spaces, respectively. From this one obtains $g_{d,\pi} \sim g_{g,\pi} \sim 1$ and $g_{d,\nu} \sim g_{g,\nu} \sim 0$ with $h = g_d - g_g \sim 0.1\ \mu_N$. At present we are investigating whether these mechanisms can provide a quantitative description of the data.

4. Future

The success of the IBA model has shown that there is a need for simple soluble models that contain the dominant degrees of freedom of the complicated nuclear many-body problem. Let me briefly mention two recent directions in which algebraic models could play a useful role.

First up to now applications of the IBA model have been restricted mostly to static problems, which can be expressed in terms of matrix elements of the infinitesimal generators Q_i of the algebra. However, there is a class of problems which requires the evaluation of the group elements $\exp(\sum_i \alpha_i Q_i)$. An example is the description of multiple excitation of collective quadrupole degrees of freedom in vibrational and/or rotational nuclei in high-energy proton scattering. There the algebraic approach seems to be a powerful alternative to the conventional coupled channel method [36].

In the impact parametrization the scattering amplitude for (p,p') can be approximated by an exponentiated transition operator [36]

$$\langle f|T(q)|i\rangle = \frac{ik}{2\pi} \int d_2 b \; e^{i\vec{q}\cdot\vec{b}} \langle f|\; 1-e^{\chi_{el}(b)} \;\; {}^{-\chi_{inel}(b)} |i\rangle \tag{4.1}$$

where $\vec{q} = \vec{k}-\vec{k}'$ is the momentum transfer, χ_{el} and χ_{inel} are the elastic and transition profile functions, respectively. Assuming that the collective nuclear degrees of freedom are described by the IBA model the latter can be expressed as a sum of multipoles

$$\chi_{inel}(b) = f_0 \frac{mi}{k} \sum_{n,\lambda} \int dz \; \rho_n^{(\lambda)}(b,z) \; Y_\lambda(\theta,\phi) \cdot \hat{Q}^{(\lambda)}(n) \tag{4.2}$$

where $\hat{Q}^{(\lambda)}(n)$ are multipole operators in the boson space, e.g. for $\lambda=2$ $\hat{Q}^{(2)}_\mu(1) = d^\dagger_\mu s + s^\dagger d_\mu$, $\hat{Q}^{(2)}_\mu(2) = (d^\dagger d)^{(2)}_\mu$. Furthermore f_0 represents the forward projectile-nucleon scattering amplitude, and $\rho_n^{(\lambda)}(r)$ are the nuclear transition densities, which can be deduced from electron scattering.

In the past the evaluation of the U(6) representation matrices, $\exp[\alpha^{(\lambda)}(b) \cdot \hat{Q}^{(\lambda)}]$, in (4.1) has been simplified by making the approximation $z = 0$ in (4.2), i.e. $\theta = \frac{\pi}{2}$. Recently a method has been developed for the evaluation of the general group elements between U(6) basis states [39]. In this way finite excitation probabilities of final states with odd angular momentum are obtained, which vanish in the $z = 0$ approximation.

Several applications of this formalism to the multiple excitation of the lowest L=0,2,4 states both in limiting cases of the IBA model (where analytic results can be derived) as well as transitional nuclei have already been discussed [36]. With increasing energy resolution of the experimental facilities experimental tests of predicted cross sections for excitation of weaker states, which are more sensitive to details of the model wave functions, become feasible. Of interest is the study of the interference between the double quadrupole and the one-step hexadecupole excitation in $0^+ \rightarrow 4^+$ transitions, the relative contribution of which varies with momentum transfer. It has also been shown that algebraic methods can be applied to a much wider class of physical problems such as electron scattering off molecular rotational and vibrational degrees of freedom.

Secondly a basic limitation of the present IBA model is the restriction to one major shell. As a consequence only low-lying collective excitations can be described. The coupling of these modes to the high frequency irrotational excitations (giant resonances) can be incorporated to a certain extent by the introduction of effective charges and renormalized interaction strengths. Still there are several phenomena which require a more explicit treatment of coupling with the E2 giant resonance. Examples are the description of transition densities for low-lying states which are much more surface peaked than can be explained in terms of a $0\hbar\omega$ shell model (e.g. see fig. 1), and the fragmentation of the giant resonances in deformed nuclei. We note that there exists an interesting algebraic unified description of both low-lying and high-lying E0 and E2 excitations in terms of the non-compact group Sp(6,R) [37]. The generators of this group which can be regarded as a generalization of the Elliott SU(3) model are basically all $\Delta N = 0$ and $\Delta N = 2$ components of the quadrupole tensor in addition to the three angular momenta. Since in this approach the pairing interaction (an essential ingredient in the IBA model) is not included applications are restricted to strongly deformed nuclei. Therefore it appears of interest to try to unify the IBA (U(6)) and Sp(6,R) approaches by starting from a larger group structure. Some work in this direction is in progress.

Acknowledgements

I would like to thank Drs. O. Scholten, H.C. Wu and B.I.M. van der Cammen for valuable contributions to various topics in this talk. This work is part of the research program of the Stichting FOM, which is supported by the Nederlandse Organisatie voor Zuiver Wetenschappelijk Onderzoek (ZWO). I also acknowledge the support by the NATO Research grant RG85/0036.

References

1. A. Arima and F. Iachello, Ann.Rev.Nucl.Sci. 31(1981)75.
2. A. Arima and F. Iachello, Adv.Nucl.Phys. 13(1984)139.
3. A.E.L. Dieperink and G. Wenes, Ann.Rev.Nucl.Part.Sci. 35(1985)77.
4. J. van der Laan, thesis 1986, University of Amsterdam (unpublished); C. de Jager et al., to be published.
5. O. Scholten, Phys.Rev. C28(1983)1783.
6. T. Otsuka, A. Arima and N. Yoshinaga, Phys.Rev.Lett. 48(1982)387.
7. D.R. Bes, R.A. Broglia, E. Maglione and A. Vittari, Phys.Rev.Lett. 48(1982)1001;
   E. Maglione et al., Nucl.Phys. A404(1983)333.
8. K. Sugawara-Tanabe and A. Arima, Phys.Lett. 100B(1982)87.
9. W. Pannert, P. Ring and Y.K. Gambhir, Nucl.Phys. A433(1985)189.
10. T. Otsuka and J.N. Ginocchio, Phys.Rev.Lett. 55(1985)276.
11. E. Maglione et al., Nucl.Phys. A411(1983)181.
12. L. de Winter, N. Walet, P.J. Brussaard, K. Allaart and A.E.L. Dieperink, to be published.
13. V.G. Soloviev, Yad.Yiz. 36(1982)1376.
14. D.G. Burke, Nucl.Phys. A442(1985)424; A445(1985)70.
15. R. Casten, Phys.Lett. 152B(1985)145; Phys.Rev.Lett. 54(1985)1991.
16. A. Wolf et al., Phys.Lett. 123B(1983)165.
17. H.C. Wu and A.E.L. Dieperink, contribution to Int.Conf. on Nuclear Structure, Reactions and Symmetries, Dubrovnik, 1986;
    H.C. Wu, A.E.L Dieperink and O. Scholten, preprint.
18. M. Samabataro et al., NuclPhys. A423(1984)333.
19. A. Wolf, D.D. Warner and N. Benczer-Koller, Phys.Lett. 185B(1985)7.
20. D.W.L. Sprung et al., Nucl.Phys. A326(1979)326.

21. R. Nojarov et al., preprint.
22. D. Bohle et al., Phys.Lett. 137B(1984)27;
23. D. Bohle et al., Phys.Lett. 148B(1984)260.
24. U.E.P. Berg et al., Phys.Lett. 149B(1984)59.
25. A.E.L. Dieperink, Prog.Part.Nucl.Phys. 9(1983)121.
26. N. LoIudice and F. Palumbo, Nucl.Phys. A326(1979)193.
27. D.R. Bes and R.A. Broglia, Phys.Lett. 137B(1984)141.
28. S. Iwasaki and K. Hara, Phys.Lett. 144B(1984)9.
29. I. Hamamoto and S. Åberg, Phys.Lett. 145B(1984)163;
R.R. Hilton, J.Phys.(Paris), C6(1984)255.
30. H. Kurasawa and T. Suzuki, Phys.Lett. 144B(1984)151.
S. Pittel et al., Phys.Lett. 144B(1984)145.
31. C. Djalabi, Phys.Lett. 164B(1985)269.
32. A. Bohr and B. Mottelson, Nuclear Structure, Vol. II.
33. A.E.L. Dieperink, O. Scholten and D.D. Warner, to be published.
34. A. Arima et al., Phys.Lett. 66B(1977)205.
35. H.C. Wu, A.E.L. Dieperink and S. Pittel, Phys.Rev.C (1986).
36. A. Amado et al., Phys.Rev. C25(1982)13; C29(1984)932;
J.N. Ginocchio et al., Phys.Rev. C23(1986)247;
G. Wenes et al., Nucl.Phys. to be published.
37. D. Rowe, Rep.Progr.Phys. 48(1985)1419.
38. W. Greiner, Nucl.Phys. 80(1966) 417.
39. B.I.M. van der Cammen, to be published.

Inst. Phys. Conf. Ser. No. 86
Paper presented at Int. Nucl. Phys. Conf., Harrogate, UK, 1986

Collective and single-particle structures at high spin

P J Nolan

Oliver Lodge Laboratory, University of Liverpool,
Liverpool, L69 3BX, ENGLAND

Abstract: The use of suppression shield arrays has extended the spin range for which measurements can be made. In ^{158}Er and neighbouring nuclei there is a change in structure near spin 40ħ arising from an interplay between collective and single-particle states. Superdeformed states have been identified in ^{152}Dy and ^{132}Ce that decay by discrete gamma rays. In ^{152}Dy states are seen to spin 60ħ corresponding to a deformation $\beta \sim 0.6$. In ^{132}Ce mean lifetimes have shown the band, seen to spin > 50ħ, to have $\beta \sim 0.5$.

1. Introduction

The study of nuclear states up to the highest angular momentum allows many features of nuclear structure to be probed. The observed properties tend to be divided into two classes: "single-particle" behaviour and "collective" behaviour. It is the influence on these two types of motion of strong centrifugal and coriolis forces that is of interest in high spin studies. At the lowest spins, pairing correlations play an important role in nuclei and it is found, for example, that a superfluid phase exists characterised by low moments of inertia. Higher spin states can be made in two ways. The orbital angular momenta of individual particles can align with the rotation axis. This type of behaviour is predominant in "spherical" nuclei near closed shells. Alternatively spin can be generated by collective rotations of a deformed nucleus. In this case the nuclear properties are often found to change abruptly at certain rotational frequencies when the coriolis force generated by the rotation is sufficient to unpair and align individual pairs of nucleons. This "backbending" effect (Johnson et al 1971, Stephens and Simon 1972) is well established in a range of nuclei.

It is the balance and interplay between these two methods of generating spin that lies at the heart of high spin studies. For states above spin 30ħ the effect of pairing correlations is expected to be much reduced so the nuclear structure may be that of a system of unpaired nucleons. When collective motion dominates it is expected that a rigid-body-like structure will be seen with properties close to those of a classical system. First observation of such behaviour at high spin was made in ^{168}Hf by Chapman et al (1983) where the relationship between spin and excitation energy was exactly that of a classical rotor in the spin range 22 - 34ħ. The study of single-particle motion when pairing correlations are not present will be more sensitive to deformation, the single-particle potential etc. and should yield information not available at low spin.

The bulk properties of deformed nuclei can be explained with a liquid drop model with shell corrections (and pairing correlations where necessary). Similar to the shell gaps for spherical nuclei (the magic numbers), others are found at various deformed nuclear shapes. In a harmonic oscillator calculation one such favoured deformation is at a 2:1 axis ratio (e.g., Bohr and Mottelson 1975). Such superdeformed states have been of interest for many years, one of the first studies being of fission isomers (Polikanov et al 1962, Specht et al 1972). At high spin the shell corrections will be somewhat different as high spin orbitals (with a small projection on the symmetry axis) are more rapidly affected by the rotation. This will lead to shell gaps in different places. Such superdeformed states at high spin have been predicted in a number of nuclei (e.g., Ragnarsson et al 1980, Bengtsson et al 1981).

Two topics have been selected for in-depth discussion. The first is the interplay between collective and single-particle structures near spin 40$\hbar$ in ^{158}Er and neighbouring nuclei. The second is the observation of superdeformed states at high spin in both ^{152}Dy and ^{132}Ce.

2. Experimental Considerations

Experiments aimed at the study of high spin states now use a combination of heavy-ion induced reactions and an array of escape suppressed spectrometers. These techniques have been described in detail by, amongst others, Herskind (1983), Twin (1983), Nolan (1984).

2.1. Heavy-ion reactions

In order to input sufficient angular momentum, it is necessary to use heavy ion beams. Typical examples are ^{124}Sn(^{48}Ca,4n)^{168}Hf, ^{100}Mo(^{36}S,4n)^{132}Ce. The experiments are generally carried out at a beam energy of ~ 4 MeV/A. The typical decay pattern following compound nucleus formation is shown in fig.1. Several nuclei are formed. The discrete line gamma transitions most often studied occur close to the yrast line, continuum studies probe the part of the decay path between the emission of statistical gamma rays and the yrast line. The spectrometers designed to study these gamma rays contain two elements. The first is a 4π sum energy-multiplicity detector which is used to aid channel selection. The second is an array of escape suppressed germanium detectors. This combination can study, in high resolution, both the discrete line and continuum gamma transitions. The suppressed array has a large number of elements to improve the efficiency of detecting γ-γ and γ-γ-γ coincidences.

2.2. Suppressed germanium detector arrays.

The successful array TESSA2 (Twin et al 1983) operated at Daresbury Laboratory for about three years until late 1985. Results from this system were so impressive that it became clear that suppressed germanium detector arrays had to be used in the study of high spin states. TESSA2 employed sodium iodide escape suppression shields. In modern systems bismuth germanate (BGO) replaces the sodium iodide. The layout of such a system is shown in fig.2. As BGO is very dense, the systems are small and many can be arranged around the target. Two such arrays are shown in figs.3 and 4. TESSA3 is the development of TESSA2 and contains a 50 element sum energy-multiplicity array with 16 BGO suppressed germanium

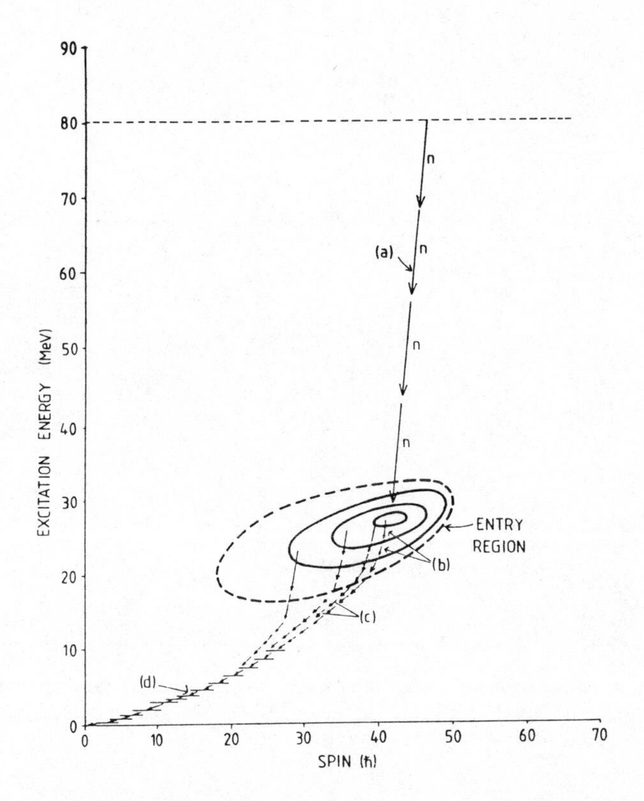

Fig.1.Typical heavy ion reaction showing (a) particle emission (4n in this case), (b) statistical γ-rays, (c) continuum γ-rays, (d) yrast (normally discrete) γ-rays.

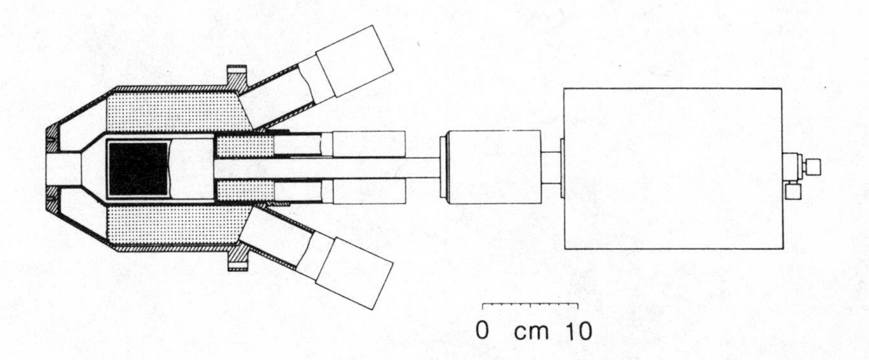

Fig.2.A bismuth germanate (BGO) shielded germanium (Ge) detector. There is a BGO catcher behind the Ge detector and a sodium iodide front cone.

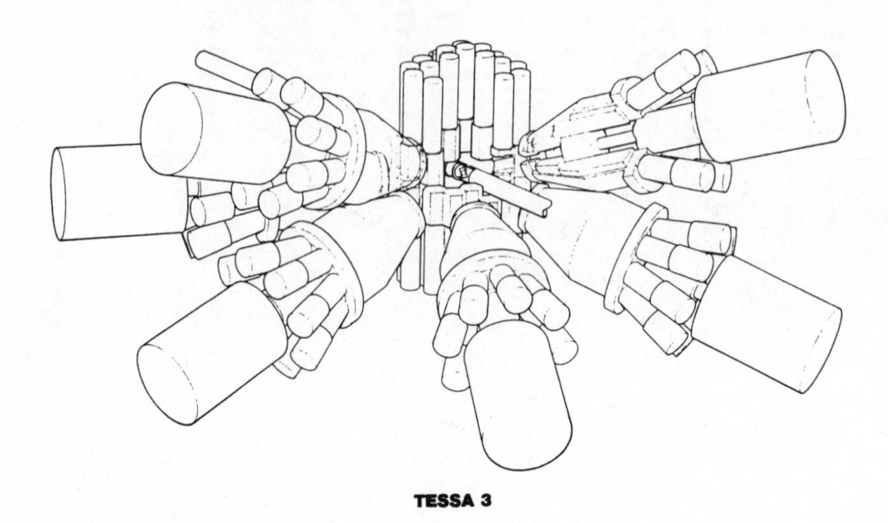

Fig.3 TESSA3 arrangement at Daresbury. A central 50 element bismuth germanate detector (Twin et al 1983) is surrounded by up to 16 escape suppressed germanium detectors.

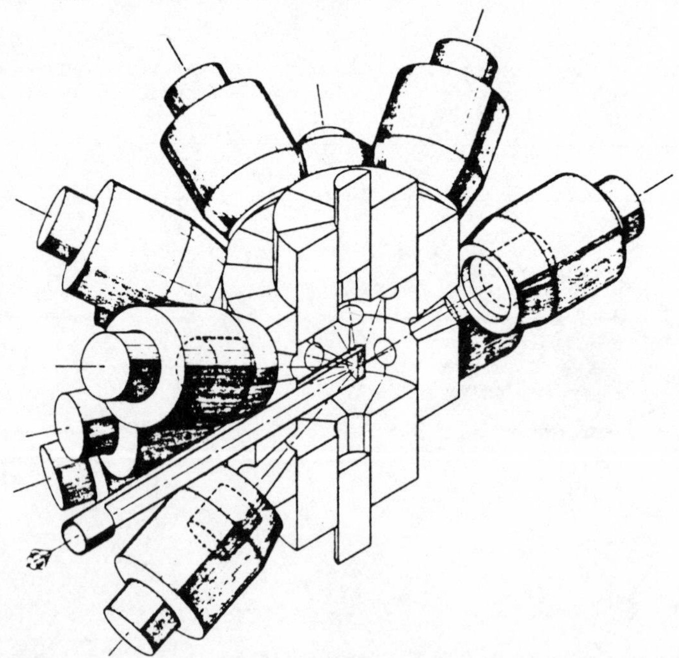

Fig.4 The HERA arrangment at Berkeley. Twenty one suppressed germanium detectors surround an inner ball.

detectors. This arrangement has operated at Daresbury since early 1986.

Many other systems are now either operational or planned. Figure 4 shows the system at Berkeley containing 21 BGO suppressed germanium detectors (Diamond and Stephens 1981) which has been operating for about 2 years. Their BGO inner ball is planned to come into operation within the next year.

The performance of these systems is measured by the photopeak-to-total ratio in each germanium spectrometer. This factor gives the fraction of useful data collected, the remainder being Compton background which severely limits the quality of the system. The importance of Compton suppression can be seen when γ-γ and γ-γ-γ coincidences are needed. The data given in Table 1 are typical for gamma rays from a ^{60}Co source. The fraction of useful data is increased by a factor of 10 for γ-γ coincidence measurements.

Table 1. Suppressed germanium detector performance.

	Fraction of photopeak events in the spectrum (%)	
	Unsuppressed	Suppressed
Single γ	18	59
γ-γ coincidence	3.2	35
γ-γ-γ coincidence	0.6	20.5

The correlated patterns of gamma-ray spectra expected in deformed nuclei for both discrete line and continuum gamma studies have been described by many authors (e.g., Herskind 1983, Deleplanque 1981). In contrast single particle structures are characterised by uncorrelated or random spectra.

In this talk I will give examples (sections 3 and 4) of both deformed and single-particle structures which have been investigated using discrete line techniques. Lack of time does not permit any in-depth discussion of continuum studies, but these will be briefly reviewed in section 5.

3. Competition between Collective and Single-Particle Structures near Spin 40$\hbar$ in ^{158}Er.

The nucleus ^{158}Er has been studied by many groups (e.g., Simpson et al 1984, Burde et al 1982, Tjøm et al 1985), at low spin it has the structure of a deformed prolate rotor with particle alignments due to $\nu(i13/2)^2$ and $\pi(h11/2)^2$. This rotational structure continues to spin 38$\hbar$. Above this there is an abrupt change in the structure which is evident in the gamma-ray spectrum shown in fig.5. The new sequence forms an irregular structure characteristic of single-particle behaviour. Further work by the group at Berkeley (Tjøm et al 1985) shows two sequences. One containing the 1203 and 1210 keV γ-rays had fast lifetimes, the remainder had slow lifetimes. The decay scheme can be seen in fig.6. These data suggest a continuation of the collective sequence (fast transitions) and a new single-particle sequence (slow transitions). This feature can be seen better in fig.7(a). The level energies are plotted relative to a smooth rotational reference. Two branches are seen, the irregular sequence being

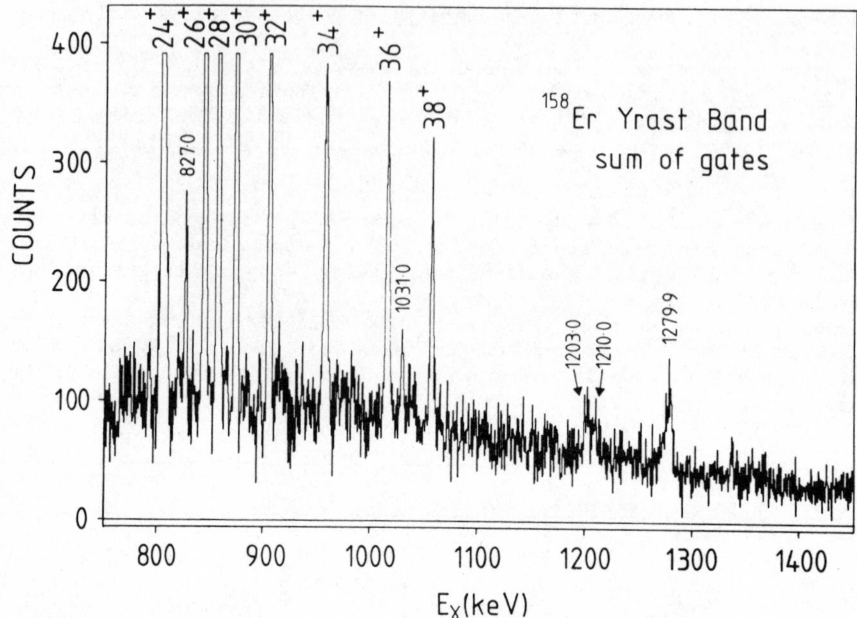

Fig.5 Spectrum of gamma rays following $^{114}Cd(^{48}Ca,4n)^{158}Er$ reaction (Simpson et al 1984).

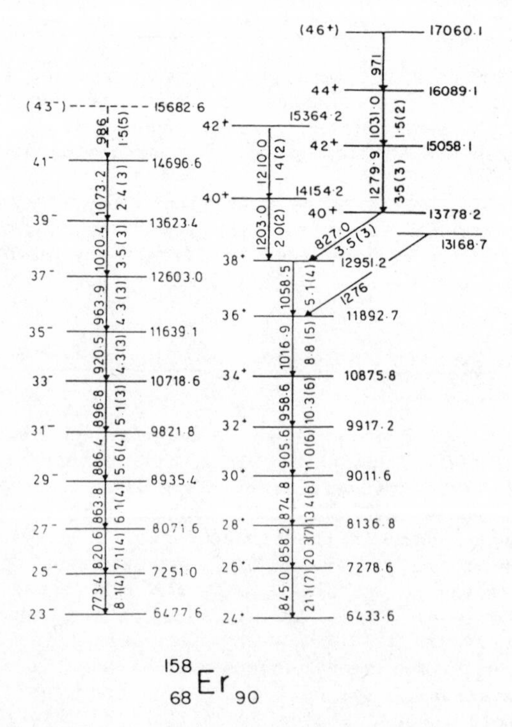

Fig.6 Partial decay shceme for high spin states in ^{158}Er (Tjøm et al 1985, Riley et al 1986).

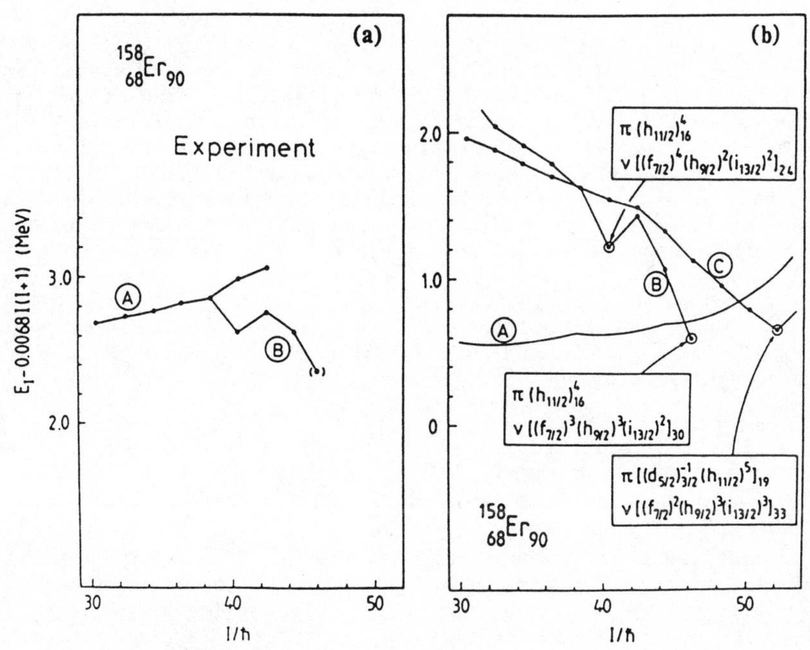

Fig.7 States in ^{158}Er plotted relative to a constant moment of inertia rotational reference (a) experiment (see fig.6), (b) calculation Bengtsson and Ragnarsson (1983) A: collective states, B: the $\pi(h11/2)^4$ ν $[(f7/2\ h9/2)^6\ (i13/2)^2]$ configuration, C: the $\pi[(d5/2)^{-1}\ (h11/2)^5]$ $\nu[(f7/2h9/2)^6\ (i13/2)^3]$ configuration.

the single-particle structure. An explanation for this behaviour lies in the existence of oblate single-particle states (Bengtsson and Ragnarsson 1983, Dudek and Nazarewicz 1985, Tanabe and Sugawara-Tanabe 1984). Figure 7(b) shows calculated energies (Bengtsson and Ragnarsson 1983) for various oblate single-particle structures. In particular consider the sequence labelled B which is a $\pi(h11/2)^4$ structure. The protons are fully aligned giving a spin 16. Two states are seen to be particularly favoured in energy. At spin 46 the state has a $\nu[(f7/2)^3(h9/2)^3(i13/2)^2]$ character and is fully aligned giving spin 30. Another favoured state occurs at spin 40 where the neutron structure has changed to a different, fully aligned configuration. Comparison with experiment shows the same pattern, the difference in the relative excitations of the collective (A) and single-particle (B) structures being the major difference. Reducing the calculated difference by 1 MeV would give reasonable agreement with experiment.

On the face of it the situation is explained by collective "prolate" states mixing with the oblate single-particle states. Closer examination of the structure of the single-particle states in the $\pi(h11/2)^4\nu[(f7/2h9/2)^6(i13/2)^2]$ configuration shows a different picture. Potential energy surface calculations (Ragnarsson et al 1986) show that the majority of states are triaxial, only the 40^+ and 46^+ have an oblate minimum. It is not clear either that the collective states are prolate. Frauendorf and May (1983) have shown that particle alignments will tend to drive the shape triaxial and finally oblate. Evidence for this can be seen experimentally from transition strength determinations which are unfortunately not yet available near spin 40ħ. Data are available (Oshima et al 1986) for states up to spin 22^+. These show a drop in collectivity consistent with a change to a triaxial shape. In all probability the two configurations that interact cannot be simply described as collective and single-particle. This illustrates that by spin 40ħ the two methods of generating spin have become less distinct and real nuclear states will be some combination of the two. The data also suggest that at spin 40, the single-particle configuration discussed i.e., $\pi(h11/2)^4\ \nu[(f7/2h9/2)^6\ (i13/2)^2]$ is a major component of the more complex structure that makes up the collective states.

Effects like the one described above are not limited to ^{158}Er. Riley et al (1986) have shown the effect to be present in the complete range of N=90 nuclei from Dy to Yb. Similar features are also observed in the N=88 isotopes (Ragnarsson et al 1986, Cranmer-Gordon et al 1986).

4. Superdeformed Structures

It has been known for some time that nuclei possess structures with deformations much larger than those typically found in ground states. These superdeformed states have been the subject of much experimental and theoretical interest for many years. Perhaps the best known example of this feature was the discovery of fission isomers (Polikanov et al 1962, Specht et al 1972). They were explained in terms of a second minimum in the potential energy surface at a large deformation.

The existence of superdeformed structures at high spin was the focus of much theoretical interest (e.g., Ragnarsson et al 1980, Bengtsson et al 1981). The first clear experimental indication of these states was found in ^{152}Dy (Nyako et al 1984, Twin et al 1985). A deformation $\beta \sim 0.6$ was deduced from the study of ridge structures in continuum gamma spectra. Evidence for superdeformed structures which decay by discrete line gamma transitions have been found recently in both ^{152}Dy and ^{132}Ce. These data will be discussed below.

4.1 A discrete line superdeformed band in ^{152}Dy.

At low spin ^{152}Dy has two distinct structures. There are a series of single-particle states which have been studied to near spin 40ħ (Styczen et al 1983, Khoo et al 1978, Merdinger et al 1979). These correspond to a spherical or slightly oblate shape. Recently a deformed structure was found corresponding to $\beta \sim 0.15 - 0.20$ (Nyako et al 1986). These states also extend to ~ 40ħ. The decay scheme summarising these states is given in fig.8.

A recent high statistics experiment has been carried out using TESSA3 (with 12 BGO suppressed detectors). The reaction used was ^{108}Pd(^{48}Ca,4n)^{152}Dy at a beam energy of 205 MeV. Over 200 million events were recorded corresponding to two (or more) suppressed germanium detectors and the BGO sum energy - multiplicity array. This experiment resulted in the observation of a superdeformed band which decays by a series of discrete line transitions. The spectrum shown in fig.9 is the classic "picket fence" spectrum expected for a rotational band. The assignment of spins to the various transitions in the band is described in detail by Twin et al (1986). The assignment is based on the feeding

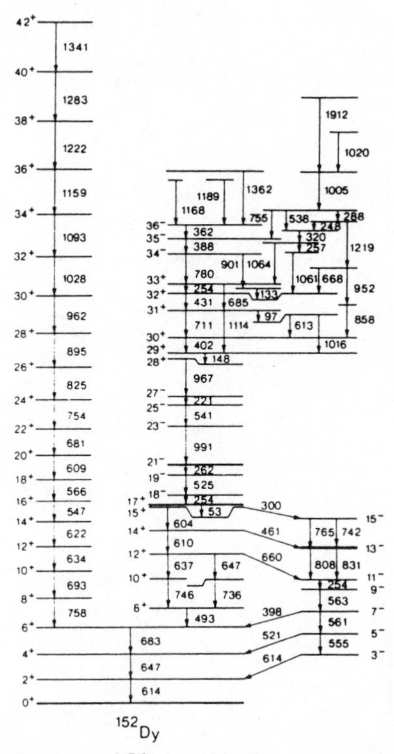

Fig.8 Partial decay scheme for ^{152}Dy. A low deformation band is seen to spin 42^+, the irregular oblate states are seen to near spin 40 (Nyako et al 1986).

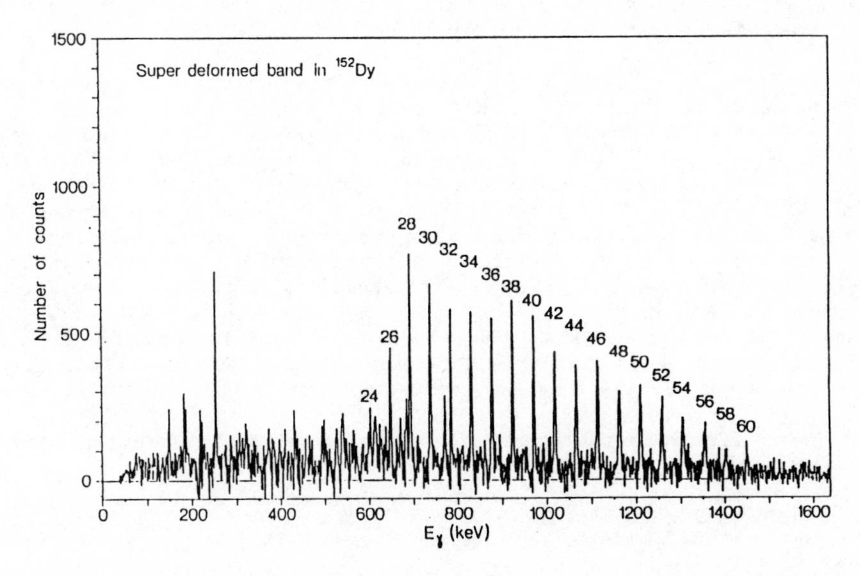

Fig.9 Gamma-ray spectrum in the superdeformed band in ^{152}Dy following the ^{108}Pd(^{48}Ca,4n)^{152}Dy reaction at 205 MeV (Twin et al 1986).

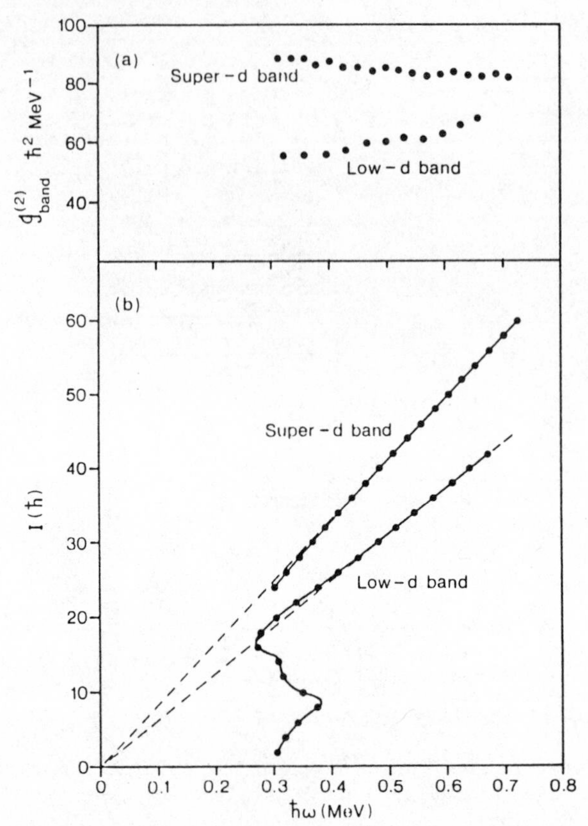

Fig.10 Spin (I) and moment of inertia ($\mathcal{J}$) against frequency ($\hbar\omega$) for the low and superdeformed bands in ^{152}Dy (Twin et al 1986).

pattern of decays into the known states and the stretched E2 character of transitions within the band. The transition spacing allows the moment of inertia to be calculated. This is shown in fig.10(a). The value found is constant at 85 MeV$^{-1}\hbar^2$. This corresponds to a superdeformed structure with $\beta \sim 0.6$. A spin versus rotational frequency plot for the band is shown in fig.10(b). Both the superdeformed and low deformation bands show the now familiar (Chapman et al 1983, Bacelar et al 1985a) straight line which extrapolates to the origin. This apparently classical rotational behaviour could be an indication of the reduction in pairing correlations, but this feature is still the subject of much theoretical work. The intensity of transitions within this band is ~ 2% of the population of the low-lying states in ^{152}Dy. This intensity accounts for the vast majority of the ridge seen in the previous gamma-gamma energy correlation studies.

4.2 Mean lifetime measurements in the superdeformed second minimum in ^{132}Ce.

A high deformation rotational band decaying by discrete line transitions was found in ^{132}Ce using the TESSA2 spectrometer (Nolan et al 1985). Based on the transition energies, the experimental moment of inertia corresponded to a deformation $\beta \sim 0.45$. These data have recently been extended using TESSA3. The reaction used was ^{100}Mo(^{36}S,4n)^{132}Ce at beam energies of 150 MeV and 160 MeV. Other experimental conditions were as for the ^{152}Dy experiment. The band was extended beyond spin 50ħ using over 300 million events collected using a thin target (500 μg cm^{-2}). The thin target ensured that all nuclei decayed at one recoil velocity and spectra could be summed by gain matching during the analysis. The resulting spectrum is shown in fig.11. Gamma transitions are seen above 2 MeV corresponding to a spin above 50ħ. The transitions linking this decay sequence to the ground state band have not been identified. The spins have been assigned in the new band by assuming 2ħ are taken by the linking transitions and that the transitions within the band are stretched E2. The intensity of the transitions in the band is given in Table 2. Transitions below spin 32ħ are essentially fed 100% by cascade within the band, above this spin there is a gradual fall in intensity. In fig.12 the data are compared to the results of calculations carried out by Bengtsson and Ragnarsson (1985). The general shape of the band is well reproduced, but the calculation predicts a spin 6 - 8ħ higher than we have assumed. As the linking transitions to the ground state sequence have not yet been seen experimentally, higher spins are a possibility.

Table 2: Gamma ray energies, relative intensities and mean lifetimes.

Eγ (keV)	I^{π}	Relative Intensity[a)]	Observed F[b)]	Apparent τ(fs)[c)]	τ(fs)[d)]
809	20	85(10)	0.50(3)	435(50)	85(30)
865	22	100(7)	0.62(1)	280(12)	90(20)
929	24	110(12)	0.74(2)	170(15)	40(17)
995	26	103(13)	0.80(2)	125(15)	<25
1060	28	105(9)	0.82(1)	108(8)	<30
1127	30	91(3)	0.85(1)	88(7)	20(10)
1195	32	94(7)	0.89(1)	62(6)	14(11)
1264	34	84(3)	0.91(1)	50(6)	<20
1335	36	62(11)	0.93(1)	37(6)	<10
1409	38	51(9)	0.93(1)	37(6)	<15
1488	40	29(7)	0.94(2)	32(11)	<15
1567	42	39(3)	0.94(2)	32(11)	<35
1652	44	46(5)	0.99(2)	<15	<10
1742	46	21(7)	0.98(3)	<25	
1836	48	25(10)			
1930	50				
2030	52				

a) Normalised to 865 keV = 100, this is equivalent to 5% of the $4^+ - 2^+$ intensity in ^{132}Ce.

b) The attenuated Doppler Shift F is the fraction of the expected full Doppler shift.

c) No feeding correction. The conversion to τ was carried out using electronic stopping powers from Northcliffe and Schilling (1970) normalised to the alpha stopping powers given by Ziegler and Chu (1974). The nuclear stopping power was claculated using the method outlined by Lindhard et al (1963) and Blaugrund (1966).

d) Corrected for feeding.

The major part of the new experiment was aimed at measuring mean lifetimes. This was carried out using a thick target (1 mg cm^{-2}) on a

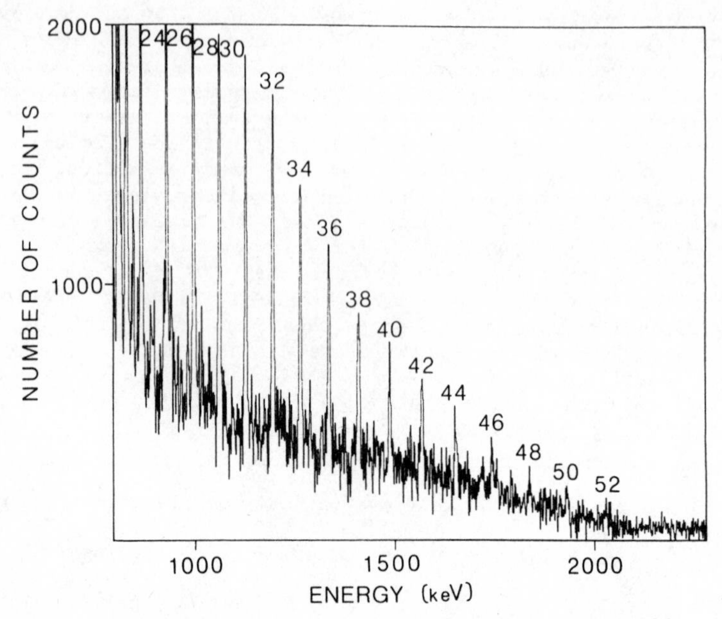

Fig.11 Gamma ray spectrum for the superdeformed band in ^{132}Ce following the ^{100}Mo(^{36}S,4n)^{132}Ce reaction.

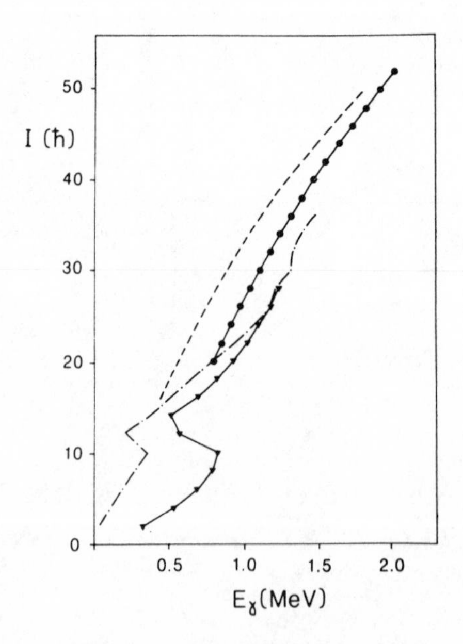

Fig.12 Spin (I) against gamma ray energy (E_γ) for states in ^{132}Ce. The experimental states are shown as: ● superdeformed band, (b) ▼ ground state sequence; the calculated states are shown by:- - - superdeformed band, -·-·- ground state sequence. The calculations (Bengtsson and Ragnarsson 1985) are without pairing.

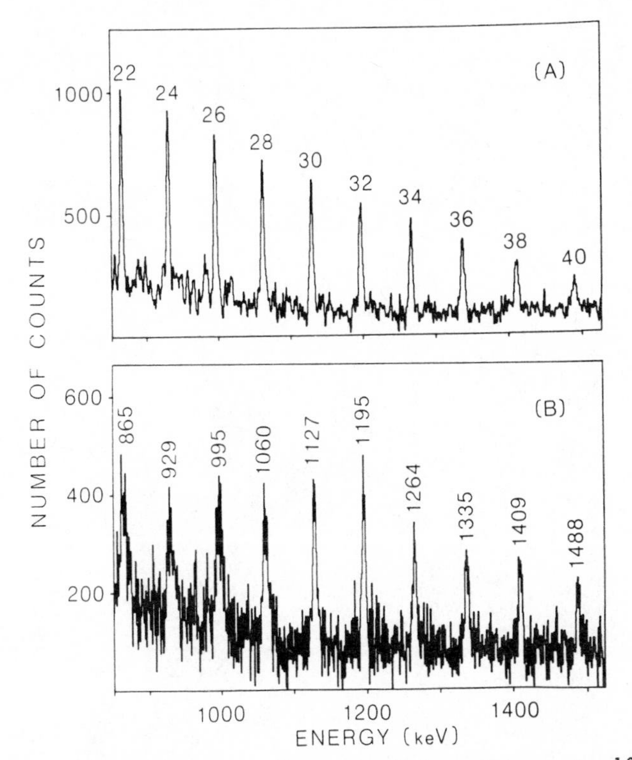

Fig.13 Gamma ray spectra for the superdeformed band in ^{132}Ce. (a) thin target labelled by spins, (b) thick target labelled by gamma ray energies.

gold backing (15 mg cm^{-2}), over 200 million events were collected. Lifetimes were deduced using the Doppler shift attenuation method (DSAM) (e.g., Nolan and Sharpey-Schafer 1979). Spectra taken with both the thin and backed targets are shown in fig.13. These spectra have been gain matched assuming the recoils decay at the initial recoil velocity. Sharp peaks are therefore seen in the thin target data. Transitions above spin 30ħ are also sharp in the backed target data showing that they arise from states with very short lifetimes. The remaining transitions show Doppler effects which implies their lifetimes must be comparable with the slowing down time of the recoils in the target and its backing. The results of the measurements are shown in fig.14. The average recoil velocity ($\langle v/c \cos\theta \rangle$) at which each state decays is plotted against spin. Also shown is the calculated initial recoil velocity (allowing for the target thickness). The gap between this line and the data points corresponds to a time of 10 - 15fs which represents the feeding time into the band. The data points show a smooth decrease in $\langle v/c \cos\theta \rangle$ as the spin decreases. This indicates that there is no significant slow sidefeeding into the band. This would be seen as a discontinuity in the curve resulting in a drop in $\langle v/c \cos\theta \rangle$. In fig.14 three curves have been included corresponding to a quadrupole moment Q_0 of 8.8 eb (curve b) and a $\pm$ 20% difference (curves a and c). These have been calculated assuming decay down a rotational band with a constant deformation. The data are best represented by Q_0 = 8.8 eb (β = 0.5). Mean lifetimes for the individual states are given in Table 2 corrected for cascade feeding

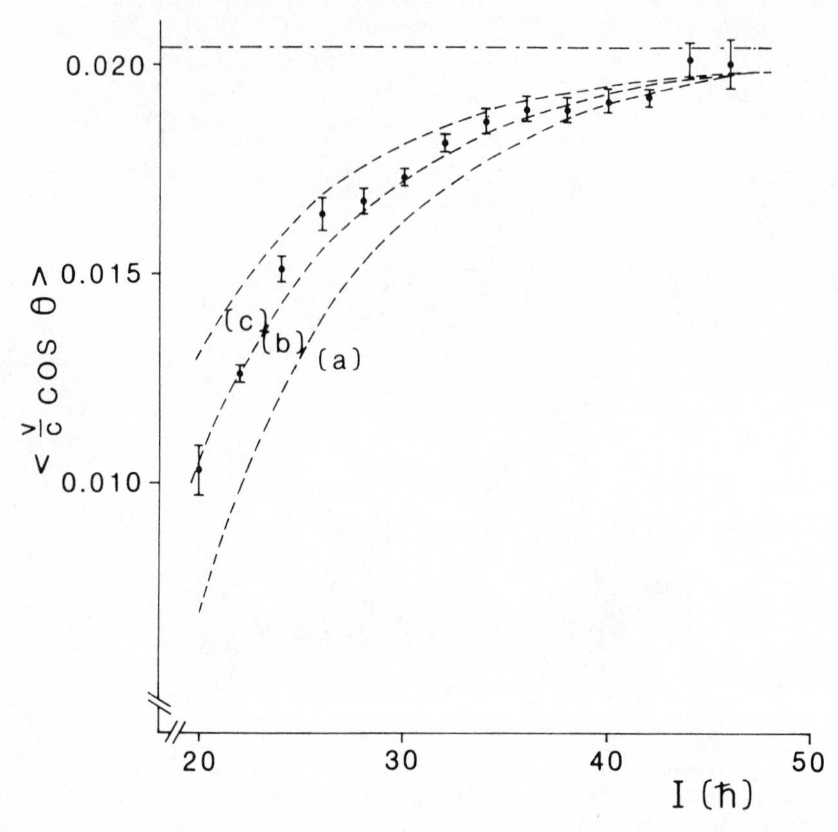

Fig.14 Average recoil velocity for the decay of each state (< v/c cos θ >) as a function of spin for the superdeformed states in ^{132}Ce. Three curves are shown (- - -) assuming decay down a constant deformation rotational band. These are normalised to the average of the top four points. Curve (b) corresponds to Qo = 8.8eb (β ~ 0.5). Curves (a) and (c) represent a change in Qo of ± 20%. Also shown is the calculated initial recoil velocity (-·-·-).

only. This will yield good results as, for the lower levels, no side feeding is seen experimentally and for the higher levels the side feeding is fast.

Hence in ^{132}Ce a superdeformed band has been observed with a quadrupole moment consistent with a deformation β ≃ 0.5. These states lie within the deformed second minimum in the potential energy surface (PES). Figure 15 shows calculated PES between spins 20ħ and 56ħ. The deformed second minimum becomes yrast above spin 30ħ. At the highest spin shown there is an oblate minimum starting to develop. This spin of 56ħ corresponds approximately to the highest spin seen experimentally in the band. Further experiments will be needed to see if the band extends further or its population terminates due to a shape change.

4.3 Summary

Superdeformed bands which decay by discrete gamma transitions have been

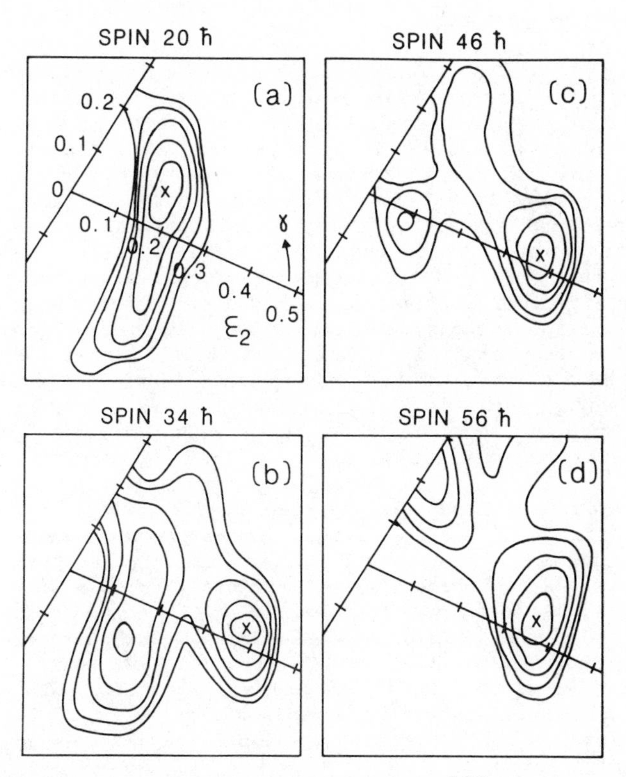

Fig.15 Potential energy surface calculations for ^{132}Ce. A constant value of $\varepsilon_4 = \varepsilon_2/6$ was used. The minimum is marked x; contours are shown at 0.2 MeV intervals.

observed to spin 60ħ in ^{152}Dy and spin >50ħ in ^{132}Ce. The energy level spacing gives a constant moment of inertia (the whole spin range in ^{152}Dy, part of the spin range in ^{132}Ce) corresponding to deformations of $\beta \sim 0.6$ (^{152}Dy) and $\beta \sim 0.45$ (^{132}Ce). Mean lifetimes measurements in ^{132}Ce are consistent with a large quadrupole moment and a deformation $\beta = 0.5$. The intensities seen in the two decay sequences are relatively strong at 2% (^{152}Dy) and 5% (^{132}Ce) of the population of the low-lying states. Bands such as these present an excellent opportunity for studying the deformed second minimum in the nuclear potential energy surface. Previous studies have been limited mainly to fission isomers whose population is many orders of magnitude lower.

There remain however a number of questions yet to be answered. In both nuclei the excitation energy of the superdeformed band is unknown and further experimental work is necessary. Twin et al (1986) have suggested that the gamma-ray transitions de-exciting the superdeformed structure will have a statistical character. This is supported by the absence of discrete linking transitions in both the ^{152}Dy and ^{132}Ce high statistics experiments. Other questions will be the subject of theoretical interest. Why is only one band observed? What determines the spin at which it decays? As this structure has only a few percent of the intensity what properties do the remaining states have? Many theoretical papers are expected in the next year.

5. Future directions

The study of nuclear structure at high spin has had many notable achievements in the past few years. Two of the major ones have been discussed in this presentation. Others have come not from one nucleus but from the results of systematic trends across many nuclei. The existence of pairing and other correlations has been discussed by Garrett et al (1986) in a series of deformed rare-earth isotopes. The data are mainly energy levels and systematics suggest that pairing is much reduced above spin 30ħ in a range of Yb nuclei (A = 162 - 169). There have been a limited number of mean lifetime measurements which can determine transition strengths and quadrupole moments. These are generally limited to ground-state sequences below about spin 24ħ. The Yb and neighbouring isotopes have been studied at Oak Ridge (Fewell et al 1985, Oshima et al 1986, Rao et al 1986). There is a general trend to reduced collectivity with increasing spin which is interpreted as a change to a triaxial shape. The data however have large uncertainties and are open to alternative explanations.

Although much work has already taken place at high spin, data are generally limited to level energies. These are not always a very exacting probe of nuclear models and often are insensitive to different predictions. Matrix element determinations provide a much more sensitive probe. Such measurements for example, will help in understanding the shapes of the states near spin 40ħ in ^{158}Er. Such data on odd nuclei and different bands within even nuclei will give a measure of the nuclear shape in different configurations and its stability when extra particles are added. These detailed measurements require high statistics and are time consuming, but will provide invaluable data to aid nuclear structure models at high spin.

In this talk I have not had time to discuss recent work on the highest spin states using continuum gamma transitions. The major step forward recently was the discovery by Love et al (1985) that the in-band transition probability within the continuum was ~ 20%. This prompted a lot of work on spreading widths (eg. Bacelar et al 1985b, Draper et al 1986) in an attempt to explain the data. Investigation of the continuum will be important in the future as this is where the highest spin states will be found. The experimental difficulties will be eased with the advent of triple gamma coincidence data. This will allow selection of particular nuclei and specific decay paths. The improved data should prove easier to analyse and give an indication of both nuclear shapes and single-particle properties (through alignments) at the highest spins. The excitation energies of the states (temperature) is another degree of freedom that can be probed.

Experiments suggested in the previous paragraphs will become possible with the large arrays now either operating or being commissioned. The Berkeley array (HERA) will be enhanced with a sum energy - multiplicity detector. The 8π spectrometer (Taras et al 1983) will operate shortly at Chalk River. This consists of 20 BGO suppressed germanium detectors surrounding a 72 element inner BGO ball. The large crystal balls at Oak Ridge and Heidelberg are being modified to incorporate a number of suppression shields. A European collaboration (ESSA30) is planned where 30 BGO suppressed germanium detectors will be used.

The technical revolution in the study of high spin states is now nearing completion. This marks the start of a new era of study of nuclear properties at extreme angular momentum. The exciting discoveries I have covered will prove to be only the start and we can look forward to many interesting and revealing phenomena.

Acknowledgements

The work described is that of many people in a number of different laboratories. I thank them all, especially those whose data have been used prior to publication.

References

Bacelar J C et al (1985a) Nucl. Phys. A442 509
Bacelar J C, Hagemann G B, Herskind B, Lauritzen B, Holm A, Lisle J C, Tjøm P O (1985b) Phys. Rev. Lett. 55 1858
Bengtsson T, Faber M, Leander G, Möller P, Ploszajczak M, Ragnarsson I and Åberg S (1981) Physica Scripta 24 200
Bengtsson T and Ragnarsson I (1983) Physica Scripta T5 165
Bengtsson T and Ragnarsson I (1985) Nucl. Phys. A436 14 and private communication
Blaugrund A E (1966) Nucl. Phys. 88 501
Bohr A and Mottelson B R (1975) Nuclear Structure Vol. II (N.Y.: Benjamin) p.592
Burde J, Dines E L, Shih S, Diamond R M, Draper J E, Lindenberger K H, Schück C and Stephens F S (1982) Phys. Rev. Lett. 48 530
Chapman R et al (1983) Phys. Rev. Lett. 51 2265
Cranmer-Gordon H W et al (1986) Proc. Int. Conf. Nucl. Phys. Harrogate 25-30 August Vol. I Contributed Papers (London: Inst. Phys.) p.126
Deleplanque M A (1981) Physica Scripta 24 158
Diamond R M and Stephens F S (1981) The High Resolution Ball, Berkeley Report
Draper J E, Dines E L, Deleplanque M A, Diamond R M and Stephens F S (1986) Phys. Rev. Lett. 56 309
Dudek J and Nazarewicz W (1985) Phys. Rev. C31 298
Fewell M P et al (1985) Phys. Rev. C31 1057
Frauendorf S and May F R (1983) Phys. Lett. 125B 245
Garrett J D et al (1986) Proc. Int. Conf. Nucl. Phys. Harrogate 25-30 August Vol. I Contributed Papers (London: Inst. Phys.) p.386
Herskind B (1983) Proc. Int. Conf. Nucl. Phys. Florence 28 Aug.-30 Sept. Vol. II Invited Papers (Bologna: Tipografia Compositori) p.117
Johnson A, Ryde H and Sztarkier J (1971) Phys. Lett. 34B 605
Khoo T L, Smither R K, Haas D, Hausser O, Andrews H R, Horn D and Ward D (1978) Phys. Rev. Lett. 41 1027
Lindhard J, Scharff M and Schiott H E (1963) K Danske Vidensk. Selsk., Mat. Fys. Meddr. 33 No. 14
Love D J G, Nelson A H, Nolan P J and Twin P J (1985) Phys. Rev. Lett. 54 1361
Merdinger J C, Beck F A, Byrski T, Gehringer C, Vivien J P, Bozek E and Styczen J (1979) Phys. Rev. Lett. 42 23
Nolan P J and Sharpey-Schafer J F (1979) Rep. Prog. Phys. 42 1
Nolan P J (1984) Fundamental Problems in Heavy Ion Collisions ed. Cindro N, Greiner W, Caplar R (World Scientific) 133
Nolan P J, Kirwan A, Love D J G, Nelson A H, Unwin D J and Twin P J (1985) J. Phys. G11 L17
Northcliffe L C and Schilling R F (1970) Nucl. Data Tables A7 233
Nyako B M et al (1984) Phys. Rev. Lett. 55 1380

Nyako B M, Simpson J, Twin P J, Howe D, Forsyth P D and Sharpey-Schafer J F (1986) Phys. Rev. Lett. 56 2690
Oshima M, Johnson N R, McGowan K K, Baktash C, Lee I Y, Schutz Y, Ribas R V and Wells J C (1986) submitted to Phys. Rev. C
Polikanov S M et al (1962) Sov. Phys. JETP 15 1016
Ragnarsson I, Bengtsson T, Leander G and Åberg S (1980) Nucl. Phys. A347 287
Ragnarsson I, Xing Z, Bengtsson T and Riley M A (1986) to be published in Physica Scripta
Riley M A, Garrett J D, Sharpey-Schafer J F and Simpson J (1986) Phys. Lett. 177B 15
Rao M N et al (1986) Phys. Rev. Lett. 57 667
Simpson J et al (1984) Phys. Rev. Lett. 53 648
Specht H J, Weber J, Konechy E and Heunemann D (1972) Phys. Lett. 41B 43
Stephens F S and Simon R S (1972) Nucl. Phys. A183 257
Styczen J, Nagai Y, Piiparinen M, Ercan A and Kleinheinz P (1983) Phys. Rev. Lett. 50 1752
Tanabe K and Sugarawara-Tanabe K (1984) Phys. Lett. 135B 353
Taras P, Waddington J C, Andrews H R and Ward D (1983) Chalk River Report
Tjøm P, Diamond R M, Bacelar J C, Beck E M, Deleplanque M A, Draper J E and Stephens F S (1985) Phys. Rev. Lett. 48 530
Twin P J (1983) Proc. Int. Conf. Nucl. Phys. Florence 28 Aug.-30 Sept. Vol. II Invited papers (Bologna: Tipografia Compositori) p.527
Twin P J, Nolan P J, Aryaeinejad R, Love D J G, Nelson A H and Kirwan A (1983) Nucl. Phys. A409 343c
Twin P J et al (1985) Phys. Rev. Lett. 55 1380
Twin P J et al (1986) Phys. Rev. Lett. 57 811
Ziegler J F and Chu W K (1974) At. Data and Nucl. Data Tables 13 464

Inst. Phys. Conf. Ser. No. 86
Paper presented at Int. Nucl. Phys. Conf., Harrogate, UK, 1986

Spin response in nuclei

Carl Gaarde

The Niels Bohr Institute, University of Copenhagen, Copenhagen, Denmark

1. Introduction

In this paper we discuss the spin response function as obtained in hadron scattering. We shall specifically make use of charge-exchange reactions at intermediate energies which have proven to be very specific as to probing the spin response. This specificity is an effect of the ratio of interactions in the spin transfer and non-spin transfer channels in the NN interaction. When we therefore measure cross sections at different momentum transfers in charge-exchange reactions we probe the spin response from the nucleus.

We shall discuss 3 regions of the response function:

i) Gamow Teller region
$|\vec{q}|$ small ; ω small

ii) quasifree peak region
$|\vec{q}| \sim 1.2\text{-}2\ \mathrm{fm}^{-1}$; $\omega \sim 50\text{-}150$ MeV

iii) Δ-region
$|\vec{q}| \sim 0.8\text{-}5\ \mathrm{fm}^{-1}$; $\omega \sim 200\text{-}400$ MeV.

Here $|\vec{q}|$ is the 3-momentum transfer and ω the energy transfer, both given in the lab. system.

2. Gamow-Teller Strength

The information on the spin strength distributions at small momentum[1,2,3] transfers is largely coming from the (p,n) studies at Indiana University. It was shown that the (p,n) reaction at intermediate energies selectively excites spinmodes in the nucleus. In Figure 1 we show spectra at 3 angles from the (p,n) reaction on ^{208}Pb at 200 MeV. The spectra correspond to the angles where the peaks dominating the spectra have the maximal cross section for angular momentum transfers of 0, 1 and 2. The more detailed analysis of spectra from the (p,n) reaction,as given in Figure 1,shows e.g. that the zero degree spectrum to a good approximation directly gives the response function for the $\sigma\tau$-operator, i.e. that the spectrum gives the Gamow-Teller strength function. The broad resonances observed at θ=4.5°and 9° are interpreted as envelopes of collective states with J = 0^-, 1^-, 2^- and 1^+, 2^+, 3^+, respectively, corresponding to the operators $r(Y_1\sigma)_{0,1,2}\tau_-$ and $r^2(Y_2\sigma)_{1,2,3}\tau_-$. It has however not been possible to resolve these broad resonances into their components. A particularly interesting case would

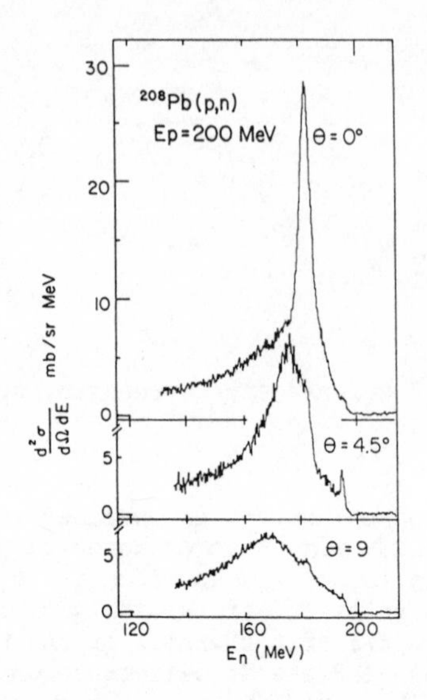

Fig.1. Spectra at 3 angles for the ^{208}Pb(p,n) reaction.

be the 0^--resonance. Correlations for 0^--states come alone from spin longitudinal interactions, so the 0^--resonance is a truly pionic mode. The 3 spectra given in the figure show that there are strong isospin-spin correlations in nuclei that form collective states. These states are pushed up in energy, i.e. the particle-hole interaction is repulsive in the isospin-spin channels at small momentum transfers.

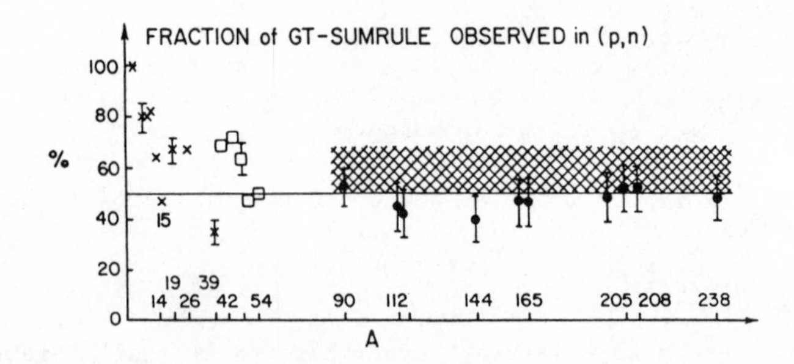

Fig. 2. Fraction of Gamow-Teller sumrule strength observed in (p,n) reactions. In the p- and sd-shell the strength is most often in a few sharp states. In the fp-shell a multipole decomposition is attempted. For heavier nuclei the dots (with error bars) represent strengths in peaks (low lying + giant), whereas the cross hatched region also includes strength under the collective state. Possible strength above (larger E_x) the collective state is not included.

Another result of the (p,n) experiments is summarized in Figure 2. Here the fraction of the Gamow-Teller sumrule strength as observed in the (p,n) studies is given as a function of mass number. A very simple sumrule applies for GT strength.

$$S_{\beta^-} - S_{\beta^+} = 3(N-Z)$$

The difference between the total β^- and β^+ strength for a given target nucleus is model independent and equal to 3(N-Z). The only assumption is that the τ and σ operators only change the direction in isospin and spin space of the nucleon. Including internal nucleon degrees of freedom, e.g. the Δ-excitation, changes the sumrule.

The figure refers to the strength in the "shell model region", i.e. under and below the collective state for the heavier systems. Above this region it is experimentally very difficult to determine how much of the cross section corresponds to Gamow-Teller strength. We cannot from the (p,n) studies exclude that ℓ=0 strength could be there. The missing GT strength would then be explained as an effect of coupling to 2p-2h states. The more detailed studies, where spin-flip probabilities are measured in ($\vec{p},\vec{n}$) [4,5] experiments do not show any GT strength in this energy region, but with the same uncertainty as above: that a small fraction of the cross section cannot be excluded as ℓ=0 strength.

Still another set of data from the particle decay of the collective Gamow-Teller state in ^{208}Bi shows that there is little 1p-1h strength above the collective state.[6]

In connection with the missing GT strength and the study of spin modes in general we mention a very interesting development at the (n,p) facility at TRIUMF.

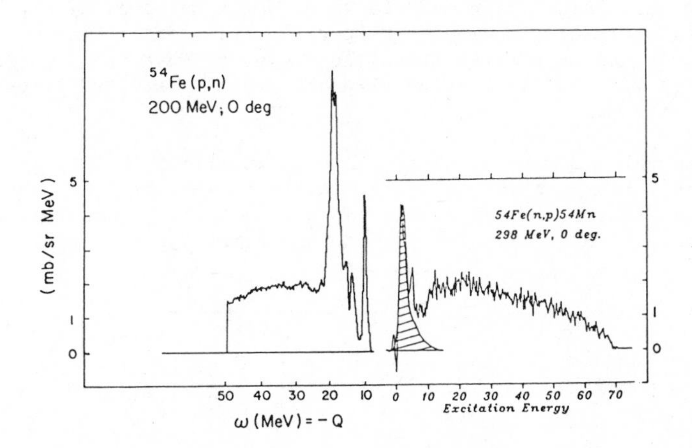

Fig.3. Spectra for (p,n) and (n,p) reactions on ^{54}Fe at zero degrees and 200 and 300 MeV bombarding energy respectively. The cross section unit is the same. The hatched area in the (n,p) spectrum refers to a shell model calculation of the β^+-strength, scaled by a factor 0.6.

In Figure 3 we show a spectrum for the (n,p) reaction on ^{54}Fe at zero [7] degrees. Also shown is a (p,n) spectrum on the same target. The spectra together therefore give the total β-decay strength function for a nucleus where both the β^-- and β^+-strength is quite large. The peaks in both spectra have been proven to have ℓ=0 angular distributions and therefore correspond to GT transitions. The sums of strength extracted are S_{β^-}=7.8±1.9 [8] and S_{β^+}=5.1±0.6. The latter number is based on a DWIA analysis, whereas the (p,n) value is based on systematics obtained from (p,n) 0° cross sections versus B(GT) values obtained from β-decay. A shell model gives S_{β^-}=15.2 and S_{β^+}=9.2 respectively, giving a difference of 6=3(N-Z). We see that the ground state correlations reduce the strength from the extreme case where ^{54}Fe would be described as $(\pi f_{7/2})^6(\nu f_{7/2})^8$. This configuration gives S_{β^-}=16.3 and S_{β^+}=10.3.

We see that the observed strength is around 50% of the strength predicted in the shell model for both the β^- and β^+-decays. We further note that

$$S_{\beta^-} - S_{\beta^+} = 2.7\pm2.0$$

is smaller than 3(N-Z) = 6, but that the errorbars are rather large. If we however compare the cross sections above the peaks in the spectra, we note that they are the same and the difference therefore zero. Before we conclude that there is no GT strength from the neutron excess in the continuum we should remember that these spectra are not obtained with the same bombarding energies.

There is, however, an interesting perspective as to the question of GT strength in the continuum in comparing (p,n) and (n,p) spectra at all angles and at the same bombarding energy. The difference spectra should be related to the strength arising from the neutron excess. It should be possible to distinguish between strength from the core and the neutron excess. Recent calculations by Desplanques et al. predict considerable[9]) strength in spin-saturated nuclei like ^{40}Ca in the continuum region. Extrapolating this calculation to ^{54}Fe we expect 4 units of GT strength up to 50 MeV of excitation energy from 2p-2h interactions in the core, i.e. in both (p,n)- and (n,p)-like transitions. We see that this contribution is comparable both to the missing strength and the observed strength in say the β^+-channel.

We summarize this section by noting that i) isospin-spin correlations in nuclei are sufficiently strong to form collective states, ii) the particle-hole interaction is repulsive, and that iii) Gamow-Teller strength is still missing. The effect of quenching due to coupling to Δ-hole states is still an open question.

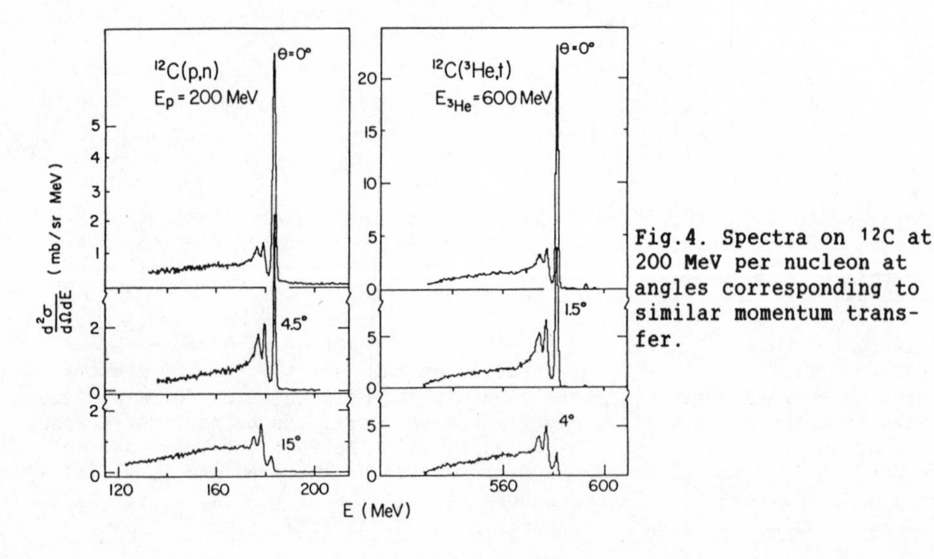

Fig.4. Spectra on ^{12}C at 200 MeV per nucleon at angles corresponding to similar momentum transfer.

3. Quasifree Peak Region

In this section we compare spectra from the (^{3}He,t) reaction at intermediate energy with spectra obtained from inelastic electron scattering. We shall compare the centroid energy and width of the quasifree peak in the two reactions.

We first note that the (^{3}He,t) reaction at intermediate energies is very similar to the (p,n) reaction. This is demonstrated in Figure 4 where spectra with ^{12}C as the target are shown: The bombarding energy per nucleon is chosen to be the same and the angles correspond to the same momentum transfer. It is seen that the spectra are indeed very similar. It is results as shown in Figure 4, that leads us to conclude that the (^{3}He,t) reaction at intermediate energies is a one-step process dominated by $\sigma\tau$-transitions.

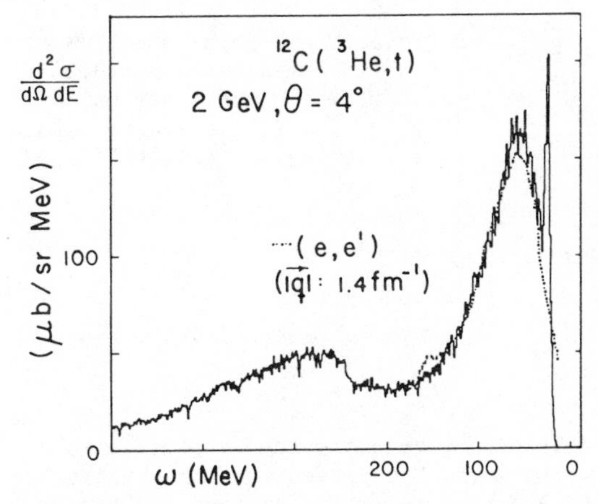

Fig.5. The (^{3}He,t) spectrum at 2 GeV and θ=4° is compared with the transverse response function (in arbitrary units) at $|\vec{q}|$=1.4.fm^{-1} 10).

In Figure 5 we show a spectrum for ^{12}C(^{3}He,t) at θ_{lab}=4° and 2 GeV bombarding energy. The sharp peak in the spectrum corresponds to the 4$^-$-state in ^{12}N (E_x~4.1 MeV). The broad structure around $\omega = T_{^3He} - T_t$ ~60 MeV, we interpret as the quasifree peak, i.e. transitions arising from charge-exchange reactions with quasifree neutrons in the nucleus.

We think of the (^{3}He,t) reaction at these energies just like the (p,n) reaction, but with a different formfactor than for the nucleon. When we therefore measure the energy loss, ω, and momentum transfer, $\vec{p}_{^3He} - \vec{p}_t = \vec{q}$ we know in a plane wave impulse approximation (PWIA) the four-momentum transfer $q_\mu = (\vec{q}, \omega)$ to the nucleon in the nucleus.

In comparing with spectra obtained in (e,e') experiments we should choose the same four-momentum transfer. This is attempted in Figure 5 where a spectrum at $|\vec{q}|$~1.4 fm^{-1} is compared with the (^{3}He,t) spectrum at θ=4°. We first note that the centroid energy and width are indeed the same for the two spectra.

Let us first comment on possible shifts of the peak in the (^{3}He,t) spectrum. In the PWIA we would write the cross section on a free nucleon as

$$\frac{d\sigma}{dt} = [|\alpha|^2 + (|\beta|^2 + |\gamma|^2 + |\varepsilon|^2) + |\delta|^2] FF^2_{^3He-t}(q_\mu)$$

where α etc. are the charge-exchange NN amplitudes (see below) and $FF(q_\mu)$ the formfactor for the ^3He-t system.

The spectrum for the quasifree reactions would therefore be distorted if the formfactor changed over the quasifree peak. The four-momentum transfer is however almost constant for ω between 0 and 100 MeV. This is illustrated in Figure 7 where curves are drawn in an ω versus $|\vec{q}|$ plot for (^{3}He,t) spectra at different angles. It is seen that the curves are almost along constant $q_\mu^2 = \vec{q}^2 - \omega^2$. The form factor is therefore almost constant in the quasifree region.

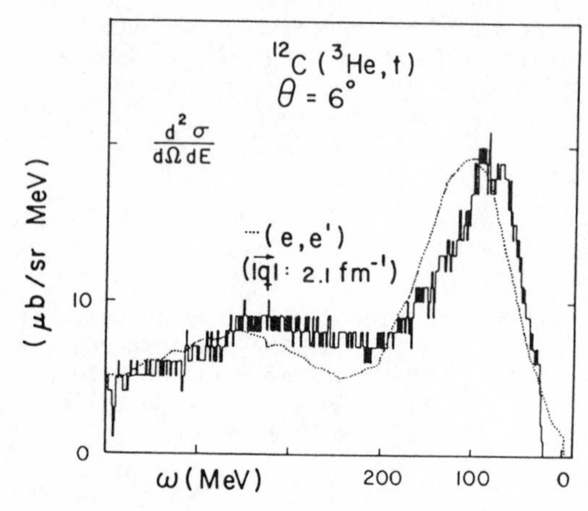

Fig.6. The (^{3}He,t) spectrum at θ=6° is compared to an (e,e') spectrum (in arbitrary units) at |q|~2.1 fm^{-1} based on data from Saclay and MIT.[10,11]).

In the (e,e') experiments it is possible to separate the response into a[10]) longitudinal and spin transverse part. The dotted curve in Figure 5 is in fact the transverse response function as extracted from the Saclay data. The dotted curve in fig.5 is therefore the $\vec{\sigma} \times \vec{q}$ response.

A very simple model seems to explain the main features of the (e,e') results on the quasifree peak.

$$\omega = \sum_i^A \left(\frac{\vec{p}_i + \vec{q}}{2M}\right)^2 - \left(\frac{\vec{p}_i^{\,2}}{2M} + B\right)$$

in which the energy of the particle hole excitation can be obtained in a plane wave limit and the Pauli-blocking is a small effect, when $|\vec{q}|$ is comparable or larger than p_F

$$\omega \simeq \frac{q^2}{2M} + |B| + \sum_i^A \frac{\vec{p}_i \cdot \vec{q}}{M}$$

The two first terms give the centroid, the last term the width.

In Figure 7 a curve with an average binding energy B=15 MeV is seen to reproduce the centroid of the peak from (e,e') in ^{12}C quite well.

In Figure 6 we compare the (^{3}He,t) at θ=6° with the (e,e') spectrum for $|\vec{q}|$~2.1 fm^{-1} (see fig.7). We now see a shift of ~15 MeV down in energy for the peak observed in the (^{3}He,t) reaction.

We interpret this shift as due to particle-hole correlations in the spin-longitudinal channel, i.e. that interactions with a spin-structure $(\vec{\sigma}_1\cdot\vec{q})(\vec{\sigma}_2\cdot\vec{q})$ are attractive at these momentum transfers and shift the quasifree peak to lower energy. The interaction from π-exchange would have this spin structure, in which case we should write

$$V_{ph} = \kappa \frac{(\vec{\sigma}_1\cdot\vec{q})(\vec{\sigma}_2\cdot\vec{q})}{\omega^2-\vec{q}^2+m_\pi^2}\,\vec{\tau}_1\cdot\vec{\tau}_2$$

This interpretation assumes that distortion effects coming from the absorption of the projectile or ejectile do not give rise to a shift of the (d, 2p) peak. It is however not expected that such a shift would be different at say $\theta=4^0$ and 6^0.

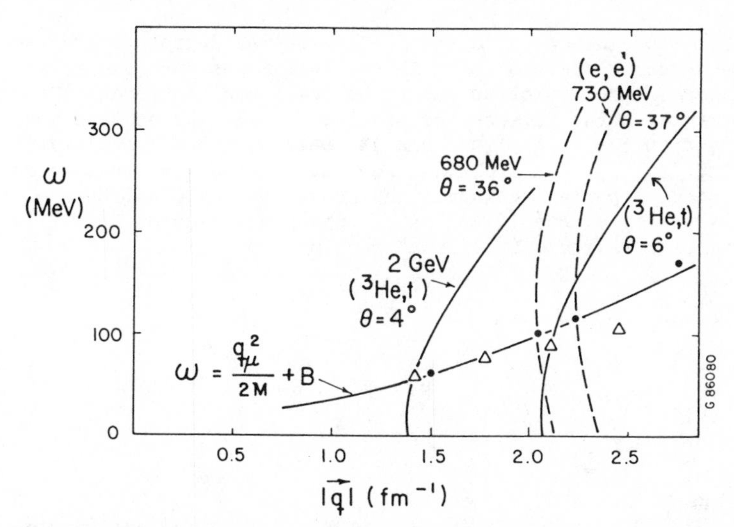

Fig.7. The centroid energies of the quasifree peak as observed in (^{3}He,t) and (e,e') is plotted in a ω versus $|\vec{q}|$ system. A given spectrum will follow one of the curves indicated in the figure.

In Figure 7 we have summarized the observed energies of the centroid of the quasifree peak. The observed shift is however not directly comparable with a possible particle-hole correlation in the spin longitudinal channel. The observed spectrum is an effect both of the probe and the nuclear medium.

The (^{3}He,t) reaction is a mixed longitudinal and transverse probe. In the above PWIA the terms $(\beta^2+\varepsilon^2+\gamma^2)$ are the transverse contributions and δ^2 the longitudinal (α is the non-spin transfer amplitude and completely negligible at these energies and q). If we look up the NN amplitudes at 515 MeV we find that $\delta^2/(\beta^2+\varepsilon^2+\gamma^2)$ is 2.5 and 1.6 at q_μ=1.7 and 2.1 fm^{-1} respectively. The (^{3}He,t) reaction is in this model a good spin longitudinal probe, but a more detailed analysis is needed to extract the effective particle-hole interaction, i.e. to separate the longitudinal from the total spin response. We note that such correlations in the nuclear medium have been discussed in the literature. We mention here specifically the work of M. Ericsson and A. Molinari and coworkers.[12])

We conclude this section by noting that i) a shift of 15-20 MeV is observed for the quasifree peak in the (^{3}He,t) relative to the (e,e') reaction; ii) a possible interpretation of such a shift would be correlations in the spin longitudinal channel. iii) The (^{3}He,t) reaction at $|\vec{q}|$~2 fm^{-1} (i.e. a short range probe) seems to probe the whole nucleus even if the reaction is a surface reaction.

4. The ($\vec{d}$,2p) Reaction

Before we go to the response function in the Δ-region we briefly discuss the (d,2p) reaction at intermediate energies. This reaction looks very promising as a probe of spin modes in nuclei.

The (d,p) reaction with the 2 protons in a relative singlet S state [$^1S_0$] is an isospin-spin transfer reaction. In the impulse approximation the (d,2p) is like an (n,p) reaction with spin transfer.

In the (d,2p) experiments performed at Saturne the 2 protons are recorded in the same magnetic spectrometer, SPES4. This means that the relative kinetic energy of the 2 protons has to be small and that means that the very experimental setup selects the singlet S state. We mention that the contribution from the ^{3}P is less than 1%. Data have been obtained at 650 MeV and 2 GeV deuteron bombarding energy and the results obtained do indeed confirm the expectations mentioned above. We shall not here go into a detailed discussion of the (d,2p) data, but show 2 figures that illustrate the spin transfer selectivity of the reaction.

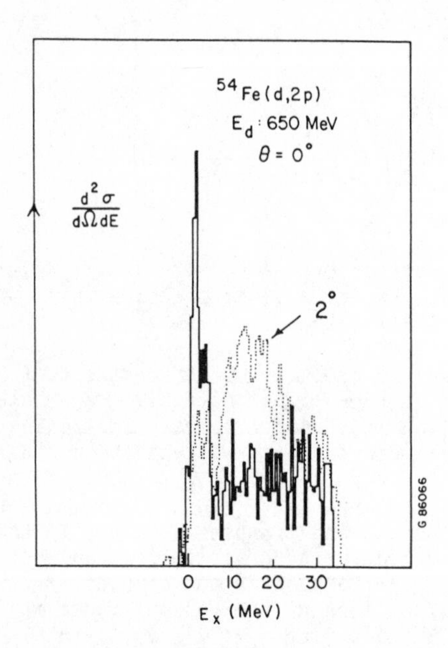

Fig.8. Spectra at 2 angles for the (d,2p) reaction on ^{54}Fe are shown. The resolution is around 1.4 MeV (FWHM).

In Figure 8 two spectra from the (d,2p) reaction on ^{54}Fe are shown. We see that the zero degree spectrum is very similar to the (n,p) spectrum shown in Figure 3, dominated by a single peak. The spectrum at θ=2° shows that the peak has a forward peaked angular distribution, characteristic of an ℓ=0 transfer. The broad structure at θ=2 we interpret as the envelope of collective $\ell=1$ states, similar to the structure in the $\theta=4.5°$ spectrum in Figure 1.

In Figure 9, spectra are given with CH_2 and C targets at 2 GeV bombarding energy. The difference spectrum then gives us the p(d,2p) data. At 2 GeV the Δ-excitation is a dominant feature of the spectrum. We see a shift in Δ-energy between the proton and ^{12}C. We shall comment on this shift below in connection with the similar effect seen in the (^3He,t) data.

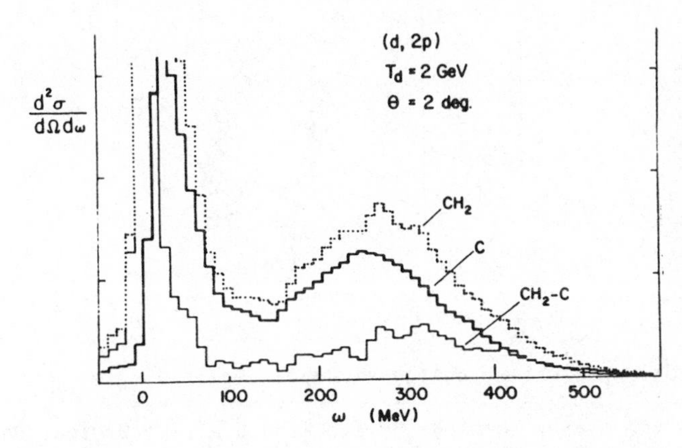

Fig.9. Spectra with CH_2 and C targets. The difference spectrum, used to obtain the p(d,2p) data, is also shown.

The data presented have in fact been obtained with a tensor polarized beam. By measuring the cross section as a function of the 3 spin projections, the tensor analysing powers T_{20} and T_{22} can be determined. That is, in a simple cross section measurement, similar information is obtained as in a $(\vec{n},\vec{p})$ experiment.

Following the notation of ref. 13 we write the unpolarized cross section (in a PWIA) on a proton, as

$$\frac{d\sigma}{dt} = \frac{1}{3}\left[\{|\beta|^2+|\varepsilon|^2+|\gamma|^2\}|S^-|^2+|\delta|^2|S^+|^2\right]$$

The only difference from the (^3He,t) case discussed above, is that the projectile-ejectile formfactor is different for the transverse and longitudinal part. This difference comes from the D-state in the deuteron wave function.

$$S^+ = S(^3S\rightarrow{}^1S) + \sqrt{2}\,S(^3D\rightarrow{}^1S)$$

$$S^- = S(^3S\rightarrow{}^1S) - \frac{1}{\sqrt{2}}\,S(^3D\rightarrow{}^1S)$$

$$F = \left(\frac{S^+}{S^-}\right)^2$$

We find further

$$T_{20} = \frac{\sqrt{2}}{2}\,\frac{\beta^2-2\varepsilon^2+\gamma^2+\delta^2F}{\beta^2+\varepsilon^2+\gamma^2+\delta^2F}$$

$$T_{22} = \frac{\sqrt{3}}{2}\,\frac{\beta^2+\gamma^2-\delta^2F}{\beta^2+\varepsilon^2+\gamma^2+\delta^2F}$$

A very interesting detail in the $(\vec{d},2p)$ reaction is this amplification factor F on the spin longitudinal amplitude. F varies from 1 for q=0 to almost 3 for q~2 fm^{-1}. It seems we have a reaction that via the tensor force in the deuteron is sensitive to spin longitudinal correlations.

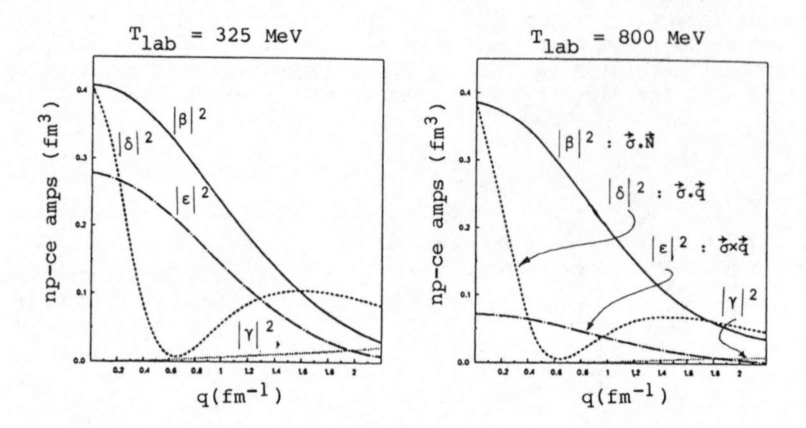

Fig.10. The squares of empirical np charge exchange amplitudes.[14])

In Figure 10 we give the charge exchange NN amplitudes at 2 energies. Only the spin-dependent terms are shown. We note a very strong energy dependence of the $\vec{\sigma}\times\vec{q}$ term.

We shall not go into further detail with the (d,2p) reaction but only mention that also for the study of the spin structure of the Δ-excitation the reaction is indeed a very useful probe. The preliminary data show a negative value for the tensor analysing power, a direct sign of the longitudinal character of the Δ-excitation.

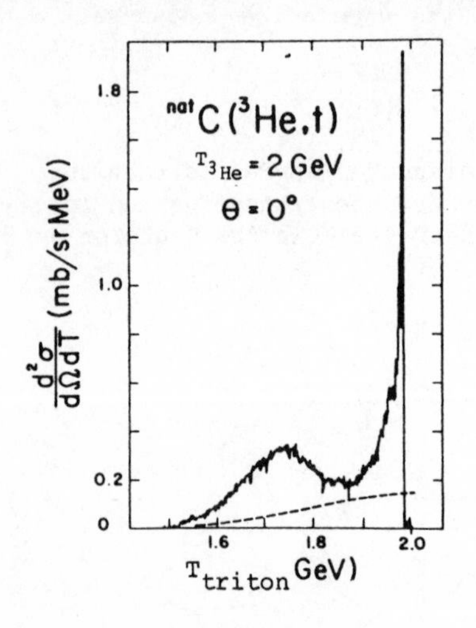

Fig.11. Triton spectrum at 2 GeV bombarding energy.

5. Δ-region

5.1 The (^{3}He,t) Reaction

In Figure 11 we show a spectrum from the (^{3}He,t) reaction on ^{12}C at 2 GeV. It is seen directly from the spectrum that the Δ-excitation is indeed a simple mode in the nucleus. The data are obtained at Saturne using the magnetic spectrometer SPES4.[15]

An important test case for our study is the p(^{3}He,t)Δ^{++} reaction. The results for this case have been discussed in a recent publication[16]. It is found that the data can be explained in terms of one pion exchange between the target proton and the ^{3}He-t system treated as a particle with spin 1/2 and a formfactor. Such a model accounts well for the absolute cross section as well as details in the resonance shape as demonstrated in Figure 12.

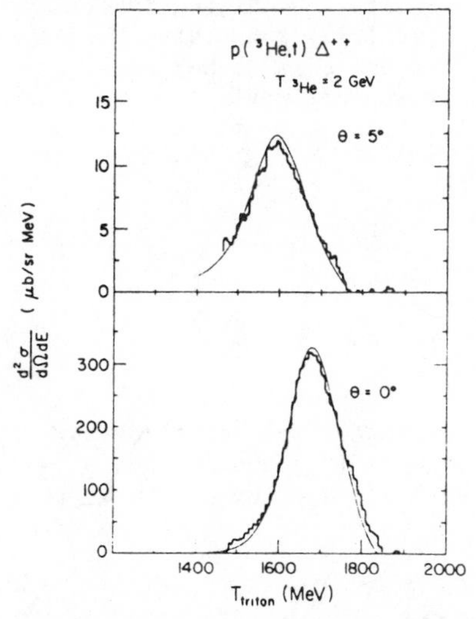

Fig.12. Spectra at 2 angles for the p(^{3}He,t)Δ^{++} reaction. The curves are calculated cross sections in an OPE approximation.

We have rather systematic data on the Δ-production in nuclei with the[15] (^{3}He,t) reaction, at higher energies from Dubna[19]. We summarize some of the results as:

i) A-dependence.
A large Δ cross section is observed for all A with a resonance energy and shape that is independent of A for A≥12. The cross section is going from 100 mb/sr for ^{12}C to 130 mb/sr for ^{40}Ca and then increasing rather slowly to 170 mb/sr for ^{208}Pb. The numbers refer to lab. cross sections at θ=0°.
ii) Angular distributions.
The angular distributions for the Δ-peak are very similar and also similar to that observed for the p→Δ^{++} reaction.
iii) Energy shift relative to quasifree Δ-production.
The Δ-peak is found to be shifted around 35 MeV relative to the energy corresponding to quasifree Δ-formation.

The findings described above are consistent with quasifree Δ-production except for the shift in energy. Several attempts have been made to account for the shift. A recent publication describes a surface-response for the (^{3}He,t) probe[17]).It is shown that ΔN^{-1} correlations of the RPA type only would give rise to energy shifts of 5-10 MeV. The observed shift is then described as an effect of the change of properties of the Δ in the nuclear medium, i.e. a smaller mass and a larger width. This is shown in Figure 13 where the spectrum at $\theta=0°$ for $^{40}Ca(^3He,t)$ at 2 GeV is given together with calculated cross sections for Δ-production.

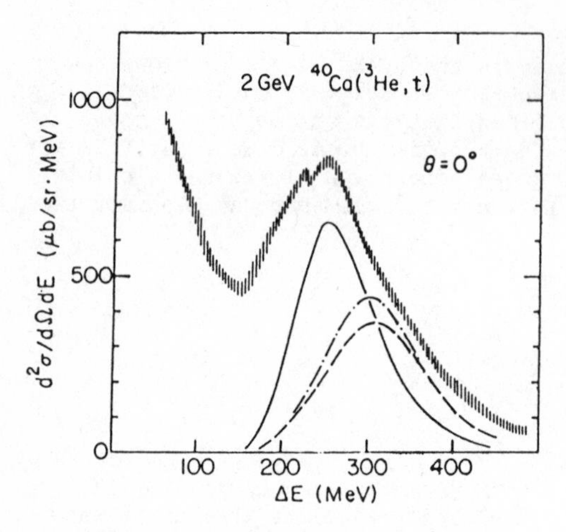

Fig.13. Triton spectrum from $^{40}Ca(^3He,t)$ is given together with calculated spectra. The dashed line is the quasifree case, the dot-dash curve includes Δ-h interactions. In the full drawn curve the mass of the Δ is 35 MeV smaller than 1232 MeV.

5.2 Heavy-Ion Charge Exchange Reactions

Charge exchange reactions with heavy ions as projectiles are also very good probes for the Δ-modes in nuclei. An example is shown in Figure 14 where spectra from the $(^{16}O,^{16}N)$ reaction with ^{12}C and the proton as targets are shown.

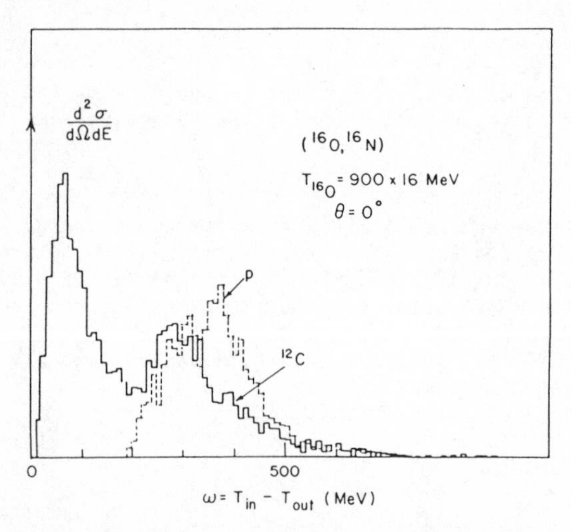

Fig.14. $(^{16}O,^{16}N)$ spectra with ^{12}C and the proton as targets (see fig.15).

A program is underway at Saturne to study such charge exchange reactions at intermediate energies[18]. The SPES4 spectrometer is used again in these experiments. The angular distributions in these reactions are extremely forward peaked, by $\theta \sim 0.5°$ the yield is down an order of magnitude. In the experiments an aperture of $\pm$ 0.85° by $\pm$1.7° is used around $\theta = 0°$ and the whole angular distribution is then contained in this angular range. Since we require definite mass and charge in our detector only bound states in the ejectile can contribute to the yield. This has some important consequences for the observed spectra. In say (^{12}C,^{12}N) only the g.s. would contribute, whereas in (^{12}C,^{12}B) several other states up to 3.4 MeV could be involved in the reactions. With the ^{20}Ne beam the 2 channels are rather different. As for the (^{3}He,t) and (d,2p) reactions the experiment on the proton is an important test case. In Figure 15 spectra are shown with the proton as the target with different projectile-ejectile combinations, but all at 900 MeV per nucleon bombarding energy.

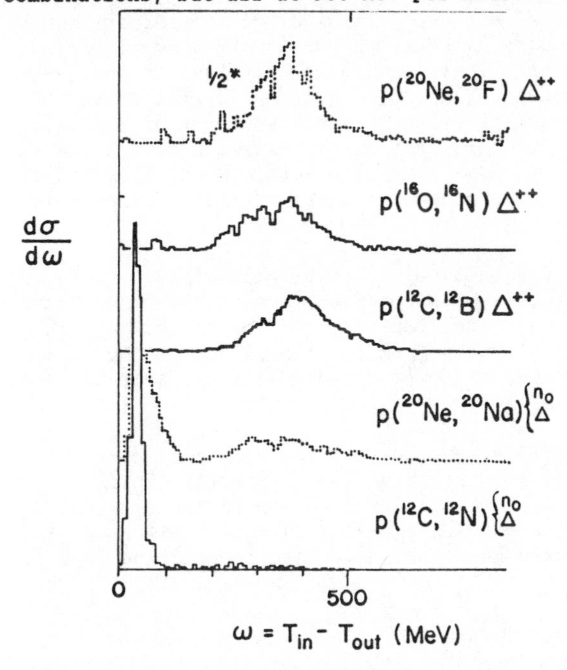

Fig.15. Spectra at 900 MeV per nucleon bombarding energy with the proton as the target. The spectrometer is at $\theta = 0°$, but with an aperture ($\pm$ 0.85° by $\pm$ 1.7°) so the spectra show in fact the total cross section.

We shall not here go into a detailed discussion of the data shown in Figure 15 or similar data on a number of nuclei from ^{12}C to ^{208}Pb. We only point to 2 cases: p(^{12}C,^{12}N)$^{n_0}_{\Delta}$ and p(^{16}O,^{16}N)Δ^{++}. In the first case only the $^{12}C \rightarrow ^{12}N$ 1^{+} g.s. transition contributes and we should know the transition formfactor fairly well. The same is true for the $^{16}O \rightarrow ^{16}N$ transitions. In the latter case several final states contribute, but with a simple shell model structure, and in contrast to the $\ell = 0$ transfer example in $^{12}C \rightarrow ^{12}N$, we have $\ell = 1$ and $\ell = 3$ transitions involved in the $^{16}O \rightarrow ^{16}N$ system. It is exactly this difference in the transition formfactor at the momentum transfer for Δ-formation ($q \sim 0.8$ fm^{-1}) for the 2 projectile-ejectile systems, we take as the main effect to explain the rather dramatic difference in Δ-yield in the 2 cases.

The picture we have used here is really like the model for the p(^{3}He,t)Δ^{++} reaction. The only difference is a different formfactor for the projectile-ejectile systems. The 2 cases discussed above, we consider as important for a more quantitative analysis of all the data.

We conclude this section by noting that also in heavy-ion charge exchange reactions, the Δ-excitation is an important feature of the spectrum. We note that the reactions look like 1-step processes, where an impulse approximation seems to apply. We see a shift in the energy of the Δ in the nuclear medium, as we have seen it in the (^{3}He,t) and (d,2p) reactions.

6. Summary

The (p,n) experiments at intermediate energies have led to a breakthrough in the understanding of spin modes in nuclei. The missing Gamow-Teller strength in the shell-model region is an intriguing problem. Is the quen ching due to coupling to Δ degrees of freedom or to 2p-2h states over a wide energy region? both interesting effects. The simplicity of the (p,n) spectra has prompted a number of other studies of charge exchange reactions at intermediate energies. It seems that also with composite particles these reactions are simple one-step processes with a striking selectivity for spin excitations.

The quasielastic peak as observed in (^{3}He,t) is shifted relative to (e,e') data on this peak. Such a shift is not surprising, but on the other hand not observed before. The findings described here are to some extent contrary to the results in the ($\vec{p}$,$\vec{p}$') data from Los Alamos, where the ratio of longitudinal to transverse response over the quasielastic peak is found to be close to 1 (at $q = 1.7\ fm^{-1}$).

We interpret the shift as an effect of pionic correlations, a reminiscent of pion-condensation in nuclei. We shall have to confirm this shift in other reactions than (^{3}He,t). Here the ($\vec{d}$,2p) reaction looks very interesting as a probe of the spin structure of low-lying as well as Δ excitations. Also in the quasielastic peak region we have interesting data at different momentum transfers.

We have seen that not only the (^{3}He,t) and (d,2p) reactions but also charge exchange with heavy ions show the Δ-excitation as a very prominent feature of the spectra. We see in all cases a shift of the Δ-peak in the medium, a real pionic effect. The future for the study of spin-response in nuclei seems very promising.

Acknowledgements

The material presented here represents the work of a large collaboration involving people from many laboratories. Most of the recent work is done at Saturne, and I am particularly indebted to M. Bedjidian, D. Contardo, J.Y. Grossiord, A. Guichard, R. Haroutunian and J.R. Pizzi from IPN, Lyon, D. Bachelier, J.L. Boyard, T. Hennino and M. Roy-Stephan from IPN, Orsay, P. Radvanyi and J. Tinsley from LN Saturne, C. Goodman from Indiana, I. Bergqvist, A. Brockstedt, and P. Ekström from Lund, and C. Ellegaard and J.S. Larsen from the Niels Bohr Institute. This work has been supported in part by the Danish Natural Science Research Council.

References

1) D.E. Bainum, J. Rapaport, C.D. Goodman, D.J. Horen, C.C. Foster, M.B. Greenfield, and C.A. Goulding, Phys. Rev. Lett. 44 (1980) 1751.

2) C. Goodman, Nucl. Phys. A374 (1982) 241c; and reference therein.

3) C. Gaarde, Nucl. Phys. A396 (1983) 127c; and references therein.

4) T.N. Taddeucci, T.A. Carey, C. Gaarde, J. Larsen, C.D. Goodman, D.J. Horen, T. Masterson, J. Rapaport, T.P. Welch, and E. Sugarbaker, Phys. Rev. Lett. 52 (1984) 1960.

5) C.D. Goodman, R.C. Byrd, I.J. van Heerden, T.A. Carey, D.J. Horen, J. Larsen, C. Gaarde, J. Rapaport, T.P. Welch, E. Sugarbaker, and T. Taddeucci, Phys. Rev. Lett. 54 (1985) 2060.

6) C. Gaarde, J.S. Larsen, A.G. Drentje, M.N. Harakeh and S.Y. van der Werf, Phys. Rev. Lett. 46 (1981) 902.

7) O. Häusser, private communication.

8) J. Rapaport, T. Taddeucci, T.P. Welch, C. Gaarde, J. Larsen, D.J. Horen, E. Sugarbaker, P . Koncz, C.C. Foster, C.D. Goodman, C.A. Goulding, an T. Masterson, Nucl. Phys. A410 (1983) 371.

9) B. Desplanques and S. Noguera, Phys. Lett. B173 (1986) 23.

10) P. Barreau, M. Bernheim, J. Duclos, J.M. Finn, Z. Meziani, J. Morgenstern, J. Mougey, D. Royer, B. Saghai, D. Tarnowski, S. Turck-Chieze, M. Brussel, G.P. Capitani, E. De Sanctis, S. Frullani, F. Garibaldi, D.B. Isabelle, E. Jans, I. Sick and P.D. Zimmerman, Nucl. Phys. A402 (1983) 515.
Z. Meziani, Thesis, 1984, Université de Paris-Sud, Orsay.

11) J.S. O'Connell, W.R. Dodge, J.W. Lightbody, Jr, X.K. Maruyama, J.O. Adler, K. Hansen, B. Schröder, A.M. Bernstein, K.I. Blomqvist, B.H. Cottman, J.J. Comuzzi, R.A. Miskimen and B.P. Quinn, Phys. Rev. Lett. 53 (1984) 1627.
B. Schröder, private communication.

12) W.M. Alberico, M. Ericson and A. Molinari, Nucl. Phys. A379 (1982) 429 and references therein.

13) D.V. Bugg and C. Wilkin, Phys. Lett. 152B (1985) 37 and 154B (1985) 243.

14) D.V. Bugg and C. Wilkin, private communication.

15) D. Contardo, M. Bedjidian, J.Y. Grossiord, A. Guichard, R. Haroutunian, J.R. Pizzi, C. Ellegaard, C. Gaarde, J.S. Larsen, C. Goodman, I. Bergqvist, A. Brockstedt, L. Carlén, P. Ekström, D. Bachelier, J.L. Boyard, T. Hennino, J.C. Jourdain, M. Roy-Stephan, M. Boivin and P. Radvanyi, Phys. Lett. 168B (1986) 331.

16) C. Ellegaard, C. Gaarde, J.S. Larsen, V. Dmitriev, O. Sushkov, C. Good man, I. Bergqvist, A. Brockstedt, L. Carlén, P. Ekström, M. Bedjidian D. Contardo, J.Y. Grossiord, A. Guichard, R. Haroutunian, J.R. Pizzi, Bachelier, J.L. Boyard, T. Hennino, M. Roy-Stephan, M. Boivin, and P. Radvanyi, Phys. Lett. 154B (1985) 110.

17) H. Esbensen and T.S.H. Lee, Phys. Rev. C32 (1985) 1966.

18) D. Bachelier, J.L. Boyard, T. Hennino, J.C. Jourdain, M. Roy-Stephan, C. Contardo, J.Y. Grossiord, A. Guichard, J.R. Pizzi, P. Radvanyi, J. Tinsley, C. Ellegaard, C. Gaarde and J.S. Larsen, Phys. Lett. B172 (1986) 23.

19) V.G. Ableev, G.G. Vorob'ev, S.M. Eliseev, S.A. Zaporozhets, V.I. Inozemtsev, A.P. Kobushkin, A.B. Kurepin, D.K. Nikitin, A.A. Nomofilov, N.M. Piskunov, I.M. Sitnik, E.A. Strokovskii, L.N. Strunov and V.I. Sharov, Sov. Phys. JETP Lett. 40 (1984) 763.

Inst. Phys. Conf. Ser. No. 86
Paper presented at Int. Nucl. Phys. Conf., Harrogate, UK, 1986

Clustering aspects of nuclear structure and reactions

R. R. Betts

Physics Division, Argonne National Laboratory, Argonne, Illinois U.S.A.

W. D. M. Rae

Department of Nuclear Physics, Oxford University, Oxford, England

Abstract. Some aspects of clustering phenomena in nuclear structure and reactions are reviewed. Particular emphasis is placed on the relationship between cluster-like states and shape-isomeric states which arise as a result of shell effects in deformed potentials. Some recent evidence in favour of this connection is presented and discussed.

The notion of clustering in nuclei dates from the earliest days of nuclear physics following the observation of the spontaneous emission of α-particles from heavy nuclei and the theoretical investigations of this phenomenon by Gamow (1930). The idea of a cluster model of light nuclei was first proposed by Hafstad and Teller (1938) and the structure of many light nuclei has subsequently been interpreted in terms of well defined cluster configurations (Morinaga 1956). This concept has an obvious appeal as a simplification of the nuclear many-body problem and the idea of well-defined clusters of nucleons interacting via some effective cluster-cluster interaction is part of our picture of many aspects of nuclear structure.

The experimental evidence supporting these general ideas is abundant. Examples of states possessing unusually large overlaps with nuclear clusters are found in both resonance scattering studies populating unbound states of the final nucleus and cluster transfer reactions to bound and unbound states. Perhaps the classic example is found in the nucleus ^{20}Ne studied by reactions such as α-particle resonance scattering on ^{16}O and α-particle transfer reactions. A spectrum (Anantaraman 1979) of the ^{16}O(^{6}Li,d) α-transfer reaction is shown in Fig. 1. The reaction is highly selective, showing strong population of relatively few levels despite the known complexity of the ^{20}Ne level scheme above 6 MeV of excitation. Particularly noticable is the population of the ground-state rotational band (0^+, 0.0; 2^+, 1.63; 4^+, 4.25; 6^+, 8.78 and 8^+, 11.95 MeV) as well as the $K^\pi=0^-$ band beginning with the 5.79 MeV $J^\pi=1^-$ level. This splitting between the positive and negative parity bands indicates the importance of the microscopic structure underlying such cluster states which in the extreme cluster model should lie on the same rotational sequence.

The location of $K^\pi=0^+$ bands in ^{20}Ne with large reduced widths for decay into an α-particle and ^{16}O in its ground and excited states are shown in

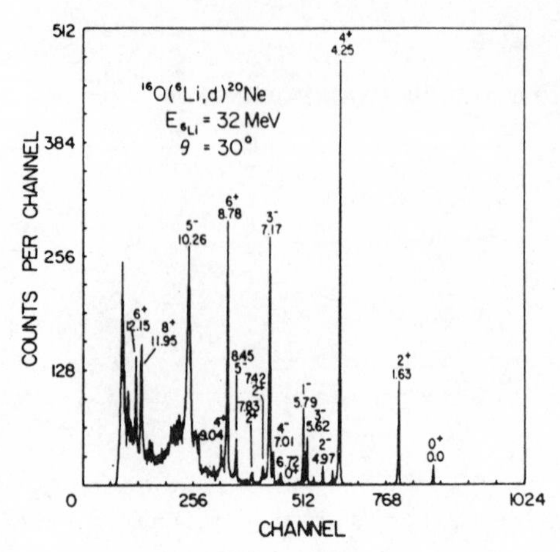

Fig 1. Spectrum of the $^{16}O(^6Li,d)^{20}Ne$ reaction (Anantaraman 1979).

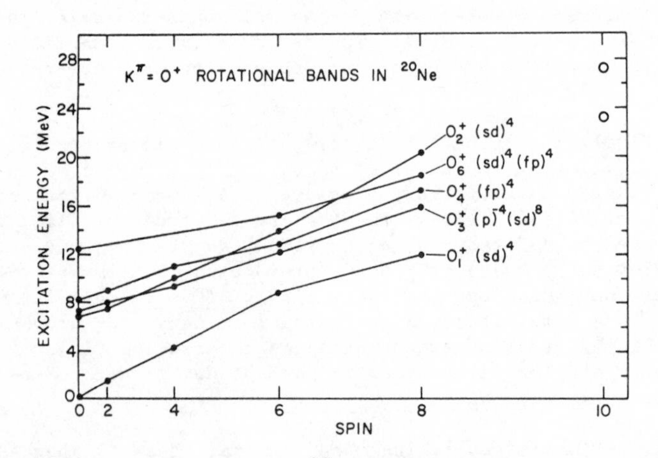

Fig. 2. $K^\pi=0^+$ cluster bands in ^{20}Ne (Hindi 1983). The location of the lowest 10^+ states (Allcock 1986) is indicated by open circles.

Fig. 2. Also indicated are the proposed dominant shell-model configurations associated with each of the bands (Hindi 1983). Of considerable interest in this respect has been the search for higher spin members of these rotational bands and, in particular, the location of the lowest $J^\pi=10^+$ level. If the extent which the cluster exists is determined by the shell-model sub-space from which its wavefunction is composed, each rotational band should terminate at the maximum spin that can be constructed for an S=0, T=0 cluster within the sub-space. In this case, for the $(sd)^4$ ground-state rotational band, we expect $J_{MAX}=8$. Much effort has been devoted to this problem over the years and in a contribution to this conference (Allcock 1986) a search for 10^+ states decaying to the ground-state of ^{16}O is reported. No $J^\pi=10^+$ states are found below 23 MeV but strength is found in a narrow peak near 23.2 MeV and in a broad peak centered at 27.4 MeV -- both these results are in accord with earlier work (Artemov 1977). The locations of these two states are shown in Fig. 2. It is clear that they do not lie on any reasonable extrapolation of the ground-state rotational band, but are most likely members of higher lying bands whose shell-model configurations allow them to have spin 10 members. It thus appears that the ground-state band of ^{20}Ne does in fact terminate at J=8 indicating the importance of the shell structure in these apparently highly clustered bands.

Information on states with large cluster decay widths involving other than α-particles came with the beginnings of the study of heavy-ion reactions. In some of the very first studies of the energy dependence of heavy-ion scattering and reactions evidence was found for the existence of long-lived resonance states with large overlaps with complex nuclear clusters. I refer of course to the experiments on ^{12}C+^{12}C and other light systems by Bromley, Kuehner and Almquist (1960). Subsequent to this, similar phenomena have been observed in much heavier systems (Braun-Munzinger 1977, Betts 1984) and, indeed, whenever accelerator and experimental technology has allowed such studies in a new region of target and projectile mass, evidence has been found for the formation of long-lived, cluster-like states at high excitation energy and in many cases at high angular momentum.

Examples of this kind of behaviour for the ^{12}C+^{12}C (Reilly 1973) and ^{16}O+^{16}O systems (Maher 1969) are shown in Fig. 3 where the the θ_{cm}=90° elastic scattering cross-sections are shown as a function of center-of-mass bombarding energy. In the case of ^{16}O+^{16}O, the broad structures observed are generally interpreted as a series of potential or shape resonances formed in pockets which occur in the ion-ion potential -- the absorption for the surface partial waves being weak enough to keep the resonance widths small. Studies of the scattering of ^{16}O+^{16}O in the

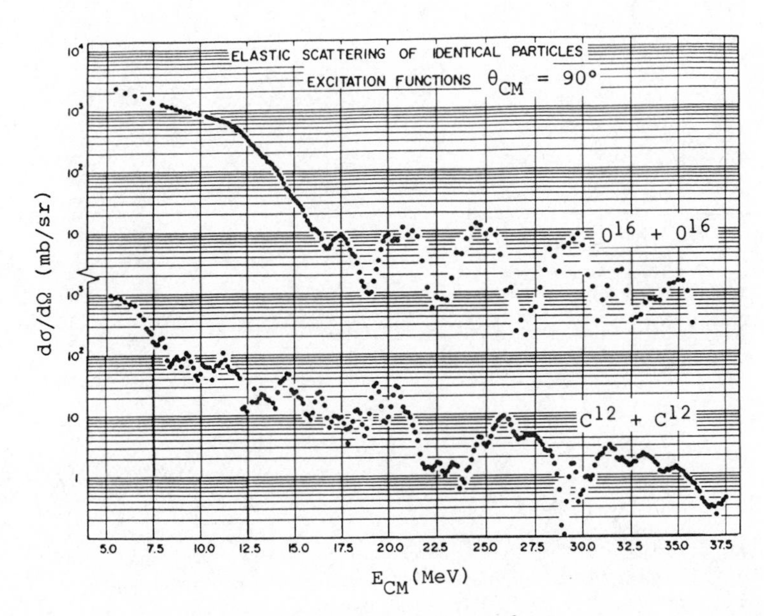

Fig. 3. Energy dependence of the θcm=90° elastic scattering of ^{12}C + ^{12}C (Reilly 1973) and ^{16}O + ^{16}O (Maher 1969).

framework of the Generator Coordinate method (Langanke 1981) have led to the identification of these broad resonances as a rotational band with a structure closely related to the lowest states allowed by the Pauli principle which can be formed in the compound nucleus ^{32}S from two ^{16}O nuclei. Consideration of this in rather general terms, according to the ideas of Harvey (1975), lead to some interesting insights into the general problem of resonance formation in heavy-ion collisions. In this picture, the two separated ^{16}O nuclei are described by simple harmonic oscillator wavefunctions as shown in Fig. 4. The two nuclei approach each other along the z axis which connects the centers of the two nuclei. As the two nuclei overlap the x and y degrees of freedom are assumed to remain unchanged. The number of oscillator quanta in the z direction, however, has to change to satisfy the Pauli principle. In the case of $^{16}O+^{16}O$ coalescing to form ^{32}S, this results in the formation of a state with four particles (2n-2p) in the N=3 shell outside a ^{28}Si core. This so-called "diabatic" configuration is that which is expected to be formed in the early stages of the collision process as it represents the configuration reached by applying the minimum changes to the wavefunctions of the colliding nuclei consistent with the Pauli principle. (These ideas are supported by investigations (Norenberg 1985) into the nature of dissipation in collisions of much heavier systems). It is now interesting to ask to which shape this diabatic configuration will relax. This may be answered by considering the distribution of oscillator quanta between the x, y and z directions and then applying an equilibrium condition (Bohr 1975) to obtain the shape. For the oscillator configuration shown in Fig. 4

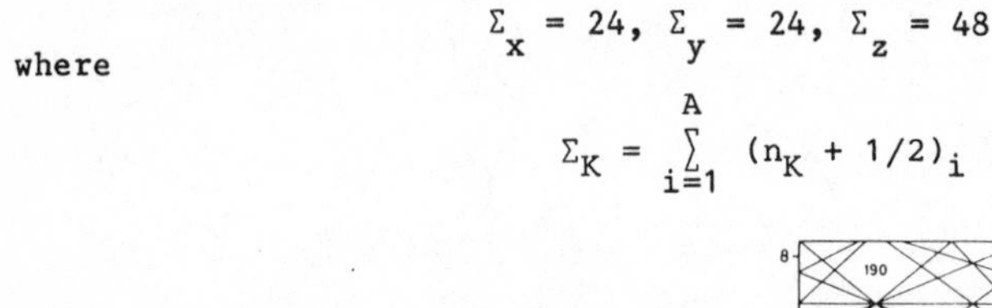

$$\Sigma_x = 24, \ \Sigma_y = 24, \ \Sigma_z = 48$$

where

$$\Sigma_K = \sum_{i=1}^{A} (n_K + 1/2)_i$$

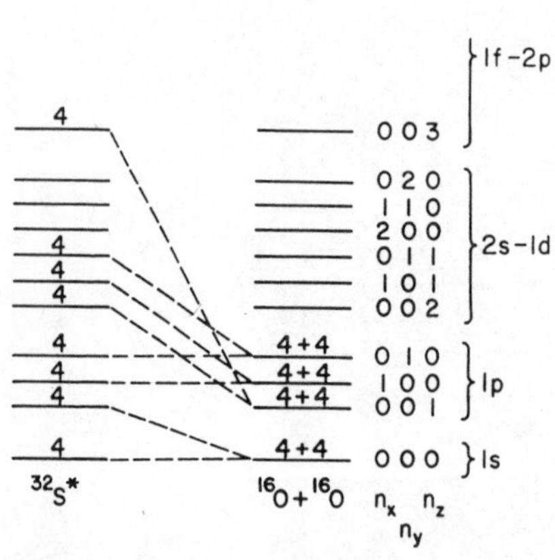

Fig. 4. Schematic drawing of the excited state in ^{32}S reached in the diabatic collision of $^{16}O+^{16}O$ (Harvey 1975).

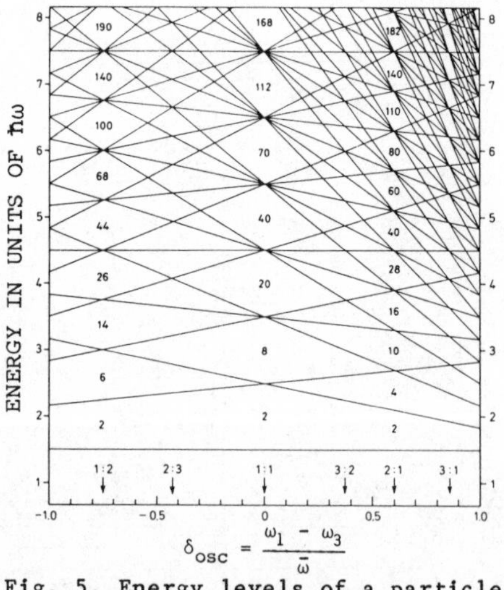

Fig. 5. Energy levels of a particle in a deformed axially symmetric harmonic oscillator. The energies are measured in units of units of $\bar{\omega} = (2\omega_\perp + \omega_3)/3$.

the equilibrium condition

$$\Sigma\omega_x = \Sigma\omega_y = \Sigma\omega_z$$

then gives

$$\omega_x : \omega_y : \omega_z = 2:2:1$$

which corresponds to an axially symmetric ellipsoid with r.m.s. axes in the ratio

$$\bar{x} : \bar{y} : \bar{z} = 1:1:2$$

This shape is of particular significance in the present case for, as shown in Fig. 5, this ratio of oscillator frequencies corresponds (Wong 1970) to a deformed shell closure which thus, as can be seen in Fig. 6, gives rise to a secondary minimum in the potential energy surface of ^{32}S (Ragnarsson 1981). These ideas, which as outlined above are for angular momentum zero, can be extended to rotating systems by cranking and, for not too high rotational frequencies or too large deformations, can be shown to lead to the same conclusions as is shown in Fig. 6 for spin 8.

The configuration discussed above corresponds to an $^{16}O+^{16}O$ cluster-like band with 2N+L=24 which has been identified in the GCM calculations of Langanke, Stademann and Timm (1981) with a rotational band of bound states in ^{32}S. The observed broad structures are associated with a band with 2N+L=28 which differs only from the lowest configuration by promotion of four nucleons to the next major shell. We see therefore that the entrance channel resonances in $^{16}O+^{16}O$ are closely related to deformed shape isomers in the compound nucleus and may be thought of as simple excitations based on the lowest state of ^{32}S in the secondary minimum with a 2:1 axis ratio.

Returning to Fig. 3, the data for $^{12}C+^{12}C$ show considerably more complexity than for $^{16}O+^{16}O$. Within the framework of the above discussion this can be seen to arise from the non-spherical nature of the ^{12}C ground-state which leads to an orientation dependence of the collision process (Harvey 1975). It is also clear from the potential energy surface (Leander 1975) for ^{24}Mg shown in Fig. 7 that this too possesses a richness of structure. Insight into the cluster structure of the potential energy minima can be obtained by comparing the results

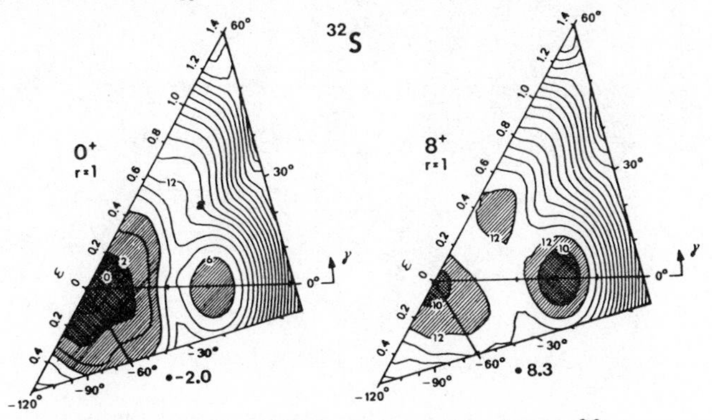

Fig. 6. Calculated potential energy surfaces for ^{32}S at spin 0 and spin 8 (Ragnarsson 1981).

of a recent calculation (Marsh 1986) of ^{24}Mg within the cranked cluster model. In this calculation, based on the cluster model of Brink (1966), the positions of six 0s α-clusters were allowed to vary without constraint and local minima in the total energy of the system searched for. The nucleon-nucleon force used was the Brink-Boeker B1 force (1967). The density contours of the solutions with large deformations are shown in Fig. 7 associated with the appropriate Nilsson-Strutinsky minima -- the association having been made using estimates of the deformation parameters from the cluster calculation density distributions and their shell model configurations. It is important to note that the stability of these deformed cluster configurations arises as a consequence of the shell structure of the deformed mean field and the spin-isospin degeneracy of the single-particle orbits and not from any inherent stability of the clusters themselves.

The triaxial minimum (ε=1.26, γ=42°) shows a distinct planar structure in which can be clearly discerned two triangular groups of three clusters. This structure of three is the same structure found for the ^{12}C ground-state in this model before parity projection, and we therefore expect this configuration to have a large overlap with two ^{12}C nuclei in their ground-states and, when rotating, two ^{12}C nuclei in the excited 2^+ state with the spins aligned parallel to the rotation axis. This connection has been pointed out by Mosel (1981) who noted the effects of the relative orientation of the two ^{12}C nuclei on the potential calculated in a two-center shell model (Chandra 1978). It was

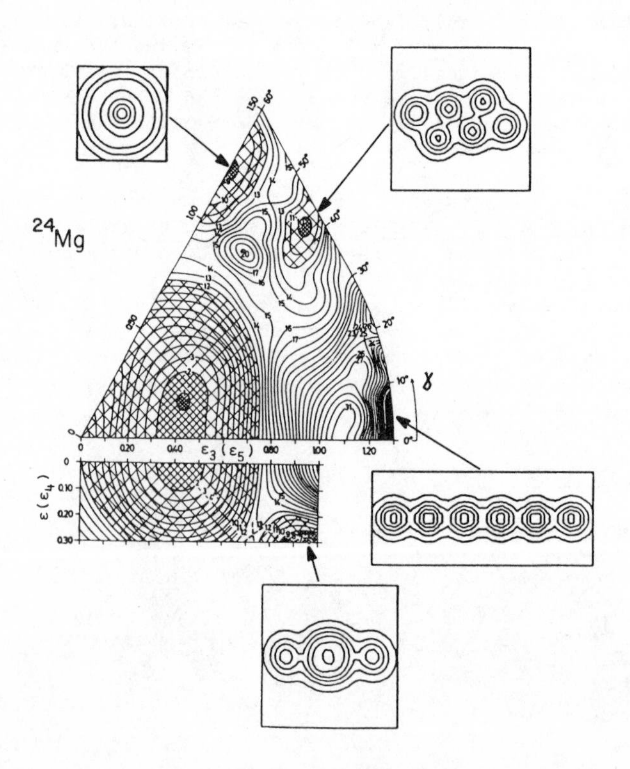

Fig. 7. Potential energy surface for ^{24}Mg (Leander 1975) together with density contours for the stable cluster configurations (Marsh 1986).

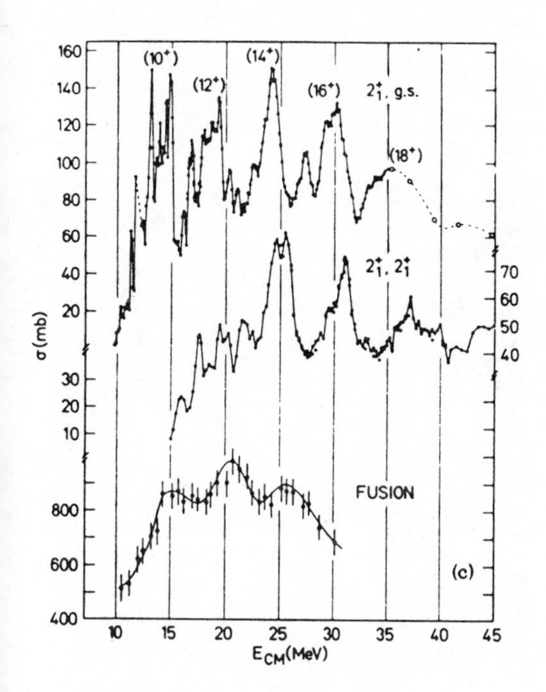

Fig. 8. Cross-sections for the single and mutual inelastic scattering of $^{12}C+^{12}C$ shown as a function of center-of-mass bombarding energy (Cormier 1978).

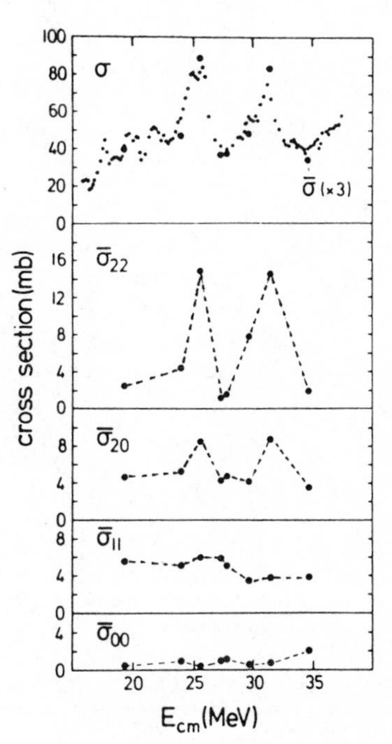

Fig. 9. Decomposition of the $^{12}C+^{12}C$ mutual inelastic scattering cross-section into contributions from different substates $\bar{\sigma}(m_1m_2)$ $\bar{\sigma}(22)$ corresponds to the aligned configuration (Konnerth 1985).

further noted by Mosel that dynamical coupling effects in the entrance channel tended to align the ^{12}C nuclei in the appropriate fashion and thus enhance the population of the triaxial isomeric shape. This configuration may therefore be associated with the band of broad resonances observed (Cormier 1977, 1978) in the elastic and single and mutual inelastic scattering of $^{12}C+^{12}C$, data for which is shown in Fig. 8, and which have been seen to decay with the spins of both ^{12}C nuclei aligned (Konnerth 1985) as suggested above and as displayed in Fig. 9.

Of particular interest is the prolate minimum with $\varepsilon=1.0$, $\gamma=0°$ and $\varepsilon_3=0.3$. The cluster model also predicts this minimum to be reflection asymmetric -- the density contours shown are those obtained after parity projection. This minimum lies somewhat below the triaxial minimum in energy and from the density contours and shell model configurations appears to have a structure based on a ^{12}C nucleus in its ground state together with another in the chain-like excited 0^+ state. A similar shape has been found in (Umar 1986) Hartree-Fock calculations and studied (Strayer 1984) in Time-Dependent Hartree-Fock calculations of $^{12}C(g.s.) + ^{12}C(0_1^+)$. The energy of this minimum suggests that this configuration be identified with the narrow resonances observed (Bromley 1960, Erb 1980) in $^{12}C+^{12}C$ in the vicinity of the Coulomb barrier (Fig. 10). Two questions arise, however, as a result of this interpretation. Firstly, the proposed cluster configuration has no obvious overlap with two ^{12}C nuclei in their ground states which leaves the ^{12}C

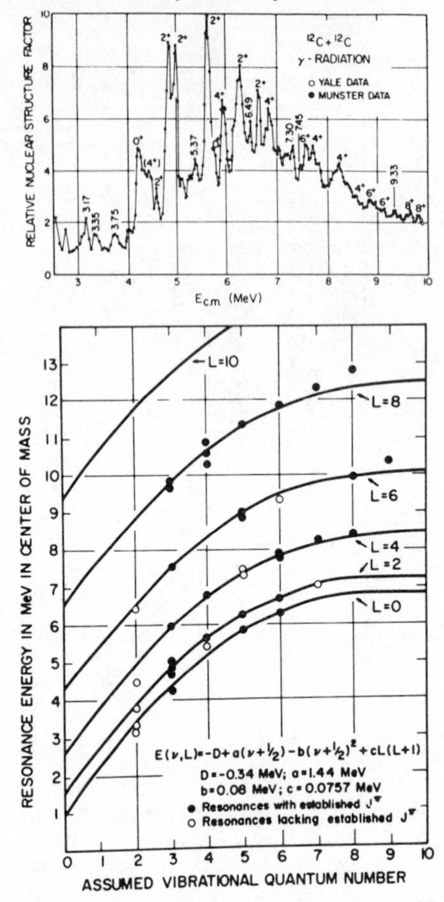

Fig. 10. Energy dependence of the low-energy $^{12}C+^{12}C$ reaction cross-section (Erb 1980) together with a rotation-vibration fit to the observed spectrum (Erb 1981).

width of the barrier resonances to be accounted for and secondly, the multiplicity of observed resonances in this energy region (see Fig. 10) require additional degrees of freedom beyond the simple rotation of this shape. A resolution of the first question comes from the work of Takigawa and Arima (1971) who found that, in order to reproduce the electromagnetic decay properties of these states in ^{12}C, a 20% mixing of the excited 0^+ chain configuration was required in the ^{12}C ground-state which would then give the barrier resonances the desired ^{12}C decay properties. The question of other degrees of freedom coupled to the rotation of this configuration has been addressed in the Hartree-Fock calculations of Umar and Strayer (1986) who find a low frequency isoscalar monopole vibrational mode ($\hbar\omega \simeq 1$ MeV) corresponding to the vibration of the excited ^{12}C through the ^{12}C in its ground-state as shown in Fig. 11. A similar result is obtained in the cranked cluster model. The energy of this mode agrees well with that extracted from a vibration-rotation model fit (Erb 1981) to the spectrum of barrier resonances shown in the lower portion of Fig. 10 and it can therefore be concluded with reasonable certainty that the observed complexity results from the coupling of this vibrational mode to the rotation.

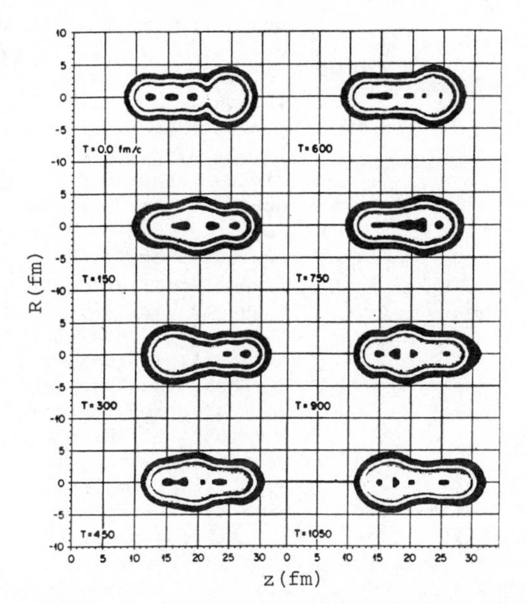

Fig. 11. Isoscalar density contours obtained in a TDHF calculation of the time evolution of a $^{12}C(g.s.)+^{12}C(0_1^+)$ configuration (Strayer 1984).

Some observations are relevant to future experiments seeking to confirm the above speculations. The wavefunction of the proposed configuration for the barrier resonances has an α-particle in the (sd) shell and an α-particle in the (fp) shell outside a closed ^{16}O core -- in agreement with that obtained in the Harvey prescription for a $^{12}C(0_1^+)$ incident along the symmetry axis of the oblate $^{12}C(g.s.)$. This configuration has not only an obviously large overlap with the ^{16}O ground-state and ^8Be but also the relationship between the $^{12}C+^{12}C$ barrier resonances and the various rotational bands in ^{20}Ne now becomes clear. In particular, we would expect to see enhanced α-decays to the ground-state, 0_4^+ and 0_6^+ rotational bands which are thought to be based on configurations which would have large overlap with the above $^{12}C+^{12}C$ isomer. This expectation is qualitatively in agreement with the suggestions of Cosman (1981) and Ledoux (1984) who, however, assigned the triaxial minimum to the $^{12}C+^{12}C$ barrier resonances. Finally, the reflection asymmetric shape found in the Nilsson-Strutinsky and cluster model calculations raises the question of the existence of negative parity members of the bands built on this configuration. These would not, of course, have been observed in the $^{12}C+^{12}C$ entrance channel due to the spin zero nature of ^{12}C but could, for example, be observable via non-symmetric entrance and exit channels such as $^{20}Ne+\alpha\rightarrow^{16}O+^8Be$.

The extremely deformed minimum at $\varepsilon=1.25$, $\gamma=0°$ in the Nilsson Strutinsky calculations also appears as an α-particle chain structure in the cluster model calculations. Similar structures appear in all A=4n nuclei. In ^{24}Mg this structure should decay into ^{12}C nuclei in the chain-like excited 0^+ state with subsequent decay into three α-particles thus giving rise to "jet" like events. The location of these states will involve the kinematic reconstruction of a multiparticle final state and their investigation is an exciting experimental challenge.

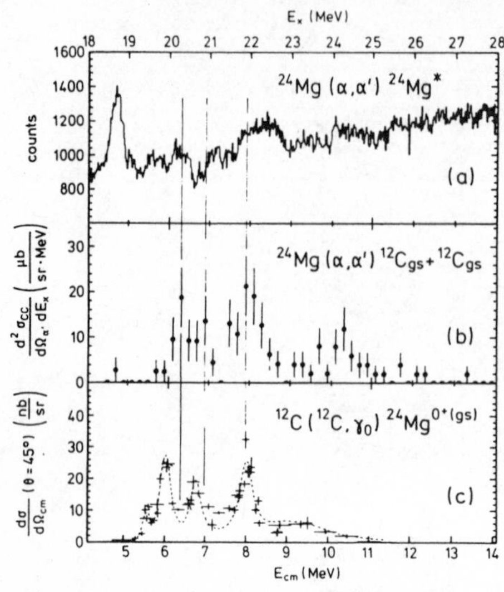

Fig. 12. Comparison of inelastic α-scattering on ^{24}Mg with coincidence events for ^{12}C(g.s) + ^{12}C(g.s.) and with the radiative capture of ^{12}C+^{12}C (Lawitzki 1986).

One important question involves the relationship of these extremely deformed states with the less deformed "normal" states of ^{24}Mg. Obviously the mixing between the two must be rather small in order to preserve the observed narrow widths. Some new insight into this question has come from studies of the "fission" of ^{24}Mg following inelastic excitation. In one of these, the results of which are shown in Fig. 12, the decay of ^{24}Mg into two ^{12}C nuclei following inelastic excitation by α-particles was measured (Lawitzki 1986). The data show some correlation with fragments of the giant quadrupole resonance seen in the inelastic scattering singles spectrum and with the results of ^{12}C(^{12}C,γ_o)^{24}Mg radiative capture (Nathan 1981) which must also proceed through 2^+ states in the continuum. Looking at the Nilsson-Strutinsky diagram in Fig. 7 we can see that this

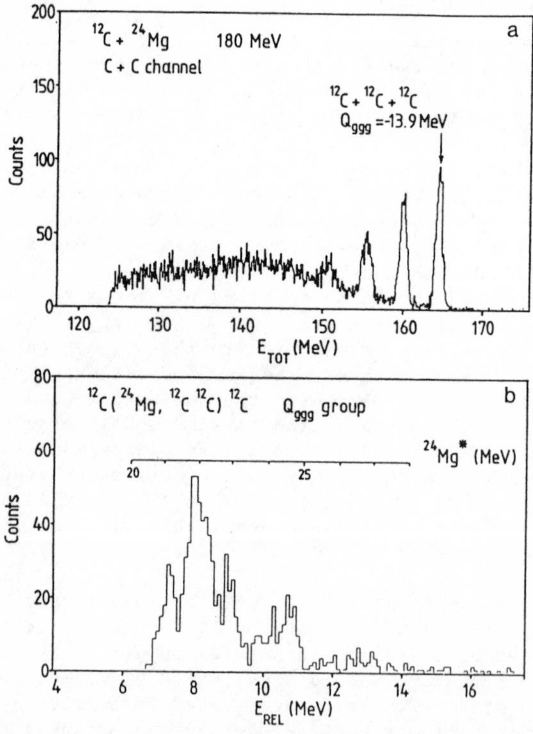

Fig. 13. (a) Total energy spectrum for fission of ^{24}Mg into two ^{12}C nuclei following inelastic scattering. (b) Excitation energy spectrum for ^{24}Mg fissioning into ^{12}C(g.s.) + ^{12}C(g.s.) (Fulton 1986).

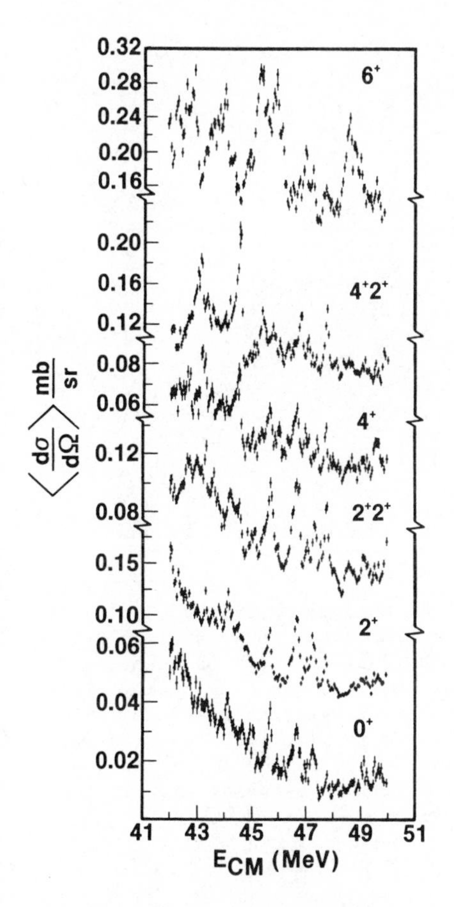

Fig. 14. Angle-average cross-sections for the large-angle scattering of $^{24}Mg+^{24}Mg$ (Zurmühle 1983).

reaction may proceed either through the K=0 component of the GQR via the prolate deformed minimum or through the K=2 component which might mix with the oblate minimum at $\varepsilon=1.23$, $\gamma=60°$. These questions remain to be explored in more detail.

An interesting new technique (Fulton 1986, Wilczynski 1986) for studying this connection has recently been reported. A ^{24}Mg beam was inelastically excited by collision with a ^{12}C target. Coincident ^{12}C nuclei were then detected at forward angles and each event kinematically reconstructed. For events in which all three ^{12}C nuclei were emitted in their ground-states the excitation spectrum of ^{24}Mg (Fig. 13) shows distinct structure similar to that observed in the "fission" following inelastic scattering and the radiative capture. This technique has many experimental advantages and its future pursuit should prove most profitable, especially as it might then prove possible to study deformed configurations which are otherwise inaccessible.

Phenomena similar to those observed in systems such as $^{12}C+^{12}C$ have also been seen in much heavier systems such as $^{24}Mg+^{24}Mg$ (Zurmühle 1983) and $^{28}Si+^{28}Si$ (Betts 1979, 1981, 1981a). An example is shown in Fig. 14, where excitation functions for the angle-averaged large-angle scattering of $^{24}Mg+^{24}Mg$ are shown. The observed narrow structures appear in a correlated fashion in all the reaction channels and have been shown (Saini 1984) to be inconsistent with statistical fluctuations and thus correspond to more or less isolated resonances in the composite system. In one case (Betts 1986) a series of measurements of elastic scattering angular distributions over one of the narrow $^{24}Mg+^{24}Mg$ resonances has been made which together with a constrained phase shift analysis, leads to a spin assignment of J=34. These results imply that the observed resonances correspond to a set of rather high spin states lying somewhat above the yrast line of the compound nucleus ^{48}Cr. A theoretical discussion along the lines of that presented above for $^{12}C+^{12}C$ and $^{16}O+^{16}O$ indicates that for two prolate ^{24}Mg nuclei incident end-on-end the diabatic configuration has an equilibrium shape which is close to the 3:1 shape for which the deformed oscillator potential has a shell gap for 24 nucleons. The effect of this shell gap on the adiabatic potential energy surface (Ragnarsson 1984) of ^{48}Cr is shown in

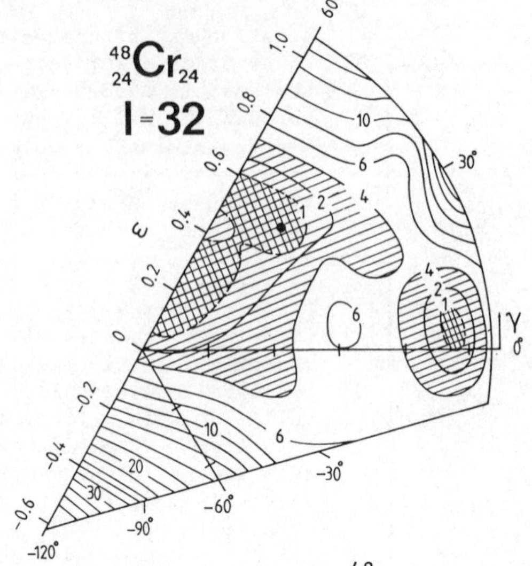

Fig. 15. Potential energy surface for ^{48}Cr at spin 32 (Ragnarsson 1984).

Fig. 15 which clearly displays the deformed minimum. Again it is likely that entrance channel interactions play an important role in aligning the two colliding nuclei thus enhancing the formation of the isomeric configuration. The experimental evidence that the two ^{24}Mg nuclei are aligned in this way comes primarily from the energy spectra (Saini 1986) at large angles, an example of which is shown in Fig. 16. The spectrum is dominated by peaks corresponding to single and mutual excitations of high spin states which, if their spins are aligned perpendicular to the

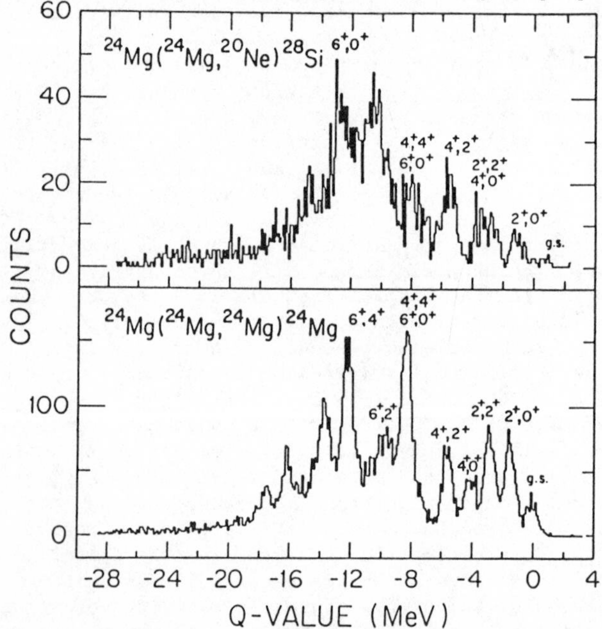

Fig. 16. Energy spectrum of inelastically scattered ^{24}Mg+^{24}Mg measured at large scattering angles (Saini 1986).

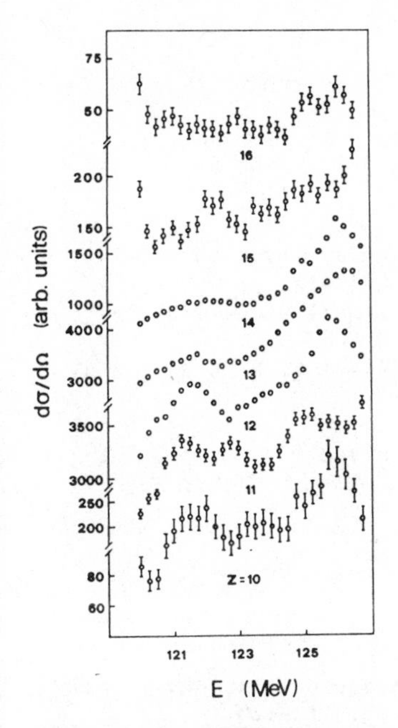

Fig. 17. Excitation functions for different Z fragments produced in the $^{28}Si+^{64}Ni$ reaction (de Rosa 1985).

reaction plane, are favoured by simple kinematic matching conditions. More definitive evidence on this point comes from some preliminary results on direct measurements of the fragment spin alignment (Mattis 1985).

Many of the features observed in the resonance data for heavy systems are similar to those of fully-damped orbiting or "deep-inelastic" collisions. Namely, complete damping of the entrance channel kinetic energy together with the formation of a relatively long-lived intermediate complex which subsequently decays, the fragments emerging with energies characteristic only of the Coulomb repulsion and orbital angular momentum of the system. These kind of processes have been studied over the range of target and projectile combinations but only recently has the detailed energy dependence of such collisions been studied (de Rosa 1985, Cardella 1986, Lucas 1986). Some of these data are shown in Fig. 17 in which excitation functions for different Z fragments produced in the $^{28}Si+^{64}Ni$ reaction are plotted. These data show a number of broad structures which appear in a similar fashion in the excitation functions for several different Z fragments. It thus appears that resonance phenomena and therefore the existence of cluster-like deformed states may occur over an even wider range of nuclei. It is an open question as to whether the diabatic approach discussed above for light systems or the adiabatic approach familiar from fission will be more appropriate for the heaviest systems. The fact that the most recent evidence for fluctuating cross-sections comes from studies of damped collisions, however, does suggest that the former may be more correct.

In summary, the connection between cluster states and shape isomeric states which arise as a consequence of shell effects in the deformed single particle potentials has been discussed. It appears that many of the features observed in the resonance scattering of heavy ions can be understood in terms ofthe interplay of reaction dynamics and the formation of deformed shape-isomers in the compound nucleus. This discussion leads us to assign the $^{12}C + {}^{12}C$ barrier resonances to a reflection asymmetric configuration which has a large parentage with $^{12}C(g.s.) + {}^{12}C(0_1^+)$ and the broad higher energy resonances to a configuration similar to two oblate ^{12}C nuclei lying in the same plane. These results, we believe, firmly establish resonances seen in heavy-ion collisions as a feature of the nuclear structure of the composite system.

This work has been supported by the U. S. Department of Energy, Nuclear Physics Division, under Contract in part W-31-109-Eng-38.

References

Allcock S et al. 1986 contributed paper to this conference.
Anantaraman N et al. 1979 Nucl. Phys. A313 445.
Artemov K P et al. 1977 Sov. J. Nucl. Phys. 26 4.
Betts R R et al. 1979 Phys. Rev. Lett. 43 253.
Betts R R et al. 1981 Phys. Lett. 100B 117.
Betts R R et al 1981a Phys. Rev. Lett. 47 23.
Betts R R 1984 Proc. 5th Adriatic Int. Conf. on Nucl. Phys. ed. N. Cindro et al. (Singapore:World Scientific) pp. 33-52.
Betts R R 1986 Proc. Second Int. Conf. on Nucleus-Nucleus Collisions Nucl. Phys. A447 257c.
Bohr Aa and Mottelson B 1975 Nuclear Structure (Reading, Mass:Benjamin) p. 77.
Bromley D A et al. 1960 Phys. Rev. Lett. 4 365.
Braun-Munzinger P et al. 1979 Phys. Rev. Lett. 38 944.
Brink D M 1966 Proc. Int. School of Phys. "Enrico Fermi" XXXVII ed. C. Bloch (New York:Academic) pp. 247-277.
Brink D M and Boeker E 1967 Nucl. Phys. A91 1.
Cardella G et al. 1986 contributed paper to this conference.
Chandra H and Mosel U 1978 Nucl. Phys. A298 151.
Cormier T M et al. 1977 Phys. Rev. Lett. 38 940.
Cormier T M et al. 1978 Phys. Rev. Lett. 40 924.
Cosman E R et al. 1981 Proceedings of Symp. on Resonances in Heavy Ion Reactions ed. K A Eberhard (Berlin:Springer) pp. 112-128.
de Rosa A et al. 1985 Phys. Lett. 160B 239.
Erb K A et al. 1980 Phys. Rev. C22 507.
Erb K A and Bromley D A 1981 Phys. Rev. C23 2781.
Fulton B R et al. 1986 Proceedings of Symp. on Many Facets of Heavy Ion Fusion, Argonne National Laboratory and contribution to this conference.
Gamow G A 1930 Proc. Roy. Soc. A126 632.
Hafstad L R and Teller E 1938 Phys. Rev. 54 681.
Harvey M 1975 Proc. 2nd Int. Conf. on Clustering Phenomena in Nuclei ed. D A Goldberg et al. (Springfield:U.S. Dept. of Commerce ORO-4856-6) pp. 549-564.
Hindi M et al. 1983 Phys. Rev. C27 2902.
Konnerth D et al. 1985 Phys. Rev. Lett. 55 588.
Langanke K et al. 1981 Phys. Rev. C24 1023.
Lawitzki S et al. 1986 Phys. Lett. 174B 246.
Leander G and Larsson S E 1975 Nucl. Phys. A239 93.
Ledoux R J et al. 1984 Phys. Rev. C30 866.
Lucas R et al. 1986 contributed paper to this conference.
Maher J V et al. 1969 Phys. Rev. 188 1665.
Marsh S and Rae W D M 1986 Phys. Rev. Lett. (in press).
Mattis A et al. 1985 Beschleunigerlabor München Annual Report p. 4.
Morinaga H 1956 Phys. Rev. 101 254.
Mosel U 1981 Proceedings of Symp. on Resonances in Heavy Ion Reactions ed. K A Eberhard (Berlin:Springer) pp. 358-371.
Nathan A M et al. 1981 Phys. Rev. C24 932.
Norenberg 1985 Proc. 2nd La Rabida Summer School on Nuclear Physics ed. M Lozano and G Madurga (Singapore:World Scientific) pp. 492-530.
Ragnarsson I et al. 1981 Physica Scripta 24 215.
Ragnarsson I 1984 (private communication).
Reilly W et al. 1973 Nuovo Cimento 13A 897.
Saini S and Betts R R 1984 Phys. Rev. C29 1769.
Saini S et al. 1986 (to be published).

Strayer M R 1984 Proc. 5th Adriatic Int. Conf. on Nucl. Phys. ed. N Cindro et al. (Singapore: World Scientific) pp. 95-108.
Takigawa N and Arima A 1971 Nucl. Phys. A168 593.
Umar A S and Strayer M R 1986 Phys. Lett. 171B 353.
Wilczynski J et al. 1986 LBL Preprint No. 21415.
Wong C Y 1970 Phys. Lett. 32B 668.
Zurmühle R W et al. 1983 Phys. Lett. 129B 384.

Inst. Phys. Conf. Ser. No. 86
Paper presented at Int. Nucl. Phys. Conf., Harrogate, UK, 1986

Sub-barrier fusion

W Reisdorf

Gesellschaft für Schwerionenforschung, D6100 Darmstadt,F R Germany

Abstract. Measured fusion excitation functions in the vicinity of the barrier reveal nuclear structure effects, due in particular to the coupling of the fusion process to direct-reaction channels.

1. Tunnelling in one dimension

Classically, if one assumes that the fusion process is irreversibly initiated at a fixed internucleus distance r_B, one obtains a simple formula for fusion cross sections (σ_{fus})

$$\sigma_{fus} = \pi r_B^2(1 - V_B/E) \qquad (1.1)$$

This equation predicts that fusion should set in sharply at a threshold energy V_B, the 'barrier', and then approach asymptotically ($E >> V_B$) towards the value πr_B^2. Underlying the derivation is that no conversion of both the orbital angular momentum and relative kinetic energy into internal degrees of freedom of the partner nuclei has taken place in the approach phase. In that case, V_B can be interpreted as the (conservative) relative potential energy V(r) of the system at the 'touching' distance r_B. It is then also an intuitively reasonable further specification of the model to state that r_B coincides with the maximum of this potential as a function of the internucleus distance r. This implies that such a maximum exists.

What do we expect when $E < V_B$? Quantum mechanics teaches us that transmission through a potential barrier can occur at classically forbidden energies via tunnelling. Using the WKB approximation we can calculate the transmission coefficient

$$T = (1 + \exp(S))^{-1} \qquad (1.2)$$

where the 'action' S is given by

$$S^2 = (8\mu/\hbar^2) \times \int_{r_i}^{r_o} [V(r)-E]dr \qquad (1.3)$$

The integration limits r_i and r_o are the inner and outer turning points,respectively, and μ is the reduced mass.

Our earliest 'practical' experience with tunnelling in nuclear physics comes from alpha decay (Gamov 1928). As we shall illustrate with figure 1 (Roeckl 1983), the WKB approximation works extremely well in this case. As a function of neutron number and for even-even nuclei, one sees the reduced quantity

$$w_\alpha \equiv N/(t_\alpha T) \qquad (1.4)$$

where t_α is the measured partial half-life for ground state to ground state ($0^+ \rightarrow 0^+$) alpha decay. N is a normalisation constant adjusted to give $w_\alpha = 1$ for ^{212}Po. The transmission probability T is calculated using eq. (1.2) (Rasmussen 1959) with a standard alpha-nucleus potential (Igo 1958) and the measured total kinetic energy of the decaying system. Chains of isotopes with the element number given in the figure are joined by straight lines to guide the eye. All w_α values can be found within a band, deviating at most by a factor of 10 from the reference value for ^{212}Po. We observe that this systematics includes ^{174}Hf with a half life of 2×10^{15} years, as well as ^{216}Ra with a half life of 0.18 μs, a difference of 29 orders of magnitude. Clearly, the WKB approach has done an excellent job in 'explaining' most of the enormous half life differences. What is left to study can be termed 'nuclear structure'.

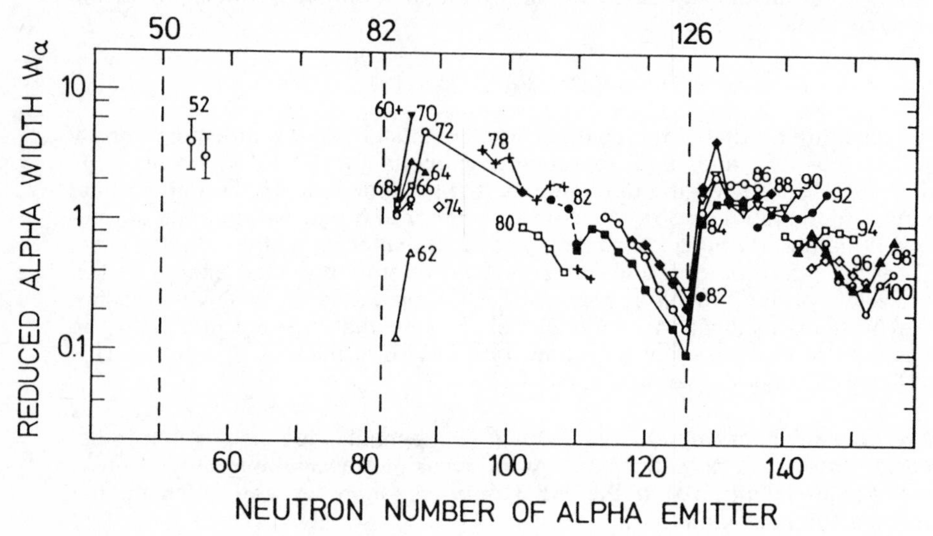

Fig. 1 Reduced alpha widths for even-even nuclei (Roeckl 1983)

In many ways, as we shall see, heavy-ion fusion probabilities can also be 'reduced' with this kind of philosophy.

In studying the fusion probability of massive nuclei below the barrier one may ask whether the WKB approximation with its treatment of only one degree of freedom (the radial distance r) and with simply the reduced mass of the system as inertial parameter is still adequate. In figure 2 we show the fusion excitation function for ^{16}O + ^{12}C (Cujec and Barnes 1976) together with a WKB calculation (solid line) and a calculation using the Hill-Wheeler (1953) analytical formula, which is a specialisation of the WKB method for the case that the potential can be described by an inverted parabola. The potential used had simply the form

$$V(r) = \frac{e^2 Z_1 Z_2}{r} - V_0 C \exp[1.333(C_1 + C_2 - r)/b] + (\hbar^2/2\mu)[L(L+1)/r^2] \qquad (1.5)$$

containing a Coulomb, a proximity (Błocki 1977) and a centrifugal term. The nuclear proximity term is assumed to be adequate for $r > C_1 + C_2 + 1$ fm, where C_1 and C_2 are the central radii, $C = C_1C_2/(C_1+C_2)$ and b is the surface width parameter (1 fm). Since the geometrical parameters can be deduced from electron scattering and other methods of measuring nuclear moments, one has in principle only the strength V_0 left as a free parameter. The value of 74 MeV/fm found necessary to describe the data is almost twice the value proposed by Błocki et al. (1977). A substantial part of the measured data refers to energies below the deduced barrier V_B (7.8 MeV)

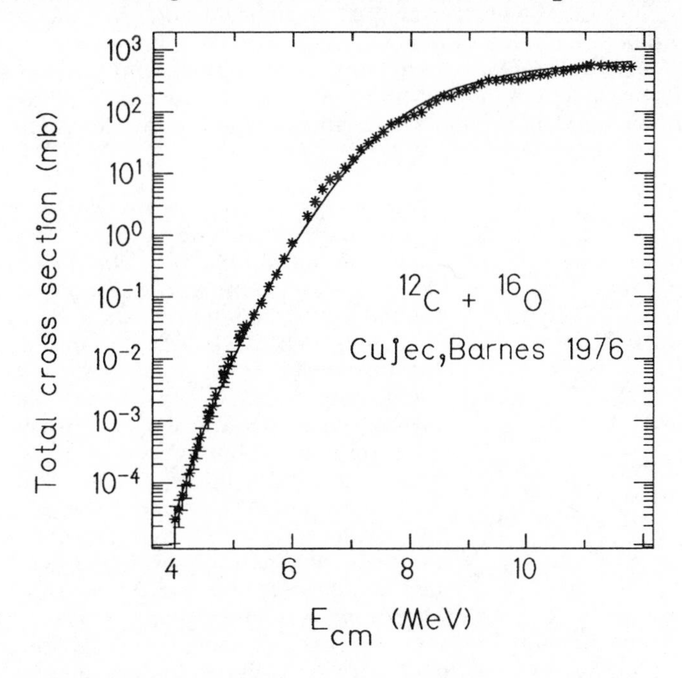

Fig. 2 Fusion excitation function for $^{12}C + ^{16}O$

The barrier transparency for such a heavy-ion system is still relatively high: the ratio of the Hill-Wheeler frequency $\hbar\omega$ to the barrier height, which is 30% in this case, is a good measure of this transparency. The 'contact' point, i.e. the barrier location, is about 3.5 fm outside the sum of the central radii. This large distance, presumably, is the secret for the apparent success of this simple calculation.

An indication that the underlying physics may be more complex comes from the fact that it is rather difficult to find a global potential that describes all 'light' ion data ($A_1, A_2 \leq 20$). Another observation that one can make, is that the $^{16}O + ^{12}C$ data contain some real structure, especially above 6 MeV, which is clearly outside the scope of WKB theory. But aside from this, it remains impressive how well, globally, the calculated cross sections follow the data over seven orders of magnitude.

The reactions just discussed play an important role in astrophysics (Barnes 1985). The systematics for p-shell nuclei are discussed by Cujec (1985).

2. Tunnelling in the presence of rotations and vibrations.

The influence of deformation on fusion excitation functions became convincingly clear with the data of Stokstad et al. (1980) for $^{16}O + ^{148,150,152,154}Sm$ and of Di Gregorio et al. (1986) for $^{16}O + ^{144}Sm$. Part of these data are shown in figure 3. The stable samarium isotopes span the transition from spherical to deformed equilibrium shapes. In figure 3 pure size effects expected from the global increase of nuclear radii with mass have been removed by plotting 'reduced' excitation functions for the heavier Sm isotopes. This procedure will be specified later. While the excitation functions merge at the higher energies, there is a remarkable isotopic trend at the lower energies.

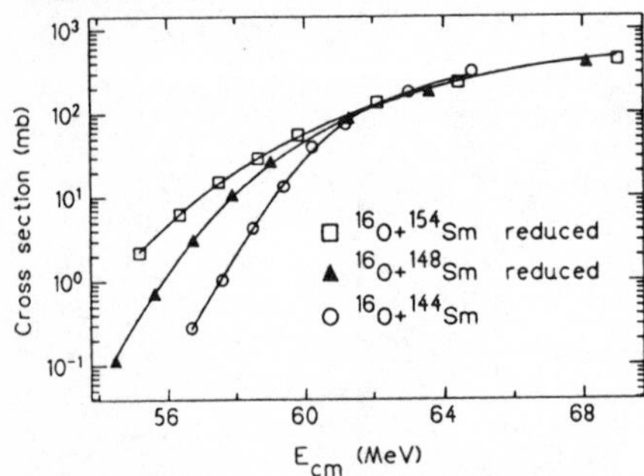

Fig.3 Comparison of fusion excitation functions for $^{16}O + ^{148,154}Sm$ (Stokstad 1980) and $^{16}O + ^{144}Sm$ (Di Gregorio 1986)

We can first fit a 'reference' potential to the 'spherical' case, $^{16}O + ^{144}Sm$ (solid curve through the ^{144}Sm data in figure 3). Then one can use this potential for 'orientation averaging' in the 'deformed' case, $^{16}O + ^{154}Sm$, using the known static quadrupole moment of ^{154}Sm. We used the procedure of Rasmussen (1971) with curvature corrections (Randrup 1978). This is shown in figure 4, solid line. As can be judged by comparison with the dashed curve (zero quadrupole moment), the bulk of the isotopic effect can be accounted for in this very simple way, although a very close look shows that the calculation tends to overestimate the cross sections above the indicated equivalent spherical barrier, while it underestimates them at subbarrier energies.

The 'subbarrier enhancement' of the ^{154}Sm cross sections can be understood qualitatively in a simple way. Because of the strongly non-linear behaviour of the fusion cross sections in the threshold region, more fusion probability is gained by lowering of the barrier in a collision with a prolate nucleus having the deformation axis parallel to the line joining the centers of the nuclei than is lost in a collision with the deformation axis rotated by 90^0 relative to this line. The proper orientation averaging introduces for typical, fully deformed prolate target nuclei a barrier fluctuation of (5-7)% roughly independent of the projectile mass. On the other hand the barrier transparency parameter, that we have mentioned earlier, $\hbar\omega/V_B$, can be shown to decrease roughly with the product $Z_1 \times Z_2$ of the projectile and target charges (Reisdorf et al. 1985b), which is another way of saying that more massive systems behave more 'classically' in terms of (one-dimensional) tunnelling. For O + Sm the quantum mechanical 'uncertainty' of the barrier has about the same magnitude as the orientational fluctuations of the barrier.

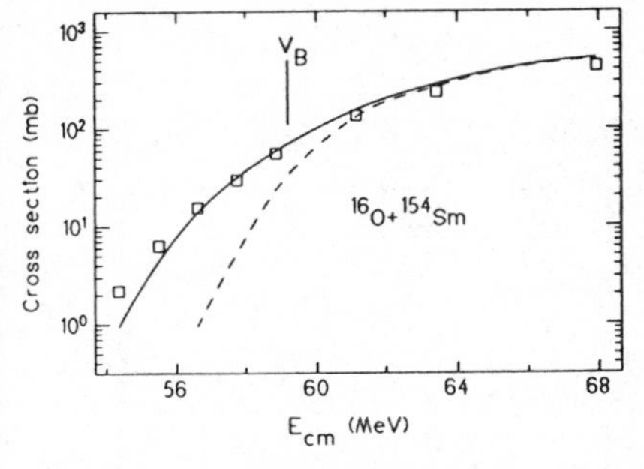

Fig.4 Fusion excitation function for ^{16}O + ^{154}Sm (Stokstad 1980). Dashed curve: WKB calculation with a potential fitted to the ^{16}O + ^{144}Sm data; solid curve: same potential, but with orientation-averaging using the known quadrupole moment of ^{154}Sm.

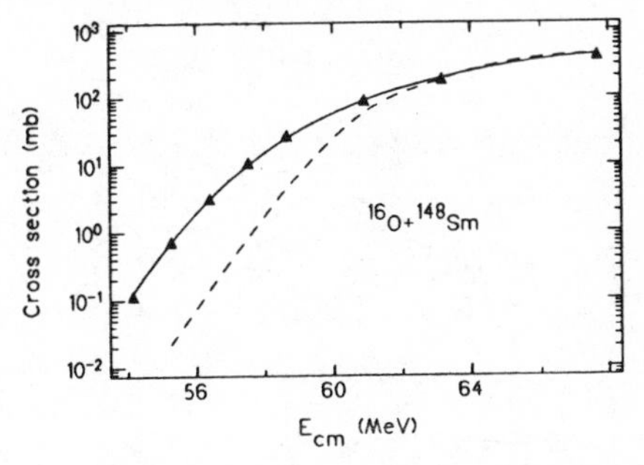

Fig.5 Fusion excitation function for ^{16}O + ^{148}Sm (Stokstad 1980). Dashed curve: WKB calculation with a potential adjusted to reproduce the data at the higher energies; solid curve: same, but including a barrier fluctuation of 4%.

Actually, the most striking effect in the O+Sm data is observed in the ^{16}O + ^{148}Sm system. In terms of nuclear spectroscopy ^{148}Sm is still considered to be a 'spherical' nucleus. Nevertheless figure 3 clearly shows that the corresponding excitation function does not coincide with that for the ^{144}Sm target, which is a nucleus with a closed neutron shell. Also, one can show (figure 5) that it is not possible to reproduce the data with a one-dimensional WKB calculation using a reasonable potential. Any calculation adjusted to reproduce the data at higher energies, fails to do so at lower energies (dashed curve in figure 5) and vice-versa. It is very tempting to interpret this as a consequence of the influence of *dynamical* deformations on the fusion process. Phenomenologically, the data can be reproduced by introducing a barrier fluctuation of about 4% (solid curve, fig.5). Efforts to invert the WKB equation (1.3) in order to extract V(r) yield unrealistically thin barriers in this and many similar cases (Balantekin et al. 1983).

From the barrier transparency arguments made above, one expects deviations from the WKB transmission calculations to be increasingly evident as still heavier systems are studied. This is demonstrated in figure 6 where a comparative study of the systems ^{40}Ar + $^{112,122}Sn$ and ^{40}Ar + $^{144,148,154}Sm$ is shown (Reisdorf 1985a). The correlation of the fusion excitation functions with collective nuclear properties is confirmed. It is the *shape* transition from ^{144}Sm to ^{154}Sm, and not the increasing neutron number (^{112}Sn to ^{122}Sn) that suggests itself as the important feature of these data. Moreover, in these systems with $\hbar\omega/V_B \approx 3.5\%$, one observes the total failure of the static potential model (dotted lines) to reproduce the data, even after adjustment in the region above 100 mb.

As for the O+Sm system (figure 3), we have again tried to remove expected macroscopic differences in the heights of the static fusion barriers. For this purpose we have defined a reference isotope ^{122}Sn, resp. ^{154}Sm. The data for the other isotopes of the same element are reduced by rescaling the abscissa -see eq. (1.1)-

$$E' = E V_B^{ref}/V_B \qquad (2.1)$$

The index 'ref' holds for the reference system.

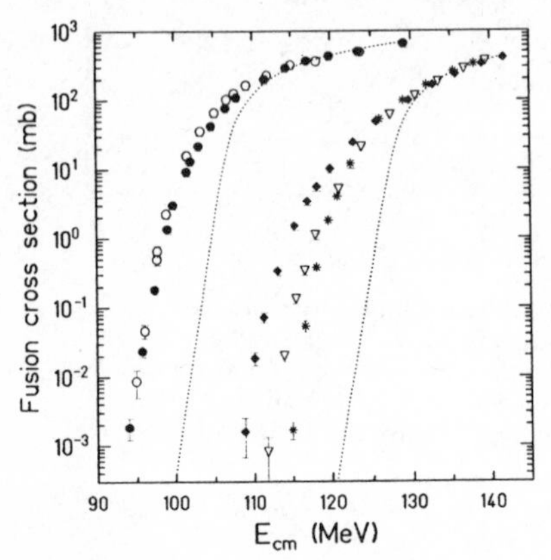

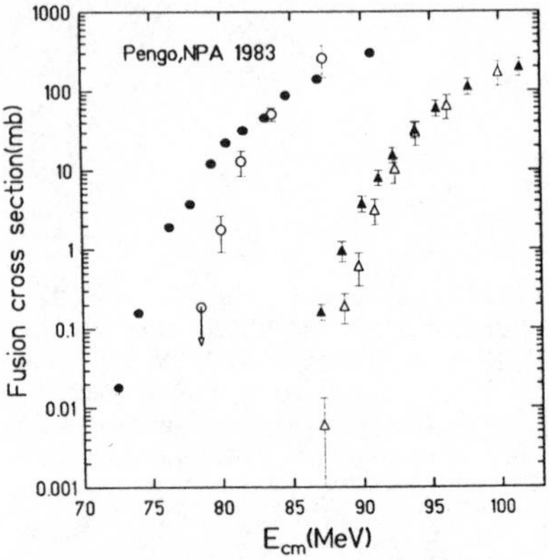

Fig.6 Comparison of fusion excitation functions for ^{40}Ar+^{112}Sn (open circles), ^{122}Sn (full circles), ^{154}Sm (full diamonds), ^{148}Sm (triangles) and ^{144}Sm (asterisks). ^{40}Ar +^{122}Sn and ^{154}Sm are the reference systems for the tin and samarium isotopes, respectively. The dotted curves are obtained from a WKB calculation with V_0 = 62.5 MeV (Reisdorf 1985a).

Fig.7 Comparison of fusion excitation functions for ^{32}S + ^{92,100}Mo (left) and ^{32}S + ^{100,104}Ru (Pengo 1983). The latter two reactions have been shifted 10 MeV to the right for better visual separation. The systems ^{32}S + ^{100}Mo and ^{32}S + ^{104}Ru are the reference systems (full symbols).

The ratio of fusion barriers was calculated with the potential eq. (1.5) for a fixed potential strength, 60 MeV/fm (it is insensitive to the choice of V_0).However, due to the small range of the attractive nuclear forces, the dependence of the height of the fusion barrier on mass A and charge Z of projectile and target is very sensitive to the way nuclear radii evolve with A and Z. The central radii have therefore been calculated using the droplet-model formulas (Myers and Schmidt 1983). It was shown that the droplet model reproduces measured root-mean-square (rms) charge radii with a standard deviation of 0.03 fm, provided the experimental or the calculated nuclear deformations are properly included. It was also shown in the same work that the only apparent influence of shell effects on the rms charge radii was via the deviations from sphericity they induce. Since we intend to remove only *macroscopic* differences between the various systems, we use the droplet-model formulas for spherical nuclei in all cases. Eq. (1.1) also suggests a re-adjustment of the ordinates in terms of the ratios of r_B^2. For isotopic series this correction is generally smaller than experimental uncertainties.

The 'geometrical' reduction of the fusion data leads to an almost perfect superposability of the Ar+Sn curves (figure 6). We note that a recent study (Jacobs et al. 1986) of the ^{16}O $+^{x}Sn$ system also suggests near superposability of the excitation functions.

The general influence of collective properties is also shown by the data of Pengo et al. (1983) for ^{32}S induced fusion. This is illustrated in figure 7. The excitation functions for the more 'collective' nuclei ^{100}Mo and ^{104}Ru (full symbols, reference systems), are less steep than for ^{92}Mo and ^{100}Ru, respectively (open symbols). We shall later see that this correlation subsists if ^{32}S is replaced by the much more massive projectile ^{86}Kr (Reisdorf 1985b).

At this point we also would like to mention the early work of Scobel et al. (1975) and Sikora et al. (1979) who observed and discussed dynamical deformation effects in subbarrier fusion of the systems $^{35}Cl, ^{40}Ca$ + ^{x}Ni.

Before presenting theoretical approaches, let me briefly illustrate in figure 8 how subbarrier fusion data can be described phenomenologically and parameterized. We have suggested as a qualitative idea that orientational and vibrational degrees of freedom introduce a dynamic barrier fluctuation which is responsible for the enhanced fusion probability with respect to the frozen-potential expectation. Phenomenologically, we can generalize the concept of barrier fluctuation and use its amplitude as a parameter to be determined by the data, irrespective of the mechanism involved. Neglecting intrinsic spins, the partial fusion probability

$$T_L(E) = \int f(V_B' p_i)\, t_L(E, V_B')\, dV_B' \qquad (2.2)$$

where the integration is over a distribution $f(V_B, p_i)$ of barriers (p_i are the parameters of the distribution). The transmission coefficients $t_L(E, V_B)$ can be generated using a one-dimensional potential of the type shown in eq. (1.5) with the strength and/or radial and/or diffuseness parameters chosen so as to give a barrier height equal to V_B. We choose for the distribution a truncated Gaussian function. Data fits using this Ansatz are shown in figure 5 (solid line) and in figure 8. The parameters are the average barrier $<V_B>$ and the standard deviation $\sigma_B(V_B)$ of the barrier distribution (technically, the average barriers were adjusted by varying the strength V_0 and the fluctuations were generated by a radius parameter distribution for the central radii in eq. (1.5)). The gaussians are truncated at $<V_B> \pm t\sigma_B$, where t is kept in the range 2-3. The necessity for this truncation comes from the empirical observation, see figures 5 and 8, that 'asymptotically', at energies well below the barrier, one finds that the logarithmic slopes of the fusion excitation functions tend to become linear and approximately equal to the slopes expected from the one-dimensional WKB calculations. This allows to define an asymptotic barrier shift that we shall call D_∞ (fig.8). The barrier value $<V_B>-D_\infty$ can be interpreted as the asymptotically lowest barrier. It has been termed 'adiabatic' barrier by some authors (Krappe 1983, Tanimura 1985, Keller 1986). Figure 8 shows that in order to determine D_∞ experimentally, one needs data below the mb level.

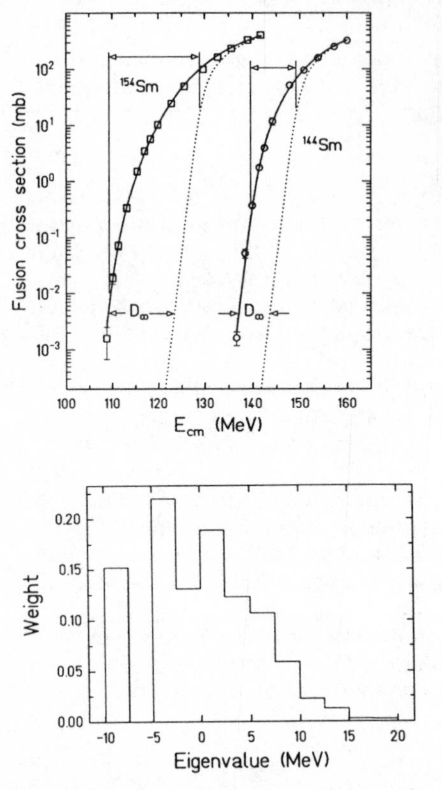

Fig.8 Parametric descriptions (solid curves, see text) of measured fusion cross sections for ^{40}Ar + ^{154}Sm and ^{40}Ar + ^{144}Sm (the latter shifted by 20 MeV). The dotted curves are calculations excluding barrier fluctuations. The double arrows in the top part of the figure span the energy region between the fitted average barrier location and the low-energy limit (within experimental accuracy) of the 'fluctuation range'. The asymptotic shifts D_∞ are shown.

Fig.10 Theoretical barrier distribution function for ^{40}Ar + ^{144}Sm. The eigenvalue zero corresponds to the unperturbed barrier location (Reisdorf 1985a).

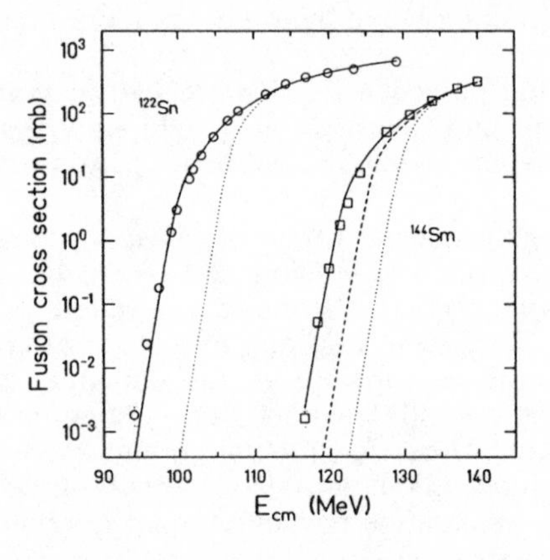

Fig.9 Fusion excitation functions for ^{40}Ar + ^{122}Sn and ^{144}Sm. Solid curves: WKB calculations including the coupling to inelastic channels; dashed: only the ^{40}Ar levels are included; dotted: no-coupling calculations.

We see the benefit of the phenomenological data fitting exercise primarily in the fact that it puts the forest of existing data and the "subbarrier cross section enhancements by many orders of magnitude" into proper perspective in a way that is similar in philosophy to the reduced alpha widths discussed in connection with figure 1. Except for systems with $Z_1 \times Z_2 \leq 1600$, the average barrier locations are reproduced with strength values V_0 = (70±20) MeV/fm with a tendency for V_0 to decrease as $Z_1 \times Z_2$ increases and with many local variations in the range indicated. What one invariably finds for fusing systems with $\hbar\omega/V_B \leq 10\%$ (where the barrier fluctuations cannot be 'swallowed' by readjustments of the potential parameters), is that the barrier fluctuations are about (4-5)% of the average barrier, diminishing to (2±0.5)% if both target and projectile are near closed shells, and increasing to (6-7)% if mid-shell nuclei with permanent deformations are involved. In absolute terms this means that the fluctuations (and D_{∞}) increase as the systems become more massive (Jahnke 1982). The correlation with nuclear shell structure simply implies a correlation with low energy nuclear spectroscopy and, in terms of nuclear reactions, a correlation with quasielastic channels. Obviously near the threshold, the decision for fusion falls at a time when the basic shell structure of the separated partners is still at least partially intact.

Among the theoretical approaches to understand subbarrier fusion, the early work of Esbensen (1981) seemed to provide an intuitively attractive picture. Fluctuations of the barrier were linked to fluctuations of the surface-to-surface distance originating from slow collective vibrational motion of the nuclear surfaces ('zero-point motion'). The standard deviations σ_λ from the spherical shape in these low-lying ($\lambda = 2^+, 3^-, \ldots$) surface modes are directly related to $B(E\lambda)$ values or deformation lengths $\delta_\lambda = R\beta_\lambda$ well known from low-energy gamma-ray spectroscopy and/or from DWBA and CCBA analyses of (p,p') and (α,α') data:

$$\sigma_\lambda = [R/Z(\lambda+3)]\,[(2\lambda+1)B(E\lambda)]^{1/2} \qquad (2.3)$$

or

$$\sigma_\lambda = \delta_\lambda/(4\pi)^{1/2} \qquad (2.4)$$

($B(E\lambda)$ is expressed in Weisskopf units)
Esbensen's approach (1981) which treated the different 'elementary' surface modes as independent oscillators (hence using quadratic addition of the σ_λ values), seemed to explain the ^{16}O + ^{x}Sm data in a simple way and was later also used with apparent success for ^{40}Ar + ^{x}Sn,^{x}Sm (Reisdorf 1982).

The dynamic limitation of this theory became soon clear, however. Jacobs and Smilansky (1983) pointed out that both the zero-point fluctuation approach (vibrations) and the 'orientation-averaging' (rotations) required that the internal vibrational or rotational spectrum was degenerate and implied a coupling to *all* members of the band. To a first approximation one has to compare the typical elementary excitation quantum to the Hill-Wheeler frequency ($\hbar\omega \cong 4$ MeV). For rotations the mean excitation energy turns out to be small compared to $\hbar\omega$ in most cases, for vibrations the degeneracy condition is less realistic.

The coupled channels theory (CCBA) as a traditional approach to direct reactions (Tamura 1965, Satchler 1983) offered the advantage of not being restricted by the above limitations. After the pioneering effort of Stokstad and Gross (1981) to explain the ^{16}O + Sm data with the CCBA approach, the disadvantage of considerable numerical effort necessary in reactions with massive projectiles was partially overcome by recent technical improvements (Rhoades-Brown et al. 1980). This allowed several groups to attempt first calculations for reactions with more massive nuclei (Rhoades-Brown and Braun-Munzinger 1984, Landowne and Pieper 1984). The CCBA approach has been reviewed recently by Steadman and Rhoades-Brown (1986).

Like the simple WKB approach, the CCBA, as used up to now, is essentially a barrier passing theory, as far as fusion is concerned. This implies that the barrier acts as a divider into two groups of reactions, the direct, quasielastic reactions and fusion. Flux penetrating inside the barrier is prevented from being reflected by either an ingoing-wave boundary condition (Landowne and Pieper 1984) or a short-ranged imaginary potential operating at radii well inside the barrier (Rhoades-Brown and Braun-Munzinger 1984). Some authors (Udagawa et al. 1985) allow for fusion 'under the barrier' by adjusting the imaginary potential to fit experimental fusion excitation functions. Conservation of flux enables the coupled-channel fusion cross section to be written

$$\sigma_{fus} = \sigma_{react} - \sigma_{dir} \qquad (2.5)$$

or

$$\sigma_{fus} = \pi\lambda^2 \sum_{L=0}^{\infty} (2L+1)(1 - \sum_i |S_{Li}|^2) \qquad (2.6)$$

where S_{Li} are the S-matrix elements in the various channels i that have been coupled to fusion and have been obtained by solving, with proper asymptotic conditions, the coupled one-dimensional equations (Tamura 1965)

$$[(\hbar^2/2\mu_i)d^2/dr^2 - (\hbar^2/2\mu_i r^2)L_i(L_i+1) - V_i(r) + E - \varepsilon_i]R_i\ (r) = \sum_j V_{ij}R_j(r) \qquad (2.7)$$

(ε_i excitation energy in channel i)
where, ideally, the optical potential $V_i(r)$ has no imaginary part outside and around the barrier if *all* couplings $V_{ij}(r)$ have been included for the calculation of the radial wave functions $R_i(r)$.

Because of the numerical complexity of the CCBA calculations several simplifying schemes have been attempted. Dasso et al. (1983) have shown that the equations can be easily decoupled if one neglects the channel dependences of the nuclear and centrifugal potentials and if one assumes that only the coupling strength in the vicinity of the 'bare' (uncoupled) potential, $V_i(r_B)$, barrier is important (constant - coupling approximation). By diagonalizing $\varepsilon_i\delta_{ij} + V_{ij}$ at the barrier with a unitary matrix U

$$\lambda_\alpha = \sum_{ij} U^+_{\alpha i}[\varepsilon_i\delta_{ij} + V_{ij}]U_{j\alpha} \qquad (2.8)$$

($\alpha = 1,2,...$)
they obtain a set of eigenchannel barriers $V(r_B)+\lambda_\alpha$ which can be put into a formula of the type eq. (2.2) each with a weight given by $|U_{0\alpha}|^2$, where the index zero stands for the ground-state channel. This approach, first used by Broglia et al. (1983), showed that the zero-point approach overestimated the fusion cross sections, as expected, but that reactions like $^{40}Ar+^{122}Sn$ and $^{58}Ni+^{58}Ni$ (Beckerman 1980) could be reasonably well reproduced if couplings to additional levels (other than the 2_1^+ and 3_1^-) were included. Figure 9 shows an application of this model by Reisdorf et al. (1985a); a typical eigenchannel distribution is shown in figure 10. The same value of V_0 (62.5 MeV/fm) was used for the two reactions shown, otherwise no parameters were adjusted. Figure 10 gives a reasonable a posteriori justification for our phenomenological Ansatz in terms of a truncated Gaussian barrier.

The physical insight that the model of Dasso et al.(1983) has given us into the barrier splitting mechanism cannot be valued high enough.

More recently, instead of the Ansatz eq.(2.8), Tanimura et al. (1985) have made an r-dependent diagonalisation that includes in principle also the channel dependences of the centrifugal and nuclear potentials:

$$\sum_{ij} U^{+}_{\alpha i}(r)\,\{[\varepsilon_i + (\hbar^2/2\mu_i r^2)L_i(L_i+1) + V_i(r)]\delta_{ij} + V_{ij}\}U_{j\beta}(r) = \zeta^J_\alpha(r)\delta_{\alpha\beta}$$

which yields a series of 'eigenpotentials' $\zeta^J_\alpha(r)$ of which the lowest ζ^0_1 can be associated to the asymptotically lowest or 'adiabatic' barrier suggested by the phenomenological data analysis. The main point is that the usual collective-model formfactors V_{ij} vary rapidly in the vicinity of the barrier due to Coulomb-nuclear competition, an effect that strongly influences the location and height of the eigenchannel barriers and hence limits the accuracy of the constant coupling approximation. For the ^{16}O + ^{x}Sm system the adiabatic barrier model seems to work well, see figure 11 (Tanimura 1985). One critique that one can make is that it is not obvious whether the 'adiabatic' barrier height converges as more couplings are added. An essential role must be played by the weight of the corresponding eigenchannel.

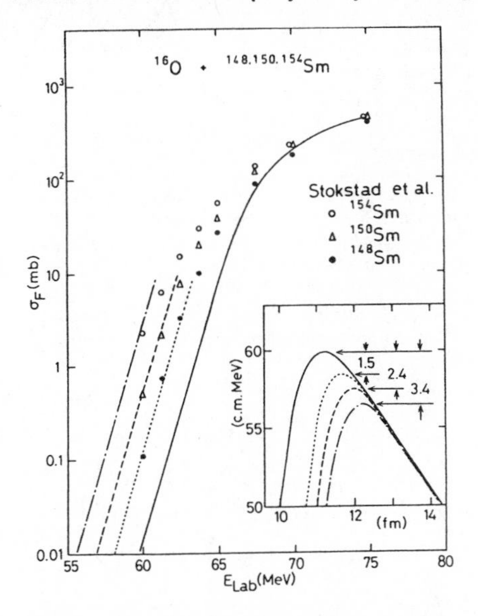

Fig.11 Cross sections for ^{16}O + ^{x}Sm (Stokstad 1980) The solid curve gives the WKB result for ^{148}Sm. The other curves represent the adiabatic-model results for the three systems, the inset shows the corresponding potential barriers. From Tanimura et al. (1985).

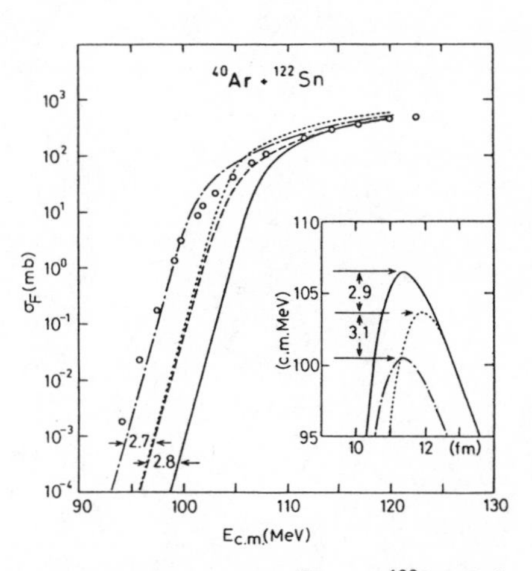

Fig.12 Calculations for ^{40}Ar + ^{122}Sn. Full CCBA: dashed curve; WKB method: solid curve; adiabatic barrier model: dotted curve; constant coupling model: dot-dashed curve. The inset shows the corresponding asymptotically lowest barriers. From Tanimura et al. (1985).

For $^{40}Ar + ^{122}Sn$ figure 12 confronts three different approximations with the 'full' coupled channels calculation (dashed line in the figure). First, the usual one-dimensional WKB calculation with a potential adjusted to reproduce the data (Reisdorf 1985a) above the 100 mb level; second the constant-coupling approach of Dasso et al. (1983), dot-dashed line, with as additional approximation (relative to the calculation shown in figure 9) the use of the Wong (1973) formula and third, a calculation using the adiabatic barrier $\zeta_1^0(r)$ (dotted line). One notices that the adiabatic barrier model agrees with the full CCBA at sufficiently low energies, that both do not reproduce the data, and that the constant coupling model overestimates the asymptotic shift D_∞ as compared to full CCBA.

3. Tunnelling in the presence of nucleon exchange.

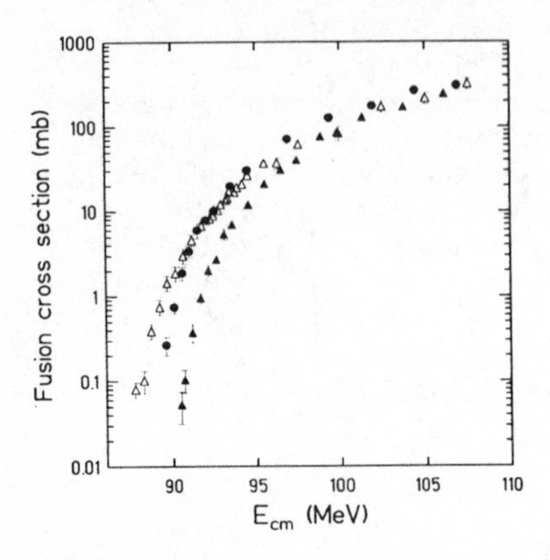

Fig.13 Reduced fusion excitation functions for $^{64}Ni + ^{64}Ni$ (full circles, reference system), $^{64}Ni + ^{58}Ni$ (open triangles) and $^{58}Ni + ^{58}Ni$ (full triangles) (Beckerman 1980,1981,1982)

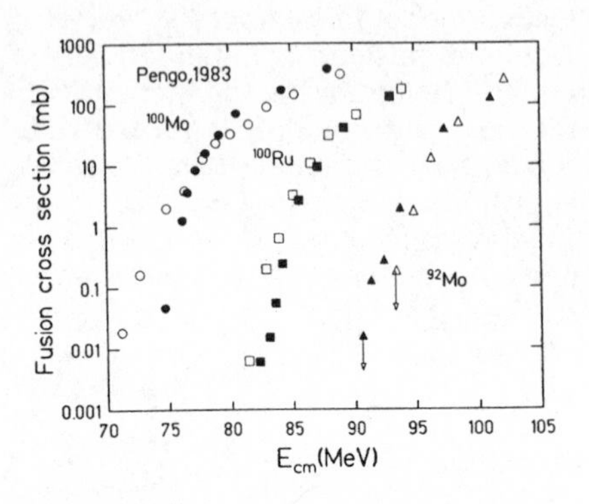

Fig.15 Comparison of reduced excitation functions for ^{32}S (open symbols) and ^{36}S (full symbols, reference systems) induced fusion reactions with the indicated targets (Pengo et al. 1983). The data for ^{100}Ru and ^{92}Mo are shifted 5 and 15 MeV, respectively, to the right for better visual separation.

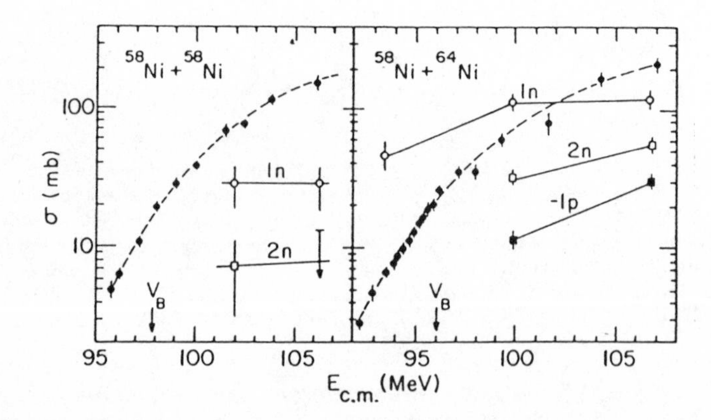

Fig.14 Energy dependence of the dominant quasielastic transfer cross sections as measured for the systems $^{58}Ni + ^{64}Ni$ and $^{58}Ni + ^{58}Ni$. From Rehm et al.(1985). The data for the fusion cross sections are from Beckerman et al.1980,1982).

One can interpret the failure of 'full' CCBA to account for the asymptotic shifts in massive systems (figure 11, see also Rhoades-Brown et al. 1985, De Castro Rizzo et al. 1985) as an indication that competition with direct channels other than purely inelastic excitation also influences the 'dressed', energy dependent (Nagarajan and Satchler 1986) potential through which tunnelling occurs. Transfer processes can be considered to be part of a microscopic description of the beginning of neck formation. Experimentally, faced with monotonically rising fusion excitation functions, it is clear that one must try to add further information by measuring all direct reaction channels in the threshold region. Signatures for the influence of pre-fusion nucleon exchanges can already be seen in some fusion data alone, however. The experimental results for the Ni+Ni system (Beckerman et al. 1980,1981,1982) are shown in figure 13. with ^{64}Ni + ^{64}Ni (full circles) chosen as reference system. The most striking effect is that ^{58}Ni +^{64}Ni (open triangles) does not behave like an intermediate system between the two symmetric systems, but crosses, in this reduced representation, the ^{64}Ni + ^{64}Ni data around the 5 mb level, a rather spectacular effect. In contrast to the symmetric systems, one can suspect that there is in the asymmetric system a strong, 'early' trend to equilibrate the neutron-to-proton ratios (Freiesleben and Kratz 1984) by either proton pickup by ^{64}Ni or by neutron stripping of^{64}Ni.

An investigation of transfer reactions for the Ni system, recently published by Rehm et al.(1985), figure 14, indeed shows that more transfer strength (mostly one and two nucleon exchange) is present in the 'mixed' system.

Another indicator of coupling of fusion to transfer reactions is shown in figure 15 using some of the data of Pengo et al.(1983). From left to right we show fusion of both ^{32}S and ^{36}S projectiles with ^{100}Mo, ^{100}Ru and ^{92}Mo respectively; the reactions with ^{36}S (full symbols) are the reference system in each case. The striking feature is the change-over in the steepness of the excitation functions; for ^{100}Mo it is ^{32}S that leads to a more gradual increase in fusion cross sections, for ^{100}Ru the two excitation functions become more similar and, finally, for ^{92}Mo it is ^{36}S which leads to the flatter excitation functions. An interchange between projectile and target, i.e. transfer reactions, could help to explain the phenomenon.

The Ca + Ca system measured by Aljuwair, Beckerman et al. (1984), has been considered by Landowne et al. (1985) to be another case strongly suggesting the influence of transfer reactions. The data are shown in figure 16, first (top panel), in an unreduced form. The striking feature is that the ^{44}Ca and^{48}Ca data "are about the same". The reduced form, lower panel, looks different. The data for the system with the mid-shell isotope, ^{44}Ca, now have the highest subbarrier cross sections. We stress that our purely macroscopic reduction does not introduce a priori nuclear structure effects. The rms charge radii of the stable Ca isotopes are known to be maximal for ^{44}Ca. Barranco and Broglia (1985) have linked both the behaviour of the charge rms radii and the fusion excitation functions to the known spectroscopy of the low-lying collective levels in the Ca isotopes.

A case that is, perhaps, more convincing, for the system S+Ni, will be presented by Stefanini et al.(1986) in a contribution to this conference.

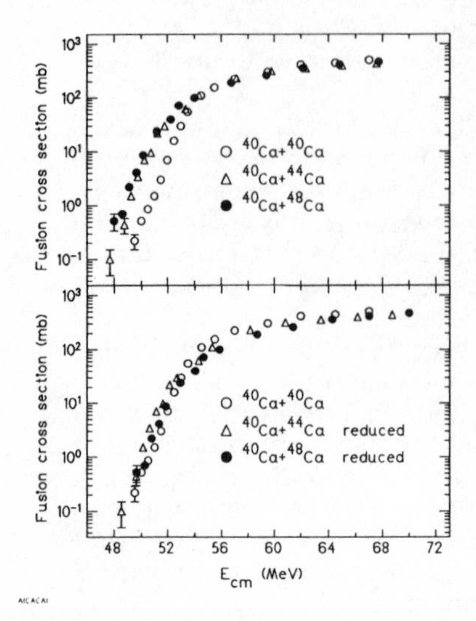

Fig.16 Top panel: Fusion excitation functions for the Ca + Ca system (Aljuwair 1984). Bottom panel: same after geometrical reduction.

Fig.18 Fusion (open circles) and evaporation residue data (full triangles) for $^{58}Ni + ^{124}Sn$ (Lesko et al. 1986). The curves are fits to the data explained in the text.

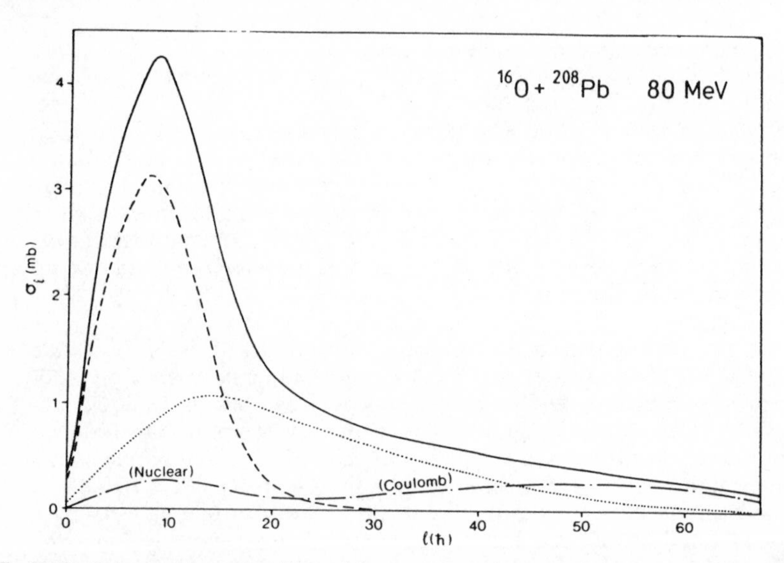

Fig.17 Partial-wave decomposition of the reaction cross sections from the full CRC calculation. The dashed curve is from fusion, the dot-dashed from inelastic excitations, and the dotted curve is from the transfer channels. The sum of these is the total reaction cross section (solid curve) (from Thompson et al.1985).

Two groups (Pieper et al.1985, Thompson et al.1985) have done coupled reaction (CRC) calculations including both transfer and inelastic channels for the system ^{16}O + ^{208}Pb. Even though they are using rather different 'bare' potentials, both groups do a rather good job in reproducing the comprehensive data of Vulgaris et al.(1986) and Videbaek et al.(1977). We learn from these calculations that a) coupling to transfer reactions can boost the fusion process. One might ask the unsophisticated question, if the effect is primarily due to the proton stripping of ^{16}O which leads in this very asymmetric system to a substantial lowering of the Coulomb barrier; b) direct channels with low cross sections, but large formfactors, can lead to strong enhancements of the fusion channel: the 3^- in ^{16}O for example (Thompson et al.1985). The matching of entrance channel and exit channel wave functions is important for the direct reaction cross section, but what matters for the lowering of the effective fusion barrier is the formfactor, rather than the cross section, associated with this channel. This may be important in judging the conjecture of Broglia et al.(1983b) that the ground-state positive-Q-value two-nucleon transfer channels are responsible for the ^{58}Ni + ^{64}Ni behaviour in figure 13; these channels, presumably because of mismatching, have not been observed by Rehm et al.(1985); c) vice-versa, large transfer cross sections (one-neutron transfer in the system ^{17}O + ^{208}Pb) do not necessarily lead to an enhancement of fusion (Thompson et al.1986). Obviously the usual conclusion that 'further investigations are needed' applies here.

Finally, let me briefly mention that the coupling of direct reactions to fusion leads to an appreciable rearrangement of the partial-wave distributions, see figure 17. The traditional subdivision of various reactions into separated zones in L-space is unrealistic. Spin distributions of fusion products at truly subbarrier energies have been published by Nolan et al.(1985), Gil et al.(1985) and Murakami et al.(1986). Space-time limitations force us to confine ourselves to mentioning the surprisingly high (beyond the expectations of CCBA) rms spin values determined by Murakami et al. for ^{16}O + ^{208}Pb, ^{232}Th and ^{12}C + ^{236}U. They have been determined from anisotropies in fission fragment angular distributions. In contrast, average spins in the ^{16}O, ^{12}C + ^{154}Sm fusion reaction (Gil et al.), determined with γ-spectroscopic methods, did not show this effect.

4. Very massive systems

One of the basic assumptions underlying our presentation so far is that fusion is treated as barrier-passing phenomenon. For systems with $Z_1 \times Z_2$ values around 1600 or higher theoretical considerations (Swiatecki 1981,1982, Davies et al. 1983) have led to the prediction that passing the barrier does not automatically lead to fusion or even to 'capture', unless a minimum extra relative kinetic energy ('extrapush') is imparted to the colliding nuclei. First experimental confirmation of this effect came from the work of Bock et al. (1982) and led to a semi-empirical parameterisation of the extrapush (Bjørnholm and Swiatecki 1982). If substantial centrifugal energy (high partial waves) is added to the Coulomb repulsion, the extrapush phenomenon can occur already in systems with $Z_1 \times Z_2 < 1600$. This is illustrated in figure 18 with the data of Lesko et al. (1986) for ^{58}Ni + ^{124}Sn. As usual, fission and evaporation residue cross sections (full triangles) have been added to obtain what is called 'fusion' cross sections (open circles). The solid curve hugging the fusion data was obtained by us using the extrapush parameters of

Björnholm and Swiatecki (1982). The dashed line running well above the data is calculated without extrapush and illustrates the amount of fusion hindrance.

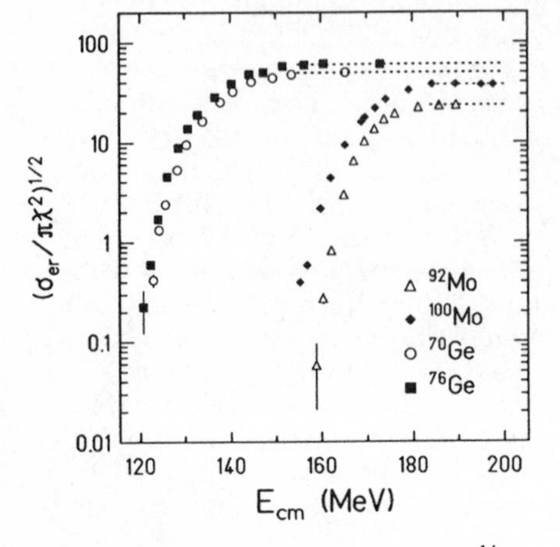

Fig.19 Excitation functions for $(\sigma_{er}/\pi\lambda^2)^{1/2}$ in ^{86}Kr induced fusion. The target nuclei are indicated.

Fig.21 Comparison of excitation functions for $^{81}Br + ^{104}Ru$ (Beckerman 1984,full triangles) and $^{86}Kr + ^{104}Ru$ (Reisdorf 1985b).

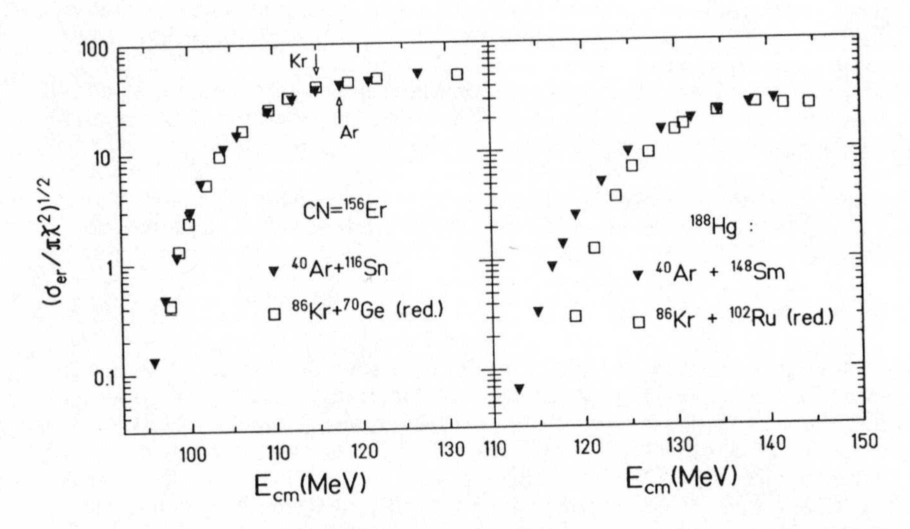

Fig.20 Comparison of excitation functions of $(\sigma_{er}/\pi\lambda^2)^{1/2}$ for pairs of reactions leading to the same indicated compound nuclei. The abscissa for the Kr-induced reactions have been rescaled by the ratio of fusion barriers calculated using the same potential strength parameter.

The phenomenon of extrapush blurs the definition of fusion, since the compound nuclear origin of 'fission' fragments is no longer obvious. For systems as light as $^{40}Ar + ^{208}Pb$ radiochemical measurements on binary fragments with given Z have shown that the angular distributions were no longer symmetric with respect to 90° in the center-of-mass, even if Z was relatively close to symmetry (see a poster contribution to this conference by Bellwied et al. 1986). For entrance channels close to symmetry the distinction between fusion-fission and deep inelastic collisions is even more difficult. Therefore methods have been developed (Sahm et al. 1985,Reisdorf et al. 1985b) to extract fusion probability information from evaporation residue data alone. The influence of fission on the excitation functions $\sigma_{er}(E)$ for evaporation residues (ER) with nuclear composition close to the compound nucleus is most readily demonstrated by plotting the dimensionless quantity $(\sigma_{er}/\pi\lambda^2)^{1/2}$, where λ is the de Broglie wave length. At sufficiently high energies one observes a saturation effect that varies characteristically with the system under investigation (figure 19). In figure 20 this saturation is shown to be a compound-nuclear property by comparing different entrance channels to ^{156}Er (left) or to ^{188}Hg (right). Figure 20 shows a comparison with similar data by Beckerman et al. (1984) for ^{81}Br induced fusion. Replacing ^{86}Kr by ^{81}Br does not affect the saturation region much (the compound nuclei have similar fissility), but the subbarrier behaviour is different: removing neutrons from the closed shell in ^{86}Kr leads to a more gradual rise of the excitation function as expected from the general systematics.

The ER data are first analysed using evaporation theory with angular momentum dependent level densities and fission barriers. For systems with a small fission probability at zero angular momentum, the evaporation calculation is adjusted to reproduce the saturation region (mostly by fitting the fission barrier) where the analysis is insensitive to the assumptions about the fusion probability as long as the lower partial waves that survive fission are assumed to be dominated by fusion (Blann and Komoto 1982). The fission probability is then extrapolated to the subbarrier region by adding theoretical information on the angular momentum dependence of the fission barriers (Sierk 1986). For very fissile systems close to symmetry, the fission probability is determined by analysing ER and fission data of sufficiently asymmetric systems leading to the same compound nuclei. Finally, with the fission probability known, fusion probability curves can be extracted from the ER data and parameterized, as before, in terms of average barriers, their dynamic fluctuations and the shifts D_{∞}. For very fissile systems the main effect of fission is that only fusion probabilities for near-central collisions can be determined, since the ER emerge selectively from low total-spin collisions.

A typical fit to ER data is shown in figure 18 (solid curve). An example of extracted central-collision fusion probabilities (P_{fus}) for ^{86}Kr induced reactions is shown in figure 22. The curves for ^{100}Mo and ^{104}Ru have been shifted along the abscissa by +0.5, and -9 MeV, in order to have barriers coincide with the data for ^{92}Mo. We identify the barrier with the energy at which P_{fus} = 1/2 (dotted lines). As can be seen, the removal of the influence of fission leads to results that are qualitatively very similar to the data of Pengo et al. (1983), already shown in figure 7, except that we have here a much more massive projectile. Again, the same nuclear structure trend prevails.

What is new with these very massive systems concerns the location of the average barrier. This is summarized in figure 23. Taking as reference fusion barriers the values V_{Bass} calculated from the global empirical potential of

Bass (1980), we have plotted V_B-V_{Bass}, the experimental excess energies versus a scaling variable proposed by Błocki et al.(1986), the 'mean fissility' x_m, which is a weighted average of the familiar compound nucleus fissility x and an 'entrance channel fissility' x_e:

$$x = (Z^2/A)/(Z^2/A)_{crit} \tag{4.1}$$

$$x_e = (Z^2/A)_e/(Z^2/A)_{crit} \tag{4.2}$$

where $$(Z^2/A)_{crit} = 50.88(1 - 1.7826I^2) \tag{4.3}$$

and $$(Z^2/A)_e = 4Z_1Z_2/[A_1^{1/3}A_2^{1/3}(A_1^{1/3}+A_2^{1/3})] \tag{4.4}$$

$$I = (A-2Z)/A \tag{4.5}$$

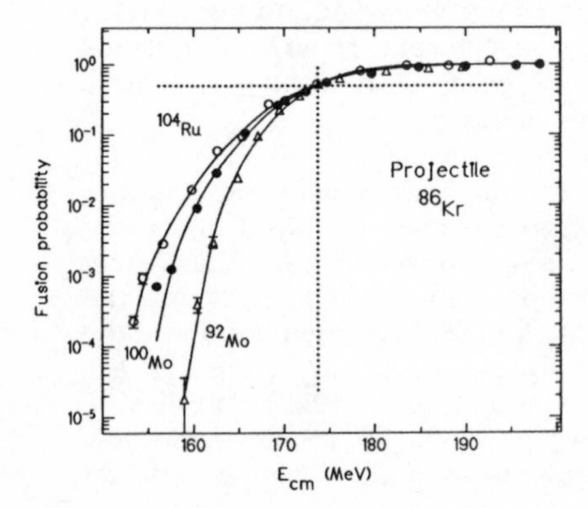

Fig.22 Central fusion probabilities for ^{86}Kr-induced reactions with the indicated targets. The dotted line indicates the fusion barrier for ^{92}Mo. The data for the ^{100}Mo and the ^{104}Ru targets have been shifted along the abscissa to make barriers coincide.

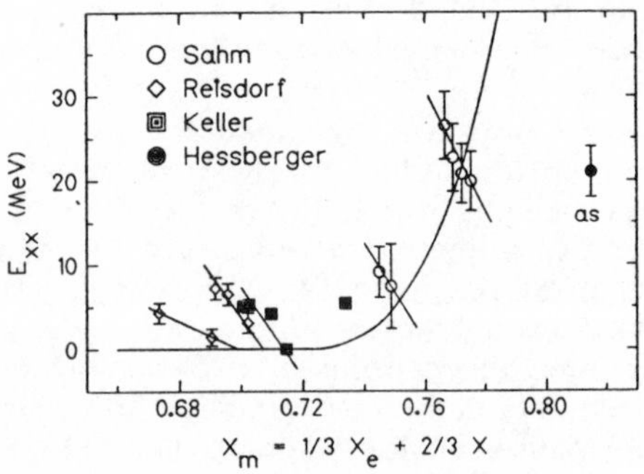

Fig.23 Extra-extra push energies for nearly symmetric systems. The data are from Sahm 1985 (open circles), Hessberger 1985 (full circle), Keller (1986) and Reisdorf (1985b). Isotopic series are joined by straight lines. The solid line is the calculation from Błocki et al.(1986).

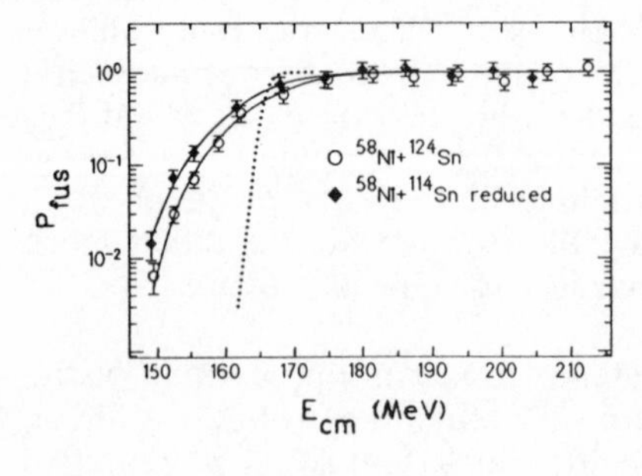

Fig.24 Central fusion probabilities for ^{58}Ni + ^{124}Sn and ^{58}Ni + ^{114}Sn. The reduction to fusion probabilities has been done using the ER data of Lesko et al.(1986). Dotted: WKB calculation for ^{58}Ni + ^{124}Sn.

The solid curve represents a recent calculation by Błocki et al. (1986) for the required excess energy, that is called 'extra-extrapush', E_{xx}, if formation of a compound nucleus (behind the saddlepoint) is required. The data, intentionally restricted to ER data, are from Sahm 1985 (open circles, Kr+Sb, Sn+Zr), Reisdorf 1985b (diamonds, Kr+Mo,Ru), Hessberger 1985 (Ti+Pb, full circle) and Keller 1986 (full squares, Zr+Y,Zr,Mo). Really high values for E_{xx} (20 MeV or more) have, so far, only been observed for the systems Sn+Zr and Ti+Pb. The relevant data do not 'scale' as predicted by Błocki (1986). The data point for Ti+Pb, the only strongly asymmetric system, is marked 'as'. Data points that refer to isotopic series are joined by straight lines to guide the eye to another effect that is sort of an 'inverse isotopic extrapush' running opposite to the theoretical trend. Essentially, one observes that the barrier is shifted towards higher energies if one adds neutrons to the target nucleus. This effect already exists for the Ni+Sn system as illustrated in figure 24. Transfer reaction data at subbarrier energies for this system are discussed by Betts et al. (1986) in a contribution to this conference. From this work and the work of van den Berg et al. (1986) one can deduce that more copious transfer is going on in systems with the heavier tin isotopes. It is therefore very tempting to associate the isotopic trends in figure 23 with the quasielastic reactions (excess energy used *before* reaching the barrier). The 'macroscopic' extrapush in the sense introduced by Swiatecki (1982), the excess energy needed *after* passing the barrier, would then be added on top of the former. The solid curve describing so well the fusion data for ^{58}Ni + ^{124}Sn in figure 18, was precisely obtained by adding the 'macroscopic' extrapush on top of the experimental barrier fitted to the ER data alone. The latter are limited to an angular momentum range which keeps them below the threshold for the macroscopic extrapush. The extrapush parameters were not readjusted from the values fitted to the data of Bock et al. (1982).

Let me close this presentation with a figure that illustrates once more in a particular striking way how fast the subbarrier behaviour changes when one starts from a closed shell system (^{90}Zr + ^{90}Zr) and adds holes or particles. In figure 25 we have plotted central fusion probabilities for several adjacent systems (Keller et al. 1986). For reference, we indicate by thin arrows the location V_{Bass} of the barrier obtained from the global potential of Bass (1980), the location of the P_{fus} = 1/2 points (thick arrows), i.e. the experimental barriers and the result of a no-coupling WKB calculation with the Bass potential (dotted lines). The dashed curves that run through the lowest energy data determine the 'adiabatic' barriers $V_{B\infty}$. Running down the panels on the left side, we notice a rapid increase of $V_{B\infty}$ - V_{Bass} with the number of added neutrons. Changing Z (right-hand side) does not change this quantity, provided the neutron number is untouched. One can conclude that it is the neutrons that are causing the shift of the adiabatic barrier. Furthermore, for the closed shell system ^{90}Zr + ^{90}Zr, the Bass barrier is equal to the experimental barrier (upper left panel). Any deviation from the shell symmetry, namely just removing one proton (upper right panel) causes an inelasticity, i.e. the P_{fus} = 1/2 point moves up in energy. A study of quasielastic reactions in these systems should soon complete these data.

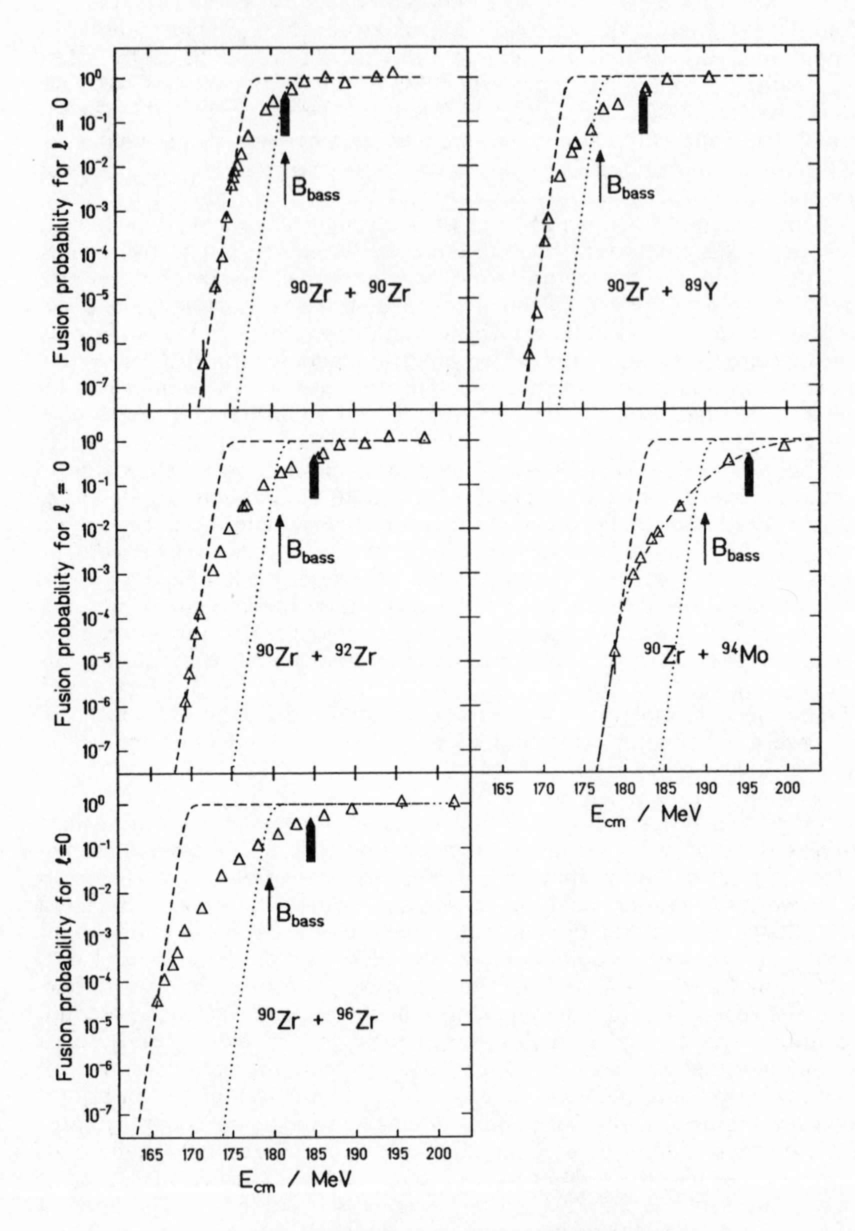

Fig.25 Measured excitation functions for P_{fus} (Keller 1986). Thin arrows: fusion barriers from the Bass potential; thick arrows: $P_{fus} = 1/2$ points. Dotted curves: WKB calculations using the Bass barriers, dashed: same but with barriers shifted to describe the lowest energy data.

References

Aljuwair H A,Ledoux R J,Beckerman M,Gazes S B,Wiggins J and Cosman E R, Betts R R,Saini S and Hansen 1984 Phys.Rev.C **30** 1223

Balantekin A B,Koonin S E and Negele J W 1983 Phys.Rev.C **28** 1565

Barnes C A 1985, see Steadman 1985 pp 70-84

Barranco F and Broglia R A 1985 Phys.Lett.B **151** 90, see also Steadman 1985 pp 196-219

Bass R 1980 Deep-inelastic and Fusion Reactions with Heavy Ions (Berlin:Springer) pp 281-293

Beckerman M,Salomaa M,Sperduto A,Enge H,Ball J,Dirienzo A,Gazes S, Yan Chen,Molitoris J D and Mao Nai-Feng 1980 Phys.Rev.Lett. **45** 1472

Beckerman M,Ball J,Enge H,Salomaa M,Sperduto A,Gazes S, Dirienzo A and Molitoris J D 1981 Phys.Rev.C **23** 1581

Beckerman M,Salomaa M K,Sperduto A,Molitoris J D, and Dirienzo A 1982 Phys.Rev.C **25** 837

Beckerman M,Wiggins J,Aljuwair H and Salomaa M K 1984 Phys.Rev.C **29** 1938

Bellwied R,Keller H,Kratz J V,Lützenkirchen, Brüchle,Moody K J,Schädel M,Sümmerer K,Wirth G,Reisdorf W 1986, poster contribution, this conference

Betts R R,Evans P M,Pass C,Poffe N,Smith A E,Stuttge L, Lilley J S,Banes D W,Connell K,Simpson J,Smith J R, Fulton B R,Bennett S,Woods P J,James A N 1986 oral contribution, this conference

Blann M and Komoto T T 1982 Phys.Rev.C **26** 1471

Błocki J,Randrup J,Swiatecki W J,Tsang C F A 1977 Ann.Phys. **105** 427

Błocki J,Feldmeier H,Swiatecki W J 1986 GSI preprint GSI-86-16

Bjørnholm S and Swiatecki W J 1982 Nucl.Phys.A **391** 471

Bock R,Chu Y T,Dakowski M,Gobbi A,Grosse E,Olmi A,Sann H,Schwalm D, Lynen U,Müller W,Bjørnholm S,Esbensen H,Wölfli W and Morenzonic E 1982 Nucl.Phys.A **388** 334

Broglia R A,Dasso C H,Landowne S and Winther A 1983 Phys.Lett.B **133** 34 , Phys.Rev.C **27** 2433

Cujec B and Barnes C A 1976 Nucl.Phys.A **266** 461

Cujec B 1985, see Steadman 1985 pp 108-128

Dasso C H,Landowne S,Winther A 1983 Nucl.Phys.A **405** 381, **407** 221

Davies K T R,Sierk A J and Nix J R 1983 Phys.Rev.C **28** 679

de Castro Rizzo D M and Alamanos N 1985 Nucl.Phys.A **443** 525

Di Gregorio D E,Fernandez Niello J O,Pacheco A J,Abriola D,Gil S, Macchiavelli A O,Testoni J E,Pascholati P R,Vanin V R,Liguori Neto R, Carlin Filho N,Coimbra M M,Gomes P R S and Stokstad R G 1986, to be published

Esbensen H 1981 Nucl.Phys.A **352** 147

Freiesleben H and Kratz J V 1984 Phys.Rep. **102** 1

Gamov G 1928 Z.Phys. **51** 204, **52** 510

Gil S,Vandenbosch R,Lazzarini A J,Lock D-K,Ray A 1985 Phys.Rev.C **31** 1752

Hessberger F P,Münzenberg G,Hofmann S,Reisdorf W,Schmidt K-H, Schött H J,Armbruster P,Hingman R,Thuma B and Vermeulen D 1985 Z.Phys.A **321** 317

Hill L D and Wheeler J A 1953 Phys.Rev. **89** 1102

Igo C 1958 Phys.Rev.Letters **71** 1

Jacobs P M and Smilansky U 1983 Phys.Lett.B **127** 313

Jacobs P M,Fraenkel Z,Mamane G and Tserruya I 1986 Phys.Lett.B **175** 271

Jahnke U,Rossner H H,Hilscher D and Holub E 1982 Phys.Rev.Lett. **48** 17

Keller J G,Schmidt K-H,Hessberger F P,Münzenberg G,Reisdorf W, Clerc H-G and Sahm C-C 1986 Nucl.Phys.A **452** 173

Krappe H J,Möhring K,Nemes M C and Rossner H 1983 Z.Phys.A **314** 23

Landowne S and Pieper S C 1984 Phys.Rev.C **29** 1352

Landowne S,Dasso C H,Broglia R A and Pollarolo G 1985 Phys.Rev.C **31** 1047

Lesko K T,Henning W,Rehm K E , Rosner G,Schiffer J P,Stephans G S F, Zeidman B and Freeman W S 1985 Phys.Rev.Lett. **55** 803, ANL preprint PHY-4776-HI-86

Murakami T,Sahm C-C,Vandenbosch R,Leach D D,Ray A and Murphy M J 1986 preprint

Myers W D and Schmidt K-H 1983 Nucl.Phys.A **410** 61

Nagarajan M A and G R Satchler 1986 Phys.Lett.B **173** 29

Nolan P J,Love D J G,Kirwan A,Unwin D J,Nelson A H,Twin P J and Garrett J D 1985 Phys.Rev.Lett. **54** 2211

Pengo R,Evers D,Löbner K E G,Quade U,Rudolph K,Skorka S J and Weidl I 1983 Nucl.Phys.A **411** 255
Pieper S C,Rhoades-Brown M J and Landowne S 1985 Phys.Lett.B **162** 43
Randrup J and Vaagen J S 1978 Phys.Lett.B **77** 170
Rasmussen J O 1959 Phys.Rev. **113** 1593
Rasmussen J O and Sugawara-Tanabe K 1971 Nucl.Phys.A **171** 471
Rehm K E,Wolfs F L H,van den Berg A M and Henning W 1985 Phys.Rev.Lett. **55** 280
Reisdorf W,Hessberger F P,Hildenbrand K D,Hofmann S,Münzenberg G, Schmidt K-H,Schneider J H R,Schneider W F W,Sümmerer K,Wirth G, Kratz J-V and Schlitt K 1982 Phys.Rev.Lett. **49** 1811
Reisdorf W,Hessberger F P,Hildenbrand K D,Hofmann S,Münzenberg G, Schmidt K-H,Schneider J H R,Schneider W F W,Sümmerer K,Wirth G, Kratz J-V and Schlitt K 1985a Nucl.Phys.A **438** 212
Reisdorf W,Hessberger F P,Hildenbrand K D,Hofmann S,Münzenberg G, Schmidt K-H,Schneider W F W,Sümmerer K,Wirth G, Kratz J-V,Schlitt K and Sahm C-C 1985b Nucl.Phys.A **444** 154
Rhoades-Brown M J,Pieper S C,MacFarlane M H 1980 Phys.Rev.C **21** 2417
Rhoades-Brown M J and Braun-Munzinger P 1984 Phys.Lett.B **136** 19
Roeckl E 1983 Nucl.Phys.A **400** 131c
Sahm C-C,Clerc H G,Schmidt K-H,Reisdorf W,Armbruster P, Hessberger F P,Keller J G,Münzenberg G and Vermeulen D 1985 Nucl.Phys.A **441** 316
Satchler G R 1983 Direct Reactions(Oxford University Press)
Sierk A J 1986 Phys.Rev.C **33** 2039
Scobel W,Mignerey A,Blann M and Gutbrod H H 1975 Phys.Rev.C **11** 1701
Sikora B,Bisplinghoff J,Scobel W,Beckerman M and Blann M 1979 Phys.Rev.C **20** 2219
Steadman S G 1985 ed Fusion Reactions Below the Coulomb Barrier, Lecture Notes in Physics, Vol. 219 (Heidelberg:Springer-Verlag)
Steadman S G,Rhoades-Brown M J 1986 Ann.Rev.Nucl.Part.Sci. **36** 649
Stefanini A M,Fortuna G,Pengo R,Meczynski,Montagnoli G,Corradi L, Tivelli A,Beghini S,Signorini C,Lunardi S,Morando M and Soramel F 1986 Nucl.Phys A in press, see also this conference
Stokstad R G,Eisen Y,Kaplanis S,Pelte D,Smilansky N and Tserruya I 1980 Phys.Rev.C **21** 2427
Stokstad R G and Gross E E 1981 Phys.Rev.C **23** 281
Swiatecki W J 1981 Physica Scripta **24** 113
Swiatecki W J 1982 Nucl.Phys.A **376** 275
Tamura T 1965 Rev. Mod. Phys. **27** 679
Tanimura O,Makowka J and Mosel U 1985 Phys.Lett.B **163** 317
Thompson I J,Nagarajan M A,Lilley J S and Fulton B R 1985 Phys.Lett.B **157** 250
Udagawa T,Kim B T and Tamura T 1985 Phys.Rev.C **32** 124
van den Berg A M,Henning W,Lee L L,Lesko K T,Rehm K E,Schiffer J P, Stephans G S F and Wolfs F L H 1986 Phys.Rev.Lett. **56** 572
Videbaek F,Goldstein R B,Grodzins L and Steadman S G, Belote T A,Garrett J D 1977 Phys.Rev.C **15** 954
Vulgaris E,Grodzins L,Steadman S G and Ledoux R 1986 Phys.Rev.C **33** 2017
Wong C Y 1973 Phys.Rev.Lett. **31** 766

Inst. Phys. Conf. Ser. No. 86
Paper presented at Int. Nucl. Phys. Conf., Harrogate, UK, 1986

Mean field physics of hot nuclei

S. Levit

Department of Nuclear Physics Weizmann Institute of Science
Rehovot, Israel, 76100

Abstract. I give a short overview of various phenomena which are presently studied in hot nuclei. I then discuss in detail the mean field theories of the temperature induced shape transitions in rotating nuclei and of the upper temperature limit of nuclear binding. The ideas of the Landau mean field theory are extremely usefull in identifying the universal features of nuclear shape transitions. The thermal Hartree-Fock approach with a proper care of continuum states allows to determine the highest possible temperature and the maximum heat a nucleus can absorb while still remaining a selfbound system.

1. Phenomena in Hot Nuclei - An Overview

Heavy-ion reactions provide an almost unique way of heating nuclei up. Although such reactions can in principle follow different reaction mechanisms it appears reasonable to expect that a substantial part of the reaction cross section goes via a formation of an equilibrated nuclear complex. Experimental efforts are presently becoming more and more elaborate and precise in the attempts to isolate and study such an equilibrated component of the nuclear cross section (Refs 1 - 4).

Microscopically equilibration means equal population of states with given excitation energy E* and possibly other conserved quantum numbers like linear and angular momentum, isospin, etc. which had fixed values in the process of equilibration. The corresponding probability distribution is therefore $\omega \sim \delta(E-H)\delta(P_0-P)\ldots$. Given this microcanonical distribution one can easily translate the discussion of the properties of equilibrated nuclei into a standard language of statistical mechanics. Unlike in infinite systems, however, such a translation involves approximations the validity of which should in principle be tested. Consider a simple example of transitions between highly excited nuclear states caused by an operator Q. If all the initial states at an energy E' were equally populated (equilibration assumption) and the only information about the final states was their energy E'', then the transition probability is

$$dP(\varepsilon=E'-E'')/d\varepsilon \sim \rho^{-1}(E')\sum_{i,f}|\langle f|Q|i\rangle|^2\delta(E'-E_i)\delta(E''-E_f)$$

which can be written as the Fourier transform

$$\int_{-\infty}^{\infty}\frac{dt}{2\pi}e^{i\varepsilon t}\mathrm{Tr}\,\delta(E'-H)Q^+(t)Q(o)/\mathrm{Tr}\,\delta(E'-H)$$

of the microcanonical response to the perturbation Q of the system at energy E'. Using $\delta(E'-H)=(2\pi i)^{-1}\int_{-i\infty}^{i\infty} d\beta \exp(\beta E'-\beta H)$ one further relates this temperature response $\mathrm{Tr}e^{-\beta H}Q^{+}(t)\bar{Q}(o)/\mathrm{Tr}e^{-\beta H}$ at the temperature $T=\beta^{-1}$. Saddle point evaluation of the β integrals gives the standard thermodynamic relatic between T and E', i.e. $E'=\partial(\beta F)/\partial\beta$, in terms of the free energy F = -Tr exp(

In this example one has explicitly constructed the bridge between the measurable quantity dP/dε and the nuclear response at finite temperature. The only approximation involved aside from the equilibration assumption is the saddle point evaluation of the β-integral which describes fluctuation of the inverse temperature β for the fixed energy E'. These fluctuations are completely negligible only in infinite systems. The experience with the nuclear level density formula which is derived using this very approximation (Ref. 5) indicates that the approximation is expected to be fairly reliable also in sufficiently large finite nuclei.

There is a number of interesting phenomena which can be studied in hot nuclei. The most obvious is the behaviour of the various nuclear collective modes. In hot systems the damping of these modes should increase with the temperature. For $T \gtrsim 1$ MeV the low frequency vibrational modes are expected to be completely damped but the giant resonance high frequency modes should still be observable for T in the few MeV range. They were indeed found in the studies of the continuum γ-rays from the deexcitation of compound nuclei formed in heavy ion reactions (Refs. 6-12). So far only the giant dipole component (GDR) has been clearly identified but the first indications of the possible isovector GQR structure were already reported [13] The systematics of the temperature dependence of the GDR parameters is slowly becoming available. The main weakness of the present generation of experiments is their inability to sharply select the energy and the angular momentum of the initial ensemble of states emitting the resonant γ-rays.

Theoretical discussions of the properties of the giant resonances in hot nuclei are based in various versions of the finite temperature RPA approach (Refs 14-20). The calculated positions of the giant dipole resonance are remarkably stable with respect to the temperature changes up to T~5 MeV (Refs 18,19). On the other hand the quadrupole strength function in closed shell nuclei is found to be considerably affected by the temperature. Within the RPA it is impossible to explain the large and increasing with the temperature width of the GDR found in experiments. As in the T=0 case much of this width is believed to arise from the coupling to the soft quadrupole degrees of freedom. Their statistical fluctuations grow significantly above T~1 MeV and in Ref. 20 the increase of the GDR width was partly related to this increase of the thermal quadrupole fluctuations.

Various phase transition-like phenomena are expected to occur in hot nuclei. In the very low temperature regime (T~0.5-1 MeV) the pairing correlations present in the ground state are expected to gradually disappear with increasing T (Refs 21,22). For higher temperatures (T~1-3 MeV) the thermal excitations wash out the shell effects. Nuclei with deformed ground states undergo transitions to spherical shapes in this region of temperatures. At finite angular momentum this tendency is counteracted by the rotational field and an interesting pattern of shape transitions may emerge [26] . Since the parameters of the giant dipole resonances depend quite sensitively on the deformation, the detailed study of such resonances in highly excited rotating nuclei is expected to open exciting possibilities to learn about the nuclear shape as a function of the temperature and the angular momentum.

First relevant results in this direction were reported in Refs 8,9 . Theoretical descriptions of shape transitions in hot nuclei are based **on the** finite temperature mean field approach (Refs 23,24) in which the equilibrium nuclear shape is found by minimizing the appropriate trial free energy at a given temperature T and angular momentum J. It is usually more convenient to work with the **intensive partner of J** - the angular velocity ω of the nuclear rotations. The trial free energy depends on variational parameters with respect to which the minimization is carried out. The choice of these parameters depends on the degree of sophistication of the theory, e.g. shape parameters of the potential in the Nilsson-Strutinsky scheme, the single particle density matrix in the Hartree-Fock approach, etc. Most of the relevant results of any such microscopic **mean-field theory** of shape transitions can be anticipated on the basis of a unified framework of the Landau theory of phase-transitions in statistical systems (cf. Section 2 and Refs. 25,26). Plotting the behaviour of the equilibrium shape in the E* vs. J plane one finds an interesting phase diagram with a **line of 1st order** transitions for J less than a certain critical J_c and a line of second order transitions for $J>J_c$. The lines meet at an analog of a tricritical point where one finds large fluctuations around the equilibrium shape.

Above the region of shape transitions one begins to approach the regime ($T \gtrsim 5$ MeV) where the binding strength of the nuclear forces may not be sufficient to overcome the tendency of the hot nuclear system to increase its entropy by simply dispersing its constituents all over the space. Examining this balance between the binding energy and the entropy, one expects that there should be a limit of maximum excitation energy (heat) which a fully equilibrated nucleus can absorb and still remain a self-bound system. Experimental evidences showing a possible detection of such a limit were presented recently (Refs 2,3,27,28). Theoretical considerations (cf. below, Section 3) indicate that the temperature T_{lim} of this limit of nuclear binding is closely related to the value T_c of the critical temperature of the liquid-gas **coexistence** in infinite nuclear matter. T_{lim} however is much below T_c, is not universal and depends on the charge and mass of the nucleus. Measuring T_{lim} in different nuclei is expected to give valuable information on the equation of state of the hot matter inside the compound nucleus and the temperature dependence of the nuclear surface tension.

Fully equilibrated nuclei can not exist above T_{lim}. However, **a partial** thermal equilibrium may possibly take place under the constraining conditions of the strong compression which presumably **exists during the initial stages** of the central collisions between two heavy ions. The hot matter thus formed expands and fragments. In the process of the expansion it "probes" different parts of its equation of state. All available calculations of the equation of state of the hot nuclear matter show (cf. Refs 29-32) that it can exist in two phases below a critical temperature $T_c \sim 15$ MeV. These phases differ only quantitatively (i.e. by their equilibrium densities at a given pressure, by their specific **entropies, etc.) and as in other condensed** materials they may be called liquid (the denser phase) and vapor or gas (the more dilute phase). The difference between the phases disappears above T_c.

Since the initial temperatures of the partial equilibrium may be as high as 10-20 MeV the expanding matter can find itself at or in the close vicinity of the critical point of the liquid-gas coexistence. At T_c large statistical fluctuations are generally expected whereas below T_c the matter can become metastable and further on mechanically unstable. Entrance into this region of instabilites was suggested as a possible mechanism of the fragmentation (Ref 33). Mass spectrum of the intermediate mass fragments was also considered as a possible candidate for a measure of the critical fluctuations in the vicinity of T_c (Refs 34-39). Several theoretical schemes were proposed for the description of the dynamics of the expansions. Hydrodynamical calculations (40), adiabatic expansion in collective monopole parameters (41), classical molecular dynamics (42) and various Monte-Carlo simulations (43 - 45) of the decay processes were reported. Summarizing these theoretical efforts it is fair to say that the question of whether or not passing the vicinity of the critical point gives an observable signal is still much under debate.

In the following sections I will discuss in some detail two regions of temperatures in hot nuclei, the shape transitions region and the approach to the limit of the nuclear binding.

2. Low Temperatures - Shape Transitions in Rotating Hot Nuclei

As in many theoretical discussions of nuclear phenomena the mean field approach provides the baseline of the analyses of shape transitions in hot nuclei. Minimizing the appropriate trial free energy at different values of the temperature T and the angular velocity ω of the rotations one finds the evolution of the equilibrium shape as a function of these parameters. The ideas of the Landau theory of phase-transitions in statistical systems are extremely useful in identifying the universal features of nuclear shape transitions. A possible way of defining the Landau theory in the context of the nuclear Hartree-Fock approach was indicated in Ref 25 . Here I follow a somewhat more general route (Ref. 26).

The application of the Landau theory begins with the identification of the symmetry which is broken in the course of the phase transition and the introduction of the relevant order parameters which acquire non-zero values when the symmetry is broken [52] . The obvious such parameters in nuclear shape transitions are the standard deformation parameters $\alpha_{2\mu}$. In the usual intrinsic frame interpretation of the nuclear shape the symmetry which is broken is the rotational invariance of the **intrinsic nuclear mean-field** Hamiltonian, which has a non-spherical equilibrium potential for deformed nuclei. In the presence of rotation the rotational symmetry in the intrinsic frame is broken explicitly by the preferred direction of the rotational axis The nuclear rotation acts in the intrinsic frame as an external field controlled by the magnitude and the direction of the rotational velocity ω. It is convenient to let it be directed along an arbitrary fixed direction, i.e. to consider vector $\vec{\omega}$.

The free energy $F(T,\vec{\omega},\alpha_{2\mu})$ is rotationally invariant and can therefore depend only upon invariant combinations of α and ω. In the absence of rotation (ω=0) these invariant combinations are $[\alpha x \alpha]_0$ and $[[\alpha x \alpha]_2 x \alpha]_0$ and assuming analyticity and expanding one can write [25] .

$$F(T,\omega=0,\alpha_{2\mu})=F_0(T)+A(T)\beta^2 -B(T)\beta^3 \cos 3\gamma+C(T)\beta^4+.... \quad (2.1)$$

where β and γ are the usual deformation parameters in intrinsic frame. The temperature dependence of the coefficients A,B,C, etc., (F_0 is irrelevant for minimization) should in principle be determined microscopically, or better still, constrained experimentally. However, for the general analysis it is sufficient to determine the main qualitative aspects of this dependence. Existing examples of the past (Refs 23-25) and ongoing (Refs 26,46 and cf. below) microscopic calculations support the following generic assumptions about the T dependence of A, B and C. For nuclei with deformed ground states the coefficient A(T) is negative at low temperatures. As the temperature rises the shell effects responsible for the deformation gradually disappear and A(T) increases and changes its sign at some $T=T_c\sim1-3$ MeV for medium heavy nuclei. The values of T_c differ for different nuclei. The coefficient B(T) is not zero reflecting the prolate-oblate asymmetry of nuclear shapes. It is positive for more frequently occurring prolate ground states. C(T) should be positive to guarantee the global stability. In the following we assume this generic behaviour of A, B and C and neglect the terms higher than $O(\beta^4)$ in (2.1). Other more exotic possibilities can also be analysed when needed.

In the presence of rotations (nonzero ω) one should add terms with invariants containing combinations of ω and $\alpha_{2\mu}$ to the expression (2.1). The lowest such invariants are quadratic in ω and sufficient for our purposes. In Cartesian coordinates one finds (Ref 26)

$$F(T,\vec{\omega},\alpha_{2\mu}) = F(T,\omega = 0,\alpha_{2\mu}) - \frac{1}{2}\sum_{ij=1}^{3} I_{ij}(T,\alpha_{2\mu})\omega_i\omega_j \tag{2.2a}$$

where I_{ij} should obviously be interpreted as the components of the nuclear moment of inertia. Its general dependence on α is [26]

$$I_{ij}=\{I_0+I_2[\alpha x\alpha]_0+\ldots\}\delta_{ij}+\{R\alpha+D[\alpha x\alpha]_2+\ldots\}_{ij} \tag{2.2b}$$

The first term is a scalar trace of I_{ij} and depends only upon β^2 and $\beta^3\cos 3\gamma$ as the terms in (2.1). The second term in (2.2b) depends, in addition to β and γ, also on the orientation angles of the deformed nucleus with respect to the rotation axis $\vec{\omega}$. Minimizing (2.2b) with respect to β,γ and these angles one finds the equilibrium deformation and orientation.

Detailed analysis (Ref. 26) shows that the equilibrium is very sensitive to the relative magnitudes of the coefficients R and D **in (2.2b). In the rigid** body moment of inertia R dominates over D whereas I_0=R=0 and D dominates in the moment of inertia for irrotational flow. Since in the temperature regime T~1-3 MeV the pairing correlations are not important, the relevant case is closer to the rigid body situation and it is therefore useful to discuss first the extreme limit of dominant R and D=0. The minimization of (2.2a) can be carried out analytically and one finds [26] that the equilibrium deformation for B>0 and D=0 depends on two combinations of the parameters τ and ω/ω_c where

$$\tau = \frac{AC}{B^2}, \quad \omega_c = \frac{9}{16}\frac{B}{C}\left(\frac{B}{R}\right)^{\frac{1}{2}}. \tag{2.3}$$

The combination τ vanishes at $T=T_c$ and is monotonic and practically linear in $T-T_c$. The results of the minimization of (2.2) are conveniently summarized on the phase diagram of τ vs. ω/ω_c, shown in Fig. 1 on which we plot the lines of constant deformations β(in units of B/C) and γ. For technical reasons it is easier to work with the convention in which the rotation is along the z axis so that $\gamma = -120^\circ$ represents a prolate shape with the rotation axis perpendicular to the symmetry axis while $\gamma = -180^\circ$ represents oblate shape with symmetry axis parallel to that of rotation. The shape is triaxial in

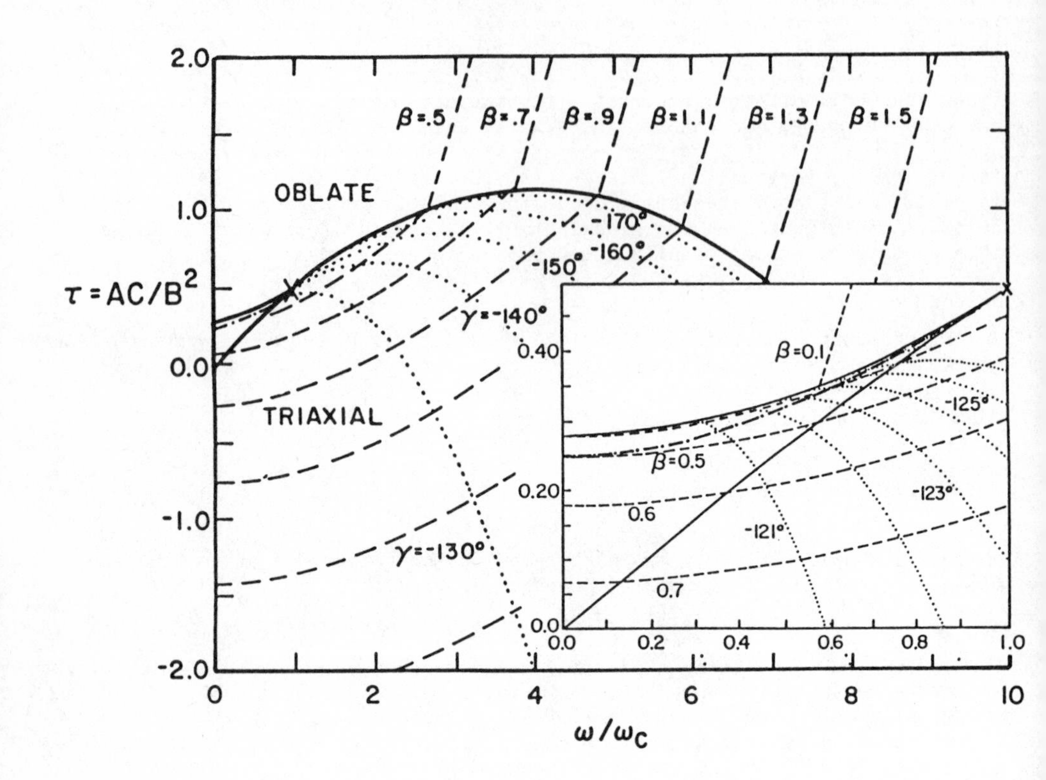

Fig. 1.
The universal phase diagram of nuclear shape transitions in the reduced temperature τ and ω/ω_c variables, eq.(2.3), for D = 0 limit. The tricritical point at $\omega/\omega_c = 1$, $\tau = 63/128$ is denoted by X. To its left the dash-dotted line denotes the first order transitions. The solid line to the right of X denotes the second order transitions. Also shown are the lines of β = const (dashed, β in units of B/C) and γ = const (dotted). Note that Z is the axis of rotation (cf. text). In the upper part of the figure γ is equal $-180°$ representing oblate shapes with the symmetry axis parallel to that of rotation. The solid lines to the left of X denote the boundary of the coexistence region of the triaxial and oblate "phases" (cf. Ref. 26).

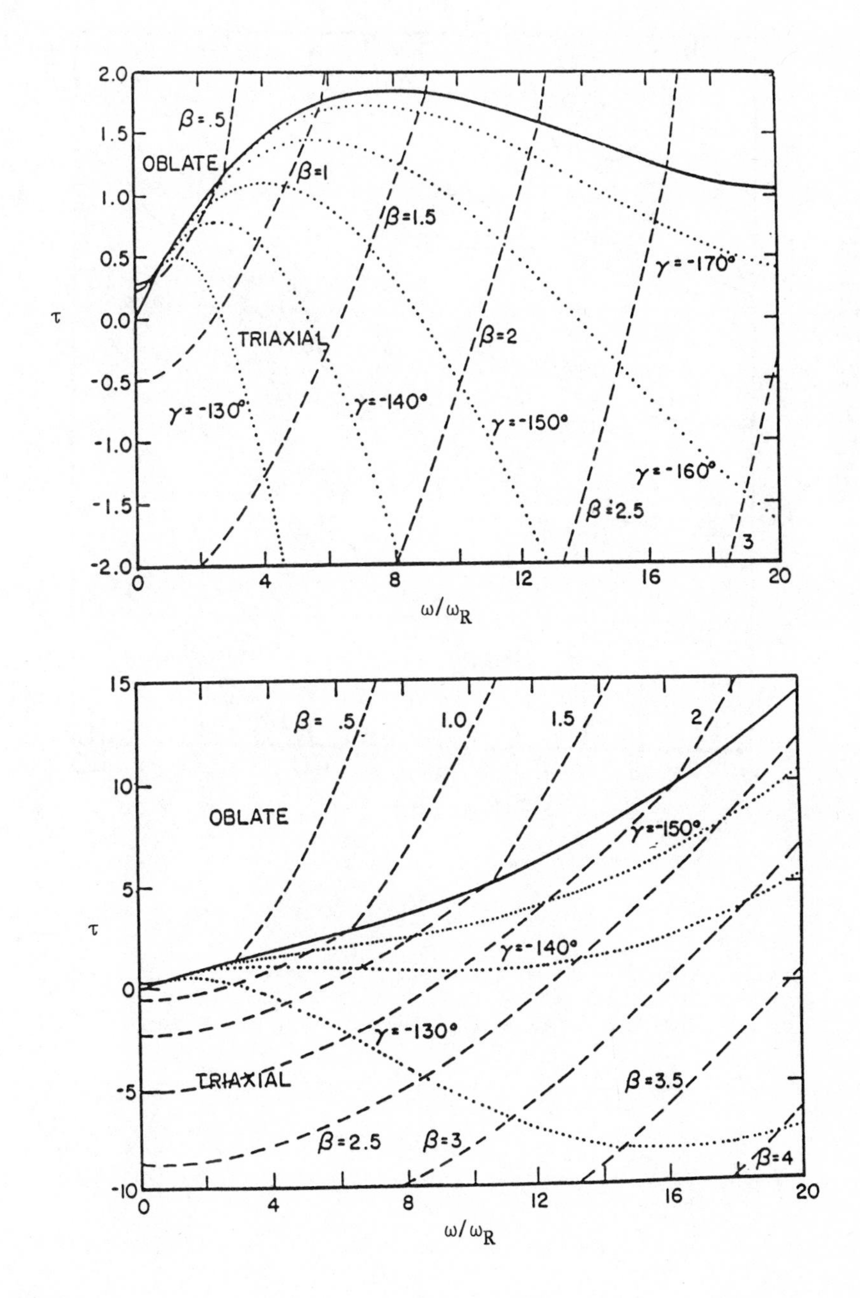

Fig. 2

Same as Fig. 1 but for $D \neq 0$ and abscissa in units of ω/ω_R. Top is for $D' \equiv \omega_R^2/\omega_D^2 = 0.01$ and bottom for $D' = 0.02$. For definitions of ω_R and ω_D cf. text. Note the sensitivity to the increase of D of the behaviour of the second order transition line at large ω.

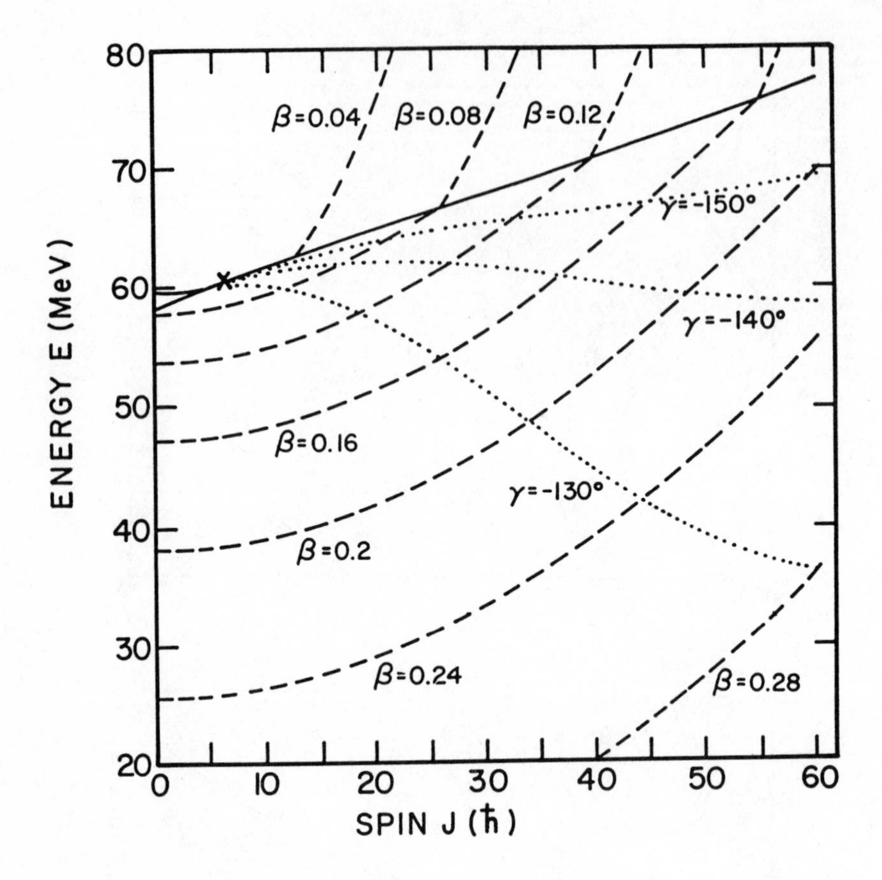

Fig. 3

The phase diagram of ^{166}Er in the plane of the excitation energy-angular momentum variables. Notation as in Fig. 1 except β is not scaled by B/C. For the details of the calculation cf. Ref. 26.

between these limits. One sees that the equilibrium nuclear shape changes rapidly from almost prolate to oblate when the temperature increases in the vicinity of T_c (i.e. $\tau=0$) and fixed ω/ω_c. The rate of the shape changes is faster for smaller values of ω/ω_c. The transition from triaxial to oblate equilibrium is not continuous for $\omega<\omega_c$ but has the abrupt first order character (dashed-dotted line). For $\omega>\omega_c$ the transitions are second order (solid line). The lines of the first and second order transitions join at the point $\tau=63/128$ and $\omega/\omega_c=1$. This point is called tricritical in the standard terminology of statistical mechanics. The phase diagram similar to Fig. 1 is found also in e.g. antiferromagnets in magnetic field or in liquid crystals in electric field[47]. The order of the transitions is probably difficult to observe in finite nuclear systems. However, the large fluctuations of the shape in the vicinity of the tricritical point should be relevant.

Departures from the D=0 limit can also be analysed analytically[26]. Now the results depend on three combinations of the parameters, $\tau=AC/B^2$, $\omega_R=(9/16)(B/C)(B/R)^{\frac{1}{2}}$ and $\omega_D=(3/2)B/(CD)^{\frac{1}{2}}$. The phase diagram in this case has similar qualitative features as in Fig. 1 but has important quantitative differences especially in the large ω region. This is shown on Fig. 2.

The transformation from the temperature and ω variables to the experimentally relevant E* (excitation energy) and J (spin) is done easily within the present approach. One must, however, know the quantitative dependence of τ, B/C, etc. on T. This dependence is presently being investigated microscopically for many nuclei[46] and one of the examples is shown in Fig. 3 where we display the results of the calculations of the phase diagram for ^{166}Er. The region J $\cong$ 10 - 20 $\hbar$, E* $\cong$ 60 MeV where ^{166}Er was recently studied experimentally (Refs. 8,9) is close to the tricritical point and the equilibrium is very sensitive to the values of E* and J offering a possible explanation of the observed rapid transition from almost prolate to oblate shape.

3. High Temperatures - Limits of Nuclear Binding

We now raise the temperature of the stove on which the nucleus is cooking. Experimentally this corresponds to forming compound nuclei at higher and higher excitation energy E*. How far up in E* can one go and still have a selfbound nuclear drop and not just a puff of dispersing vapour? How do the nuclear properties change on the way to losing the binding effects? Is the disappearance of the binding a gradual process or a sudden phase-transition-like limit? The temperature-dependent Hartree-Fock calculations provide this information. The first generation of such calculations[23,29] went to temperatures $T\sim3$-4 MeV and showed that the mean field potential changed very little from its cold shape in this temperature region and that nuclei were still safely bound. In order to extend these calculations to even higher temperatures it was necessary to understand how to consistently deal with the single-particle states in the continuum which become significantly populated at $T\gtrsim5$ MeV. As will be seen shortly the problem is not just technical but its resolution is necessary in order to understand how the concept of the equilibrated compound nucleus can be extended to very high excitation energies.

Consider an example of the Hartree-Fock calculations at T=7 MeV shown in Fig. 4. The population of the continuum s.p. states in this example leads to a nonvanishing density of nucleons extending outside the nuclear volume up to the walls of the quantization volume in which the calculations are performed. This is not surprising since in a static statistical theory one necessarily considers the nucleus in a thermal equilibrium with the particles which it evaporates, i.e. with its vapor.

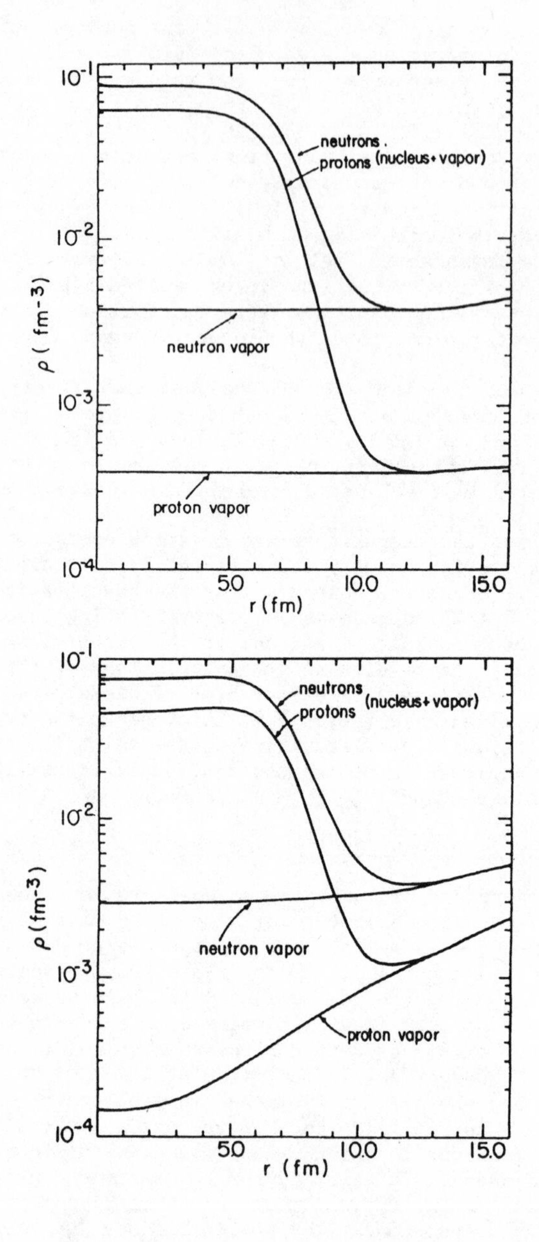

Fig. 4.
Radial dependence of the neutron and proton densities of nucleus + vapor and vapor solutions of the thermal Hartree-Fock equations at T = 7 MeV for uncharged (above) and normal, charged (below) ^{208}Pb. The particle numbers in both solutions for a constant chemical potential μ are volume dependent. μ was fixed to give the volume independent number of protons (82) and neutrons (126) in the nucleus which was defined using the "nucleus + vapor minus the vapor" subtraction procedure discussed in the text and Refs. (48-50).

The extensive quantities for such a system of nucleus+vapor, e.g. the energy, the entropy, the nucleon number, etc. all depend on the volume in which the system is placed. In Fig. 4 this means that the density of the outside nucleons will extend to larger and larger distances if one moves the wall further away from the nucleus. For the physical properties of the compound nucleus itself this volume dependence is spurious and one must learn how to extract the information about the nucleus from the properties of the nucleus-vapor system which emerges as a solution of Hartee-Fock equations.

This problem is not peculiar to the Hartree-Fock method only. In fact it was always present in the standard statistical model of the compound nucleus and resolving it was an important part of the definition of the model. The compound "levels" consist of the bound as well as the continuum part (i.e. have escape width), the latter corresponding to the components of the wave function with nucleons at large separations from the nucleus. In defining the statistical model the bound part is isolated by employing some separation procedure, e.g. R-matrix boundary conditions, projection operators, etc. Even the definition of such a common notion as "compound nucleus level density" implies such separation since the true density of nuclear states above the neutron threshold is, of course, infinite (i.e. quantization volume dependent). The finite temperature Hartree-Fock theory offers a physically satisfactory and probably least ambiguous separation procedure of the continuum. This procedure is based in the observation [48] that the Hartree-Fock equations at a given temperature and chemical potential possess two distinct solutions, the nucleus+vapor mentioned above, and another solution representing vapor alone (cf. Fig. 4). Using the differences of the extensive quantities (i.e. energy, entropy, etc.) of these solutions one obtains volume independent and, moreover, physically meaningful (cf. Ref.49) results required by the standard statistical model.

This subtraction procedure was extended and used in Refs 49,50 to calculate the properties and upper limiting temperatures of several nuclei. The sensitivity of the results to different effective interactions which are commonly used in the Hartree-Fock calculations was also tested. It was found in Refs 49,50 that despite the significant dependence of the single-particle potential on the temperature for T>5 MeV the deviations of the dependence of the entropy, energy, chemical potentials, etc. on T from the simplest Fermi gas relations is not significant (Fig. 5). This is explained by the delicate cancellation of several counteracting effects.

The existence of the upper limiting temperature of nuclear binding is demonstrated in Fig. 6 where the temperature dependence of the density profiles for the normal ^{208}Pb and the ^{208}Pb with artificially switched off Coulomb interaction is represented [50]. In this example the normal ^{208}**Pb becomes** unstable above T=8 MeV whereas its uncharged partner continues to exist up to T=12.5 MeV but becomes less and less dense with very diffuse surface spreading over **larger radii.** The uncharged nucleus disappears above 12.5 MeV and only the uniform vapor solution exists at T=13 MeV. This merging of the nucleus and the vapor solution in the uncharged system corresponds to reaching the critical temperature T_c of the liquid-gas phase coexistence found in the nuclear matter calculations[29-32]. Indeed the Hartree-Fock solution for the nuclear+vapor system at $T<T_c$ explicitly exhibits this coexistence of the more dense, liquid-like phase inside the nucleus and the less dense gas-like phase, i.e. vapor outside. As the temperature is raised towards T_c,

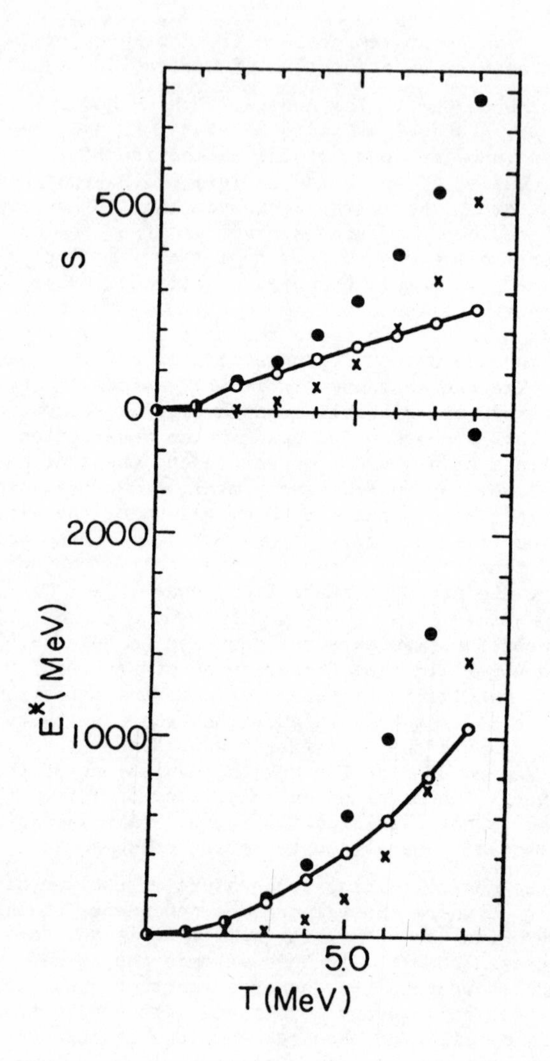

Fig. 5

Temperature dependence of the entropy S and the excitation energy E* of the nucleus+vapor (dots) and vapor (crosses) solutions and the subtracted quantities (circles) for ^{208}Pb (cf. Ref. 50). Despite significant changes of the single-particle potential for T> 5 MeV the subtracted quantities continue to show the Fermi gas behaviour, **$S \sim T$, $E^* \sim T^2$. Above T = 8 MeV** the nucleus becomes unstable (cf. Fig. 6 and Ref. 50)

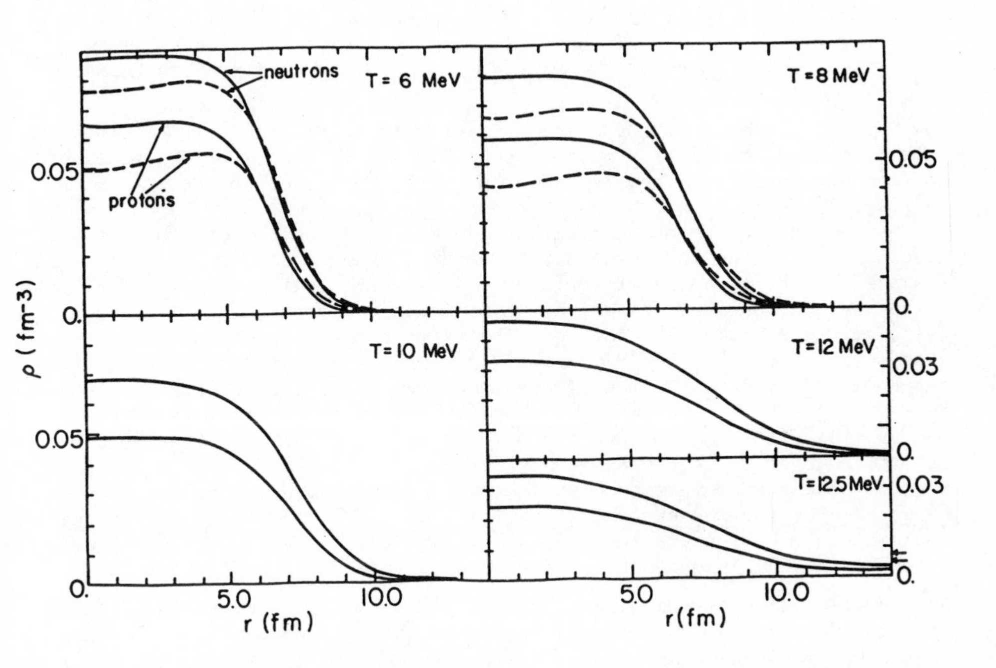

Fig. 6.
Subtracted proton (lower curves) and neutron (upper curves) density distributions at different temperatures for uncharged ^{208}Pb (full lines) and normal, charged ^{208}Pb (dashed lines) calculated with SKM interaction in R = 16 fm box (Ref.50).

Normal ^{208}Pb is unstable above T = 8 MeV. Above T = 12.5 MeV there is only a uniform solution of the Hartree-Fock equations for uncharged ^{208}Pb signifying the regime above the critical temperature of the liquid-gas coexistence. The arrows show the proton and neutron density of the uniform solution.

the difference between the phases decreases and disappears at and above T_c.

Such behaviour, however, is featured only in the uncharged nuclear system. The normal charged nucleus becomes electrostatically unstable due to the Coulomb repulsion long before the critical temperature T_c is reached (Fig. 6). The value T_{lim} of the temperature of this Coulomb instability depends on the charge-to-mass ratio of the nucleus and is, moreover, very sensitive to the choice of the effective interaction used in the Hartree-Fock equations (cf. Ref.50).

The Coulomb instability and its sensitivity to the choice of the effective interaction can be better understood within the framework of a simple finite-temperature version of the liquid-drop model which incorporates the main macroscopic features of the Hartree-Fock approach (Ref. 51). In this model the matter inside the hot nuclear drop obeys the equation of state of the liquid phase of the bulk nuclear matter. The drop is uniformly charged and has a temperature-dependent surface tension which vanishes at T_c. The conditions at the surface are chosen in accordance with the Hartree-Fock calculations by demanding the equilibrium with the surrounding vapor of evaporated particles. For simplicity the vapor is assumed to be completely screened. The main equations of the model are the coexistence conditions $P_n = P_v$, $\mu_n = \mu_v$ for the pressures and the chemical potentials of the matter inside the nucleus and the outside vapor at the same temperature. For the charged nuclear drop $P_n = P_{on} + P_{coul} + P_{surf}$ and $\mu_n = \mu_{on} + \mu_{coul}$, where the subscript zero denotes the bulk quantities given by the liquid phase part of the nuclear matter equation of state. The surface and the Coulomb contribution are found in a standard way. The vapor P_v and μ_v are given by the gas phase part of the bulk matter equation of state.

An example of the graphical solution of the coexistence equation is shown in Figs. 7,8. As the temperature increases the differences $P_{on} - P_v$, $\mu_{on} - \mu_v$ and P_{surf} decrease and vanish at T_c. The Coulomb contributions, P_{coul} and μ_{coul}, on the other hand, do not vanish as $T \to T_c$. At some limiting temperature T_{lim} the coexistence point reaches the upper boundary of the vapor phase and no coexistence is possible above T_{lim}. The value of T_{lim} (6.57 MeV in this example) is much below T_c (17.22 MeV). It is obvious that contrary to the universal character of T_c, the limiting temperature is not universal and depends on the nucleus under consideration. The theoretical dependence of T_{lim} on atomic weight A, charge to mass ratio Z/A, the equation of state and on the temperature dependence of the surface tension was studied in detail in Ref 51. The stiffer equation of state leads to higher values of T_{lim}. This dependence can be correlated with the dependence of T_{lim} on the effective interaction in the microscopic Hartree-Fock calculations of Refs 49,50.

The physical significance of T_{lim} is simple. Above the excitation energy which corresponds to T_{lim} a given nucleus cannot reach complete equilibrium and will decay in a more violent way than by slow particle evaporation. In recent analyses (Refs 2,3,27,28) evidence was found for the existence of the maximum energy that a nucleus can receive without breaking. In view of the theoretical picture discussed above studies of this energy and its dependence on the charge-to-mass ratio of formed nuclei may provide valuable information on the basic characteristics of nuclear matter - its equation of state and the temperature dependence of its surface tension.
It is interesting to address the question of the nature of the instability which occurs at T_{lim}. On Fig. 9 the comparison is given of the free energies of the bound Hartree-Fock solution and another solution which simulates the nuclear material dispersed in space. The picture is typical of the first order transitions where the second solution has initially higher free energy for low temperatures but becomes the global minimum above the limiting temperature T_{lim}.

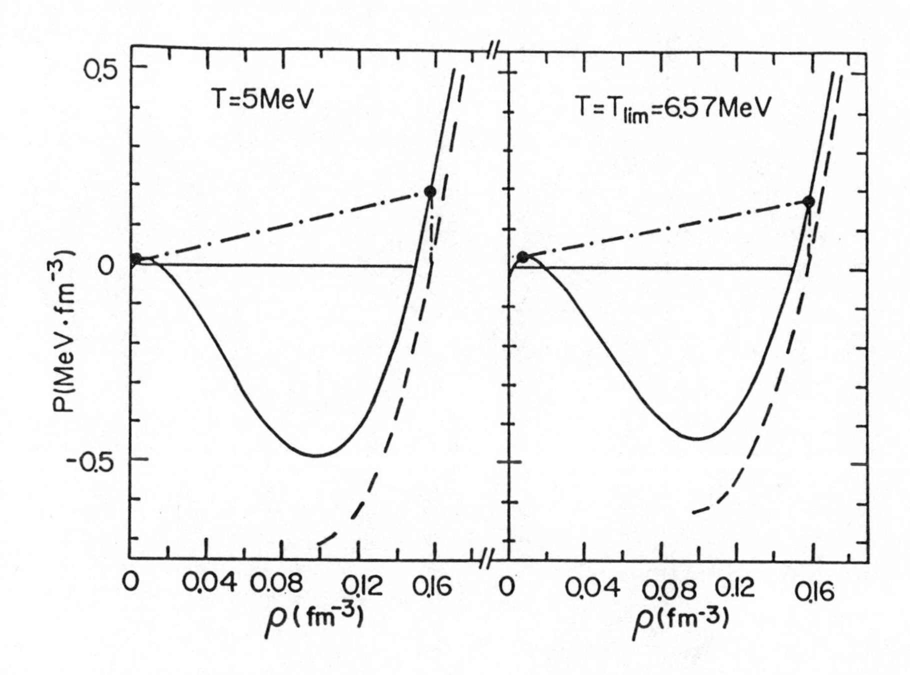

Fig. 7

Modified Maxwell construction for the nuclear charged drop-vapor **coexistence** (heavy dots connected by dashed-dotted lines) for ^{109}Ag for temperatures below (T = 5 MeV) and at the limiting value (T=6.57 MeV). Full drawn curves are the bulk isotherms with horizontal lines depicting the conditions of the **coexistence** of the bulk materials (normal Maxwell construction). Dashed lines show the full pressure $P_{on} + P_{surf}$ vs. the density inside the charged drop (ref. 51).

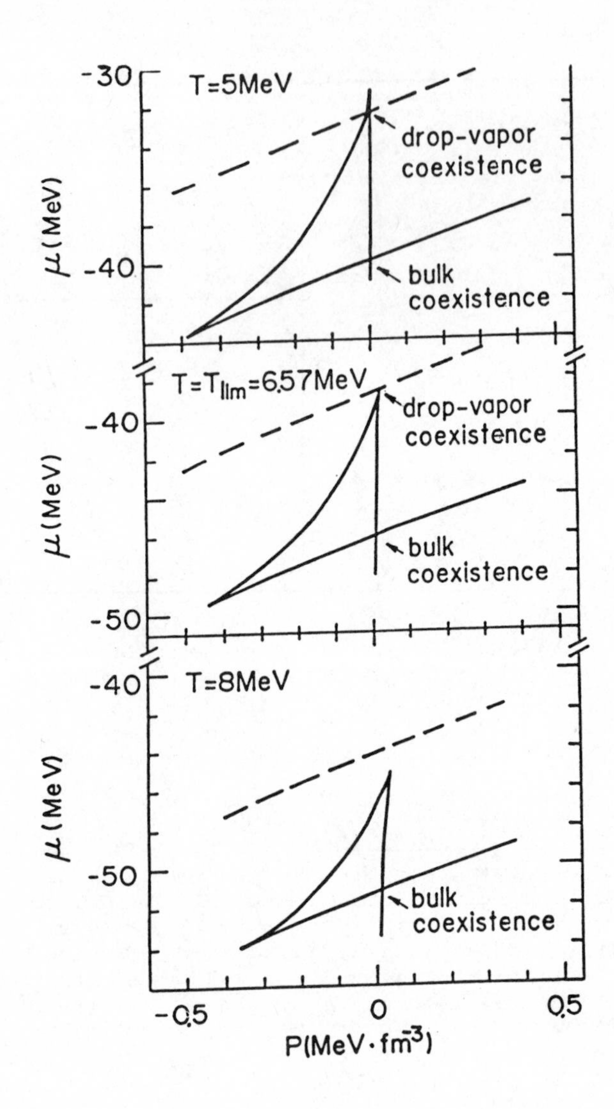

Fig. 8

Dependence of the chemical potential on the pressure for temperatures below (T= 5 MeV), at (T = 6.57 MeV) and above (T = 8 MeV) the limiting temperature. Full drawn curves denote the dependence for the bulk quantities, dashed curve – the dependence of $\mu_{on} + \mu_{coul}$ on $P_{on} + P_{coul} + P_{surf}$ inside a charged drop representing ^{109}Ag. No **coexistence** is possible above T_{lim} (Ref. 51)

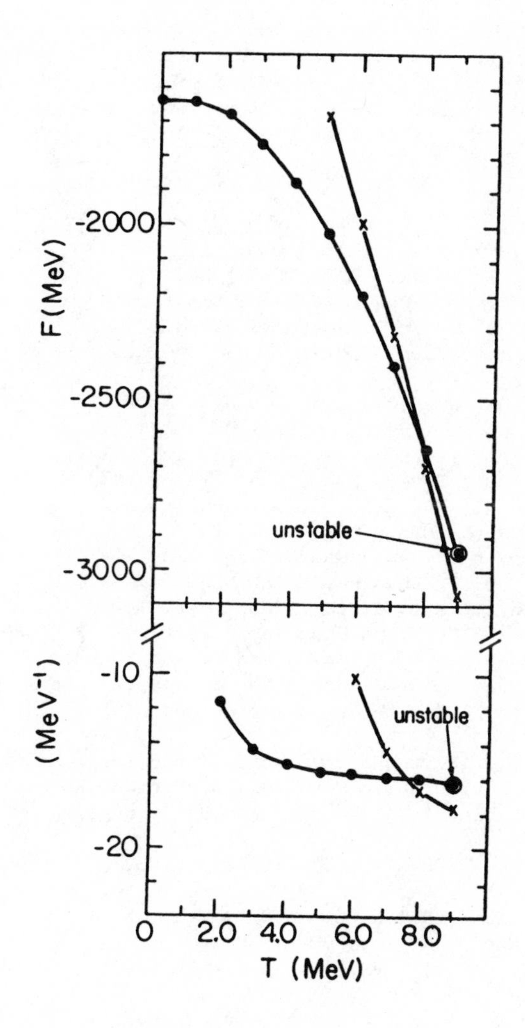

Fig. 9

Comparison of the free energies of the ^{208}Pb nucleus (dots) and of the same nuclear material dispersed, i.e. pressed against the walls of the quantization box (R=16 fm). The lower graph shows the same quantities but plotted as $(F - E_{g.s.})/T^2$. The behaviour is typical for the first order transition of the bound into dispersed situation.

Acknowledgement. Much of my knowledge about the phenomena in hot nuclei was aquired during my collaborations with Arthur Kerman (on general properties of the thermal Hartree-Fock), Paul Bonche and Domenique Vautherin (properties and stability of hot nuclei) and Yoram Alhassid, Jon Zingman and John Manoyan (shape transitions), to all of whom I am sincerely **grateful**.

References

1. S. Song et al, Phys.Lett. 130 B (1983) 14
2. B. Borderie et al, Z.Phys. Z316 (1984) 243
3. D. Jacquet et al, Phys.Rev.Lett.53 (1984) 2226
4. R.J. Charity et al, Phys.Rev.Lett.56 (1986) 1354
5. A. Bohr and B. Mottelson, Nuclear Structure, vol. 1 (Benjamin, New York, 1969) ch.2, app. 2
6. J.O. Newton et al, Phys.Rev.Lett. 46 (1981) 1383
7. J.J. Gaardhøje et al, Phys. Lett.139 B (1984) 273
8. J.J. Gaardhøje et al, Phys.Rev.Lett. 53 (1984) 148
9. C.A. Gossett et al, Phys.Rev.Lett. 54 (1985) 1486
10. W. Hennerici et al, Nucl.Phys. A396 (1983) 329 c
11. B. Haas et al, Phys.Lett. 120 B(1983) 79
12. A.M. Sandorfi et al, Phys. Lett. 130 B (1983) 19
13. J.J. Gaardhøje et al, Phys.Rev.Lett. 56 (1986) 1783
14. K. Neergard, Phys. Lett. 110 B (1982) 7
15. D. Vautherin and N.V. Vinh Mau, Phys. Lett. 120 B (1983) 261
16. P. Ring, L.M. Robledo, J.L. Egido and M. Faber, Nucl. Phys. A 419 (1984) 261 and the references therein for the earlier work of this group
17. O. Civitarese, S. Furni, M. Ploszajczak and A. Faessler, Nucl. Phys. A 408 (1983) 61
18. H. Sagawa and G.F. Bertsch, Phys.Lett. 146 B (1984) 138
19. O. Civitarese, R.A. Broglia and C.H. Dusso, Ann. Phys. 156 (1984) 142
20. M. Gallardo, M. Diebel, T. Dossing and R.A. Broglia, Nucl.Phys. A 443 (1985) 415
21. A.L. Goodman, Nucl. Phys. A 352 (1981) 30, 45, A 369 (1981) 365, A 370 (1981) 90
22. K. Tanabe, K. Sagawara-Tanabe and M.J. Mang, Nucl.Phys. A 357 (1981) 20,
23. L.G. Moretto, Nucl. Phys. A 182 (1972) 641
   M. Brack and P. Quentin, Phys.Scripta A 10 (1974) 163, Phys. Lett. B 52 (1974) 159
   P. Quentin and H. Flocard, Ann.Rev.Nucl.Sci. 28 (1978) 523
24. A.K. Ignatiuk, I.N. Mikhailov, L.H. Molina, R.G. Nazmutdinov and K. Pomo Nucl. Phys. A346 (1980) 191
25. S. Levit and Y. Alhassid, Nucl. Phys. A 413 (1984) 439
26. Y. Alhassid, S. Levit and J. Zingman, Phys. Rev. Lett. 57 (1986) 539 and Weizmann Institute preprint WIS-11/86
27. E.C. Pollacco et al, Phys. Lett 146 B (1984) 29
28. X. Campi, J. Desbois and E. Lipparini, Phys.Lett. 138 B (1984) 353, 142 B (1984) 8, Nucl.Phys. A 428 (1984) 327 c
29. U. Mosel, P.-G. Zint and K.M. Passler, Nucl. Phys. A 236 (1974) 252,
   G. Sauer, M. Chanda and U. Mosel, Nucl. Phys. A 264 (1976) 221
30. W.A. Kupper, G. Wegmann and E.R. Hilf, Ann.Phys. 88 (1974) 454
31. M. Jaqaman, A.Z. Mekjian and L. Zamick, Phys. Rev. C 27 (1983) 2782
32. M.W. Curtin, M. Toki and D.K. Scott, Phys. Lett. 123 B (1983) 289
33. G. Bertsch and P.J. Siemens, Phys. Lett. 126 B (1983) 9,
   J.A. Lopez and P.J. Siemens, Nucl. Phys. A 431 (1984) 728

34. J.E. Finn et al, Phys.Rev. Lett. 49 (1982) 1321
35. R.W. Munich et al, Phys. Lett. 118 (1982) 458
36. M.M. Gutbrod et al, Nucl. Phys. A 387 (1982) 177 c
37. C.B. Chitwood et al, Phys. Lett. 131 B (1983) 289
38. A. D. Pangiotou et al, Phys. Rev. Lett. 52 (1984) 496
39. P.J. Siemens, Nature 305 (1983) 410
40. H. Schulz et al, Phys. Lett. 147 B (1984) 17
41. H. Sagawa and G.F. Bertsch, Phys. Lett. 155 B (1985) 11
42. A. Vicentini, G. Jacucci and V.R. Pandharipande, Phys. Rev. C31 (1985) 1783
43. W. Bauer, D.R. Dean, U. Mosel and U. Post, Phys. Lett 150 B (1985) 53
44. J.P. Bondorf et al, Phys. Lett. 150 B (1985) 57
45. D.M.E. Gross, Zhang Xiao-ze and Xu Shu-yan, Phys. Rev. Lett. 56 (1986) 1544 and the references therein
46. Y. Alhassid, J. Manoyan and S. Levit, in preparation
47. A. Aharony , in "Critical Phenomena", Lecture Notes in Physics, 186,207 (Springer Verlag, New York 1983);
R. Hornreich, Phys. Lett. 109 A (1985) 232
48. A.K. Kerman and S. Levit, Phys. Rev. C 24 (1981) 1029,
A.K. Kerman, S. Levit and T. Troudet, Ann. Phys. 148 (1983) 436
49. P. Bonche, S. Levit and D. Vautherin, Nucl. Phys. A 427 (1984) 278
50. P. Bonche, S. Levit and D. Vautherin, Nucl. Phys. A 436 (1985) 265
51. S. Levit and P. Bonche, Nucl. Phys. A 437 (1985) 573
52. L.D. Landau and E.M. Lifshitz, Statistical Physics (Pergamon, N.Y. 1970) Sect. 139

Inst. Phys. Conf. Ser. No. 86
Paper presented at Int. Nucl. Phys. Conf., Harrogate, UK, 1986

Intermediate energy heavy ion reactions

B. Tamain

Laboratoire de Physique Corpusculaire, University of Caen,
14032 Caen Cedex, France

Abstract. Intermediate energy heavy ion physics has now been intensively studied for 3 years. Various observed reactions are described and show that very hot or exotic nuclei can be produced and studied. The limits of stability of hot nuclear matter are reached and typical times governing nuclear matter evolution can be determined. Special emphasis is laid on the open problems.

1. Introduction

For many years, heavy ion physics has been extensively studied at low bombarding energy, i.e. below 10 MeV/nucleon (Lefort 1978) and in the relativistic incident energy range, i.e. around 1 GeV/nucleon (Babinet 1986). In both cases the reactions can be divided in two classes, peripheral and central collisions, depending on the initial impact parameter. At low energy an important process observed in central collisions is fusion whereas peripheral reactions lead mainly to transfers. In the relativistic case central collisions lead to explosion whereas fragmentation is observed for more peripheral reactions. These features can be understood by using general physical concepts: at low energy the reaction mechanisms are governed by mean field or collective properties, and in the relativistic region nucleon-nucleon collisions are dominant.

In between these two regimes is intermediate energy heavy-ion physics which concerns projectiles with an incident energy ranging from 10 to 100 MeV/nucleon. In this region, one expects that mean-field effects and nucleon-nucleon collisions are competing processes. An important aim would be to observe and understand this competition. In the case of central collisions a further topic of interest is that very hot nuclei can be built and one is then able to study for the first time the properties of very hot and, eventually, compressed nuclear matter.

From an experimental point of view very few facilities are able to deliver intermediate energy heavy-ion beams and this physics is still relatively new and open. As in the low and high energy domains it is possible to classify most of the experiments performed into two classes depending on their topic of interest: peripheral or central collisions. In this paper we will keep this classification to give an overview of the available data. We will see that some interesting features have already been clearly established whereas several questions are still quite unclear.

2. Central Collisions and the Fusion Limits

Below 10 MeV/nucleon central collisions leading to fusion can be considered as two step processes: the fusion process itself (entrance channel) and the decay of fusion products (exit channel). The fusion process leads generally to a compound nucleus which is characterized by its excitation energy and angular momentum. The decay step can be either fission or evaporation depending mainly on compound nucleus mass. Above 10 MeV/nucleon, many experiments have been performed in order to detect either fission fragments or evaporation residues. In both cases information has been obtained both on the limits to the fusion process and on the properties of hot nuclei.

The simplest quantity which gives information on the fusion process is the fusion nucleus linear momentum (or recoil velocity). In the case of light systems this quantity can be directly measured by detecting the evaporation residues (Bizard 1985, Blachot 1985). For heavy systems the fusion nucleus undergoes fission and its velocity is determined by measuring the folding (correlation) angle between the two coincident fission fragments.

An example is shown in Fig. 1 (Fatyga 1985). For the smallest bombarding energy (7.4 MeV/nucleon) the correlation angle distribution exhibits one single bump, the mean value of which corresponds to a full momentum transfer of the projectile to the target (complete fusion). The width around this mean value is simply due to sequential decay of the fission fragments. For increasing bombarding energies the correlation angle distributions exhibit two bumps: one is located around 180° and is clearly associated with peripheral collisions. The second peak (smaller correlation angles) is the fusion peak (central collisions). Its position indicates that fusion is less and less complete as the incident energy is increased.

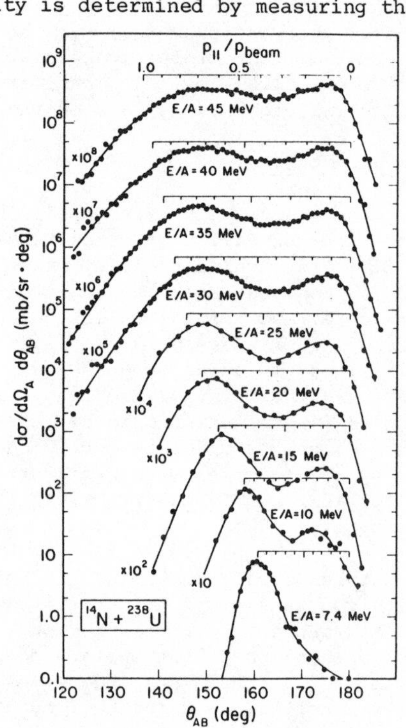

Fig. 1 Evolution of the fission fragment correlation angle distribution as the bombarding energy is increased from 7.4 to 45 MeV/nucleon. For each spectrum, the upper scale ($p_{\parallel}/p_{beam}$) indicates the proportion of projectile linear momentum which has been transferred to the target (Fatyga 1985). The data are for the $^{14}N + ^{238}U$ system.

This last feature is quite general (for a review see Grégoire 1986) and has also been deduced from experiments concerning evaporation residue detection. Therefore it reflects an entrance channel property rather than an exit channel effect. In fig. 2 it appears that a lot of experimental results can be brought together on the same universal curve whatever the nature of the projectile: the abscissa is the relative velocity of the interacting nuclei at contact and the ordinate is the proportion of the projectile linear momentum which is transferred to the fusion nucleus. The relevant entrance channel parameter is then a relative velocity rather than the total incident energy. On the other hand, the ordinate quantity is a proportion rather than the absolute value of the transferred momentum.

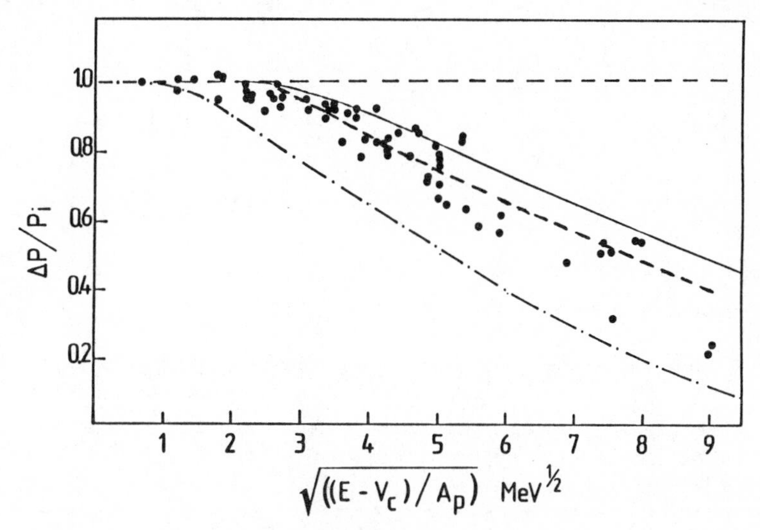

Fig. 2 Average proportion of the initial projectile linear momentum transferred to the target versus the square root of the incident energy per nucleon above the Coulomb barrier. The points have been compiled from both fission data and residue data. The curves are explained in the text.

This means that, on the average, each projectile nucleon transfers a linear momentum $\langle p \rangle$ to the target which does not depend on the projectile mass. Moreover, it turns out that the $\langle p \rangle$ value is rather constant over a wide incident energy range (Cassagnou 1986). The dashed curve of fig. 2 has been calculated simply by assuming $\langle p \rangle$ = 180 MeV/c. This result is reproduced in dynamical calculations (Grégoire 1984) involving both mean field effects and nucleon-nucleon collisions. It turns out that both mechanisms have to be considered (solid curve in fig. 2) in order to reproduce the data. The dot-dashed curve (fig. 2) indicates the mean field contribution.

A very interesting result for fusion cross sections is shown in fig. 3. The fission correlation angle distributions are plotted for the $^{40}Ar + ^{232}Th$ system at four bombarding energies (Conjeaud 1985). The two bumps associated with peripheral collisions and fusion are clearly visible at 31 MeV/nucleon, but the fusion peak decreases dramatically for higher bombarding energies: it is no longer visible at 44 MeV/nucleon. Quite similar behaviour has been obtained in experiments concerning evaporation residue detection (Blachot 1985, Auger 1986, Bizard 1985, Borderie 1984a). This rather fast transition is then observed whatever the exit channel. In the case of argon projectiles, the corresponding bombarding energy is about 40 MeV/nucleon and does not depend significantly on the target mass. On the other hand, this projectile velocity value is not universal since a fusion bump has been recognised at 60 MeV/nucleon in the case of lighter projectiles (Cassagnou 1986, Galin 1982).

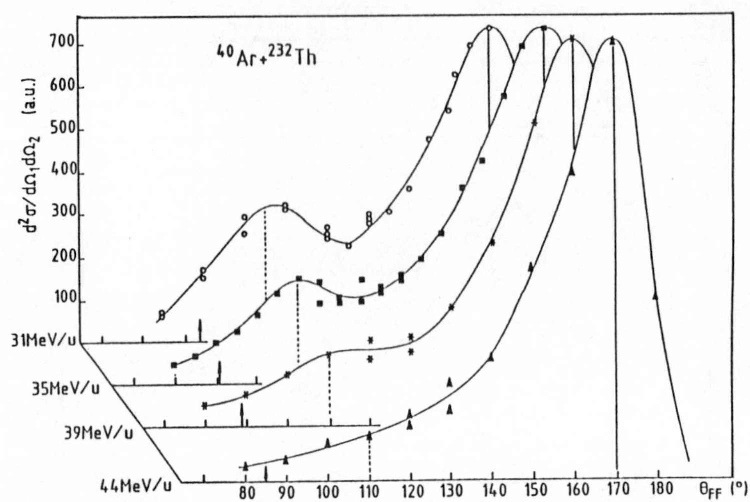

Fig. 3 Folding angle angular distributions for the Ar + Th system at four bombarding energies. The peripheral cross section has the same magnitude in all cases but the fusion part has vanished at 44 MeV/u. Taken from Conjeaud (1985).

The detailed interpretation of the above features is difficult but it turns out that the most important reason for the observed transition is connected with the maximum excitation energy bearable by an equilibrated nucleus. Figure 4 is an illustration of this statement: the ordinate quantity is the calculated maximum Ar projectile energy for which fusion can take place. This calculation takes into account the entrance channel effect summarised in fig. 2, and the maximum excitation energy bearable by the fusion nucleus has been obtained in a Hartree-Fock calculation (Levit 1985). The calculated argon energy is found to be independent of the target mass, in agreement with experiment, and its value is close to the experimental 40 MeV/nucleon value quoted above. It then appears that the main assumptions of the model are valid.

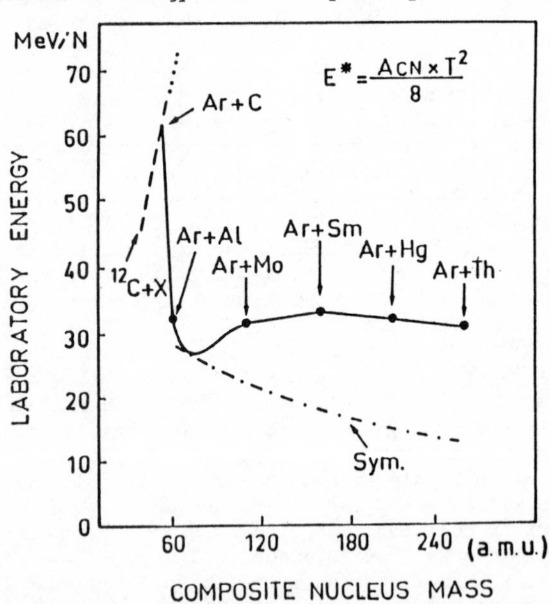

Fig. 4 Calculated threshold laboratory energies per nucleon for the observation of fusion in Ar induced reactions versus the fusion nucleus mass (Auger 1986).

However, such a reasonable agreement between the model and experimental results should not mask several questions:

i) The disappearance of the fusion bump noted in fig. 3 does not mean that fusion really vanishes at a given bombarding energy. Indeed, even at 44 MeV/nucleon, some events of fig. 3 can be attributed to fusion and the disappearance of fusion would not then be very abrupt.

ii) If fusion nucleus stability was the only concept governing fusion disappearance, one would observe that the transition bombarding energy would correspond to an excitation energy of the fusion nucleus which would be projectile independent. This does not seem to be the case (Cassagnou 1986): higher excitation energies can be reached when the projectile mass is larger. Angular momentum effects in the fusion nuclei cannot explain this feature and one is rather constrained to believe that very hot nuclei stability limits are not the only reason for the disappearance of the fusion bump. Entrance channel effects also play some role.

iii) Another difficulty has to be stressed: the measured quantities (fig. 2) are generally fusion nuclei linear momenta rather than excitation energies or temperatures. The relations between these quantities are not straightforward and are model dependent. Generally, it is assumed that one part of the projectile fuses with the whole target, the remaining part of the projectile flying at zero degrees with the beam velocity. This assumption, which is clearly wrong for symmetric systems, seems to reproduce detailed results for asymmetric ones rather well (Nifenecker 1986). However, some doubts remain about the corresponding data interpretation.

iv) Finally, the fundamental question concerns the degree of thermalization of the fusion nuclei. Are we dealing with equilibrated or non-equilibrated fusion nuclei? This fundamental question will be discussed extensively in section 4. Section 3 also sheds some light on this matter.

3. Peripheral Collisions

From an experimental point of view peripheral reactions are recognized by detecting a final product which resembles the projectile both in mass and velocity. At low bombarding energies, quasi-elastic transfers (in which one or few nucleons of nucleus A jump to quantum states of nucleus B) belong to this class of events. They are a clear indication of mean field effects. In the case of relativistic energies, peripheral reactions lead to fragmentation processes which can be described by the participant spectator model (Goldhaber 1974): nucleons of nucleus A which overlap with nucleus B interact strongly with it (participant zone) whereas the non-overlapping part of A (spectator) keeps its initial velocity. Mean field effects seem to be negligible in such processes.

At first glance (but we will see that this is incorrect) intermediate energy heavy-ion peripheral collisions look roughly like relativistic energy ones. Figures 5 and 6 are two illustrations of this feeling. Figure 5 shows typical isotopic distributions for projectile-like fragments resulting from peripheral reactions. They look quite similar for 44 MeV/nucleon and 213 MeV/nucleon, at variance with the 5.5 MeV/nucleon result. In this last case the results indicate a rather fast isospin degree of freedom equilibration: the projectile-like fragment isospin is

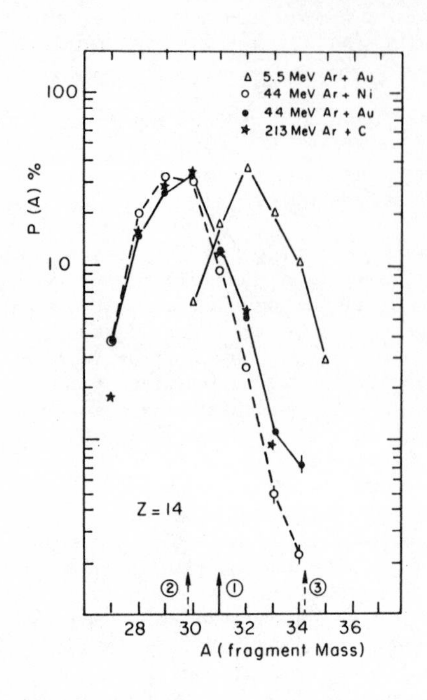

Fig. 5 Comparison of silicon isotopes production in peripheral reactions for Ar induced reactions at various bombarding energies. The 44 MeV/nucleon results look like the 213 MeV/nucleon ones, at variance with the 5.5 MeV/nucleon data. From Guerreau (1983).

strongly influenced by the target one. At 213 MeV/nucleon, the reaction time is so short that this collective degree of freedom is not equilibrated: the projectile-like fragment isospin is equal to the projectile one. In Guerreau (1983) a similar explanation is suggested for the intermediate energy (44 MeV/nucleon) case. The high energy regime would then already be reached and mean field effects would already have vanished at 44 MeV/nucleon.

A similar conclusion may be reached in studying the projectile-like fragment velocity distributions. They are peaked roughly at the beam velocity and their widths can be understood (Barette 1984a, Dayras 1986) in the participant-spectator model framework as in the relativistic energy case. In fig. 6 the experimental points are rather well fitted with such a model. Again one could be tempted to conclude that the high energy regime is already reached.

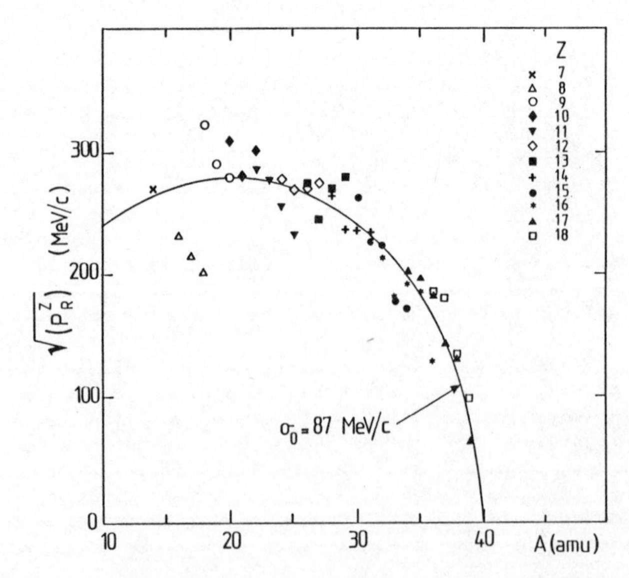

Fig. 6 Variance of the projectile-like fragment linear momentum distribution versus the projectile-like fragment mass. The points are experimental. The curve has been calculated in the participant spectator model approach: the projectile is assumed to be geometrically cut out by the target; its momentum distribution simply reflects the abraded nuclei Fermi momenta. System Ar 44 MeV/nucleon + Al. From Barette (1984).

Such a conclusion is in fact quite exaggerated (Dayras 1986) and many experimental results indicate clearly that mean field effects do play a determinant role even at 50 or 60 MeV/nucleon. First of all it is necessary to stress that the results of figs. 5 and 6 can be misinterpreted. In the case of fig. 5, a sequential decay of the primary projectile-like fragments can modify significantly the primary distributions that one would like to understand. Moreover, target isospin has been found to influence the projectile-like fragment isospin in many experiments (Dayras 1986). This feature which is extensively used to form and study very exotic nuclei (Detraz 1986) is a first indication of mean field effects in peripheral (fast) reactions. In the case of fig. 6, it is rather important to note that the participant spectator model predicts that the projectile-like fragment velocity distributions should be gaussian, which is definitely not the case in the intermediate energy region. Figure 7 is a nice illustration of this feature: if most of projectile-like fragments have velocities close to the beam velocity one can note that: i) the beam velocity peaks exhibit strong low velocity tails; this component has been subtracted in order to get the experimental points of fig. 6: ii) a small velocity (relaxed ?) component exists: in the case of projectile-like fragment atomic numbers close to the projectile one this component is clearly separated from the previous one. These more or less relaxed quasi-projectiles clearly result from peripheral reactions for which the mean field has played a determinant role. In order to support this statement let us consider in more detail the low velocity tails quoted above. In fig. 8 (Chomaz 1986, Frascaria 1986) they have been carefully analysed for projectile-like fragments close to the projectile. They exhibit several bumps which have been already observed and understood below

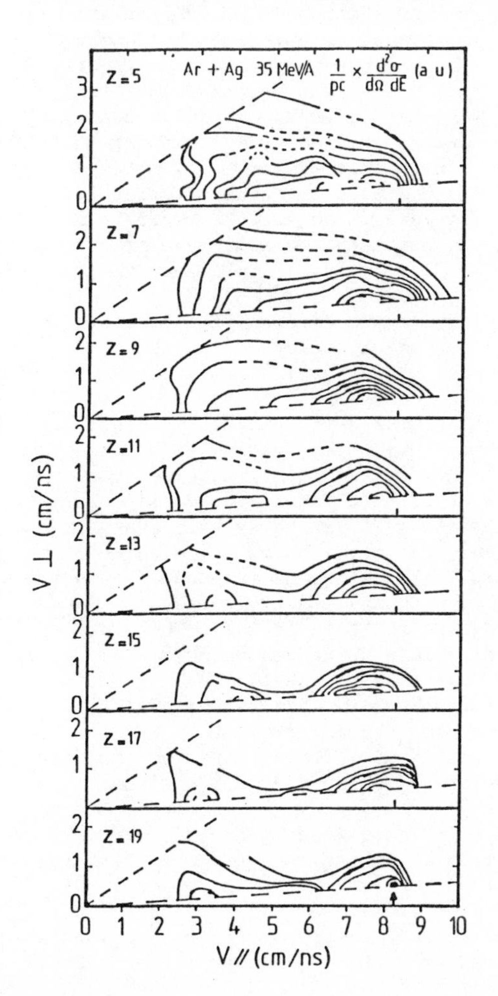

Fig. 7 Contour plots for the quantity $1/p_c\ d^2\sigma/d\Omega p$ (invariant quantity under galilean transformation) versus the velocity vector of detected projectile-like fragments. $V_\parallel$ and $V_\perp$ are the velocity components in the parallel and perpendicular directions to the beam. The various plots correspond to various atomic numbers of projectile-like fragments. System Ar + Ag at 35 MeV/nucleon (Bizard 1986a).

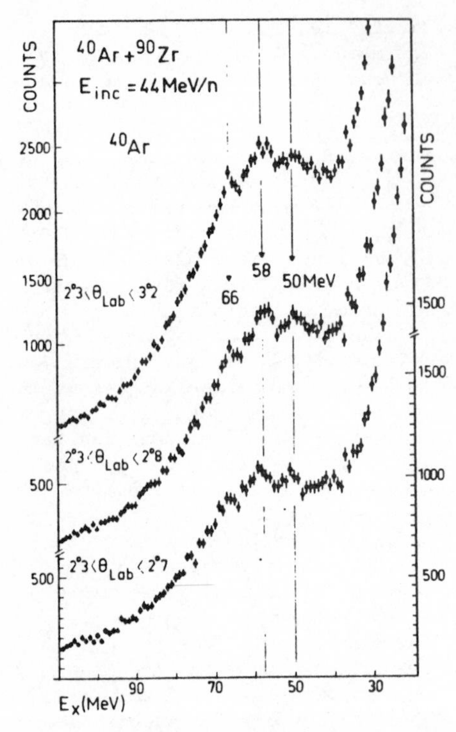

Fig. 8 Tails of Ar projectile-like fragment kinetic energy spectra obtained in the case of the reaction Ar + Zr at 44 MeV/nucleon. The abscissa has been converted to an excitation energy scale. A few bumps are clearly visible. From Chomaz (1986).

10 MeV/nucleon. They are not due to transfer product decay (Blumenfeld 1985). Their positions depend neither on the detection angle (fig. 8), nor on the incident energy (Chomaz 1986), nor on the nature of the projectile (Chomaz 1984a). Instead they are directly connected with target properties and have been successfully interpreted (Chomaz 1984b) as target multiphonon excitations. They are then strong evidence of collective, or mean field effects, which could be responsible for energy damping in the entrance channel. Recently they have been observed again with the SPEG spectrometer installed at GANIL.

Further experimental evidence of the persistence of mean-field effects in intermediate energy heavy-ion induced peripheral reactions may also be found in the existence of an unexpectedly strong quasi-elastic transfer reaction component. This mechanism has been unambiguously recognized either in inclusive experiments where definite exit channels were selected and analysed in a spectrometer (Barette 1986), or in exclusive measurements performed with large solid angle multidetector devices (see the review by Bizard 1986b). One example of such multidetector devices is the plastic wall (Bizard 1986c), installed at GANIL, which detects every light charged particle emitted between 3° and 30° to the beam. If such a device is used in coincidence with a projectile-like fragment telescope transfer reactions may be recognized when no light charged particle has been detected in the wall. A typical result is shown in fig. 9. The ordinate quantity is the percentage of projectile-like fragments which can be attributed to transfer (the remaining part can reasonably be attributed to fragmentation). It is plotted versus the number of transferred charges. The percentages quoted in fig. 9 are astonishingly large at 35 MeV/nucleon. Recent results (Bizard 1986d) indicate that the transfer component is still important even for argon beams of 60 MeV/nucleon.

Let us now consider the small velocity (relaxed ?) component observed in fig. 8. Its origin is not yet understood. It has been proposed that it is due to target shattering (Fields 1984) or participant-zone decay (Borderie 1984b). However it appears that, whatever the projectile nature, its mass distribution extends up to (but not above) the projectile mass (Bizard 1986a, Bonnafond 1986). This feature excludes at least the

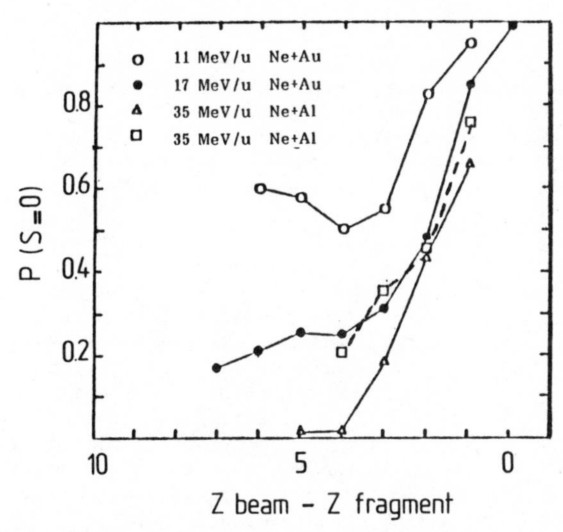

Fig. 9 Percentage of transfer reactions for various projectile-like fragments. The abscissa is the difference between the projectile and projectile-like fragment atomic numbers. Results corresponding to various beams are plotted.

first interpretation and the existence of a deep-inelastic-like component cannot be ruled out. Anyhow, strong energy relaxation is observed.

It then appears that, depending on the observed experimental features, intermediate energy heavy-ion peripheral reactions exhibit either nucleon-nucleon collision effects or mean field properties, or both. As expected, we are dealing with a transition region between low and high energy regimes; however, the mean field contribution remains unexpectedly strong. This is a first indication that the energy relaxation time could be comparable to, or shorter than, the collision time (less than 10^{-22} s) even if other evidence is needed in order to support or weaken such a feeling (see sections 4 and 5). From a theoretical point of view the Landau-Vlasov equation which includes both mean fields and nucleon-nucleon collision effects is now extensively used in order to reproduce the data (Grégoire 1985; and a review by Grégoire 1986).

4. To What Extent is Energy Relaxation Achieved?

The results discussed in section 2 indicate that hot nuclei can be formed in central collisions above 10 MeV/nucleon. In section 3, we have seen that mean-field effects still play a determinant role in this energy range. It is reasonable to describe the reaction as a two contribution process. In the early stage of the collision, nucleon-nucleon collisions and projectile break up lead to fast particle emission. These features are responsible for the fact that fusion is incomplete. The remaining part of the incident energy is transformed into the excitation energy of the outgoing products. We have seen that the energy thermalization time is probably small, but from the results discussed up to now, we do not know to what extent thermalization of outgoing products is achieved. In other words, we do not know if it is possible to extract from the data a temperature which characterizes the outgoing products. This question holds of course for fusion nuclei but also for final products resulting from a peripheral collision. One may answer it by trying to measure these excitation energies or temperature values (if any). In this section we discuss the corresponding experiments. A difficulty will be to make sure that the temperatures extracted from the data are true thermodynamical temperatures.

4.1 Excitation energy measurements

Very few excitation energy measurements have been performed up to now. Their principle is to count in a 4π device the number of neutrons evaporated from the fusion nuclei. In the case of heavy systems (mass larger than 200) 85 to 90% of the excitation energy is carried away by neutrons and their multiplicity is then strongly correlated with excitation energy (Galin 1986).

A typical result is shown in fig. 10 (Galin 1986, Jahnke 1986). It appears that the neutron multiplicity is roughly a linearly increasing function of the linear momentum transferred to the target. This result is in good agreement with the simple incomplete fusion model described in section 2 and it is a strong indication that the transferred excitation energy is really thermalized in the fusion nucleus. However, no result is available for bombarding energies exceeding 20 MeV/ nucleon and experiments in this energy range are strongly needed. They will be performed very soon at GANIL.

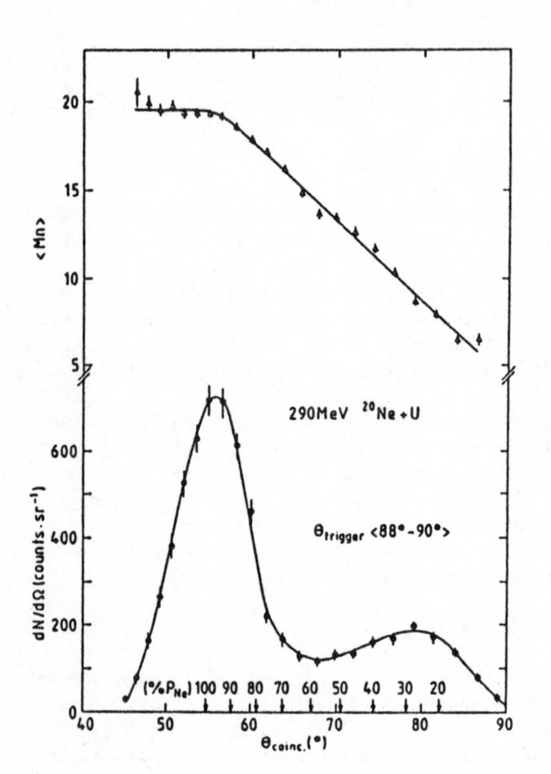

Fig. 10 The folding angle distribution of the fission fragments (lower curve) and the associated neutron multiplicity (upper curve). System 14.5 MeV/nucleon Ne + U.

4.2 Temperature measurements

At low bombarding energy (below 10 MeV/nucleon) it is rather easy to deduce a temperature from the exponential shapes of light particle kinetic energy spectra because these particles are generally evaporated from one or two well-known thermalized sources. Above 10 MeV/nucleon this prescription has also been extensively used, but it is now well known that the corresponding results are often unreliable for at least two reasons: i) the growing importance of direct emission (i.e. non-evaporated particles); ii) The multiplicity and the various recoil velocities of the evaporative sources. It is possible to get reliable temperature values (if any) only if i) the direct particle influence is strongly reduced; ii) evaporative sources are selected. These two aims can be achieved only in coincidence experiments which select definite mechanisms and definite recoil velocities of evaporative sources. Two examples are given below:

a) Coincidences between particles and fusion decay products

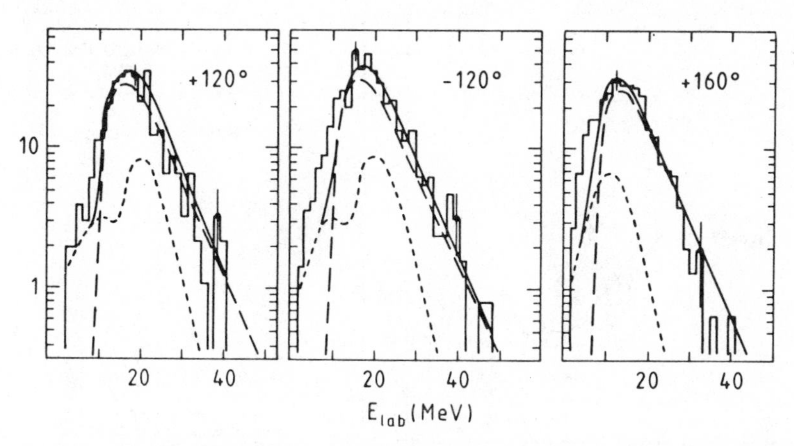

Fig. 11 Kinetic energy distributions for α particles detected at ± 120° and 160° in coincidence with fission fragments resulting from fusion nucleus decay. Ar + U at 27 MeV/nucleon. See text for further detail. From ref. Galin (1986), Jacquet (1986).

Figure 11 shows (Galin 1986) kinetic energy spectra for particles detected in triple coincidence with two fission fragments resulting from a fission following fusion (selection of small folding angles). In such an experiment the direct particle contribution is minimized by choosing a backward detection angle (120° or 160°). On the other hand, fission following complete fusion is selected by a proper choice of the fission fragment correlation angle. In fig. 11, it turns out that the measured kinetic energy spectra (histograms) exhibit exponential shapes. From their shapes one can extract a temperature value of 4.2 MeV (long dashed curve) in agreement with what is expected if these particles are evaporated before fission from thermalized nuclei. Such an agreement (which has also been obtained for 60 MeV/nucleon C projectiles (Song 1983) is a good indication that thermalization of excitation energy has been reached and is possible even for large bombarding and excitation energies.

b) Particle-particle coincidences at small relative angles

We will see in section 5 that a very important property of very hot nuclei would be that they can decay by cluster evaporation. Let us assume that such clusters are evaporated from a thermalized fusion nucleus: for instance a ^{6}Li cluster. From general statistical thermodynamical arguments, the probability of observing the ^{6}Li cluster in a given quantum state of excitation energy E* is proportional to $e^{-E^*/T}$ if T is the temperature of the heat bath from which the ^{6}Li is evaporated. Now, the excited ^{6}Li states are unstable resonances, the decay of which leads to ^{4}He and ^{2}H unexcited nuclei. One can determine the ^{6}Li excitation energy E* by measuring the relative kinetic energy E_{rel} of ^{4}He and ^{2}H final nuclei. In fig. 12 (Pochodzalla 1985a) is given the E_{rel} distribution (left side). Two peaks are clearly visible: they correspond to two excited states of ^{6}Li clusters. Their abundance ratio can be used to extract the temperature T. A value of 5 MeV leads to the best agreement with the data.

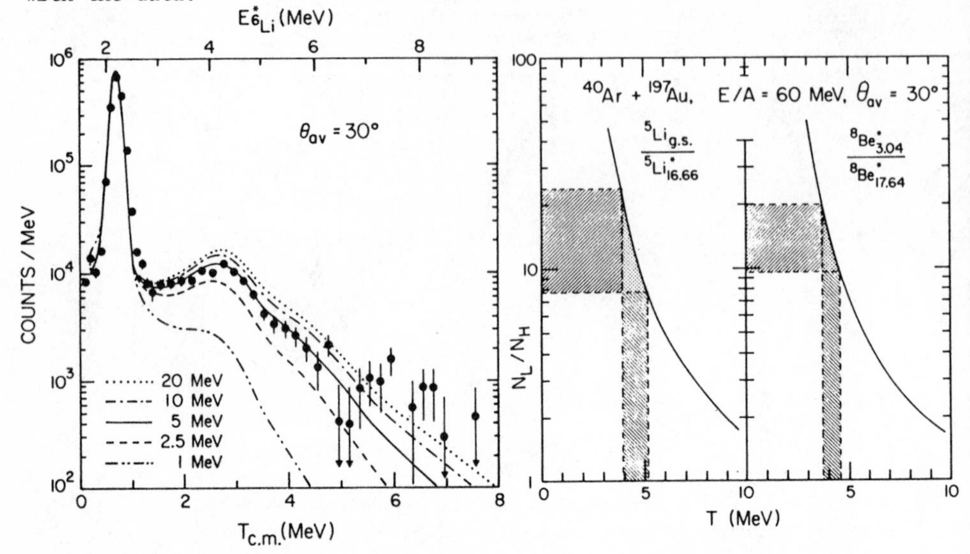

Fig. 12 Left: Relative kinetic energy distribution for two coincident ^{4}He-^{2}H nuclei. One recognizes two bumps which are correlated with two excited states of ^{6}Li. The population ratio of these two states may be used to deduce the emitting source temperature. Right: Population ratios for two states of ^{7}Li on the one hand and ^{8}Be on the other hand versus the temperature of the emitting source. The curve is the theoretical law. The hatched areas correspond to the experimental results (including uncertainties). The deduced temperature is the same in both cases. From Pochodzalla (1985 a,b).

One can make several criticisms of such experimental studies. First, the detected light nuclei can result from independent emissions rather than from excited clusters; secondly, some excited clusters can result from previous cluster decays rather than directly from the hot source; thirdly, such experiments do not select definite processes (for instance fusion); last but not least, data interpretation is performed <u>assuming</u> that thermalization was achieved.

Concerning the first criticism, it turns out that the measured relative kinetic energy distributions of the detected nuclei (left part of fig. 12) can be decomposed in two contributions: i) background, taking into account the sequential evaporations; ii) the peaks themselves which are due to two particles emitted at the same time and having no further interaction with the emitting source; this second component has all the characteristics of a cluster decay. The population ratios are calculated by subtracting the background component.

The second criticism is severe only if the considered excited states are close one to the other (Pochodzalla 1985 a,b) which is not the case for the results of fig. 12b.

The third criticism is less fundamental if clusters are detected at large angles to the beam, which excludes very peripheral reactions. However, various mechanisms are still mixed up and the measured temperatures (if any) are perhaps temperatures of subsystems in the process of expanding rather than temperatures of final equilibrated nuclei (Chitwood 1986). An improvement of the method will be to select central collisions in detecting coincident fusion decay products (evaporation residues or fission fragments).

The most serious criticism is the fourth one concerning the thermalization assumption. An answer may be found in the right part of fig. 12 (Pochodzalla 1985b): for a given system, whatever the considered excited clusters, the deduced temperature is the same (slightly below 5 MeV). This is a strong justification of the basic hypothesis. However, in similar studies concerning similar systems in which the state populations have been deduced from gamma ray-cluster coincidence measurements (Morrissey 1985), the deduced temperatures have been found to be much smaller (< 1 MeV). This discrepancy, which is probably due to sequential feeding of the levels involved in gamma-ray measurements (Gelbke 1986), remains open to discussion.

c) Indirect measurements

Temperature values have also been extracted from careful evaporation residue analysis. The main limitation of the method is that the conclusions are strongly model dependent and are only indications (but not proofs) that thermalization is really achieved.

An example concerning the system Ar + Al at 19.5 MeV/nucleon (Auger 1985b) is shown in fig. 13. The kinetic energy distributions of the detected fragments exhibit two peaks. The left one can be interpreted as a fusion peak. It can be theoretically reproduced in a simple model of incomplete fusion where the whole heavy partner (argon nucleus) has fused with part of the light one (see section 2). The subsequent decay is then simulated

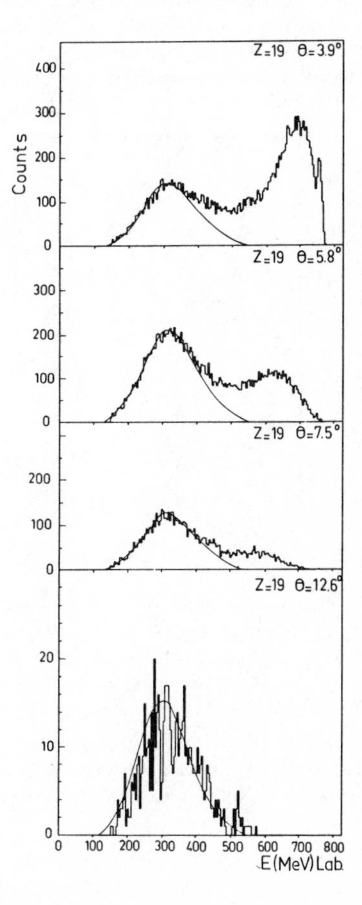

Fig. 13 Kinetic energy distributions of Z = 19 nuclei detected at various angles for the reaction Ar + Al at 19.5 MeV/nucleon. The left peak is the fusion peak. The right peak is due to peripheral collisions. From Auger (1985b).

by a Monte-Carlo calculation. The fusion nuclei excitation energy is deduced from its recoil velocity. It is then converted into a temperature in order to perform the evaporation calculation. In the above example the deduced temperature is about 7 MeV. Its reliability lies in the agreement between calculated and experimental spectra (see section 5).

4.3 Conclusion

It is rather difficult to answer the thermalization question. What we know is that very hot nuclei can be formed far above 10 MeV/nucleon. For bombarding energies below 20 MeV/nucleon they are thermalized; for larger bombarding energies we have only good indications that thermalization could be approximately achieved up to 60 MeV/nucleon (at least for light projectiles) but a more precise answer will be obtained only in very exclusive experiments in which most of the hot nuclei decay products will be detected.

5. Properties of Very Hot Nuclear Matter

In the preceding sections we have shown that very hot equilibrated nuclei are probably formed in intermediate energy heavy-ion reactions. It is then rather interesting to try to understand their decay properties. At low temperature this is achieved by using the statistical theory in which the main assumption lies in the microcanonical description. When the temperature is increased (Tamain 1985) the most important evolution of this theory can be understood in looking at fig. 14 (Delagrange 1986a).

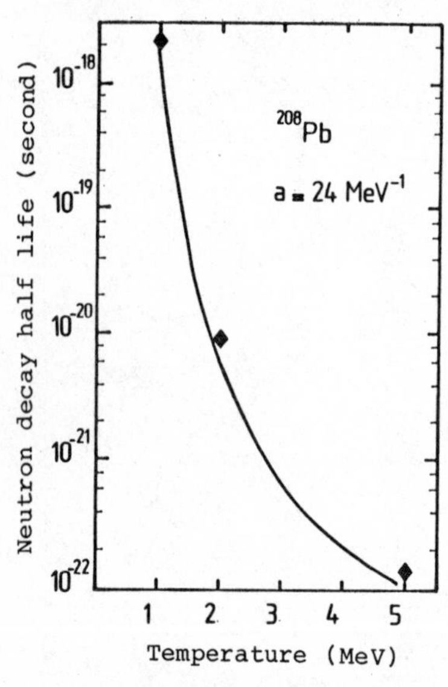

Fig. 14 Mean time between two successive neutron evaporations from a Pb nucleus characterized by a temperature T.

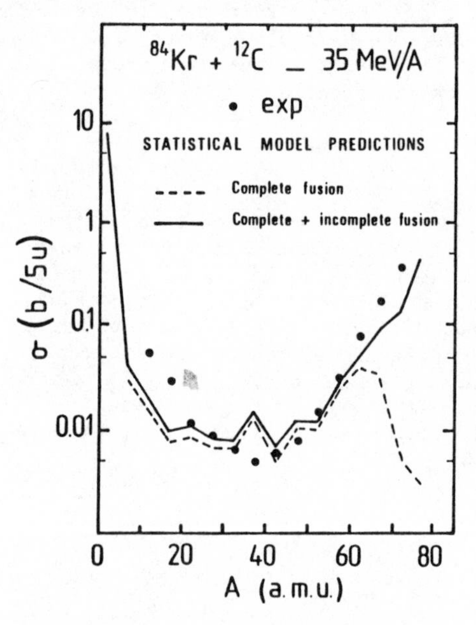

Fig. 15 Final mass distribution for the reaction Kr+C at 35 MeV/nucleon. The solid curve is calculated assuming complete + incomplete fusion processes followed by particle and cluster evaporation. From Auger (1985a).

The ordinate quantity is the calculated mean time Δt between two successive neutron evaporations from a lead nucleus. The abscissa is the corresponding temperature T. It appears that Δt is a strongly decreasing function of T. For a temperature value of 5 MeV it is about 10^{-22} s, which may be compared with the thermalization time estimated in section 3. It turns out that full thermalization is probably difficult to achieve for temperatures exceeding 5 to 7 MeV: in Delagrange (1986b) it is, for instance, calculated that evaporation may occur <u>during</u> the incident energy relaxation which would reduce the maximum temperature of outgoing nuclei significantly when the total available energy is large. On the other hand, Δt can also be compared with the typical fission time of heavy fusion nuclei (about 10^{-21} s, Nix (1984)). It turns out that the fission time is relatively so long that evaporation becomes possible during the fission process itself. In other words, fission-evaporation competition cannot be reproduced any more in a microcanonical representation: dynamical effects due to the fact that fission is a slow process become dominant.

An interesting feature is that these expectations have really been observed experimentally (Jacquet 1986, Kildir 1982, Bizard 1986e). For instance, in the case of results quoted in fig. 11, the usual statistical theory is unable to explain why the backward detected particles are pre-fission evaporated. It turns out that this kind of experiment gives good information about nuclear characteristic times.

Let us now consider the evaporation process itself. An interesting feature concerning the statistical theory at large temperatures (Grégoire 1986) is that cluster evaporation becomes a widely open channel (cluster means nuclei with atomic numbers between 3 and ~ 10). Indeed some experimental results concerning cluster detection have been fully understood in this way. For instance in fig. 15 (Auger 1985a) the overall mass distribution is reproduced assuming incomplete fusion followed by evaporation (including cluster emission).

However, in many other examples, such an analysis is not sufficient and one is rather constrained to assume that clusters have various origins (Borderie 1986). Some of them could result from peripheral reactions (see discussion of fig. 7 in section 3) or target shattering. However, their most exciting origin would be a phase transition which would take place when the excitation energy of the system exceeds a limiting value. In this picture

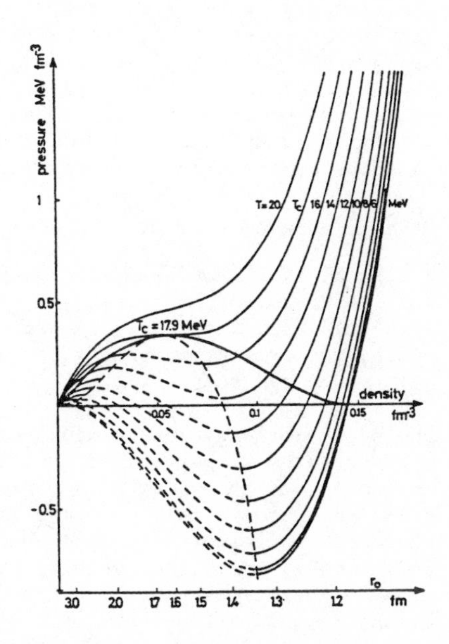

Fig. 16 Nuclear matter equation of state considering nuclear matter as a Van der Waals fluid. The curves are isotherms. For a given pressure, at a given temperature, two density values are possible corresponding to gas and liquid phase. These two solutions are mixed up at the critical temperature.

nuclear matter is described as a Van der Waals fluid and its calculated equation of state is given in fig. 16 (Sauer 1976) in the case of infinite nuclear matter. It appears that the critical temperature T_c of this fluid is about 17 MeV. A more realistic calculation performed for finite nuclear matter leads to much smaller numbers (for instance 5 to 8 MeV from Levit (1985)) in good agreement with maximum temperature values deduced from experiment in section 4 (see also fig. 4). Above this limiting value phase instabilities would lead to droplets (clusters), the mass distribution of which would follow an $A^{-\tau}$ law (A = mass number; $\tau = \sim 2.7$) in agreement with the data (Borderie 1986). Quite similar patterns can be obtained in more microscopic models: for instance, Bondorf et al (1985) reproduce quite well the above power law in a simple statistical model where the system is described by a microcanonical ensemble. An interesting result is that cluster emission is responsible for a saturation effect of the mean system temperature when the excitation energy is between 3 and 7 MeV/nucleon (fig. 17). This plateau may be regarded as a phase transi-

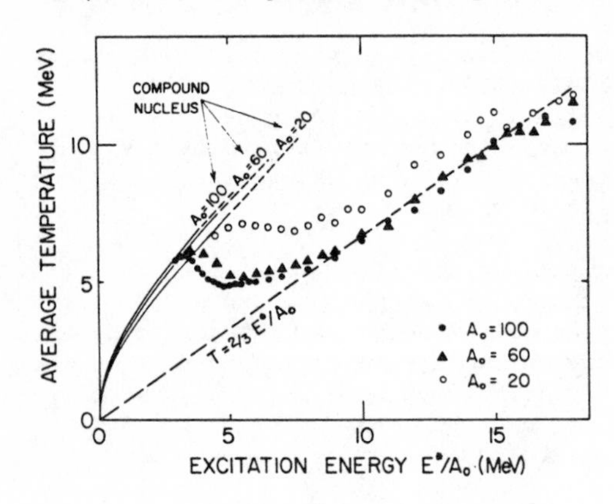

Fig. 17 Average temperature of outgoing clusters in a multi-fragmentation model of a very hot equilibrated nucleus of 100 nucleons. The abscissa is the excitation energy in MeV/nucleon; from Bondorf (1985).

tion signature. However, the above power law is also reproduced in a variety of models (Grégoire 1986, Borderie 1986) and cannot be used to recognize a phase transition. On the other hand, one has to keep in mind that finite number effects certainly weaken this transition (if any) as is clearly shown in percolation theories (Campi 1985) and the situation is still rather unclear. In order to progress one needs experiments in which clusters are detected in coincidence with other reaction products.

6. Conclusion

Intermediate energy heavy-ion reactions are irreplaceable probes to form and study nuclear matter in very exotic states. Either one is dealing with cold nuclei very far from the stability line or one is dealing with nuclei the excitation energy of which is close to or above the maximum value they can support.

The cold exotic nuclei studies have been extensively discussed at this conference by C. Detraz and that is why they have been to a large extent excluded from this review (see section 3). Let me only recall that exotic cold nuclei are mainly produced in peripheral reactions. We have seen that both nucleon-nucleon and mean field effects are observed in these collisions. Mean field effects have been used to reach the neutron drip line in collisions involving neutron rich targets. On the other hand, direct fragmentation-like reactions (nucleon-nucleon collision effects) have been used to reach the proton drip line for low Z nuclei.

Very hot nuclei are produced in central collisions. The disappearance of the fusion process when the bombarding energy is increased (section 2) is probably connected with the maximum bearable excitation energy of fusion products. However, we have seen that it is often rather difficult to be sure that the excitation energy is really thermalized and the validity of the temperature concept is questionable in extreme cases.

From a theoretical point of view a major difficulty is then to choose the relevant variables. For example, Vautherin et al (1985) have described the evolution of a hot nucleus without assuming the validity of the temperature concept. It turns out that bubble nuclei could be formed and that the maximum excitation energy of a nucleus could exceed its binding energy at variance with what is expected if thermodynamical equilibration is always achieved.

From an experimental point of view it is often a challenge to try to select between such various theroetical approaches. An example has been presented in section 5 concerning cluster emission and its interpretation in terms of a phase transition. In every case one needs coincidence experiments as exclusive as possible in which most of the outgoing products are detected. They are just now under development at GANIL, MSU and around future facilities. One may expect that they will lead to more definite answers concerning open questions:

- what are the primary processes responsible for energy relaxation around 40 MeV/nucleon (section 3)?

- to what extent thermalization is achieved above 30 MeV/nucleon (section 4)?

- to what extent phase transition effects could be recognized in the intermediate energy domain (section 5)?

- what are the properties of very hot equilibrated nuclei? For instance fission barriers are expected to decrease strongly for temperatures exceeding 4 MeV (Brack 1985). This effect has never been checked experimentally;

- to what extent compression effects or more generally coherent processes could be recognized in the intermediate energy domain? Such effects are looked for in large-angle particle-particle correlations (Ardouin 1985) or in subthreshold pion production (for a review, see Grégoire 1986) but no definite conclusions have yet been obtained.

Acknowledgements

The author thanks C. Grégoire for valuable discussions and M. de Claverie for typing this manuscript.

References

Ardouin D, Bizard G, Delagrange H, Doubre H, Gelbke K, Grégoire C, Kyanowski A, Lefebvres F, Lynch W, Maier M, Mittig W, Peghaire A, Peter J, Pochodzalla J, Quebert J, Saint-Laurent F, Tamain B, Viyogi Y P and Zwieglinski B 1985 Nucl. Phys. A447 585c.

Auger G, Berthier G, Cunsolo A, Foti A, Mittig W, Pascaud J M, Plagnol E, Quebert J and Wielesko J P 1985a Proc. XXIII Int. Winter Meeting on Nucl. Phys. Bormio p.459

Auger G, Jouan D, Plagnol E, Pougheon F, Naulin F, Doubre H and Gregoire C 1985b Z. Phys. A321 243

Auger G, Plagnol E, Jouan D, Guet C, Heuer D, Maurel M, Nifenecker H, Ristori C, Schussler F, Doubre H and Gregoire C 1986 Phys. Lett. 169B 161

Babinet R and Cugnon J 1986 Ann. Phys. France in press

Barette J, Berthier B, Chavez E, Cisse O, Dayras R, Legrain R, Mermaz M C, Pagano A, Pollaco E, Delagrange H, Mittig W, Heusch B, Lanzano C and Palmeri A 1984 Proc. XXII Int. Winter Meeting on Nucl. Phys. Bormio 561

Barette J, Berthier B, Gastebois J, Gillibert A, Lucas R, Matuszek J, Mermaz M C, Miczaika A, Van Renterghem E, Suomijarvi T, Boucenna A, Disdier D, Kraus L, Linck I, Lott B, Rauch V, Reibmeister R, Scheibling F, Sens J C, Schulz N, Grunberg C and Mittig W 1986b Proc. Hicofed Conference, Caen

Bizard G, Brou R, Doubre H, Drouet A, Guilbault F, Hanappe F, Harasse J M, Laville J L, Le Brun C, Oubahadou A, Patry J P, Peter J, Ployart G, Steckmeyer J C and Tamain B 1985 Proc. XXIII Int. Winter Meeting on Nucl. Phys. Bormio p.472, and 1986 Z. Phys. A323 459

Bizard G, Brou R, Doubre H, Drouet A, Guilbault F, Hanappe F, Harasse J M, Laville J L, Lebrun C, Oubahadou A, Patry J P, Peter J, Ployart G, Steckmeyer J C and Tamain B 1986a Proc. XXIV Winter Meeting on Nucl. Phys. Bormio

Bizard G 1986b Proc. Hicofed Conference, Caen

Bizard G, Drouet A, Lefebvres F, Patry J P and Tamain B 1986c Nucl. Inst. Meth. Phys. Res. A244 483

Bizard G, Brou R, Doubre H, Guilbault F, Hanappe F, Laville J L, Le Brun C, Patry J P, Peter J, Rudolf G, Scheibling F, Steckmeyer J C and Tamain B 1986d to be published

Bizard G, Brou R, Doubre H, Drouet A, Guilbault F, Hanappe F, Harasse J M, Laville J L, Le Brun C, Oubahadou A, Patry J P, Peter J, Ployart G, Steckmeyer J C and Tamain B 1986e Nucl. Phys. A456 173

Blachot J, Crançon J, De Goncourt B, Gizon A, Lleres A, Nifenecker H, 1985 Proc. XXX Int. Winter Meeting on Nucl. Phys. Bormio p.598

Bondorf J P, Donangelo R, Mishustin I N, Schutz H 1985 Nucl. Phys. A443 321 and Phys. Lett. 162B 30

Bonnafond C 1986 Thesis Grenoble

Blumenfeld Y, Roynette J C, Chomaz P, Frascaria N, Garron J P and Jacmart J C 1985 Nucl. Phys. A445 151

Borderie B, Rivet M F, Cabot C, Fabris D, Gardes D, Gauvin H, Hanappe F, Peter J 1984a Z. Phys. A318 315

Borderie B, Rivet M F, Cabot C, Fabris D, Gardes D, Gauvin H, Hanappe F, Peter J 1984b Z. Phys. 316 243

Borderie B 1986 Proc. Hicofed Conference, Caen

Brack M, Guet C and Hakansson H B 1985 Phys. Rev. 123 275

Campi X, Desbois J and Lipparini E 1985 Proc. XXIII Int. Winter Meeting on Nucl. Phys. Bormio

Cassagnou Y, Conjaud M, Dayras R, Harar S, Legrain R, Mostefai M, Pollaco E C, Volant C, Klotz Engmann G and Oeschler H 1986 Preprint Saclay DPh-N No. 2344

Chitwood C B, Gelbke C K, Pochodzalla J, Chen Z, Fieds D J, Lynch W G, Morse R, Tsang M B, Boal D H and Schillcock J C 1986 Phys. Lett. 172 27

Chomaz P, Frascaria N, Blumenfeld Y, Garron J P, Jacmart J C, Roynette J C, Bohne W, Gamp A, Von Oertzen W, Buenerd M, Lebrun D and Martin P, 1984a, Z. Phys. A318 41
Chomaz P, Blumenfeld Y, Frascaria N, Garron J P, Jacmart J C, Roynette J C, Bohne W, Gamp A, Von Oertzen W, Nguyen Van Giai and Vautherin D 1984b Z. Phys. A319 167
Chomaz P 1986 Proc. Hicofed Conference, Caen
Conjeaud M, Harar S, Mostefai M, Pollaco E C, Volant C, Cassagnou Y, Dayras R, Legrain R, Oeschler H and Saint-Laurent F 1985 Phys. Lett. 159B 244
Dayras R 1986 Proc. Hicofed Conference Caen
Delagrange H, Gregoire C, Scheuter F and Abe Y 1986a Z. Phys. A323 437
Detraz C 1986 Contribution to this conference
Fatyga M, Kwiatkowski K, Viola V E, Chitwood C B, Fields D J, Gelbke C K, Lynch W G, Pochodzalla J, Tsang M B and Blann M 1985 Phys. Rev. Lett. 55 1376
Fields D J, Lynch W G, Chitwood C B, Gelbke C K, Tsang M B, Utsunomiya H and Aichelin J 1984 Phys. Rev. C30 1912
Fields D J, Lynch W G, Nayak T K, Tsang M B, Chitwood C B, Gelbke C K, Morse R, Wilczynski J, Awes T C, Ferguson R L, Plasil F, Obenshain F E and Young G R 1986 Preprint MSU CL-553
Frascaria N 1986 Proc. XXIV Int. Winter Meeting on Nucl. Phys. Bormio
Galin J, Oeschler H, Song S, Borderie B, Rivet M F, Forest I, Bimbot R, Gardes D, Gatty B, Guillemot H, Lefort M, Tamain B and Tarrago X 1982 Phys. Rev. Lett. 48 1787
Galin J 1986 Nucl. Phys. A447 519c
Gelbke K 1986 personal communication
Goldhaber A S 1974 Phys. Lett. 53B 306
Gregoire C and Scheuter F 1984 Phys. Lett. 146B 21
Gregoire C, Vinet L, Remaud B, Sebille F and Schuck P 1985 GANIL Preprint 85-12
Gregoire C and Tamain B 1986 Ann. Phys. France in press
Guerreau D, Borrel V, Jacquet D, Galin J, Gatty B and Tarrago X 1983 Phys. Lett. 131B 293
Jahnke U 1986 Proc. Hicofed Conference, Caen
Jacquet D, Galin J, Borderie B, Gardes D, Guerreau D, Lefort M, Monnet F, Rivet M F, Tarrago X, Duek E and Alexander J 1986 submitted to Phys. Rev.
Kildir M, Logan D, Kaplan M, Zisman M S, Guerreau D, Alexander J and Vaz L 1982 Z. Phys. A306 323
Lefort M and Ngo C 1978 Ann. Phys. (Paris) 3 5
Levit S and Bonche P 1985 Nucl. Phys. A437 426
Morrissey D J, Benenson W, Kashy E, Bloch C, Lowe M, Blue R A, Ronningen R M, Sherril B, Utsunomiya H and Kelson I 1985 Phys. Rev. C32 877
Nifenecker H, Blachot J, Crançon J and Lleres A 1986 Nucl. Phys. A447 533c
Nix J R and Sierk A J 1984 Proc. Int. Conf. on Nucl. Phys. Bombay
Pochodzalla J, Friedmann W A, Gelbke C K, Lynch W G, Maier M, Ardouin D, Delagrange H, Doubre H, Gregoire C, Kyanowski A, Mittig W, Peghaire A, Peter J, Saint-Laurent F, Viyogi Y P, Zwieglinski B, Bizard G, Lefebvres F, Tamain B and Quebert J 1985a Phys. Rev. Lett. 55 177
Pochodzalla J, Friedmann W A, Gelbke C K, Lynch W G, Maier M, Ardouin D, Delagrange H, Doubre H, Gregoire C, Kyanowski A, Mittig W, Peghaire A, Peter J, Saint-Laurent F, Viyogi Y P, Zwieglinski B, Bizard G, Lefebvres F, Tamain B and Quebert J 1985b Phys. Lett. 161B 275

Sauer G, Chanda M and Mosel U 1976 Nucl. Phys. A264 221
Song S, Rivet M F, Bimbot R, Borderie B, Forest I, Galin J, Gardes D, Gatty B, Lefort M, Oeschler H, Tamain B and Tarrago X 1983 Phys. Lett. 130B 14
Tamain B 1985 Proc. Conf. de la Société Française de Physique Nice
Vautherin D and Veneroni M 1985 Orsay Preprint IPNO-Th-85-44

Inst. Phys. Conf. Ser. No. 86
Paper presented at Int. Nucl. Phys. Conf., Harrogate, UK, 1986

Relativisitic proton scattering

S J Wallace

University of Maryland, College Park, Maryland 20742 USA

Abstract. Recent developments in the theoretical interpretation of spin effects in proton-nucleus elastic scattering are reviewed. Emphasis is placed on use of the Dirac equation with an optical potential constructed from nucleon-nucleon scattering amplitudes and the nuclear density. Recent results based on a complete set of Lorentz invariant nucleon-nucleon amplitudes are presented.

1. Introduction

In 1981, some important experimental measurements of spin observables were performed at Los Alamos by Hoffmann et al (1981) using a polarized proton beam and the High Resolution Spectrometer. Beautiful data were obtained for the diffraction scattering of protons by ^{40}Ca at 500 MeV laboratory kinetic energy. At the time, this was the highest energy where sufficient nucleon-nucleon scattering data were available to characterize accurately both the proton-neutron and proton-proton scattering amplitudes in terms of phase shifts. The nucleon-nucleon experimental work was done at TRIUMF and phase shifts were analyzed by Bugg et al. 1978 and Bugg et al. 1980. More recent phase shift analyses have been performed by Arndt et al (1983). Also the nuclear density was rather tightly constrained by electron scattering data and Hartree-Fock predictions (Negele and Friar 1975). Therefore, the stage was set for a test of the "$t\rho$" impulse approximation in which one predicts the proton optical potential from two ingredients: i) free t-matrices for pp and np scattering, and ii) the nuclear density. According to multiple scattering theory, the "$t\rho$" approximation is the leading term in the multiple scattering expansion of the optical potential. Subsequent terms involving correlations were believed to be small because estimates of the correlation effects generally found small corrections (Lambert and Feshbach 1973, Wallace and Alexander 1980, Ray 1979). Moreover, it was established that the $t\rho$ approximation predicted cross sections in good agreement with earlier Los Alamos data obtained with an unpolarized beam (Wallace 1981). Quasifree multiple scattering was thought to be the dominant reaction mechanism at intermediate energy and this was consistent with general success of distorted wave impulse approximation predictions for inelastic proton scattering and with observations of inclusive (p,p') scattering (Alexander et al. 1980).

Using the formalism of Kerman, McManus and Thaler (1959) for the optical potential, Hoffmann et al. (1981) compared their data with the $t\rho$ prediction. As expected the cross section was in good agreement with theory as shown in Fig. 1. However, the analyzing power was off the mark, as shown in Fig. 1 and no reasonable corrections were found to improve matters since the t-matrix and density were tightly constrained. The impulse approximation seemed to have broken down when it came to describing the spin data at small angles.

In elastic scattering of protons by spin-zero nuclei, there are two complex scattering amplitudes:

$$F(\theta) = f(\theta) + i\vec{\sigma}\cdot\hat{n}\; g(\theta) \qquad (1)$$

where $\hat{n}$ is a unit vector normal to the scattering plane. From these one can form three observables: the cross section,

$$\sigma = |f|^2 + |g|^2 , \qquad (2)$$

the analyzing power,

$$A_y = \frac{|f+ig|^2 - |f-ig|^2}{\sigma} = \frac{\mathrm{Re}\{2f(ig)^*\}}{\sigma} , \qquad (3)$$

and the spin rotation function,

$$Q = \frac{\mathrm{Im}\{2f(ig)^*\}}{\sigma} . \qquad (4)$$

The cross section is $\sigma_\uparrow = |f+ig|^2$ for a proton which is polarized parallel to $\hat{n}$. The cross section is $\sigma_\downarrow = |f-ig|^2$ for proton which is polarized antiparallel to $\hat{n}$. These

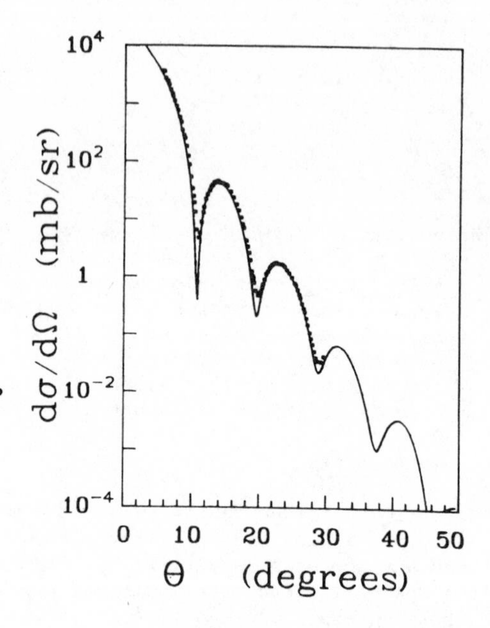

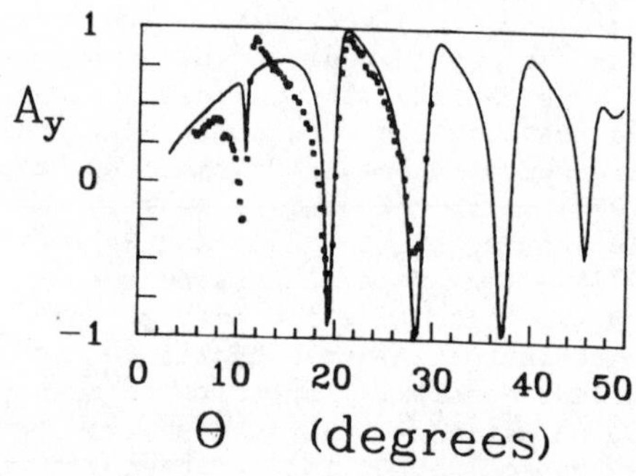

Fig. 1 Cross section and analyzing power results for proton scattering by ^{40}Ca at 500 MeV.

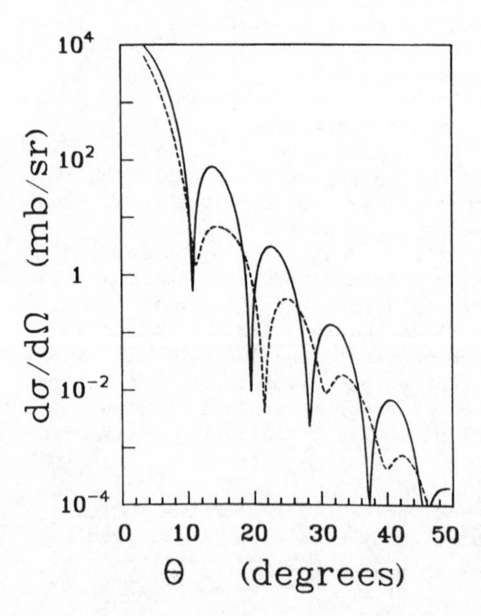

Fig. 2 Solid line shows $\sigma_\uparrow$ and dash line shows $\sigma_\downarrow$ for proton scattering by ^{40}Ca at 500 MeV.

differ due to the spin-orbit interaction, and the analyzing power A_y is directly proportional to the difference. Figure 2 shows $\sigma_\uparrow$ and $\sigma_\downarrow$ and one sees that each of these is a simple diffraction pattern. Due to the fact that $\sigma_\uparrow$ and $\sigma_\downarrow$ have pronounced minima, the analyzing power A_y is a very sensitive measure of the spin-orbit interaction. A third spin observable, Q, completes the set, except for an overall phase which cannot be observed. Due to the $\vec{\sigma}\cdot\hat{n}$ term in the scattering amplitude (1), the scattering process rotates the proton's spin. Therefore if a longitudinally polarized proton is scattered, its spin will be rotated about the normal to the scattering plane. After scattering, there is a component of spin along the third orthogonal direction, called sideways.

This spin rotation can be measured in a second scattering. A nice way to characterize the spin rotation, suggested by Glauber and Osland (1979), is to combine two measurements; one with L (longitudinal) polarization of the beam and a second using S(sideways) polarization of the beam such that one forms the quantity Q of (4). The point is that Q is orthogonal to A_y in the sense that Q measures the imaginary part and A_y measures the real part of one complex quantity. Measurement of both A_y and Q fixes the relative phase of spin-independent and spin-dependent scattering amplitudes. A polarimeter was installed in the High Resolution Spectrometer at Los Alamos and measurements of Q were performed for 500 MeV proton scattering by ^{40}Ca (Rahbar et al. 1981). Thus a complete set of measurements was obtained, and the Q data further constrained the theorists who were eager to explain the new spin data using the impulse approximation with no free parameters.

Calculations of proton scattering using an optical model and the Dirac equation had been performed for some years at Ohio State University by B. C. Clark, L. Arnold, R. L. Mercer and their collaborators (Arnold et al. 1976, 1979a, 1979b, 1981). In principle this Dirac phenomenology was equivalent to a Schroedinger phenomenology in the sense that one can define a Schroedinger potential which is equivalent to any Dirac potential. Thus by solving the Schroedinger equation with the Schroedinger equivalent potential, one gets the same answer as by solving the Dirac equation.

The Dirac phenomenology was used by Clark et al. (1983a) to fit the 500 MeV data on σ and A_y for proton scattering by ^{40}Ca and this could be done quite accurately using the 12 parameters of the Dirac potential. Then it turned out that the Dirac phenomenology <u>predicted</u> the correct results for Q without any further adjustment of parameters. The $t\rho$ potential used in the Schroedinger analyses provided no such linkage between the A_y and Q observables and therefore it became apparent that something interesting and evidently nontrivial was coming out of the Dirac phenomenology. The main assumption of this approach was that the Dirac optical potential consisted of just a scalar term S and a vector term V. To add to the puzzle these potentials were much larger ($\approx$ 300 MeV) than any potentials predicted by the $t\rho$ formalism ($\approx$ 50 MeV).

2. Relativistic Impulse Approximation: IA1

In order to clarify the connection of the Dirac approach to the impulse approximation, it was necessary to develop a Lorentz invariant description of NN scattering. This was done by McNeil, Ray and Wallace (1983) who chose to represent the NN amplitude as an expansion in the Fermi covariants as follows:

$$\hat{F}(q) = F_1 S + F_2 V + F_3 T + F_4 P + F_5 A \tag{5}$$

where S is the unit operator, $V=\gamma_1\cdot\gamma_2$, $T=\sigma_1^{\mu\nu}\sigma_{2\mu\nu}$, $P=\gamma_1^5\gamma_2^5$ and $A=\gamma_1^5\gamma_1^\mu\gamma_2^5\gamma_{2\mu}$. The five amplitudes F_1 to F_5 are Lorentz scalar functions of the Mandelstam invariants s, t and u, which represent the squares of total four-momentum, the momentum transfer, and exchange momentum transfer, respectively. One has the form (5) for pp and np amplitudes with different invariant amplitudes F_1 to F_5. An elementary analysis then showed that the optical potential for the Dirac equation could be deduced from the same $t\rho$ idea as was used in nonrelativistic multiple scattering theory. Thus it was suggested by McNeil, Shepard and Wallace (1983) that the Dirac optical potential should be connected to NN amplitudes and nuclear densities. One should solve the Dirac equation,

$$(E\gamma^0-\vec{\gamma}\cdot\vec{p}-m-\hat{U})\psi = 0 \ , \tag{6}$$

using a potential given in momentum space by the rule

$$\hat{U}(q) = -\frac{1}{4}\,\mathrm{Tr}_2\{\,\hat{M}(q)\ \hat{\rho}(q)\} \tag{7}$$

where

$$\hat{M}(q) = \frac{4\pi i P_L}{m}\,\hat{F}(q) \tag{8}$$

is a Feynman invariant amplitude, P_L is the laboratory momentum and m is the nucleon mass.

A relativistic density $\hat{\rho}$ can be constructed from wave functions of the relativistic Hartree approach to the nuclear shell model. Miller and Green (1972) developed this approach and it was later clarified further by Horowitz and Serot (1981) within the quantum hadrodynamics model of Walecka (1974). Shell-model wave functions are calculated using the Dirac equation with a potential obtained as in Eq. (7) except that a one-meson exchange interaction is used. For example, the Hartree potential for scalar and vector mesons is based on the interaction

$$\hat{M}_H(q) = \frac{g_S^2}{\vec{q}^{\,2}+m_S^2} - \gamma_1\cdot\gamma_2\,\frac{g_V^2}{\vec{q}^{\,2}+m_V^2}\ . \tag{9}$$

The nuclear density is formed out of the occupied state wave functions by

$$\hat{\rho}(\vec{q}) = \int d^3r\, e^{i\vec{q}\cdot\vec{r}}\,\{\sum_\alpha \psi_\alpha(\vec{r})\bar{\psi}_\alpha(\vec{r})\}\ , \tag{10}$$

where $\overline{\Psi}_\alpha = \psi_\alpha^\dagger\gamma^0$, and α denotes quantum numbers of the bound orbitals. In the case of a closed-shell nucleus, one finds scalar, vector and tensor contributions as follows:

$$\hat{\rho}(q) = \rho_S(q) + \gamma_2^0\rho_V(q) - \frac{\vec{\alpha}_2\cdot\vec{q}}{2m}\rho_T(\vec{q}) \ . \tag{11}$$

Given the nuclear form factors and the ansatz that (5) is an adequate representation of the nucleon-nucleon amplitudes in the Dirac space of two nucleons, we were able to replace the free parameters of the Dirac phenomenology by the impulse approximation. Equations (5)-(11) yield scalar, vector and tensor terms,

$$\hat{U}(q) = S(q) + \gamma^0 V(q) - \frac{\vec{\alpha}\cdot\vec{q}}{m}T(q) \ , \tag{12}$$

where

$$S(q) = \kappa F_1(q)\rho_S(q) \ , \tag{13}$$

$$V(q) = \kappa F_2(q)\rho_V(q) \ , \tag{14}$$

$$T(q) = \kappa F_3(q)\rho_T(q) \ , \tag{15}$$

with $\kappa = -\dfrac{4\pi i P_L}{m}$ being a kinematic factor. For short reference, this

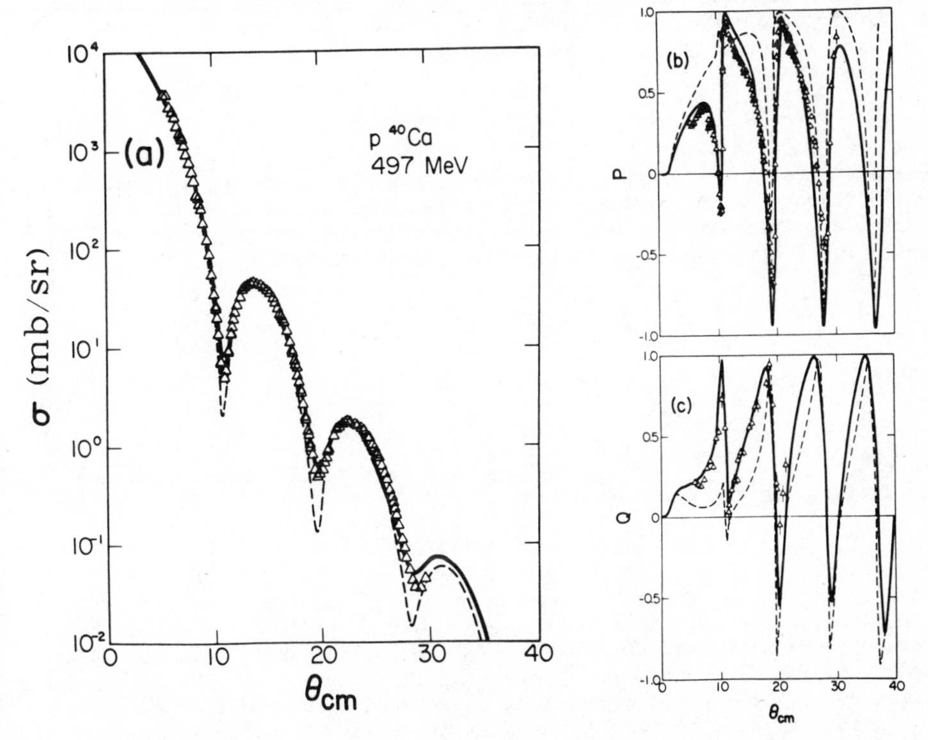

Fig. 3 Solid lines show IA1 results and dash lines show nonrelativistic tρ results in comparison with data for p+^{40}Ca at 500 MeV.

form of the impulse approximation will be called IA1. The scalar amplitude $\kappa F_1(q)$ turns out to be strongly attractive, consistent with the scalar meson interaction of (9). Similarly, the vector amplitude is strongly repulsive, also consistent with the vector meson interaction of (9). Due to a q/m factor in (12), the tensor term is not very important. Moreover the potentials essentially agree with the phenomenological ones deduced by Clark et al. (1983a) from fits to the 500 MeV data.

First calculations for proton-nucleus scattering were done independently by Shepard, McNeil and Wallace (1983) and by Clark et al. (1983b). The IA1 result is shown in Fig. 3 along with the nonrelativistic result. With no free parameters, IA1 produces an excellent description of the three observables in $p+^{40}Ca$ scattering at 500 MeV. Spin observables exhibit a clear preference for the relativistic approach. Subsequent calculations based on IA1 generally found superior fits to data than were found using the nonrelativistic impulse approximation. However, calculations at low proton energy were not as successful.

Some theoretical puzzles associated with IA1 soon became obvious. Non-relativistic and relativistic optical potentials are based on the same input, namely the nuclear density plus positive energy NN amplitudes which are fixed by experimental measurements. One needs to explain why the predictions for scattering differ sharply. A momentum space calculation by Hynes et al. (1984, 1985) pointed out that the implicit incorporation of virtual pair effects in the Dirac approach is the essential reason for the difference. One may expand the Dirac wave function in terms of positive and negative energy eigenfunctions of the free Dirac equation in momentum space,

$$\psi = u^+(\vec{p})\psi^+ + u^-(\vec{p})\psi^- \tag{16}$$

where $\psi^\pm$ are Pauli spinors. Moreover, the negative energy component, which represents the virtual pair contribution, may be eliminated to arrive at an equation involving just the positive energy component, ψ^+, which looks like the relativistic Schroedinger equation,

$$\{E - \sqrt{\vec{p}^2 + m^2} - U^{++} - U_{pair}\}\psi^+ = 0, \tag{17}$$

where

$$U_{pair} = U^{+-} \frac{1}{E + \sqrt{\vec{p}^2 + m^2} - U^{--}} U^{-+} . \tag{18}$$

Four potentials $U^{\rho'\rho}$ for $\rho = \pm$ and $\rho' = \pm$, are matrix elements of the potential $\hat{U}$ in the Dirac equation,

$$U^{\rho'\rho}(\vec{p}'\vec{p}) = \bar{u}^{\rho'}(\vec{p}')\hat{U}(\vec{p}'\vec{p})u^{\rho}(\vec{p}) . \tag{19}$$

in Eq. (17) it is straightforward to check the importance of virtual pair contributions by switching off U_{pair}. Hynes et al (1984) conclude that when virtual pair contributions are omitted, the nonrelativistic impulse approximation is reproduced. When U_{pair} is included, the characteristic difference of the Dirac approach is reproduced. The puzzling point is how U_{pair} is predicted correctly from just the positive energy NN data.

To clarify how the pair potential arises in Dirac phenomenology it is helpful to estimate the potentials U^{++} and U_{pair} assuming the Dirac potential to be composed of a scalar and vector term, i.e., $\hat{U} = S+\gamma^0 V$. One finds

$$U^{++} \simeq S + V - \frac{\vec{\sigma}\cdot\vec{p}'}{E+m}(S-V)\frac{\vec{\sigma}\cdot\vec{p}}{E+m} . \qquad (20)$$

$$U_{pair} \simeq \frac{\vec{\sigma}\cdot\vec{p}'}{E+m}\frac{(S-V)^2}{(E+m)}\frac{\vec{\sigma}\cdot\vec{p}}{E+m} \qquad (21)$$

A spin-orbit term involving $\sigma\cdot\vec{p}'\times\vec{p}$ lies in the spin-dependent part of U^{++} and this involves the difference S-V. Since S ~ -400 MeV is attractive and V ≃ +300 MeV is repulsive this is a large term in Dirac phenomenology. Correspondingly the U_{pair} term is not negligible even through it is higher order in $(E+m)^{-1}$. The size of U_{pair} is directly linked to the size of the spin-orbit term in U^{++} since each is fixed by S-V. Dirac phenomenology generally predicts S-V to be large and the origin of this effect goes back to the scalar attraction and vector repulsion inherent in the one-meson-exchange interaction (9).

A more puzzling point concerns the energy dependence of the S and V potentials which are predicted by using the NN amplitudes. Figure 4 shows values of S and V appropriate to a constant nuclear density of 0.16 fm^{-3}, i.e., nuclear matter density. At low energy, the potentials are predicted to become very large. One sees that this will make U_{pair} much larger than one expects from scalar plus vector meson exchange. This feature of the impulse approximation can be traced to one-pion exchange contributions which automatically enter when antisymmetric NN amplitudes are represented by use of the Fermi covariants in (5) (Tjon and Wallace 1985a). Figure 5 shows the one-pion exchange contributions to the impulse approximation

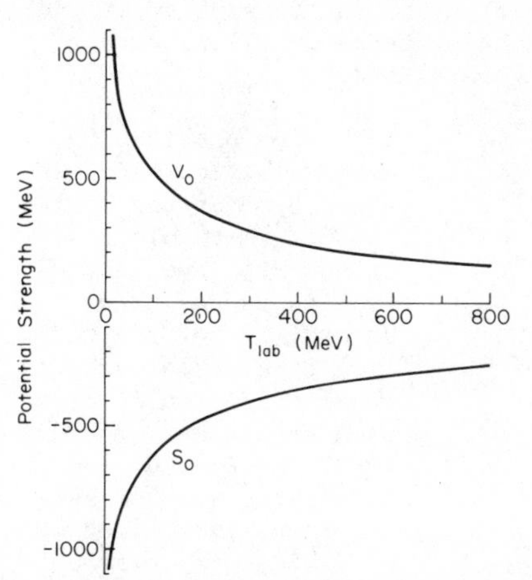

Fig. 4 Scalar and vector potential strengths for nuclear matter density based on IA1.

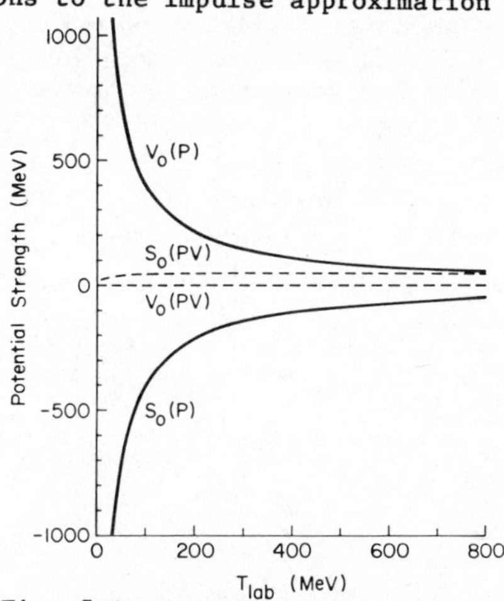

Fig. 5 One-pion exchange contributions to scalar and vector strengths. P is pseudoscalar and PV is pseudovector πN coupling.

potentials for both pseudoscalar and pseudovector πN coupling. In the pseudoscalar case, there is a very large effect at low energy while this is essentially absent in the pseudovector case. The use of Fermi covariants as in (5) can be shown to force the one-pion exchange contribution to enter as pseudoscalar πN coupling predicts. Meson exchange models of the NN interaction generally require pseudovector πN coupling to suppress overly large virtual pair contributions. Evidently this is also needed in the impulse approximation.

3. Generalized Impulse Approximation: IA2

The difference between pseudoscalar and pseudovector contributions points out a serious ambiguity in the representation of NN amplitudes in Lorentz invariant form. The possibility of ambiguities was noted first by Adams and Bleszynski (1984). A little thought convinces one that more than the positive energy scattering data is needed to eliminate all ambiguity in the relativistic amplitude F. Indeed, one needs the complete NN amplitude in the Dirac space of two particles and the five-term representation of Eq. (5) is not sufficient. Elementary counting shows that each Dirac particle has four states and thus there are 16 states available to two Dirac particles. Scattering transitions involve a 16×16 matrix, i.e., 256 matrix elements. In contrast, Eq. (5) has just five terms. As discussed by Tjon and Wallace (1985b), there are symmetries which reduce the number of amplitudes one needs. Parity invariance shows that 128 terms can be independent, and time-reversal invariance further reduces the number to 72. Charge symmetry holds to a good approximation and this reduces the number of independent amplitudes to 56. Finally, if all particles are on-mass-shell, there are only 44 independent amplitudes. This number is not small and therefore IA1 does not have a secure foundation unless one can show that all amplitudes other than those in (5) are unimportant. Therefore it is necessary to specify a complete set of NN amplitudes and to see if the very simple impulse approximation based on (5) is either theoretically consistent, or needs to be generalized.

It is now possible to answer this question as shown by Tjon and Wallace (1986). A complete set of Lorentz invariant NN amplitudes has been determined for the first time in a meson exchange model of the NN interaction (Fleischer and Tjon 1980, van Faassen and Tjon 1983, van Faassen and Tjon, 1984). The model is Lorentz invariant

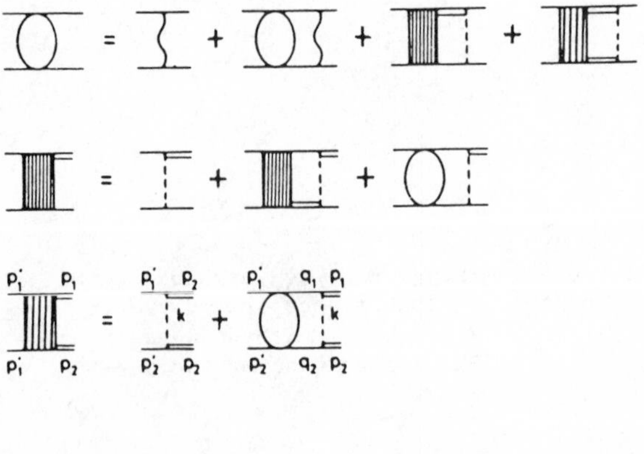

Fig. 6 Diagrammatical representation of coupled equations solved by van Faassen and Tjon (1984) for NN, $N\Delta$ and $\Delta\Delta$ scattering processes.

and it provides a description of all the basic features seen in the past two or three decades of experimentation on NN scattering. Figure 6 illustrates the physical processes included by van Faassen and Tjon (1984) to obtain a coupled-channel description of NN scattering, NΔ scattering and ΔΔ scattering. The latter two channels are needed to model the dominant absorption mechanisms in order to extend the usual boson-exchange dynamics in a unitary fashion to energies above 300 MeV where pion production becomes possible. The extension works very well. A reasonably accurate reproduction of the NN phase shifts and inelasticities is obtained in the 0 to 1000 MeV range where sufficient data for phase shift analysis exist. Moreover the model is relativistic in nature. Tjon and collaborators solve the coupled-channel, relativistic Bethe-Salpeter equations! Therefore the meson exchange model provides a prediction for the matrix elements of the NN amplitude in the space of two Dirac particles, using meson couplings consistent with the known NN interaction over a broad energy region.

Two important features of the meson exchange model are i) the assumption of pseudovector πN coupling, and ii) the use of the cutoffs at high momentum transfer of the form $\Lambda^2/(\Lambda^2-q^2)$ at each meson-baryon vertex. The cutoff mass is Λ = 1.15 GeV. Recent progress in electromagnetic physics has shown that similar assumptions provide a successful description of meson exchange currents (MEC) in ^{3}H and ^{3}He form factors (Lina and Goulard 1986). Predictions of MEC based on pseudoscalar πN coupling do not reproduce the recent electron scattering measurements from ^{3}H. Thus pseudovector πN coupling is preferred although it should also be noted that ambiguities exist in the use of relativistic MEC with nonrelativistic Faddeev wave functions which inhibit a clear answer.

Complete sets of Lorentz invariant amplitudes for on-shell NN scattering have been calculated from the meson exchange model of van Faassen and Tjon. The formalism needed to do this is given by Tjon and Wallace (1985b, 1986). I will outline the general idea since details may be found elsewhere. As in (5), one expands the NN amplitudes in terms of invariant amplitudes times kinematic covariants. The form of this expansion is

$$\hat{M} = \sum_{\rho'_1=\pm}\sum_{\rho'_2=\pm}\sum_{\rho_1=\pm}\sum_{\rho_2=\pm} \Lambda_1^{\rho'_1}(\vec{p}'_1)\,\Lambda_2^{\rho'_2}(\vec{p}'_2)\Big(\sum_{n=1} M_n^{\rho'_1\rho'_2\rho_1\rho_2}\,K_n\Big)$$

$$\times\,\Lambda_1^{\rho'_1}(\vec{p}_1)\,\Lambda_2^{\rho'_1}(\vec{p}_2)\,, \qquad (22)$$

where

$$\Lambda_i^{\pm}(\vec{p}_i) = \frac{\pm\,(\gamma_i^0\sqrt{\vec{p}_i^{\,2}+m^2} - \vec{\gamma}_i\cdot\vec{p}_i) + m}{2m} \qquad (23)$$

are covariant projection operators for positive (+) and negative (-) energy states of each Dirac particle of either the initial or the final states. Thus the full Dirac space scattering amplitude is divided into 16 ρ-spin sectors. Physical NN scattering amplitudes are in the $\rho'_1=\rho'_2=\rho_1=\rho_2=+$ sector. All other ρ-spin sectors involve one or more negative energy states and these must be determined from the relativistic meson exchange model. Kinematic covariants K_n are the same for each ρ-spin sector, however the invariant amplitudes $M_n^{\rho_1'\rho_2'\rho_1\rho_2}$ differ. Recall that only 44 invariant amplitudes are independent and so the others used in (22) are determined by symmetry relations. Tjon and Wallace (1986) give the needed symmetry relations and also provide analytic expansions of each independent amplitude in terms of Yukawa functions, as follows:

$$M_n^{\rho_1'\rho_2'\rho_1\rho_2} = \sum_k B_{nk} \{ M_k(t) \pm M_k(u)\} , \qquad (24)$$

where B is a simple matrix and

$$M_k(t) = \sum_{\ell=1}^{4} \frac{g_\ell^2}{\mu_\ell^2 - t} \frac{\Lambda^2}{\Lambda^2 - t} \qquad (25)$$

is the typical sum of Yukawa terms with a cutoff factor. The coupling constants g_ℓ vary depending on the ρ-spin labels $\rho_1'\rho_2'\rho_1\rho_2$, however a fixed set of masses μ_ℓ is used in the Yukawa fit.

Logically, this is a straightforward procedure. Given the Yukawa parameters, μ_ℓ, g_ℓ and Λ in (25), the matrix B, and the symmetry relations which determine dependent amplitudes in terms of independent ones, one constructs a complete set of Feynman amplitudes using (23) and (24). Invariant amplitudes eliminate the need to worry about how one boosts the NN amplitude from one Lorentz frame to another, for example, from the nucleon-nucleon cm frame to the proton-nucleus cm frame.

Given the general form of M, it is straightforward to calculate the Dirac optical potential of Eq. (7). Omitting the details of this calculation, we find the following results. The optical potential contains, in general, eight terms as pointed out by Celenza and Shakin (1983). The potential is formulated in momentum space, however it is possible to approximate nonlocal factors such as $\sqrt{\vec{p}^2+m^2}$ by the asymptotic energy E in order to obtain a sufficiently local form to perform calculations using a coordinate space analysis of the Dirac equation. We find

$$\hat{U}(\vec{r}) = S(r) + \gamma^0 V(r) + C(r)\{\frac{-E\gamma^0+\vec{\gamma}\cdot\vec{p}+m}{m}\} - i\vec{\alpha}\cdot\vec{r}\, T(r)$$

$$- \{S_{LS}(r)+\gamma^0 V_{LS}(r)\}\, \vec{\sigma}\cdot\vec{L} - i\vec{\gamma}\cdot\hat{r}\, D(r) + F(r)\, \vec{\alpha}\cdot\vec{p} \, . \qquad (26)$$

In principle, each function of r depends also on the energy E. Due to time-reversal invariance, the last two terms involving D and F vanish when the on-shell NN amplitudes are used to construct the potential. Thus there are six non-zero terms in (26) and it is straightforward to incorporate each of these into the coordinate space Dirac equation. Equation (26) represents a generalized impulse approximation in which the implicit virtual pair contributions embodied in U_{pair} of (18) are consistently determined from the meson exchange model of van Faassen and Tjon (1984) and the relativistic nuclear densities of Horowitz and Serot (1981). For short reference, this generalized form of the optical potential will be called IA2. There are no free parameters in IA2 nor are any unjustified symmetries assumed for the Lorentz invariant NN amplitudes. All 44 independent terms are present. Therefore it is of interest to compare the predictions with experimental data.

In general, the scalar and vector terms in (26) are the largest contributions to IA2, however these are smaller in magnitude and much less energy dependent than in the case of IA1. Figure 7 illustrates this difference in potential strengths and also offers an explanation. It is possible to reformulate IA1 in such a way that one-pion exchange contributions are forced to be pseudovector. Note, however, that this does not remove the ambiguity of using less than 44 amplitudes. The pseudovector reformulation is shown in Fig. 7 and one sees that the bulk of the difference between IA1 and IA2 is consistent with the changeover from pseudoscalar to pseudovector πN coupling. We note that inspection of the complete sets of NN amplitudes show many large contributions not associated with this changeover. Therefore the real test of IA2 is comparison with scattering data. Figures 8, 9 and 10 present results for scattering of polarized protons by ^{40}Ca. In general, IA2 is seen to be very successful. At 200 MeV, there is a substantial improvement over IA1 and this is primarily due to the changeover from pseudoscalar to pseudovector πN coupling with attendant reduction of the scalar and vector potential strengths. Spin observables A_y and Q agree remarkably with experiment. At 500 MeV, IA2 produces improvements in the description of the differential cross section, however IA1 could be preferred with regard to the description of spin observables A_y and Q. At 800 MeV, either IA2 or IA1 yields good results. Overall, IA2 succeeds to describe the p+^{40}Ca data over a broad energy range and this is a nontrivial result. The puzzles which were associated with IA1 are eliminated with IA2 and the successful description of spin data is not lost in the process. Virtual pair contributions are given a theoretical foundation in meson theory similar to the foundation which we have for meson exchange corrections in electromagnetic reactions.

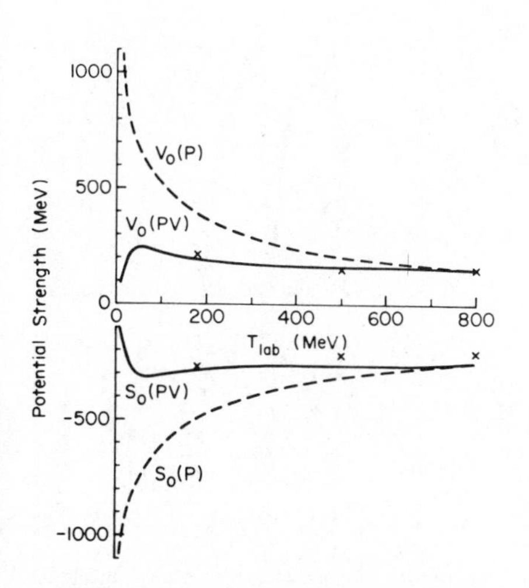

Fig. 7 Dashed lines show potential strengths for IA1 and solid lines show results for a variant of IA1 in which pseudovector πN coupling is implemented. x's show results based on the complete set of Lorentz invariant amplitudes.

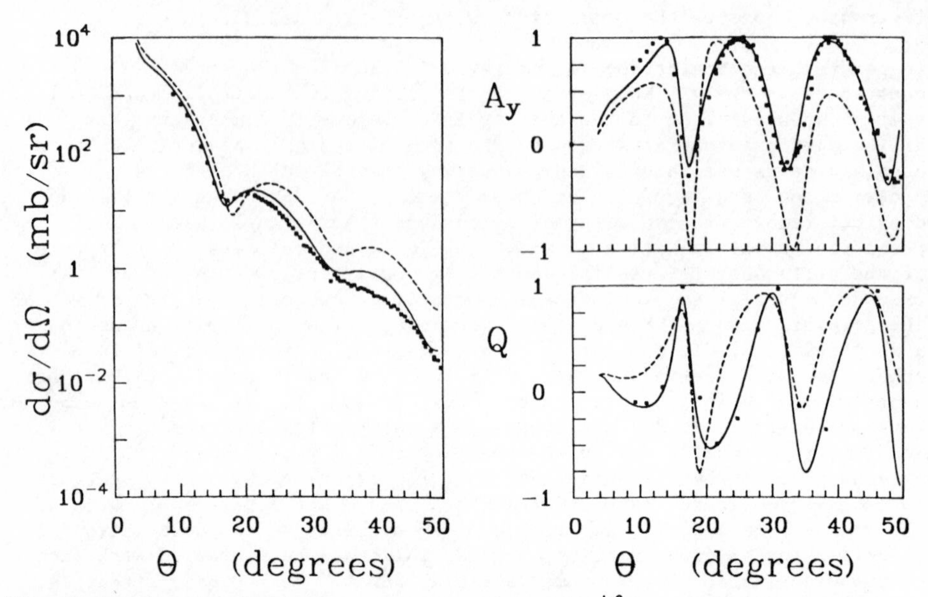

Fig. 8 Results for 200 MeV proton scattering by ^{40}Ca. Solid lines show IA2 and dash lines show IA1 results. Experimental data from Schwandt (1985) and Stephenson (1986).

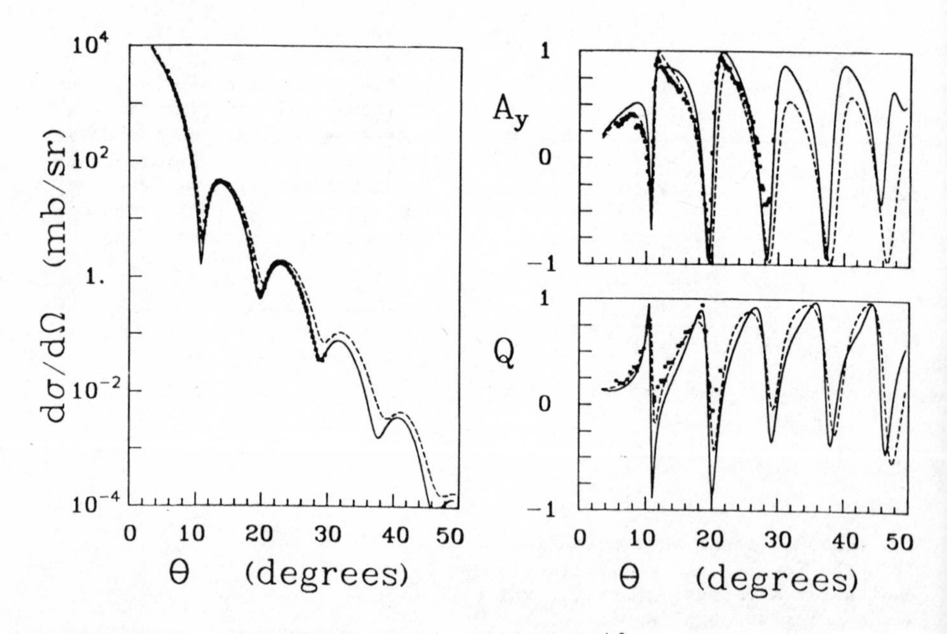

Fig. 9 Results for 500 MeV proton scattering by ^{40}Ca. Solid lines show IA2 and dash lines show IA1 results. Experimental data from Hoffmann et al (1981) and Rahbar et al. (1981).

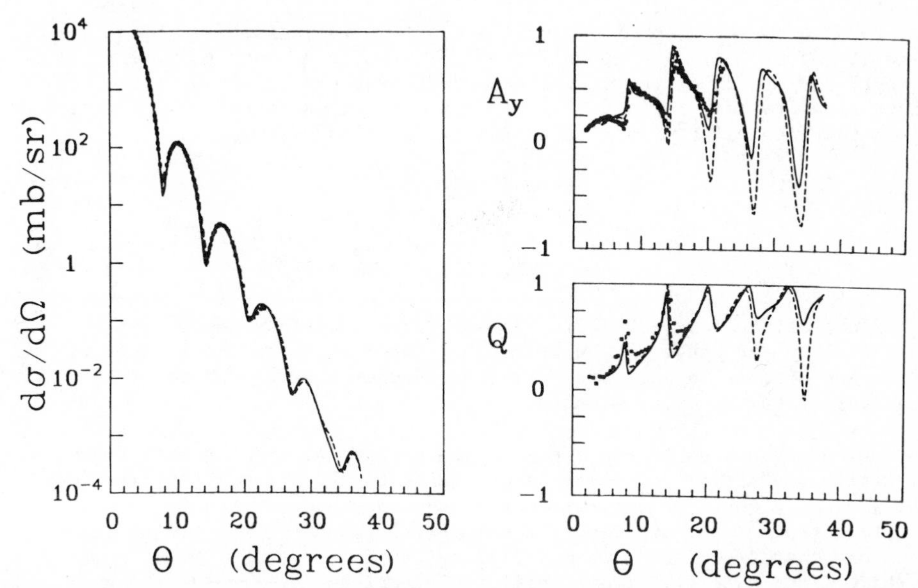

Fig. 10 Results for 800 MeV proton scattering by ^{40}Ca. Solid lines show IA2 and dash lines show IA1 results. Experimental data from Ray and Hoffmann (1985) and Fergerson et al. (1986).

4. Virtual Pairs or Lorentz-Lorenz Corrections?

Recently, there has been speculation that the Dirac approach can be reinterpreted to diminish the importance of virtual pair contributions (Theis 1986). The idea is to replace U_{pair} of (17) by a Lorentz-Lorenz correction which, under certain circumstances, takes the same mathematical form as U_{pair}. Suppose that the nucleon-nucleon t-matrix in the Pauli spin representation takes the nonlocal form

$$\langle \vec{p}'|t|\vec{p}\rangle = t_C(q) + \frac{\vec{\sigma}\cdot\vec{p}'}{E+m}\, t_{LS}(q)\, \frac{\vec{\sigma}\cdot\vec{p}}{E+m} + \ldots \qquad (27)$$

where $q = |\vec{p}-\vec{p}'|$ and the omitted terms represent spin-spin effects which do not contribute to the optical potential. The structure and nonlocality assumed in (27) is exactly the same as one would obtain from an underlying relativistic amplitude as in (5) where scalar and vector terms are dominant and they have opposite signs. On the other hand, nonrelativistic calculations usually assume a spin-orbit term of the form $t_{LS}\,\vec{\sigma}\cdot\vec{q}\times\vec{\kappa}$ where $\vec{\kappa} = (\vec{p}+\vec{p}')/2$ and thus (27) involves some new insight into how the t-matrix should depend on off-shell momenta.

The $t\rho$ potential based on (27) is essentially the U^{++} term of the Dirac approach since $t_C(q)\rho(q) = S(q)+V(q)$ and $t_{LS}(q)\rho(q) = S(q)-V(q)$ where the only change from Eq. (26) is the use of a nonrelativistic density $\rho(q)$. In addition, some care is required to avoid multiple scattering processes which require two nucleons to be very nearly in the same place (Ericson and Ericson 1966). Possible short-range contributions arise from iterating the $\vec{\sigma}\cdot\vec{p}$ nonlocality in the t_{LS} term of (27). For example, if dependence of $t_{LS}(q)$ on momentum transfer, q, is negligible, then iterations of the t_{LS} term will produce δ-function contributions which should be removed by use of the Lorentz-Lorenz prescription. Thus for zero-range forces, Theis (1985) finds a correction to the U^{++} potential such that the relativistic Schroedinger equation becomes

$$(E-E_p-U^{++}-U_{LL})\psi = 0, \tag{28}$$

where

$$U_{LL} = \frac{\vec{\sigma}\cdot\vec{p}}{E+m}\,\frac{(S-V)^2}{E+m}\,\frac{\vec{\sigma}\cdot\vec{p}}{E+m}, \tag{29}$$

is identical to U_{pair} of (21). In effect, assumptions about the nonlocality and range of the spin-orbit interaction can be invoked to deduce a Lorentz-Lorenz corrected Schroedinger equation which is equivalent to the Dirac equation in the form (17).

A logical extension of the argument shows that one ends up with the Dirac equation whether or not there is quenching of short-range contributions. To see this, consider the case where quenching of short-range contributions is incomplete, for example, because t_{LS} has finite range. Suppose that U_{LL} of (28) is modified to the form $(1-\lambda)U_{LL}$ where λ takes a value from 0 to 1. If quenching is incomplete (i.e., $\lambda \neq 0$) there must also be some contribution due to virtual pairs, because they take exactly the same form as U_{LL}. A careful analysis shows that U_{LL} and U_{pair} are the same on rather general grounds. Therefore we must add to the relativistic Schroedinger equation a term $\lambda\, U_{pair}$ to account for the incomplete quenching of NN pairs and this yields

$$\{E-E_p - U^{++} - \lambda U_{pair} - (1-\lambda)U_{LL}\}\psi = 0\ . \tag{30}$$

Since $U_{LL} = U_{pair}$, the quenching factor λ is irrelevant and one is led back to a form of the Dirac equation, whether or not quenching takes place, provided the underlying relativistic dynamics produces the nonlocal t-matrix of (27). Without a relativistic dynamics as a guide, the form $t_{LS}(\vec{q})\sigma\cdot\vec{q} \times \vec{k}$ has usually been assumed and this interaction does not generate a Lorentz-Lorenz correction.

The schematic analysis above can be generalized by formulating a "no pairs" theory of proton multiple scattering along lines proposed by Sucher (1980) for atomic physics. In place of the Dirac propagator, one has only positive energy propagation described by

$$G^+(\vec{p}) = \frac{m}{E_p}\,\frac{\Lambda^+(\vec{p})}{E - E_p + i\eta} \tag{31}$$

Virtual pair contributions are left out of this approach, however they may be introduced perturbatively. A key point is that the price for eliminating virtual pairs is the nonlocal Green's function (31) which contains intrinsic short-range propagation effects like exp(-mr)/r which vanish outside a range $r \sim m^{-1} \sim 0.2$ fm. For consistency, these short-range effects should also be removed by the Lorentz-Lorenz reorganization, on the same grounds that justify omission of virtual pair contributions. A straightforward analysis shows that the long-range contributions which are left involve the Dirac propagator $[E\gamma^0-\vec{\gamma}\cdot\vec{p}-m]^{-1}$. In general the process of removing short-range parts of the positive energy propagator in the "no pairs" theory brings one back to a theory based on the Dirac propagator.

In order to understand whether the Dirac equation succeeds because virtual pair contributions are important, or because all short range effects are quenched, or for a combination of these reasons, it is essential to calculate correlation corrections to the optical potential. An estimate by Cooper and Jennings (1986) indicates that quenching of virtual pair processes is about 20%, i.e. $\lambda = 0.8$ in (30). Standard NN correlations are combined with standard meson ranges and cutoffs to evaluate the leading-order pair diagram by Cooper and Jennings. Although quenching would be complete for zero-range forces, realistic ranges allow two nucleons which are separated by 0.5 fm or more to contribute to a scattering process which involves a virtual pair, according to available nuclear correlation functions. Thus the Cooper and Jennings (1986) estimate indicates that the straightforward Dirac approach is more consistent than the Lorentz-Lorenz reinterpretation.

Lumpe and Ray (1986) have performed complementary analyses of the second order optical potential in a Dirac multiple scattering approach. If the Dirac approach is consistent, this should turn out to be much less important than the first order optical potential. Such is the case according to the work of Lumpe and Ray (1986). Although there are some significant corrections to the imaginary parts of the scalar and vector components in the optical potential, calculations of cross sections and spin observables show very small corrections. More work along these lines is clearly welcome. The work of Lumpe and Ray is based on using just the scalar and vector NN amplitude terms of (5). These are the most important isoscalar terms and they may be expected to provide the major second-order correction for $N = Z$ nuclei, however, the pionic term should also be included.

5. Summary

Relativistic analyses of proton-nucleus elastic scattering based on the Dirac equation have produced interesting predictions, particularly for spin observables. The original form of the impulse approximation, IA1, invoked a minimal five term representation of NN amplitudes in an effort to eliminate parameters used previously in Dirac phenomenology. Predictions of IA1 were successful for protons of energy 300 MeV or higher. However low-energy predictions were found to be unreasonable and this could be understood in terms of the implicit incorporation of pseudoscalar pion-nucleon coupling. To resolve the problems at low energy, and to overcome an inherent ambiguity in the extension of NN amplitudes to the full Dirac space of two nucleons, a theoretical model of NN scattering has been adopted. A complete set of Lorentz invariant amplitudes has been calculated for on-shell kinematics and these have been used to determine a generalized impulse approximation, IA2, by Tjon and Wallace (1986). The relativistic meson exchange model of van Faassen and Tjon (1984) is used to predict all couplings to negative-energy states which are implicit in the Dirac approach. Pion-nucleon coupling is assumed to be pseudovector and cutoffs are inserted to regulate high momentum behaviour.

Nucleon-nucleon scattering data, in the form of phenomenological phase shifts, are successfully described from 0 to 1000 MeV by the model. Results for proton scattering demonstrate considerable success for IA2 in the 200 MeV to 800 MeV energy range. Therefore IA2 unifies the description of elastic scattering over the entire intermediate energy range. The meson exchange model provides a clear theoretical basis for the impulse approximation which is consistent with the theoretical basis for meson exchange currents in nuclear physics.

The success of IA2 clearly indicates that a relativistic meson-exchange model of the nuclear force provides a good starting point for the description of the Dirac optical potential. Various interesting extensions of this approach can be envisaged. For example, relativistic nuclear theory provides a very simple and attractive mechanism for the explanation of nuclear saturation as shown by Walecka (1974). More recently, Anastasio et al (1983) have found that the relativistic explanation of saturation carries over to a Brueckner-Hartree-Fock analysis which is based on an NN interaction with relativistic meson exchange. Brockmann and Machleidt (1984) and Malfliet and ter Haar (1986) obtain similar results and improve upon the technique of the earlier work by Anastasio et al. A similar meson exchange dynamics also succeeds in intermediate energy proton scattering, as shown in this paper. Effects of the nuclear medium on the NN scattering, such as Pauli blocking and self-consistency, are not incorporated into the optical potential work as yet. Otherwise IA2 has the same dynamical ingredients as the NN interaction used in the Brueckner-Hartree-Fock approaches to nuclear saturation. Therefore it is very likely that incorporation of nuclear medium effects within one consistent framework will succeed to unify the description of nucleon-nucleus interactions for all energies.

Spin observables in inelastic proton scattering provide the most fertile ground for further experimental tests of the relativistic approach. Shepard et al (1984) have developed a relativistic distorted-wave impulse approximation for inelastic scattering. Particularly for spin observables one expects to find differences between the standard nonrelativistic analysis and a relativistic one. However relativistic wave functions for nuclear states are too primitive at present to describe the nuclear structure correctly.

Correlation effects are also a subject of considerable interest. The work of Theis has shown that nonlocalities in the NN interaction can, in the limit of zero-range forces , lead essentially to the Dirac equation. Lorentz-Lorenz corrections provide the mechanism for this transition. According to this scenario, one may use the Dirac equation to avoid short-range effects and it is unnecessary to include virtual pair effects. On the other hand, calculations of Cooper and Jennings (1986) and those of Lumpe and Ray (1986) suggest otherwise. Use of realistic force ranges shows that the corrections to the Dirac optical potential due to short range correlation are of minor importance. According to this scenario, the Dirac approach succeeds because it organizes the important physics into the first order term with rather small corrections. Obviously more work will be forthcoming to clarify these opposing points of view. However it is worth noting that the nonlocality needed to motivate the Theis interpretation is natural to an underlying relativistic dynamics and has not been suggested previously as a feature to be expected from nonrelativistic potential models of the nuclear force.

Acknowledgement

The support of the United States Department of Energy for this work is gratefully acknowledged. A portion of this work was performed at the Lewes Center for Physics.

References

Adams D and Bleszynski M 1984 Phys. Lett. 136B 10
Alexander Y, Van Orden J W, Redish E F and Wallace S J 1980 Phys. Rev. Lett. 44 1579
Anastasio M R, Celenza L, Pong W S and Shakin C M 1983 Phys. Rep. 100 327
Arnold L G, Clark B C, Mercer R L, Ravenhall D G and Saperstein A M 1976 Phys. Rev. C14 1878
Arnold L G, Clark B C and Mercer R L 1979a Phys. Rev. C19 917
Arnold L G and Clark B C 1979b Phys. Lett. 84B 46
Arnold L G, Clark B C, Mercer R L and Schwandt P 1981 Phys. Rev. C23 1949
Arndt R A, Roper L D, Bryan R A, Clark R B, Ver West B J and Signell P 1983 Phys. Rev. D28 97
Brockmann R and Machleidt R 1984 Phys. Lett. 149B 283
Bugg D V, Edgington J A, Amsler C, Brown R C, Oram C J, Shakarchi K, Stewart N M, Ludgate G A, Clough A S, Axen D, Jaccard S and Vavra J 1978 J. Phys. G4 1025
Bugg D V, Edgington J A, Gibson W R, Wright N, Stewart N M, Clough A S, Axen D S, Ludgate G A, Oram C J, Robertson J R and Amsler C 1980 Phys. Rev. C21 1004
Celenza L S and Shakin C M 1983 Phys. Rev. C28 1256
Clark B C, Hama S and Mercer R L 1983a Proc. Workshop on the Interaction between Medium Energy Nucleons in Nuclei (New York: Amer. Inst. Phys.) p 260
Clark B C, Hama S, Mercer R L, Ray L and Serot B D 1983b Phys. Rev. Lett. 50 1644
Cooper E D and Jennings B K 1986 Phys. Rev. C to be published
Ericson M and Ericson T E O 1966 Ann. Phys. (NY) 36 323
van Faassen E and Tjon J A 1983 Phys. Rev. C28 2354
van Faassen E and Tjon J A 1984 Phys. Rev. C30 285
Fergerson R W, Barlett M L, Hoffmann G W, Marshall J A, Milner E C, Pauletta G and Ray R L 1986 Phys. Rev. C33 239
Fleischer J and Tjon J A 1980 Phys. Rev. D21 87
Glauber R J and Osland P 1979 Phys. Lett. 80B 401
Hoffmann G W, Ray L, Barlett M L, Fergerson R, McGill J, Milner E C, Seth K K, Barlow D, Bosko M, Iverson S, Kaletka M, Saha A and Smith D 1981 Phys. Rev. Lett. 47 1436
Horowitz C and Serot B D 1981 Nucl. Phys. A368 503
Hynes M V, Picklesimer A, Tandy P C and Thaler R M 1984 Phys. Rev. Lett. 52 978
Hynes M V, Picklesimer A, Tandy P C and Thaler R M 1985 Phys. Rev. C31 1438
Kerman A K, McManus H and Thaler R M 1959 Ann. Phys. (NY) 8 551
Lambert E and Feshbach H 1973 Ann. Phys. (NY) 76 80
Lina J M and Goulard B 1986 Phys. Rev. C34 714
Lumpe J D and Ray L 1986 University of Texas preprint
Malfliet R and ter Haar B 1986 Phys. Rev. Lett. 56 1237
McNeil J A, Ray L and Wallace S J 1983 Phys. Rev. C27 2123
McNeil J A, Shepard J R and Wallace S J 1983 Phys. Rev. Lett. 50 1439
Miller L D and Green A E S 1972 Phys. Rev. C5 241

Negele J and Friar J L 1975 Advances in Nuclear Physics (New York:Plenum) Vol. 8 pp 219-376
Rahbar A, Aas B, Bleszynski E, Bleszynski M, Haji-Saeid M, Igo G J, Irom F, Pauletta G, Wang A T M, McClelland J, Amann J F, Carey T A, Cornelius W D, Barlett M L, Hoffmann G W, Glashausser C, Nanda S and Gazzaly M M 1981 Phys. Rev. Lett. 47 1811
Ray L and Hoffmann G W 1985 Phys. Rev. C31 538
Ray L 1979 Phys. Rev. C19 1855
Schwandt P 1985 private communication
Shepard J R, McNeil J A and Wallace S J 1983 Phys. Rev. Lett. 50 1443
Shepard J R, Rost E and Piekarewicz J 1984 Phys. Rev. C30 1604
Stephenson E 1986 J. Phys. Soc. (Japan) Suppl. 55 316
Sucher J 1980 Phys. Rev. A22 348
Theis M 1985 Phys. Lett. 162B 255
Tjon J A and Wallace S J 1985a Phys. Rev. C32 267
Tjon J A and Wallace S J 1985b Phys. Rev. C32 1667
Tjon J A and Wallace S J 1986 submitted to Phys. Rev. C.
Walecka J D 1974 Ann. Phys. (NY) 83 491
Wallace S J 1981 Advances in Nuclear Physics (New York: Plenum) Vol. 12 pp 135-253
Wallace S J and Alexander Y 1980 Phys. Lett. 90B 346

Inst. Phys. Conf. Ser. No. 86
Paper presented at Int. Nucl. Phys. Conf., Harrogate, UK, 1986

The relativistic nuclear many-body problem

John Dirk Walecka

Institute for Theoretical Physics, Department of Physics
Stanford University, Stanford, California 94305

INTRODUCTION

This talk is based on a book entitled " The Relativistic Nuclear Many-Body Problem" written with Brian Serot which recently appeared as a separate volume in the series Advances in Nuclear Physics edited by John Negele and Erich Vogt [1]. The purpose of the book is pedagogical. The aim is to introduce research workers in nuclear physics to the techniques of quantum field theory as applied to the relativistic nuclear many-body problem. We also summarize results of recent research in this area. At Stanford, we have had many collaborators in these investigations, and I would like to mention W.Bardeen, A.Bielajew, S.Chin, R.Freedman, R.Furnstahl, C.Horowitz, E-J.Kim, T.Matsui, R.Rosenfelder, and F.Serr. The book contains an extended list of references: I will not repeat them here.

I want to spend some time motivating what I am going to talk about:

1) Let me first define what I call "the traditional non-relativistic nuclear many-body problem." In this approach [2], one starts with static, two-body potentials fit to free nucleon-nucleon scattering data. These potentials are inserted in the non-relativistic many-particle Schrödinger equation, and that equation is solved in some approximation. The nuclear currents are then constructed from the properties of free nucleons, and they are used to probe the system. Although this approach to nuclear physics has had many successes, it is clearly inadequate for a more detailed understanding of nuclear structure.
2) A more appropriate set of degrees of freedom for the nuclear problem are the observed strongly interacting particles, the hadrons, baryons and mesons. It is known that the long-range part of the force between two nucleons is mediated by meson exchange. The Paris potential, for example, probably the most accurate now available, consists of a long-range part arising from the exchange of the following mesons: $\pi(0^-,1)$, $\sigma(0^+,0)$, $\omega(1^-,0)$, and $\rho(1^-,1)$. Explicit hadronic effects are now observed as exchange current contributions to elastic magnetic electron scattering at high momentum transfer q^2 . Intermediate-energy meson reactions with nuclei are dominated by the internal excitations of the nucleon. As we probe the nucleus at shorter and shorter distance scales, explicit hadronic degrees of freedom play a more and more central role.
3) We are also interested in nuclear physics in describing the properties of nuclear matter under extreme conditions. For example, we want to know

[1]Supported in part by NSF grant PHY 85-08735.

what happens at high pressure, at high temperature, and at high flow velocities. We are also interested in describing the nuclear response to high q^2 probes.
4) In order to obtain an appropriate description of the nuclear system under these conditions, it is essential to retain general principles of physics such as quantum mechanics, special relativity, and causality. The only consistent theory we have of such an interacting, relativistic, many-body system is relativistic quantum field theory based on a local lagrangian density [3-5]. I like to refer to relativistic quantum field theories of the nuclear system based on hadronic degrees of freedom, the baryons and mesons, as "quantum hadrodynamics (QHD)."
5) We shall also ask that the theory be renormalizable, there are several reasons for this. First and foremost, this gives us a consistent theoretical framework within which we can calculate and compare with experiment. The theory may or may not provide an accurate description of nature; however, we are now in a position to answer this question, as with any question in physics, by comparing theoretical predictions with experimental results. Second, we can now summarize our ignorance, as with any field theory, in terms of a minimal number of phenomenological coupling constants and masses determined from experiment. Third, the condition of renormalizablity places severe restrictions on the class of interactions we need to consider. Finally, this condition makes one as insensitive as possible to the short-distance behavior of the field theory.

2. A Simple Model

Consider first the simple model that Brian and I call QHD-1. It is composed of a baryon field $\psi=\binom{P}{n}$, a neutral scalar meson field ϕ coupled to the scalar density $\bar{\psi}\psi$, and a neutral vector meson field V_λ coupled to the conserved baryon current $i\bar{\psi}\gamma_\lambda\psi$ [6]. (See also refs.[7-8].) This model reproduces the main qualitative features of the force between two-nucleons. In the static limit (which we do not assume), there is a long-range Yukawa attraction coming from scalar meson exchange and a short-range Yukawa repulsion from vector meson exchange. It is now also known empirically that there is a large Lorentz scalar attraction and a large Lorentz four-vector repulsion in the observed nucleon-nucleon scattering amplitude. Here we are primarily concerned with developing a description of the bulk and single-particle properties of nuclear matter, and we concentrate on the hadronic degrees of freedom that predominantly determine these bulk properties. The effects of the pion, for example, largely cancel in these bulk properties because of the strong spin and isospin dependence of the interaction. The lagrangian density for QHD-1 is

$$\begin{aligned}\mathcal{L} = &-\bar{\psi}\left[\gamma_\mu\left(\frac{\partial}{\partial x_\mu}-ig_vV_\mu\right)+(M-g_s\phi)\right]\psi \\ &-\frac{1}{2}\left[\left(\frac{\partial\phi}{\partial x_\mu}\right)^2+m_s^2\phi^2\right]-\frac{1}{4}F_{\mu\nu}F_{\mu\nu}-\frac{1}{2}m_v^2V_\mu V_\mu\end{aligned} \tag{2.1}$$

The field tensor for the vector meson is given by

$$F_{\mu\nu}=\frac{\partial V_\nu}{\partial x_\mu}-\frac{\partial V_\mu}{\partial x_\nu} \tag{2.2}$$

and the conserved baryon current is defined by

$$B_\mu = i\bar{\psi}\gamma_\mu\psi \quad (2.3)$$

The equations of motion derived as the Euler-Lagrange equations from this lagrangian density are perhaps more transparent

$$\frac{\partial}{\partial x_\nu} F_{\mu\nu} + m_v^2 V_\mu = ig_v\bar{\psi}\gamma_\mu\psi \quad (2.4)$$

$$(\Box - m_s^2)\,\phi = -\,g_s\bar{\psi}\psi \quad (2.5)$$

$$\left[\gamma_\mu\left(\frac{\partial}{\partial x_\mu} - ig_v V_\mu\right) + (M - g_s\phi)\right]\psi = 0 \quad (2.6)$$

Equations (2.4) are just Maxwell's equations where the vector meson field is given a mass and the source is now the conserved baryon current rather than the conserved electromagnetic current of QED. Equation (2.5) is just the Klein-Gordon equation with the Lorentz scalar density $\bar{\psi}\psi$ as source. Equation (2.6) is the Dirac equation for the baryon field with the scalar and vector fields included in a minimal fashion. This theory is renormalizable since it is analogous to massive QED with a conserved current and an additional scalar interaction.

3. Nuclear Matter-Mean Field Theory (MFT)

We have gained very little by simply writing down a lagrangian and a set of field equations since the coupling constants g_s and g_v are large. We have not made much progress unless we can find a reasonable first approximation to the problem. Fortunately, one exists [6]. Imagine nuclear matter confined to a large box and now squeeze on it to increase the baryon density. At high baryon density the source terms on the right hand side of the meson field Eqs.(2.4-2.5) get large. The quantum meson fields may then be replaced by their expectation values, which are classical fields (analogous to the classical fields $\underset{\sim}{E}$ and $\underset{\sim}{B}$ of electrodynamics). Thus, in this limit, we can talk about the classical, condensed meson fields

$$\langle\, \phi \,\rangle \equiv \phi_o \quad (3.1)$$

$$\langle\, V_\lambda \,\rangle \equiv i\delta_{\lambda 4}V_o \quad (3.2)$$

Uniform nuclear matter offers the great advantage that these fields are constants, independent of space and time. The Dirac Eq.(2.6) for the baryon field now becomes

$$\left[\gamma_\mu \frac{\partial}{\partial x_\mu} + g_v\gamma_4 V_o + (M - g_s\phi_o)\right]\psi = 0 \quad (3.3)$$

Since this equation is linear, it may be solved exactly. Note that V_o just serves to change the frequency of the solutions, and the field ϕ_o changes the mass to a new value which we shall denote by M^*

$$M^* \equiv M - g_s\phi_o \quad (3.4)$$

The baryon field in Eq.(3.3) may be expanded in terms of the positive and negative energy normal mode solutions of the new linearized Dirac equation. The canonical anticommutation relations for the field then imply that the coefficients $A_{\underset{\sim}{k}\lambda}$ and $B_{\underset{\sim}{k}\lambda}$ in this expansion are creation and destruction operators for the new baryon states, just as in conventional field theory. The hamiltonian density in this MFT then can be written in the following form

$$\hat{\mathcal{H}} = \hat{\mathcal{H}}_{MFT} + \delta\mathcal{H} \tag{3.5}$$

where

$$\hat{\mathcal{H}}_{MFT} = g_v V_o \hat{\rho}_B + \frac{1}{V} \sum_{\underset{\sim}{k}\lambda} (\underset{\sim}{k}^2 + M^{*2})^{1/2} (A^+_{\underset{\sim}{k}\lambda} A_{\underset{\sim}{k}\lambda} + B^+_{\underset{\sim}{k}\lambda} B_{\underset{\sim}{k}\lambda})$$
$$+ \frac{1}{2} m_s^2 \phi_o^2 - \frac{1}{2} m_v^2 V_o^2 \tag{3.6}$$

In this expression $\hat{\rho}_B$ is the baryon density operator

$$\hat{\rho}_B = \frac{1}{V} \sum_{\underset{\sim}{k}\lambda} (A^+_{\underset{\sim}{k}\lambda} A_{\underset{\sim}{k}\lambda} - B^+_{\underset{\sim}{k}\lambda} B_{\underset{\sim}{k}\lambda}) \tag{3.7}$$

Since the baryon current is conserved, the baryon number, and hence the baryon density for a uniform system, is a constant of the motion. This MFT hamiltonian is now <u>diagonal</u>. All of the eigenvalues and eigenstates are determined.

The additional term in Eq.(3.5) comes from normal ordering the creation and destruction operators in Eq.(3.6) and takes the form

$$\delta\mathcal{H} = -\frac{1}{V} \sum_{\underset{\sim}{k}\lambda} \left[(\underset{\sim}{k}^2 + M^{*2})^{1/2} - (\underset{\sim}{k}^2 + M^2)^{1/2}\right] \tag{3.8}$$

This term can be readily interpreted in terms of Dirac hole theory. We shall return to this "vacuum fluctuation" correction later in the talk. For now, we concentrate on the MFT with hamiltonian density given by Eq.(3.6).

4. Nuclear Matter

Consider uniform nuclear matter [6,9]. The ground state of the hamiltonian (3.6) is obtained by filling levels up to the Fermi wave number k_F. We assume a spin-isospin degeneracy of γ. For nuclear matter composed of protons and neutrons with spin up and spin down, $\gamma=4$. The baryon density simply counts the number of occupied levels

$$\rho_B = \frac{\gamma}{(2\pi)^3} \int_o^{k_F} d\underset{\sim}{k} \tag{4.1}$$

The energy density is given by

$$\mathcal{E}(\rho_B;\ \phi_o V_o) = g_v V_o \rho_B + \frac{\gamma}{(2\pi)^3} \int_o^{k_F} (\underset{\sim}{k}^2 + M^{*2})^{1/2}\, d\underset{\sim}{k}$$
$$+ \frac{1}{2} m_s^2 \phi_o^2 - \frac{1}{2} m_v^2 V_o^2 \tag{4.2}$$

The first term in this expression arises from the shift in frequency caused by the vector field V_o. The second term is the relativistic energy of a Fermi gas of baryons of mass M^*. The final two contributions are the mass terms of the constant, condensed meson fields. Since we have a lagrangian, we can calculate the stress tensor $T_{\mu\nu}$ and identify the pressure

$$P = -\frac{1}{2} m_s^2 \phi_o^2 + \frac{1}{2} m_v^2 V_o^2 + \frac{1}{3} \frac{\gamma}{(2\pi)^3} \int_o^{k_F} \frac{\underset{\sim}{k}^2}{(\underset{\sim}{k}^2 + M^{*2})^{1/2}}\, d\underset{\sim}{k} \tag{4.3}$$

There are two parameters in the energy density (4.2), the fields V_o and ϕ_o. We may determine V_o in terms of conserved quantities directly from the field equation (2.4); for constant V_o this takes the form

$$V_o = \frac{g_v}{m_v^2} \rho_B \tag{4.4}$$

V_o is thus given directly by the baryon density. In contrast, ϕ_o is a dynamical quantity. The most satisfying way to determine ϕ_o is to use a little thermodynamics. A system at constant volume and baryon number will minimize its energy. Thus we demand

$$\left(\frac{\partial \mathcal{E}}{\partial \phi_o}\right)_{\rho_B} = 0 \tag{4.5}$$

This leads immediately to the following equation

$$\phi_o = \frac{g_s}{m_s^2} \rho_s \tag{4.6}$$

which is identical to the scalar meson field equation (2.5) in the case of a constant field. The quantity on the right hand side of Eq.(4.6) is the scalar density given by

$$\rho_s \equiv \frac{\gamma}{(2\pi)^3} \int_o^{k_F} \frac{M^*}{(\underset{\sim}{k}^2 + M^{*2})^{1/2}} d\underset{\sim}{k} \tag{4.7}$$

It is the sum over the occupied levels of the quantity M^*/E^*. This is the Lorentz contraction factor relating the Dirac wavefunctions $\bar{U}U = (M^*/E^*)U^+U$. Equation (4.6) is a self-consistency relation for the scalar field.

There are two parameters in this MFT equation of state, the ratios $(g_s/m_s)^2$ and $(G_v/m_v)^2$. We choose to determine them by fitting the equilibrium binding energy and density of observed nuclear matter. By choosing to fit our coupling constants and masses to these quantities, we are putting minimal demands on the extrapolation of the equation of state, and thus we can have more confidence in our analysis. The resulting nuclear saturation curve is shown in Fig.1, and the values of the parameters used to fit the mimimum of this curve are

$$\begin{aligned} C_s^2 &\equiv g_s^2\,(M^2/m_s^2) = 266.9 \\ C_v^2 &\equiv g_v^2\,(M^2/m_v^2) = 195.7 \end{aligned} \tag{4.8}$$

The solution to the self-consistency relation (4.6) for the effective mass at each density is shown in Fig.2. The effective mass falls quickly with increasing density, and at high baryon density the nucleons are massless. At equilibrium nuclear matter density the effective mass of the baryons is $M^*/M=0.56$. On the right hand side of Fig.2 the self-consistent value of the scalar field is indicated. The scalar and vector fields are very large, several hundred MeV, in this theory, and they introduce a new energy scale into the nuclear matter problem. The equilibrium binding energy arises from a cancellation of large quantities. Furthermore, the property of nuclear saturation is inherently a relativistic effect in this model; a variational estimate based on the previously discussed potential limit proves that the non-relativistic many-body system is <u>unstable</u>

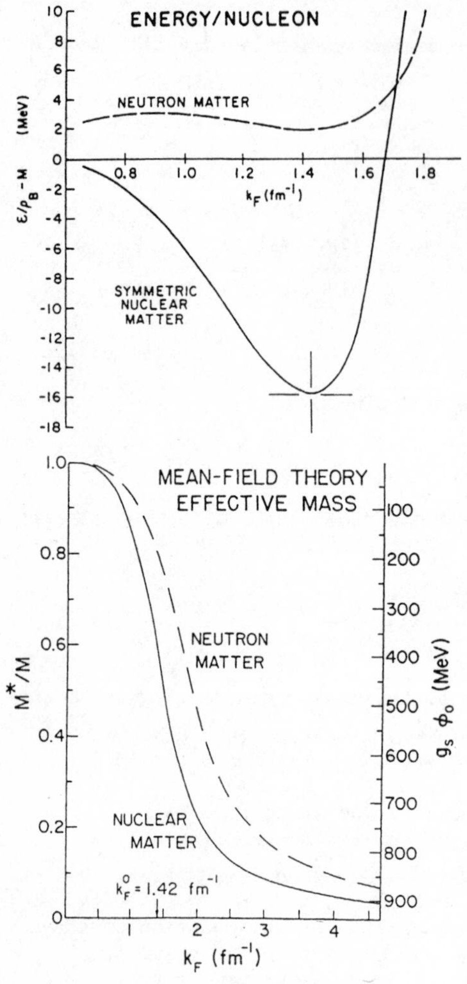

Fig.1 Saturation curve for nuclear matter calculated in MFT. The minimum is fit. The prediction for neutron matter is also shown [1,9].

Fig.2 Effective mass as a function of density for nuclear and neutron matter based on Fig.1 [1,9].

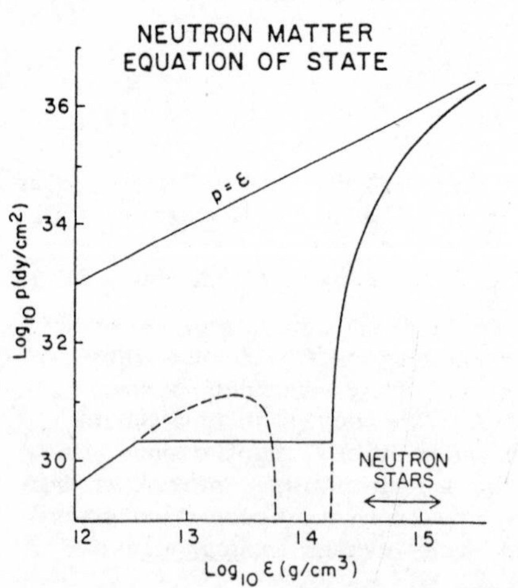

Fig.3 Predicted equation of state for neutron matter at all densities [1,9].

against collapse.

Once the two parameters in Eq.(4.8) have been determined, everything else can be calculated in this MFT of nuclear matter [9]. Neutron matter, for example, is obtained merely by setting $\gamma=2$. The resulting effective mass is shown in Fig.2 and the binding energy in Fig.1. Neutron matter is unbound, but the minimum in the binding energy curve leads to a phase separation (a liquid-gas phase separation similar to that in the Van der Waal's equation of state) and the properties of the two phases can be determined by a Maxwell construction. The equation of state of neutron matter at all density is shown in Fig.3 in more conventional units. Note the asymptotic approach to the "causal limit" $P=\mathcal{E}$ at high density. In this limit, the thermodynamic velocity of sound is equal to the speed of light. The velocity of sound always lies below the causal limit in this relativistic theory.

5. Finite Temperature

We now have a model with a diagonal hamiltonian (3.6), and it is interesting to investigate other properties of the system. For example, we can do some statistical mechanics and calculate the properties of the relativistic many baryon system at finite temperature [10]. Imagine first that our system is in contact with heat and particle reservoirs, and consider uniform nuclear matter at chemical potential μ, volume V, and temperature T. The thermodynamic potential of such a system is obtained as

$$\Omega\ (\mu,V,T;\phi_o,V_o) = -\frac{1}{\beta}\ln\ (G.P.F) \tag{5.1}$$

$$(G.P.F) \equiv \sum_{(n)} \langle n_1 \ldots n_\infty |\ \exp\left[-\beta\ (\hat{H}_{MFT} - \mu\hat{B})\right] |n_1 \ldots n_\infty\rangle \tag{5.2}$$

where $\beta \equiv 1/k_B T$. The hamiltonian and baryon number operators appearing in the grand partition function are given by Eqs.(3.6, 3.7).[2] Because these operators are diagonal, the trace in Eq.(5.2) can be evaluated analytically. There are still two parameters in the thermodynamic potential. The vector field V_o can again be determined in terms of the baryon density by taking the thermal average of the vector meson field Eq.(2.4). The result is

$$V_o = \frac{g_v}{m_v^2}\ \rho_B\ (\mu,V,T;\phi_o,V_o) \tag{5.3}$$

The scalar field is still dynamical. It may be obtained with the aid of a little thermodynamics. We learn from Gibbs that a system at fixed μ,V, and T will minimize its thermodynamic potential. Thus

$$\left(\frac{\partial\Omega}{\partial\phi_o}\right)_{\mu,V,T} = 0 \tag{5.4}$$

This leads to a self-consistency equation for M^* that must be solved at each T. The energy density and pressure are calculated to be

---

[2] Here we suppress additive contributions from the non-interacting scalar and vector mesons.

$$\varepsilon = \frac{g_v^2}{2m_v^2}\rho_B^2 + \frac{m_s^2}{2g_s^2}(M-M^*)^2 + \frac{\gamma}{(2\pi)^3}\int d\underset{\sim}{k}(\underset{\sim}{k}^2+M^{*2})^{1/2}\,[n(T) + \bar{n}(T)] \tag{5.5}$$

$$P = \frac{g_v^2}{2m_v^2}\rho_B^2 - \frac{m_s^2}{2g_s^2}(M-M^*)^2 + \frac{1}{3}\frac{\gamma}{(2\pi)^3}\int d\underset{\sim}{k}\,\frac{\underset{\sim}{k}^2}{(\underset{\sim}{k}^2+M^{*2})^{1/2}}\,[n(T) + \bar{n}(T)] \tag{5.6}$$

The baryon density is given by

$$\rho_B = \frac{\gamma}{(2\pi)^3}\int d\underset{\sim}{k}\,[n(T) - \bar{n}(T)] \tag{5.7}$$

The scalar density appearing in the scalar meson field Eq.(4.6) is given by

$$\rho_s = \frac{\gamma}{(2\pi)^3}\int d\underset{\sim}{k}\,\frac{M^*}{(\underset{\sim}{k}^2+M^{*2})^{1/2}}\,[n(T) + \bar{n}(T)] \tag{5.8}$$

and the effective mass is defined by Eq.(3.4). In contrast to the corresponding expressions at T=0 (Eqs.(4.1, 4.2, 4.3, 4.7)) which sum contributions from the degenerate Fermi gas of baryons, these expressions at finite T receive contributions from both baryons and antibaryons with thermal distribution functions given by

$$n(T) = [\exp\beta(E^*-\mu^*) + 1]^{-1} \tag{5.9}$$

$$\bar{n}(T) = [\exp\beta(E^*+\mu^*) + 1]^{-1} \tag{5.10}$$

Here $E^* \equiv (\underset{\sim}{k}^2+M^{*2})^{1/2}$ and μ^* is the effective chemical potential defined by

$$\mu^* \equiv \mu - g_v V_o \tag{5.11}$$

Note the important feature that all the antibaryon contributions are positive, except for the baryon density (5.7).

These relations must be solved numerically, and the resulting self-consistent baryon mass at zero baryon density is illustrated in Fig.4. It quickly drops to zero with increasing temperature T, and at high temperature (just as at high density) the baryons are massless in this theory.

The calculated isotherms of the equation of state for neutron matter are shown in Fig.5. Note the presence of a critical region and a critical temperature above which the liquid-gas phase separation disappears. Note also the approach to the causal limit from below. One interesting feature of these results is that the <u>isotherms</u> <u>terminate</u> at high enough temperature (right angles in the figure). The system exhibits behavior analogous to black-body radiation due to the presence of pairs. There is a minimum pressure and energy density at a given T, corresponding to a vanishingly dilute gas of baryons in a sea of pairs. One then moves up along an isotherm as the baryon density is increased.

6. Relativistic Hartree Theory of Finite Nuclei

Consider next real, finite nuclei. It is an immediate extension of the preceeding analysis to allow the static meson fields V_o and ϕ_o to acquire a <u>spatial</u> <u>dependence</u>. The field equations become

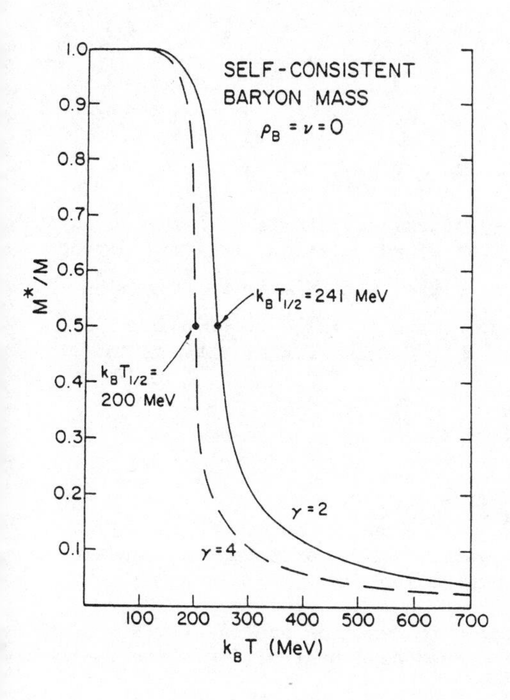

Fig.4 Self-consistent nucleon mass as a function of temperature at vanishing baryon density calculated in MFT. Results are indicated for both neutron and nuclear matter [1,10].

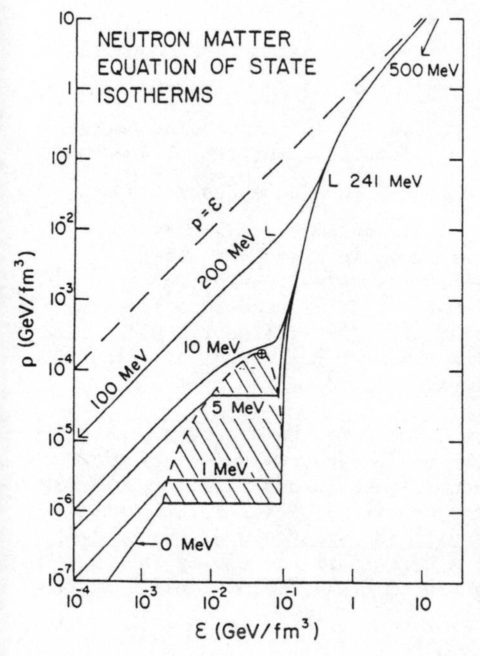

Fig.5 Isotherms of the neutron matter equation of state at finite temperature calculated in MFT. The curves are labelled by k_BT, and the left-hand endpoint corresponds to zero baryon density. The shaded area shows the region of phase separation, and the critical temperature is $k_BT = 9.1 \pm 0.2$ MeV [1,10].

$$(\nabla^2 - m_v^2)\, V_o(\underset{\sim}{x}) = -g_v \rho_B(\underset{\sim}{x}) \qquad (6.1)$$

$$(\nabla^2 - m_s^2)\, \phi_o(\underset{\sim}{x}) = -g_s \rho_s(\underset{\sim}{x}) \qquad (6.2)$$

$$[\gamma_\mu \frac{\partial}{\partial x_\mu} + g_v \gamma_4 V_o(\underset{\sim}{x}) + (M - g_s \phi_o(\underset{\sim}{x}))]\, \Psi = 0 \qquad (6.3)$$

The source terms in the meson field equations are the local baryon density $\rho_B(\underset{\sim}{x})$ and the scalar density $\rho_s(\underset{\sim}{x})$. The Dirac equation, in turn, depends on the local fields $V_o(\underset{\sim}{x})$ and $\phi_o(\underset{\sim}{x})$. In the relativistic Hartree approximation, we assume single baryon orbitals filled up to a level F, and evaluate the source terms in Eqs.(6.1, 6.2) by simply summing the individual contributions

$$\rho_B = \sum_\alpha^F \mathcal{U}_\alpha^+(\underset{\sim}{x})\, \mathcal{U}_\alpha(\underset{\sim}{x}) \qquad (6.4)$$

$$\rho_s = \sum_\alpha^F \bar{\mathcal{U}}_\alpha(\underset{\sim}{x})\, \mathcal{U}_\alpha(\underset{\sim}{x}) \qquad (6.5)$$

Equations (6.1-6.5) are local, non-linear, coupled differential equations. Fortunately, iteration procedures converge rapidly.

This system of equations can be extended to describe nuclei with $N \neq Z$ if one borrows from QHD-II and includes a condensed neutral field for the rho meson

$$(\nabla^2 - m_\rho^2)\, b_o(\underset{\sim}{x}) = -g_\rho\, \rho_3(\underset{\sim}{x}) \qquad (6.6)$$

$$\rho_3 = \sum_\alpha^F \mathcal{U}_\alpha^+(\underset{\sim}{x})\, \frac{1}{2}\tau_3\, \mathcal{U}_\alpha(\underset{\sim}{x}) \qquad (6.7)$$

as well as the Coulomb potential. There are now 4 parameters (coupling constants and masses) in this model: g_s, g_v, g_ρ, and m_s.[3] We show the results of Serot and Horowitz [11] who choose to fit the binding energy (E/B), density k_F, and symmetry energy a_4 of nuclear matter. Since one length is required to determine m_s, the root-mean-square charge radius of ^{40}Ca is (arbitrarily) also fit. The resulting calculated charge densities of ^{208}Pb, ^{40}Ca, and ^{16}O are shown in Figs.6-8. They are compared with experimental values obtained from elastic electron scattering. The height of the Pb curve at 3-5 fm is used to define the density of nuclear matter, and the radius of ^{40}Ca is also fit as stated; the remaining shapes are predicted. The shape of the charge distribution in ^{16}O in Fig.8, for example, comes free.

We get something else for free. Figure 9 exhibits the occupied single-particle orbitals in ^{208}Pb. The shell closures and spacings of the nuclear shell model are clearly seen. Thus, in addition to providing a minimal theoretical description of the densities, these relativistic Hartree calculations based on QHD-I <u>predict the existence of the nuclear shell model</u>. Its origin is apparent. A Dirac particle moving in spatially varying vector $V_o(|\underset{\sim}{x}|)$ and scalar $\phi_o(|\underset{\sim}{x}|)$ fields feels a strong

---

[3] It is assumed that $m_v = m_\omega$.

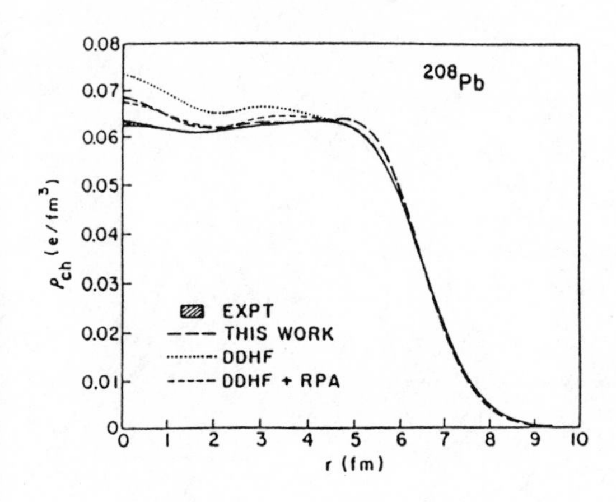

Fig.6 Relativistic Hartree change density of ^{208}Pb (heavy dashed line). The solid curve and shaded area represent the fit to the experimental electron scattering data [11,1].

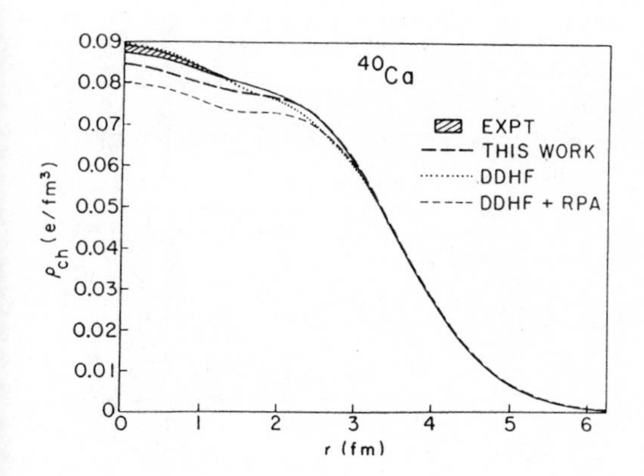

Fig.7 Same as Fig.6 for ^{40}Ca.

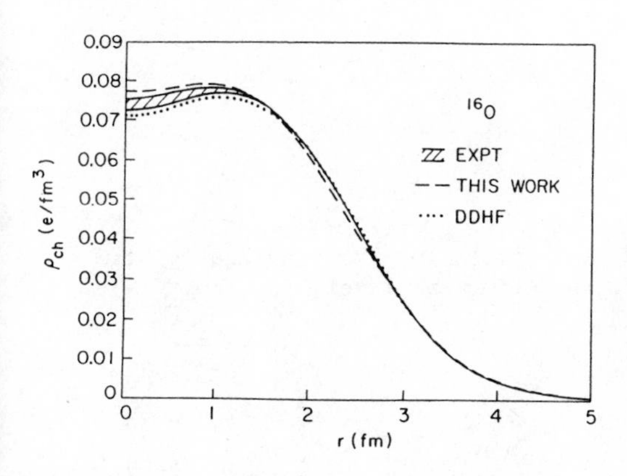

Fig.8 Same as Fig.6 for ^{16}O.

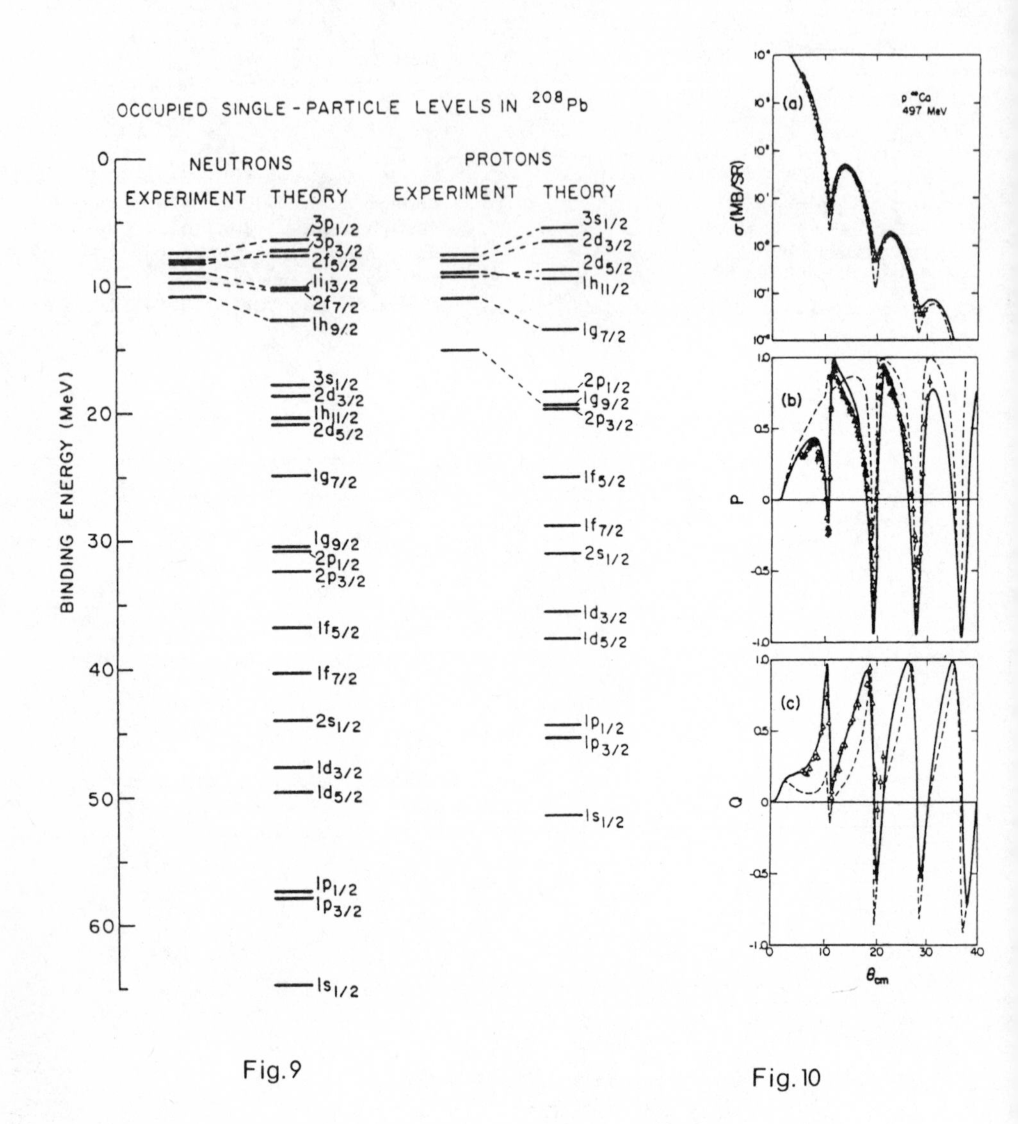

Fig.9 Predicted spectrum for occupied levels in ^{208}Pb in relativistic Hartree calculation [11,1]. Experimental levels are from neighboring nuclei.

Fig.10 Calculated cross section (a), analyzing power (b), and spin-rotation function (c) for p + ^{40}Ca at T_L = 497 MeV using the Dirac impulse approximation (RIA)-solid curve [1]. The authors would like to thank B. Clark for preparing this figure.

spin-orbit interaction. (Recall the origin of the atomic spin-orbit force as due to a Dirac electron moving in a spatially varying Coulomb potential.) In contrast to the binding energy of nuclear matter which arises from a strong cancellation, the spin-orbit interaction gets additive contributions from the vector and scalar fields; it is thus a very stable element of the model.

And there is more that comes essentially for free. Consider the scattering of an energetic, polarized proton from the nucleus as measured, for example, at LAMPF. This process may be described in the relativistic impulse approximation (RIA). Since simple scalar and vector exchange will not reproduce the detailed spin-dependence of the nucleon-nucleon (N-N) scattering amplitude, we first agree to compromise and use the empirical N-N scattering amplitude at the appropriate energy. This amplitude may be written in Lorentz invariant form as

$$f_{NN} = f_s\, 1^{(1)} \cdot 1^{(2)} + f_v\, \gamma_\mu^{(1)} \cdot \gamma_\mu^{(2)} + \dots \qquad (6.8)$$

There are five terms. It is an empirical fact, however, that under the conditions where it is used, this amplitude is dominated by large Lorentz scalar f_s and vector f_v terms. This amplitude now has both real and imaginary parts. It may be folded in with the previously discussed vector (baryon) and scalar densities to construct an optical potential. When this potential is inserted in the Dirac Equation, the scattering problem can be solved by numerical integration. The names associated with this RIA development are Clark, Hama, Mercer, Ray, Serot [12], and McNeil, Shepard and Wallace [13]. The calculated cross section for the scattering of 497 MeV protons on ^{40}Ca is shown in Fig.10. A lovely description of the differential cross section is obtained. Figure 10 also compares the calculated and measured polarization P; it is very well described within the same framework. Furthermore, the spin rotation function Q, another independent spin observable and the last quantity shown in Fig.10, is predicted with remarkable accuracy in this approach. The dominant feature of this spin physics is the same spin-orbit interaction that gives rise to the nuclear shell model; as remarked above, this is a stable element of the theory.

In summary, the relativistic Hartree calculations based on QHD-I provide a minimal unifying theoretical framework which allows us to understand nuclear densities, the nuclear shell model, and intermediate energy proton-nucleus scattering, three essential aspects of nuclear structure.

7. Feynman Rules for QHD-I

The phenomenology of this model is thus very appealing. The proposed goal, however, is to develop a theory within which it is possible, in principle to systematically calculate modifications of the MFT results. Such a theory may be summarized in a set of Feynman rules, from which all the Green's functions can be calculated [1,6]. For illustration, we state the Feynman rules for the S-matrix in QHD-I:

1) Draw all topologically distinct connected diagrams
2) Include a factor $(-i)^n$ for each order
3) Include a scalar $-g_s \cdot 1$ and vector $-ig_v \cdot \gamma_\mu$ vertices
4) Include a factor $(2\pi)^4\, \delta^{(4)}$ (sum of four-momenta) at each vertex
5) Integrate over internal momenta $\int d^4q$

6) Include a factor (-1) for each closed fermion loop
7) Include wavefunctions for the external particles
8) Use the following propagators
scalar meson

$$\frac{-i}{(2\pi)^4} \frac{1}{k^2 + m_s^2 - i\eta} \tag{7.1}$$

vector meson

$$\frac{-i}{(2\pi)^4} \frac{1}{k^2 + m_v^2 - i\eta} \left[\delta_{\mu\nu} + \frac{k_\mu k_\nu}{m_v^2}\right] \tag{7.2}$$

baryon [4]

$$\frac{i}{(2\pi)^4} (i\not{p} - M) \left[\frac{1}{p^2 + M^2 - i\eta} - 2\pi i \delta(p^2 + M^2)\, \theta(p_0)\, \theta(k_F - |\mathbf{p}|)\right] \tag{7.3}$$

The additional terms in $k_\mu k_\nu$ in the vector meson propagator will not contribute to the S-matrix because the vector mesons couple to the conserved baryon current. The proof is the same as that in QED which demonstrates the vanishing of the gauge-dependent parts of the photon propagator.

It is the baryon propagator which manifests the complexity of the finite-density, relativistic, many-body theory. The first term in Eq.(7.3) is the free Feynman propagator. The second term in Eq.(7.3) arises because of the presence of the stationary, finite-density nuclear medium. Its role is to raise a certain number of poles above the real axis. When expectation values are computed by closing contours in the upper-half p_0 plane, a contribution from the filled Fermi sea of baryons will then be obtained; in addition, one also picks up the negative-frequency contributions from the free Feynman propagator. This aspect of the many-body problem is completely absent in the non-relativistic theory where the negative frequency poles are essentially pushed out to infinity and antiparticle propagation is neglected completely.

8. Extensions

As one example of a calculation of the modification of the MFT results, consider the "vacuum fluctuation" correction mentioned previously [1]. In the Hartree theory of nuclear matter, the scalar and vector tadpole contributions are retained in the baryon proper self-energy and then summed with Dyson's equation to give the Green's function; they serve to shift the mass and frequency. When the ground-state energy is calculated, the positive-frequency particle contribution leads precisely to the MFT result in Eq.(4.2). In this calculation, however, the negative-frequency poles are also modified, and the ground-state energy receives an additional contribution which is just that in Eq.(3.8).

Now in the renormalizable theory QHD-I, there is actually a counterterm $\delta\mathcal{L}_{CTC}$ in the lagrangian density (2.1) which we have so far suppressed. This term contains up through quartic powers of the scalar field and is to be chosen so that the one-, two-, three-, and four-scalar meson amplitudes

---

[4] We suppress isospin

in the vacuum sector take certain <u>renormalized</u> values which are prescribed by experiment. Given these phenomenological values, all other physical quantities can then be calculated. This extra piece of the lagrangian will also contribute to the ground state energy of the many-baryon system. The result of this RHA calculation is an energy density of the form

$$\mathcal{E}_{RHA} = \mathcal{E}_{MFT} + \Delta\mathcal{E}_{VAC} \tag{8.1}$$

The first term is the MFT result in Eq.(4.2) and the additional "vacuum fluctuation" term takes the form

$$\Delta\mathcal{E}_{VAC} = -\frac{1}{V}\sum_{\underset{\sim}{k}\lambda}\left[(\underset{\sim}{k}^2+M^{*2})^{1/2} - (\underset{\sim}{k}^2+M^2)^{1/2}\right] - \sum_{n=1}^{4}\frac{c_n}{n!}\phi_0^n \tag{8.2}$$

The coefficients c_n have now been determined (for simplicity, we assume that the three and four-point scalar meson amplitudes vanish at vanishing four-momenta). The resulting expression in Eq.(8.2) is <u>finite</u>, and is given by [1,14-16]

$$\begin{aligned}\Delta\mathcal{E}_{VAC} = -\frac{1}{4\pi^2}\Big[&M^{*4}\,\ell n\,\frac{M^*}{M} + M^3(M-M^*) - \frac{7}{2}M^2(M-M^*)^2 \\ &+ \frac{13}{3}M(M-M^*)^3 - \frac{25}{12}(M-M^*)^4\Big]\end{aligned} \tag{8.3}$$

The mass M^* is obtained by minimizing the energy density. In Figs.11 and 12 we show the modification of the previous MFT results (solid lines) for the binding energy and equation of state of nuclear matter when the contribution of Eq.(8.3) is retained (long-dashed lines). These calculations are due to Horowitz and Serot and use the MFT coupling constants of Eq.(4.8). The shift in binding energy is not large on the scale of scalar and vector energies in the nuclear matter problem (several hundred MeV). Indeed, it would be distressing if these "vacuum fluctuation" effects arising from antibaryons were to play an important role in nuclei as we know them. On the other hand, such effects are <u>completely absent</u> in non-relativistic potential calculations of nuclear matter, and one has to at least speculate on the inherent limitation of such an approach for obtaining a truly quantitative description of nuclear saturation. Although differing at observed nuclear matter densities, the equation of state shown in Fig.12 does rapidly approach the MFT result at high density. This serves as a partial justification of the previous claim that the MFT should provide the correct equation of state in QHD-I at high baryon density.

9. Electroweak Interactions

We have seen that QHD provides a relativistic framework for describing nuclear structure. The static Hartree properties have been determined from the interactions with neutral mesons (QHD-I). We would like to probe this structure with electron scattering (e,e'). In order to have a realistic description of this process, however, it is necessary to enlarge our theoretical framework to include charged mesons (QHD-II). There is an <u>internal</u> electromagnetic structure of the nucleon in QHD-II coming, for example, from the following process:

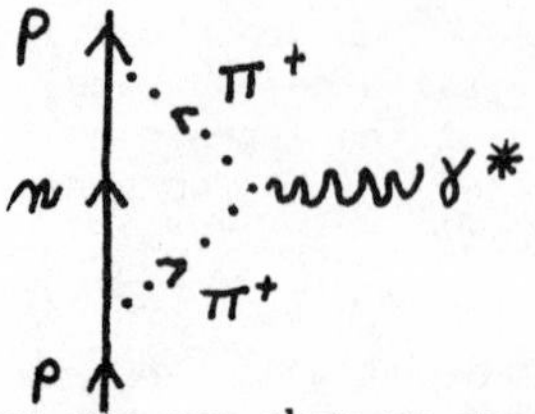

p ↑ π⁺ ; n ↑ γ* ; π⁺ ; p ↑

$$F(q^2) = \frac{1}{\pi}\int_{(2m_\pi)^2}^{\infty} d\sigma^2 \frac{\rho(\sigma^2)}{\sigma^2 + q^2} \tag{9.1}$$

Now one can show on general grounds that the electromagnetic form factors of the nucleon have a spectral representation as indicated in Eq.(9.1). The lowest-mass part of the spectral weight function $\rho(\sigma^2)$ comes from the lightest-mass intermediate states in the absorptive part (which consists of two pions as illustrated above). A simple Fourier transform of the static limit of Eq.(9.1) shows that the lowest-mass meson spectrum gives rise to the longest-range part of the ground-state charge and current density of the nucleon. Let us make the simplifying assumption that the two-pion contribution dominates the spectral weight function everywhere. The resulting magnetic properties of the nucleon calculated under this assumption are then equivalent to those calculated directly from the Feynman diagram in Eq.(9.1); they are shown in Table 9.1

TABLE 9.1 MAGNETIC PROPERTIES OF THE NUCLEON CALCULATED FROM THE TWO-PION CONTRIBUTION IN QHD-II [1] (*)

	$\frac{1}{2}(\lambda_p+\lambda_n)$	$\frac{1}{2}(\lambda_p-\lambda_n)$	$(\langle r^2\rangle_{2p})^{1/2}$
Theory	0	1.60	0.49 fm
Experiment	-0.06	1.853	0.8 fm

(*) This calculation gives $F_{2n}(q^2) = - F_{2p}(q^2)$

A detailed calculation of the internal electromagnetic properties of a free nucleon is as yet prohibitive in QHD-II, but these arguments indicate that we may hope at least to be in the right ballpark.

In order to probe the Hartree structure of nuclei, we shall for now be content to use an effective current, which includes this internal electromagnetic structure. We first assume, consistent with observations over the range of momentum transfers of interest to nuclear physics, that there is a common single-nucleon form factor. We thus define an effective Møller potential $f_{s.n.}(q^2)/q^2$. We next define an effective electromagnetic current operator which is to be used in lowest order in QHD-I to describe electromagnetic interactions. We write

$$J_\mu^\gamma(x) = i\bar{\psi}\gamma_\mu Q\psi + \frac{\partial}{\partial x_\nu}\frac{1}{2M}(\bar{\psi}\sigma_{\mu\nu}\lambda\psi) \tag{9.2}$$

where the charge and anomalous moments are defined by

$$\underline{Q} \equiv \frac{1}{2}(1 + \tau_3)$$

$$\underline{\lambda} \equiv \frac{1}{2}(1 + \tau_3)\lambda_p + \frac{1}{2}(1 - \tau_3)\lambda_n \tag{9.3}$$

Although oversimplified, this current does have the following essential features to recommend it:

1) It is local
2) It is covariant

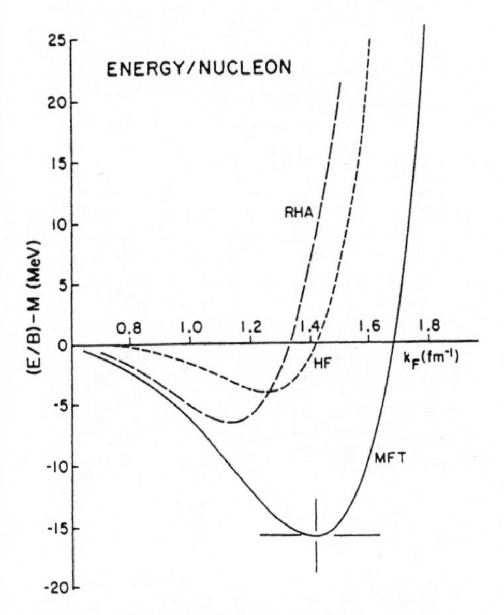

Fig.11 Energy/nucleon in nuclear matter calculated in the relativistic Hartree approximation (RHA) which includes vacuum fluctuations (long-dashed curve). The coupling constants of Eqs.(4.8) are used, and the previous MFT results are also shown (solid curve) [1].

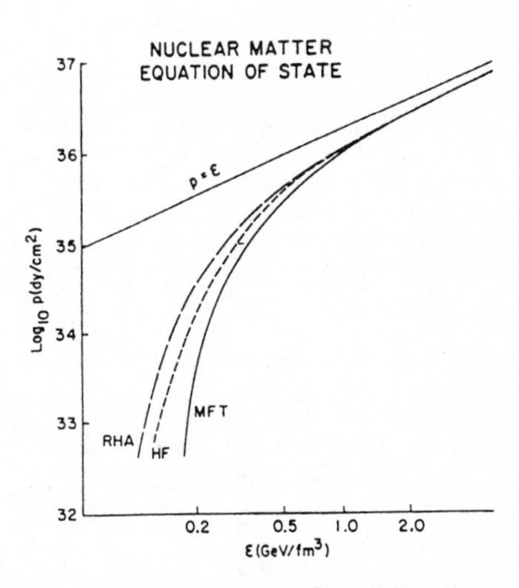

Fig.12 Nuclear matter equation of state. The curves are labelled as in Fig.11 [1].

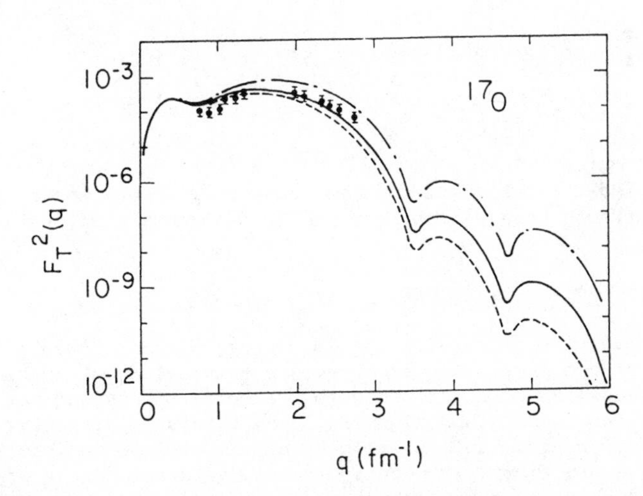

Fig.13 Relativistic calculation of the transverse elastic magnetic form factor for electron scattering from the $1d_{5/2}$ neutron in ^{17}O. The dash-dot line omits $f_{s.n.}(q^2)$ and the dashed-line omits the center-of-mass correction $f_{c.m.}(q^2)$ [17].

3) It is conserved in QHD-I

$$\frac{\partial}{\partial x_\mu} J^\gamma_\mu(x) = 0 \tag{9.4}$$

4) It gives the correct result for a free nucleon.

As one application, we show the calculations of E.-J. Kim for elastic magnetic scattering from ^{17}O in Fig.13 [17]. The valence neutron is in a $(1d_{5/2})$ orbit and the wave function is found by solving the Dirac equation in the ^{16}O core. The transverse form factor has been calculated out to large q^2 since the calculation is now fully relativistic and there is nothing that inherently limits the theoretical results to low q^2.[5]

An effective weak current can be similarly constructed. We define the nuclear axial-vector current by

$$J^{(\pm)}_{\mu 5}(x) = \left[\delta_{\mu\nu} - \frac{1}{\Box - m_\pi^2}\frac{\partial}{\partial x_\mu}\frac{\partial}{\partial x_\nu}\right] j^{(\pm)}_{\nu 5}(x) \tag{9.5}$$

where the basic axial current is given by

$$j^{(\pm)}_{\mu 5}(x) = i\bar{\psi}\gamma_5\gamma_\mu \underline{\omega}^{(\pm)}\psi \tag{9.6}$$

and the weak charge by

$$\underline{\omega}^{(\pm)} \equiv F_A(0)\tau_\pm \tag{9.7}$$

As justification, we note that the axial-vector current in Eq.(9.5) has the following properties:

1) It is covariant
2) It satisfies PCAC (it is partially-conserved)

$$\frac{\partial}{\partial x_\mu} J_{\mu 5} = O(m_\pi^2) \tag{9.8}$$

3) It gives the correct result for a free nucleon.

The full electroweak currents take the form

$$\mathcal{J}^{(\pm)}_\mu = J^{(\pm)}_\mu + J^{(\pm)}_{\mu 5} \tag{9.9}$$

$$\mathcal{J}^{(0)}_\mu = \mathcal{J}^{V_3}_\mu - 2\sin^2\theta_W J^\gamma_\mu \tag{9.10}$$

We now have a model relativistic nuclear many-body theory where the electroweak currents have all the general transformation properties of the Standard Model.

10. Quantum Chromodynamics (QCD)

One of the reasons the discussion of QHD was motivated in such detail is that a glance at the particle tables reveals a multitude of hadrons. Hadrons may themselves be described with a simpler underlying set of degrees of freedom, quarks and gluons. There exists an elegant gauge theory of the strong interactions binding quarks and gluons into hadrons [18-20]. Quarks are first given an additional internal degree of freedom called color which takes three values; this is analogous to isospin for the nucleon. Quantum chromodynamics (QCD) is then a Yang-Mills gauge

---

[5] Except for the harmonic oscillator center-of-mass correction which here does not play a dominant role [Fig. 13].

theory [21] based on this $SU(3)_c$ internal symmetry group. The gluons are the massless vector mesons introduced to maintain local gauge invariance. Certain non-linear couplings of the gluon fields are required to maintain local gauge invariance; they are present because the generators of the symmetry group do not commute (it is non-abelian).

QCD has two remarkable properties. The first is "asymptotic freedom." Asymptotic freedom essentially states that when all the momenta entering into a process are very large, the renormalized coupling constant for that process becomes very small. This is just the opposite to QED where the bare charge is shielded by vacuum polarization, and the charge gets larger and larger as one goes to shorter and shorter distances, that is, to higher and higher momentum transfer. QCD corresponds to antishielding, where the color charge gets smaller at shorter distances. Asymptotic freedom arises because of the non-linear gluon couplings in this Yang-Mills theory. The second property is "confinement." QCD is a theory where the underlying degrees of freedom, colored quarks and gluons, are not observed as asymptotic free states in the laboratory. It is a hypothesis of QCD that only color singlets correspond to observed hadrons. Lattice gauge theory calculations provide strong indications that confinement is indeed a dynamical property of QCD.

We are faced with the problem of reconciling QHD and QCD. QHD is presumably a "correct" limit of the underlying QCD theory at large distances, where we know that meson exchange governs the N-N interaction. One approach, therefore, is to work in coordinate space and assume a radius outside of which one can use QHD and inside of which one can use asymptotically free QCD. This is the basis of bag models of the hadrons. Another approach is to work in momentum space using dispersion relations. The contribution of nearby singularities can be expressed in terms of observed hadrons, while that of distant singularities can be evaluated using QCD and asymptotic freedom. Still another approach is to assume that QHD provides the most direct description of the baryon/meson phase of nuclear matter, while asymptotically free QCD provides the most direct description of a quark/gluon phase. One can then look for a phase transition [22]. I would like to conclude my talk with a very simple model calculation of this phase transition [23-25,1]

We restrict the discussion to the "nuclear domain" where only up and down quark fields are retained

$$q \doteq \binom{u}{d} \qquad ;\text{nuclear domain} \tag{10.1}$$

They are assumed massless. The states can still contain any number of $(\bar{q}q)$ pairs, and the nuclear domain encompasses the hadrons relevant for nuclear structure. The confinement property will be modeled by assuming it takes a finite amount of energy/volume to hollow out a region in the vacuum into which the quarks and gluons are confined

$$(E/V)_{vac} = +\, b \tag{10.2}$$

The quantity b will be treated as a parameter. Equation (10.2) is the essential ingredient of the M.I.T. bag model [26]; however, only the volume properties of the uniform quark/gluon system are required in the present study, and it is unnecessary to model the more complex hadron surface confinement region at this stage.

Consistent with asymptotic freedom, it is now assumed that this vacuum "bubble" is filled with a non-interacting gas of quarks and gluons. Quarks are spin-1/2 particles which come in 3 colors and 2 flavors. Gluons are massless spin-1 particles (like photons) which come in 8 colors. The degeneracy factors for these gases are therefore

$$\gamma_Q = 3\cdot2\cdot2 = 12$$
$$\gamma_G = 8\cdot2 \quad = 16 \tag{10.3}$$

It is now a simple exercise in statistical mechanics to obtain the equation of state for this uniform quark/gluon system. The energy density and pressure are given by (compare Eqs.(5.5-5.7))

$$\varepsilon = b + \frac{\gamma_G}{(2\pi)^3}\int d\underset{\sim}{k}\,\frac{k}{e^{\beta k}-1} + \frac{\gamma_Q}{(2\pi)^3}\int d\underset{\sim}{k}\; k\,[n(T) + \bar{n}(T)] \tag{10.4}$$

$$P = -b + \frac{1}{3}\left\{\frac{\gamma_G}{(2\pi)^3}\int d\underset{\sim}{k}\,\frac{k}{e^{\beta k}-1} + \frac{\gamma_Q}{(2\pi)^3}\int d\underset{\sim}{k}\; k\,[n(T) + \bar{n}(T)]\right\} \tag{10.5}$$

and the baryon density takes the form

$$\rho_B = \frac{1}{3}\,\frac{\gamma_Q}{(2\pi)^3}\int d\underset{\sim}{k}\,[n(T)-\bar{n}(T)] \tag{10.6}$$

The chemical potential of the gluons has been set equal to zero because gluons are not conserved, and each quark is assigned a baryon number of 1/3. The thermal distribution functions for the quarks and antiquarks are given by (compare Eqs.(5.9-5.10))

$$n(T) = [\exp\beta\,(k-\mu/3)+1]^{-1} \tag{10.7}$$
$$\bar{n}(T) = [\exp\beta\,(k+\mu/3)+1]^{-1} \tag{10.8}$$

Here μ is the chemical potential of the baryons, and there is no self-consistency equation to be satisfied.

A combination of Eqs.(10.4) and (10.5) immediately yields the equation of state of the quark/gluon system at all temperatures and baryon densities

$$3\,(P+b) = \varepsilon - b \tag{10.9}$$

At vanishing baryon density where $\rho_\beta = \mu = 0$, all integrals can be evaluated analytically to give

$$3\,(P+b) = \frac{37}{30}\,\pi^2(k_BT)^4 \tag{10.10}$$

This expression is analogous to that for black-body radiation; it is the thermal pressure of quark-antiquark pairs and gluons in a vanishingly dilute sea of baryons. It is now possible to look for a <u>self-sustaining</u> quark/gluon plasma where the thermal pressure of the constituents just balances the inward pressure for the vacuum bubble; simply set P=0

$$b = \frac{37}{90}\,\pi^2\,(k_BT_o)^4 \tag{10.11}$$

This expression relates the temperature T_o at which the plasma becomes self-sustaining to the parameter b. To get some feel for b, the energy/baryon for the quark/gluon system at zero temperature is plotted in Fig.14 (solid line). Here it is the Fermi pressure of the quark gas at T=0 that gets played off against the bubble pressure, and the quark/gluon system evidently saturates. Also reproduced in Fig.14 is the previous MFT result of QHD-I (dashed curve). In order to assure that observed nuclear matter is indeed in the baryon/meson phase and is not a quark/gluon plasma, we <u>arbitrarily</u> chose a value

$$\mathfrak{R} \equiv 3(2\pi^2b)^{1/4} = 1.2M \tag{10.12}$$

in illustrating this and the remaining results. The mimimum in the solid

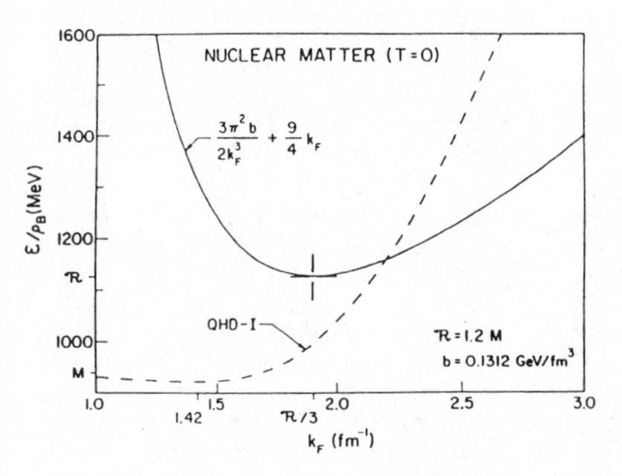

Fig.14 Saturation curves at T=0 for nuclear matter. The solid curve denotes the quark/gluon result in the present model. The quantity R is defined in Eq.(10.12), and the baryon density is parametrized by k_F through Eq.(4.1). The long-dashed curve is the previous MFT result for nuclear matter. A value R = 1.2M is used here for illustration [1].

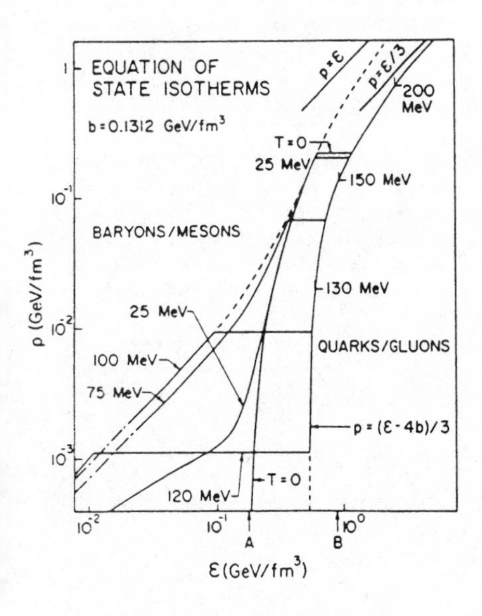

Fig.15 Equation of state isotherms for nuclear matter for the indicated values of $k_B T$. Phase equilibrium exists along the horizontal segments; elsewhere the stable phases, baryon-meson or quark/gluon, are indicated. The left-hand endpoints correspond to zero baryon density [1].

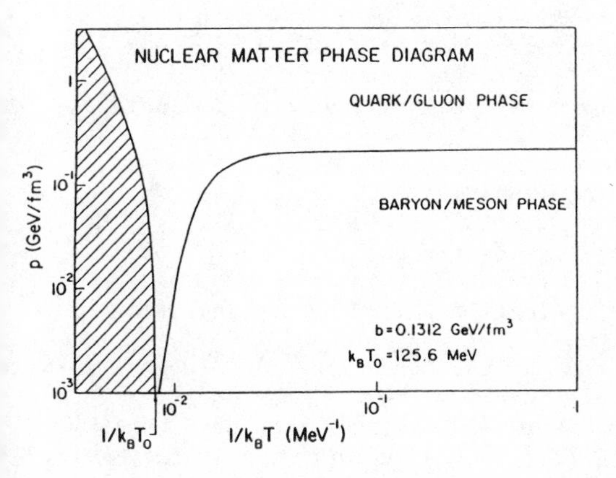

Fig.16 The equilibrium vapour pressure is plotted against $1/k_B T$. The boundary of the shaded region indicates the minimum obtainable pressure at each temperature [1].

curve in Fig.14 now occurs at an energy/baryon = $\mathcal{R}$ = 1.2M and at a density $k_F = \mathcal{R}/3 = 0.4M$.

We now have two distinct models for two distinct phases of nuclear matter. A meson/baryon phase described by the analysis of Eqs.(5.5-5.11) and a quark/gluon phase described by Eqs.(10.4-10.8). How can we combine these analyses? The answer was again given by Gibbs. The conditions for phase equilibrium are

$$P_1 = P_2$$
$$\mu_1 = \mu_2$$

at constant T. (10.13)

When two systems have the same pressure and chemical potential at a fixed temperature, they can coexist in phase equilibrium. (Elsewhere it is the phase with the lowest free energy which is stable.) Thus we simply look for solutions to Eqs.(10.13) using the two different descriptions. The resulting isotherms for the equation of state of nuclear matter are shown in Fig.15. Let us follow one of them. At low density we start with the baryon/meson phase. As the density is increased, we move up the isotherm until a pressure is reached at which the quark/gluon phase begins to form. The two phases coexist at this "vapor pressure" until, by a slight increase in pressure, the system is entirely converted to the quark/gluon phase. Increased pressure then moves us up along the equation of state (10.9). The resulting equilibrium vapor pressure curve for nuclear matter is plotted against 1/T in Fig.16. At high P or at high T, it is always the quark/gluon phase which is stable. The shaded forbidden region in Fig.16 simply reflects the black-body nature of the equation of state.

Although this calculation is a very simple one, it does have several non-trivial properties:

1) The models used for the two phases are both consistent with the principles of special relativity
2) The model used for the meson/baryon phase describes the properties of observed nuclei
3) The model used for the quark/gluon phase satisfies asymptotic freedom
4) The statistical mechanics has been done exactly

11. Summary

Quantum Hadrodynamics (QHD) is a relativistic quantum field theory of nuclear structure based on hadronic degrees of freedom; it provides a theoretical basis for the relativistic nuclear many-body problem. It correlates and explains many features of nuclear structure, including

- Nuclear densities
- The shell model
- Intermediate energy (p,p) scattering

QHD is evidently an approximation to quantum chromodynamics (QCD).
QHD is in some sense "correct" at large distances.

Within the framework of QHD there are many outstanding problems, for example:

1) The lightest mass quantum of the nuclear force, the pion, presents the biggest challenge. A renormalizable theory which includes additional fields for the charged pi and rho mesons (QHD-II) has been developed by Serot [27]. Calculations indicate [28] that some form of chiral symmetry

must be imposed on this lagrangian to correctly describe both the properties of nuclear matter and the low-energy optical potential for pions in the nuclear medium. We have not yet been able to accomplish this. Furthermore, the $(3/2^+,3/2)$ first excited state of the nucleon plays a major role in intermediate-energy physics. Within a renormalizable theory, this state must arise from the dynamics of the meson-nucleon interaction. Fortunately, Chew and Low have shown how this can come about [29-30]. The theoretical analysis within QHD of the behavior of this state in the nuclear medium is a topic of current investigation.

2) QHD is a strong-coupling theory at short distances. One would like to have an exact solution to the theory with which to compare in this regime. A lattice calculation here is an interesting possibility.

3) The big question is can one derive QHD from QCD? Can one demonstrate the existence of condensed scalar and vector fields ϕ_o and V_o in the many baryon system starting from QCD?

Quantum chromodynamics(QCD) is the underlying theory of the strong interactions binding quarks and gluons into hadrons. It is formulated in terms of a completely new set of degrees of freedom for the nuclear problem. The theory is simple at short distances (asymptotic freedom), and is successful at describing short-distance phenomena.

At large distances QCD is a complicated strong-coupling theory (confinement). To be a correct theory, QCD must reproduce nuclear structure including

- Meson exchange
- Baryon dynamics in nuclei
- Meson dynamics in nuclei

It will be a long time before one will be able to describe these phenomena starting from the lagrangian of QCD. The following is my favorite example in this regard: imagine being given the lagrangian of QED, the underlying theory of atomic structure, and being asked to predict the phenomenon of superconductivity!

As for the <u>future</u>, in my opinion the most exciting possibility is to look for interesting, qualitatively new nuclear phenomena based on this new underlying set of degrees of freedom, quarks and gluons. The phase transition we discussed is just one example. There will undoubtedly be new collective phenomena, and new macroscopic color configurations in the many-baryon system. Where should one look? What are their properties? It is an exciting time in nuclear physics.

What one will most likely do in the future is to compute nuclear properties as accurately as possible in both QHD and QCD and then compare the calculations with each other and with experiment in order to study

- Quarks in nuclei
- The inadequacies of QHD
- Approximations to QCD

It is a long, tough program, and it will take a lot of hard work; however, the physics payoff should be high.

I would like to close with three quotations which I find very interesting and thought-provoking:

> "We have been doing nuclear physics for 50 years
> without quarks, why do we need them now?"

(H.L.Anderson, LAMPF II Workshop (1983))

This is actually a profound question for nuclear physics and nuclear physicists. I ask you to think about it very carefully.

"The single most important practical application of the recent advances in particle physics may be the revolution in our picture of the nucleus" (R.R.Wilson, private communication (1984))

And finally, an appropriate new definition of the field

"Nuclear physics is the study of the strong interaction aspects of QCD" (N.Isgur, CEBAF Workshop (1984))

REFERENCES

[1] B.D.Serot and J.D.Walecka "The Relativistic Nuclear Many-Body Problem." Advances in Nuclear Physics, Vol.16, eds. J.W.Negele and E.Vogt, Plenum Press, New York (1986)
[2] A.L.Fetter and J.D.Walecka, Quantum Theory of Many-Particle Systems, McGraw-Hill, New York (1971)
[3] J.D.Bjorken and S.D.Drell, Relativistic Quantum Mechanics, McGraw-Hill, New York (1964)
[4] J.D.Bjorken and S.D.Drell, Relativistic Quantum Fields, McGraw-Hill, New York (1965)
[5] C.Itzykson and J-B.Zuber, Quantum Field Theory, McGraw-Hill, New York (1980)
[6] J.D.Walecka, Ann. Phys. 83, 491 (1974)
[7] M.H.Johnson and E.Teller, Phys.Rev.98, 783 (1955)
[8] H.P.Duerr, Phys. Rev.103, 469 (1956)
[9] S.A.Chin and J.D.Walecka, Phys. Lett.52B, 24 (1974)
[10] J.D.Walecka, Phys.Lett.59B, 109 (1975)
[11] C.J.Horowitz and B.D.Serot, Nucl. Phys.A368, 503 (1981)
[12] B.C.Clark,S.Hama,R.L.Mercer,L.Ray, and B.D.Serot, Phys. Rev. Lett.50, 1644 (1983); Phys. Rev. C28, 1421 (1983)
[13] J.A.McNeil, J.R.Shepard, and S.J.Wallace, Phys. Rev. Lett. 50, 1439, 1443 (1983)
[14] S.A.Chin, Ann. Phys. 108, 301 (1977)
[15] T.D.Lee and M.Margulies, Phys. Rev. D11, 1591 (1975)
[16] W.Bardeen, unpublished
[17] E.-J.Kim, Phys. Lett. 174B, 233 (1986)
[18] H.Fritzsch and M.Gell-Mann, in Proc. XVI Int. Conf. on High Energy Physics, eds. J.D.Jackson and A.Roberts, Vol.II, p.35, Fermi Nat'l. Acc. Lab., Batavia Ill. (1972)
[19] H.D.Politzer, Phys. Reports C14, 129 (1974)
[20] F.Wilczek, Ann. Rev. Nucl. Part. Sci. 32, 177 (1982)
[21] C.N.Yang and R.Mills, Phys. Rev 96, 191 (1954)
[22] J.C.Collins and M.J.Perry, Phys. Rev. Lett. 34, 1353 (1975)
[23] G.Baym and S.A.Chin, Phys.Lett. 62B, 241 (1976)
[24] S.A.Chin, Phys. Lett. 78B, 552 (1978)
[25] J.Kuti et al., Phys. Lett. 95B, 75 (1980); 98B, 199 (1981)
[26] A.Chodos, R.L.Jaffe, K.Johnson, C.B.Thorn, and V.F.Weiskopf, Phys. Rev. D9, 3471 (1974)
[27] B.D.Serot, Phys. Lett. 86B, 146 (1979);Err. 87B, 403 (1979)
[28] T.Matsui and B.D.Serot, Ann. Phys. 144, 107 (1982)
[29] G.F.Chew and F.E.Low, Phys. Rev. 101, 1570 (1956)
[30] S.C.Frautschi and J.D.Walecka, Phys. Rev. 120, 1486 (1960)

Inst. Phys. Conf. Ser. No. 86
Paper presented at Int. Nucl. Phys. Conf., Harrogate, UK, 1986

Ultrarelativistic heavy-ion collisions and the quark–gluon plasma

Gordon Baym

Department of Physics, University of Illinois at Urbana-Champaign
1110 W. Green St., Urbana, Illinois 61801

Abstract. The expected form of nuclear matter under extreme conditions of temperature or baryon density is a quark-gluon plasma. The basic physics of such a plasma and the transition from hadronic to deconfined matter are reviewed. Ultrarelativistic heavy-ion collisions allow one to study properties of nuclear matter under extreme conditions in the laboratory. The formation of plasmas in such collisions and their expected evolution are outlined.

1. Introduction

The basic question I would like to address in this talk is: how does one go about learning the properties of extended matter at extremely high energy densities, an order of magnitude or more beyond that of normal nuclear matter? The energy density of normal nuclear matter is essentially the rest mass density, of order 0.15 Gev/fm^3, large compared with the scale of low-energy spectroscopy, of order Mev/fm^3. How do we expect nuclear matter to act when we raise its energy density to the range of 1-10 Gev/fm^3 say? One of the most interesting possibilities then is that matter will form a new state, the quark-gluon plasma.

To carry out such studies will require colliding heavy nuclei together at energies well above 1 Gev per nucleon in the center-of-mass frame. A program of fixed-target experiments with lighter nuclear projectiles will begin shortly at the CERN SPS at lab energies of 60 and 225 GeV per nucleon, and at the Brookhaven AGS at lab energies of 12-14 GeV per nucleon. These experiments will be reviewed in detail in the following talk by Hans Specht. In addition, the Brookhaven Relativistic Heavy Ion Collider (RHIC) is quite far along on the drawing boards, and will provide the capability of colliding nuclei as heavy as Au on Au at 100 GeV per nucleon c.m. (equivalent to 20 TeV per nucleon lab).[1]

Studying the behavior of matter at high energy densities in the laboratory offers many important opportunities for new physics. Formation of a deconfined quark-gluon plasma will permit one to study quantum chromodynamics (qcd) and quark phenomena over distances very large compared with those available in light hadron systems. In addition, nuclear matter

1. Useful general references on relativistic nucleus-nucleus collisions are the proceedings of the ongoing conferences on quark matter (Ludlam and Wegner 1984, Kajantie 1985, Gyulassy *et al.* 1986).

under extreme densities is of great interest in astrophysics, a question I will return to later.

The exploration of the behavior of nuclear matter under extreme conditions gives one a rather complementary approach to the study of quark-gluon degrees-of-freedom, and the related question of relativistic effects, in nuclei. High precision electron scattering from nuclei, which will be carried out over the coming years, will provide information on quark properties in individual nuclear states, the effects of quarks in nuclear wave functions. Hadronic probes will provide further detailed information. Ultrarelativistic heavy-ion collisions on the other hand answer a different question: what are the highly excited thermodynamic states, the thermodynamic role of quark and gluon degrees of freedom? As Erich Vogt's comment -- that simply by knowing about Coulomb forces one does not necessarily know the properties of water -- emphasizes, these are rather different aspects of the problem. Heavy-ion collisions will primarily address the problem of the gross statistical behavior of matter, rather than the details of individual wave functions.

Our principal region of interest, in the phase diagram of nuclear matter in the temperature-baryon density plane, Fig. 1, is that at high temperature or baryon density, where one can reach deconfinement of quarks and gluons. In the low temperature-baryon density region the basic degrees-of-freedom are hadronic, those described by Dirk Walecka in the previous talk. Between these two regions may or may not be a sharp phase transition. Later we will come back to the question of how one can learn about the phase diagram from various experiments, but first let us consider the elementary properties of quark matter.

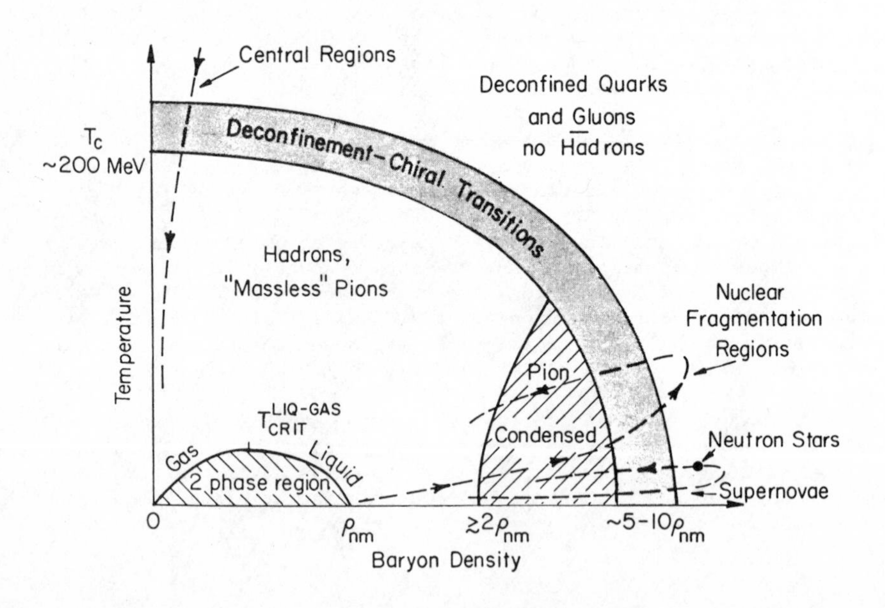

Fig. 1 Phase diagram of nuclear matter in the baryon density, temperature plane showing regions of hadronic and deconfined matter. Normal nuclear matter density ρ_{nm} is 0.16 fm^{-3}.

2. Quark-gluon plasma

As matter is heated or compressed its degrees of freedom change from composite to more fundamental. For example, by heating or compressing a gas of atoms, one eventually forms a plasma in which the nuclei become stripped of the electrons, which go into continuum states forming an electron gas. Similarly, when nuclei are squeezed, as happens in the formation of neutron stars in supernovae where the matter is compressed by gravitational collapse, the matter merges into a continuous fluid of neutrons and protons. Since nucleons themselves are made of quarks, one further expects that a gas of nucleons, when squeezed or heated, turns into a gas of uniform quark matter, composed of quarks, and at a finite temperature, antiquarks and gluons as well, which are no longer confined in individual hadrons but are free to roam over the entire volume of the deconfined region.

Before turning to the properties of quark matter, let us recall a few relevant features of qcd. In quantum electrodynamics, photon exhange produces the basic force between charges, which between static point charges is simply the Coulomb interaction, $\sim e^2/r$. The force between charges of opposite sign is opposite to that between like charges; thus in qed one can form electrically neutral systems, such as positronium or hydrogen, which do not give rise to long-range Coulomb fields. Qcd has a similar structure, in that the forces between quarks arise from exchange of gluons, and the color degree of freedom functions as a three-valued charge, rather than simply + or −, as in qed. Again one can form color singlet or neutral systems that do not give rise to long-range color Coulomb fields. [If one very naively pictures the forces between different colored quarks to be −1/2 that between similar colored quarks, then a nucleon made of three different colored quarks would have no long-range Coulomb interaction with another quark.] To have such a charge scheme requires eight gluons, rather than a single photon, themselves having color and hence coupling directly to themselves producing the rich non-linear structure of qcd. Because qcd allows color neutrality, quark matter in equilibrium in its state of lowest energy, or free energy at finite temperature, will on average have no long-range color Coulomb fields, as in an ordinary electrically neutral plasma.

Qcd is also asymptotically free. In qed an electron gathers around it a polarization cloud of electron-positron pairs in the vacuum which decreases the net charge seen at large distances; the effective charge of an electron at large distances is given by $e^2/\hbar c = 1/137$. At short distances, inside the polarization cloud, the effective charge on the electron grows, diverging at zero distance -- one of the troublesome divergences of qed. In qcd a rather different behavior occurs. Because the gluons themselves carry color, they also screen the bare charges, but their net effect is opposite that of quark-antiquark pairs; the result is that as one goes close to a quark, the effective charge does not become infinite, but rather goes to zero -- the property of asymptotic freedom. At short distances, corresponding to large momentum scales, interactions become arbitrarily weak [with the effective coupling $\alpha(p) = g^2/4\pi \approx 6\pi/[(33-2N_f)\ln(p/\Lambda)]$ where N_f is the number of quark flavors that are relevant, Λ the qcd scale parameter is of order 100-200 MeV, and p is the momentum scale]. At large distances however quite the opposite happens; if one tries to separate colored particles, the forces become larger and larger, giving rise to confinement.

Imagine then highly compressing or heating nuclear matter, so that many quark-gluon degrees of freedom are excited. To a first approximation, one can at very high densities treat the system as a non-interacting gas of relativistic quarks, antiquarks and gluons, since as in an ordinary plasma, any small region of the matter will, at high densities, be on average color neutral and not produce long-range (color) Coulomb fields, while the residual short distance forces in the region become weak as the interparticle separation becomes small, due to asymptotic freedom. [See Svetitsky 1986 and Polonyi 1986 for a more careful discussion of interaction effects in the high density limit.]

Although for the temperatures and baryon densities of plasmas realistically expected in laboratory collisions, interactions will in fact be important, the non-interacting limit provides a useful first handle on the quark-gluon plasma. While ordinary nuclear matter has 4 helicity states, 2 for spin times 2 for isospin, a quark-gluon plasma has many more internal degrees-of-freedom; the quarks have from 24 to 36 helicity states, composed of 2 spin, 3 color, 2 particle-antiparticle, and 2 to 3 flavor degrees-of-freedom, depending on whether strange (s) quarks are also present in addition to the light up (u) and down (d) quarks; the massless gluons have in addition 16 helicity states (2 spin and 8 color). From a thermodynamic point of view a quark-gluon plasma at a given energy density has a high entropy. Hot free quark matter looks very much like ordinary black-body radiation with energy density $E \sim T^4$, where T is the temperature; in a system with equal number of u, $\bar{u}$, d, and $\bar{d}$, as well as gluons,

$$T \approx 160 \text{ MeV } E^{1/4} \tag{1}$$

and the total density of excitations per fm^3 is

$$n_{exc} \approx 2.25 \ E^{3/4} \tag{2}$$

with E measured in GeV/fm^3. Since the qcd phase transition to a quark-gluon plasma is believed to occur at T of order 200 MeV, we see from (1) that the scale of energy densities that must be deposited in collisions to excite a plasma is of order several GeV/fm^3. Because of the slow dependence of T on E, it will not be easy to heat a plasma in a nuclear collision much beyond hundreds of MeV.

In order to neglect interactions, temperatures (or momenta or chemical potentials) should be large compared with the qcd scale parameter Λ. The temperatures produced in collisions are expected, however, to be at most on the order of a few times Λ; one must generally take interactions into account. They are clearly always important near the deconfinement transition. At first one is tempted to use perturbation theory. Writing down all the Feynman diagrams for the thermodyamic free energy one is capable of doing in an afternoon takes one to order $\alpha^2 \ln \alpha$. Taking into account terms of order α^2 is considerably harder. The eventual result, an asymptotic expansion of the free energy in α, is absolutely useless at the coupling strengths of interest. For example, the first correction to the entropy density, calculated by differentiating the free energy with respect to T, is given by $s = s_o[1 - (54/19\pi)\alpha]$, where s_o is the non-interacting entropy density. This first order expression for the entropy turns negative for $\alpha \sim 1$, while coupling constants can become larger at low densities. Now this is impossible -- entropies must be positive. What we see is a signal that perturbation theory breaks down early, and is not a satisfactory way to calculate.

The only useful approach so far to calculating effects of interactions is Monte Carlo lattice gauge theory, where by putting the theory of qcd on a lattice one becomes capable of dealing with all strengths of interaction. (See Satz 1985 for a general review.) Lattice gauge theory requires rather large computers to get adequate statistics, but recently, as Cray type supercomputers with substantial megaflop rates have become available to the community, the ability to compute with good statistics on large lattices has improved enormously, and lattice gauge theory is really at the point where it will be able to give quantitatively good information on the properties of quark matter over large ranges of temperature and also baryon density. Calculations with finite baryon density are just in their beginning, and I shall in the limited time available review only the finite temperature, zero baryon density results.

Figure 2 shows early Monte Carlo calculations of the energy density, plotted in units of the ideal non-interacting system (Stefan-Boltzmann) energy, $\sim T^4$, in a system with just gluons, and no quarks whatsoever -- pure Yang-Mills theory -- in a) for SU(2), with just 2 colors (Engels et al. 1981), and in b) for SU(3) (Çelik et al. 1983). The temperature is measured in units of the lattice qcd scale parameter Λ_L. In SU(2), the energy density exhibits a clean second order phase transition. At low temperature the system behaves as a gas of massive glueballs, and is confined, while at the transition temperature, T_c, the system turns smoothly into a gas of deconfined gluons, rather rapidly approaching the high temperature T^4 limit. In SU(3), by contrast, one sees a rather sharp first transition, like the boiling of water, with a large latent heat, of order a few GeV/fm^3. Indeed the Monte Carlo calculations in this case exhibit considerable hysteresis effects, which are a good signal of a first order transition.

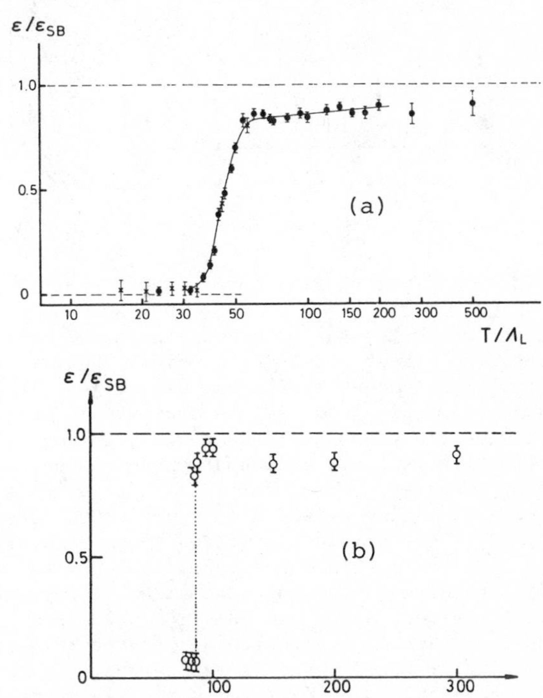

Fig. 2 Energy density of pure gauge theory in units of the ideal gas energy: (a) SU(2), (b) SU(3).

In a pure gluon theory one can measure whether the system is confined or not by adding a massive quark-antiquark pair of test particles to the system, and asking how much energy is required to separate the pair to infinity. If the system is confining, it is impossible to separate them and the energy of separation ε diverges. [Color octet gluons cannot screen the force between color triplet quarks.] Thus the "Wilson line," defined by $W = \exp(-\varepsilon(R\to\infty)/T)$, goes to zero as the separation R goes to infinity, while in the deconfined state it should be non-zero. The Wilson line functions in pure gluon theory as a useful order parameter to distinguish the confined from the deconfined phase. Calculations of W for the pure gluon theory are shown in Fig. 3; in a) for SU(2), W begins to rise above the transition point, corresponding to the onset of deconfinement, while in b) it jumps discontinuously, consistent with the behavior of E.

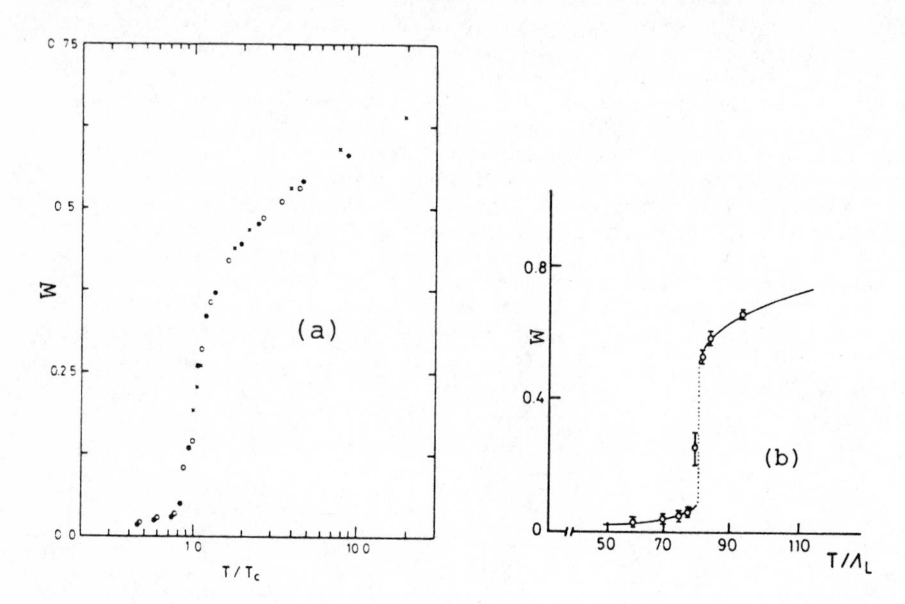

Fig. 3 The Wilson line for (a) SU(2) and (b) SU(3) pure gauge theory.

When one begins to include finite mass quark degrees of freedom, q and $\overline{q}$, the simple test of adding a pair of heavy test quarks $Q\overline{Q}$ runs into trouble, since at sufficient separation, it becomes energetically favorable to create a $q\overline{q}$ pair in the system, which screens out the interaction between the test pair; the q binds to the $\overline{Q}$, and the $\overline{q}$ to the Q, creating effectively a pair of mesons which can be separated to infinity with finite energy. The point is that once one has light quarks in the system there no longer exists a good measure of whether the system is in a confined or deconfined state, and there need not be a sharp transition between the confined and deconfined phases. The transition between the two phases can be smooth, as occurs for example in ionization of a gas as it is heated, where one goes gradually from gas molecules to electrons and nuclei; the two states are qualitatively different and there is a reasonably rapid onset of ionization, but it is not sharp. Alternatively, the transition may be first order, as in the boiling of water.

Qcd matter with light quarks turns out technically to be analogous to a ferromagnet in the presence of an external magnetic field, H, with the quark mass playing a similar role to H^{-1}. Let us consider the case of a ferromagnet in which the transition in zero external field happens to be first order. Then with increasing H the transition becomes weaker and weaker, and finally beyond a critical H_c, the transition becomes smoothed out; in enormous H, the spins are all aligned and there is no transition. In the T-H plane, one finds a line of first order transition points, terminating in a critical point at H_c. Similarly a quark-gluon plasma, at large quark mass, has a well-defined transition, which weakens as the mass decreases; it is not immediately apparent whether for realistic quark masses, analogous to large H, there is a sharp transition or not. One problem is that to take light mass quarks into account accurately in lattice gauge theory requires quite extensive computing, and only very recently has one begun to achieve a reasonable understanding of the nature of the transition (at zero baryon density).

One further ingredient which must be taken into account is chiral symmetry, the nearly exact SU(2)xSU(2) symmetry of the strong interactions in low nuclear physics generated by the conserved vector current

$$\vec{V} = \int d^3r\, \overline{\psi}(r)\gamma^o\vec{\tau}\psi(r)$$

together with the partially conserved axial vector current (PCAC)

$$\vec{A} = \int d^3r\, \overline{\psi}(r)\gamma^o\gamma_5\vec{\tau}\psi(r) \;.$$

Because the axial current is not precisely conserved, chiral symmetry is not exact; the level of violation is measured by the smallness of the pion mass: $(m_\pi/m_n)^2 \sim 1/50$. From the point of view of the underlying quark structure, chiral symmetry is exact only for zero mass u and d quarks; the violation of chiral symmetry is a reflection of the fact that the light quark are not precisely massless, but have masses on the order of 10 MeV.

Symmetries can be realized in a physical situation in two ways: the Wigner mode in which the states can be classified according to the representations of the rotation group, as in atoms and nuclei, and the Goldstone mode, in which the equilibrium state picks out a given direction in the group space, analogous to the situation in a ferromagnet (or more precisely an anti-ferromagnet), where the magnetized state chooses a special spatial direction for the magnetization, breaking the overall rotational symmetry of the state. In nuclear physics chiral symmetry is in fact spontaneously broken. The long wavelength small oscillations of the spins in an aligned ferromagnet are low-lying modes, the spin waves; in the case of broken chiral symmetry, the analogous low-lying excitations, or Goldstone bosons -- the oscillations of the spontaneously selected direction -- are the physical pions.

The implications for qcd are the following. At low temperatures and baryon densities chiral symmetry is spontaneously broken. On the other hand, if at the high momentum scales of very high temperatures or densities the system becomes asymptotically free, then one expects chiral symmetry to be fully restored. Between these two limits a chiral symmetry restoring phase transition should occur. Associated with this transition is a well-defined order parameter $\langle\overline{\psi}\psi\rangle$ (where ψ is the quark field) which is non-zero in the spontaneously broken phase and zero where chiral symmetry is fully restored. Finite quark mass can wash out the chiral transition; however as m becomes sufficiently small one expects the appearance again of a sharp transition associated with the chiral symmetry.

Figure 4 shows several recent calculations of the Illinois Monte Carlo group (Kogut and Sinclair 1986, and earlier references therein) in SU(3) with finite quark mass m, on a 6x10x10x10 lattice. The horizontal scale is the effective coupling $\beta = 6/g^2$, which increases monotonically with temperature. The mass is given in terms of the transition temperature T_c (generally of order twice the qcd scale parameter Λ) by 6 times the values in the figures (there in units of a the lattice spacing). Fig. 4a, the Wilson line for a system with relatively heavy quarks -- over an order of magnitude more massive than the u and d quarks -- nicely illustrates how matter with heavy quarks has a first order phase transition. The Wilson line shows a fairly sharp onset of deconfinement; hysteresis in the transition provides good evidence that it is actually first order. Figure 4b, for intermediate mass ma = 0.05, shows a smooth onset of the deconfined phase; the behavior is more like a second order or smoothed-out phase transition. Also plotted is the chiral order parameter $\langle\bar{\psi}\psi\rangle$, which goes rapidly from a finite to a small value (not exactly to zero, since the quark mass is finite in these calculations) as deconfinement sets in.

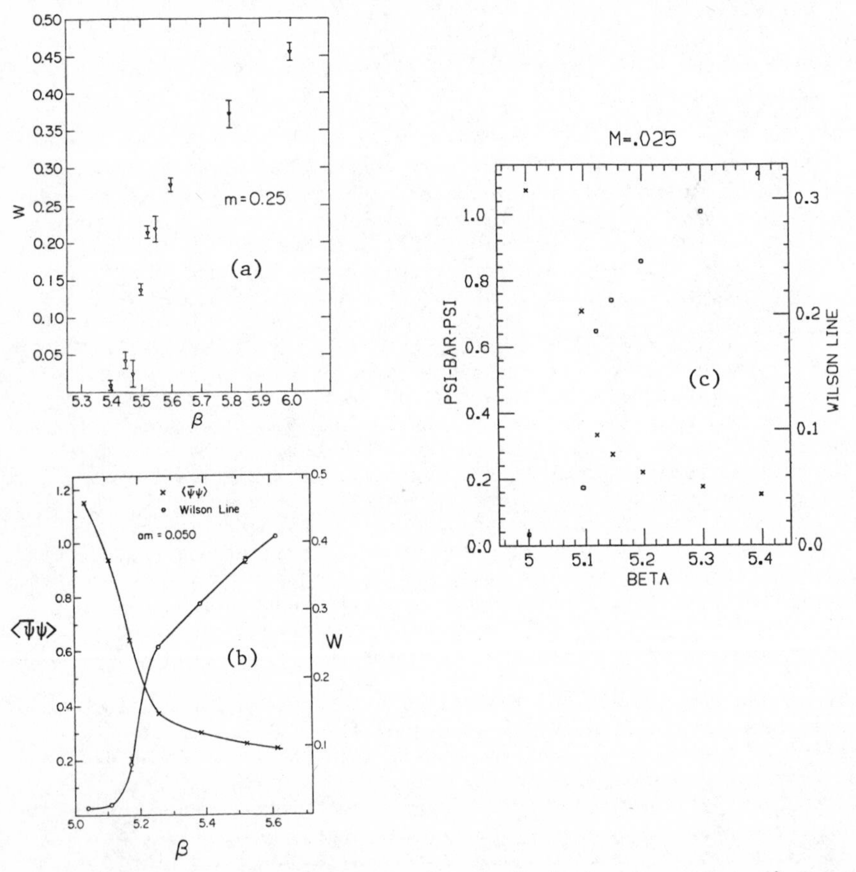

Fig. 4 Wilson line in SU(3) with a) relatively heavy quarks, b) intermediate mass; c) light mass; b) and c) show the chiral behavior as well.

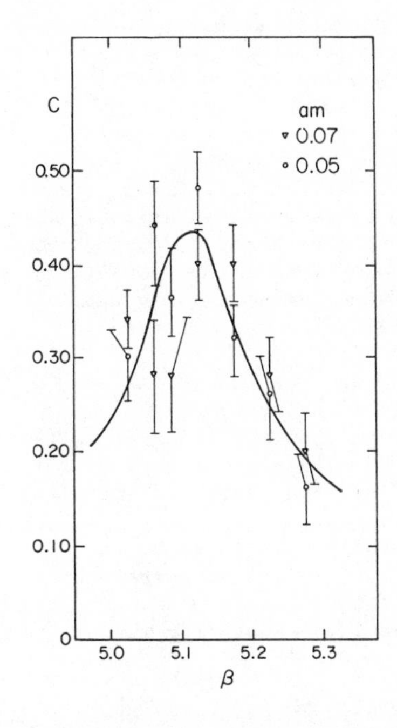

Fig. 5 Specific heat with intermediate mass quarks .

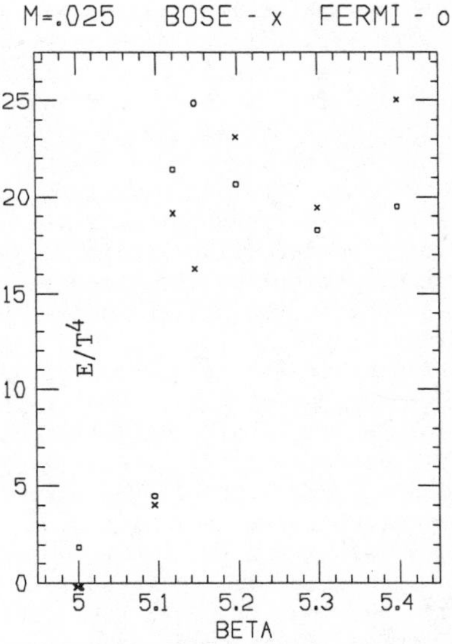

Fig. 6 Energy density in units of T^4 for light quarks.

Another measure of the sharpness of the transition is the specific heat, which indicates the temperature range over which energy must be put into the system to go from one phase to the other; Fig. 5 (Kogut 1986) shows, for the intermediate mass cases ma = 0.05 and 0.07, the specific heat versus β (or effectively the temperature) in the neighborhood of the transition. We again see not a sharp transition in this case, but rather a large but smooth bump; the total energy under the curve is of order a few GeV/fm^3.

However for even lower mass, in Fig. 4c (Kogut and Sinclair unpublished) -- ma = 0.025, m/T_c = 0.15, closer to the realistic case -- we see evidence that the restoration of chiral symmetry, and deconfinement, is now via a sharp transition. Figure 6 (Kogut and Sinclair, to be published) shows the energy density divided by T^4, versus temperature for this light mass case. Above the deconfinement transition, the energy goes rapidly to the Stefan-Boltzmann limit (19.1 for fermion quarks with 4 flavors, as considered here).

To summarize, present lattice gauge theory calculations with finite mass quarks indicate that at large m the transition between the confined and deconfined phases is first order, while with decreasing m one enters an intermediate regime where the transition is somewhat washed out; at still lower quark masses, when the chiral behavior is accurately taken into account the transition sharpens again, driven by restoration of chiral symmetry, and is likely (weakly) first order.

3. Exploring the phase diagram

The physics of matter at high temperature and density can be explored in astrophysical as well as laboratory situations. Let us first briefly discuss the applications to supernovae, neutron stars and the early universe. In a (Type II) supernova explosion, a massive star which has burnt out its fuel at the end of its evolution can no longer support itself against gravitational collapse, and begins to implode. In the infall the matter is crushed to very high densities, several times that of normal nuclear matter. The core of the star bounces back, as shown by the trajectory in the phase diagram; the matter in the core may or may not cross into the deconfined region. The energy output in a supernovae, which should depend on the strength of this bounce, may provide a handle on the nature of the matter in the interior reached in the collapse.

In neutron stars the properties of matter under extreme conditions play a particularly crucial role. One quickly learns in trying to construct models of neutron stars how little is known about the properties of matter at densities beyond $\sim 2\rho_{nm}$. The lack of such knowledge is reflected, for example, in uncertainty in the maximum mass of neutron stars, an important quantity in trying to identify black holes unambiguously. Measured masses of neutron stars are ~ 1.4 solar masses, with radii calculated to be ~ 10 km. Neutron stars are giant nuclei, with $A \sim 10^{57}$. Typical temperatures are very low, less than one MeV. The central conditions in a neutron star are indicated on the phase diagram. The matter in the interior may possibly be deconfined.

One can in fact study the interior of neutron stars by observing their cooling. In their early years ($<10^5$y), cooling is governed primarily by neutrino emission. For phase space reasons, cooling via the nucleonic URCA process, $n \rightarrow p + e + \bar{\nu}$, and $e + p \rightarrow n + \nu$, is considerably slower than it would be via the corresponding process with light mass deconfined quarks, $d \rightarrow u + e + \bar{\nu}$, and $u + e \rightarrow d + \nu$. Combining knowledge of the ages of astrophysical objects containing neutron stars with measurements of neutron star surface temperatures, taken with X-ray telescopes, gives one a measure of how rapidly neutron stars cool. Particularly rapid cooling would provide evidence for unusual states of matter in the interior. Present observations, which generally provide an upper bound on surface temperatures, are so far consistent with the interiors being normal nuclear matter, although future satellite observations should sharpen these bounds and provide a more definitive answer to the nature of the matter in the interiors of neutron stars.

In the first microseconds of the early universe, the temperature falls as

$$T \approx 0.5\ (t_{seconds})^{-1/2}\ \text{MeV}\ ,$$

so that prior to ~ 6 microseconds after the big bang, when the temperatures are $>$ hundreds of MeV, matter is in the form of a quark-gluon plasma. The matter of the early universe has a relatively small net baryon density, of order 1 part in 10^9 (which appears as the present photon/baryon ratio). As the universe expands it cools and matter hadronizes, following a downward trajectory practically along the vertical axis of the phase diagram. Matter emerging from the transition is primarily in the form of pions, with a slight baryon excess. Possible astrophysical consequences of the transition in the early universe from deconfined plasma to hadrons will be touched upon by Dave Schramm in his talk.

4. Ultrarelativistic heavy-ion collisions

Coming back down to earth, let us consider how one can study the phase diagram in the laboratory via very energetic collisions of heavy nuclei. Imagine colliding two nuclei together at energies from 1 to 100 GeV per nucleon energy in the center-of-mass. At low energies, the regime that will be studied in the forthcoming SPS and AGS experiments, one can picture the two Lorentz-contracted colliding nuclei as nearly stopping each other, with reasonable probability of forming, to a crude first approximation, a fireball. [In reality, parts of the nuclei will generally pass through the collision rather than remain in a fireball.] Such collisions may achieve energy densities of order a few Gev/fm^3 and baryon densities several times ρ_{nm}; the matter may indeed cross into the deconfined region, as shown by the curve in Fig. 1 labelled "fragmentation regions," and then expand out. The high density matter produced in such collisions will be relatively baryon rich.

As the beam energy is increased, the time τ_o it takes for excitations to form as the nuclei collide, measured in the center-of-mass, becomes, as a consequence of Lorentz time dilation, effectively longer than the time it takes for the nuclei to pass through each other -- the phenomenon of nuclear transparency. In this case the nuclei pass through each other, become highly excited internally, and at the same time, leave the vacuum between them in a highly excited state containing quarks, antiquarks and gluons as illustrated in Fig. 7. Nuclear transparency is very important in the ultrarelativistic regime, above ~ 10 GeV per nucleon (c.m.). The nuclear fragmentation regions, which recede from each other at the speed of light, contain essentially all the baryons of the original nuclei; the central region, to a first approximation, has no baryon excess, and resembles the hot vacuum of the early universe. [In fact, the baryons of the colliding nuclei will spread somewhat into the central region; predicting how much is an important problem on which the forthcoming heavy-ion experiments at CERN and Brookhaven are expected to shed light. However, the energy per baryon will in general be very high.]

In ultrahigh energy collisions, one expects a useful strong correlation between the spatial structure of the collision region, and the final rapidities of the detected particles emerging from the collision.

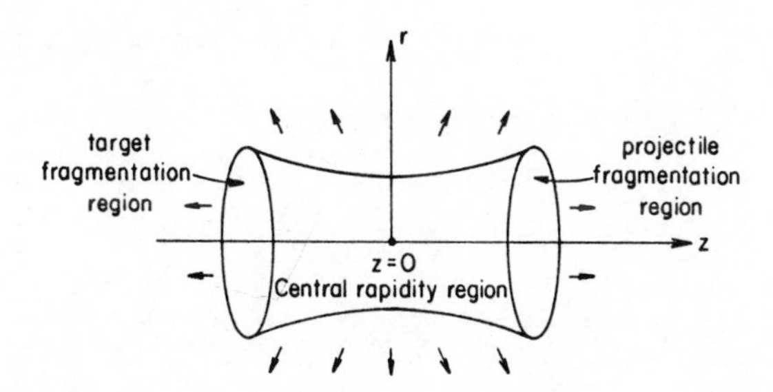

Fig. 7 Nuclear fragmentation and central rapidity regions in an ultrarelativistic central heavy-ion collision.

[Relativistic collisions are conveniently described in terms of the rapidity y of the beams and collision products, defined by $y = (1/2)\ln[(E+p_z)/(E-p_z)]$, where E is the particle energy, and p_z its momentum along the beam axis. For motion purely along the z axis, the velocity is given by $v/c = \tanh y$. Rapidities, unlike ordinary relativistic velocities, have the nice property of being additive; under a Lorentz transformation along z by velocity $u = c \tanh y_u$, rapidities transform by $y \rightarrow y + y_u$.] To see how this correlation works, think of the fragmentation regions after the collision as travelling at c away from each other, with the central region being uniformly stretched out in between, so that its velocity increases linearly with distance, from −c at the left fragmentation region to +c at the right fragmentation region (Fig. 7). Thus, particles observed at large positive rapidities come primarily from the right fragmentation region, those at large negative rapidities from the left fragmentation region, while the intermediate rapidity particles arise from the central regions. The intrinsic motion of the particles with respect to the local average motion will, of course, blur this correspondence somewhat, on the order of one unit of rapidity.

The total spread in rapidity in a collision is given by $\Delta y = 2 \ln(2E/m_n)$ where E is the beam energy per nucleon in the center-of-mass. Thus ultrarelativistic collisions provide sufficient total rapidity spread -- for example 10.6 units at 100 GeV on 100 GeV -- that this correlation enables one to sort out the different collision regions from the rapidities of their final state products; this ability to distinguish different regions experimentally is one of the principal reasons for going to high energies. The baryons will appear predominantly at large rapidities, while in the central rapidity region, which has little excess of baryons over anti-baryons, one will see primarily mesons. Indeed, such a structure emerges in proton-proton scattering, as carried out at the CERN ISR, and in $\bar{p}p$ collisions at the S$\bar{p}$pS. The charged particle multiplicity vs. rapidity for 30 GeV on 30 GeV pp collisions is shown schematically in Fig. 8; the meson spectrum (unshaded) is spread out over the central region, while the net baryon density (shaded) is peaked near the rapidities of the two incident colliding beams, with a width of order 2 units in rapidity.

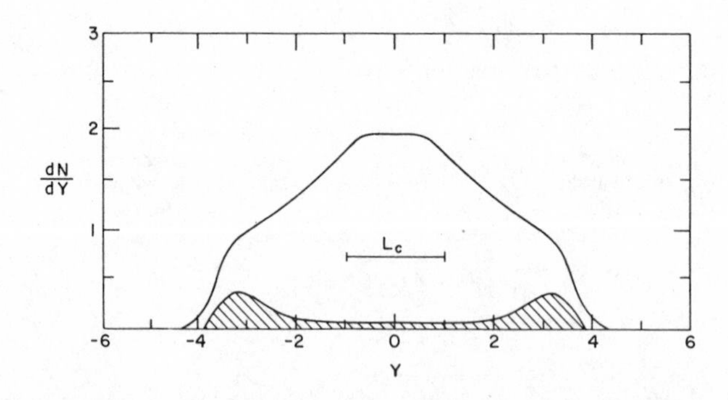

Fig. 8 Charged particle multiplicity in pp collisions at 30 GeV on 30 GeV.

The scale of energy densities expected in ultrarelativistic collisions can be estimated from the observation that pp or $\overline{p}p$ collisions in this energy range produce ~ 2-3 charged particles, predominantly pions, per unit of rapidity, with typical transverse energy, ~ 400 MeV. Adding in neutral pions as well, we find an energy density of ~ 2 GeV per unit of rapidity. A very conservative extrapolation to a nucleus-nucleus collision is to multiply this produced energy by a factor ~ (1-2) A, i.e., assume that each nucleon of one of the nuclei makes only one or two collisions going through the other nucleus. The resulting estimate of the energy per unit rapidity in an AA collision is > (2-4)A GeV, and the corresponding energy density is

$$E \gtrsim 0.4A^{1/3}/t \text{ GeV/fm}^3$$

where t is the time in fm/c. At a time of 1 fm/c the energy density is at least of order 2.5 GeV/fm^3 in the average central collision.

The actual energy density can be much larger, as can be seen from a similar argument based on extrapolating from particle multiplicities (Von Gersdorff et al. 1986). If we assume that entropy is conserved in the evolution of the collision, then at early times, $tT^3 \sim (dN/dy)/\pi R^2$, where T is the temperature, R is the nuclear radius, and dN/dy is the final multiplicity density in rapidity. Clearly the earlier the time, the greater the temperature and hence energy density. The uncertainty principle implies that the initial formation time, τ_o, when one can first begin to describe the system in terms of well-defined interacting excitations (quarks, antiquarks and gluons), and the initial temperature, T_o, at that time, obey $T_o\tau_o \gtrsim 1$. Let us assume that the mean multiplicity in an AA collision is of order A times that in a pp collision. Combining with the uncertainty principle relation we deduce that the formation time is $\sim A^{-1/6}$ in fm/c, that the initial temperature is $\sim 200\ A^{1/6}$ MeV, and most importantly, the initial energy density one expects to achieve, proportional to T_o^4, is $E \sim A^{2/3}$ in GeV/fm^3, which is of order 30 GeV/fm^3 for Au or U collision partners. Furthermore, in rare events, on which one can certainly trigger, one expects even larger energy depositions. Thus we have good reason to believe that the energy densities in collisions will be sufficiently high in many events to form interesting states of matter.

A further feature of pp and $\overline{p}p$ scattering observed at CERN collider energies is that the multiplicity distributions in the central regions are roughly flat, as can be seen in Fig. 8. Since changing rapidity is equivalent to making a Lorentz transformation, the lack of change under a shift of the horizontal rapidity scale implies that conditions in the central region are approximately Lorentz invariant. The assumption of Lorentz invariance in the central region provides a very simple first picture of the evolution of the central region, as is illustrated in the space-time diagram, Fig. 9, which shows a slice of the collision along the central axis, z, versus time t. At negative times the projectile and target approach each other along the light cone and collide. The first event in the collision is production of excitations, after a finite formation time, which stream out from the collision point, the origin. In heavy nucleus-nucleus collisions, unlike in pp collisions, sufficiently large numbers of excitations are made, and the system sizes are correspondingly large, that after a further finite time the excitations come into local thermodynamic equilibrium; this means that the system enters a regime where it can be described by hydrodynamics. Two interesting phenomena occur now. One is that the matter passes through the

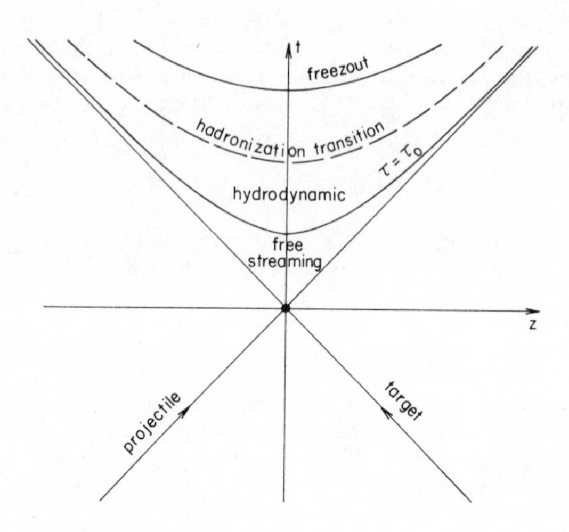

Fig. 9 Space-time picture of the evolution of a central collision.

hadronization transition, as shown in the phase diagram by the curve labelled "central regions," and emerges from the transition in the form of hadrons. Eventually the system expands sufficiently that the interactions among the hadrons cease -- "freezeout" -- and the system becomes a collection of freely streaming particles, which are eventually detected. The lines separating the regions in the space-time diagram are essentially hyperbolae, since as a consequence of the approximate Lorentz invariance in the central region, one expects the same conditions at all z in the central region at the same proper time, $(t^2 - z^2)^{1/2}$ (corresponding to local velocity z/t along the central axis).

One very important point that this diagram illustrates is that the matter undergoes considerable processing from the initial quark-gluon plasma phase to the finally observed hadronic products; tracing back from the observations to the properties of the plasma will clearly be a challenging problem. In the collision, the two fragmentation regions move away from each other longitudinally, and the central region undergoes a longitudinal stretching and cooling, as illustrated in Fig. 9. In addition, the system begins to undergo transverse expansion, which initially occurs hydrodynamically with a rarefaction wave propagating inward from the outer edge at the speed of sound.

To summarize, the basic picture of ultrarelativistic collisions, while clearly a large extrapolation from measured pp and $\bar{p}p$ collisions, is likely rather well understood on the whole. The situation is reminiscent of the recent discovery in New Mexico of some 8 fragments of data -- which turned out to be dinosaur tail bones -- in a cliff. With further digging

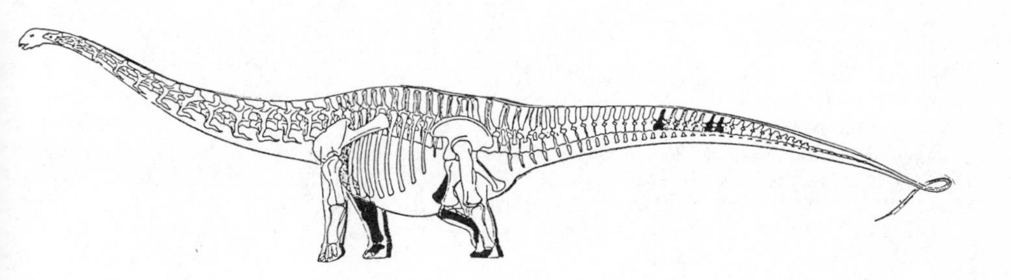

Fig.10

several leg bones were discovered, from which a very beautiful extrapolation was made in Fig. 10 -- the world's largest dinosaur. [The bones that were found are shaded.] The extrapolation is really remarkable, even to the smile on the face and the curl of the tail. These details may be wrong: the neck may go straight up and the dinosaur may face the other way and be frowning, but the basic outlines of the picture, like those of ultrarelativistic collisions, are reasonable. In ultrarelativistic collisions, we not only have the possibility of extracting the physics of the states of matter produced in collisions from the forthcoming data (albeit a difficult task), but we have the possibility of discovering major surprises, which together make the study of ultrarelativistic collisions very exciting.

I would like to thank John Kogut and Don Sinclair for making available their recent Monte Carlo results in Figs. 4 and 6, prior to publication.

* Supported in part by U.S. National Science Foundation Grant PHY84-15064.

Çelik T, Engels J, and Satz H 1983 Phys. Lett. 129B 323
Engels J, Karsch F, Montvay I and Satz H. 1981 Phys. Lett. 101B 89
Gyulassy M et al. 1986 eds Quark Matter '86, Proc. 5th Int. Conf. on Ultra-relativistic Nucleus-Nucleus Collisions, Nucl. Phys. A (in press)
Kajantie K 1985 ed Quark Matter '84, Proc. 4th Int. Conf. on Ultra-relativistic Nucleus-Nucleus Collisions, Lect. Notes in Phys. 221, (Berlin: Springer)
Kogut J 1986 Phys. Rev. Lett. 56 2557
Kogut J and Sinclair D 1986 preprint ILL-(TH)-86-#46, Nucl. Phys. B [FS] (in press)
Ludlam T W and Wegner H E 1984 eds Quark Matter '83, Proc. 3rd Int. Conf. on Ultra-relativistic Nucleus-Nucleus Collisions, Nucl. Phys. A418
Polonyi J 1986 Quark Matter '86, Proc. 5th Int. Conf. on Ultra-relativistic Nucleus-Nucleus Collisions, Gyulassy M et al. eds Nucl. Phys. A (in press)
Satz H 1985 Ann. Rev. Nucl. Part. Phys. 35 245
Svetitsky B 1986 Quark Matter '86, Proc. 5th Int. Conf. on Ultra-relativistic Nucleus-Nucleus Collisions, Gyulassy M et al. eds Nucl. Phys. A (in press)
Von Gersdorff H, McLerran L, Kataja M and Ruuskanen P V 1986 Phys. Rev. D34 794

Inst. Phys. Conf. Ser. No. 86
Paper presented at Int. Nucl. Phys. Conf., Harrogate, UK, 1986

Nuclear collisions at very high energies

Hans J. Specht
Physikalisches Institut der Universität Heidelberg
D-6900 Heidelberg

Abstract: Nuclear collisions at very high energies will permit experimental studies of the transition to a plasma of deconfined quarks and gluons expected to occur in strongly interacting matter at high energy densities. We discuss the suggested signatures, their relation to the phenomenology of hadron collisions, and the forthcoming experimental program with beams of 200 GeV/A $^{16}O/^{32}S$ at CERN and 15 GeV/A ^{32}S at BNL.

1. Introduction

Statistical Quantum Chromodynamics predicts that strongly interacting matter will undergo a transition to a plasma of deconfined quarks and gluons at energy densities exceeding $\varepsilon_{crit} \sim$ 2.5 GeV/fm^3. Nucleus-nucleus collisions at very high energies (> 10 GeV/A) appear to be the only laboratory tool able to reach such energy densities over sufficiently large space-time volumes to permit experimental studies of this new state of matter. The preceding talk (Baym 1986) has discussed the theoretical basis for these expectations. The present paper concentrates on the experimental aspects, dealing with the accelerator situation, the conditions expected in nuclear collisions, the signatures for quark matter formation, and the way the signals are covered by the 5+3 major electronic experiments which are scheduled to start in the late autumn of 1986 at CERN and at BNL, respectively. The written version of the talk is considerably condensed compared to the oral one. The reader interested in more details is referred to earlier reviews (Jacob and Tran Thanh Van 1982) and, in particular, to the proceedings of the four Quark Matter Conferences in Bielefeld (1982), Brookhaven (1983), Helsinki (1984) and Asilomar (1986).

2. The Accelerator Scene

At CERN, a dedicated preaccelerator for heavy ions has been installed. It has been prepared by a GSI Darmstadt / LBL Berkeley collaboration and consists of an ECR source built by Geller (Grenoble) to provide the necessary high ionic charge states like O^{6+} or S^{12+}, and an RFQ accelerator plus rebuncher etc. It injects into the existing accelerator complex to reach 12.5 MeV/A after Linac I, 400 MeV/A after the Booster Synchrotron, 7 GeV/A after the PS and a maximum energy of 225 GeV/A after the SPS. About 10^8 ions circulate per cycle of about 14 s and are extracted from the SPS during a flat top period of about 3 s. The beam emittance is < 1 π mm mrad. External beams are simultaneously available in several beam lines in the North and West experimental areas. The 1986 program foresees a 17 day period with ^{16}O at 60 and 200 GeV/A. For 1987, following an upgrade of the ion source, ^{32}S (possibly ^{40}Ca) will become available for a further 17 day period. No formal approval exists yet for running beyond 1987, but depending on the outcome of the first exploratory round, a continuation may be anticipated. A proposal is also being prepared for extension of the program up to Au- or Pb-beams. This requires a new preaccelerator / linac system together with an improvement of the Booster vacuum (Stock and Bock 1986) and could be operational by 1991. Hadron acceleration in the LEP tunnel in the later 1990's would then open up still more dramatic possibilities.

At BNL, a transfer line has been installed between the Tandem Van de Graaff and the AGS. This combination allows to accelerate ions up to ^{32}S with energies up to 14 GeV/A and intensities comparable to those at CERN. The experimental program will start nearly simultaneously towards the end of 1986, but offers somewhat more generous running time during the following years. By around 1989/1990, following the completion of a booster synchrotron, beams up to Au will be provided from the AGS. A much more ambitious proposal exists moreover to build a Relativistic Heavy Ion Collider (RHIC) with c.m. energies of 100+100 GeV/A (20 TeV/A fixed target equivalent) for ions up to Au+Au, using superconducting magnets in the unused ISABELLE tunnel. If approved soon, operation could start in 1993/94.

3. Conditions in Nuclear Collisions

Here, we shortly comment on three issues, (i) baryonic vs. mesonic matter, (ii) the energy densities attainable in nuclear collisions, and (iii) the problem of sufficient spatial size.

From a physics point of view, the three energy regimes at AGS, SPS and RHIC are well matched and somewhat complementary. As schematically illustrated in Fig. 1, 15 GeV/A ($\sqrt{s}$ = 5.6 GeV/nucleon-nucleon pair in the c.m. system) corresponds to the "stopping" regime: as a result of the collision, the baryons from the two colliding nuclei experience a rapidity shift

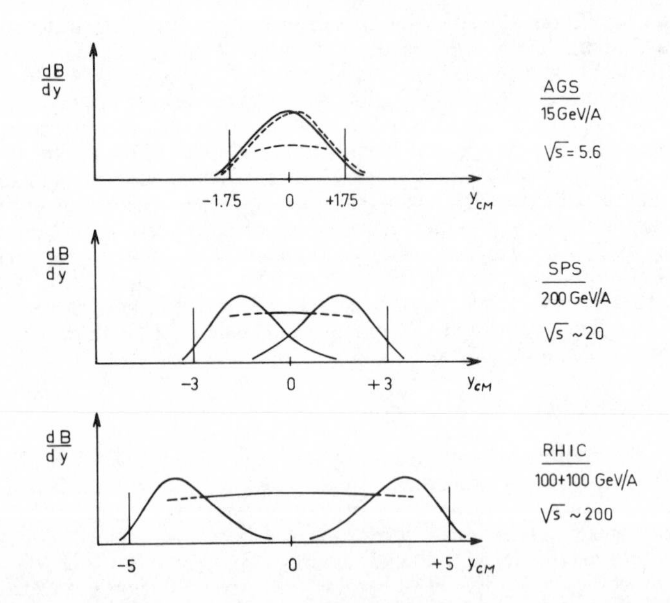

Fig. 1 Rapidity density of baryon number vs. c.m. rapidity at AGS, SPS and RHIC energies. The vertical bars mark the beam- and target rapidities before the collision.

such that, on average, they pile up on top of each other in rapidity space, creating a region of maximum baryon density (baryon chemical potential μ_B or quark minus antiquark density

$\neq 0$). At SPS energies (scaling regime for pp), some "transparency" sets in: the rapidity shift is insufficient for a complete overlap, and a central region with a possibly decreased value of μ_B may emerge between the fragmentation regions. Average amount and width of the rapidity shift in nucleus-nucleus collisions are actually a matter of great controversy and one of the important bread- and butter-results to be expected from the first round of experiments. While the median rapidity shift $\Delta y \sim 1$ in pp, some evidence exists for $\Delta y \sim 2$ in pA interactions (Busza and Goldhaber 1984), but this is not yet settled and the extrapolation to the case of AA collisions is totally unclear (see Quark Matter Conferences 1983, 1984, 1986). Triggering on particularly high transverse energy E_T may, moreover, select events with rapidity shifts larger than average, very probably filling in the $y_{cm} = 0$ region with baryons even at SPS energies (where the much larger energy reservoir still is extremely attractive). RHIC energies, on the other hand, should be sufficient to guarantee a baryon-free central region of pure mesonic matter (μ_B=0), irrespective of the present uncertainties in nucleus-nucleus stopping. Only μ_B=0 (quark- and antiquark densities equal) is, of course, equivalent to the early universe situation, while certain signals for quark matter formation are peculiar to $\mu_B \neq 0$.

At energies high enough to be in the scaling regime, the initial energy densities attainable in nucleus-nucleus collisions can readily be estimated (in the central region) by considering the production of secondary particles (essentially pions) in a local, longitudinally expanding volume at rapidity y. Following Bjorken (1983), the comoving energy density ε is then obtained as

$$\varepsilon_A = \left(\frac{dn}{dy}\right) \cdot \langle m_T \rangle \cdot \frac{1}{\pi (r_0 A^{1/3})^2 (\tau c)}$$

where dn/dy denotes the number of secondaries per rapidity interval, $\langle m_T \rangle$ the mean energy $(p_T{}^2 + m_\pi)^{1/2}$ per secondary, $r_0 \sim 1.1$ fm and (τc) the formation length, i.e. the longitudinal extension of the interaction region producing the secondaries considered. With dn/dy ~ 3, $\langle m_T \rangle$=0.4 GeV and $(\tau c) \sim 1$ fm for pp, $\varepsilon_p \sim 0.3$ GeV/fm^3 for a minimum bias pp collision. Assuming (dn/dy) = 3A for AA, $\varepsilon_A \sim 0.3\ A^{1/3}$ GeV/fm^3, equivalent to a superposition of $A^{1/3}$ pp collision. For Pb+Pb, this yields about 2 GeV/fm^3 , close to but not larger than the critical value. However, two effects may very much increase ε_A. The enhanced nuclear stopping mentioned before points to a smaller formation length $(\tau c) < 1$ fm in pA, decreasing even further in AA (McLerran 1984). Still more important, triggering on the tails of the multiplicity- or transverse energy distribution ($dE_T/dy = dn/dy \cdot \langle m_T \rangle$) may yield up to nearly an order-of-magnitude increase in ε_A relative to minimum bias. Thus, values of $\varepsilon_A \sim$ 5-50 GeV/fm^3 >> ε_{crit} appear to be achievable. As an example for the importance of tails, Fig. 2 shows the transverse energy distribution for 200 GeV/c p+Pb recently measured by the HELIOS collaboration (Åkesson et al. 1986). The spectrum extends to 40 GeV, twice the kinematic limit for pp ($\sqrt{s}$ = 20 GeV) in the full rapidity interval ($\Delta\eta = 6$). The slope is too flat to be accounted for by convoluting a number of independent pN collisions; in fact no theoretical model exists at present which is able to describe these data. Calculating the energy density at E_T = 40 GeV with the expression given above (somewhat questionable in this case), a value of ~ 6 GeV/fm^3 is obtained for (τc) = 1 fm.

In order to justify the use of a thermodynamical description with equilibrium quantities like temperature etc., the interaction volume has to be sufficiently large. Apart from the $A^{2/3}$ increase relative to the typical pp interaction volume, a further increase (by $\varepsilon_A/\varepsilon_{crit}$) arises during the expansion stage as long as the system is still in the plasma phase. Volumes larger by about a factor of 100 than in pp may therefore be expected for a ^{32}S collision with a heavy nucleus.

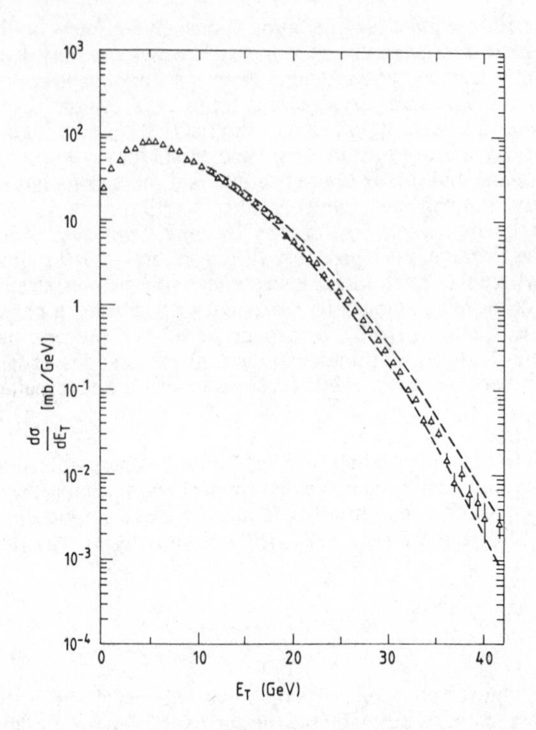

Fig. 2 Cross section of transverse energy measured for 200 GeV/c p+Pb in the pseudorapidity interval $0.6 < \eta < 2.4$. The dashed lines indicate the systematic error.

4. Signatures for Quark Matter Formation

The questions of how to know whether a plasma was formed, and how to quantify its properties are obviously of crucial importance. Two classes of problems have to be distinguished. The first is related to the links between plasma properties or parameters like temperature T, energy density ε, pressure P, entropy density σ etc. and the experimental observables like the final state hadrons, leptons, photons, their y (η-) and p_T-distribution, their flavor content etc. The second arises from the finite volume of the collision complex in space time: stages of compression, excitation, thermalization, (chemical) equilibration, expansion and rehadronization occur; every observable thus results from the folding over a complicated history, receiving in general contributions both from the plasma- and the hadronic world. A deeper theoretical understanding of most of these problems is still in its beginning. We will not pursue this complicated discussion any further, but refer instead again to the series of Quark Matter Conferences. In the following, we reproduce the standard list of relations between variables of interest and experimental observables and comment on a few selected items in more detail.

Variable	Measurement
impact parameter	energy in forward calorimeter; energy- or multiplicity distribution in general; projectile fragments
participant energy	distribution of dE_T/dy or dn/dy

temperature	p_T (or E_T-) distribution of hadrons or (better) leptons and photons
expansion velocity	boost of thermal p_T-distribution, resulting in a dependence on the mass of the secondaries (π, K, p)
pressure	induces expansion and p_T-boost; also acceleration of spectator matter by pressure of participants: distribution of spectators (p, d, α) in fragmentation region
energy density	total transverse energy / volume
entropy density	total multiplicity / volume
chemical potential	flavor content : baryon number, strangeness, antimatter
shocks, deflagrations	fluctuations in dE_T/dy or dn/dy for individual events
spatial extension	identical particle correlations (Hanbury-Brown-Twiss); speckle interferometry
chiral symmetry	disappearance of the ρ-meson.

The combined effects of temperature and collective transverse expansion of the fireball (induced by the pressure difference between the matter and the vacuum) determine the transverse momentum- (or energy-) distribution of the secondaries. Reducing this information to a single number, e.g. the slope parameter from an exponential fit or the average value $<p_T>$, one generally has (Shuryak 1980, van Hove 1982)

$$<p_T> = <p_T>_{thermal} + <p_T>_{collective}$$

An important difference exists between the emission of hadrons and photons or lepton pairs. Hadrons can only emerge when the plasma is cool enough to hadronize, i.e. they sample the surface of the fireball and in general the late stages of the development, reflecting essentially the critical temperature $T \approx T_c$. Photons and lepton pairs, on the other hand, can be created through quark and gluon processes at any stage; the Stefan-Boltzmann behaviour of the production rates ($\sim T^4$) and the absence of final state interactions in fact emphasize the hot interior and the early stages of the development with $T >> T_c$, although contributions from the hadronic stage will also occur.

A particular situation arises if, as generally expected, the phase transition is steep, i.e. has a large change of energy density ε and entropy density σ for a small change of temperature T and pressure P (not necessarily a first order transition). A characteristic relation between $<p_T>$ and ε or σ (measured by dn/dy) then arises (van Hove 1982): as the energy density increases, T, P and thus $<p_T>$ increase, then reach a plateau in the regime of coexistence between plasma and hadronic matter at $T = T_c$, and finally continue to rise (less for hadrons than for photons and leptons).

A qualitative behaviour like this may already have been observed in cosmic ray events (Burnett et al 1985). Fig. 3 shows a compilation of data on the dependence of $<p_T>$ on energy density ε where ε has been computed from the measured dn/dy as discussed above. The constant behaviour of the ISR pp data has historically been interpreted as indicating a limiting temperature for the strong interaction, but in Hagedorn's more modern interpretation reflects instead the critical temperature; the 30% rise plus the final flattening in the SPS $p\bar{p}$ data has also been discussed in terms of a phase transition (van Hove 1982). The significance of a correlation of $<p_T>$ vs. ε in the cosmic ray events is surely less evident in a plot vs. the directly measured quantity dn/dy, but the sole existence of very large values of $<p_T>$ (seen before in older data) is still extremely interesting. The first round of experiments at accelerators will obviously settle this issue soon.

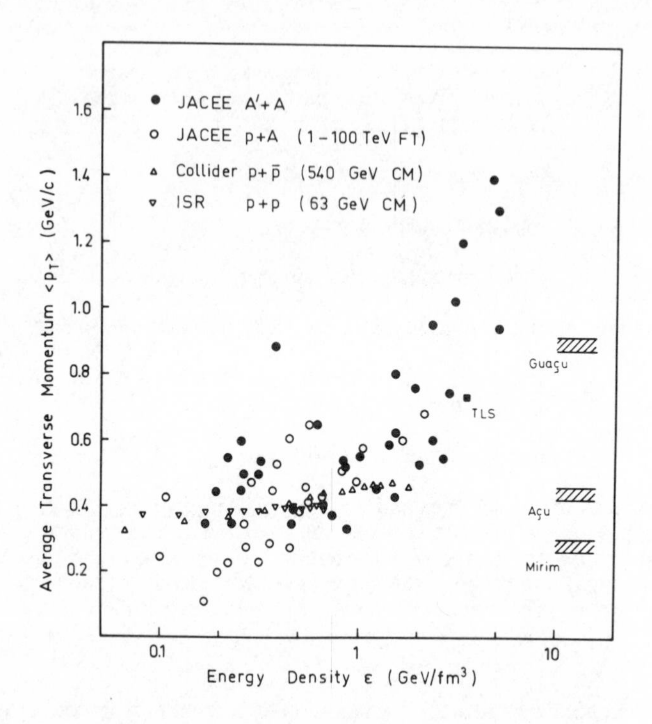

Fig. 3 The dependence of $<p_T>$ on ε for the JACEE cosmic ray events (Burnett et al 1985), $p\bar{p}$ collisions at the CERN SPS (Arnison et al 1982), and pp collisons at the CERN ISR (Breakstone et al 1983). The $<p_T>$ positions of certain classes of older cosmic ray events with unknown abscissa are also indicated (Lattes et al 1980)

If the phase transition is steep, the release of the large latent heat could also occur irreversibly by plasma deflagration (van Hove 1983) or detonation. This may lead to collective fluctuations in the production of secondaries. Fluctuations in the rapidity density in individual events may again have been seen in cosmic ray interactions (Burnett et al 1983, Burnett et al 1985, Ivai et al 1982), but the statistical significance is far from being clear.

The use of photons and lepton pairs as thermometers meets a particular problem in the sense that the phenomenology of their production in high energy hadron collisions is hardly understood in that region of transverse momenta or transverse masses (200-1000 MeV/c or MeV/c^2) which is of special relevance for thermal production. Since we have recently reviewed this field elsewhere including a complete list of references (Specht 1984), we discuss it only shortly here; more recent references on photon and lepton pair production include McLerran et al (1985), Kajantie et al (1985) and Kajantie et al (1986), on the J/ψ (and its disappearance as a signal for quark matter formation) Matsui et al (1986).

The order-of-magnitude of the inclusive direct photon cross sections in high energy hadron collisions is illustrated in Fig.4. In the range 10 MeV/c < p_T<10 GeV/c, a decrease of about 12 orders is observed. The high-p_T part is reasonably understood and interpreted as radiation created in a single hard scattering process of a quark on a gluon or antiquark; of the lowest order ($\alpha\alpha_s$) QCD diagrams, the QCD quark-gluon Compton graph dominates. Only one experiment exists on very soft photons, with $p_T \leq 50$ MeV/c (Chliapnikov et al 1984). The observed yield is higher by a factor of ~ 5 compared to what one expects from incoherent hadronic bremsstrahlung, presenting an intriguing puzzle. Nothing at all is empirically known in the region in between, where direct photons, presumably produced here on a level of order 1/137 relative to π^0's, are completely masked by the photons from π^0-decays.

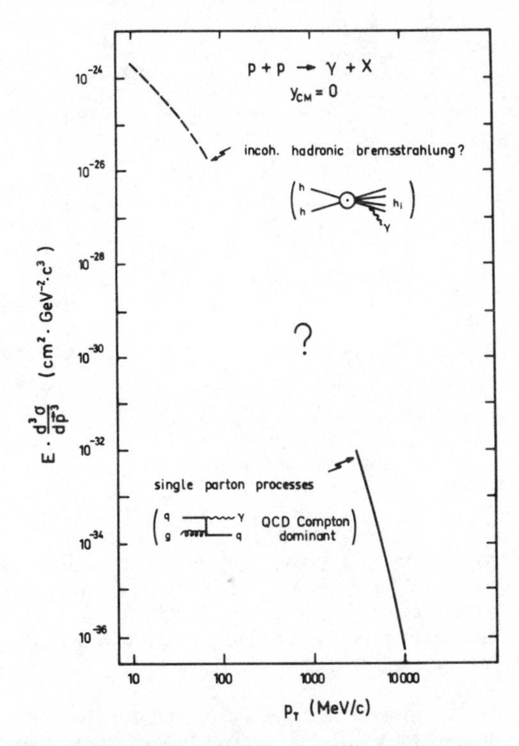

Fig. 4 Illustration of inclusive direct photon cross sections. The high-p_T part has been measured for pp at $\sqrt{s} = 63$ GeV (Anassontzis et al 1982), the low-p_T part for K^+p at $\sqrt{s} = 12$ GeV (Chliapnikov et al 1984)

Photons in the nucleus-nucleus case can occur as (coherent) bremsstrahlung in the fragmentation regions ($p_T \lesssim 100$ MeV/c), and as thermal black-body radiation from the hot plasma ($p_T \gtrsim 100$ MeV/c). With so little known on the comparison basis pp, the situation may seem disastrous. However, more has to be and surely will be learned on pp through a continuing experimental program including the HELIOS experiment. Two physics arguments also help. Whereas the π^0's are emitted from the surface of the interaction region, the direct photons are created in the hot interior, leading to a volume/surface gain for the γ/π^0 ratio by one order of magnitude or even more ("gammaization", Feinberg 1976). In addition, the spectral shape of the π^0-photons reflects T_c, whereas yield and shape of the direct photons greatly profit from the T^4-dependence of the production at the higher interior temperature $T >> T_c$.

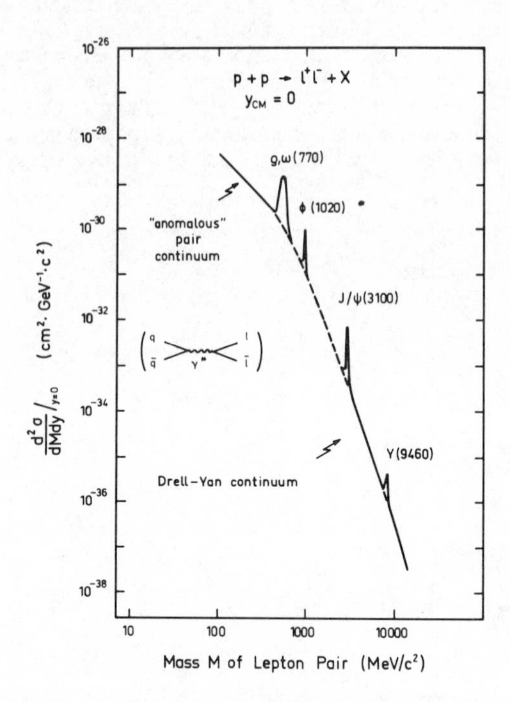

Fig. 5 Order-of-magnitude illustration of inclusive lepton pair cross sections. The high-mass part applies to $\sqrt{s}$ = 28 GeV, the low-mass part has been taken from a compilation (Specht 1984) and is rather insensitive to $\sqrt{s}$. The vector mesons are only drawn in a qualitative way.

In analogy to photons, Fig. 5 illustrates the order-of-magnitude of the inclusive lepton pair cross section in high energy hadron collisions, covering the whole mass range of interest. A structured continuum is observed which sharply falls with mass and spans again many decades. The superposed structures reflect resonant production of the vector mesons with the same quantum numbers as the photon (1^{--}). The high-mass continuum is reasonably well understood; it is thought to be entirely due to the Drell-Yan mechanism (Drell and Yan 1970) in which a valence quark from one hadron annihilates with a sea quark from the other to create an e^+e^-- or $\mu^+\mu^-$ pair via an intermediate virtual photon. The low mass part (< 1 GeV/c^2) on the other hand, where most of the cross section lies, exceeds the yield expected from Drell-Yan by 2 orders of magnitude, presenting a puzzle since many years.. The most popular interpretation of these "anomalous pairs" is based on quark-antiquark annihilation of the numerous soft quarks produced by gluon processes during the collision (Bjorken and Weisberg 1976). Recent experimental evidence on a quadratic dependence of the pair cross section on the associated hadron multiplicity (Jarlskog et al 1986) seems to support this picture, but the issue is far from being settled. Again, further progress has to be and will be achieved by a continuing pp program (including HELIOS). The relevance for the nuclear case is best documented by noting that these very pairs have been interpreted by Shuryak (1978) as evidence for quark gluon plasma formation in pp collisions.

Fig. 6 shows an example of a recent theoretical result on the expected lepton pair signal in nucleus-nucleus collisions (Kajantie et al 1986 I). This particular calculation appears to give a comfortably large excess over the (scaled) pp low-mass and Drell-Yan continua. First results also exist on the incorporation of transverse flow effects (Kajantie et al 1986 II). Further characteristic features related to the p_T-distribution of the pairs and the dependence on the hadron multiplicity density can be found in the same references.

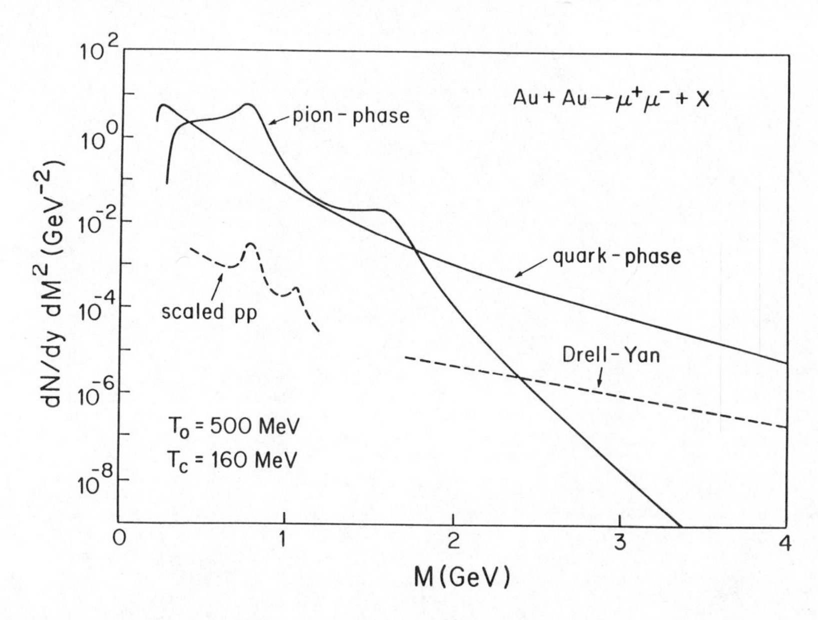

Fig. 6 The calculated dilepton mass distribution for Au+Au at central rapidities, assuming an initial temperature of 500 MeV and a critical temperature of 160 MeV (Kajantie et al 1986 I)

If the quark plasma is not only in thermal equilibrium, but also in equilibrium with respect to the production of different quark flavors ("chemical" equilibrium), the observable flavor content of the final state hadrons may be quite different from that in pp collisions. In particular, as repeatedly pointed out by Rafelski (Koch, Müller and Rafelski 1986) and others, the abundance of s and $\bar{s}$ quarks in the plasma may significantly exceed that of $\bar{u}$ and $\bar{d}$ quarks due to the Pauli-principle, i.e. the existence of a chemical potential of order > 200 MeV for u and d quarks hindering additional $u\bar{u}$ and $d\bar{d}$ production (in the baryon-rich region). If this large relative amount of strangeness survives all expansion effects (a somewhat open issue at present), intensity ratios like K^+/π^+ or, better, $\bar{\Lambda}/\bar{p}$ ($\triangleq \bar{s}/\bar{u}$), $\bar{\Xi}/\bar{p}$ ($\triangleq \overline{ss}/\overline{uu}$) etc may be very much enhanced, possibly up to > 1. Since antihyperon relative to antiproton production in pp collisions is only of order 10^{-1}, falling by about a factor of 4 for each additional unit of S (Bourquin et al 1979), this would present a dramatic signal indeed. Fig. 7 shows an example for a theoretical calculation of the density ratios ρ_s/ρ_q and $\rho_{\bar{s}}/\rho_{\bar{q}}$ in the plasma, demonstrating values > 1 for the latter (Koch et al 1986).

5. Overview of Experimental Program

At present, 5(3) major electronic experiments and 8(9) smaller experiments with nuclear emulsions, plastic detectors etc. are approved at CERN and at BNL, respectively. The majority of these experiments is nearly ready to run. The creation of such a huge program within a period of only two years is truely impressive and surely unprecedented in the history of nuclear physics. It was made possible by a joint effort from particle physics and nuclear physics (more so at CERN), and by the recuperation of a large amount of equipment used in

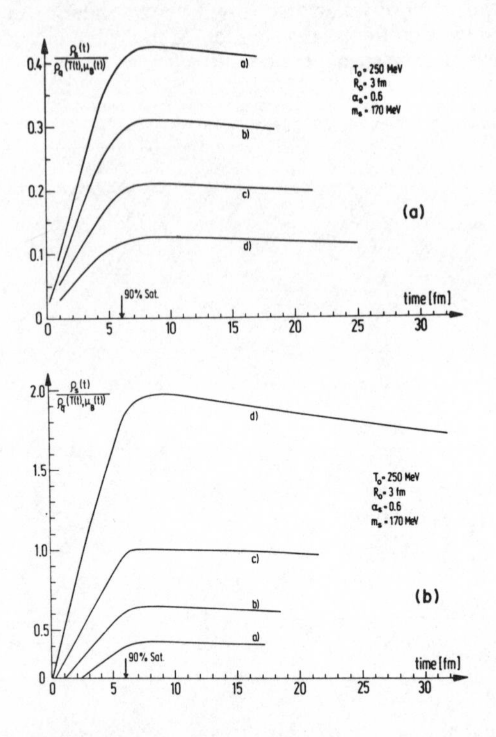

Fig. 7 Calculated density ratios ρ_s/ρ_q (upper part) and $\rho_{\bar{s}}/\rho_{\bar{q}}$ (lower part) of strange and antistrange quarks in the plasma (Koch et al 1986)

previous high energy experiments. The total number of physicists directly involved in the program is presently about 500 (~ 330 at CERN).

The general strategy of the experiments takes full account of the exploratory nature of the field:

(i) coverage of several observables within the same set-up to be able to cross-correlate them in individual events;

(ii) systematic comparison of data from nucleus-nucleus collisions with those from hadron-hadron and hadron-nucleus collisions, if possible measured in the same apparatus and at the same beam energy/nucleon.

Total charged-particle multiplicities are expected to be about 800 in an average central collision of a 200 GeV/A ion such as ^{32}S with a heavy nucleus. Such prohibitive numbers, together with the need to select events with more than average energy density has led to a rather consistent use of calorimetry as the common element in all electronic experiments (at least at CERN).

In the following, we restrict the discussion to the capabilities of the eight major experiments. We take the liberty of treating one (NA34) at more length than the others, but it is rather clear from the outset that indeed only the combination of several experiments with rather different virtues offers a chance for success in this first exploratory stage of a new (and extremely complicated) field.

NA34 (HELIOS)/CERN

Consistent with the need for a better understanding of some features of (soft) photon and lepton production in hadron collisions before dealing with the nuclear case (apart from other motivations), this experiment consists nominally of two separate parts, one (Gordon et al 1983) devoted to hadron-nucleon interactions, the other (Gordon et al 1984) to nucleus-nucleus interactions; the case hadron-nucleus is common to both. The overall lay-out, shown in Fig. 8, contains five major components: a target plus vertex detector system, a 4π-calorimeter with separate sections for electromagnetic and hadronic energy, a compact forward electron spectrometer, a forward muon spectrometer, and an external hadron spectrometer at sideward angles. Several smaller components exist, in addition, for the detection of real photons. A sixth major component, an electron spectrometer at larger angles specifically for the nuclear case (not shown in Fig. 8), is under development.

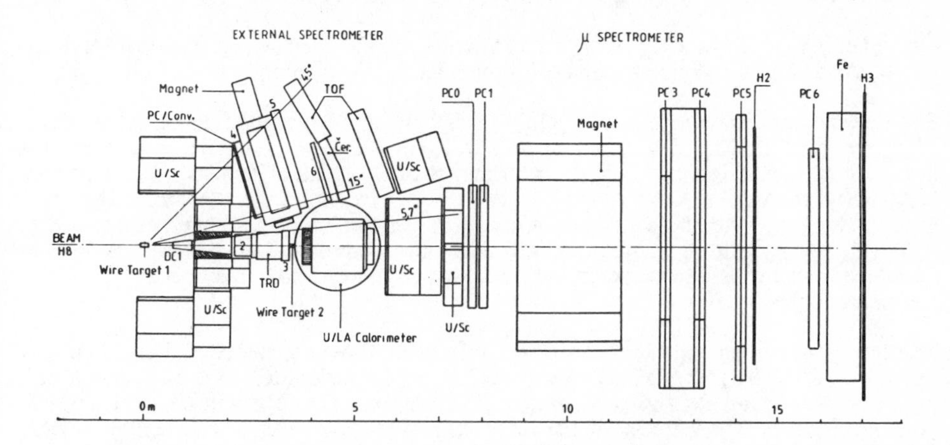

Fig. 8 Top view of the HELIOS experiment

The hadron part of the experiment is devoted to a high-accuracy study of the production of electrons, muons and neutrinos. The main physical issues concern the problem of e/μ universality in charm decay (not relevant for the purpose of this talk), and the anomalous low-mass lepton pairs (200-1000 MeV/c^2) and low p_T inclusive leptons (< 500 MeV/c).

The calorimeter, guaranteeing full hermeticity, re-uses in its box- and wall-parts the uranium/scintillator modules (U/Sc) developed and used before in R807/808 at the ISR. The very forward part consists of a new high-granularity uranium/liquid argon calorimeter (U/LA), followed by further uranium/scintillator modules for complete energy containment. The calorimeter fulfils a dual purpose. It allows for precise missing energy determination (resolution < 2% at 450 GeV/c) in the study of semi-leptonic decay channels, and it also allows the investigation of the (hardly known) E_T- dependence of inclusive lepton- and lepton-pair production.

The forward electron spectrometer utilizes special drift chambers with high multitrack capability and a calorimetrized weak field (dipole) magnet. Hadron rejection on a level > 10^5 is achieved by combining a set of transition radiation detectors (TRD) with the high-granularity electromagnetic part of the uranium/liquid argon calorimeter. Photon conversions and low-mass Dalitz decays can be vetoed on the trigger level: the low field integral allows a straight-line interpolation between a downstream hit and the target (a beryllium wire with a diameter of 50 μm and a length of 20 mm in the beam direction); a

Si-pad detector at a distance of 15 cm then recognizes close pairs by double dE/dx at the respective intersection point. Several Si-strip detectors, also installed close to the target, serve for accurate vertex determination in connection with charm decays.

The muon spectrometer uses major components of the previous NA3 dimuon experiment, but rearranged in a more compact form. The forward calorimeter also serves the purpose of the necessary hadron filter. Tracks in the muon spectrometer can be matched with upstream tracks inside the electron spectrometer to achieve an excellent mass resolution. The acceptance covers the mid- and forward rapidity part and the whole mass regime of interest (from 250 MeV/c^2 to beyond the J/ψ).

The nuclear part of the experiment emphasizes the exploratory nature of this new field. It aims at (i) measuring as many observables as possible for a given event, and (ii) comparing, with high precision, to hadron-nucleus collisions under identical instrumental conditions, drawing from the unique advantage of an independent hadron progam.

Calorimetry now plays a key role, both in terms of self-interest (energy flow, rapidity structure) and trigger function (central collision- high-E_T selection).

The external spectrometer at wide angles (15-45°), containing particle identification with silica aerogel Cerenkov detectors and time-of-flight hodoscopes and viewing the target through a narrow horizontal slit in the calorimeter, allows to study inclusive hadron spectra (strangeness production, $\langle p_T \rangle$ versus E_T, etc.) as well as identical particle correlations, and to search for long-lived unusual charge or mass particles. An active multiwire target (36 wires with different A) is used to minimize secondary interactions and photon conversions, to detect the vertex with high accuracy, and to recognize and eliminate interactions from beam fragments.

The spectra of real photons are investigated with two different approaches. The first uses a small array of 36 BGO crystals, either directly, or in a converter-method in conjunction with proportional wire planes and two crossed drift chambers for obtaining the best possible pointing accuracy. The whole arrangement is easily movable to scan a wide rapidity region. The second method also uses conversions of the photons in a special converter-wire-chamber set-up, in this case placed immediately in front of the external hadron spectrometer for magnetic analysis of the electron pair. It thus covers the angular region of 15°-45°, which is of particular interest because of maximum particle production in asymmetric nuclear collisions.

The muon spectrometer, situated behind the hadron filter and thus not subject to charged particle loads in the hundreds, is used for the study of lepton pairs. It is served by a second active wire target immediately in front of the U/LA calorimeter to minimize the random combinatorial background from $\pi \rightarrow \mu\nu$ decays. Matching to upstream tracks becomes impossible, due to the prohibitive particle flux in the (unusable) electron spectrometer, resulting in some loss of mass resolution (limited now by multiple scattering in the calorimeter). The acceptance of the instrument allows, for the running conditions foreseen with the 200 GeV/A ^{16}O-beam at CERN, > $3 \cdot 10^5$ central collision events to be accumulated in the region of anomalous lepton pairs, decreasing steeply with mass from thereon, but still yielding about 10^4 J/ψ's (assuming no new physics to occur).

As an approach complementary to the muon spectrometer, a new electron spectrometer is also under development. It is motivated by desires to extend the mass range of dileptons down to 100 MeV/c^2, to extend the rapidity coverage to larger angles, to obtain improved mass resolution (complete ρ, ϕ separation), and to allow the first target to be used in connection with lepton pairs (more complete event information). The present concept foresees an azimuthally symmetric magnetic spectrometer with an acceptance Θ = 12°-25°, consisting of a superconducting very short (ring-like) solenoid, a Ring Image Cerenkov counter, and a sectored gas drift chamber, all arranged within the box calorimeter. The RICH, operated in a threshold mode, provides the necessary hadron rejection. The detector components are in the prototype stage.

Finally, for a short fraction of the runs, nuclear emulsions will be used as a target. Events can be tagged by a continuous movement of the stack; they can thus be associated with any information recorded synchronously in the electronic detector. Vice versa, 4π-tracking of charged particles can be done for preselected events.

NA35/CERN

This experiment (Aris et al 1985), shown in Fig. 9, consists of a 2 m streamer chamber within a vertex magnet (previously NA5), and a set of four calorimeters: a highly-segmented photon position detector (PPD), a segmented hadron calorimeter (both covering the mid-rapidity region), an intermediate calorimeter at more forward angles, and a veto calorimeter at very forward angles for trigger purposes. These components have been used before in NA5 and NA24.

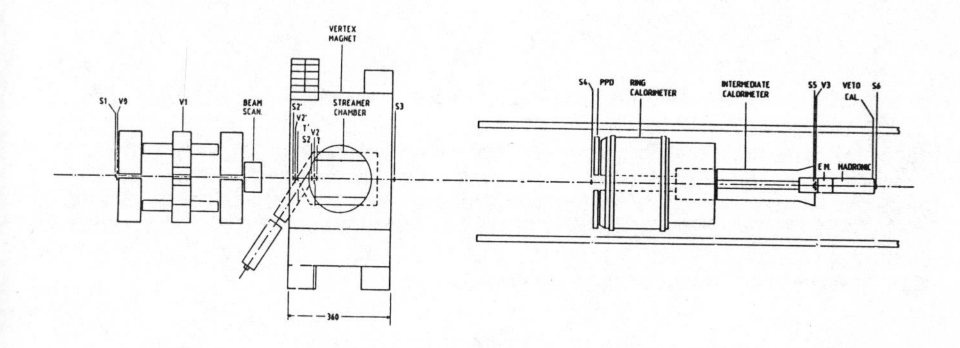

Fig. 9 The NA35 Experiment

This is the only experiment in the CERN/Brookhaven program which provides full (visual) track information in a magnetic field, with the added bonus of being triggerable in conjunction with calorimetry. The goal is twofold: (i) extensive information on the gross properties of hadron production, i.e., single particle observables (proton and pion rapidity and transverse momentum distributions), particle correlations (rapidity distributions event by event, two-particle correlations etc.); (ii) detailed information on strangeness production (Λ and K_S^0 spectra, total $\bar{\Lambda}$ cross-sections, Λ polarization, etc.).

NA36/CERN

This experiment (Elefteriades et al 1984) places its major emphasis on the possible enhancement of strangeness production as a signal for quark matter formation. As shown in Fig. 10, it consists essentially of a new three-dimensional time projection chamber (TPC), coupled to a somewhat modified configuration of the previous European Hybrid Spectrometer EHS (including electromagnetic and hadronic calorimeters in the mid- and forward rapidity regions). For the same reasons as quoted above, an active target is used. A magnet surrounding the TPC sweeps the majority of the produced particles away from the tracking chambers, allowing the investigation of the production of neutral kaons, Λ, Ξ, Ω and their antiparticles with good identification and high statistics.

NA38/CERN

In this experiment (Iversen et al 1985), the emphasis is placed on the detection of muon pairs as one of the prime signals. The experiment re-uses the essentially unmodified NA10 dimuon spectrometer (Fig. 11); an active target and an electromagnetic calorimeter for the measurement of the neutral transverse energy flow are added. The calorimeter consists of

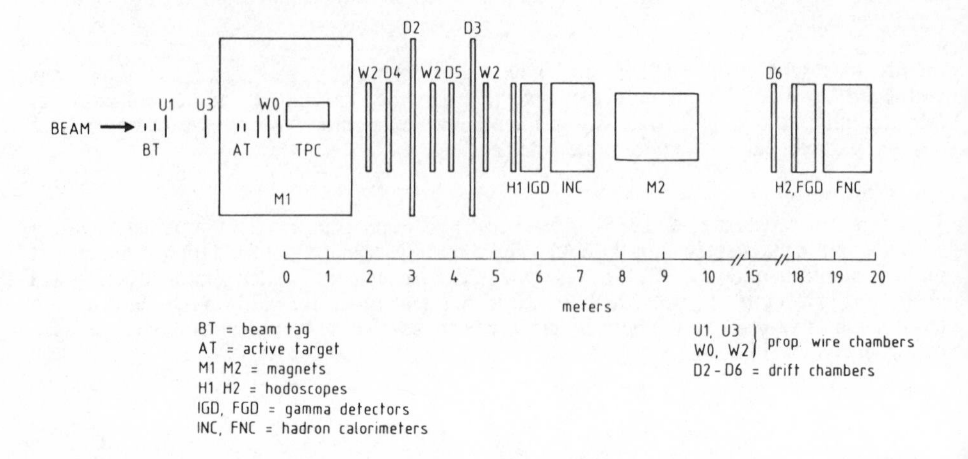

Fig. 10 The NA36 Experiment

segmented Pb/scintillating fibre structures and covers a pseudorapidity range of $1 < \eta_{lab} < 3.5$ with a resolution of $\Delta\eta = \pm 0.2$. The acceptance of the dimuon spectrometer is optimal for transverse masses ≥ 1 GeV/c^2, but the rapidity range is restricted to $3 \leq \eta_{lab} \leq 3.8$ (3-6 for NA34/2). The major advantage of NA38 is the capability to run at the maximum expected luminosity ($2 \cdot 10^7$ interactions per pulse), guaranteeing an excellent sensitivity in the high mass region even in the case of more subtle effects.

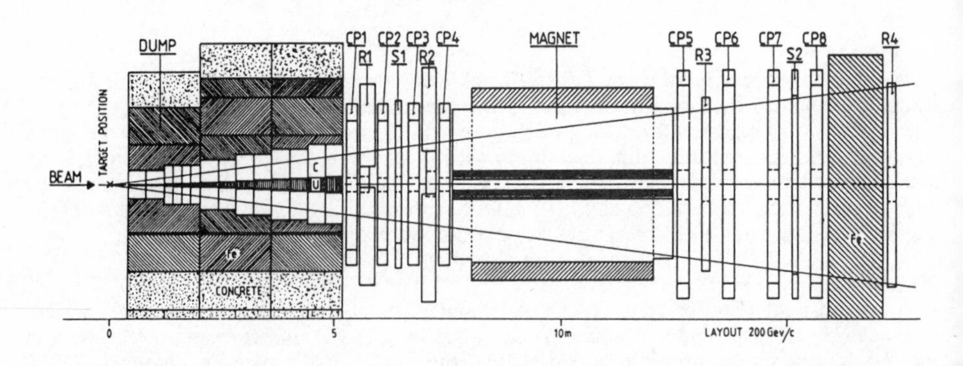

Fig. 11 The NA38 Experiment

WA80/CERN

This experiment (Albrecht et al 1985) presents still another example for the effective re-use of equipment operated successfully before. The layout, shown in Fig. 12, centres around the plastic ball, an array of 655 (double dE/dx) scintillator modules with identification and kinetic energy measurement for the slow protons, deuterons, tritons, alphas, etc., in the target rapidity region; the ball has been in use for years in the nuclear beam program of the Berkeley BEVALAC accelerator. The mid- and forward rapidity region is fully covered by calorimeters with separate electromagnetic and hadronic sections (design analogous to the U/Sc modules in NA34/2), and charged-particle multiplicity information (via Iarocci-type streamer tubes

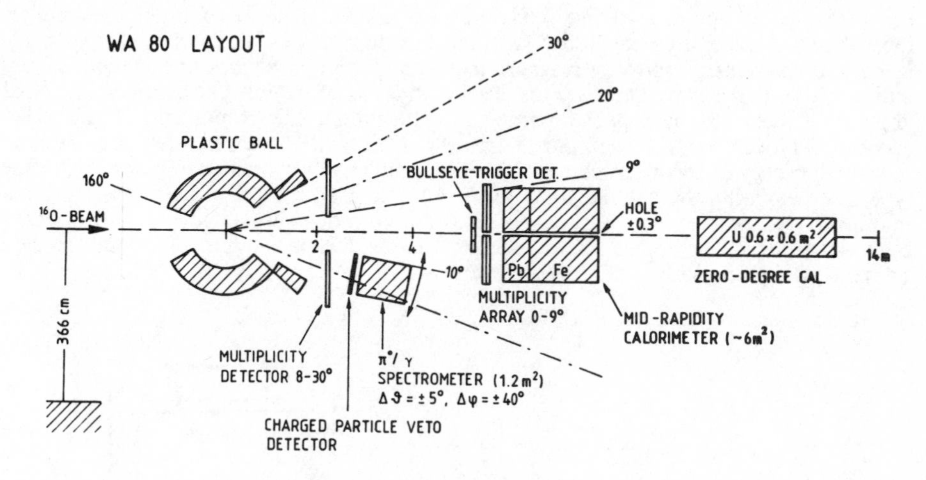

Fig. 12 The WA80 Experiment

with pad read-out for 25000 pads) is available for all angles < 30°. A further zero degree calorimeter exists for trigger purposes, and still another, highly-granular calorimeter (1350 segments of Pb glass) can be used at variable angles for the study of high p_T π^0's and direct photons. The major strength of this experiment is thus rather complete information on hadron production in conjunction with the target spectator decay pattern.

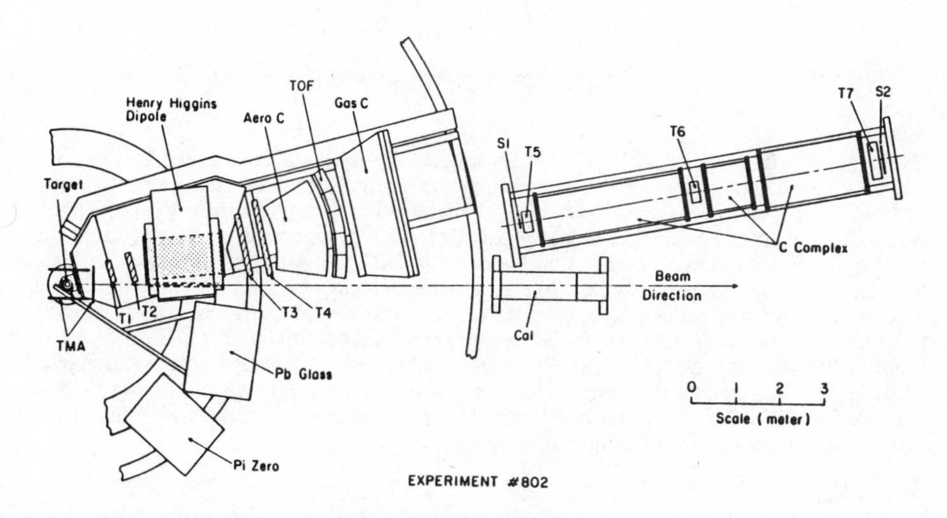

Fig. 13 The E802 Experiment

E802/BNL

This experiment (Alburger et al 1984) will, somewhat earlier than the others, open the BNL relativistic heavy ion program. As shown in Fig. 13, its basic components are a movable 25 msr single arm magnetic spectrometer, a further small-angle forward spectrometer, a Pb-glass array at y_{cm} ~ 0, a multiplicity array surrounding the target, and a small-angle forward calorimeter. The magnetic spectrometer uses drift chambers for tracking and achieves particle identification for p, π, K up to 4 GeV/c with a combination of highly segmented time-of-flight and silica aerogel and gas Cerenkov counters. Its acceptance is large enough in both dimensions to not only measure inclusive particle spectra, but to record a large number of particles simultaneously for the investigation of particle correlations. The small angle forward spectrometer pushes particle identification to higher momenta. The Pb-glass array (336 blocks, ex R110 equipment from the CERN ISR) measures (electromagnetic) transverse energy distributions plus individually resolved π^0's and direct photons. A streamer tube array covering the angular range of 5°-135° is used to obtain the charged-particle multiplicity associated with each event, and the zero-degree calorimeter with separate electromagnetic and hadronic sections presents the usual handle to trigger on central collisions or events with particularly small forward energy.

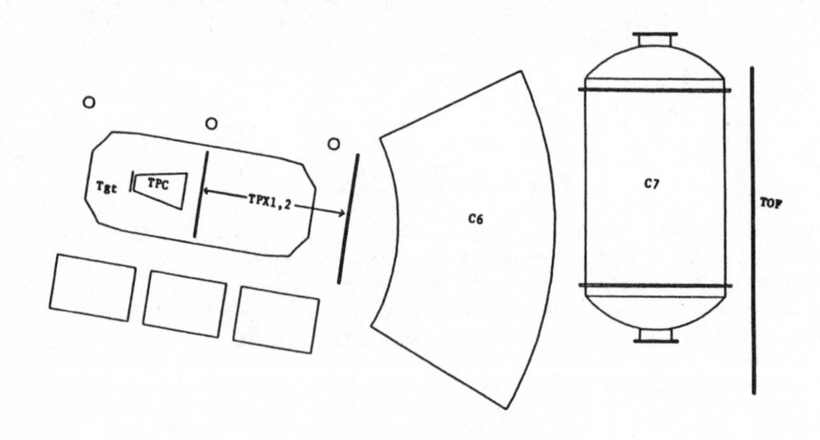

Fig. 14 The E810 Experiment

E810/BNL
Experiment 810 (Etkin et al 1984) centers around the multi-particle spectrometer (MPS) which has, as a general facility, served a number of particle physics experiments over many years (Fig. 14). The insertion of modern drift chambers will help to handle the huge particle multiplicities, but the existing Cerenkov- and time-of-flight systems are less suitable. In its first phase, the experiment focusses on the detection of the more energetic secondaries and the identification of hyperon decays in a way analogous to NA36 at CERN, using the same TPC and displacing beam and target such that the very soft particles (and most of the positives) will be swept away by the MPS magnetic field and bypass the chambers. Particle multiplicities and pseudorapidity distributions are measured by a nearly hermetic arrangement of several CCD's around the target. This is a technologically very appealing development, suitable for multiplicities in the thousands, but with the present readout schemes the data taking rate is reduced to ~ 1 central collision/sec.

E814/BNL
Experiment 814 (Lissauer et al 1985) is different from all the others in the sense that the forward region is optimally covered by a high resolution forward spectrometer (Fig. 15). This allows, in conjunction with 4π compact calorimetry around the target, to follow two quite different physics objectives, (i) the completely exclusive investigation of projectile fragmentation processes at large impact parameters, in particular those induced by the Coulomb force through multiple giant dipole excitation (with the aim of searching for exotic decay products like the tetraneutron or bound states of neutrons and negative pions), and (ii) a high resolution study of the leading baryons together with complete energy flow information for more central collisions. The technique of the spectrometer consists of

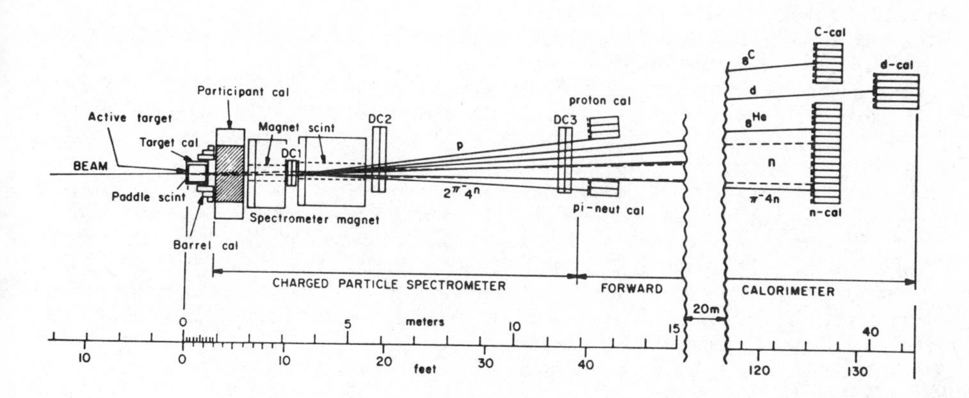

Fig. 15 The E814 Experiment

combining accurate momentum and dE/dx information from a set of drift chambers with calorimeter modules (U/Sc from NA34 at CERN) arranged at such a large distance that individual particles (including neutrons) can be separated and precisely identified. The set-up is obviously particularly sensitive to negatively charged objects. Although the general energy scale is many GeV, the Q-value resolution for peripheral reactions is expected to be ~ 10 MeV.

6. Conclusion

Nuclear and particle physics are entering a joint new field. The physics goal is well defined. The preparation of a dedicated first generation program of experiments is nearly finished. The lack of reliable quantitative guidance may be regretted, but it enhances the smell of a venture into the unknown, something rather rare in present day physics. One will surely learn something - whether it will just be on some ill-understood features of soft hadronic collisions, or on the deconfinement transition, or on something completely unexpected, I do not dare to predict.

References

Åkesson T et al 1986 Proc. 33'rd Int. Conf. on High Energy Physics Berkeley

Albrecht R et al 1985, Proposal P190 (1982) to the PSCC and M406 to the SPSC, CERN-SPSC/85-35; accepted as WA80:
Albrecht R, Bock R, Claesson G, Gutbrod H H, Kolb B, Schmidt R, Schulze R (Darmstadt GSI); Doss M, Kristiansson P, Poskanzer A M, Ritter H G (LBL); Garpmann S, Gustafsson H A, Oskarsson A, Otterlund I, Persson S, Soderstrom K, Stenlund E, (Lund Univ.); Beckmann P, Berger F, Dragon L, Glasow R, Kampert K H, Loehner H, Peitzmann T, Purschke M, Santo R, Wienke R (Münster Univ.); Awes T, Baktash C, Beene J, Ferguson R, Gross E, Johnson J, Lee I Y, Obenshain F, Plasil F, Young G, Sorensen S (Oak Ridge Nat. Lab.); Siemiarczuk T, Stepaniak Y, Zielinski I (Warsaw Univ.)

Alburger D et al 1984, Proposal 802 to the AGS Program Committee; accepted as E802:
Alburger D, Bond P D, Chasman C, Chu Y Y, Cumming J B, Duck E, Hansen O, Haustein P, Katcoff S, LeVine M J, Ludlam T, Olness J, Pfoh A, Remsberg L P, Shor A, Sunyar A, Tanaka M, Tannenbaum M J, Thieberger P, van Dijk J M, Vincent P, Wegner H (BNL); Sugigate T (Hiroshima Univ.); Greiner D, Mulera T, Perez-Mendez V (LBL Berkeley); Enge H A, Grodzins L, Ledoux R J, Steadman S G, Woodruff D (MIT), Akiba Y, Hamagaki H, Hashimoto O, Homma S, Miake Y, Nagamiya S, Tanihata I,

Torikoshi X (Tokyo Univ.)
Anassontzis E et al 1982, R806, Z Phys. **C13** 277
Aris A et al 1985, Proposal P190 (1982) to the PSCC and M 407 to the SPSC, CERN SPSC/85-36; accepted as NA35:
Aris A, Kakoulidou M, Margetis S, Panagiotou A (Athens Univ.); De Marzo C, De Palma M, Favuzzi C, Lapova P, Maggi G, Nappi E, Posa F, Ranieri A, Selvaggi G, Spinelli P (Bari Univ.); Bangert D (CERN); Bartke J, Kowalski M (Cracow Inst. Nucl. Phys.); Bock R, Brockmann R, Guerra C, Humanic T, Sandoval A, Stroebele H, Wensveen M (Darmstadt GSI); Stock R (Frankfurt/Main Univ.); Bamberger A, Heck W, Runge K (Freiburg Univ.); Renfordt R, Schall D (Heidelberg Univ.); Harris J, Odyniec G, Pugh H G, Schröder L, Ticknell M (LBL); Puehlhofer F, Rauch W, Roehricht D, Wenig S (Marburg Univ.); Derado I, Eckardt V, Fendt J, Freund P, Gebauer H J, Pretzl K P, Schmitz N, Seyboth P, Seyerlein J (Munich MPI); Wolf K (Texas A&M Univ.); Bialkowska H, Gazdzichi M, Szwed R, Skrzypczak E (Warsaw Univ.); Kadija K, Paic G, Vranic D (Zagreb Rudjer Boskovic Inst.)
Arnison G et al 1982, UA 1 Collaboration, Phys. Lett. **118B** 167
Baym G 1986, this Conference
Bjorken J D and Weisberg H 1976 Phys. Rev. **D13** 1405
Bjorken J D 1983 Phys. Rev. **D27** 140
Bourquin M et al 1979 Nucl. Phys. **B153** 13 and 1980, Z. Phys. **C5** 275/
Breakstone A et al 1983, R420, Contrib. to the Int. Conf. on High Energy Physics Brighton
Burnett T H et al 1983 JACEE Collaboration, Phys. Rev. Lett. **50** 2062
Burnett T H et al 1985 JACEE Collaboration, Int. Conf. on Nucleus-Nucleus Collisions Visby, Nucl. Phys. **A447** 189c
Busza W and Goldhaber A S 1984 Phys. Lett. **139B** 235
Chliapnikov P V et al 1984 Phys. Lett. **141B** 276
Drell S D and Yan T M 1970 Phys. Rev. Lett. **25** 316 and 1971 Ann. Phys. **66** 578
Elefteriades G et al 1984, Proposal P196 to the SPSC, CERN-SPSSC/84-13; accepted as NA36:
Elefteriades G, Gazis E N, Papadopoulos C, Vlastroy R (Athens Nat. Tech. Univ.), Lovhoiden G, Thorsteinsen T F (Bergen Univ.); Nelson J M, Morrison G C (Birmingham Univ.); Etkin A, Foley K, Hackenburg R, Morris T, Longacre R, Lone W, Platner E, Saulys A (Brookhaven Nat. Lab.); Barnes P D, Franklin G, Hertzog D, Quinn B (Carnegie-Mellon Univ.); Montanet L, Powell B (CERN); Bhatia V S, Kohli J M, Mittra I, Singh J (Chandigarh, Punjab. Univ.); Chang C, Kramer M, Lindenbaum S (City College of New York); Natkaniec Z, Rozanska M, Rybicki K, Sakrejda I, Turnau J (Cracow Inst. Nucl. Phys.); Geist W M, Greiner D, Gruhn C R, Heiden M, Pugh H G, (LBL); Blaes R, Braun H, Huss M, Michelon A, Michelon M E, Riester J L, Voltolini C (Strasbourg Univ.); Hrubec J, McNaughton J, Neuhofer G, Porth P, Rohringer H (Vienna Akad. Wissensch.)
Etkin A et al 1984, Proposal 810 to the AGS Program Committee; accepted as E810:
Etkin A, Foley K J, Hackenburg R W, Longacre R S, Love W A, Morris T W, Platner E D, Saulys A C (BNL); Lindenbaum S J (BNL/CC New York); Chan C S, Kramer M A (CC New York); Geist W M, Gruhn C R, Heiden M (LBL Berkeley)
Feinberg E L 1976 Nuovo Cim. **34A** 391
Gordon H et al 1983, Proposal P189 to the SPSC, CERN-SPSC/83-51; accepted as NA34
Gordon H et al 1984, Proposal P203 to the SPSC, CERN-SPSC/84-53; accepted as NA34/2:
Gordon H, Ludlam T, Olson L H, Polychronakos V, Rahm D C, Stumer I, Woody C (Brookhaven Nat. Lab.); Akesson T, Atherton H, Chesi E, Dederichs K, Esten M J, Fabjan C, Mjornmark U, Piuz F, Rudge A, Schukraft J, Seman M, Vanuxem J P, Wigmans R, Willis W J (CERN); Glaessel P, Goerlach U, Kroh V, Pfeiffer A, Ries H, Soltani J, Specht H J (Heidelberg Univ.); von Hecke H, DiGiacomo N J, McGaughey P L, Sondheim W E, Sunier J W (Los Alamos Nat. Lab.); Almehed S, Haglund R, Hedberg V, Jarlskog G, Johansson S, Lorstad B (Lund Univ.); Corriveau F, Hamel L A, Leroy C, Sirois Y (Montreal McGill Univ.); Beaudoin G, Beaulieu J M, Depommier P, Jeremie H, Lessard L, Lounis A (Montreal Univ.); Gavrilenkov I, Mayburov S, Shmeleva A (Moscow Lebedev Phys. Inst.); Cherniatin V, Dolgoshein B,

Golubkov Yu, Kalinovsky A, Kantserov V, Nevsky P, Sumarakov A (Moscow Eng. Phys. Inst.); Sidorov V, Tikhonov Yu (Novosibirsk Inst. Nucl. Phys.); Cleland W, Clemen M, Collick B, Murray M, Thompson J (Pittsburgh Univ.); Gaidot A, Gibrat F, London G W, Pansart J P (Saclay CEN DphPE); Erlandsson B, Hellman S, Nielsson S, Selldon B (Stockholm Univ.); Roosen R (Brussels IIHE); Bettoni D (Syracuse Univ.)

Ivai J et al 1982, Nuovo Cim. **69A** 295

Iversen P S et al 1985, Proposal P211 to the SPSC, CERN-SPSC/85-2; accepted as NA38: Iversen P S (Bergen Univ.); Sonderegger P (CERN); Baldit A, Castor J, Fargeix J, Force P, Landraud G, Peyrard G (Clermont Ferrand Univ.); Barreira G P, Bordalo P, Casaca A, Gago J M, Maio A, Pimenta M, Ramos S, Valera J (Lisbon Nat. Inst. Sci. Res.); Bedjidian M, Descroix E, Guichard A, Haroutunian R, Pizzi J R (Lyon Univ.); Perrin D (Neuchatel Univ.); Gerschel C (Orsay IPN); Borenstein S, Busson P, Chaurand B, Kluberg L, Romana A, Salmeron R (Palaiseau Ecole Poly. LPNHE); Britz J, Gorodetzky P, Kraus L, Linck I (Strasbourg/CRN); Velasco J (Valencia Univ.)

Jacob M and Tran Thanh Van J (editors) 1982 Quark Matter Formation and Heavy Ion Collisions Phys. Rep. **88** No.5

Jarlskog G, Hedberg V, Schuhkraft J et al, 1986 R807/808, Proc. 33'rd Int. Conf. on High Energy Physics Berkeley

Kajantie K and Hwa R C 1985 Phys. Rev. **D32** 1109

Kajantie K, Kapusta J, McLerren L, Mckjian A 1986, preprint RHIC-PH-11 (I)

Kajantie K, Kataja M, McLerran L, Ruuskanen P V 1986, preprint RHIC-PH-12 (II)

Koch P, Müller B and Rafelski J 1986 Phys. Rep. **142** 167

Lattes C M G, Fujimoto I and Hasegawa S 1980, Phys. Rep. **65** 151

Lissauer D et al 1985, Proposal 814 to the AGS Program Committee; accepted as E814: Lissauer D, Stumer I (BNL); Willis W J (CERN); Boissevain J, Britt H C, DiGiacomo N J, Gavron A, Jacak B V, McGaughey P L, Sondheim W E, Sunier J W, Van Hecke H (Los Alamos Nat. Lab.); Bassalleck B, Kominos N Wolfe D (Univ. of New Mexico); Alamanos N, Braun-Munzinger P, Paul P, Stachel J (SUNY Stony Brook); Benary O, Dagan S, Oren Y (Tel Aviv Univ.)

Matsui T and Satz H 1986, preprint RHIC-PH-14

McLerran L D 1984 in Quark Matter 1984 p.1

McLerran L D and Toimela T 1985, Phys. Rev. **D31** 545

Quark Matter Formation and Heavy Ion Collisions 1982 Bielefeld, Jacob M and Satz H (editors) (Singapore : World Scientific)

Quark Matter 1983 Brookhaven, Ludlam T W and Wegner H E (editors) Nucl. Phys. **A418** 1984

Quark Matter 1984 Helsinki, Kajantie K (editor) Lecture Notes in Physics **221** (Heidelberg:Springer)

Quark Matter 1986 Asilomar, Gyulassi M (editor) to appear

Shuryak E V 1978 Phys. Lett. **78B** 150; Sov. J. Nucl. Phys. **28** 408

Shuryak E V 1980 Phys. Rep. **61** 71; Shuryak E V and Zhirov O V 1980 Phys. Lett. **89B** 253

Specht H J 1984, in Quark Matter 1984 p.221

Stock R and Bock R 1986 draft document

van Hove L 1982 Phys. Lett. **118B** 138

van Hove L 1983 Z. Phys. **C21** 93

Inst. Phys. Conf. Ser. No. 86
Paper presented at Int. Nucl. Phys. Conf., Harrogate, UK, 1986

Chromodynamics in nuclear physics

Nathan Isgur

Department of Physics, University of Toronto, Toronto, Canada M5S 1A7

Abstract. An overview of chromodynamics and its application to multiquark systems is presented. While some of the main results which have been obtained are mentioned, the emphasis of the discussion is to explain why QCD must be used as the framework for fundamental questions in nuclear physics.

1. Introduction

While it is now generally accepted that quantum chromodynamics (QCD) is the correct theory of the strong interaction, there has been an understandable reluctance to abandon the phenomenologically successful traditional (nucleons-interacting-via-meson-exchange) picture of nuclear physics. This stance is possible because very little is rigorously known about QCD in the long-distance (confinement) regime relevant to nuclear physics. Thus although quark-based models of aspects of nuclear physics have produced some potentially revolutionary conclusions on the nature of the nucleus and of nuclear forces, any particular calculation can be dismissed as being a model-dependent aberration.

The main purpose of these brief remarks will be to describe not the details of any particular quark-based calculation, but rather the general framework on which such calculations are anchored. I will argue that it is an inescapable conclusion that fundamental questions in nuclear physics must be addressed within this framework. I will also conclude that there is almost certainly new QCD-based physics in operation in nuclei that must at least coexist with, but may even radically alter, the traditional view of the nucleus.

I will begin with a brief review of chromodynamics, followed by an explanation of why perturbative QCD cannot be used to study confinement-regime physics. This will lead us to adopt the strong-coupling language of lattice QCD with its quark and flux-tube degrees of freedom. I will then describe a simple picture of the quark model which follows from this point of view, and its ramifications for multiquark systems. I conclude with a brief discussion of the main conclusions which follow from studying the nucleus in this light.

2. A Review of QCD

QCD is a non-abelian generalization of QED in which local U(1) phase invariance is replaced by a local SU(3) invariance acting on a new three dimensional internal degree of freedom called colour. If $\psi_\alpha(x)$ (α=1,2,3) is a Dirac field, then under a local (i.e., x-dependent) SU(3) transformation,

$$\psi_\alpha(x) \to \psi'_\alpha(x) = \{\exp(-ig\sum_{i=1}^{8} \theta^i(x) \frac{\lambda^i}{2})\}_{\alpha\beta}\psi_\beta(x) \tag{1}$$

where the λ^i are the usual eight 3x3 generators of SU(3) and $\theta^i(x)$ are "phases". If the Dirac Lagrangian is to be invariant under such transformations, then there must be a set of eight minimally coupled compensating vector fields G_i^μ (i=1,...8) (the gluons); by this reasoning one is led to

$$L_{QCD} = \sum_{flavour} \bar{q}\{\gamma_\mu[i\partial^\mu - g\frac{\lambda^i}{2}G_i^\mu] - m_q\}q - \frac{1}{4}F_{i\mu\nu}F_i^{\mu\nu} \tag{2}$$

where $F_i^{\mu\nu}$ is the QCD field strength tensor

$$F_i^{\mu\nu} = \partial^\mu G_i^\nu - \partial^\nu G_i^\mu - gf_{ijk}G_j^\mu G_k^\nu \tag{3}$$

in which the f_{ijk} are the structure constants of SU(3). Notice that the "extra term" in (3) arises here and not in QED because the gauge group is non-abelian. Thus under the gauge transformation which leads to (1) the gluon fields are both "shifted" and "rotated" so that the QED gauge shift $A^\mu(x) \to A'^\mu(x) = A^\mu(x) + \partial^\mu\theta(x)$ becomes here (in the form appropriate to differential phases $\delta\theta^i(x)$)

$$G_i^\mu(x) \to G_i'^\mu(x) = G_i^\mu + \partial^\mu\delta\theta^i + gf_{ijk}\delta\theta_j G_k^\mu \ .$$

It is this extra non-linear term which is responsible for the remarkable properties of confinement and asymptotic freedom; it is also this innocent-looking term which makes QCD so difficult to solve.

Although it is not the regime in which we are most interested here, we briefly review the phenomenon of asymptotic freedom, which has allowed QCD to be applied with success using perturbative techniques to certain processes involving short-distance physics. At first sight it seems paradoxical that a theory of strong interactions, in which, naively, g would seem required to be large, could ever be treated by perturbation theory. The first step in understanding this is to realize that the coupling constant g appearing in (2) is the "bare" coupling constant and, as such, is not necessarily simply related to the couplings one will observe in various physical processes. Quantum chromodynamics, like any quantum field theory, only makes sense if it is defined (regulated) in terms of some renormalization scheme since the point-like operators appearing in L_{QCD} are highly divergent in character. In the next section we will define QCD in terms of a lattice regularization scheme, but it is more normal to cut the theory off at some very large mass scale M. A renormalizable theory, like QCD, has the property that its predictions for processes with intrinsic scales much less than M are independent of M. Consider, in this light, the quark-gluon coupling at some scale $Q^2 \ll M^2$. We would calculate

$$\alpha_s(Q^2) = f(g,Q^2,M^2) \equiv \alpha_s(g,Q^2,M^2)$$

where f is, for example, obtained by the sum of all relevant Feynman diagrams with integrations cut off at the scale M. Now for a given bare g and cutoff M we see that we can calculate $\alpha_s(Q^2)$ at any Q^2; consequently, any measurement of α_s can be used to fix the bare value of g. This choice of g will, however, depend on M since, crudely speaking, g is acting not as the bare coupling of the point-like L_{QCD} but rather as the bare coupling of the M-regulated theory. Thus, given an observed $\alpha_s(Q^2)$, we must choose

$g=g(M^2)$; it is this observation that leads to the renormalization group equations. In QED such an analysis indicates that the bare electric charge $e(M^2)$ must be <u>increased</u> with increasing M^2 to maintain α at its observed value of 1/137 at low Q^2; this fact can be attribute to the screening of the bare charge by vacuum polarization, much as an ordinary charge is screened in a dielectric medium. In QCD one finds that the bare colour charge $g(M^2)$ must be <u>decreased</u> with increasing M^2 in order to maintain the physical coupling $\alpha_s(Q^2)$ at fixed values. This observation then leads to the conclusion that for sufficiently small scale phenomena, QCD will behave like a theory with a small coupling and so may be studied by perturbation theory. Conversely, analysis leads to the conclusion that $\alpha_s(Q^2)$ inevitably grows as Q^2 is decreased, until at some characteristic scale Λ_{QCD} one finds $\alpha_s(\Lambda^2_{QCD})\sim O(1)$, at which point the perturbative analysis fails and along with its failure the possibility that QCD confines emerges.

We are now in a position to understand why the confinement regime of QCD cannot be studied by ordinary weak coupling perturbation theory. Imagine trying to calculate the energy per unit length (called the string tension b) between a static quark and antiquark at large separation. Since there are no intrinsic dimensionful parameters in QCD (we can ignore for now the possible existence of dynamical quarks), the calculation of the dimensionful parameter b must yield an expression of the form

$$b=M^2F(g^2)$$

where F is a function of the bare coupling $g(M^2)$. Thus, since b is an observable, it must be that as $M^2\to\infty$, $F(g(M^2))\to 0$ so as to keep their product constant:

$$\frac{db}{dM^2} = F(g^2)+M^2\,\frac{dF}{dg^2}\,\frac{dg^2}{dM^2} = 0$$

However, the perturbative analysis described above gives for the bare coupling constant the famous expression

$$g^2(M^2) = \frac{g^2(Q^2)}{1+\frac{g^2(Q^2)}{48\pi^2}(33-2N_f)\ln\left(\frac{Q^2}{M^2}\right)} \tag{4}$$

in lowest order so that with the number of dynamical fermions $N_f=0$

$$M^2\,\frac{dg^2}{dM^2} = -\,\frac{11g^4}{16\pi^2}\quad . \tag{5}$$

It follows that

$$F(g^2)\propto\exp\left(-\,\frac{16\pi^2}{11g^2}\right)\,, \tag{6}$$

a function which obviously cannot be found by perturbating about g=0! In order to understand QCD in the confinement regime we therefore need to abandon perturbative quark-gluon field theory in favour of some other approach. It turns out that this strong interaction physics part of QCD may be studied, both numerically and also analytically as a function of 1/g, if it is latticized.

3. Strong Coupling Hamiltonian Lattice QCD

In the Hamiltonian version of lattice QCD, space (but not time as in most numerical studies) is discretized. In this formulation the lattice spacing "a" will, without reference to a perturbative expansion, play the role of the regulator mass M. Latticizing the theory also has another advantage: it will allow us to set up a strong coupling perturbation expansion in which the expansion parameter for lattice QCD is 1/g instead of g. We may expect to be able to learn more about the strongly coupled regime of the theory in terms of such an expansion, and indeed this seems to be the case: for example, confinement is an automatic property of the $g\to\infty$ limit of lattice QCD. Moreover, the natural degrees of freedom of the strong coupling regime are not quarks and gluons, but rather quarks and flux tubes, the latter being more in accord with various qualitative ideas on the nature of confinement in QCD. Of course, space is not coarse-grained (at least not on the scale of 10^{-15} metres), so that to relate lattice QCD to real QCD we must consider the limit $a\to 0$. In this limit, as we have discussed, $g\to 0$ so that a strong coupling expansion must fail; this is just the other side of the failure of the weak coupling expansion for small Q^2. If, however, it can be shown that the two regimes "match" around g=1, thereby proving that lattice QCD as $a\to 0$ is QCD, then one would nevertheless expect the strong coupling expansion to be useful in many situations where large scales dominate, just as the weak coupling expansion is useful for short distance physics.

A simple analogy may be useful. Consider approximating a continuous one dimensional harmonic oscillator by a particle hopping along a one dimensional lattice of points $x=na(n=\ldots,-2,-1,0,1,2,\ldots)$ with lattice spacing "a". The lattice Hamiltonian can be chosen to be

$$H_{mn} = \left(\frac{1}{ma^2} + \frac{1}{2}ka^2n^2 \right)\delta_{mn} - \frac{1}{2ma^2} \left(\delta_{m,n+1}+\delta_{m,n-1}\right) \tag{7}$$

since then the Schrödinger equation

$$i\frac{\partial\psi_m}{\partial t}(t) = H_{mn}\psi_n(t) \tag{8}$$

becomes

$$i\frac{\partial\psi(x,t)}{\partial t} = \left[-\frac{1}{2m}\frac{\partial^2}{\partial x^2} + \frac{1}{2}kx^2\right] \psi(x,t) \tag{9}$$

as $a\to 0$. Now for $a\to 0$ with k and m fixed, the potential energy term $\frac{1}{2}ka^2n^2\delta_{mn}$ dominates and the eigenstates correspond to the particle sitting on single lattice sites; corrections to this limit are of relative order $x=1/kma^4$ and one can proceed to systematically do perturbation theory in this hopping strength. Since the characteristic scale of the harmonic oscillator is $\alpha^{-1}=(km)^{-1/4}$, one will not get realistic wave functions or eigenenergies for the harmonic oscillator for $a >>\alpha^{-1}$ where lowest order perturbation theory applies, but for $x\sim 1$ one will begin to get good approximations to the solutions of the continuum problem if one works to sufficiently high order in x. By contrast, starting with free particle solutions to the continuum Hamiltonian and treating $\frac{1}{2}kx^2$ as a perturbation is hopeless. (The difference, of course, is that the hopping parameter expansion for the ground state, for example, will be accurate if a matrix of dimension of order $1/a\alpha$ is diagonalized.)

We now turn to the formulation of QCD on a (cubic) spatial lattice. In this formulation the quark degrees of freedom of the theory "live" on the lattice sites while the gluonic degrees of freedom "live" on the links between these sites (see Figure 1). Let's consider first the theory without quarks: we describe this theory in terms of link variables U_ℓ which (before quantization) are 3x3 SU(3) group elements. The pure gauge field Hamiltonian is then the sum of two parts, one involving only the U's and one which has non-trivial commutation relations with the U's:

$$H_{glue} = \frac{g^2}{2a} \sum_\ell C_\ell^2 + \frac{1}{ag^2} \sum_p \mathrm{Tr}[2-(U_{\ell_4}U_{\ell_3}U_{\ell_2}U_{\ell_1}+h.c.)] \tag{10}$$

with "a" the lattice spacing and g the corresponding coupling constant. Here C_ℓ^2 is defined in terms of the eight generators $E^a{}_{\ell\pm}$ of SU(3) transformations of U_ℓ at the beginning (-) or the end (+) of the link ℓ

$$[\, E^a_{\ell+}\, , U_\ell\,] = -\frac{\lambda^a}{2} U_\ell \tag{11}$$

$$[\, E^a_{\ell-}\, , U_\ell\,] = +\, U_\ell \frac{\lambda^a}{2} \tag{12}$$

by $C_\ell^2 = \Sigma_a (E^a{}_{\ell+})^2 = \Sigma_a (E^a{}_{\ell-})^2$. In the second term the product of the U's is taken in order around the plaquettes p. To complete lattice QCD one simply adds to (10) a lattice Hamiltonian H_{quark} for the quarks interacting with the glue. With the quark fields as site variables we have

$$H_{quark} = \sum_{flavour} m_q \sum_{\vec{n}} q^\dagger_{\vec{n}} q_{\vec{n}} + \frac{1}{a} \sum_{\substack{flavour \\ \ell_{ji}}} q^\dagger_j U_{\ell_{ji}} \alpha_{\ell_{ji}} q_i \tag{13}$$

where $\alpha_{\ell ji}$ is the Dirac matrix in the direction of the link ℓ_{ji}.

Our complete Hamiltonian $H^{lattice}{}_{QCD} = H_{glue} + H_{quark}$ has H_{QCD} in $A^0=0$ gauge as its naive continuum limit; it is, furthermore, invariant under arbitrary gauge transformations at the lattice sites. Gauss' law takes the form of a constraint in the theory that the only physically relevant states are those which are gauge invariant.

We are now ready to consider the properties of $H^{lattice}_{QCD}$. We note first that in the strong coupling limit where "a" (and as we shall see, therefore g) is large

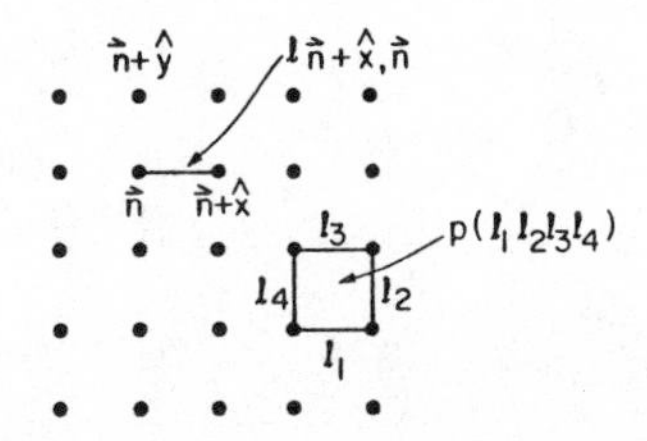

FIG. 1. A two dimensional (x,y) slice of the lattice showing a typical lattice point $\vec{n}=(n_x,n_y,n_z)$, a typical link $\ell_{\vec{n}+\hat{x},\vec{n}}$ from $\vec{n}$ to $\vec{n}+\hat{x}$, and a typical plaquette $p(\ell_1\ell_2\ell_3\ell_4)$

$$H_{QCD}^{lattice} \rightarrow H_{sc} = \frac{g^2}{2a} \sum_{\ell} C_{\ell}^2 + \sum_{flavour,\vec{n}} m_q q_{\vec{n}}^{\dagger} q_{\vec{n}} \tag{14}$$

The eigenvalues of C_{ℓ}^2 are just those of the square Casimir of SU(3): 0 for the singlet, 4/3 for 3 or $\bar{3}$, 10/3 for 6 or $\bar{6}$, 3 for the octet, etc. The quark part of H_{sc} is, on the other hand, diagonalized by an arbitrary number of quarks and antiquarks at arbitrary lattice sites. Since, however, the only physically relevant eigenstates are those which are gauge invariant, the strong coupling eigenstates may be classified as follows:

1) <u>the strong coupled vacuum:</u> In this case all links are unoccupied ($C_{\ell}^2 = 0$) and there are no fermions: the total energy is zero.

2) <u>the pure glue sector</u>: There are still no quarks, but links are excited in such a way that gauge invariant states are produced. The simplest such pure glue states ("glueloops") have a closed path of links in the 3 (or $\bar{3}$) representation. These have energy $(2g^2L)/(3a^2)$ where L is the length of the path; the simplest such state just has the links around the perimeter of an elementary plaquette excited: Tr $[U_{\ell_4}U_{\ell_3}U_{\ell_2}U_{\ell_1}]|0\rangle$, where $|0\rangle$ is the vacuum. Of course more complicated configurations are allowed, including those with non-triplet flux and those with more complicated topologies. For example, three flux links can emerge from a single lattice site since a gauge invariant combination can be formed with the ϵ_{ijk} invariant tensor. See Figure 2.

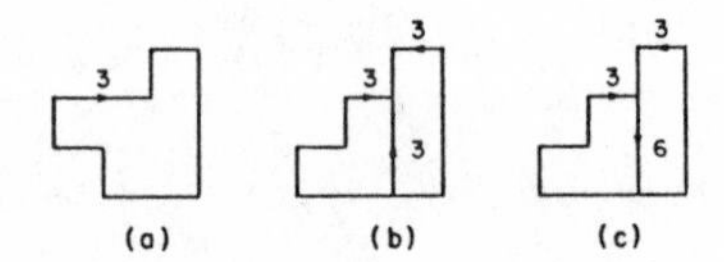

FIG. 2. some primitive pure glue states

3) <u>the meson sector</u>: The simplest quark-containing state consists of a quark and antiquark on the lattice joined by a path of flux lines (for gauge invariance). These will have energy $m_q + m_{\bar{q}} + (2g^2L)/(3a^2)$ so that we automatically have quark confinement in strong coupling. See Figure 3.

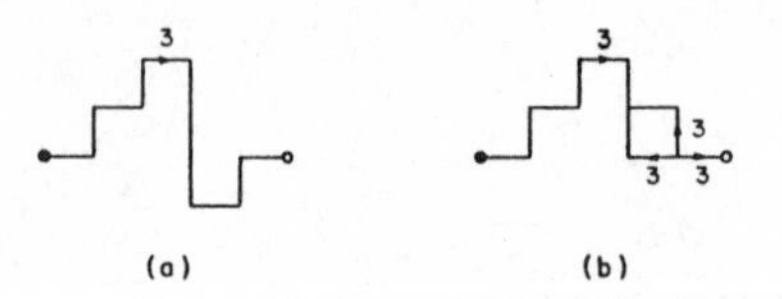

FIG. 3. some primitive meson states

4) <u>the baryon sector</u>: The next simplest quark-containing state consists of three quarks connected by an ϵ_{ijk}-type flux junction. Such quarks will also be confined. See Figure 4.

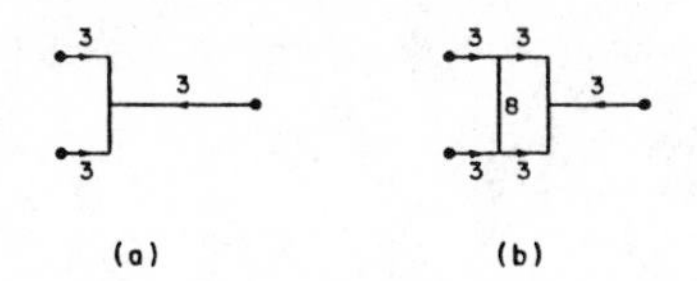

FIG. 4. some primitive baryon states

5) multi-quark sectors: When there are more quarks than those required for a meson or baryon, then in general the system will not be completely confined. The simplest such system consists of two quarks and two antiquarks. See Figure 5.

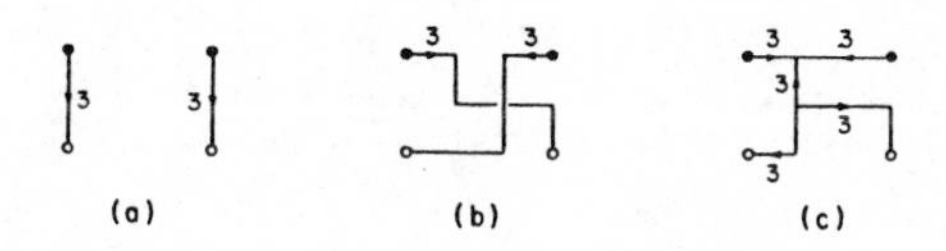

FIG. 5. some primitive $qq\bar{q}\bar{q}$ states

With these examples, the general structure of the eigenstates of the strong coupling limit is clear: it consists of "frozen" gauge invariant configurations of quarks and flux lines. Of course these are not the eigenstates of QCD, but they do form a complete basis (in the limit $a\to 0$) for the expansion of the true strong interaction eigenstates.

The full eigenstates of QCD can be found (in principle!) by considering corrections to the strong coupling limit from the terms we have neglected so far. These terms can induce a variety of effects. Consider first of all the $q^+U\alpha q$ term. It can, among other things,

1) annihilate a quark at one point and recreate it at a neighbouring point with an appropriate flux link (Figure 6a);

2) break a 3-flux line and create a pair (Figure 6b).

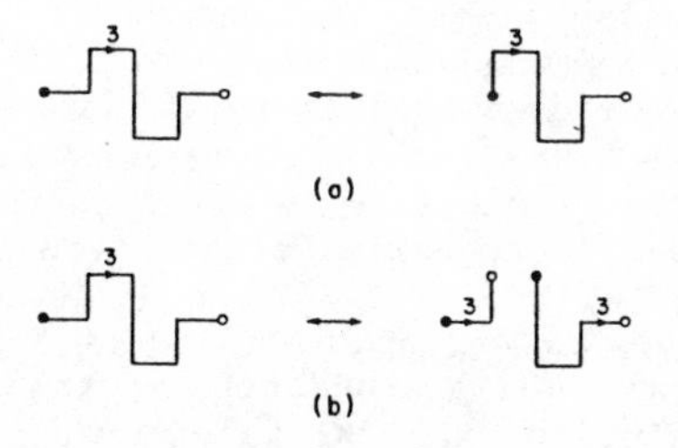

FIG. 6. some effects of $q^\dagger U\alpha q$: (a)quark hopping, (b)flux breaking pair creation.

This term thus plays a role analogous to both the usual quark kinetic energy term and quark-gluon coupling term of the weak coupled theory. Next consider the $Tr[2-U_{\ell_4}U_{\ell_3}U_{\ell_2}U_{\ell_1} + h.c.)]$ term. It can, among other things,

1) allow flux to hop across plaquettes (Figure 7a)
2) change flux topology (Figure 7b).

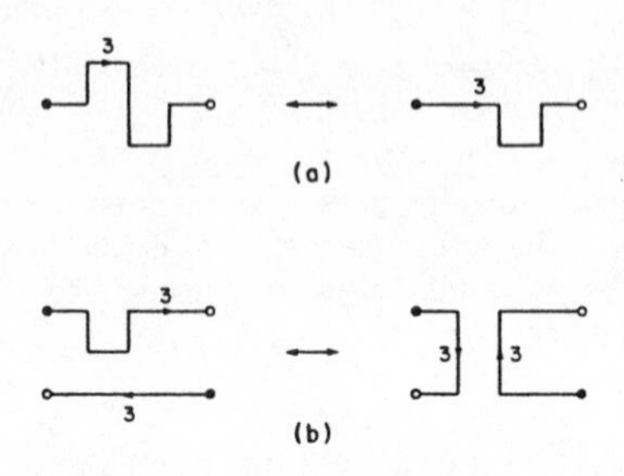

FIG. 7. some effects of Tr[2-(UUUU+h.c.)]: (a)flux tube hopping, (b)flux tube topological mixing by rearrangement.

The full diagonalization of this Hamiltonian problem as $a \to 0$ would constitute an exact solution of QCD. Unfortunately, this diagonalization represents a numerical problem which is of a magnitude well beyond presently available computing capacity.

4. <u>The Flux Tube Model</u>

Even after successful numerical calculations within QCD are possible, it will still be useful to have models which summarize the very complex structure of this theory. Many such models have already been proposed, including potential models, bag models, string models, etc. We will base our discussion instead on a simple model which emerges from the considerations of the previous section on Hamiltonian lattice QCD.

To illustrate this model in the simplest possible context, consider first a heavy $Q\bar{Q}$ system and ignore light quarks. Departures from the strong coupling limit will have a number of effects, but among them will be flux hopping terms which will convert the frozen flux tube lines of the strong coupling limit into dynamical strings which can sustain transverse oscillations. (The conversion from frozen to fluid strings corresponds to the "roughening transition" seen in numerical lattice work). We accordingly first imagine summing all such terms so that we can deal with the $Q\bar{Q}$ pair connected by a dynamical string. Next we consider departures from strong coupling that involve "topological mixing" either to other connected string states or to states with disconnected vacuum fluctuations. If L is large we can hope to approximate the state of the system in terms of a lattice of scale "a" with $L \gg a \sim \lambda_0$, where λ_0 is the scale where g=1 and topological mixing becomes important. Our model for this system is thus that of a discrete string. The ground state of this system for infinitely heavy quarks will consist of the $Q\bar{Q}$ pair with a ground state string stretched between them, the first excited state will be doubly degenerate with either a right-handed or left-handed phonon excited in the lowest string mode, etc. As the distance R between the Q and $\bar{Q}$ is varied slowly, the eigenenergy of the string eigenstate S will trace out an adiabatic potential $V^{(S)}(R)$; we associate such potentials with an adiabatic approximation to the physics of mesons. When the $Q\bar{Q}$ pair move in the adiabatic potential bR of the ground state of this QCD string, one recovers the usual spectrum of mesons in the quark model. When the pair moves in the adiabatic potential of an excited string, the resulting hadrons correspond to a new species not contained in the usual quark model: hybrid mesons with both quark and gluonic degrees of freedom in evidence.

The baryon sector, while more complicated, is analogous to the mesons: the ordinary baryons of the quark model correspond to three quarks moving in the adiabatic potential of the ground state of the three junction (Y) string, while excited strings will lead to hybrid baryons.

In multiquark systems we are forced for the first time to go beyond the simple vibrating string picture to consider topological mixing. This is because in such systems adiabatic surfaces will always cross in the absence of mixing, as can be seen by considering Figure 7(b) for the case when the $qq\bar{q}\bar{q}$ system is arranged at the corners of a square. Such systems are consequently considerably more complicated than those we have already considered. They are the main subjects of the next sections.

Of course the simple classification scheme we have presented here is both incomplete and inaccurate. For example, one must expect to find (at high masses) "topological meson hybrids" in which the $Q\bar{Q}$ system moves in the potential of more complicated string topologies. There will also be mixing between topologies via the $Tr[U_{\ell_1}U_{\ell_2}U_{\ell_3}U_{\ell_4}$ +h.c.] and $q^{\dagger}U\alpha q$ terms and deviations from the adiabatic limit that mix states with different degrees of phonon excitation. While all such effects can be described within the context of this model, its utility will depend on the degree to which they can be ignored, at least for low-lying states. Perhaps the very success of the naive quark model for mesons and baryons provides some evidence in favour of the viability of such a separation.

5. Multiquark Systems in Chromodynamics

We should not be misled into believing that we understand hadron dynamics in QCD on the basis of our being able to understand mesons and baryons. The $q\bar{q}$ and qqq systems are very special cases since for these states the lowest adiabatic potential (corresponding to the gluonic ground state in the presence of fixed quark sources) is very well isolated from excited glue adiabatic potentials. In this sense the whole of the ordinary quark model spectrum is analogous to the vibrational and rotational bands of a molecule moving in the lowest adiabatic surface of the electronic ground state of the molecule. We have yet to find experimentally the "hybrid" mesons and baryons which are the analogs of the vibrational and rotational molecular bands built on excited electronic (versus gluonic) states.

While the hybrid states depend on the properties of the gluonic fields, their main characteristics should be quite similar to those of ordinary mesons and baryons. In contrast, we should expect multiquark states to behave in entirely new ways: <u>multiquark states characteristically have many low-lying adiabatic surfaces which cross</u>. Figures 8 and 9 illustrate this characteristic behaviour in the case of the $qq\bar{q}\bar{q}$ system. Figure 8 illustrates very clearly the fact that the state of a $qq\bar{q}\bar{q}$ system depends not only on the positions of the quarks and antiquarks, but also on the state of the flux which flows between them. This is also true for simpler

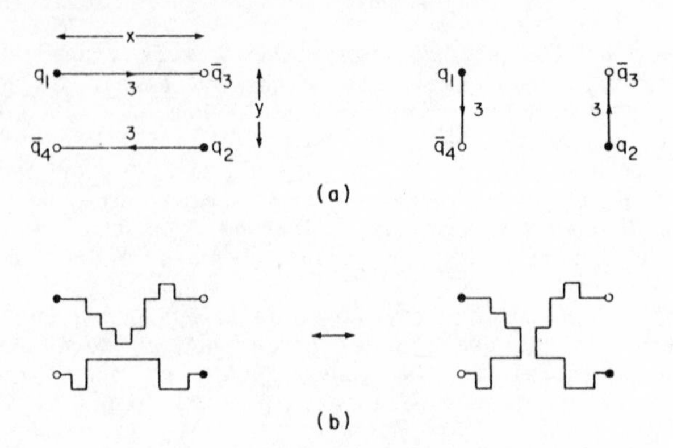

FIG. 8. (a)two low-lying $qq\bar{q}\bar{q}$ configurations, (b)topological mixing between these two configurations

systems like $q\bar{q}$ (where the state of the glue distinguishes the ordinary mesons of the quark model from hybrid mesons), but in multiquark systems we cannot expect to be able to segregate the dynamical states into sectors with an approximately fixed (i.e, adiabatically evolving) glue state. Figure 8(b) shows why this is true: the two flux configurations of Figure 8(a) will be dynamically mixed and this mixing <u>will always be strong</u> in those regions of the lowest adiabatic surface of the $qq\bar{q}\bar{q}$ system where unmixed adiabatic surfaces of a given topology would have crossed. This effect is illustrated more explicitly in Figures 9(a) and 9(b), which show slices through the three lowest adiabatic potentials of the $qq\bar{q}\bar{q}$ system for

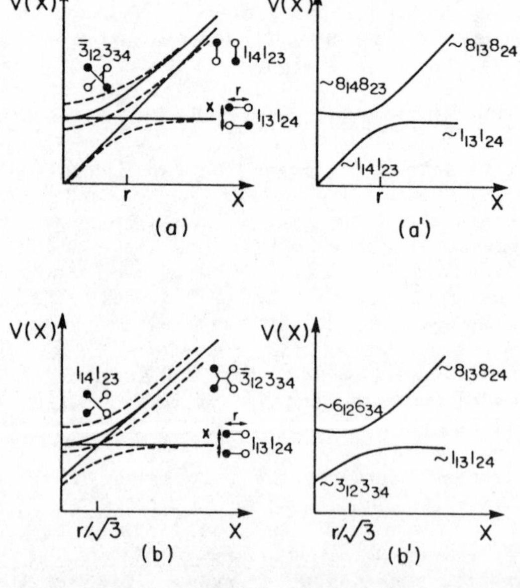

FIG. 9. the adiabatic potentials of the flux tube model and of the $\vec{F}_i \cdot \vec{F}_j$ potential model for two $qq\bar{q}\bar{q}$ geometries: (a) and (b) show the flux tube potentials before (solid curves) and after (shown schematically by dashed curves) topological mixing, while (a') and (b') show the related $\vec{F}_i \cdot \vec{F}_j$ potentials

two different rectangular arrangements of the quarks and antiquarks. They are shown as a function of the length x of one of the sides of the rectangle, with the other side kept at fixed length r. (We will return to Figures 9(a') and 9(b') below when we discuss possible models for this complex behaviour.) We note that the lowest adiabatic surface in this prototypical multiquark system has completely different flux tube topologies dominant in different regions of $qq\bar{q}\bar{q}$ configuration space. In a dynamical situation, we cannot expect to be able to describe this system in terms of only a quark wavefunction: for a given $qq\bar{q}\bar{q}$ configuration we will in general have a quantum state that is a strong coherent superposition of the possible states of the glue.

The physics of this complex situation is, as indicated in Figure 8, controlled by flux tube topological mixing. Since this is an effect that becomes important for $g<1$ where strong coupling perturbation theory is not useful, we cannot provide a very quantitative description of this process. Nevertheless, there are some qualitative features of this mixing which we believe we can safely expect and which are sufficient to form a basis for discussing most of its effects. Consider the example of $qq\bar{q}\bar{q}$ depicted in Figure 8(a) once again. We would like to know the amplitude for conversion between these two topologies as a function of x and y. To estimate this function we first imagine that both x and y are much greater than λ_0 so that a lattice spacing $a \sim \lambda_0$ where $g \sim 1$ can be used to describe the system. In this region topological mixing is not especially weak, but we can expect that a lowest order treatment will give a qualitatively sensible picture of the physics. Such lowest order mixing can occur when two (discrete) strings find themselves, in the tail of their vibrational wavefunctions, in a situation where they each have an element occupying a link on opposite sides of an elementary plaquette (see Figure 8(b)). They can then mix with an amplitude of order $1/ag^2 \sim 1/\lambda_0$. The amplitude that their wavefunctions will allow them to mix is, however, damped like a Gaussian in the required displacements of the flux tubes from equilibrium so that, even after summation over all plaquettes, the mixing will be very small if x and y are both much larger than the transverse size $b^{-1/2}$ of the string wavefunctions. Conversely, if there is good overlap between all four relevant string wavefunctions, then this mixing amplitude will be of order $1/\lambda_0$ and will be strong.

There is an immediate application of these ideas to the question of residual forces between hadrons. If y corresponds to typical hadronic dimensions but $x \gg y$, then flux tube topological mixing between the ground states of the two short (vertical) flux tubes to any states of the two long (horizontal) flux tubes will be suppressed. We conclude that the residual forces between colour singlet hadrons will have no long range (power law) tail, in accord with experiment. Conversely, for distances $x \sim y \sim 1$ fermi we should expect strong topological mixing between colour singlets leading to medium range van der Waal's forces which could play a role in nuclear binding. Indeed, we will see below that such effects are important in explicit calculations, and they may indeed be responsible for the bulk of nuclear binding energies!

A full-fledged treatment of the gluon dynamics of multiquark systems along the lines imagined above would obviously be a formidible task. In the meantime much recent work on multiquark dynamics has been based on simpler models. I would say that the minimum requirements for such a model are:

1) If it is to simplify calculations significantly, it should specify adiabatic potentials that depend only on the quark coordinates (including

space, spin, and colour);
2) It should confine colour;
3) It should reduce to the known $q\bar{q}$ and qqq potentials in the appropriate limits;
4) It should produce flux-tube-like potentials between well separated quarks or clusters of quarks with, for example, the correct ratio of the string tension between 3-$\bar{3}$ clusters and 6-$\bar{6}$ clusters, and
5) It should incorporate the known short distance interquark forces from one gluon exchange.

The only potential I know which meets these minimum requirements is the one which assumes that confinement develops smoothly from Coulomb's law so that

$$V = \sum_{i<j} V_{ij} \tag{15}$$

where

$$V_{ij}(r_{ij}) = -[-\frac{\alpha_s(r_{ij})}{r_{ij}} + \frac{3}{4}c + \frac{3}{4}br_{ij}]\ \vec{F}_i \cdot \vec{F}_j \tag{16}$$

where $\vec{F}_i = \vec{\lambda}_i/2$ for a quark and $\vec{\lambda}_i^c/2 = -\vec{\lambda}_i^*/2$ for an antiquark. Aside from meeting the above "minimum requirements", this potential seems to do a remarkable job of imitating the low-lying adiabatic surfaces of the full flux tube picture. This is illustrated in Figures 9(a') and 9(b') which show the two adiabatic surfaces (since there are two distinct colour couplings in this system, it has exactly two adiabatic surfaces instead of the infinite number of QCD) of (15) corresponding to the same $qq\bar{q}\bar{q}$ configurations as Figures 9(a) and 9(b). Unfortunately, the resulting imitation is imperfect. The $\vec{F}_i \cdot \vec{F}_j$ potential does not know about the limited transverse extension of the flux lines which form the potential and so does not cut off the analog of topological mixing fast enough to avoid a long range power law tail to hadron-hadron potentials. This flaw, however, by no means makes the $\vec{F}_i \cdot \vec{F}_j$ potential useless: it just means that one must proceed with due caution for possible spurious effects that it might introduce. In many cases in practice this is easy to do.

6. The $qq\bar{q}\bar{q}$ System: A Multiquark Primer

I can think of no better introduction to multiquark systems than to discuss the simplest multiquark system of two quarks and two antiquarks in the $\vec{F}_i \cdot \vec{F}_j$ model. In early applications of this model, illegitimate approximations were made that led to erroneous predictions of a rich discrete spectrum of multiquark states. We can begin to educate ourselves in the complexity of multiquark systems by learning from these old mistakes. Recall as well in what follows that the $\vec{F}_i \cdot \vec{F}_j$ model is a simple approximation to a much more complicated structure.

In fact, let us begin by considering an even simpler theoretical model which exposes most clearly the basic issues: imagine that the interquark potential were just

$$V_{ij}(r_{ij}) = \frac{1}{2}kr_{ij}^2\ \vec{F}_i \cdot \vec{F}_j \quad , \tag{17}$$

a purely "harmonic" force. (At first one might even imagine that this simple problem could be solved analytically, but it cannot be.) With the labels and coordinate system shown in Figure 10, and taking as a colour

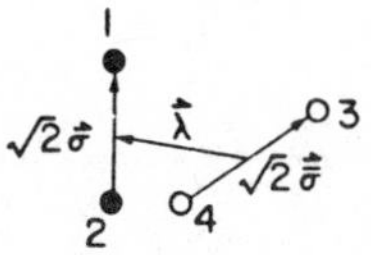

FIG. 10. the relative coordinates of the system $q_1 q_2 \bar{q}_3 \bar{q}_4$

basis the states $|\bar{3}_{12} 3_{34}\rangle$ and $|6_{12} \bar{6}_{34}\rangle$ (where C_{ij} means that particles i and j are coupled in the colour representation C and in each case the system is in overall colour singlet), we have

$$|\psi\rangle = \psi_{\bar{3}3}(\vec{\sigma},\vec{\bar{\sigma}},\vec{\lambda})\,|\bar{3}_{12}3_{34}\rangle + \psi_{6\bar{6}}(\vec{\sigma},\vec{\bar{\sigma}},\vec{\lambda})\,|6_{12}\bar{6}_{34}\rangle \tag{18}$$

and (with "1" $\leftrightarrow$ $|\bar{3}_{12}3_{34}\rangle$ and "2" $\leftrightarrow$ $|6_{12}\bar{6}_{34}\rangle$ in matrices)

$$H = \frac{1}{2m}(p_\sigma^2+p_{\bar{\sigma}}^2+p_\lambda^2)\begin{pmatrix}1 & 0\\ 0 & 1\end{pmatrix} + \frac{1}{2}k\begin{pmatrix} 2\sigma^2+2\bar{\sigma}^2+\frac{4}{3}\lambda^2 & -2\sqrt{2}\,\vec{\sigma}\cdot\vec{\bar{\sigma}} \\ -2\sqrt{2}\,\vec{\sigma}\cdot\vec{\bar{\sigma}} & \sigma^2+\bar{\sigma}^2+\frac{10}{3}\lambda^2 \end{pmatrix} \tag{19}$$

If we now assume that mixing between the $|\bar{3}_{12}3_{34}\rangle$ and $|6_{12}\bar{6}_{34}\rangle$ sectors can be neglected (or at least treated perturbatively) then we will erroneously conclude that both the $\bar{3}3$ and $6\bar{6}$ sectors of the system are confined in all three relative coordinates, so these systems will be analogous to ordinary mesons and baryons. On this basis we would expect to find two separate infinite towers of discrete levels (in the absence of mixing), or, if mixing is treated by perturbation theory, a single mixed tower of discrete states. However, this conclusion must be qualitatively wrong since one can easily show that this Hamiltonian contains solutions corresponding to free mesons in situations where the system is split into two $q\bar{q}$ clusters in that mixture of $|\bar{3}_{12}3_{34}\rangle$ and $|6_{12}\bar{6}_{34}\rangle$ corresponding to having each of these mesonic clusters in colour singlets. We can thus prove that in at least one simple case the $-2\sqrt{2}\,\vec{\sigma}.\vec{\bar{\sigma}}$ terms cannot be treated perturbatively: at least an infinite number of the states of the (erroneously deduced) discrete tower of states must collapse into a continuum of free mesons. Indeed, matters are even worse: there are an infinite number of continuous spectra, corresponding to the infinite number of possible internal excitations of the separated meson clusters. From this perspective, one might expect that if non-trivial multiquark states exist in this system (i.e., states that are not essentially two mesons), then they are likely to correspond to either weakly bound states, or to meson-meson resonances. Certainly, it would seem, the picture of a tower of discrete states analogous to the spectrum of ordinary mesons is very inappropriate.So far as I am aware, we still do not know the full spectrum of this harmonic multiquark model. However, a recent study of its ground state using variational methods strongly suggests that it does not even have any weakly bound states or resonances; rather, it appears to be best viewed as a theory of two mesons with weak residual interactions.

It would take us too far afield to discuss in detail what happens in more realistic models of $qq\bar{q}\bar{q}$ with an $\vec{F}_i.\vec{F}_j$ potential. Suffice it to say that there are strong suggestions that the physics of such meson-meson systems is roughly analogous to that of the nucleon-nucleon system. Indeed, it may even turn out that the S*(980) and δ(980) mesons are $K\bar{K}$ bound states roughly analogous to the deuteron.

7. The Nucleon-Nucleon System

Given the success of the valence quark model (in which mesons are approximated by $q\bar{q}$ and baryons by qqq), and armed with our experience from the simpler $qq\bar{q}\bar{q}$ system, it is natural to ask what the baryon number two system looks like in the same approximation. Notice that from this point of view it is not even clear that the simplest nucleus can be approximated by a two nucleon system: we must set out in an unprejudiced fashion to solve the six quark problem with (a model for) chromodynamics. In this six quark (or valence) approximation, we will be ignoring the effects of pair creation which are responsible for meson exchange. Such a calculation thus will examine the nature of the nucleus in an approximation in which the system is controlled by interquark forces and gluon dynamics alone. This is roughly analogous to studying molecular hydrogen without taking into account positronium exchange. In the molecular case this approximation is certainly justifiable to a very high level of accuracy; its validity in the case at hand will be discussed below.

There have been many calculations based on this valence approximation to the nucleon-nucleon problem. As mentioned already, it is not my intention to review this large and growing body of work. I will, however, mention the results of the one calculation with which I am most familiar. This calculation found that the six quark system with S-wave pn quantum numbers (but not the general six quark system) was to an excellent approximation describable as an interacting proton and neutron; it moreover extracted the effective nucleon-nucleon S-wave potentials shown in Figure 11. In

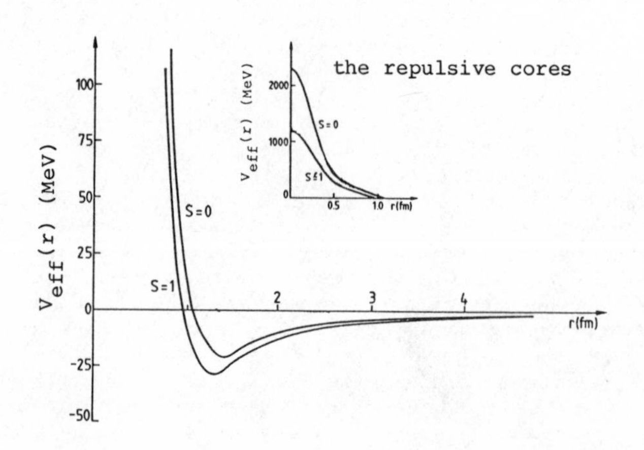

FIG. 11. the effective nucleon-nucleon potentials in the $^3S_1$ and $^1S_0$ channels arising from residual colour forces

this calculation the repulsive core is due to what Lipkin has dubbed the "Pauli, Fermi, Heisenberg Effect" since it arises from an interplay of the Pauli exclusion principle, the Fermi contact spin-spin interaction, and the Heisenberg uncertainty principle. It should be stressed that this effect is a quite general property of this system which, as its name implies, can be understood in a qualitative, parameter-independent, way. Equally interesting is the fact that this calculation gives an attraction in the 1-2 fm range which has its origin almost entirely in the gluonic rearrangement effect discussed in Section 5.

One might wonder at this point whether this molecular-like picture of the NN force is consistent with the empirical evidence for the existence of potentials with meson exchange quantum numbers . The fact that the observed potentials can be associated with the quantum numbers of the low-lying $q\bar{q}$ states is, indeed, completely expected in the molecular picture. Quark exchange potentials will have the quantum numbers of a quark-quark hole pair, which are identical to those of $q\bar{q}$. Moreover, the attractive gluonic rearrangement potential will have vacuum quantum numbers, identical to those of the hypothetical σ.

While molecular physics-like elements of the NN force could therefore be dominant so far as we know, a complete calculation must obviously allow for $q\bar{q}$ pair creation. It is, however, completely unclear that pair creation effects can be adequately described in terms of meson exchange. This is crudely illustrated in Figure 12 which strongly suggests that meson

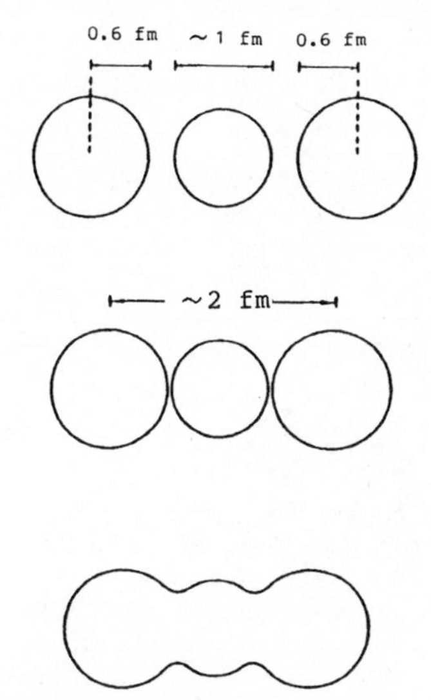

FIG. 12. a cartoon illustrating that for $r \lesssim 2$ fm meson exchange is unlikely to be appropriate to the description of the inter-nucleon potential

exchange can be used reliably only if the meson exchange range $m_M^{-1} \gg 2(r_N+r_M)$, where r_N and r_M are the nucleon and meson radii and m_M is the meson mass. In QCD this condition is unnatural since m_H and $r^{-1}{}_H$ are both normally of order Λ_{QCD} for all hadrons H. There is only one exception to this rule: $m_\pi \simeq 0$ due to its special status as the (almost) Goldstone boson of spontaneously broken chiral symmetry. This suggests that a useful model of the NN potential might be one which is the simple sum of the six quark molecular-like potential and a cut-off one-pion-exchange potential.

8. Conclusions

I have tried to make the case that QCD is the solid foundation on which nuclear physics (as well as the rest of strong interaction physics) must build. I have explained that simple models for chromodynamics in the confinement regime, which should be reliable for qualitative conclusions, indicate

1) nuclear physics is approximately the physics of nucleons,

2) not all six quark systems exhibit such clustering,

3) the repulsive core is at least in large part a "Pauli-Fermi-Heisenberg Effect", and

4) there is a large contribution to NN binding from glue dynamics.

suggesting, respectively, that

1) nuclear physics did not need quarks for the last 50 years,

2) there are many (difficult) experimental tests of the molecular-like picture in exotic six quark systems,

3) the role of non-pion-exchange potentials should be critically reexamined, and

4) nuclear binding itself may be a very old clue for the existence of glue.

Of course it goes without saying that there is tremendous scope for possible improvements in the models and in the complex calculations based on them. In addition, we can look forward to rigorous checks from numerical lattice gauge theory calculations of the basic physics of the models. Nevertheless, I am confident that it is not too soon to assert that QCD will become a permament and important ingredient of the nuclear physics of the future.

10. Acknowledgements and Apologies

Consistent with foregoing a review of this subject, I have avoided attempting a list of references. However, fairly complete references can be found in my recent papers with Jack Paton, John Weinstein, and Kim Maltman whose collaboration in much of the work described in Sections 3, 4 and 5, Section 6, and Section 7, respectively, of this article is gratefully acknowledged*. I also thank the Department of Theoretical Physics of the University of Oxford for their hospitality during the period when this manuscript was prepared.

---

*Corporate references referred to are:

N Isgur and J Paton 1983 Phys. Lett. 124B 247 (More recently Phys. Rev. D same author)
J Weinstein and N Isgur 1983 Phys. Rev. D27 588
K Maltman and N Isgur 1984 Phys. Rev. D29 952

Inst. Phys. Conf. Ser. No. 86
Paper presented at Int. Nucl. Phys. Conf., Harrogate, UK, 1986

Quarks in nuclei

Anthony W. Thomas

Department of Physics, University of Adelaide,
G.P.O. Box 498, Adelaide S.A. 5001

Abstract. Over the past few years theoretical attempts to use the quark model to obtain a deeper microscopic understanding of nuclear physics have become quite sophisticated. We review some highlights of this work, bearing in mind the constraints of existing data and mentioning possible future experimental tests.

1. Introduction

In view of the generality of the topic assigned, it is worthwhile to spend a little time explaining the philosophy adopted here. We begin from the premise that QCD is the correct theory for the strong interactions. However we accept that for the present we do not have reliable techniques for calculating even the properties of single hadrons - let alone multi-nucleon systems. (Although noble efforts are, of course, being made on this problem.)

Given these circumstances, the most effective approach would appear to be an inspired blend of experimental work, guided by phenomenology, linked as closely as possible to the underlying theory. Thus we hope for a system with positive feedback, simultaneously refining our models of nucleon and nuclear structure in a manner consistent with all the relevant data. Of course there will always be good and bad phenomenology. We should be suspicious of models which have more parameters than predictions, just as we should discard those models in conflict with known properties of QCD (e.g. its symmetries).

The structure of this review is as follows. In the next section we describe the most commonly used, QCD-inspired models of hadron structure. We present arguments for preferring those which explicitly keep the quarks, rather than replacing them by effective Bose fields (as in the Skyrme model, for example). Some comments are also made concerning the phenomenological constraints on key parameters of such models - notably the size of the confinement region. In section 3 we deal with systems of baryon number two. This includes calculations of specific components of the N-N force, the possible relationship to N-$\bar{N}$ and hyperon-N scattering, as well as various reactions involving the deuteron.

In section 4 we deal with heavier nuclei. We first clarify some confused thinking about the relationship between quark models and conventional meson exchange current (MEC) calculations. Next we review some recent work on the phase transition from nucleon to quark matter. Then we turn to explicit calculations of the interactions of weak and electromagnetic probes with nuclei, including the so-called EMC effect. Finally, section 5 contains some remarks on important directions for research in the near future.

2. The Structure of the Nucleon

In discussing QCD-motivated models of hadron structure in the present context, we shall avoid those models which, like the Skyrme model[1)], bosonise QCD[2)]. There are three reasons for this. First, while Witten's arguments[3)], based on a 1/Nc expansion[4)], are thought-provoking, they do not constitute a proof that hadrons are topological solitons. Second, by excluding explicit inclusion of colour degrees of freedom, such models eliminate even the possibility of investigating new physics in the nucleus. Third, by construction such models are not suited to describing deep-inelastic scattering[5)], and therefore cannot play a useful role in the consistent interpretation of data from new facilities, like CEBAF and the hadron facilities[6)], which will extend the kinematic boundaries of conventional nuclear physics.

Let us therefore turn to QCD-motivated, explicit quark models. The rather sad lack of constraint from QCD is evident in the incredible variety of such models, ranging from the potential models (non-relativistic,[7)] or otherwise [8)]), through bag models [9)], non-topological bag models[10)], chiral bag [11)] or soliton models, and finally topological, chiral bag models [12,13)]. It is pointless to go over these models here, because each has been the subject of more than enough review articles. Instead we concentrate on the essential differences.

On the basis of very general arguments we know that the pion plays a special role in nature.[11,14)] By incorporating the pion in a manner consistent with chiral symmetry the chiral bag/soliton and the topological-bag/soliton models (hereafter called class A) certainly have a stronger claim to be considered realistic than the other models (class B). The first major difference between them is whether or not the Lorentz scalar field which confines the quarks[15)] is the chiral partner of the pion (as in the linear σ-model), or not. In the former case we are led to topological models[3,12,13)], in the latter we are led to the cloudy bag[11)].

A second difference arises over whether or not heavier mesons should be included in addition to the pion. For example, the Maryland group argues strongly that gluons should not appear explicitly in their topological soliton calculations, rather one should add heavier mesons (ω, ρ, A_1...)[16)]. We find this extremely unlikely, particularly in the nuclear context where one might expect some colour polarisation. Furthermore, if the effective bag radius (or size of the confinement region) is relatively large (say greater than 0.8 fm), one can make a strong case against including heavier mesons[17)]. On the other hand, we do not find any purely theoretical arguments, based upon our present understanding of QCD, sufficiently compelling to distinguish between models in class A. For this purpose we need to examine the experimental evidence.

Amongst the static properties of the nucleon the axial form-factor, $g_A(q^2)$, is the one most closely linked to the quark distributions. In a model like the cloudy bag, which includes only quarks and pions and has no discontinuities in the pion field, the quarks alone determine $g_A(q^2)$. Experimentally this form-factor is fitted by a dipole with mass less than 1GeV[18], which corresponds to a bag radius a little larger than 1 fm.[19] Since one can also establish a link between the πNN form-factor and $g_A(q^2)$ in a relatively model independent way,[19] one finds that $g_{\pi NN}(q^2)$, if parameterised as a dipole, would have a mass parameter (Λ) of order 0.9 GeV. Of course this is far smaller than what seems to be needed phenomenologically - e.g. the Bonn potential uses $\Lambda \sim 1.5$ GeV,[20] and similar values are required for the deuteron d-state wave-function.[21]

The naive response to this is to adopt a smaller bag radius. If one is willing to add an A_1 meson, with a range one fifth of the size of the bag it is supposed to dress, this can probably be made consistent with $g_A(q^2)$. On the other hand, in discussing the σ-term (a direct handle on chiral symmetry breaking in QCD, with a major contribution from pion loops[22,23]) Gasser[24] refers to $\Lambda = 1$ GeV as "somewhat exotic" - essentially because he considers it too large! Clearly what is natural to a nuclear theorist can appear absurd to a particle theorist, and vice versa. When this occurs it usually means there is something deeper which needs to be understood. We feel it most likely that Λ is of order 0.9 GeV for a single, free nucleon, but that when <u>two</u> nucleons begin to overlap there are more interesting physical processes which may simulate a harder form-factor.[25]

The discussion just concluded regarding $g_A(q^2)$ and $g_{\pi NN}(q^2)$ would be even more model dependent for any other form-factor. Quite clearly anyone with a favourite model can emerge from such a discussion with prejudice unshaken. We are firmly convinced that while agreement with low energy hadron properties is a requirement for any model of hadron structure, one will never be able to draw unambiguous conclusions about the size of the confinement region (hereafter loosely called a bag radius) from such a comparison.

It has recently been suggested that the bag radius may in fact have no physical significance[26] - the so-called "Cheshire cat" picture. Indeed, for the 2D toy model where this claim was demonstrated (a fermion theory with an exactly equivalent boson theory), it is quite true that the boundary where one switches from fermion to boson fields is irrelevant. However, we find this extremely unlikely for QCD. Figure 1 shows the dependence of the Σ^- and Σ^+ magnetic moments on the bag radius within the cloudy bag model.[27] At least for $R \gtrsim 0.8$ fm the CBM should be fairly accurate. Clearly while the Σ^- behaves as the Cheshire-cat hypothesis would suggest, the Σ^+ does not.

Ultimately the only really unambiguous tool available to us is deep inelastic scattering,[28] which in principle can probe the quark distributions directly. An early attempt in this direction used the SU(3) flavour breaking in the sea to put a lower limit of 0.85 fm on the bag radius.[29] Recent work on the strange sea has given a similar result.[30] However, both of these calculations put limits on baryon form-factors, rather than directly on the valence quarks. It is the latter distributions which hold the promise of directly determining R. Although early work on bag structure functions[31] did suffer from lack of momentum conservation, Jaffe and Ross[32] were able to conclude that the bag is consistent with data on the valence distribution of the nucleon at $Q^2 \sim 0.7$ GeV^2. More recently Bickerstaff and Thomas[33] have been

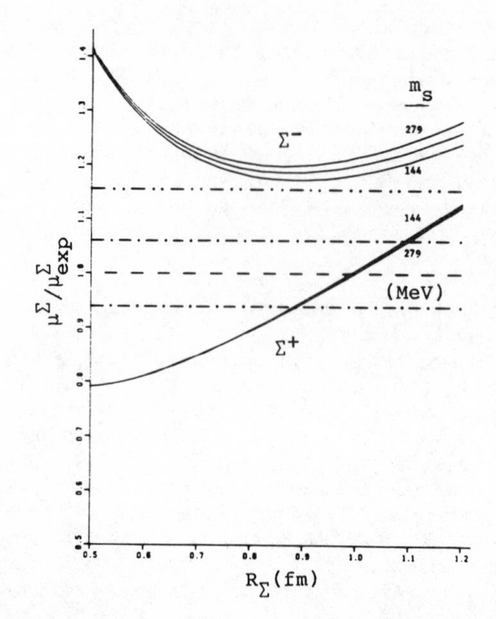

Figure 1: An illustration of the explicit dependence on bag radius of the Σ^+ and Σ^- magnetic moments in the cloudy bag model, Théberge and Thomas (1983)

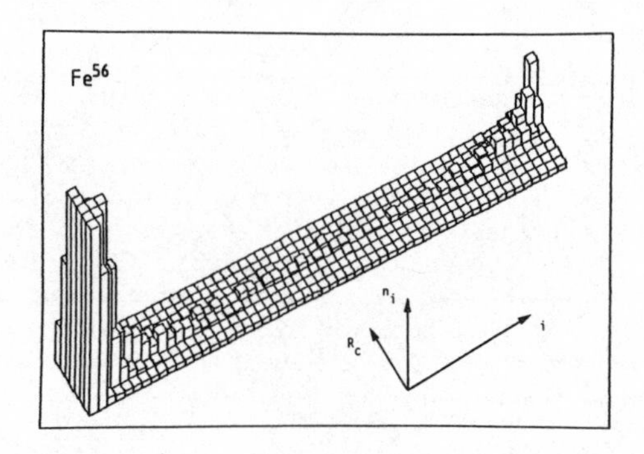

Figure 2: Relative probability n_i, of finding clusters of i nucleons in ^{56}Fe, as a function of the critical distance ($2R_c$) between nucleons where they behave as 6-quarks. R_c is in steps of 0.25 fm - from Güttner and Pirner (1985)

able to improve the support of the bag structure function, by restoring its translational invariance. We have hopes that this may finally give unambiguous limits on the bag size.

For the present we shall adopt the working hypothesis that R is of order 1 fm, and ask about the consequences for conventional nuclear physics. Since the average separation distance between nearest neighbours is about 2 fm in nuclear matter, one would naively expect that the consequences could be dramatic. Indeed in the late 70's there was speculation that quark models would imply some crystalline structure in nuclei,[34] or at the very least some percolation.[35] That is, it seemed likely that some nucleons in a nucleus might overlap sufficiently for quarks to move freely along a chain. Very recently Güttner and Pirner[36] have made a more quantitative estimate of this effect using Monte-Carlo techniques. Under the assumption that quarks can move freely between nucleons whenever they are within a distance $2R_c$ of each other, Fig. 2 shows (for ^{56}Fe) the relative number (n_i) of clusters containing i nucleons, for various values of R_c - in steps of 0.25 fm. With a bag size of 1 fm one might expect R_c to be between 0.5 and 0.75 fm. Clearly the probability of finding quarks percolating between three or four nucleons is not negligible!

Of course the previous discussion is very naive because there is almost no dynamical information going in. In sections 3 and 4 we shall present some much more sophisticated calculations. For the present we merely conclude that this way of looking at Nature presents nuclear physicists with some very challenging questions.

3. The nucleon-nucleon system

The idea that the quark model might contribute to our understanding of the N-N interaction is far from new. Immediately after the invention of the MIT bag model Fairley and Squires showed that in the simplest possible approximation one would get several hundred MeV attraction at separations of 1-2 fm.[37] De Tar showed that the colour magnetic interaction would lead to short-range repulsion.[38] Matveev and Sorba took great care to emphasise the important role of colour in a 6-quark bag with deuteron quantum numbers. Indeed the probability of finding two 3-quark clusters with non-singlet colour is some 80% in the bag model.[39] In many ways the possible appearance of this new degree of freedom in systems with more than one nucleon is the key distinguishing feature of the quark-level (rather than meson-exchange current, MEC) description of short-distance physics.

Because human beings are as they are, every one of the models of nucleon structure mentioned in section 2 has been applied to the N-N system. In many ways the cleanest approach involves the least detailed dynamics. In the boundary condition models one has an interior region which contains only quarks, and an exterior region containing nucleons.[40-45] The interior region has been given many colourful names, including "soul" and "primitive". Clearly, if one were only concerned with elastic scattering, the model of the interior would be irrelevant. However we are also concerned with nuclear reactions involving weak and electromagnetic probes, as well as hadrons. In these reaction calculations the clean separation of interior and exterior regions is a real simplification, and we shall cite a number of such calculations below.

In the quark compound bag model[45,46] the deuteron wave-fuction contains two pieces, ψ_q and ψ_h. The former is a confined, primitive 6q-state, while the latter describes the relative motion of two 3q-clusters. The Schrödinger equation for ψ_h includes an effective potential resulting from the elimination of ψ_q. In this model ψ_h is not necessarily zero in the interior region. Kalashnikova et al. report that $P_q \lesssim 1\%$ and the corrections to μ_d and Q_d were less than 1% and 6% respectively.[45]

At the other end of the spectrum, some groups have attempted to use resonating-group techniques to calculate N-N scattering completely within a model. For this purpose the non-relativistic quark model (NRQM) is undoubtedly the simplest, and not surprisingly the most used.[47-54] (Indeed the bag model is terribly complex to handle this way.[55]) We shall first comment on the central, spin-orbit and tensor pieces of the force generated in this model, and then comment on some deeper questions about the whole approach.

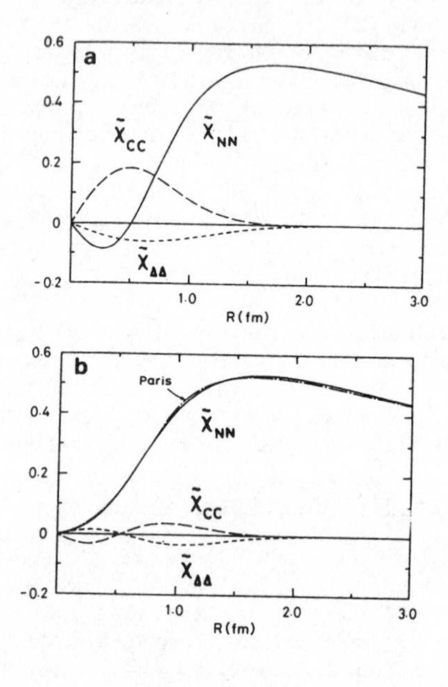

Figure 3: Two equivalent representations of the relative wave-functions of the N-N, Δ-Δ and C-$\bar{C}$ components of the deuteron, in a resonating group calculation.[56]

The apparent repulsive core in the N-N interaction is understood in this approach as a node in the relative wave function associated with the necessary [42] symmetry component of the wavefunction. This wave function is illustrated in Fig. 3(a), which is taken from Yamauchi and Wakamatsu.[56] In this case there is a 2.4% $(0s)^6$ component in the deuteron wavefunction. An alternate representation in which all the $(0s)^6$ components are put into the resonating group N-N wavefunction is shown in Fig. 3(b). In this case the similarity to the Paris wavefunction is remarkable. Furthermore, because the short distance wavefunctions are suppressed, the effect of the quark-level anti-symmetrizer is not so great. Physical observables calculated from either representation should be the same, but clearly the second form (Fig. 3(b)) is much closer to conventional nuclear physics.

Considerable effort has also been put into calculating the spin-orbit component of the N-N force arising from one-gluon-exchange (OGE) with quark interchange. The consensus appears to be that this mechanism can give similar results to conventional OBEP.[51,53] However we have some lingering theoretical doubts. Firstly the quark-gluon coupling constant needed is larger than the original MIT value (which was itself considered large). Secondly the size of the quark confinement region is quite small. Thirdly, the origin of the whole effect, namely the L.S component of the OGE interaction is quite mysterious. It is in fact usually discarded in single baryon spectroscopy.[57] One very interesting proposal, which may shed some light on this problem, is to compare the predictions for the ΛN, ΣN and NN spin-orbit potentials in the quark models with those of conventional OBEP. (Note, however, the considerable differences between Yin He et al.[58] and Pirner.[59])

The tensor piece of the N-N interaction has received less attention in these quark level calculations. In part this is because pion exchange should probably play a more important role than gluon exchange. Of course, in view of the controversy over the πNN form-factor mentioned in section 2, there is an urgent need for more sophisticated calculations to follow up the work of Guichon and Miller.[25]

At present it must be said that these resonating group calculations raise more questions than they answer. For example, Fujiwara and Hecht[60] have shown that the inclusion of a 3q-$q\bar{q}$ component in one of the two colliding nucleons can lead to significant changes in the effective N-N interaction. In particular the effective $q\bar{q}$ exchange potential becomes attractive for N-N separations less than 0.8 fm - unlike conventional ω-exchange. Perhaps the most disagreeable aspect of this calculation is its essential complication. There are some 25 possible exchange terms to include. Unless some convergence criterion can be established one shudders at the thought of allowing both the incident and outgoing nucleons to be a linear combination of 3q and 3q-$q\bar{q}$ configurations. Furthermore, only OGE is included in this calculation, whereas explicit pion exchange should probably be included,[61] plus two gluons, etc.

Models based on the NRQM with (say) harmonic oscillator confining potentials also suffer from spurious van der Waal's forces. Thus one can never be sure whether a success or failure at intermediate N-N separations is spurious or not. Such considerations have led many authors back to introduce explicit links[62] or flux tubes[63] between the quarks. Alternatively one may resort to bag models.[64] In any case, once one takes this problem seriously it seems inevitable that many-body forces must be included at the quark-level.[64,65]

To end this rapid tour of N-N elastic scattering, we would like to stress the possibility of using the phenomenological evidence concerning N-$\bar{N}$ annihilation as a guide to what physics should be included for the N-N system. As Dover has discussed at length,[66] the conventional quark rearrangement mechanism for $p\bar{p}$ annihilation into mesons cannot explain several important selection rules - see also ref. (67). For example, the ASTERIX group[68] has found a sizeable branching ratio for $^{33}P_1 \rightarrow \pi^0 f$. In order to understand this from the phenomenological point of view, it is essential that the contributions R_2 and A_2 shown in Fig. 4 (where a $q\bar{q}$ pair pops into or out of the air - this is the "$^3P_0$ mechanism") are of the same order of magnitude. We are aware of no argument why this mechanism for $q\bar{q}$ pair creation and annihilation should not play an important role in the N-N system.

To conclude this section we turn to reaction calculations in the A=2 system, beginning with the simplest and cleanest, namely deep-inelastic scattering. If the deuteron has a non-zero 6q-component, Kobushkin pointed out some ten years ago[69] that the structure function would extend into the region of Bjorken x beyond that allowed for a free nucleon. Indeed he was able to fit some preliminary SLAC data with a 6q probability about 6%. Unfortunately the calculation of the competing mechanism, namely fermi motion of the nucleon, is not yet under control - as we mention in section 4 in connection with the EMC effect. This deserves more theoretical work, particularly in view of the possibility of new data from the NPAS facility.

Another suggestion by Mulders and Thomas overcomes the objections about fermi-motion corrections, but does require new ν and $\bar{\nu}$ data on hydrogen to compare with e-d and e-p data.[70] By comparing the true ratio of down to up quark distributions extracted from the former with the effective distributions coming from the latter we can get a handle on the 6q component. Unfortunately the $\nu,\bar{\nu}$ experiments are time-consuming and, to our knowledge, no study is yet planned. We hope this will be remedied soon!

Considerable emphasis has been put on the asymptotic behaviour of the elastic, deuteron form-factor ($F_d(Q^2) \sim (Q^2)^{-5}$ for 6 quarks), as a signal of a 6q component.[71] At the present time the data is not convincing (up to 6 GeV^2), and a behaviour more like $Q^{-5.5}$ has been suggested as a best fit.[72] An alternate suggestion is to turn instead to polarisation observables like P_x/P_{xz}, at very high Q^2, as a signature of the quark components of the deuteron.[73] This proposal is being taken seriously in the plans for the first measurements at CEBAF.[73]

The reaction d(e,e')np, with a relatively low energy for the final np system, has been cited for many years as an example of the success of conventional nuclear physics. The impulse approximation plus MEC can explain the data very well. In a recent study based on the boundary condition model (i.e. an interior, bag region and an exterior, N-N region), Kisslinger was able to put this in better perspective.[74] Figure 5 shows his results which make it quite clear that this is not the reaction with which to test for a 6q component in the deuteron. (The dotted contribution is negligible at all values of the momentum transfer considered.) Incidentally, for the same parameters, Kisslinger does find a sizeable 6q contribution to the elastic charge form-factors of ^3H and ^3He. Indeed this contribution is essential to his good fit to the data. (The relatively large 6q probability that he finds, namely 14%, is rather spectacular however.)

The BCM has also been used to calculate pion production,[75] while the quark compound bag has been applied to the problem of parity violation in deuteron photodisintegration.[76] The latter raises the interesting question of whether quark model calculations offer the hope of a better understanding of symmetry breaking in nuclear physics. For example, charge symmetry has been the subject of a recent high precision test at TRIUMF.[77]

Looked at from the quark level, where $m_u \neq m_d$, charge symmetry does not appear to be very fundamental. Indeed it is only because confinement, which leads (in bag/soliton models) to light quark energies of (300-400) MeV which are much greater than m_u or m_d, and <u>not</u> very sensitive to the latter, that charge symmetry is good at the hadronic level. (This is also the reason why the colour dielectric model[78], where the confinement mechanism is more sensitive to the quark masses, has received so much attention recently.[79-81])

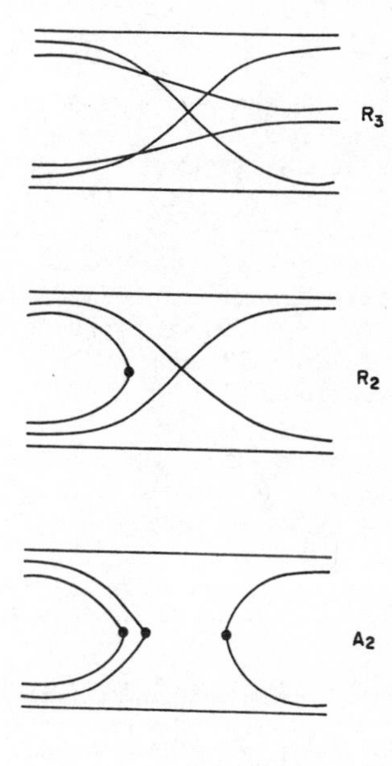

Figure 4: Some mechanisms proposed to explain $p\bar{p}$ annihilation, which may also be relevant to N-N scattering - from Dover.[66]

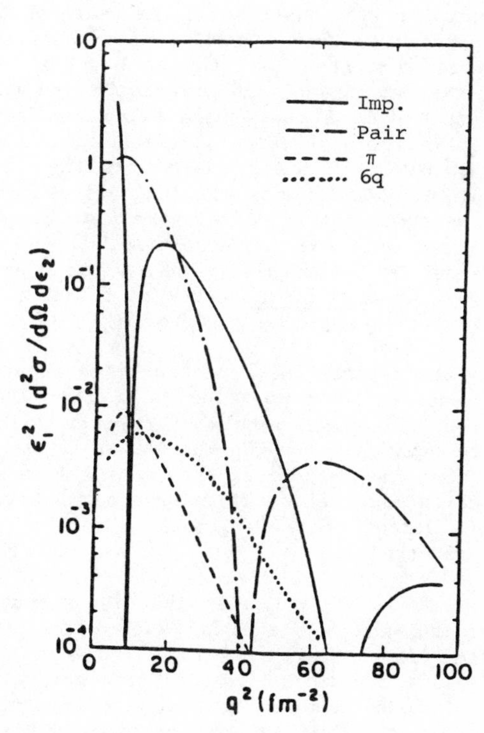

Figure 5: Various theoretical contributions to the reaction d(e,e')np, calculated by Kisslinger.[74] The 6q component of the deuteron wavefunction (dotted line) makes a negligible contribution.

The TRIUMF measurement found a tiny difference between the n and p analysing powers in n-p elastic scattering at 477 MeV : $(A_n - A_p) = (37 \pm 17 \pm 8) \times 10^{-4}$. At first sight the theoretical treatment[82] suggests that quark physics plays a negligible role. Indeed Miller et al., using the boundary condition model, found a contribution of only 1×10^{-4} from OGE (because $m_u \neq m_d$).[82] On the other hand, the calculation was very sensitive to the form-factor in the OPEP term, so that one may have a new test (for a completely new piece of the N-N force) of the interplay between short-distance physics and the effects of form-factors. This should complement the information obtained from studies of η.[21,25] Finally, recent work by Holzenkamp et al.[83] suggests that the dominant OPEP contribution is also very sensitive to the intermediate range N-N wavefunction - with the Bonn potential yielding significantly different results from the Reid potential at 188 and 350 MeV. This deserves more work, both theoretical and experimental.

Finally we mention the fact that a great deal of effort has been put into the search for exotic states with A=2.[84] Thus far we do not find any of the evidence compelling. However, there can be no doubt that the discovery of a stable, or at least narrow, A=2 state in addition to the deuteron (e.g. the H or di-lambda of Jaffe[85]) would be a tremendous help in formulating our quark level descriptions of the N-N system.

4. Quark models of real nuclei

Before turning to some actual calculations we deal first with a matter of principle. The question at the heart of the debate over the quark model, as opposed to conventional nuclear physics with its nucleons, mesons and isobars, is often lost in the heat of argument. This question is what are the key degrees of freedom in the many-body system. We shall soon see that one of the issues being keenly debated is whether the nucleon gets bigger, or "swells", inside a nucleus. Quite clearly this question is unanswerable until one states the framework within which the issue should be decided. In the language of isobars and MEC, a swollen nucleon may be equivalent to a relatively small admixture of Δ or Roper resonance into the many-body wavefunction.[86] In the 1980's this would not be regarded as particularly new. On the other hand, the necessity to include an additional degree of freedom, namely colour, into the nuclear wavefunction would be quite new.

From the mathematical point of view the question is even easier to state and answer. The question is whether the set of all baryon number one hadrons, together with the observed mesons, constitute a complete set. Quite obviously for QCD the answer is no, in principle. Of course it may be that in practice the corrections associated with the explicit appearance of colour are always negligible. It is unfortunate that there is no direct probe for colour which we could use to answer the question experimentally.

In section 2 we mentioned that quark models of hadronic structure led to speculation some years ago about the occurrence of exotic, multiquark states in nuclei. We also showed the naive estimates of cluster formation due to Güttner and Pirner (Fig. 1). These authors (along with many others) assumed that an exotic state would be formed whenever two nucleons came within some critical distance. We now discuss some more sophisticated calculations which relate to this matter.

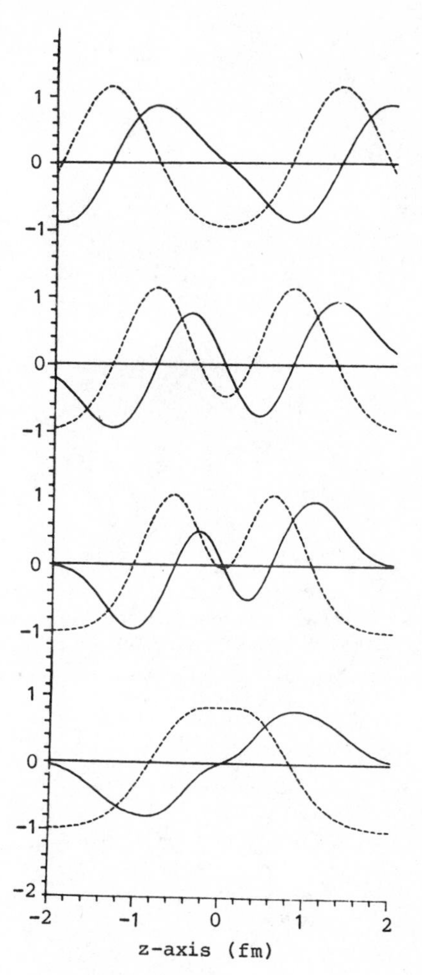

Fig. 6 The scalar field (dotted line) and pion field (solid line) for the collision of two solitons in the Birse-Banerjee model. In (a)-(d) the separations are 2.7, 1.5, 1.0 and 0.5 fm respectively - from Pineda.[87]

Figure 6 shows the σ-field and the pion field for two nucleons, in the Born-Oppenheimer approximation, at several separations.[87] This calculation is based on the soliton model of Birse and Banerjee,[13,16] where the quarks are "confined" by a scalar field which is the chiral partner of the pion (an assumption with which we disagree). Pions are included to guarantee chiral symmetry, and the hedgehog approximation[88] has been made to "simplify" the numerical work. (Even so, the calculation required an array processor.) For our purposes the key point is that the two clusters of quarks ("nucleons") remain relatively distinct until about 0.8 fm, at which point the bags suddenly merge. It obviously makes more sense to speak of the system in terms of six quarks for $r_{NN} < 0.8$ fm!

In Fig 7 we show the quark density in a unit cell in an idealisation of nuclear matter as a crystal with cubic symmetry.[89] The model used is the non-topological soliton model of Friedberg and Lee,[10] which (in the absence of gluons - as here) does not actually confine the quarks. Because of numerical difficulties, the solution shown corresponds to a parameter set for which the free nucleon has a fairly diffuse surface. Nevertheless the result is quite remarkable - the phase transition to quark matter occurs at about 7/10 nuclear matter density. This sort of investigation should clearly be carried further, with different parameter sets, including glue, eventually relaxing the unrealistic, regular lattice.

One non-topological soliton model which is of considerable interest in this regard is the colour dielectric model of Chanfray, Nachtmann and Pirner.[78] Motivated by the work of Neilsen and Patkos,[90] this model confines the quarks by a coupling of the form $(m/\chi)\bar{\psi}\psi$, with m the quark mass and χ the confining scalar field. As $\chi \to 0$ outside the hadron the quark becomes infinitely massive. With a suitable choice of parameters this model can produce solutions which look quite bag-like[80,91] - see Fig. 8. Unfortunately the nuclear crystal-soliton calculation has not

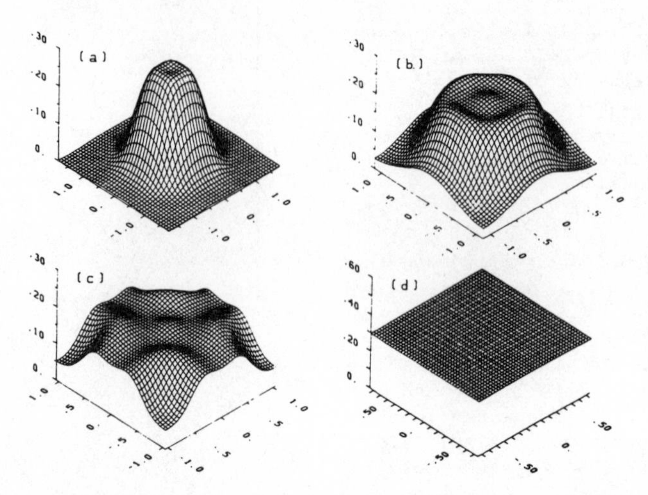

Figure 7: The quark density in the (100) plane of a unit cell for a crystalline approximation to nuclear matter in the Friedberg-Lee model. Figs. (a)-(d) correspond to densities ρ/ρ_o = 0.1, 0.5, 0.7 and 2.0, respectively - from Achtzehnter et al.[89]

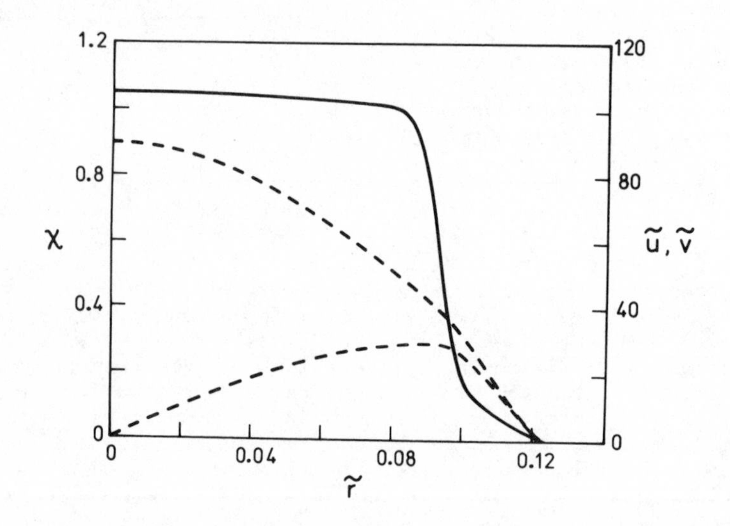

Figure 8: An example of a bag-like soliton in the colour dielectric model. The upper and lower components are $\tilde{u}$ and $\tilde{v}$ respectively, the radius $\tilde{r} = r\sigma_v$, with σ_v = 15 MeV. Finally, χ is the confining, scalar field - from Williams et al.[79]

yet been made in this model.

However, the colour dielectric model has been applied, at least in an approximate analytic form, to the problem of charge symmetry breaking in nuclei.[79] In fact, it is easily shown[78,79] that while $\chi \to 0$ outside an isolated nucleon, the lowest energy configuration inside a nucleus involves a non-zero value, χ_N, between the nucleons. Thus the quarks can move between the nucleons - in principle. Exactly how much leakage of quarks occurs depends on the density of the many-body system and the soliton parameters. Because of the way the quark mass enters the Lagrangian density - i.e. as m/χ, with $\chi \to \chi_N$ which is small between the nucleons - the effects of small u-d mass differences can be relatively large. While Williams and Thomas[79] show that there are reasonable parameter sets which allow one to reproduce the Nolen-Schiffer anomaly,[92] the key point is not the agreement with data. The key point is that if this model is at all realistic such effects, which are really new many-body forces, must be considered. In conclusion, we note that important work on the Nolen-Schiffer anomaly has also been carried out in the boundary condition model.[93]

Other applications of the quark model in nuclear systems abound. We have mentioned Kisslinger's work on the ^3H-^3He form-factors,[74] but similar results have also been obtained by Burov et al.[94] Karl et al.[95] have estimated the effect of quite large 6q admixtures in the A=3 wave-function, which they interpret as an increased nucleon size. However the main claims in regard to the latter have been related to the EMC effect[96,97] and recent quasi-elastic electron scattering data.[98] We deal with the latter first.

In the past couple of years, careful separation of the longitudinal and transverse response functions in nuclear (e,e') scattering at $q \sim (300$-$400)$ MeV/c,[98] have given theorists a headache. Until recently the claim was that the transverse response was not anomalous, but that there was some 50% of the strength missing from the longitudinal response. Since the longitudinal response is obtained by dividing data by the free nucleon charge form-factor, the anomaly could be explained by assuming a decrease in the nucleon form-factor in the medium (at the same q^2) - that is, an increase in size. This idea was attractive to many people in view of claims that such a size increase could also explain the EMC effect.[99]

This superficially attractive suggestion is clouded by three developments. First, Sick has shown that y-scaling prohibits a "size change" of more than about 5% in ^3He and ^{20}Ne. (However, as Mulders has observed[101] y-scaling involves total cross-sections and is therefore sensitive to both the magnetic and the charge form-factors.) Second, Stroth et al.[102] have been able to fit the ^{12}C longitudinal response at 300 MeV/c using the RPA, with the Gogny force, by taking anti-symmetry into account. While the results were not so good for ^{56}Fe they argue that 2p-2h excitations should be more important there. Third, Mulders has shown that if the short-range N-N system is treated as a 6q-bag, and one includes isobar excitation, both the longitudinal and transverse response functions can be fitted.[101] The price is not a change in nucleon size but a change in its effective magnetic moment - he finds $\mu^*/\mu_p^{free} \simeq 1.25$-$1.30$ for a variety of nuclei from ^{12}C to ^{56}Fe.

Inconclusive though these claims of a change in nucleon size may be, a number of theoretical calculations have emerged recently which prove that such a thing must happen. Noble originally suggested that this might happen in a model with point-like nucleons plus σ- and ω-mesons.[103]) Much more recently Oka and Amado claim to have shown, on very general grounds, that the size of nucleon must increase when it is placed in an attractive potential [104]) - see however the results of Oka et al in the Skyrmion model.[105]) This would seem to support claims based on the City University soliton model,[106]) the linking model of Lenz and co-workers,[62]) the crystal model of Goldman and Stephenson[107]) and many others.

Even though one can find objections to each one of these claims individually, their unanimity would be impressive: were it not for the fact that they are not uncorrelated. With the exception of Noble's work, all of this speculation about an increase in size followed the interpretation of the EMC effect as an increase in confinement size inside a nucleus.[99,108]) It is to the EMC effect, and to these claims that we now turn.

The data on the EMC effect from CERN[96]) and SLAC[97]) shows a definite depletion of the fraction of the momentum of the nucleon carried by its valence quarks. This effect increases roughly as ℓn A. There may or may not be a corresponding increase in the sea. Close and collaborators discovered that this effect could be parameterised by shifting the value of Q^2 (the 4-momentum of the virtual photon) at which the nucleon structure function was evaluated for a given nucleus:

$$F_3{}^A(x,Q^2) = F_3{}^N(x,\xi Q^2),$$

with $\xi \simeq 2$ for ^{56}Fe. This observation struck a responsive chord for three reasons. First the Q^2-dependence is directly given by QCD. Second, Close and collaborators were able to construct an argument based on the MIT bag model, which related a change of momentum scale to a change in confinement size.[99]) Third, practical calculations of the A-dependence of the effect, using realistic nuclear wavefunctions to estimate the probability of nucleon overlap, were able to reproduce the systematics of the SLAC data with a single free parameter.[108])

Unfortunately we do not find these arguments convincing. Despite some fascinating speculation from the group at Rutherford laboratory recently (including the requirement that the pion should contain more constituent glue than valence quarks),[109]) we have seen no physics arguments linking perturbative QCD to the clearly non-perturbative process of nuclear binding. It has also been demonstrated that the argument linking a change of momentum scale to a change in confinement scale is not consistent with anything like the 15% increase in size required in Fe (both for EMC and quasi-elastic scattering).[110]) In fact the valence quark correlation functions in the two targets are identical (by assumption) at their respective scales. On the practical side the 6q-probabilities needed in ref. (108) to explain the EMC data for ^{56}Fe were unbelievably large.

After all this, it has also been shown that a more careful treatment of binding and fermi-motion corrections for the nuclear data can explain most of the EMC effect.[111,112]) Figure 9 shows the calculations of Adelaide University group, using standard single-particle potentials and removal energies, together with an ansatz for the off-mass-shell behaviour of the nucleon structure function.[112] Rather than a change of distance scale, the EMC effect seems to reflect a change of mass scale due to nuclear binding. This has been discussed at much greater length by Bickerstaff and Thomas.[113]

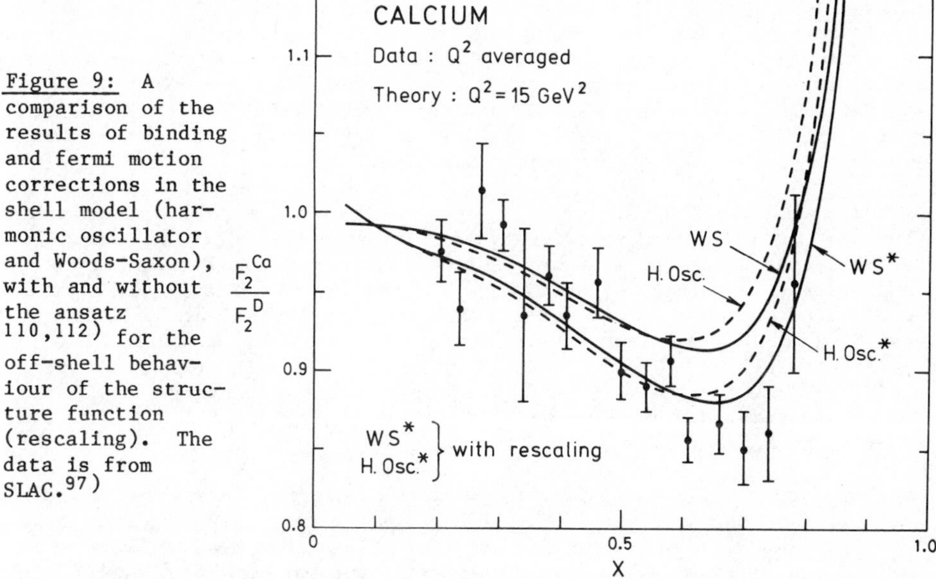

Figure 9: A comparison of the results of binding and fermi motion corrections in the shell model (harmonic oscillator and Woods-Saxon), with and without the ansatz [110,112] for the off-shell behaviour of the structure function (rescaling). The data is from SLAC.[97]

To conclude this section we repeat our belief that the concept of an increase of the nucleon size inside a nucleus, even when properly formulated, is not supported by any unambiguous experimental evidence. Defining this question more sharply, and designing questions to answer it, must have a high priority in the next few years.

5. Future directions

Hanging over all discussions of quarks in nuclei are the undisputed successes of the shell model. These are probably best symbolised by the beautiful electron scattering experiments from Saclay which have established the existence of a $3S_{1/2}$ orbital in the Pb region,[114] with a spectroscopic factor of 70%. Any quark model must respect the successes of the shell model and for this reason its limits must continue to be mapped out by careful experimental work. It will be crucial to push to higher-Q^2 and to deeper shell-model states using (e,e'p) and (e,e'n) - although this may need to wait for CEBAF to come on line in 1992. In the mean time further tests of y-scaling, along the lines already proposed[100], would be most useful.

The high-energy community has a duty to pursue systematic studies of the Q^2, x and A-dependence of the EMC effect. Some important tests of the various models using the Drell-Yan process[115] also need to be followed up. On the theoretical side, a great deal remains to be done on rather mundane, but fundamental issues like treating binding and fermi motion in a consistent way - e.g. without the need to resort to an ansatz for the off-mass-shell behaviour of the nucleon structure function. Until this can be done we shall not be sure whether there is evidence of new physics in the EMC effect. The relationship between models involving binding and an excess of pions[116,117] also needs to be explored. Direct tests of an enhancement of the virtual pion field of the nucleus[118] also need to be tested less ambiguously.[119]

Amongst the tests for multi-quark configurations in nuclei, we find the cumulative region, $x > 1$, which is forbidden for a free nucleon, very promising.[120] However binding and fermi motion corrections give even worse theoretical problems here. The deuteron is much simpler and we recall once again the suggestion of Mulders and Thomas - see section 2 and ref. (70). Of course we should also have our eyes open for anomalies - either in accurate tests of low energy parameters of the N-N system,[121] or in tests of symmetry breaking.[77,82]

Although we have not mentioned it previously we also hold considerable hope for insights from experiments with hypernuclei. It has long been realized that while the Λ viewed as an elementary particle experiences no Pauli exclusion principle, at the quark level its non-strange quarks will be affected. Indeed it has been suggested that the effects of this may have already been seen[122] in the existence of a narrow state in ${}^{6}_{\Sigma}H$ (possibly a (1 s_p^{-1}, 1 s_{Σ^-}) configuration), but not in heavier nuclei. Experiments with doubly strange nuclei, possibly in higher density systems, may need to wait for future hadron facilities, but are eagerly anticipated.

Finally, we do look to QCD itself for theoretical insights to both hadronic and nuclear structure. This may necessarily involve large numerical calculations on a lattice.[123] We also need to return to basics, and re-examine how to formulate the nuclear many-body problem in the absence of a sensible definition of a N-N potential - as seems to be the case in the quark model.

The days of innocence are past. Establishing a decent, microscopic understanding of nuclear physics at the quark level will not be easy. Nevertheless, no-one can doubt that we shall learn a great deal from the endeavour.

Acknowledgements

It is a pleasure to acknowledge important discussions with my colleagues in Adelaide and elsewhere. I would particularly like to thank Mrs. Elaine Gregory for her assistance in the production of this paper under considerable time pressure.

This work was supported by the University of Adelaide and the ARGS.

References

1. T.H.R. Skyrme, Proc. Roy. Soc. A260 (1961) 127.
2. R.T. Cahill and C.D. Roberts, Phys. Rev. D32 (1985) 2419.
3. E. Witten, Nucl. Phys. B160 (1979) 57; B227 (1983) 433.
4. G. t'Hooft, Nucl. Phys. B72 (1974) 461.
5. M.B. Wise and H. Sonoda, Caltech preprint, CALT-68-1157; G.V. Dunne and A.W. Thomas, Nucl. Phys. A446 (1985) 437.
6. Report of the 1985 and 1986 Summer Study Workshops, CEBAF Va.; TRIUMF KAON FACTORY proposal (1985); Los Alamos preprint LA-10720-MS (1986)

7. N. Isgur and G. Karl, Phys. Rev. D18 (1978) 4187; ibid. D19 (1979) 2653.
8. W. Weise, Prog. Part. Nucl. Phys. 11 (1984) 123.
9. C. De Tar and J.F. Donoghue, Ann. Rev. Nucl. and Part. Sci. 33 (1983) 235.
10. R. Friedberg and T.D. Lee, Phys. Rev. D15 (1977) 1694; R. Goldflam and L. Wilets, Phys. Rev. D25 (1982) 1951; R. Saly and M.K. Sundaresan, Phys. Rev. D29 (1984) 525; Th. Köppel and M. Harvey, Phys. Rev. D31 (1985) 171.
11. G.A. Miller et al., Comm., Nucl. Part. Phys. 10 (1981) 101; A.W. Thomas, Adv. Nucl. Phys. 13 (1984) 1; G.A. Miller, Int. Rev. Nucl. Phys. 2 (1984).
12. M. Rho et al., Phys. Rev. Lett. 51 (1983) 747; J. Goldstone and R.L. Jaffe, Phys. Rev. Lett. 51 (1983) 1518.
13. M.C. Birse and M.K. Banerjee, Phys. Lett. 136B (1984) 284.
14. H. Pagels, Phys. Rep. 16 (1979) 219.
15. S.J. Adler, Princeton preprint (1986).
16. W. Broniowski and M.K. Banerjee, U. Maryland preprint #85-99.
17. R.F. Alvarez - Estrada and A.W. Thomas, J. Phys. G9 (1983) 161.
18. E. Amaldi et al., Pion Electroproduction at low energy... (Springer, Berlin, 1979).
19. P.A.M. Guichon et al., Phys. Lett. 124B (1983) 109.
20. R. Machleidt et al., Phys. Rep., to appear (1986).
21. T.E.O. Ericson and M. Rosa-Clot, Nucl. Phys. A405 (1983) 497.
22. J. Gasser, Ann. Phys. 136 (1981) 62.
23. G. Chanfray and A.W. Thomas, U. Lyon preprint LYCEN/8574 (Nov. 1985).
24. Ref. (22), op. cit. p.85.
25. P.A.M. Guichon and G.A. Miller, Phys. Lett. 134B (1984) 32.
26. R. Perry and M. Rho, U. Washington preprint 40048-07-N6 (1985).
27. S. Théberge and A.W. Thomas, Nucl. Phys. A393 (1983) 252.
28. F.E. Close, An Introduction to Quarks and Partons (Academic, N.Y., 1979)); E. Leader and E. Predazzi, An Introduction to Gauge Theories and the New Physics (Cambridge, 1982).
29. A.W. Thomas, Phys. Lett. 126B (1983) 97; Prog. Part. Nucl. Phys. 11 (1984) 325.
30. A. Signal and A.W. Thomas, (1986) to be published.
31. R.L. Jaffe, Phys. Rev. D11 (1975) 1953; Nucl. Phys. B229 (1983) 205.
32. R.L. Jaffe and G.G. Ross, Phys. Lett. 93B (1980) 313.
33. R.P. Bickerstaff and A.W. Thomas, (1986) to be published.
34. D. Robson, Nucl. Phys. A308 (1978) 381.
35. G. Baym, Physica 96A (1979) 131.
36. F. Güttner and H. Pirner, CERN. TH. 4287 (1985)
37. G.T. Fairley and E.J. Squires, Nucl. Phys. B93 (1975) 56.
38. C. De Tar, Phys. Rev. D17 (1978) 323; (E) D19 (1979) 1028.
39. V. Matveev and P. Sorba, Nuovo Cim. 45A (1978) 257.
40. R.L. Jaffe and F. Low, Phys. Rev. D19 (1979) 2105.
41. P.J. Mulders, Phys. Rev. D25 (1982) 3039.
42. E.M. Henley et al., Phys. Rev. C28 (1983) 1277.
43. E. Lomon, A.I.P. Conf. Proc. 97 (1983) 79.
44. Y.E. Kim and M. Orlowski, Phys. Lett. 140B (1984) 275.
45. Yu. S. Kalashnikova et al., Zeit. f. Phys. A323 (1986) 205.
46. Yu. A. Simonov, Nucl. Phys. A416 (1984) 109.
47. D.A. Liberman, Phys. Rev. D16 (1977) 1542.
48. M. Oka and K. Yazaki, Phys. Lett. 90B (1980) 41; O. Morimatsu et al., N.P. A420 (1984) 573.
49. M. Harvey et al., Nucl. Phys. A424 (1984) 428.
50. M.B. Kislinger, Phys. Lett. 79B (1978) 474.

51. F. Wang and C.W. Wong, N.P. A438 (1985) 620; Yin He et al., Nucl. Phys. A448 (1986) 652.
52. K. Holinde, Phys. Lett. 157B (1985) 123.
53. Y. Suzuki and K.T. Hecht, Nucl. Phys. A420 (1984) 525.
54. K. Maltman and N. Isgur, Phys. Rev. D29 (1984) 952.
55. S. Furui and A. Faessler, Nucl. Phys. A397 (1982) 413.
56. Y. Yamauchi and M. Wakamatsu, Phys. Lett. 172B (1986) 161.
57. N. Isgur, Proc. Baryon 80 (U. Toronto Press, 1980).
58. Yin He et al., Nucl. Phys. A451 (1986) 653.
59. H.J. Pirner, Phys. Lett. 85 (1979) 190.
60. Y. Fujiwara and K.T. Hecht, Nucl. Phys. A444 (1985) 541; ibid (1986) to appear.
61. X.-H. Yang and Z.-Y. Ma, Peking University preprint, September 1985.
62. F. Lenz et al., SIN report 85-09.
63. J. Paton and N. Isgur, U. Oxford Preprint (1985).
64. L. Heller, in "Quarks and Nuclear Forces", Vol. 100, Springer Tracts in Modern Physics (Springer, Berlin, 1982) p. 145; P. Hasenfratz et al., Phys. Lett. 94B (1980) 401.
65. Th. Pfenninger and A. Faessler, Nucleon-nucleon interaction with a new model for confinement, Tubingen preprint (July, 1986).
66. C.B. Dover, Brookhaven preprint BNL 37502 (1986).
67. A.M. Green, Acta. Physica Austriaca, Suppl. XXVII (1985) 129; A.M. Green and J.A. Niskanen, Ann. Rev. Nucl. Physics 1 (1985).
68. S. Ahmad et al., Proc. LEAR Workshop, Tignes, France (1985).
69. A.P. Kobushkin, ITP-76-I45E (November 1976).
70. P.J. Mulders and A.W. Thomas, Phys. Rev. Lett. 52 (1984) 1199.
71. S. Brodsky and B. Chertok, Phys. Rev. D14 (1976) 3003; S. Brodsky in D.C. Fries and B. Zeitnitz, "Quarks and Nuclear Forces" (Springer, Berlin, 1982).
72. F. Gross, CEBAF Spectrometer Workshop, Williamsburg (Oct. 1983).
73. V.V. Burov et al., Zeit. Phys. A315 (1984) 205; F. Gross and F. Khanna, private communication.
74. L.S. Kisslinger, Carnegie-Mellon preprint (1986).
75. G.A. Miller and L.S. Kisslinger, Phys. Rev. D27 (1983) 1669; W.-Y. P. Hwang, contribution to CEBAF Summer Workshop (1986).
76. V.M. Dubovik and I.T. Obukhovsky, Zeit. Phys. 10C (1981) 123.
77. W.T.H. van Oers et al., Phys. Rev. Lett. 56 (1986) 2571.
78. G. Chanfray et al., Phys. Lett. 147B (1984) 249.
79. A.G. Williams and A.W. Thomas, Phys. Lett. 154B (1985) 320; Phys. Rev. C33 (1986) 1070.
80. L.R. Dodd et al., U. Adelaide preprint ADP-360/T26; ADP-356/T24, to appear Phys. Lett.
81. G. Chanfray and H. Pirner, Lyon preprint, LYCEN 8610 (1986).
82. G.A. Miller et al., Phys. Rev. Lett. 56 (1986) 2567.
83. R. Holzenkamp et al., contribution to Int. Conf. Few Body Physics (Tokyo, 1986); and to be published.
84. A. Yokosawa, Argonne preprint ANL-HEP-CP-85-93.
85. R.L. Jaffe, Phys. Rev. Lett. 38 (1977) 195.
86. M. Ericson, CERN preprint, CERN-TH. 4485/86.
87. F.J. Pineda, U. Maryland preprint (July 1986).
88. A. Chodos and C.B. Thorn, Phys. Rev. D12 (1975) 2733.
89. J. Achtzehnter, W. Scheid and L. Wilets, Phys. Rev. D32 (1985) 2414.
90. H.B. Nielsen and A. Patkos, Nucl. Phys. B195 (1982) 137.
91. W. Broniowski et al., Maryland preprint #86-081.
92. J.A. Nolen and J.P. Schiffer, Ann. Rev. Nucl. Sci. 19 (1969) 471; N. Auerbach, Phys. Rep. 98 (1983) 273.

93. V. Koch and G.A. Miller, Phys. Rev. C31 (1985) 602; J.M. Greben and A.W. Thomas, Phys. Rev. C30 (1984) 1021.
94. V.V. Burov et al., Z. Phys. A318 (1984) 67.
95. G. Karl, G.A. Miller and J. Rafelski, Phys. Lett. 143B (1984) 326.
96. J.J. Aubert et al., Phys. Lett. 123B (1983) 275.
97. R.G. Arnold et al., Phys. Rev. Lett. 52 (1984) 727.
98. P. Barreau et al., Nucl. Phys. A402 (1983) 515; Z.E. Meziani et al., Phys. Rev. Lett. 52 (1984) 1233; C. Marchand et al., Phys. Lett. 153B (1985) 29.
99. F.E. Close et al., Phys. Lett. 129B (1983) 727.
100. I. Sick, Nucl. Phys. A434 (1985) 677.
101. P.J. Mulders, NIKHEF preprint, P-8 (1986).
102. H. Stroth et al., Phys. Lett. 171 (1986) 339.
103. J.V. Noble, Phys. Rev. Lett. 46 (1981) 412.
104. M. Oka and R.D. Amado, Pennsylvania preprint UPR-0295-T (1986).
105. M. Oka et al., Penn. preprint, UPR-0289-T (1985).
106. C. Shakin, in Workshop on Nuclear Chromodynamics (World Sci., Singapore, 1986) p. 507.
107. T. Goldman and G.J. Stephenson Jr., Phys. Lett. 146B (1984) 143.
108. O. Nachtmann and H.J. Pirner, Zeit. Phys. C21 (1984) 277; F.E. Close et al., Phys. Rev. D31 (1985) 1004.
109. F.E. Close et al., Phys. Lett. 168B (1986) 400.
110. G.V. Dunne and A.W. Thomas, Phys. Rev. D33 (1986) 2061.
111. S.V. Akulinichev et al., Phys. Lett. 158B (1985) 727.
112. G.V. Dunne and A.W. Thomas, Nucl. Phys. A455 (1986) 701; S. Bass, E. Duff and A.W. Thomas, to be published.
113. R.P. Bickerstaff and A.W. Thomas, Adelaide preprint, ADP-352/T20 (1986).
114. J.M. Cavendon et al., Phys. Rev. Lett. 40 (1982) 978.
115. R.P. Bickerstaff et al., Phys. Rev. Lett. 53 (1984) 2532; M. Ericson and A.W. Thomas, Phys. Lett. 148B (1984) 191.
116. C.H. Llewellyn Smith, Phys. Lett. 128B (1983) 107; M. Ericson and A.W. Thomas, Phys. Lett. 128B (1983) 112.
117. E. Berger et al., Phys. Rev. D24 (1984) 398.
118. T.A. Carey et al., Phys. Rev. Lett. 53 (1984) 144.
119. W. Alberico et al., Nucl. Phys. A379 (1982) 429; LYON preprint LYCEN/8575 (1986); G. Chanfray, LYCEN/8641 (1986).
120. I. Lovas, Budapest preprint, KFKI-1986-05/A.
121. D. Sprung, "Determination of the Deuteron Mean Square Radius", McMaster preprint (1986).
122. E.V. Hungerford and L.C. Biedenharn, Phys. Lett. 142B (1984) 232.
123. H. Markum and M. Meinhart, Vienna preprint (1985); H. Pirner, invited talk at the CEBAF Workshop (July 1986); H. Matsuoka and D. Sivers, ANL-HEP-PR-85-89.

Inst. Phys. Conf. Ser. No. 86
Paper presented at Int. Nucl. Phys. Conf., Harrogate, UK, 1986

Electron scattering—from meson to quark degrees of freedom

Ingo Sick
Department of Physics
University of Basel
CH-4056 Basel, Switzerland

Meson degrees of freedom of nuclei are studied via exchange current contributions and the effects of a 3-body force, quark degrees of freedom via changes of form factors at large momentum transfer. The change of nucleons due to the nuclear medium is investigated.

1. Introduction

Nuclear physics traditionally has dealt with nuclei in terms of nucleons and the nucleon-nucleon interaction. Many concepts and models have been developed to deal with a strongly interacting many-body system. For light nuclei and nuclear matter, ab-initio predictions are fairly successful, while for nuclei $A > 4$ calculations of nuclear properties involve much phenomenological input.

The limitations of this nucleon-only approach become visible as soon as one seeks quantitative predictions; these very rarely can be obtained. For instance, the binding energy and density of nuclear matter are off by 20%. The short-range properties of nuclei, even for $A \lesssim 4$, are not understood. Some of these deficiencies can be cured by taking into account meson exchange currents, by introducing three-body forces that reflect the truncation of the space to nucleons only, etc. At this level, however, one again is forced to introduce considerable phenomenology.

Nuclear physics in the future can be expected to increasingly deal with the nucleus as a system of nucleons, mesons, deltas, ... and, on the more microscopic level, quarks and gluons. When treating explicitly these subnucleonic degrees of freedom we can hope to calculate the short-range properties, meson exchange currents, and eliminate phenomenological ingredients like a three-body force. We then can hope to understand the nucleus in terms of a fundamental theory, QCD.

The interest in QCD results from two distinct motivations: We want to understand the nucleus in terms of fundamental constituents and interactions; by eliminating (or deriving) the effective degrees of freedom we can hope for a more quantitative description. We want to find out whether QCD is indeed the correct theory; by exploiting the partial deconfinement of quarks that occurs in nuclei, we can investigate the most intriguing aspect of QCD, confinement, and the strong-interaction limit.

In these studies of nuclei in terms of QCD, the electromagnetic probe plays an important role. Quark effects will be particularly important at short range, and these short distance phenomena can be probed at large momentum transfer q. Contrary to more strongly interacting probes, the relation between range and transfer, $\Delta x \sim 1.5/q$, is not destroyed by multistep reactions. At short range we can best increase the ratio of signal to "noise", i.e. the ratio of quark to nucleonic effects. Accordingly, already now mesonic and quark degrees of freedom do play an important role in electromagnetic observables.

In this talk, I want to give a number of examples that deal with the question of mesonic and quark degrees of freedom of nuclei. I should emphasize that some of the ideas presented are still exploratory, or even speculative. When discussing the non-nucleonic degrees of freedom, I will concentrate on A = 2-4, simply because for these nuclei the nucleonic "background" can be calculated with good accuracy. In this case, we have the best chance to identify genuine quark effects.

2. Meson exchange currents

When addressing the question of meson exchange currents (MEC), I will restrict the discussion to the three-body system, ^{3}He and ^{3}H. For these nuclei, we can calculate the nucleonic sector "exactly". For 3 nonrelativistic nucleons bound by a modern nucleon-nucleon interaction (Paris, Reid soft core, ..), the Schrödinger equation can be solved numerically. Different ways to calculate the wave function agree within differences that are much smaller than the ones to experiment. These differences to experiment then can teach us something on non-nucleonic degrees of freedom. In addition, the 3-nucleon system is the only case where both partners of an isospin doublet are (almost) stable; the separation of form factors into their T = 0/T = 1 components is very valuable given the different sensitivity to ingredients of the calculation.

This separation into T = 1,0 previously was not possible since for ^{3}H no data of quality comparable to the ones[1] for ^{3}He were available. An experiment on ^{3}H is difficult, for obvious reasons. For electron scattering at large momentum transfers one needs thick targets (grams of tritium) and intense beams (tens of μA); such beams can weld holes, and pose safety problems. These difficulties have been overcome by an experiment we performed[2] in Saclay.

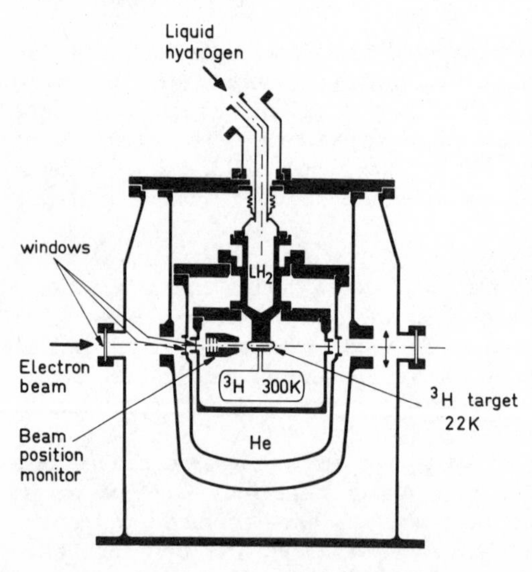

Fig. 1 Tritium target used in Saclay experiment[2].

The Saclay experiment used electrons up to 700 MeV and 25μA, and a 10K-Curie target shown in Fig. 1. The target, containing 5cm^3 of liquid ^{3}H, is surrounded by 3 containment vessels with thin windows, and a transport container which is closed in the presence of personnel. The integrity of the containment vessels is permanently tested via the leak rate of helium present between vessels 2 and 3. Two microprocessors continuously monitor the operating conditions of the target, and each other.

As an example of the data obtained in this experiment, Fig. 2 shows the magnetic form factor. The data show a clear diffraction minimum of $q^2 \simeq 23\text{fm}^{-2}$; the diffraction maximum could only be partly explored due to the limited maximum electron energy available.

The calculations shown in Fig. 2 have been performed by Hajduk et. al.[3]. The nucleon wave function is obtained by solving the Faddeev equations for the Paris potential. The resulting impulse approximation (IA) form factor, the dashed line, strongly deviates from experiment. In order to account for the longest range mesonic degrees of freedom, the MEC diagrams of Fig. 3 have been included. The resulting form factor (dotted line Fig. 2) agrees much better with experiment. The solid curve differs by terms of relativistic order (q^2/M^2), which are not consistently treated in the MEC calculation. The use of G_e for the electromagnetic vertex form factor respects current conservation, while F_1 does not. Calculations such as the one shown in Fig. 2 still suffer from a number of other ambiguities such as the choice of the πNN vertex form factors and the treatment of shorter-range MEC.

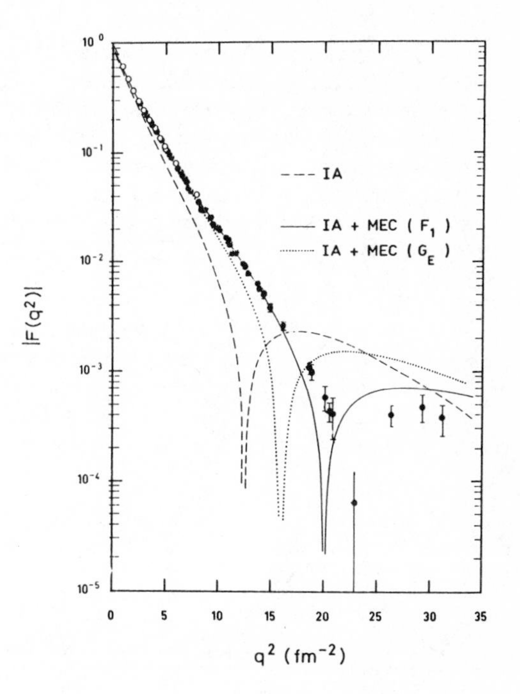

Fig. 2 ^{3}H magnetic form factor[2] compared to calculation of Hajduk and Sauer[3].

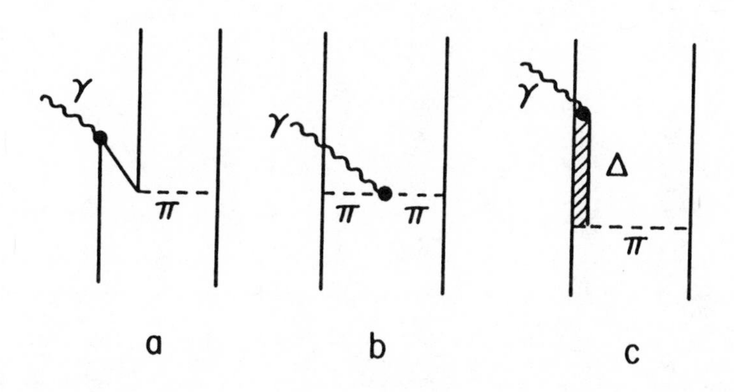

Fig. 3 Exchange current diagrams.

During the past year significant progress has been made in the calculation of these exchange currents. Modern NN potentials are based on the meson exchange picture, and the parameters governing these exchange processes have been fitted to NN scattering data. Identifying in V_{NN} the π-like and ρ-like exchange terms allows to derive[4] these exchange terms. Thus a consistent calculation can be done where both the nucleonic wave function and the MEC are based on the nucleon-nucleon interaction. This eliminates a number of ambiguities.

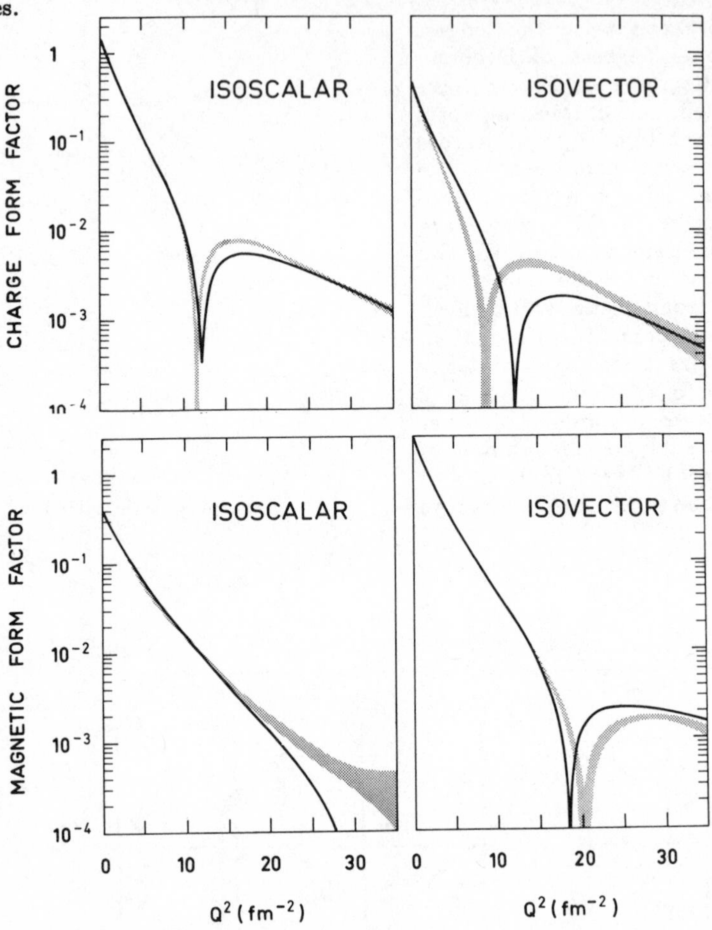

Fig. 4 Experimental A = 3 form factors[5] (hatched area) component to calculation of Hajduk and Sauer[3].

In order to give a more complete representation of the A = 3 form factors, we have separated the charge and magnetic form factors into their T = 0,1 components[5]. The data (Fig. 4, dotted band) is compared to the calculations of the Hannover group[6] which include MEC. A detailed comparison shows that the largest differences occur in those observables where the MEC contribution is largest. We thus may suspect that MEC are not accounted for correctly. Remaining uncertainties include the relativistic terms mentioned above, and the Δ-diagram which does not follow directly from V_{NN}. At large q, degrees of freedom other that π, Δ, $\bar{N}$ must be expected to play a role.

3. Three-body force

Under this heading I want to discuss another area of mesonic degrees of freedom of nuclei. As long as we treat the full spectrum of nuclear constituents, N, Δ, π,.. only two-body forces occur. As soon as we truncate the space to e.g. nucleons only, we need to include three-body forces (3BF). The diagram with the largest range responsible for a 3BF is shown in Fig. 5.

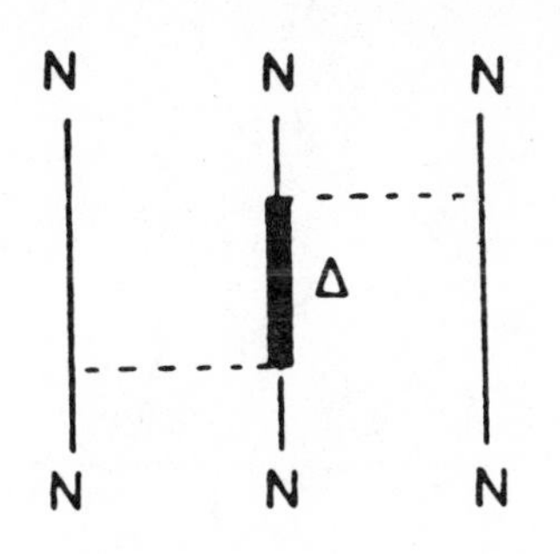

Fig. 5 Dominant diagram responsible for three-body force.

During the past year several calculations that include the 3BF have been published[7]. Using the Tucson-Melbourne force together with one of the standard 2BF in a Faddeev calculation gives a remarkable result: the binding energy of the A = 3 system is increased by ~ 1.5 MeV. This seems to provide the ~ 1 MeV that has been missing in the nucleons-only calculations. The 3BF also leads to a small improvement of the charge form factor[8]. This is related to the appearance of a central depression in the point density, a feature derived[9] some time ago by Fourier transforming the experimental ^{3}He charge form factor (corrected for the contribution of MEC). The tendency toward a central depression is show in Fig. 6 which displays the result of a Faddeev calculation of the Los Alamos group[10].

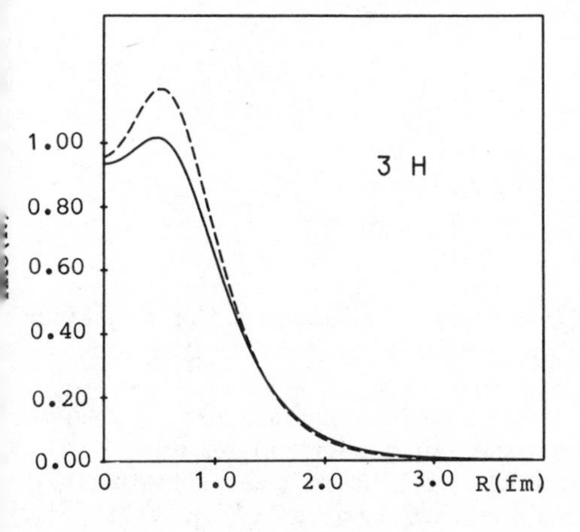

Fig. 6 ^{3}H point proton density without (solid) and with (dashed) 3BF.

These calculations show that π, Δ degrees of freedom are very important for a quantitative understanding of nuclei. For ^{4}He, equivalent calculations provide an increase of the binding energy of ~ 6 MeV. For nuclear matter, changes of 20% are obtained with similar 3BF; the Tucson-Melbourne force actually overbinds nuclear matter to the point of collapse. This indicates that at very short range the π, Δ degrees of freedom still are difficult to treat using a 3BF.

Some of the difficulties that occur can be identified by comparing with the calculation of the Hannover group[3]. In these calculations pions and deltas initially are kept in the wave function. Pions as explicit constituents are shown to play a minor role, in which case the A = 3 nuclei can be treated using the coupled channel N, Δ Faddeev approach. Such a calculation requires that the NN interaction be redetermined by allowing the same Δ degrees of freedom in NN scattering.

The results of this calculation give a much smaller effect of the π, Δ degrees of freedom than the 3BF used above to simulate their effect. The increase in the binding energy is 0.3 MeV only, partly because dispersion effects are included, partly because the repulsive contribution of ρ-exchange is accounted for as well. Again, this calculation finds a tendency of the point density towards a central depression. The density of Δ's, which exhibits a pronounced depression (Fig. 7), polarizes the nucleonic density.

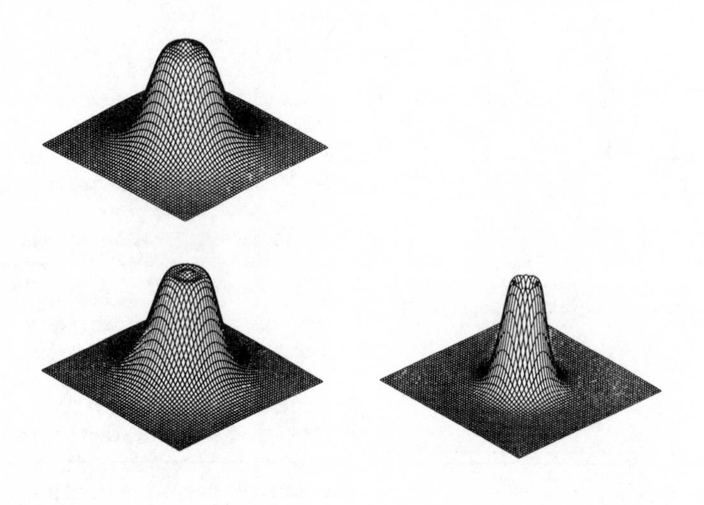

Fig. 7 ^{3}H point proton density[3] wit hout (top) and with (bottom left) π,Δ degrees of freedom. Δ-density is shown on lower right.

It will be a task for the future to completely understand the difference between the two approaches to include the π and Δ degrees of freedom. A treatment via a 3BF obviously can be generalized to A > 3 much more easily. One therefore will have to "calibrate" this procedure in the A = 3 system where an explicit treatment of π, Δ in the wave function is still feasible.

The 3BF related to the diagram shown in Fig. 5 has a clear signature: it energetically prefers the triangular over the linear arrangement of nucleons (Fig. 8). This is the feature that leads to the central depression in $\rho_p(r)$. It is obvious that such a change of the wave function will be most clearly visible when looking at $l \neq 0$ components of the wave function. Only for $l \neq 0$ can the spatially different arrangement of 3 nucleons (Fig. 8) be differentiated. This point emphasizes that in the future it will be important to study $l \neq 0$ states, and D-states (which are preferably coupled in N-Δ transitions) in particular.

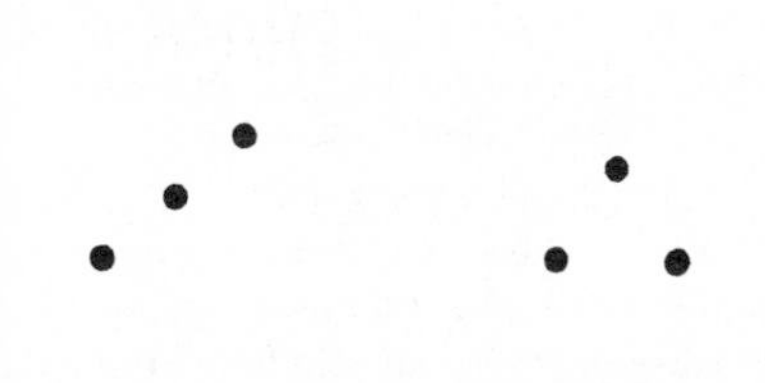

Fig. 8 Triangular and aligned configuration of 3 nucleons.

The D-states in the A = 3 system have a small probability (~ 8%), and are therefore difficult to isolate. It can be done, however, as show by Jourdan, et. al.[11]. The tensor polarization observable T_{20} in $p + \vec{d}$ radiative capture at E ~ 15 MeV selectively picks out the D-state wave function, and allows to measure its medium-range behavior without complications due to MEC. If the p + d system is treated exactly, as is done by Jourdan, et. al., who perform a coupled channel Faddeev continuum calculation, T_{20} can be used to give accurate information on the D-state.

The lesson from these explorations of the effects of the 3BF: the large effects obtained from the truncation of space indicates that with a nucleon-only picture of nuclei we cannot expect to achieve a quantitative description. It appears that the 3BF gives contributions of the right sign and magnitude to explain some of the longstanding discrepancies in the A =3,4,∞ nuclei. Even at low energy and momentum transfers one cannot always neglect the mesonic degrees of freedom, a fact to be kept in mind when we consider the role of quark degrees of freedom.

4. Quark degrees of freedom

These degrees of freedom of the nucleus are fairly well hidden, given the confining nature of QCD. In order to see them, we will have to look at the nucleus with a spatial resolution Δx appropriate for the degree of freedom. Present experiments on light nuclei typically reach a momentum transfer of $5fm^{-1}$, i.e. a resolution of $\Delta x \simeq 1.5/q = 0.3fm$. At this resolution we are in the region where quark degrees of freedom may play a role (the nucleon radius in ~ 0.8fm). In order to get quarks to dominate the observables, we will need significantly larger transfers.

A number of calculations have been performed in order to find how to best describe quarks in nuclei. They seek a framework which conserves as much as possible from QCD, and which is simple enough to treat many-quark systems. At the same time, the calculations seek for the observables where quark effects are best visible.

The magnetic form factor B(q) of the deuteron turns out to be reasonably sensitive to quark contributions, the nucleonic wave function can be calculated with confidence, and the standard MEC are small since mostly of T = 1 nature. Below, I discuss three representative examples of calculations that have been done for B(q).

Nyman and Riska[12] use the Skyrme model, in which the nucleons are formed as topological solitons of pions and scalar meson fields. This model is used to calculated the exchange current contribution to B(q), a quantity that is poorly determined in the classical MEC calculations which provide little guidance for the isoscalar transverse current. For the impulse approximation part, the standard form factor calculated for the Paris NN interaction is used.

Kisslinger and collaborators[13] use a hybrid quark cluster model, where the nucleonic and quark degrees of freedom are allowed for in different regions of space. For nucleon separations $r > r_o$, the standard nucleonic wave function, NN interaction and pionic exchange currents are used. For $r < r_o$, the nucleons are considered to overlap so much that a description in terms of a six quark bag is more appropriate. For the description of the 6q-bag a variant of the MIT bag model is employed.

Yamauchi et. al.[14] use also a hybrid quark cluster model. The coexistence of quarks and nucleons is treated via the resonating group technique, which allows to antisymmetrize the quark wave functions in the two regions $r \gtrless r_o$. From past studies of light nuclei ($A \simeq 6$) in terms of both d,α clusters and nucleons, we know that the resonating group approach is a powerful tool in describing systems in terms of two different sets of constituents.

The results of above calculations are shown in Figs. 9-11. The impulse approximation contribution of Nyman and Riska (Fig. 9) fails to explain the data at large q; the exchange contribution obtained from the Skyrme model provides an increase of B(q) at large q. Near perfect agreement with the data is obtained if the isoscalar nucleon form factor rather than the Skyrme model is used to determine the chiral angle.

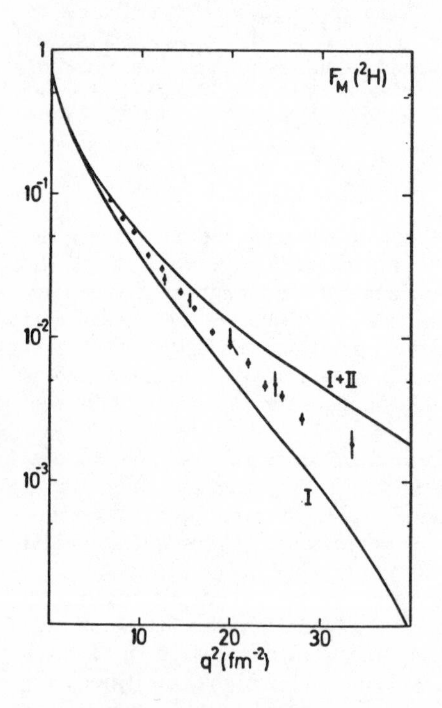

Fig. 9 Deuteron magnetic form factor, with impulse approximation (I) and exchange contribution (I + II) (ref. 12).

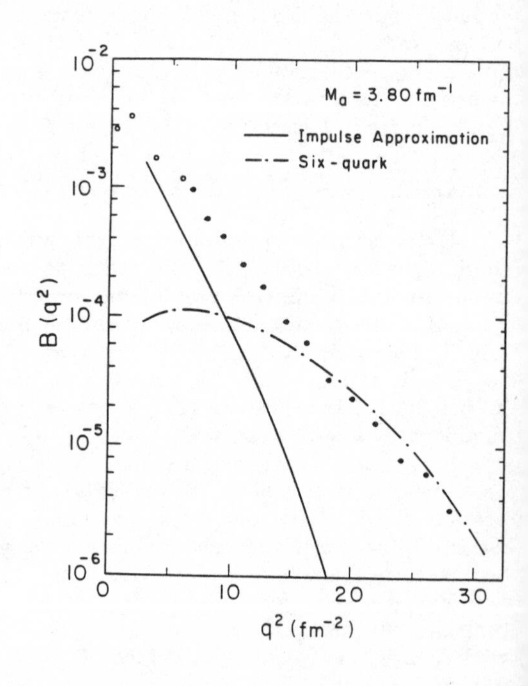

Fig. 10 Deuteron magnetic form factor, with calculation[13] of IA (solid) and six-quark contribution (dashed).

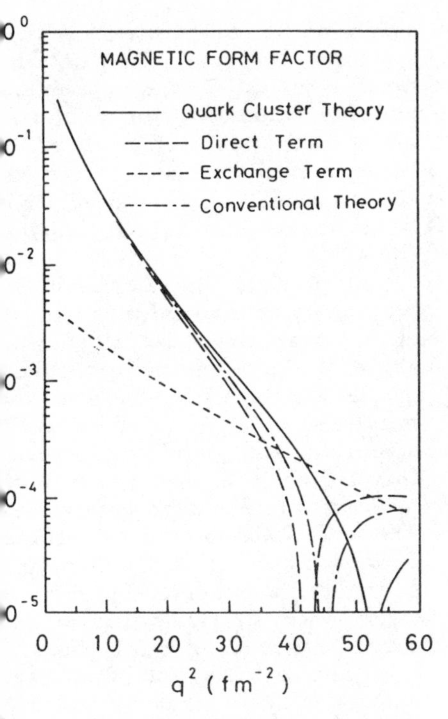

Fig. 11 Deuteron magnetic form factor, with calculation of Yamauchi, et. al.[14].

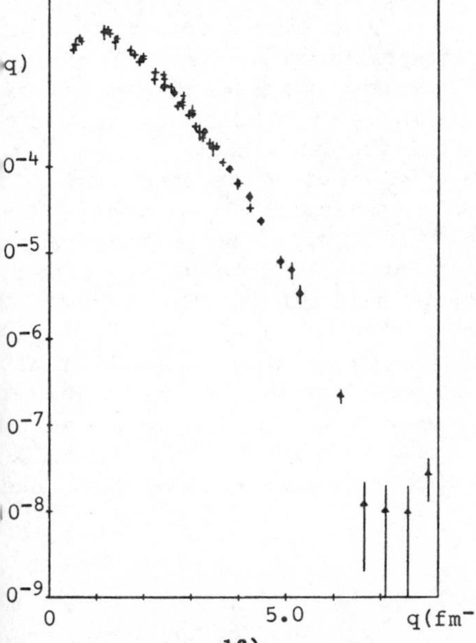

Fig. 12 New Data[16] on deuteron B(q) of Auffret et. al. (dots) and Arnold et. al. (triangles).

The hybrid model calculations of Kisslinger, performed with parameters taken from fits to the A = 2,3 charge form factors, predicts a large contribution of quarks at large q, and a large 6 quark percentage (> 5%) in the wave function. Both quantities depend sensitively on the choice of r_o.

The resonating group calculation of Yamauchi et. al. predict a much smaller effect of quarks, as (approximately) shown by the "exchange" terms in Fig. 11. Due to the antisymmetrization of the quark wave functions this calculation allows to better identify the genuine 6q-contribution. This calculation predicts a shift of the diffraction minimum of B(q) to 7.3fm^{-1}, a feature observed by recent experimental data[16] (Fig. 12), which show a minimum at 7.2fm^{-1}. A similar shift due to quark exchange was found in the resonating group calculation of Chemtob and Furui[15].

The above studies of quark effects on B(q) should be taken with a grain of salt. Past calculations[17] have shown that B(q) receives important contributions from the ΔΔ configuration in the deuteron wave function. This is emphasized again in the contribution of Lomon et. al.[18], who obtain a perfect fit to B(q) when adjusting ΔΔ percentages and the $\pi\rho\gamma$ coupling constant. It would be very desirable to identify in the quark model calculations the term corresponding to a ΔΔ configuration, and to account for it explicitly. This term probably is better associated with the longer-range π exchange terms, while the genuine quark effects presumably are associated mostly with the shorter-range processes.

What should we conclude from the above discussion of quark effects on B(q)? At present, calculations still are exploratory, and it is not clear whether the deuteron magnetic form factor is indeed the best observable,

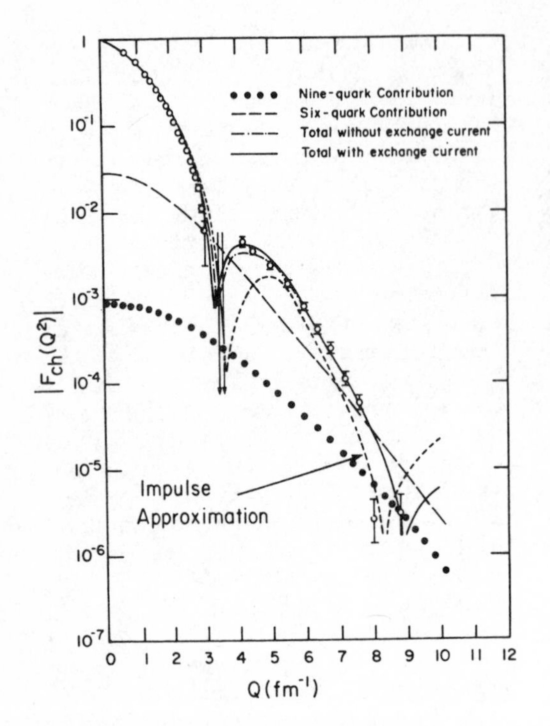

Fig. 13 ^{3}He charge form factor, with resonating group calculation[19]; nucleonic (short-dash) and six-quark (long-dash) contributions.

simply because the deuteron is such a dilute system. ^{3}He or ^{4}He would seem to be more appropriate, and the resonating group cluster model calculation of Kim et. al.[19] (done in a spirit similar to the one of Yamauchi et. al.) indeed finds large effects due to 6q clusters. (Fig. 13).

It is clear that at present we do not have available a set of electromagnetic observables that extend to large enough q to make quarks dominate over nucleons and mesons. Only a GeV CW accelerator will allow us to explore this region at large q, for both inclusive and exclusive observables. The 4 GeV CW accelerator CEBAF to be built in the US will permit us to push experiments into a region where we can expect to find cleaner signatures of quark effects.

The calculations discussed above deal with light nuclei only. The question of quarks in heavy nuclei remains to be explored. It is not yet clear how to deal with the problem of a nucleus where > 90% of the wave function is described in terms of nucleons. Standard models assume that quark degrees of freedom become active in short-range NN interactions, basically. More radical proposals, like the one of Pirner and Nachtmann[20], postulate quarks to be confined only as long as studied with bad spatial resolution. At good spatial resolution total deconfinement of the quarks could occur, with modes that extend over the entire nucleus. This (extreme) model would allow for N-clusters near the Fermi surface (where experiments do indeed observe nucleons), but deeply bound "nucleons" could be dissolved into their quark content.

The concept of how to deal with different layers of constituents in the nucleus is not yet very clear. It is not obvious how to distinguish nuclear properties attributable to specific constituents, and it is not even clear whether in a nucleus the notion of a "nucleon" identical to a free nucleon is tenable. This last question has received considerable attention during the past years, and is discussed below.

5. Nucleon size in medium

Nucleons in nuclei are separated by an average distance that exceeds the nucleon diameter by 20% only. Nucleons could overlap a fair fraction of the time, and could change their shape and size. Given the small average distance of nucleons, it actually would be astonishing if nucleons would remain inert in the medium.

Various pieces of evidence for an increase in nucleon size due to the nuclear medium have been advanced:

- The EMC experiment[21] finds that the momentum distribution of quarks in nucleons is changed due to binding in nuclei. At large momenta, x ~ 0.7, the density is reduced. Based on the uncertainty principle, this is interpreted as an increase in confinement volume. One particular model for that is a general increase of the bound-nucleon size.

- The q-dependence of the (e,e´p) cross section on ^{12}C, at fixed initial and final proton momentum, shows a small deviation from the q-dependence of the free e-p cross section[22]. An increase of the bound-nucleon size could be responsible.

- The longitudinal response function in quasielastic electron-nucleus scattering is smaller than expected[23]. The integral over $S_L(q,\omega)$, which at $q >> 2k_F$ should approach $2 \cdot \sigma_{ep}$, is too small by 20-40%. Taken at face value, this result could be interpreted as being due to a smaller σ_{ep}, again due to an increase of nucleon size due to binding.

None of the above observations provides clear evidence, however. There are many interpretations of the EMC effect (see talk of K. Rith). The interpretation of (e,e´p) ignores processes like (e,e´n) plus subsequent (n,p) charge exchange, which could produce effects of the size observed. The longitudinal response function is measured at too small momentum transfer, $q < 2\ k_F$, where the sum rule does not hold and where FSI and Pauli blocking is not negligible. In addition, the data show a somewhat unphysical behavior, a response function that goes to zero too quickly at large energy loss ω; this indicates the presence of systematic errors, or difficulties with radiative corrections at low final electron energy. Adding the different reasons for reduction - 10% for FSI, 10-15% for short-range NN correlations that move strength to very large ω, 10-20% for experimental problems - easily accounts for the lack of strength observed.

A number of theoretical studies of medium effects have been published. The soliton model of Shakin and collaborators[24] yields a ~ 20% increase of the nucleon charge radius, the Skyrmion model of Oka[25] predicts a decrease of the nucleon radius. Exploiting the link between total photon cross sections and polarizability, Ericson et. al.[26] interpret the increase of the gamma-nucleus cross section - commonly attributed to quasideuteron mechanism related to MEC - to nucleon intrinsic properties.

From the above studies, no clear picture emerges on what to expect. The work of Amado et. al.[27] provides a more general result. A quantum mechanical system, described by a fairly general Hamiltonian, shows an increase in size if placed in an attractive external field. This increase depends on the depth of the potential, and the excitation energy of the monopole resonance. Inserting numbers for nuclear binding energies and the energy of the Roper resonance yields a radius increase of ~ 2%. This change is small compared to the values of ~ 20% advocated in some of the models discussed above.

The cleanest experimental evidence presently available comes from inclusive electron scattering at large q, analyzed in terms of y-scaling[28,29]. The process considered is again quasielastic electron scattering off a nucleon in the medium.

In the limit of $q \to \infty$ the inclusive cross section $\sigma(q,\omega)$ is expected to scale, $\sigma(q,\omega)/\sigma_{ep}(q)\ d\omega = F(y)dy$, i.e. depend on a single variable y only, and not on q,ω separately. The physical meaning of y is the component of the nucleon momentum parallel to q, F is the momentum distribution. This scaling is derived using the kinematics of the e-p scattering only.

Scaling gives information on the reaction mechanism. If processes like MEC, FSI dominate the cross sections do not scale. In the region where these processes give small contribution (typically $y < o$), one can exploit the scaling property to learn something on the q-dependence of σ_{ep}, hence nucleon size. Use of an improper q-dependence of $\sigma_{ep}(q)$ leads to a q-dependence of F(y), i.e. loss of scaling.

The use of scaling to determine the q-dependence of the constituent form factor is analogous to the use in deep inelastic electron-nucleon scattering. The q-dependence of the constituent form factor derived via x-scaling provides the most convincing evidence for the presence of point-like constituents, quarks, in nucleons.

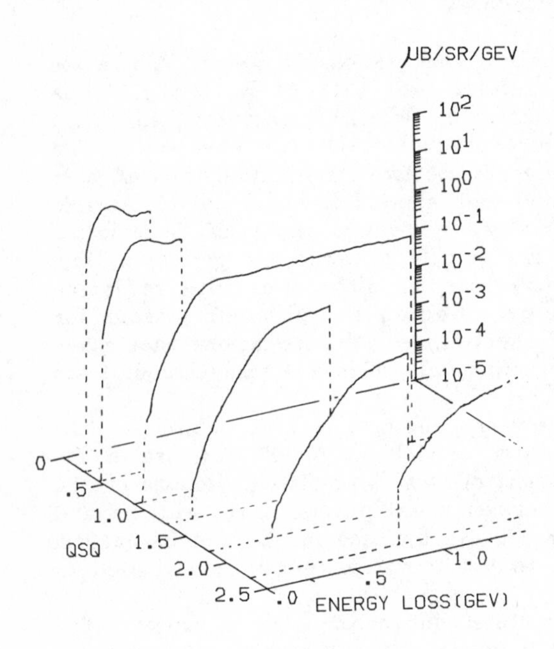

Fig. 14 Cross sections for $^{56}Fe(e,e)$ (drawn as continuous line) as function of q^2 and ω.

In a recent experiment[30] at NPAS we have measured inclusive data for nuclei A = 4-197 and large $q = 2\text{-}12\text{fm}^{-1}$. Fig. 14 gives an impression of the data for iron, which cover many orders of magnitude in σ at a given y. This data is measured at small scattering angles, 15-39°, and the cross section receive contributions from both charge and magnetization density. Plotting the same data in terms of the scaling variable (Fig. 15) leads to spectacular scaling; for $y < o$ the data define a unique curve.

The scaling of Fig. 15 is obtained using for σ_{ep} the q-dependence of the free nucleon cross section. A change of nucleon size, and q-dependence of σ_{ep}, leads to a different scaling behavior. In order to give a quantitative measure of the quality of scaling, we have fitted F(y) with a flexible parametrization. The χ^2 of this fit determines how well all the data defines a unique curve.

Fig. 16 shows χ^2 as a function of the assumed change of the nucleon radius. The minimal value of χ^2 is found for a radius 2 ± 20% larger than the free nucleon radius; the error bar includes an estimate

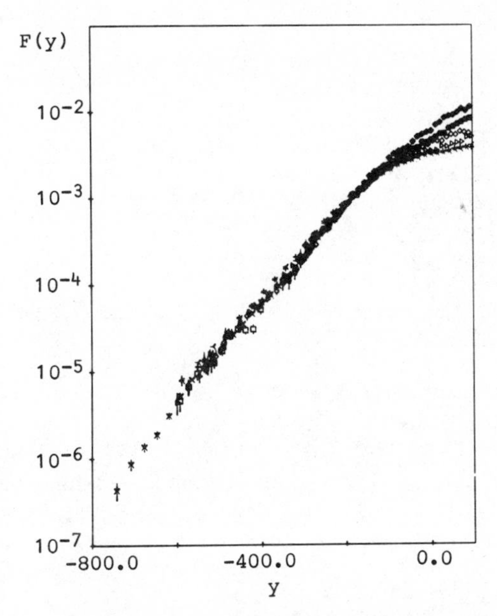

Fig. 15 Scaling function F(y) as a function of y for the Fe(e,e) data of Fig. 14.

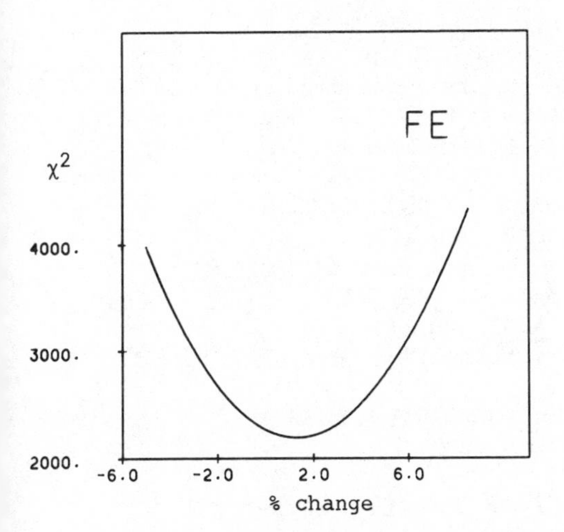

Fig. 16 Chi-squared of fit to F(y) as function of assumed change of nucleon radius.

for lack of convergence of F(y) (since $q \neq \infty$) and nucleon form factor uncertainties.

From Fig. 16 we conclude that the nucleon size changes little as one puts nucleons into the medium. A general change of size probably is not a useful concept anyway. As long as nucleons are far apart from each other, they have little reason to change. As they enter short-range NN collisions, a description in terms of 6-quark system (of larger confinement volume) is more appropriate.

6. Outlook

From the above discussion it becomes clear that at present our knowledge on quark degrees of freedom of nuclei still is very limited. We are still looking for more reliable ways to treat nuclei in terms of quarks while conserving as many features of QCD as possible. We do need to increase the momentum transfer in experimental observables, both inclusive and exclusive, such as to enhance the ratio of quark to nucleonic and mesonic effects. Once we have facilities like the CEBAF 4GeV/CW accelerator, we can measure these observables. We then are in a better position to address one of the major questions for nuclear physics of the future, the understanding of nuclei in terms of fundamental constituents and interactions.

References

1) J.S. McCarthy, I. Sick, R.R. Whitney, Phys. Rev. C15 (77) 1396.
R.G. Arnold et. al., Phys. Rev. Lett. 40 (78) 1429.
P.C. Dunn et. al., Phys. Rev. C27 (83) 71.
C.R. Ottermann et. al., Nucl. Phys. A436 (85) 688.
J.M. Cavedon et. al., Phys. Rev. Lett. 49 (82) 986.
2) F.P. Juster et. al., Phys. Rev. Lett 55 (85) 2261.
3) C. Hajduk, P.U. Sauer, Nucl. Phys. A369 (81) 321.
P.U. Sauer, Prog. Part. Nucl. Phys. 16, 35.
4) D.O. Riska, Physica Scripta 31 (85) 471.
A. Buchmann, W. Leidemann, H. Arenhovel, Nucl. Phys. A443 (85) 726.
5) J.M. Cavedon et. al., to be published.
6) P.U. Sauer, priv. com.
7) R.B. Wiringa et. al., Phys. Lett. 143B (84) 273.
C.R. Chen et. al., preprint LA-UR 85-1472.
S. Ishikawa et. al., Phys. Rev. Lett. 53 (84) 1877.
8) I. Sick, Proc. Conf. Three-Body Int. in Three-Body System, Washington, 1986.
9) I. Sick, Lect. Notes in Physics, 82 (78) 236
10) J. Friar, priv. com.
11) J. Jourdan, et. al., Nucl. Phys. A453 (86) 220.
12) E.M. Nyman, D.O. Riska, preprint HU-TFT-85-47
D.O. Riska, Proc. CEBAF Workshop, 1986, to be published.
13) L.S. Kisslinger, Phys. Lett. 112B (82) 307.
T.S. Cheng, L.S. Kisslinger, to be published.
14) Y. Yamauchi et. al., Nucl. Phys. A443 (85) 628.
15) M. Chemtob, S. Furui, Nucl. Phys. A454 (86) 548.
16) S. Auffret, et. al., Phys. Rev. Lett. 54 (85) 649
R.G. Arnold, et. al., Contr. paper to this conference.
17) For review see T.W. Donnelly, I. Sick, Rev. Mod. Phys. 56 (84) 461.
18) E. Lomon, et. al., Contribution to this conference.
19) M.A. Maize, Y.E. Kim, Phys. Rev. C31 (85) 1923.
20) O. Nachtmann, H.J. Pirner, preprint HD-THEP-84-1
21) J.J. Aubert, et. al., Phys. Lett. 123B (83) 123.
22) G. van der Steenhoven, et. al., Phys. Rev. Lett. 57 (86) 182
23) R. Altemus, et. al., Phys. Rev. Lett. 44 (80) 965.
R. Barreau, et. al., Nucl. Phys. A402 (83) 515
24) L.S. Celenza, A. Rosenthal, C.M. Shakin, Phys. Rev. Lett. 53 (84) 892.
25) M. Oka, Phys. Lett. 165B (85) 1.
26) M. Ericson, M. Rosa-Ust, preprint CERN-TH-4420/86
27) M. Oka, R.D. Amedo, preprint UPR - 0295-T
28) I. Sick, Phys. Lett. 157B (85) 13
29) I. Sick, D. Day, J.S. McCarthy, Phys. Rev. Lett. 45 (80) 871
30) D. Day, et. al, to be published.

Inst. Phys. Conf. Ser. No. 86
Paper presented at Int. Nucl. Phys. Conf., Harrogate, UK, 1986

The EMC effect

Klaus Rith

Max-Planck-Institut für Kernphysik, D-6900 Heidelberg, FRG

1. Introduction

The challenge for todays nuclear physics is to develop a fundamental theory of nuclei and of nuclear forces based on quarks, gluons and their interactions.

In this connection there are a lot of basic questions to be answered, which are partially illustrated in fig. 1: are the properties of the nucleon, like its mass or size, affected by the presence of other nucleons in a nucleus? Do nucleons swell, but still keep their quark and gluon content? Do they (or at least some of them) overlap and form multiquark clusters or even one big bag, where there is free colour flow and the quarks can no longer be assigned to individual nucleons while due to the strong colour forces it is not possible to liberate them from free hadrons? Are the confinement conditions in a nucleus different from the free nucleon case and could therefore the nucleus be used as a laboratory to study and to try to understand confinement?

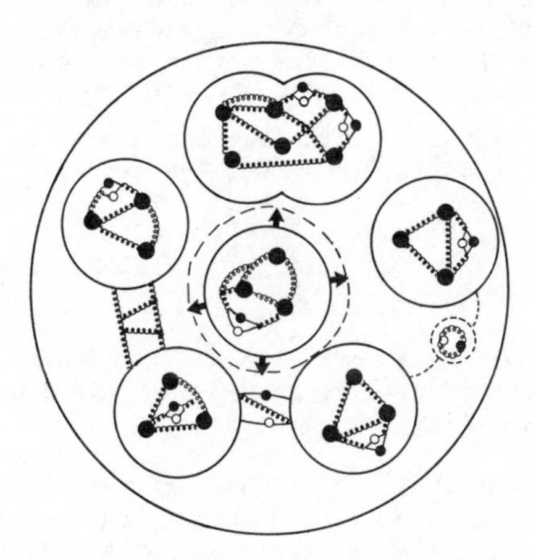

Fig. 1

A speculative view of the nucleus

Furthermore, how are these colourless objects bound together? Is it by the exchange of other colourless composite objects like pions or generally speaking meson exchange currents, a mechanism which is currently not understood in the framework of QCD? Can at least the short range part of the nuclear forces be directly described by the fundamental strong force between the fundamental particles, that means by the exchange of quarks and gluons or some multigluon states? Can QCD be extended into a region where the strong coupling constant is large and is it possible to find a unique description of the strong colour forces and the nuclear forces?

2. The Tool

In my opinion the best way to attack these questions is to look at quark and gluons directly in deep inelastic lepton-nucleus scattering experiments, which allow to probe the nucleus at distances which are two to three orders of magnitude smaller than the nucleon size, and to study the modifications of quark and gluon distributions $q(x,Q^2)$ and $g(x,Q^2)$ due to the nuclear environment.

For charged leptons these are obtained from the measurement of the structure functions $F_1^N(x,Q^2)$ and $F_2^N(x,Q^2)$ per nucleon. In the quark parton model F_2^N is given by

$$F_2^N(x,Q^2) = x\sum_f z_f^2(q_f(x,Q^2) + \bar{q}_f(x,Q^2))$$

where the sum runs over the different quark flavours f, z_f is the charge of the quark (in units of $|e|$) and $q_f(x,Q^2)$ ($\bar{q}_f(x,Q^2)$) is the probability that a quark (antiquark) of flavour f carries the fraction x of the nucleon light cone momentum. Seaquarks contribute to F_2^N only at values of x below 0.35–0.4. x is given by $x=Q^2/2M\nu$, where Q^2 is the negative square of the four-momentum carried by the virtual photon exchanged between lepton and nucleon, M is taken to be the proton mass and $\nu=E-E'$ is the energy transferred by the virtual photon from the lepton to the nucleon.

F_1^N and F_2^N are linked by $R=\sigma_L/\sigma_T$, the ratio of the longitudinal to transverse virtual photon cross sections, via the expression

$$2\,x\,F_1^N(x,Q^2) = F_2^N(x,Q^2)(1+\frac{4M^2x^2}{Q^2})/(1+R(x,Q^2))$$

Apart from QCD and quark mass corrections R is equal to zero if the virtual photon interacts with a point-like spin-½ particle and equal to infinity for scattering on point-like spin-0 objects.

It should be pointed out that only if both structure functions show the same dependence on the nuclear mass number A, or equally if R is independent of A, then the ratio of cross sections for two different nuclei is equal to the corresponding structure function ratio.

The gluon distribution $g(x,Q^2)$ can only be obtained indirectly from an analysis of the Q^2 evolution of the structure functions or for example from the production of open or closed charm (see section 3.1.2).

Nuclear effects in F_2^N were first observed by the European Muon Collaboration (EMC) at CERN (Aubert et al 1983), therefore the name EMC effect, and then by a series of other deep inelastic electron, muon and neutrino experiments. In this contribution I will discuss the main features of the existing data and some implications due to recent results and will briefly summarise the large variety of possible theoretical explanations.

3. The Data

3.1 EMC Results

3.1.1 Quark distributions

Figure 2 shows the original data of the EMC effect, the ratio of the nucleon structure functions for iron, $F_2^N(Fe)$, and deuterium, $F_2^N(D)$, plotted against x, together with the expectation of previous Fermi motion model calculations (Bodek and Ritchie 1981). The data have been taken at muon energies between 200 and 280 GeV. The mean Q^2 of the datapoints at low x is about 16-20 GeV^2, it increases to about 80 GeV^2 for x=0.65. The shaded area indicates the range for the error on the slope of a straight line fit to the data, the systematic errors for each point are somewhat larger. In addition there is an overall normalisation uncertainty of ±7%. It is obvious that these errors are too large for detailed conclusions about the underlying physics and that dedicated simultaneous measurements on pairs of nuclei are needed to reduce normalisation errors and systematic errors to a minimum.

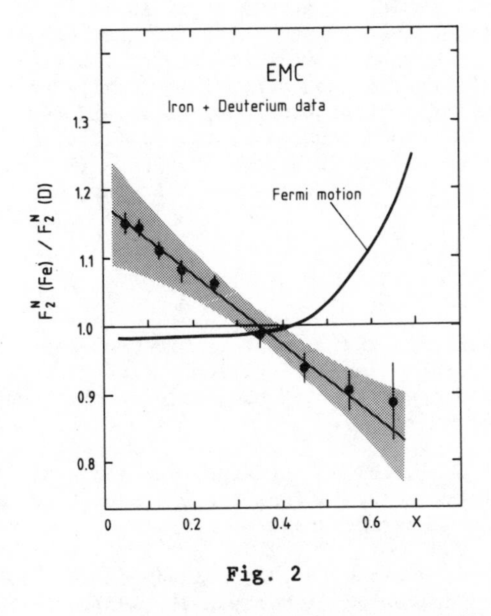

Fig. 2

Q^2 averaged ratio of the nucleon structure function F_2^N for iron and deuterium

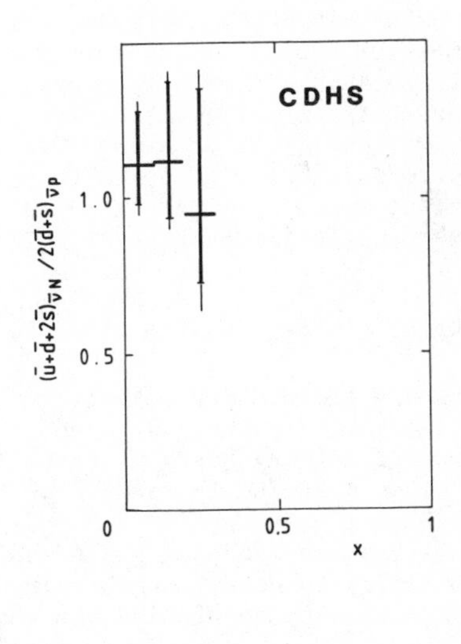

Fig. 3

Ratio of the seaquark distribution in iron and hydrogen as measured by CDHS

Keeping in mind these restrictions the result tells us that at x>0.3 (valence quark region) less quarks with high momentum are found in a nucleus than in a free nucleon. At x=0.65 the difference to the expectation for iron is about 30-40%. At low x (sea quark region), where only little difference had been predicted, an enhancement is observed, suggesting an increase of the momentum fraction carried by seaquarks.

This behaviour of the data immediately provoked the question already formulated in the first theoretical paper on the EMC effect by Jaffe (1983): does one observe an additional sea component in the nucleus? Are those the sea quarks responsible for the short range forces? Or does one observe the presence of extra pions (meson exchange currents) in the nucleus as proposed by Llewellyn Smith (1983), Ericson and Thomas (1983), Berger et al (1984, 1985) and other authors (Alonso and Sanchez-Gomez 1985, Titov 1983, Sapershtein and Shmatikov 1985).

To answer this important question obviously better experimental information is needed. Due to their large normalisation uncertainty the EMC data allow any value for an enhanced sea between 0 and 45%. For these studies, in principle, neutrino/antineutrino experiments are superior since they allow to extract the sea quark distributions directly, but unfortunately they will always suffer from too low statistics. The CDHS experiment (Abramowicz et al 1984), for example, obtained for the ratio of the antiquark distributions in iron, $\frac{1}{2}(\bar{u}+\bar{d}+2\bar{s})_{\bar{\nu}Fe}$, and in hydrogen, $(\bar{d}+\bar{s})_{\bar{\nu}p}$, a mean value of 1.10±0.11 (stat.) ±0.07 (syst.), where the systematic error, in my opinion, is still underestimated. The data are shown in fig. 3. The results of the other neutrino experiments (Parker et al 1984, Cooper et al 1984, Ammosov et al 1984) are even more inconclusive. In a contribution to this conference the WA25/WA59 collaboration (1986) presented a full analysis of approximately 40000 neutrino/antineutrino events taken with the BEBC bubble chamber with neon and deuterium filling. From these fits they conclude that there is a (16±8%) decrease of the sea in neon. Hopefully Drell-Yan experiments on different nuclei, with incident proton energies around 800 GeV, which will be soon possible at FNAL, as discussed by Garvey (1986) at this conference, will prove to be better suited for these studies. (For a detailed discussion and an extensive list of references see also Berger (1986) and Bickerstaff et al (1986)).

3.1.2 Gluon distribution

Quark and gluon distributions are related by the elementary QCD processes of gluon emission by quarks and the quark-antiquark creation out of gluons. Therefore, if the quark distributions are modified by nuclear effects, then the gluon distribution must also be affected.

Apart from the QCD analysis of the Q^2 evolution of structure functions a possibility to determine the gluon distribution is from the cross section for J/ψ production studied via the $\mu^+\mu^-$ decay of the J/ψ.

The EMC found that the cross section per nucleon for J/ψ production is much larger for iron than for hydrogen-deuterium (Aubert et al 1985). In the framework of the photon-gluon fusion model (Leveille and Weiler 1979, Glück and Reya 1979) this implies that the gluon distribution is enhanced in iron compared to hydrogen and deuterium. The data are shown in fig. 4. They cover a range $0.026<x<0.086$. The average ratio of the cross sections, or the gluon distributions respectively, is 1.45±0.12(stat.) ±0.22(syst.). However, these data have too low statistics to allow any conclusion about the magnitude and the x and A dependence of the change of the gluon distribution. Obviously more and better data are needed.

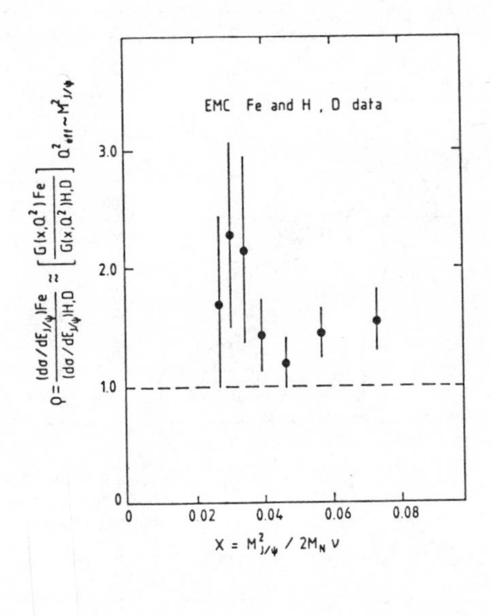

Fig. 4

The ratio of the cross sections per nucleon for J/ψ production for iron and hydrogen-deuterium measured by EMC

3.2 SLAC Electron Data

Figure 5 shows the results of the SLAC experiment E139 of Arnold et al (1984a), which has measured cross sections per nucleon for a series of nuclei with incident electron energies between 10 and 20 GeV. The data cover a Q^2range of $2 \leq Q^2 \leq 15$ GeV². Also plotted are the recovered target wall data for aluminium and steel from the experiments E49B and E87 (Bodek et al 1983a,b), the results of the shadowing experiment E61 (Stein et al 1975) and the EMC data.

For all nuclei one observes that the nuclear cross section σ^A is reduced compared to the 'free nucleon' one, σ^D, in the region $0.3<x<0.8$. Obviously there is little Q^2 dependence of the effect, since there is good agreement between these data and the EMC result in shape as well as normalisation (see fig. 5a) although their Q^2 range differs by more than a magnitude. The rapid increase of the ratio at larger x is dominantly a kinematic effect since the free nucleon cross section vanishes for $x \to 1$. It is however, not clear whether this increase is entirely due to Fermi motion of nucleons or whether it is caused by collective phenomena such as nucleon-nucleon correlations or, for instance, the formation of multiquark clusters. The reduction in the region $0.3<x<0.8$ has a very characteristic shape which is similar for all nuclei, with a minimum around $x \sim 0.65$. The effect is already present for helium, its magnitude increases <u>logarithmically</u> with the atomic <u>weight A</u>, as can be seen from fig. 6.

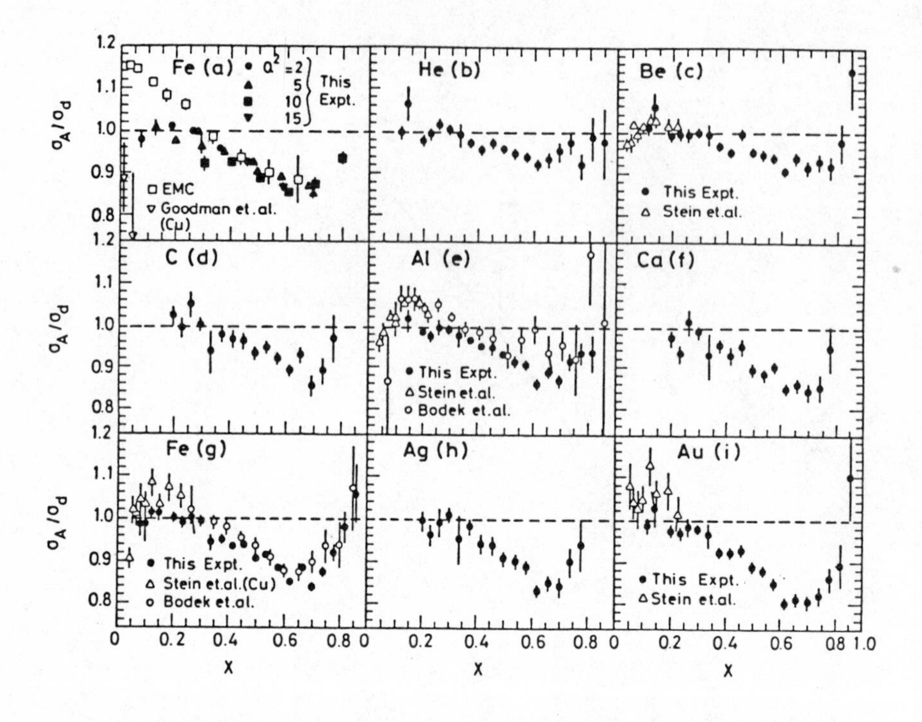

Fig. 5

A dependence of the EMC effect measured by SLAC experiment E139

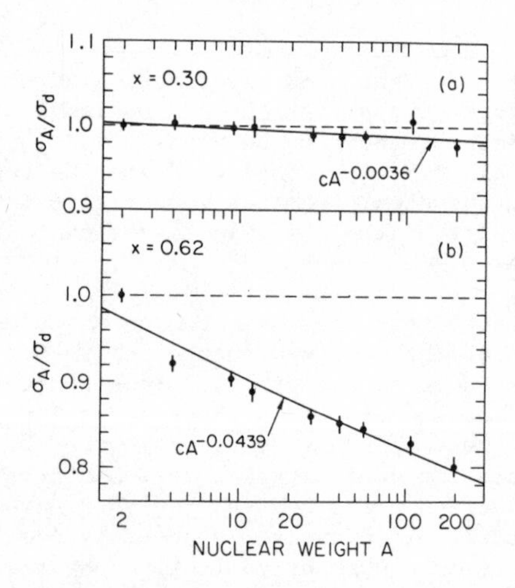

Fig. 6

The ratio σ^A/σ^D as a function of nuclear weight A

The average nuclear density $\bar{\rho}(A)$ shows a similar A dependence, therefore all models which relate the effect in some way to this quantity (increase of confinement size of nucleons, fraction of multiquark clusters, ..) are able to reproduce this A dependence. Also the number of extra pions per nucleon, as obtained from standard nuclear physics calculations by Friman et al (1983) increases approximately like log A. Consequently, pion models can, in principle, describe as well an enhancement at low x as the A dependence of the cross section reduction at higher x, simply by assuming that the momentum carried by the pions has been taken from the valence quarks of the nucleons. Another quantity showing the same A dependence and which has therefore, been used by some authors e.g. Close et al (1986) to show the agreement between their model and experimental data, is the shift of the quasi-elastic peak $\epsilon_{meas.}$ measured in A(e,e')A-1 reactions, for instance by Moniz et al (1971). One should however, be aware of the fact that after coulomb corrections, which, following Rosenfelder (1980), should be applied to the experimental data points, ϵ_c is nearly constant for all A. It is also interesting to note that the shift of the quasi-elastic peak is different for the transverse and longitudinal part of the cross section (Mulders 1986). Similarly the mean separation energy $\bar{E}_s$ experimentally has an almost constant value of around -25 MeV for all nuclei heavier than carbon (Frullani and Mougey 1984). This quantity has been used as input for the so called x-rescaling models to explain the EMC effect as a standard nuclear physics phenomenon caused by nuclear binding. Because of the saturating behaviour of this quantity these models, which will be discussed in more detail in chapter 4, fail to reproduce the measured A dependence.

While in the medium x range the experimental situation seems to be reasonably clear (although details such as the Q^2 dependence have still to be clarified), this is not the case for low x. First of all, there are for most nuclei no data at x<0.2 from E139, secondly the results are controversal. This is most clearly seen for iron (fig.5 a,g): the structure function ratio measured by EMC at high Q^2 shows a continuous rise between x=0.3 and x=0.05, the low Q^2 cross section ratios from E139 are compatible with unity while the data from E61 show a clear enhancement with a maximum at x around 0.1-0.15 which seems to increase (fig. 5c,e,g,i) with A. Part of the discrepancy could be caused by a strong Q^2 dependence of the nuclear effects at low x, but it could also be due to systematic problems in one or several of the data sets. Therefore, much more detailed experimental information is needed especially at low x.

Another possible explanation for the difference between the SLAC cross section ratios and the EMC structure function ratio could be an A dependence of $R=\sigma_L/\sigma_T$. R could change with A if higher twist contributions are different for bound and free nucleons (Shuryak 1985) or if there is a substantial amount of spin-0 objects present in the nucleus, which could lead to a decrease of F_1^N. Indeed there is an indication of such an A dependence of R from the measurements of E139 (Arnold 1984b, Gomez 1985). The results for R are shown in fig. 7 as a function of A at various x values for Q^2=5 GeV^2 for D, He, Fe and Au. There is a trend in the data to increase with A, but due to the large error bars the results are not conclusive enough to decide whether that is really the case.

If one assumes that $R^{Fe}=R^D+0.15$, as suggested by the data, and then converts the cross section ratios to structure function ratios the apparent difference between the EMC and the SLAC data disappears and there is reasonable agreement between the two data sets as can be seen from fig. 8.

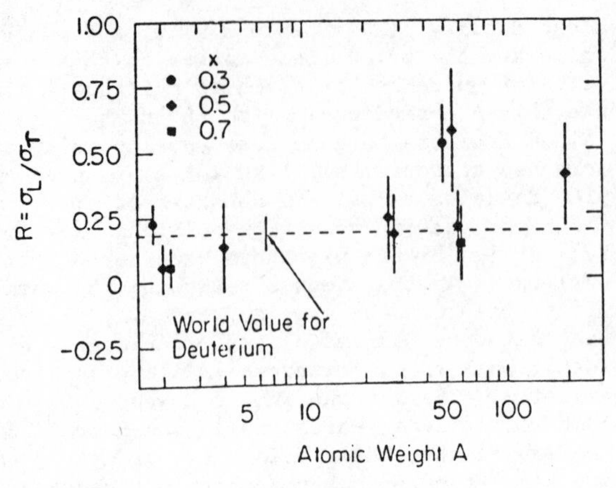

Fig. 7

SLAC E139 results for $R=\sigma_L/\sigma_T$ at $Q^2=5$ GeV² and various x values versus nuclear weight A

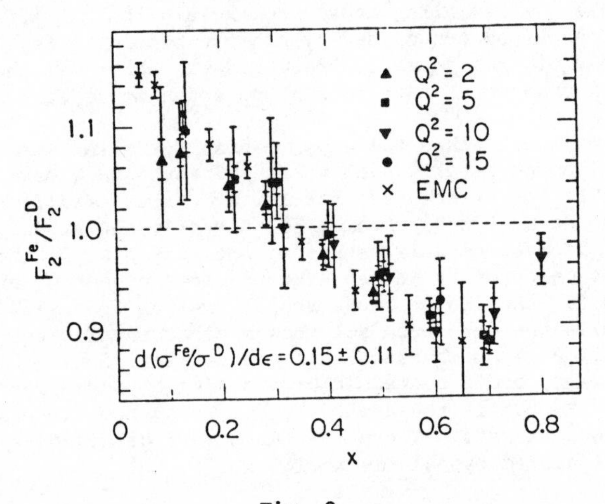

Fig. 8

SLAC E139 results for the ratio of structure functions F_2^{Fe}/F_2^D extracted under the assumption $R^{Fe}=R^D+0.15$

A dedicated experiment to measure R(A) has been performed at SLAC (Arnold 1984b), preliminary data (Rock 1986) indicate that the A dependence, if it exists at all, might be very small.

3.3 Neutrino Data

The data from the CDHS counter experiment (Abramowicz et al 1984) and the three bubble chamber experiments BEBC WA25/59 (Cooper et al 1984), BEBC-TST (Parker et al 1984) and FNAL 15ft (Ammosov et al 1984) are summarised in fig. 9. The mean Q^2 values of these data are typically a factor of ten smaller than those of the EMC data. The results of all four experiments are, within their large error bars, compatible with the EMC data for $x \geq 0.1$ showing some enhancement below $x \simeq 0.3$ and some depletion at higher x. At $x < 0.1$ however, they decline below one. On the other hand data for neon/deuterium obtained by the experiment E545 (Hanlon et al 1985) at the 15ft bubble chamber at FNAL do not show the drop at low x and preliminary data for iron/deuterium, which have been obtained by the same group from a comparison of events originating from the steel walls of the bubble chamber and the deuterium filling show a clear rise at low x (Kitagaki 1986). Since these data, which are shown in fig. 10, have been taken simultaneously, many of the systematic errors should cancel.

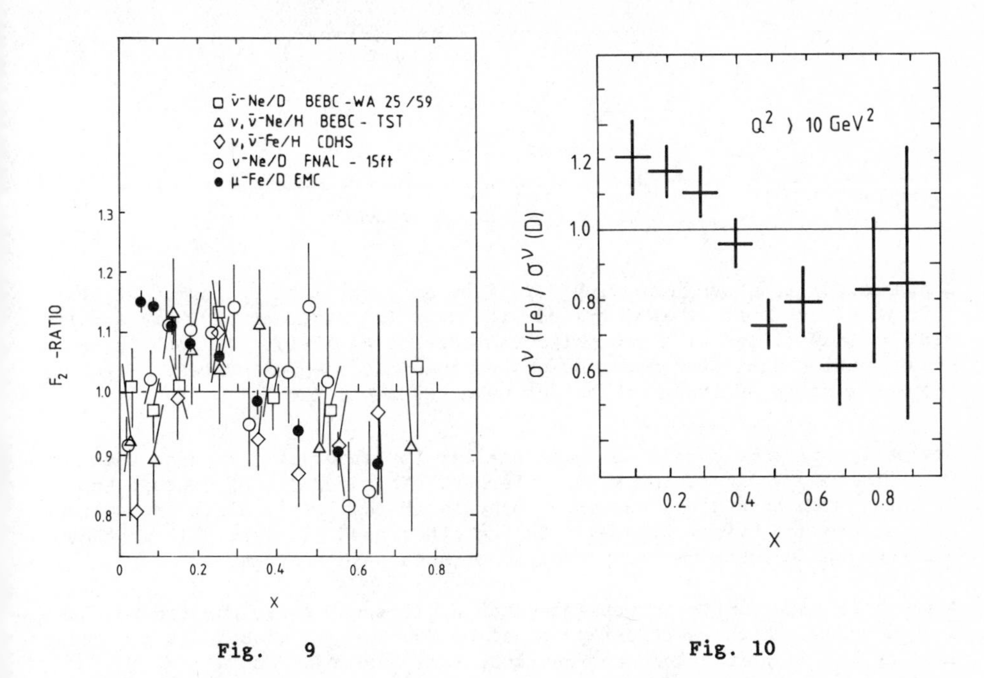

Fig. 9

Neutrino results for the EMC effect

Fig. 10

Cross section ratios for Fe/D from FNAL neutrino experiment E545

There is another interesting bubble chamber result which might give some answer to the question, whether nuclear effects are the same for proton and neutron. This must not necessarily be the case since neutrons and protons have, for instance, different separation energies and Fermi momenta and might also occupy different areas of space inside the nuclear volume.

In neutrino bubble chamber experiments one is able to distinguish whether the interaction took place on a proton or a neutron by looking at the summed charge of the hadronic final state. With incident antineutrinos for instance one obtains

$$\bar{\nu}p \to \mu^+ + X^\circ , \qquad \bar{\nu}n \to \mu^+ + X^- .$$

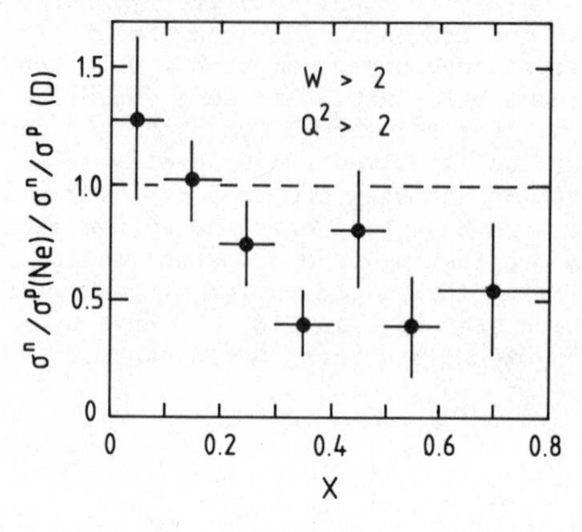

Fig. 11

The ratio of σ^n/σ^p for neon and deuterium from antineutrino bubble chamber experiments.

A comparison has been made with data from an experiment performed at the 15ft bubble chamber at FNAL filled with neon (Asratyan et al 1985) and from BEBC at CERN filled with deuterium (Allasia et al 1981). In both cases the cross section ratio for neutrons and protons has been extracted. Figure 11 shows the ratio of these ratios for neon and deuterium.

Obviously at large x σ^n/σ^p is much smaller (by about 50%) in neon than in deuterium, while at low x it is larger. This result, of course, has to be taken with care since two experiments with completely different systematic errors are being compared. On the other hand it might tell us that neutrons are affected more by nuclear effects than protons.

A possible reaction to investigate such a different behaviour for protons and neutrons, which was first proposed by Von Harrach (1985), is the comparison of the nuclear structure functions for $^{40}_{18}Ar$ and $^{40}_{20}Ca$.

Since the two nuclei have the same mass number and their density and mean square radius are very similar, one would expect that their structure function per nucleon $F_2^N(A)$ is the same, provided the argon data are properly corrected for the neutron excess.

$F_2^N(A)$ for a nucleus with N neutrons and Z protons is given by

$$F_2^N(A) = \frac{F_2^n+F_2^p}{2}\left(1-\frac{N-Z}{N+Z}\,\frac{1-F_2^n/F_2^p}{1+F_2^n/F_2^p}\right)$$

$$= \frac{F_2^n+F_2^p}{2}\cdot K ,$$

where F_2^n and F_2^p are the structure functions for neutron and proton (modified by nuclear effects). For ^{40}Ca the correction factor K is zero, for ^{40}Ar it depends on F_2^n/F_2^p in this nucleus. Assuming that neutron and proton are affected in the same way for the two nuclei one could determine $F_2^n/F_2^p(A)$ from the cross section ratios and the comparison with the measurements on deuterium would allow to study differences in the EMC effect for neutron and proton.

3.4 New Muon Data From BCDMS and EMC

There are recent data from the BCDMS and EMC muon experiments at CERN. The BCDMS results for N/D and Fe/D (Bari et al 1975, Milsztajn 1986) are shown in fig. 12. The Fe/D ratio shows a clear enhancement at x<0.3 with a maximum of ~1.06 at x=0.1. Taking into account the systematic errors, indicated by the two curves, and a possible normalisation difference of 3-4%, there is globally good agreement between this data set and the original EMC data. The point below x=0.1 indicates a possible drop towards low x. Only a small enhancement of 2-3% below x=0.3 is seen in the N/D ratio. The drop at larger x is very similar also in magnitude to that of the Fe/D data indicating that the A dependence at high values of Q^2 might be different from that extracted from the SLAC data. This in turn would point to a sizable Q^2 dependence of the effect.

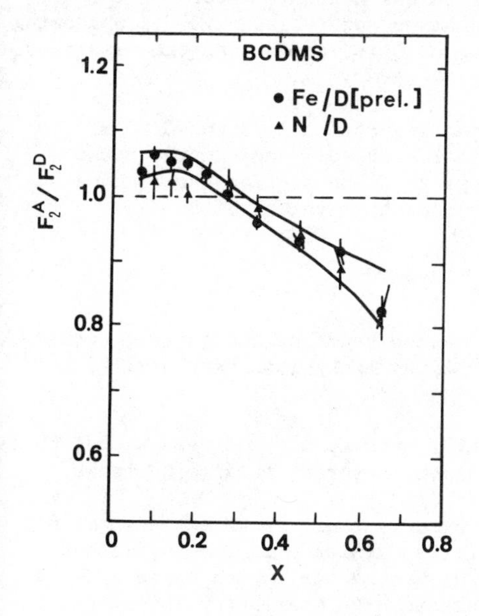

Fig. 12

Preliminary results for N/D and Fe/D from BCDMS. The curves indicate the range of systematic errors for the iron data.

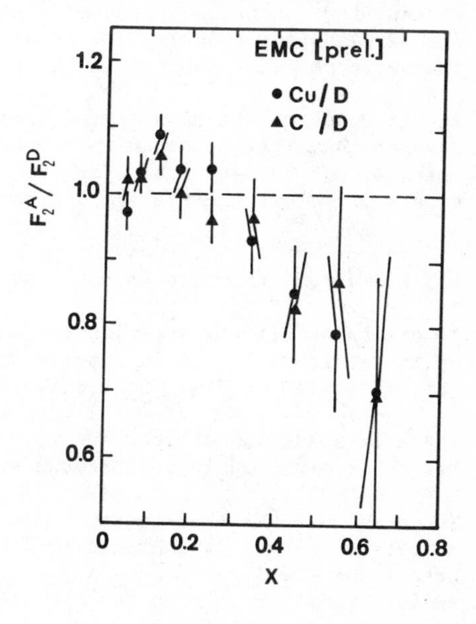

Fig. 13

Preliminary results for C/D and Cu/D from EMC.

During the measurements of the polarised proton structure functions in 1984/85 the EMC has performed a parasitic measurement from several nuclear targets (D, He, C, Cu, Sn), each approximately 8g/cm² thick. These were placed downstream of the polarised target, one at a time, and exchanged every few hours, thus minimising systematic errors due to variations in the acceptance and apparatus performance. The data are still being analysed. Preliminary results for C/D and Cu/D, based on approximately 10K events for each target are presented in fig. 13. These data also show an enhancement at $x<0.3$ which is more pronounced for Cu/D than for C/D. The Cu/D results are compatible with the original Fe/D data if one takes into account the point to point systematic errors and allows for a normalisation shift of the Fe/D results downwards by about 3-4%. Only the point at $x=0.05$ differs by more than one standard deviation.

The Cu/D ratio shows a drop towards very small x which could be an indication that shadowing, undoubtedly observed in photoproduction ($x=Q^2=0.0$), persists to relative high Q^2 values around 10 GeV^2. This would favour the quark parton model explanations of shadowing (Nikolaev and Zakharov 1975, Müller 1982, Nachtmann and Pirner 1984, Nachtmann 1986) in which shadowing/antishadowing is a nuclear effect caused by fusion of low momentum quarks and gluons of different nucleons which are able to 'extend in space' over the whole nuclear volume. In the other approaches, which assume that the quark-gluon structure of the nucleons is not affected by the nuclear medium and relate shadowing to the hadronic nature of the photon (for a detailed discussion see Grammer and Sullivan 1978), shadowing should disappear quickly with Q^2 as soon as Q is larger than the masses of the vector mesons involved in the interaction.

At higher x the decrease of the cross section ratios Cu/D and C/D is steeper than that observed in the original Fe/D ratio and in the BCDMS results, but the error bars are too large to allow any further conclusion especially about a different A dependence at high values of Q^2.

3.5 Summary of the Experimental Results

If one takes all the experimental information together the general trend of the nuclear effects in quark distributions might look very similar to that indicated in figs. 5c,e,g,i.

There is shadowing at very low x, probably even at high Q^2 values, but this has to be explored in a dedicated experiment covering a large Q^2 range.

There is no doubt that in the range $0.05<x<0.3$ there is an enhancement of σ^A over σ^D with a maximum around $x=0.1$, its magnitude increasing with A. Details of the A dependence and of the Q^2 dependence, which appears to be small, are still unknown and require further high statistics investigations before one can relate it to effects like antishadowing, extra pions or an enhanced sea. Neither the large enhancement at $x\leq0.08$ in the original EMC data nor the nearly constant ratio around unity below $x\simeq0.3$ in the E139 data has been confirmed by recent experiments. Therefore there is little need for an A dependence of $R=\sigma_L/\sigma_T$, nevertheless this question has to be clarified in its own right to find out whether there are more pointlike spin-0 objects in a nucleus than in a free nucleon.

At $0.3<x<0.8$ there is a reduction of the cross section ratio which has the same shape for all nuclei. The effect has its maximum at $x \simeq 0.65$ and increases approximately with log A, while nuclear binding corrections have nearly the same magnitude for all A. There are indications that this A dependence might look different at high Q^2. The Q^2 dependence of the effect in this region has to be measured with a precision in the level of a few per mille to find out whether the effect can be related to perturbative QCD (see next chapter) or whether quark-quark correlations between different nucleons in the nucleus (higher twist effects) play an important role (Shuryak 1985).

At even higher x σ^A/σ^D rises again above one; essentially nothing is known about the behaviour of σ^A at $x>1$, the region where multiquark clusters and other nucleon-nucleon correlation effects may play the dominant role (Baldin 1983, Garsevanishvili and Mentheshashvili 1984, Kondratyuk and Shmatikov 1984, Vary 1984, Frankfurt and Strikmann 1985). The exploration of this kinematic regime might be essential to discriminate between models.

Very little is known about the modification of the gluon distribution in nuclei. The EMC results for J/ψ production indicate a substantial increase of the gluon distribution at low x, but much more precise data are needed to clarify the magnitude and the x and A dependence of this enhancement.

I have not discussed the influence of the nuclear medium on multiplicities, energy and transverse momentum distributions of hadrons created in the deep inelastic process. These quantities can be modified by nuclear effects by several processes: in the pre-EMC effect era it was considered that in the nucleus the struck quark can lose energy and change direction due to multiple scattering and bremsstrahlung through the strong force before it fragments and/or that the produced hadrons can undergo internuclear cascading dependent on their energy and fragmentation length. Generally this leads to an attenuation of the number of energetic hadrons and a broadening of the angular distribution. For a more detailed discussion of the different models and experimental data I refer to Arvidson et al (1984).

After the advent of the EMC effect one has to take into account in addition, that already the parent quark distribution is different for bound and free nucleons and that the fragmentation function can be different due to the modified gluon field inside the nucleus (Nachtmann and Pirner 1984b, Dias de Deus 1986). Since the process is very complex and different mechanisms can contribute, very high statistics over a wide kinematic range is needed and many different parameters have to be studied to find out which mechanisms influence the production of hadrons in nuclei.

4. Interpretations

The discovery of the EMC effect caused a lot of excitement in the nuclear physics community since, for the first time, the interplay between nuclear and quark-gluon degrees of freedom had been clearly observed. Consequently there was and still is a flood of publications explaining the effect with all the different ideas mentioned in the introduction, which go beyond the simple picture of the nucleus as a dilute gas of nucleons (and pions).

It is beyond the scope of this review to outline and discuss the details of all the different models which have been proposed to explain the EMC effect.

(At present my incomplete collection of publications about this subject contains nearly 200 papers and the number is still increasing). Instead, I will summarise the main key ideas which are fortunately not so many. More detailed discussions can be found in several reviews (Nachtmann 1984, Rith 1984, Savin 1984, Dias de Deus 1985, Jaffe 1985, Llewellyn Smith 1985, Krzywicki 1985, Berger 1985, Rith 1985) and a recent publication by Bickerstaff et al (1986) in which different aspects of several approaches are quantitatively compared.

4.1 Convolution Models

The basic idea of most of the models is to describe the structure function of a nucleus as a convolution of the probability $f_c^A(y)$ to find a certain cluster c of momentum y in the nucleus and the structure function of this cluster $F_2^c(x/y)$ and to take the incoherent sum over contributions of all clusters c:

$$F_2^A(x,Q^2) = \sum_c \int_x^A dy f_c^A(y) F_2^c(\frac{x}{y}) \ .$$

Examples for such clusters are the nucleon itself, undisturbed or with a reduced mass due to nuclear binding or with an increased size due to different boundary conditions in the nuclear environment, extra pions which are responsible for nuclear binding, Δ-isobars, multiquark clusters like 6q, 9q or 12q bags or the whole nucleus as one big bag with free quark and colour flow throughout the whole nuclear volume, as already proposed by Krzywicki in 1976, to explain the anomalous enhancement of inclusive hadron spectra at large transverse momenta.

It has been argued by Jaffe et al (1984) that the assumption of incoherence at the cluster level is highly suspect and wrong in most cases. In addition there is a lot of freedom in these approaches, concerning as well the choice of f_c^A as that of the parameterisation of F_2^c, which are, if at all, badly known.

One example of these convolution models is the pion model (Llewellyn Smith 1983, Ericson and Thomas 1983, Berger et al 1984, 1985, Alonso and Sanchez-Gomez 1985, Titov 1983, Sapershtein and Shmatikov 1985), where the nucleus is considered as a bound system of pions and nucleons only, with quarks and gluons confined within these hadrons. The quark and antiquark distributions are assumed not to be affected by the nuclear medium and identical to those of the free particles. This model can, in principle, describe the behaviour of the nuclear structure functions over the whole x range. The additional pion contribution causes the excess at low x, the imposition of momentum balance between pions and nucleons causes the depression at medium x and the rise at $x \to 1$ is due to Fermi motion of the nucleons. If the number of excess pions is chosen to be approximately proportional to the nucleon density, as for instance calculated by Friman et al (1983), also the A dependence at medium x can be reproduced. Therefore the authors claim that there is no need for extra components beyond standard nuclear physics.

This result by itself would already be exciting since then the deep inelastic scattering experiments would be the first ones which have directly 'seen' pions in the nucleus.

I tend, however, to some scepticism. First of all, as pointed out in the previous paragraph, the experimental information at low x is incomplete and not sufficient to deduce any reliable A dependence. I would prefer to see the pion model tested in this kinematic region where pions directly contribute. If in addition the present indications that shadowing persists to very high Q^2 turn out to be correct it may be very difficult to describe this behaviour in the framework of this model.

Furthermore, the choice of f_π^A and of F_2^π is model dependent and very much disputed (Bickerstaff et al 1986). It is also not at all clear to me why the on-shell pion structure function, measured in the Drell-Yan process $\pi A \to \mu^+ \mu^- X$, should be sufficient to describe all the contributions to nuclear binding due to the exchange currents of off-shell mesons.

Most of the authors of the other convolution approaches, using other clusters than pions, also succeed to reproduce the data rather well, at least in the medium x range, which might be due to the fact that their input parameters are less constrained than in the pion model. To be fair, the experimental information is far too incomplete to discriminate between these models and I find it very exciting that all these different aspects are investigated to learn more about the true picture of the nucleus.

4.2 Rescaling Models

In the second class of approaches the EMC effect is explained by a change of either the Q^2-scale or the x-scale for the nuclear structure function compared to the free nucleon.

In the 'Q^2-rescaling' models, first proposed by Close et al (1983) and Nachtmann and Pirner (1984a), the EMC effect is related to a change of confinement size inside the nucleus. The A dependence of quark and gluon distributions for bound nucleons and the Q^2 evolution of these distributions for free nucleons both have the same origin. They are caused by the colour forces between quarks and gluons which ensure confinement. The qualitative argument goes as follows: the strength of the strong force between quarks is not just determined by the resolution $1/\sqrt{Q^2}$ at which they are probed, but also by the size λ_A of the volume in which they are confined. Therefore, the relevant parameter in the strong coupling constant α_s is not just Q^2 but $(Q \cdot \lambda_A)^2$. If the confinement size is modified inside the nucleus, either due to a 'swelling' of nucleons, the formation of multiquark bags or free quark and colour flow throughout the whole nucleus then, as a consequence, quark and gluon distributions obtained for different nuclei A and B, at the same value of Q^2, are related by

$$q_A(x,Q^2) = q_B(x,\xi \cdot Q^2), \quad g_A(x,Q^2) = g_B(x,\xi \cdot Q^2)$$

Note that the distributions for A and B are only different at a fixed value of Q^2 since they are evolving with Q^2. This scale breaking behaviour of the quark and gluon distributions is completely neglected in the other approaches.

ξ is a rescaling parameter determined by the two confinement scales λ_A and λ_B involved. In the so called 'dynamical rescaling' models it is given by

$$\xi = (\lambda_A/\lambda_B)^{2\alpha_s(\mu_A^2)/\alpha_s(Q^2)}$$

where μ_A is a low momentum cut-off for radiating gluons. In this class of models the enhancement of the quark and gluon distributions at low x and the depression at medium x is naturally explained, the only free parameters being the A dependent confinement size and the cut-off parameter μ_A. On the other hand there are also some problems, especially the fact that the x value where the structure function ratios cross one (x>0.3) is not the same one where the scaling violations disappear (x~0.15) and it is also unclear what the relevant confinement size really is.

A good test will be a careful study of the Q^2 dependence of the EMC effect at x>0.35. The New Muon Collaboration at CERN will perform such measurements on the few per mille level. These data will help to decide whether the EMC effect is really a QCD phenomenon.

Independent of this question I find the concept of a modified confinement size inside the nuclear medium fascinating. If quark and gluon distributions are modified by the fact that due to different boundary conditions in the nuclear medium the volume in which they are confined by the strong force changes, then in turn the study of this change in the momentum distributions as a function of A could provide us with information about these boundary conditions. This would offer another possibility, than just the study of hadron spectroscopy, to attack the 'confinement problem' which is one of the basic questions in todays theory of elementary particles.

The '<u>x-rescaling</u>' model is based on the observation of Garcia Canal et al (1984) and Staszel et al (1984), that the depression of the nuclear structure functions at medium x can be very well reproduced if for a nucleus the scaling variable x is replaced by a modified one

$$x^* = (M/M_A^*),$$

where M_A^* is an effective mass of the nucleon which is smaller than the free nucleon mass M due to binding effects.

This idea has been refined by several authors (Akulinichev et al 1985 a-d, Krzywicki 1985, Levin 1985, Birbrair et al 1986, Dunne and Thomas 1986, Nakano and Raffelski 1986) who have recalculated Fermi motion and binding effects in the single-particle approach. It has been discussed in some detail by Thomas (1986) at this conference.

In the single-particle approach the virtual photon with four momentum q interacts with an off-mass-shell nucleon which has three momentum $\vec{p}_i$ and energy $p_\circ=M_A-E_{A-1}^i=M+E_s^i-|\vec{p}_i^{\,2}|/2M_{A-1}$, where A-1 is the recoiling excited nucleus and E_s^i is the nucleon's separation energy from a given shell state i as measured in quasi-elastic knock-out reactions. The scaling variable $x=Q^2/2pq$ therefore depends on the mean separation energy $\bar{E}_s$ and on the momentum distribution of nucleons in the nucleus. For large masses A and not too high momenta $\vec{p}_i$ one obtains

$$x^* \simeq \frac{1}{1+\bar{E}_s/M} \cdot x$$

and for the structure function ratio in addition to momentum (kinetic energy) dependent Fermi motion corrections (as for instance calculated by Bodek and Ritchie 1981) an extra term which is proportional to the mean separation energy $\bar{E}_s$

$$F_2^A/F_2^N = (F_2^A/F_2^N)_{\text{Fermi motion}} + C,$$

$$C = (\bar{E}_s/M) \cdot x \cdot (dF_2^N/dx)/F_2^N \ .$$

Since at large x F_2^N falls approximately like $(1-x)^3$ this term explains naturally the declining of the ratio with x.

This observation is the basis of the claim that the EMC effect can be completely explained within the framework of conventional nuclear physics.

This conclusion however, goes much too far. As outlined in detail by Strikman and Frankfurt (1986), these calculations are not unambiguous. Care must be especially taken in the choice and normalisation of the nucleon momentum distribution not to violate either the baryon number conservation or the energy-momentum sum rule. They also argue that an additional term proportional to $\langle\vec{p}_i^{\,2}\rangle/M$ has to be taken into account which is of nearly the same magnitude as the term C but of opposite sign, resulting in a much smaller correction than that calculated by the other authors.

Furthermore, this model is <u>not</u> able to describe the experimental data. As mentioned in section 3.2 and outlined in more detail by Gupta (1986), the measured mean separation energies $\bar{E}_s(A)$ do not increase continuously with log A but saturate already very early at a value around -25 MeV (Frullani and Mougey 1984). Without further ingredients such as nucleon-nucleon correlations, multiquark clusters or other effects the measured A dependence of the EMC effect cannot be reproduced, the model fails one of the main criteria used to judge the quality of the approaches.

The enhancement at low x, which is undoubtedly experimentally confirmed, can also not be reproduced since in that region F_2^N is nearly independent of x and therefore C vanishes. This defect could probably be cured by taking the objects, responsible for the binding, (e.g. pions) into account.

My conclusion is that there might be a grain of truth in this ansatz and that it might explain a fraction of the depression at medium x, but I am convinced that additional components which go beyond standard nuclear physics are necessary to describe the whole effect.

5. <u>OUTLOOK</u>

Many of the proposed theoretical approaches can describe some features of the existing data, which have just scratched the surface of this exciting field. However, much more detailed experimental information is needed until the final picture with all its facets emerges and one will find a unique explanation for the observed nuclear effects. Nearly all of the open experimental questions, indicated in the previous paragraphs, will be investigated during the following years by the New Muon Collaboration, NA37 at CERN (NMC 1985).

For the study of the A and Q^2dependence of F_2^A/F_2^D at low x and nuclear effects in hadron distributions, targets of relatively low weight (100 g/cm²) are being used. Foreseen are measurements on D, He, Li, Be, C, Al, Si, Ca, Nb and Ho. In parallel to the normal trigger, which accepts muons with scattering angles larger than ~10 mrad, a special trigger has been installed to detect also scattered muons at even smaller angles. This setup allows to cover the whole kinematic range $x>0.005$ and and $1<Q^2<200$ GeV². Data are taken from pairs or triples of nuclei simultaneously. The targets are split into several alternating short sections and the whole setup is frequently exchanged by its complementary one with the different target materials having interchanged their positions along the beam. Due to this procedure overall normalisation errors and point to point systematic errors will be kept below a level of 1%. With $7\cdot10^{11}$ muons per target pair the resulting statistical accuracy in the F_2 ratios will be better than 2.5% for most (x,Q^2) bins in the region $x<0.3$ and better than 1% per x bin when averaged over Q^2. In addition a total of about one million events usable for hadron analysis are expected.

For the study of the A dependence of $R=\sigma_L/\sigma_T$ at high Q^2, the Q^2 dependence of F_2^A/F_2^D at $x>0.3$, the behaviour of F_2^A/F_2^D at $x>1$ and the A dependence of the gluon distribution high luminosity measurements are needed. Therefore, an active target will be used with a core of alternating target pieces of C or Be and Nb of 300 g/cm² each in the beam region. With a total flux of 10^{13} muons, which can be accumulated in about 100 days, these measurements will allow to determine A dependent changes in R with an accuracy of $\Delta R=0.02$ for several kinematic bins in the range $0.05<x<0.25$ and $7<Q^2<50$ GeV².

The statistics at high x will be sufficient to measure the Q^2 dependence of the EMC effect with an accuracy of $d(F_2^{Pb}/F_2^C)/d\ln Q^2 \simeq 2-4\cdot10^{-3}$ (for $0.3<x<0.6$). Many events will also be collected at $x>1.0$.

Furthermore about 7000 J/ψ events are expected from each nuclear target plus about 1700 (800) from a measurement on hydrogen (deuterium) which is also being done. These statistics will allow a precise determination of the magnitude and x dependence of the <u>change</u> of the gluon distribution with A, even if one restricts the analysis to the 15-20% inelastic events where the theoretical uncertainties for the interrelation of J/ψ cross section and gluon distribution are assumed to be much smaller than for the elastic ones. The J/ψ events from hydrogen and deuterium will also provide the additional information whether nuclear effects are already present for deuterium.

The aspect of the enhanced sea will be explored by the Drell-Yan experiment E772 at FNAL (Garvey 1986).

My hope is that in a few years we will understand better how and why the nuclear environment influences quark and gluon distributions and confinement and that the data will help to understand nuclear forces in a fundamental way in terms of quarks, gluons and their interactions.

References

Abramowicz H et al 1984 Z. Phys. C25 29
Akulinichev S V et al 1985a Phys. Lett. 158B 485
Akulinichev S V et al 1985b Phys. Rev. Lett. 55 2239
Akulinichev S V et al 1985c J. Phys. G, Nucl. Phys. 11 L245
Akulinichev S V et al 1985d JETP Lett. 42 127
Allasia D et al 1981 Phys. Lett. 107B 148
Alonso J A and Sanchez-Gomez J L 1985 Phys. Lett. 154B 441
Ammosov V V et al 1984 PISMA v. ZHETF 39 327
Arnold R G et al 1984a Phys. Rev. Lett. 52 1431
Arnold R G et al 1984b SLAC proposal E140
Asratyan A E et al 1985 preprint ITEP-115
Arvidson A et al 1984 Nucl. Phys. B246 381
Aubert J J et al 1983 Phys. Lett. 123B 275
Aubert J J et al 1985 Phys. Lett. 152B 433
Baldin A M 1983 Dubna preprint E2-83-415
Bari G et al 1985 Phys. Lett. 163B 282
Berger E L et al 1984 Phys. Rev. D29 398
Berger E L and Coester F 1985 Phys. Rev. D32 1071
Berger E L 1985 Argonne preprint ANL-HEP-CP-85-70 and Proceedings of the Topical Seminar on Few and Many Quark Systems, San Miniato, Italy, March 1986
Berger E L 1986 Nucl. Phys. B267 231
Bickerstaff P P et al 1986 Phys. Rev. D33 3228
Birbrair B I et al 1986 Phys. Lett. 166B 119
Bodek A and Ritchie J L 1981 Phys. Rev. D23 1070; D24 140
Bodek A et al 1983a Phys. Rev. Lett. 50 1431
Bodek A et al 1983b Phys. Rev. Lett. 51 534
Close F E et al 1983 Phys. Lett. 129B 346
Close F E et al 1985 Phys. Rev. D31 1004
Close F E et al 1986 Phys. Lett. 168B 400
Cooper A M et al 1984 Phys. Lett 141B 133
Dias de Deus J 1985 Proceedings of the Int. Europhysics Conference on High Energy Physics, Bari 1985, Nitti L and Preparata G ed. p 571
Dias de Deus J 1986 Phys. Lett. 166B 98
Dunne G V and Thomas A W 1986 Phys. Rev. D33 2061
Ericson M and Thomas A W 1983 Phys. Lett. 128B 112
Frankfurt L L and Strikman M I 1985 Nucl. Phys. B250 143
Friman B L et al 1983 Phys. Rev. Lett. 51 763
Frullani S and Mougey J 1984 Advances in Nuclear Physics 14 1
Garcia Canal C H et al 1984 Phys. Rev. Lett. 53 1430
Garvey G T 1986 these proceedings
Garsevanishvili V R and Menteshashvili Z R 1984 Dubna preprint E2-84-314
Glück M and Reya E 1979 Phys. Lett. 83B 98
Gomez J 1985 SLAC Pub 3552
Grammer G and Sullivan J 1978 in Donnachie A and Shaw G "Electromagnetic Interactions of Hadrons" Vol II Plenum Press
Gupta S 1986 Bombay preprint TIFR/TH/86-17
Hanlon J et al 1985 Phys. Rev. D32 2441
Jaffe R L 1983 Phys. Rev. Lett. 50 228
Jaffe R L et al 1984 Phys. Lett. 134B 449
Jaffe R L 1985 MIT preprint 1261
Kitagaki T 1986 E545 data presented at the 12th Int. Conf. on Neutrino Physics and Astrophysics, Sendai, Japan, June 1986
Kondratyuk L and Shmatikov M 1984 ITEP preprint -13
Kondratyuk L and Shmatikov M 1985 Z. Phys. A321 301

Krzywicki A 1976 Phys. Rev. D14 152
Krzywicki A 1985 Nucl. Phys. A446 135C
Levin E M 1985 Leningrad preprint 85-1147
Leveille J P and Weiler T 1979 Nucl. Phys. B147 147
Llewellyn Smith C H 1983 Phys. Lett. 128B 107
Llewellyn Smith C H 1985 Nucl. Phys. A434 35C
Milsztajn A 1986, to appear in the proceedings of the "Int. Symposium on Weak and Electromagnetic Interactions in Nuclei" Heidelberg 1-5 July 1986
Moniz E J et al 1971 Phys. Rev. Lett. 26 445
Müller A H 1982, proceedings of the XVIIth Rencontre de Moriond ed. J. Tran Thanh Van, Vol. 1 p 13 and private communication
Mulders P J 1986 NIKHEF preprints P-8, P-10 and contribution to this conference
Nakano K and Rafelski J 1986 University of Cape Town preprint UCT-TP 50
Nachtmann O 1984 Proc. 11th Conf. on Neutrino Physics and Astrophysics, Nordkirchen near Dortmund, p 204
Nachtmann O and Pirner H J 1984a Z. Phys. C21 277
Nachtmann O and Pirner H J 1984b Heidelberg preprint HD-THEP-84-7 to appear in Ann. Phys. (Leipzig)
Nachtmann O 1986 to appear in the proceedings of the "Int. Conf on Weak and Electromagnetic Interactions in Nuclei" Heidelberg 1-5 July 1986
Nikolev N N and Zakharov V J 1975 Phys. Lett. B55 397
NMC 1985 Proposal CERN/SPSC/85-18 SPSC/P210
Parker M A et al 1984 Nucl. Phys. B232 1
Rith K 1984 Proceedings of the 3rd Klaus Erkelenz Symp. ed. Bleuler K Lecture notes in physics 197 (Springer 1984) p 214
Rith K 1985 Proceedings of the "Zuoz Spring School on Strong Interactions", Zuoz, Switzerland, April 1985 SIN 1985 p 201
Rock S 1986 Data presented at the XXIII Int. Conf. on High Energy Physics, Berkeley July 1986
Rosenfelder R 1980 Ann. Phys. 128 188
Sapershtein E E and Shmatikov M Kh 1985 JETP Lett. 41 53
Savin I 1984 Proceedings of the XIIth Int. Conf. on High Energy Physics, Leipzig 1984 Meyer A and Wieczovek ed. Vol II p 241
Shuryak E V 1985 Nucl. Phys. A446 259C
Staszel M et al 1984 Phys. Rev. D29 2638
Stein S et al 1975 Phys. Rev. D12 1884
Strikman M I and Frankfurt L L 1986 Leningrad preprint LINR 1197
Thomas A W 1986 these proceedings
Titov A I 1983 Sov. Journ. Nucl. Phys. 40 50
Vary J 1984 Proceedings of the 7th Int. Conf. on High Energy Physics Problems, Dubna 1984, p 147
Von Harrach D 1985 private communication
WA25/59 collaboration 1986 paper submitted to this Conference

Inst. Phys. Conf. Ser. No. 86
Paper presented at Int. Nucl. Phys. Conf., Harrogate, UK, 1986

Nuclear tests of symmetries

W. C. Haxton

Institute for Nuclear Theory, Department of Physics, University of Washington, Seattle, Washington 98195

1. Introduction

Years of careful spectroscopy have produced a great deal of information on the properties of nuclear levels. Today this knowledge is being exploited in new tests of fundamental symmetries. The nucleus, with its capacity to filter and amplify special interactions, has proven to be a rich testing ground for fundamental physics. Twenty-five years ago elegant nuclear experiments revealed the nature of the weak interaction and prepared the way for the standard model. Today there is great hope that some subtle violation of low-energy symmetry laws might provide a glimpse of physics beyond the standard model. We nuclear physicists, with our tradition of precision experiments, find ourselves well positioned to contribute to this search for new interactions.

My talk today, a summary of symmetry tests in nuclear physics[1], will be flawed by many errors of omission. I apologize to the many physicists whose deserving work in this field may not be cited. Unfortunately, in view of the varied backgrounds of the physicists present, I've concluded that it is wise to cover a few topics adequately rather than many superficially. The topics I have chosen, parity-mixed nuclear levels, time-reversal-odd nuclear moments, and the Mikheyev-Smirnov mechanism, have generated a great deal of recent excitement, though they are not unique in this respect.

I hope, however, that Table 1 may at least suggest the breadth of present activities in nuclear tests of symmetries. The topics listed in the table are generally of interest to both nuclear physicists and the broader physics community. The topics include both space-time symmetries (parity, time reversal) and internal symmetries (isospin, charge conjugation), and encompass many possibilities for new interactions and new particles. One aspect of the first topic, parity violation in the hadronic weak interaction, will be discussed in some detail later in this talk. Parity-violating observables in the two-nucleon system and parity mixing in nuclei provide our only practical tests of the neutral weak interaction between hadrons.[2] The weak interaction between the electron and the nucleus is being probed in polarized electron scattering experiments[3] at Bates and Mainz, and through the parity mixing of levels in heavy atoms.[4] Unlike parity nonconservation, our knowledge of CP violation is very limited, as CP violation has so far been observed only in the neutral kaon system.[5] However important constraints on theories incorporating CP violation have been imposed by recent measurements of CP-violating triple correlations in β decay[6] and by sensitive atomic measurements of time-reversal-odd nuclear moments.[7] In addition, rather stringent constraints on T-odd but P-even nuclear forces have been obtained from statistical analyses of nuclear level densities.[8]

Double beta decay[9] provides our most sensitive test of lepton number conservation and the charge conjugation properties of the electron neutrino. In a parallel session Dr.

Table 1. A partial list of nuclear physics tests of symmetries and conservation laws.

Symmetry/conservation laws	Nuclear observables
parity nonconservation	scattering of polarized nucleons;[13] parity-mixed nuclear levels;[2] helicity dependence of electron-nucleus scattering;[3] parity-mixed atomic levels[4]
CP/T nonconservation	triple correlations in β decay;[6] complex phases in electromagnetic mixing ratios;[14] nuclear electric dipole moments;[7] statistics of nuclear energy levels[8]
lepton number nonconservation	double beta decay[9]
muon number nonconservation	$\mu \rightarrow e$ conversion in the nuclear field[12]
charge symmetry, charge independence of the NN interaction	^{3}He-^{3}H mass difference;[20] NN scattering lengths;[15] isospin-forbidden reactions[19]
neutrino mass, mixing	neutrino oscillations; tritium β decay;[21] double beta decay[9]
searches for exotic particles, interactions	axion production in nuclear reactions;[24] the Eötvös experiment:[25] nuclei exhibiting unusual properties

Tomoda has discussed neutrinoless $\beta\beta$ decay limits on the Majorana neutrino mass and right-handed couplings. A number of extraordinary experiments are presently underway. The limit on the 0ν $\beta\beta$ decay halflife of ^{76}Ge now exceeds $4 \cdot 10^{23}$ years[10], while the TPC ^{82}Se experiment of the Irvine group[11] may soon yield the first laboratory measurement of the lepton number-conserving 2ν decay mode. Programs to test the conservation of separate lepton number have been pursued with enthusiasm at LAMPF, TRIUMF, and SIN. Searches for the muon number-violating decays $\mu \rightarrow e\gamma$ and $\mu \rightarrow 3e$ and for $\mu \rightarrow e$ conversion in the nuclear field yielded limits on the branching ratios of $1.7 \cdot 10^{-10}$, $2.4 \cdot 10^{-12}$, and $1.4 \cdot 10^{-11}$, respectively.[12] Major efforts to improve these limits are currently underway.

In view of the large deviation of the current quark mass ratio $m_d/m_u = 1.79$ from unity, isospin invariance is an intriguing approximate symmetry of the strong NN interaction. In talks given in the parallel sessions, the results of the TRIUMF n-p elastic scattering experiment[15] and the Los Alamos $\pi^+/\pi^- - {}^4$He experiment[16] were presented. The former is in good agreement with the predictions of Miller. Thomas, and Williams[17], while the latter found no evidence of the strong isospin mixing deduced from the photonucleon ratio $\sigma(\gamma,p)/\sigma(\gamma,n) = 1.7$.[18] Recently Coon and Preedom[19] have calculated the $\pi\eta$-mixing contribution to the charge symmetry-forbidden reaction $d + d \rightarrow \pi^0 + {}^4$He. The resulting cross section ($\approx 0.1\ pb/sr$) may be within the reach of experiment. Similar meson-mixing estimates do not account for the full 80 keV non-electromagnetic binding energy difference between ^{3}H and ^{3}He, considered one of the best measures of charge symmetry breaking.[20]

Nuclear physics experiments are a major component of any particle physics conference on neutrinos. I've already mentioned double beta decay tests of the charge conjugation properties of the neutrino. Kundig will review direct measurements of the neutrino mass in tritium β decay[21] and electron capture[22] experiments. One of the most remarkable results of this year, the discovery by Mikheyev and Smirnov of enhanced solar neutrino oscillations, will be discussed in some detail later in this talk.

The final entry in Table 1 is the use of nuclear physics and nuclear chemistry techniques in searches for exotic particles and interactions. The enthusiasm for axions as an explanation of the GSI e^+e^- events was intense[23] though short-lived (as discussed by Schwalm at this meeting). This explanation was inconsistent with constraints from other experiments, including searches for nuclear axion decay and nuclear "multipole-meter" experiments.[24] A reanalysis of the Eötvös experiment by Fischbach and collaborators provided evidence for a long-range force coupling to baryon number.[25] This result depends on a fundamental property of nuclei, that binding energy differences lead to significant variations in the ratio of mass to baryon number. Many other experiments have been performed or proposed in which exotic nuclear properties (unusual chemistry, mass spectrometry, or isotopic distribution) are the anticipated signatures of free quarks, heavy charged leptons, or other new stable particles.

Table 1 could be expanded by a factor of two or three and yet still be incomplete. Nevertheless, I hope this brief introduction provides some sense of the activity in nuclear tests of symmetries. I'll now discuss three of these topics in more detail.

2. Parity Mixing of Nuclear Levels

While elegant experiments have been performed to test the predictions of the standard model for leptonic and semileptonic interactions, the hadronic weak interaction has been more elusive. At low energies it can be studied only when the strong and electromagnetic interactions are forbidden by a symmetry principle, such as flavor conservation. However, in the standard model the neutral-current contributions to $\Delta S = 1$ and $\Delta C = 1$ weak processes are greatly suppressed. Therefore neutral weak interactions can only be probed in flavor-conserving processes, where parity nonconservation must serve as the filter to isolate the relevant observables. The NN interaction is the single practical example.

The weak NN interaction is a sum of isoscalar, isovector, and isotensor terms. The charged current contribution to the $\Delta I = 1$ amplitudes is suppressed by $\sin^2\theta_c$, where θ_c is the Cabibbo angle, while the neutral current contribution is unsuppressed. Thus this component of the weak NN interaction should provide a direct measure of the neutral current interactions of quarks. In the conventional single-meson-exchange description of the weak NN interaction, where one meson-nucleon vertex is governed by the strong interaction and the second by the weak, the isovector amplitude is dominated by pion exchange. The neutral current is expected to increase the weak pion-nucleon coupling f_π from the Cabibbo value (~ 0.5) to approximately 12.

The experimental task of isolating the pion exchange amplitude is a difficult one. At low energies there are six independent S-P amplitudes (if one considers the long-range pion-exchange and short-range vector-meson exchange contributions to the $^1S_0$ - $^3P_0$

amplitude to be distinct). Ideally one would make six independent measurements i the NN system to determine these amplitudes. To date only one definitive result, th helicity dependence of the $\vec{p} + p$ cross section[13], has been obtained.

Another approach to this problem is the extraction of the weak meson-nucleon coupling from measurements of the parity mixing of nuclear states. The success of this approac depends in part on our ability to determine the many-body matrix elements of the wea NN potential. I will discuss only a single example, ^{18}F. A more complete discussion o parity violation in nuclei can be found in Ref. 2.

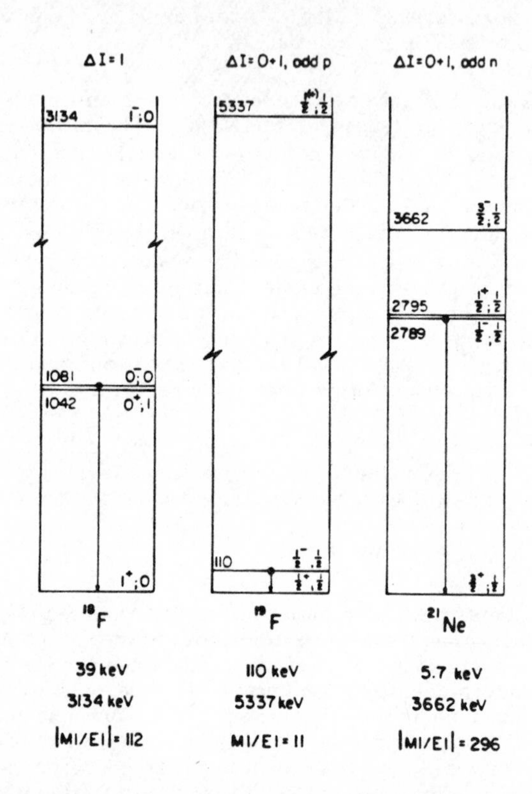

Figure 1. Parity-mixed doublets in light nuclei. The transitions displaying the amplified PNC effect are indicated. The quantities ΔE and $\Delta E'$ are the smallest and next smallest energy denominators governing the parity mixing. The quantities shown in the bottom row are "amplification factors."

A diagram of the relevant levels in ^{18}F (and of the parity doublets that have been studied in the neighboring nuclei ^{19}F and ^{21}Ne) is shown in Fig. 1. Since the separation ΔE of the $J^\pi T = 0^-0$ and 0^+1 levels is only 39 keV, one can reasonably assume that the parity admixture in the 0^-0 state is due to mixing with the 0^+1 state. This admixture can be measured by observing the circular polarization P_γ of the gamma ray emitted in the $0^-0 \rightarrow 1^+0$ (g.s.) decay:

$$P_\gamma = \frac{2}{\Delta E} Re \left[\frac{< 1^+0 \| M1 \| 0^+1 >}{< 1^+0 \| E1 \| 0^-0 >} < 0^+1\, V_{PNC}^{\Delta I=1} | 0^-0 > \right] \tag{1}$$

This formula illustrates a number of attractive features of the doublet. The definite isospin of the nuclear states isolates a single component, $V_{PNC}^{\Delta I=1}$, of the weak NN potential. While the parity-allowed E1 transition is isospin-forbidden, the parity-forbidden M1 transition is unusually strong, 10.3 ± 1.5 W.u. Thus the matrix element ratio $< M1 > / < E1 > \approx 110$ amplifies the effect of the parity mixing. A further amplification occurs because of the small energy denominator separating the doublet states. Thus one expects $P_\gamma \sim 10^{-3}$, four orders of magnitude greater than the scale for parity nonconserving observables in the NN system.

A final remarkable aspect of the ^{18}F doublet involves a relation[26] between the PNC matrix element in Eq. (1) and the β decay of ^{18}Ne. The ^{18}Ne ground state is the isospin analog of the 1.042 MeV 0^+1 state in ^{18}F. Thus the first-forbidden transition leading to the 1.081 MeV 0^-0 state in ^{18}F links the same states that appear in Eq. (1). In the long wavelength limit this transition proceeds through the axial charge operator. In addition to the familiar one-body axial charge operator

$$F_A \sum_{i=1}^{A} \frac{\vec{\sigma}(i) \cdot \vec{p}(i)}{M} \tau_- \tag{2}$$

there is an important two-body contribution that is also of order (v/c). This operator, which is determined by current algebra and PCAC, is identical, apart from an isospin rotation and overall coupling constants, to the pion-exchange piece of $V_{PNC}^{\Delta I=1}$.

More importantly, the exchange current, when averaged over core nucleons, becomes an effective one-body operator whose form is also given by Eq. (2). That is, the exchange current, to an excellent approximation, renormalizes the one-body coupling constant F_A. Thus, even though the matrix elements of the one- and two-body operators are difficult to calculate, the ratio of these matrix elements should be very insensitive to nuclear physics uncertainties. Thus an estimate of $< V_{PNC}^{\Delta I} >$ can be obtained from the measured[27] first-forbidden β decay rate and the calculated ratio of one- and two-body matrix elements. The more detailed arguments given in Ref. 26 indicate that this procedure is remarkably reliable.

This result permits one to extract from an experimental measurement of P_γ (and thus $< V_{PNC}^{\Delta I=1} >$) a reliable value for the weak pion-nucleon coupling f_π. The beautiful experiments of the Queen's University/Princeton/Caltech[28] and Florence[29] groups, when combined with earlier measurements[30], yield $P_\gamma = (0.8 \pm 3.9) \times 10^{-4}$ and $| < V_{PNC}^{\Delta I=1} > | < 0.09$ eV. The effort expended on these experiments is extraordinary: the total running time to date is approximately 8000 hours. The corresponding limit on f_π is a factor of five smaller than the standard model best value calculated by Desplanques, Donoghue, and Holstein.[31]

This result is consistent with other measurements of PNC ($A_L(\vec{p}+p)$, $A_L(\vec{p}+{}^4\mathrm{He})$, $A_L(\vec{p}+d)$, $A_\gamma({}^{19}\mathrm{F})$), with the exception of the circular polarization measurement in ^{21}Ne. However, the interpretation of the ^{21}Ne result depends entirely on shell model estimates of the nuclear matrix element. The constraints imposed by these experiments on effective isoscalar and isovector couplings are shown in Fig. 2. (The $p+\vec{p}$ results do not appear on this figure because they depend strongly on the isotensor weak NN potential.) Even if one accepts the ^{21}Ne constraint, a global fit[2] of weak meson-nucleon couplings to PNC observables indicates that the neutral current enhancement of f_π is less than that predicted by theory.

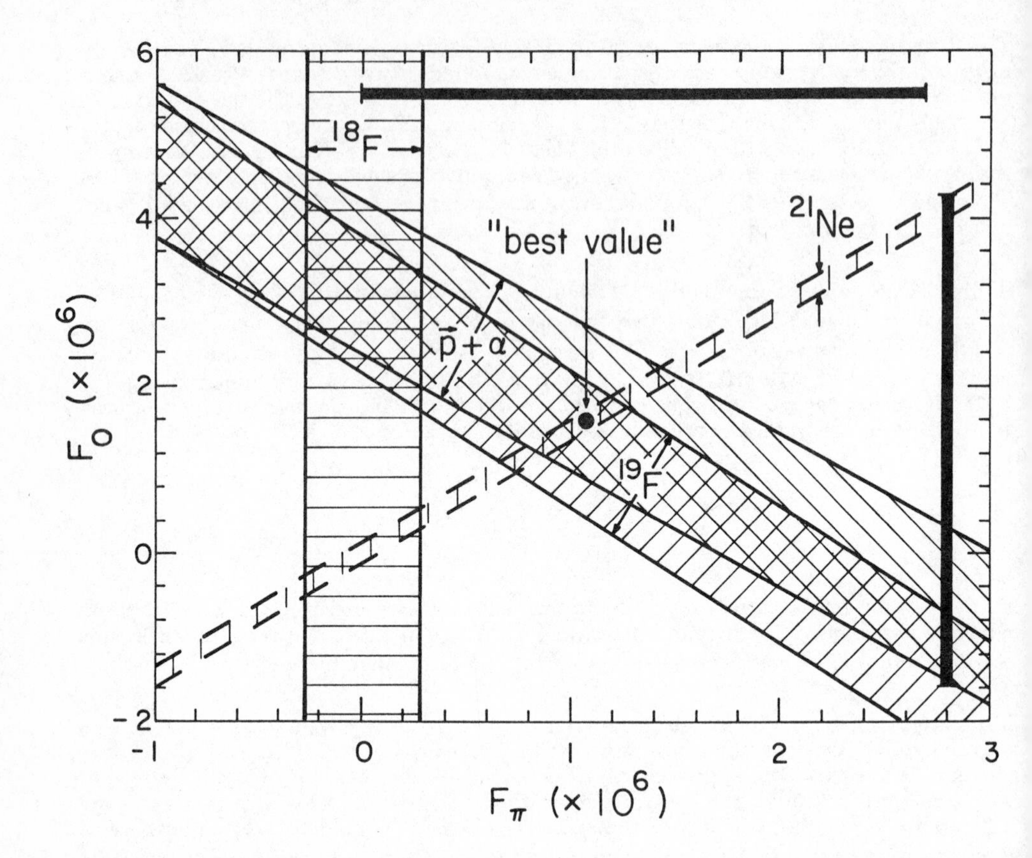

Figure 2. The results of a simplified two-parameter analysis of parity mixing in nuclear systems in which the weak NN potential is approximated by isovector pion exchange (F_π) and isoscalar ρ exchange (F_0). The point marked by the solid circle is the best value of Desplanques, Donoghue, and Holstein, and the heavy solid lines are their reasonable ranges. The results for $A_L(\vec{p}+p)$ are not shown because this observable depends strongly on the isotensor component F_2 of the weak NN force. However, the experimental results are quite consistent with the Desplanques, Donoghue, and Holstein best values for F_0 and F_2. (The isotensor interaction in complex nuclei tends to zero as an average is taken over core nucleons.)

The discrepancy indicated by the ^{18}F experiment is similar in magnitude (after strong interaction corrections are made) to the $\Delta I=1/2$ rule in $\Delta S=1$ interactions. There is clearly great interest in possible additional experiments on parity nonconservation that could confirm the surprising isospin dependence of V_{PNC}.

3. T-odd Nuclear Moments

The observation of CP-violation in neutral kaon decays[5], coupled with the CPT theorem, suggests elementary particles may have non-zero "odd" static moments (C1/E1, M2, C3/E3,...) forbidden by time reversal symmetry. Although no non-zero odd moments have been detected, the sensitive limit on the electric dipole moment (edm) of the neutron[32] ($d_n < 4 \cdot 10^{-25} e \quad cm$) has provided important constraints on theoretical models incorporating CP violation.

Recently very powerful techniques for measuring the edm's of neutral atoms have been developed, yielding, in the case of the ^{199}Hg experiment of Fortson *et al.*, a limit of approximately 10^{-26} e-cm.[7] The importance of atomic experiments as tests of the electron edm has long been appreciated. However, with the exquisite precision of recent measurements, these experiments are also emerging as a competitor to the neutron edm measurements in probing possible CP-violating interactions between quarks.

As a probe of nuclear edm's, atomic experiments must contend with Schiff's theorem: for a point nucleus, there is no term in the interaction energy of a neutral atom in an external field that is linear in the edm.[33] The perfect shielding limit is not realized in Nature due to the finite nuclear size: the interaction energy is proportional to the r^2-weighted moment of the difference between the normalized nuclear charge and edm distributions. Thus heavy nuclei are favored. There are also terms in the interaction energy that depend on the M2 nuclear moment and on relativistic corrections to the electron-nucleus Coulomb interaction.

Henley and I discussed the possibility that the sensitivity of atomic edm experiments to the parameters governing CP violation might, in favorable cases, be much greater than one would expect from single-particle estimates.[34] This suggestion was originally made by Feinberg[33], who pointed out that nearly degenerate ground-state opposite-parity doublets are found in certain rare earth and actinide nuclei. In such cases enhanced T-odd nuclear moments can result from the mixing of the doublet states by the CP-nonconserving NN interaction, just as occurs for favorable parity-mixed doublets. We calculated the moments that would result if the QCD θ parameter were the dominant source of CP-violation. In this case the CP-violating NN interaction is due to single pion exchange, where one πNN vertex is CP-violating (and proportional to θ) and the other is governed by the strong interaction. The edm's due to this many-body effect are, in favorable cases, 100 times that due to the unpaired valence nucleon. In one extraordinary case, the 200 eV parity doublet in ^{229}Pa, the enhancement is 10^4. I will not discuss this work further in the written version of my talk because adequate references already exist.[34]

Similar work was done by Khriplovich[35] and his collaborators for the Kobayashi Maskawa CP-violating phase in the quark mass matrix. Khriplovich found that the CP-violating KM NN interaction was dominated by a kaon-exchange diagram. The KNN CP-violating vertex was estimated by evaluating a penguin diagram. (The πNN vertex for the θ parameter is fixed by current algebra.) These calculations yield the surprising result that the KM NN interaction, for a fixed value of the neutron edm, is roughly two orders of magnitude stronger than the θ-parameter NN interaction. The calculations of Ref. 35 suggest that the Seattle result for $d_A(^{199}\mathrm{Hg})$ may be about an order of magnitude more restrictive than the neutron edm measurement in limiting the size of the KM phase. No special enhancement due to a ground-state degeneracy occurs for this nucleus. Clearly there exists the possibility that a significant increase in sensitivity to CP-violation would result if the experimental techniques could be adapted for a more favorable nucleus. A recent current algebra estimate of the KM CP-violating NN interaction by McKellar and Choudhury[36] agreed well with Khriplovich's direct term, but yielded a much weaker exchange term. Clearly considerable effort should also be spent on improving the nuclear and atomic theory needed to interpret experiment.

4. Matter-enhanced Oscillations of Solar Neutrinos

Mikheyev and Smirnov[37] have discovered a mechanism by which electron neutrinos produced in the solar core could be efficiently converted into neutrinos of a different flavor. This mechanism depends on an effective density-dependent electron neutrino mass arising from the weak charged-current interactions with solar electrons, a phenomenon first discussed by Wolfenstein.[38] The Mikheyev-Smirnov-Wolfenstein mechanism provides a plausible particle physics solution to the solar neutrino puzzle: neutrino oscillation governed by small vacuum mixing angles can produce the needed factor-of-three suppression of the ^{37}Cl counting rate.[39]

Numerical studies of the effect of matter-enhanced oscillations on the counting rate of the ^{37}Cl and ^{71}Ga experiments have been carried out by Mikheyev and Smirnov[37] Rosen and Gelb[40], Hampel[40], and others. However, most of the physics underlying these detailed calculations can be understood in a simple model that exploits the analogy between the MSW mechanism and the phenomenon of adiabatic level crossing in atomic collisions. I would like to briefly describe this approach. More detailed discussions can be found in the papers of Bethe[41], Messiah[42], Haxton[43], and Parke.[44]

In the adiabatic approximation one describes the evolution of the neutrino wavefunction in terms of the stationary eigenstates of a time-independent Hamiltonian evaluated for the appropriate instantaneous electron density. Consider the case of two-state mixing where the ν_e, in vacuum, is composed primarily of the lighter mass eigenstate and the "ν_μ" of the heavier. Then adiabatic propagation of the neutrino through the sun will lead to a large $\nu_e \rightarrow \nu_\mu$ amplitude provided the ν_e, when produced in the solar core was initially the heavy eigenstate of the instantaneous core Hamiltonian. Such level crossing occurs if

$$\beta\rho(t=0) > \cos 2\theta \tag{3}$$

where β and ρ are the dimensionless quantities $\beta = 2E/\delta m^2 R_s$ and $\rho(t) = \sqrt{2}\, G_F\ \eta_e(t) R_s$, with E the neutrino energy, $\delta m^2 = m_2^2 - m_1^2 > 0$, R_s the sun's radius, and $\eta_e(t)$ the instantaneous electron density at time t; t is measured in units of R_s and thus runs from 0 to 1 for a ν_e produced at the sun's center. An additional constraint comes from the requirement of adiabatic propagation. For small vacuum mixing angles θ, the oscillation frequency moves through a pronounced minimum at a critical density that determines the level-crossing point

$$\beta\rho(t_c) \equiv \beta\rho_c = \cos 2\theta \tag{4}$$

The adiabatic condition relates the allowed mixing angles to the rate of change of the density

$$\tan^2 2\theta \gtrsim \frac{\sin^3 2\theta}{\left[(\beta\rho - \cos 2\theta)^2 + \sin^2 2\theta\right]^{3/2}} \left| \frac{1}{\rho_c^2} \frac{d\rho}{dt} \right| \tag{5}$$

a constraint that becomes particularly stringent at t_c

$$\gamma \equiv \frac{\tan^2 2\theta}{\left| \frac{1}{\rho^2} \frac{d\rho}{dt} \right|_{t_c}} \gtrsim 1 \tag{6}$$

The term adiabatic conversion will be used to describe neutrino oscillations when both the level-crossing condition (Eq. (3)) and the adiabatic condition (Eq. (6)) are satisfied.

The constraints imposed by Eqs. (3) and (6) are plotted in Fig. 3 as a function of the two relevant parameters, $\delta m^2/E$ and $\sin^2 2\theta$. Above the horizontal line no level crossing occurs: even in the solar core the electron neutrino is dominantly the lighter of the two instantaneous mass eigenstates. Below the diagonal line (defined by $\gamma = 1$) the propagation of the neutrino in the vicinity of the crossing point is nonadiabatic.

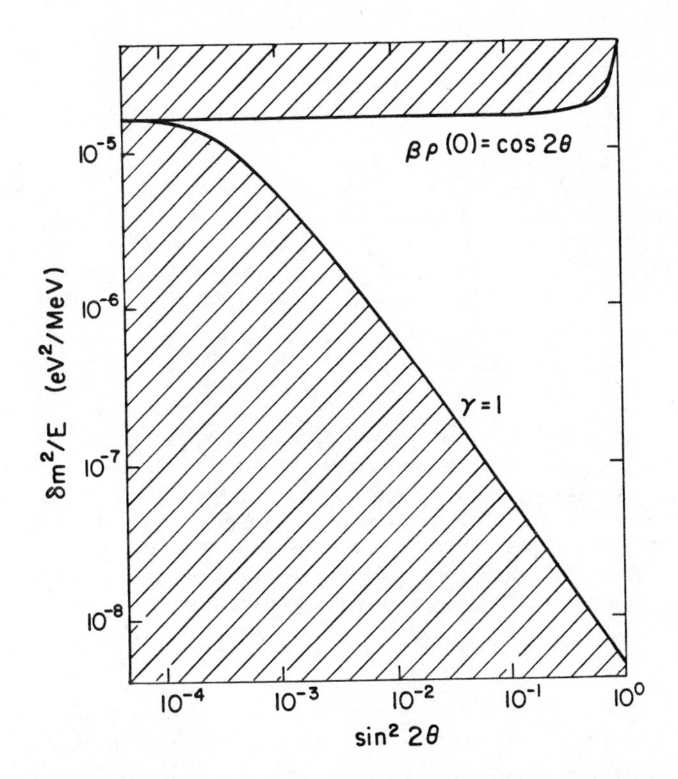

Figure 3. The shaded areas define those values of $\delta m^2/E$ and $\sin^2 2\theta$ that fail to satisfy the constraints imposed by Eq. (3) (upper region) and by the condition of adiabatic propagation (Eq. (6), lower region). The conversion of solar $\nu_e \to \nu_\mu$ is highly efficient in the unshaded region. Solutions consistent with Davis's experiment lie along the boundaries for an appropriate choice of an average E.

The region of adiabatic conversion lies entirely inside these boundaries. (The adiabatic condition will be discussed more fully below.)

As Davis[39] has measured a *nonzero* signal of 2.0 ± 0.3 SNU, some of the solar neutrinos that contribute to the ^{37}Cl counting rate are not converted to ν_μ. Bethe[41] noted that Eq. (3) permits one to exclude low-energy neutrinos from adiabatic conversion for an appropriate choice of δm^2. Bethe found $\delta m^2 \sim (0.008\text{eV})^2$, corresponding to a critical neutrino energy $E_c^B \sim 6$ MeV, provides the correct suppression of the experimental signal. Eq. (6) then requires $\sin^2 2\theta \gtrsim 8 \cdot 10^{-4}$. This solution is one of two found in the numerical studies of the MSW mechanism. Bethe's analysis identifies this solution with the boundary in the $\delta m^2 - \sin^2 2\theta$ plane where the level-crossing requirement for adiabatic conversion fails for low-energy neutrinos.

The second solution found in the numerical studies can be identified with the adiabatic boundary ($\gamma \approx 1$) in Fig. 3. For a fixed angle θ, the condition $\gamma = 1$ defines a density that I will denote by $\rho^{\gamma=1}(\theta)$. Now note that the denominator on the right-hand side of Eq. (6) is a monotonically increasing function of t, running from 0 at $t = 0$ to 1 at $t \sim 0.9$ for a neutrino produced at the sun's center. Thus, for fixed θ, Eq. (6) is satisfied by any δm^2 and E that lead to $\rho_c > \rho^{\gamma=1}(\theta)$; i.e.,

$$\frac{\delta m^2 R_s}{2E} > \frac{\rho^{\gamma=1}(\theta)}{\cos 2\theta} \,. \tag{7}$$

For our present "back-of-the-envelope" purposes we will assume that the condition $\gamma = 1$ defines a sharp boundary between the adiabatic and nonadiabatic regions. (This approximation will be improved below.) Thus, for fixed δm^2, neutrinos of energy E *above* a critical value will not satisfy Eq. (7) and thus not undergo adiabatic conversion. Davis's result is consistent with an absence of neutrinos with energies below 10.4 MeV. Thus a second solution to the solar neutrino problem is obtained from Eq. (7)

$$\delta m^2 = 5.9 \cdot 10^{-9} \mathrm{eV}^2 \, \frac{\rho^{\gamma=1}(\theta)}{\cos 2\theta} \tag{8}$$

so that δm^2 is determined as a function of θ. For the spectrum of neutrinos sampled by the ^{37}Cl detector, this solution connects smoothly to Bethe's solution at $\delta m^2 \sim (0.008 \text{ eV})^2$ and $\sin^2 2\theta \sim 8 \cdot 10^{-4}$, but runs to much smaller δm^2 and large mixing angles. Values satisfying Eq. (8) for a range of mixing angles are given in the third column of Table 2 and are very similar to the results of detailed numerical calculations. Thus solutions to the ^{37}Cl puzzle are uniquely defined as those δm^2 and $\sin^2 2\theta$ that lie along the two boundaries of the region of adiabatic conversion, an intuitively appealing result.

Table 2. Values of δm^2, $\rho_c^{\gamma=1}(\theta)$, and $\sin^2 2\theta$ of the adiabatic-boundary solution for Eq. (8) (δm^2) and for the calculation that includes the width of the adiabatic boundary and presence of hybrid solutions for small $\sin^2 2\theta$ (δm^2_{LZ}). The right-hand side of Eq. (3) was taken from a smooth fit to the tabulated density profile of Ref. 45. $\rho_c^{\gamma=1}$ can be compared to the sun's central density, $\rho(0) = 2.8\text{E}4$. (EX $= 10^x$).

$\sin^2 2\theta$	$\rho^{\gamma=1}$	$\delta m^2(\mathrm{eV}^2)$	$\delta m^2_{LZ}(\mathrm{eV}^2)$	$\sin^2 2\theta$	$\rho_c^{\gamma=1}$	$\delta m^2(\mathrm{eV}^2)$	$\delta m^2_{LZ}(\mathrm{eV}^2)$
8E-4	1.1E4	6.4E-5	7.3E-5	3E-2	3.3E2	2.0E-6	9.8E-7
1E-3	9.8E3	5.5E-5	6.1E-5	6E-2	1.6E2	9.9E-7	3.2E-7
3E-3	3.6E3	2.1E-5	1.1E-5	1E-1	8.4E1	5.2E-7	1.8E-7
6E-3	1.9E3	1.1E-5	5.4E-6	3E-1	2.0E1	1.4E-7	8.6E-8
1E-2	1.1E3	6.3E-6	3.2E-6				

The level-crossing solution discussed by Bethe leads to a reduced flux of higher energy neutrinos and thus, in various radiochemical experiments, mimics the signal expected of a nonstandard solar model. The critical densities for this solution are achieved in the solar core. For the adiabatic-boundary solution the strongest suppression occurs (except for the small-angle hybrid solutions discussed below) for the low-energy neutrinos. Thus a distinctively low counting rate will result in the gallium experiment. The critical densities become progressively lower (i.e., the crossing point occurring nearer the surface) as one moves away from the level-crossing boundary to smaller δm^2.

It is rather easy to improve the simple arguments above to account for the fact that the boundaries represent transition regions. The behavior near the level-crossing boundary is governed by the overlap of ν_e with the heavy-mass eigenstate $\nu_H(t)$

$$< \nu_e|\nu_H(t) >= \frac{1}{\sqrt{2}}\left[1 \pm \frac{1}{\sqrt{1+\alpha^2}}\right]^{1/2} \tag{9}$$

where $\alpha(t) = \sin 2\theta/(\beta\rho(t) - \cos 2\theta)$ and the + (-) sign is taken for α positive (negative). At the level-crossing boundary $\alpha(t = 0)$ is infinite. However, for small $\sin 2\theta$, modest variations in $\delta m^2/E \propto \beta^{-1}$ drive $\alpha(t = 0)$ to small values and the overlap to ~ 0 (above the boundary) or to ~ 1 (below boundary). The representation of the transition region by a sharp boundary in $\delta m^2/E$ is thus a good approximation for small $\sin 2\theta$, and Bethe's E_c^B is well defined.

The behavior near the diagonal (adiabatic) boundary is determined by the probability of remaining on the heavy mass trajectory while crossing the critical point. This is given approximately by the Landau-Zener[46] factor $1 - e^{-\pi\gamma/2}$. This result, which is familiar from level-crossing problems in atomic physics, is derived by approximating the off-diagonal mixing matrix element (and therefore $n_e(t)$) by a linear function of t near the crossing point. The Landau-Zener factor is an excellent smooth approximation to the numerical boundary profiles given, effectively, as a function of $\delta m^2/E$ in Figs. 1 of the paper by Rosen and Gelb. A single parameter γ governs the physics of the adiabatic region, the boundary region, and the highly nonadiabatic region below the diagonal line in Fig. 3.

There is an explicit quadratic dependence of γ on $\delta m^2/E$ and an implicit dependence through $d\rho/dt_c$. This dependence is much more gentle than that of Eq. (9), so that the notion of sharp cutoff energy at 10.4 MeV is an oversimplification. However, by folding the neutrino spectrum with the appropriate Landau-Zener factors, the arguments leading to Eq. (3) can be generalized to account for the width of the adiabatic boundary. The results are given in Table 2 as δm^2_{LZ}. Near the intersection with Bethe's solution (i.e., for small $\sin^2 2\theta$) low-energy neutrinos (^{7}Be, pep, CNO) fail to satisfy the level-crossing condition and thus contribute substantially to the ^{37}Cl capture rate. Excluding these small $-\sin^2 2\theta$ hybrid solutions, one must reduce the naive coefficient in Eq. (8) from 5.9 to approximately 2.5 to reproduce on average the last column of Table 2.

Efficient adiabatic conversion of solar ν_e's to ν_μ's will occur throughout much of the unshaded region in Fig. 3. One quickly realizes the implications of the MSW mechanism in view of the *nonzero capture rate* for the ^{37}Cl experiment. Davis's result excludes, in the case of two-state mixing, a substantial portion of the $\sin^2 2\theta - \delta m^2$ plane corresponding, roughly, to the region bounded by the two solutions to the ^{37}Cl puzzle and by the line $\sin^2 2\theta \sim 0.8$ (a requirement imposed by Eq. (9) as $\rho \to 0$). The excluded masses and mixing angles are totally unexplored in terrestrial experiments.

Clearly the proposed Ga experiment will provide a crucial test of neutrino physics. The Ga detector is sensitive to a flux that is distinctively solar, the low energy pp neutrinos (0.42 MeV endpoint) produced copiously in the ppI cycle. The predicted standard-solar-model capture rate is 122 SNU. The minimum astronomical rate consistent with the assumption of steady-state hydrogen burning is 78 SNU.[47] For $\sin^2 2\theta \gtrsim 10^{-2}$ the adiabatic-boundary solution yields a very suppressed pp neutrino flux. Counting rates far below the minimum astronomical value would result, providing a compelling argument for neutrino masses and mixing. As one continues along this solution to somewhat smaller angles so that the pp neutrinos begin to cross the level-crossing boundary in Fig. 3, the Ga counting rate increases. These hybrid solutions are special in that both $\sin^2 2\theta$ and δm^2 could be determined from the results of the Ga and Cl experiments. For still smaller values of $\sin^2 2\theta$ the entire pp flux lies above the region of adiabatic conversion, guaranteeing that the Ga counting rate exceeds the minimum astronomical value. As in the case of the level-crossing solution, an experiment capable of measuring the spectrum of ^{8}B neutrinos would be needed to distinguish oscillations from nonstandard solar physics.

Any substantial counting rate in the Ga experiment will unambiguously rule out a large region in δm^2 and $\sin^2 2\theta$ where adiabatic conversion of the pp neutrinos must take place. This region is similar in size and shape to that tested by the ^{37}Cl experiment but is shifted, as the effective E in $\delta m^2/E$ is about an order of magnitude smaller, to lower δm^2. Adiabatic conversion for δm^2 as small as 10^{-8} eV^2 can take place. For those larger δm^2 that are also tested in the ^{37}Cl experiment, the adiabatic region for the Ga experiment will extend to somewhat smaller mixing angles.

References

1. A more comprehensive review of this subject has been given by Denys Wilkinson, J. Phys. Soc. Japan 55, 347 (1986) Suppl.
2. E.G. Adelberger and W.C. Haxton, Ann. Rev. Nucl. Part Sci. 35, 501 (1985).
3. P.A. Souder, Proc. Conf. on the Intersections between Particle and Nuclear Physics, AIP Conf. Proc. #123, pg. 490.
4. For a general review see E.N. Fortson and L.L. Lewis, Phys. Rpts. 113, 289 (1984).
5. J.H. Christenson, J.W. Cronin, V.L. Fitch, and R. Turley, Phys. Rev. Lett. 13, 138 (1964).
6. A.L. Hallin *et al.*, Phys. Rev. Lett. 52, 337 (1984); R.I. Steinberg *et al.*, Phys. Rev. D 13, 2469 (1976); B. Erozolimski *et al.*, Sov. J. Nucl. Phys. 28, 48 (1978).
7. T.G. Vold, F.J. Raab, B. Heckel, and E.N. Fortson, Phys. Rev. Lett. 52, 2229 (1984); E.N. Fortson, B. Heckel, J. Jacobs, S. Lamoreaux, F.J. Raab, and T.G. Vold, to be published.
8. E.P. Wigner, SIAM Rev. 9, 1 (1967); J.B. French, V.K.B. Kota, A Pandey, and S. Tomsovic, Phys. Rev. Lett. 54, 2313 (1985); D. Boosé, H.L. Harney, and H.A. Weidenmüller, Phys. Rev. Lett. 56, 2012 (1986).
9. W.C. Haxton and G.J. Stephenson, Jr., Prog. Part. Nucl. Phys. 12, 409 (1984); M. Doi, T. Kotani, and E. Takasugi, Prog. Theor. Phys. Suppl. 83, 1 (1985); M.G. Schepkin, Sov. Phys. Usp. 27, 555 (1984); H. Primakoff and S.P. Rosen, Ann. Rev. Nucl. Part. Science 31, 145 (1981).
10. D.O. Caldwell *et al.*, Phys. Rev. D 33, 2737 (1986); F.T. Avignone *et al.*, Phys. Rev. Lett. 54, 2309 (1985); E. Bellotti *et al.*, Phys. Lett. 146B, 450 (1984); P. Fischer, Proc. of the Sixth Moriond Workshop on Massive Neutrinos, ed. O. Fackler and J. Trân Thanh Vân (Edition Frontières, 1986), pg. 615; H. Ejiri *et al.*, Nucl. Phys. A448, 271 (1986).
11. S.R. Elliott, A.A. Hahn, and M.K. Moe, Phys. Rev. Lett. 56, 2582 (1986).
12. W.W. Kinnison *et al.*, Phys. Rev. D 25, 2846; M. Blecher *et al.*, Bull. Am. Phys. Soc. 30, 727 (1985); W. Bertl *et al.*, SIN Newsletter 17, 7 (1985).
13. D.E. Nagle *et al.*, Proc. 3rd Int. Symp. on High Energy Physics with Polarized Beams and Polarized Targets, AIP Conf. Proc. 51, pg. 24 (1979); R. Balzer *et al.*, Phys. Rev. C 30, 1409 (1984); P. von Rossen, U. von Rossen,, and H.E. Conzett, Proc. 5th Int. Symp. on Polarization Phenomena in Nuclear Physics, AIP Conf. Proc. 69, pg. 1442 (1981); D.M. Tanner *et al.*, Proc. Int. Conf. on Nuclear Physics, pg. 697 (1983); V. Yuan *et al.*, submitted to Phys. Rev. Lett. (1986); M. Simonius, to be published in the Proc. of the Second Conference on the Intersections between Particle and Nuclear Physics (1986).

14. J.L. Gimlett, H.E. Henrikson, N.K. Cheung, and F. Boehm, Phys. Rev. Lett. 42, 354 (1979); R.J. Blin-Stoyle and F.A. Bezerra Coutinho, Nucl. Phys. A211, 157 (1973).
15. R. Abegg *et al.*, Phys. Rev. Lett. 56, 2571 (1986); S.E. Vigdor, in *Current Problems in Nuclear Physics* (Hellenic Physical Society Conference Series, 1986) pg. 193.
16. C.L. Blilie, Phys. Rev. Lett. 57, 543 (1986).
17. G.A. Miller, A.W. Thomas, and A.G. Williams, Phys. Rev. Lett. 56, 2567 (1986).
18. J.R. Calarco, B.L. Berman, and T.W. Donnelly, Phys. Rev. C 27, 1866 (1983).
19. S.A. Coon and B.M. Preedom, Phys. Rev. C 33, 605 (1986).
20. R.A. Brandenburg, S.A. Coon, and P.U. Sauer, Nucl. Phys. A294, 305 (1978); S.A. Coon and M.D. Scadron, Phys. Rev. C 26, 562 (1982).
21. W. Kundig, these Proceedings; papers by V. Lubinov, J.F. Wilkerson, and H. Kawakami, Proc. Sixth Moriond Workshop on Massive Neutrinos, ed. O. Fackler and J. Trân Thanh Vân (Editions Frontières, 1986).
22. A. De Rujula, Nucl. Phys. B188, 414 (1981); J.A. Anderson *et al.*, Phys. Lett. 122B, 461 (1983); B. Janson *et al.*, Nucl. Phys. A396, 479C (1983).
23. T. Cowan *et al.*, Phys. Rev. Lett. 54, 1761 (1985); A.B. Balantekin, C. Bottcher, M.R. Strayer, and S.J. Lee, Phys. Rev. Lett. 55, 461 (1985); L. Krauss and F. Wilczek, Yale preprint YTP-86-03; R.D. Peccei, T.T. Wu, and T. Yanagida, DESY preprint (1986).
24. F.P. Calaprice, A.L. Hallin, R.W. Dunford, and A.B. McDonald, 1985 Progress Report of the Princeton Nuclear Physics Group; R.D. Peccei *et al.*, DESY preprint (1986); M.J. Savage *et al.*, Caltech preprint (1986).
25. E. Fischbach, D. Sudharsky, A. Szafer, C. Talmadge, and S.H. Aronson, Phys. Rev. Lett. 56, 3 (1986); F.D. Stacey and G.J. Tuck, Nature (London) 292, 230 (1981).
26. W.C. Haxton, Phys. Rev. Lett. 46, 698 (1981); C. Bennett, M.M. Lowry, and K. Krien, Bull. Am. Phys. Soc. 25, 486 (1980).
27. E.G. Adelberger *et al.*, Phys. Rev. C 27, 2833 (1983).
28. H.C. Evans *et al.*, Phys. Rev. Lett. 55, 791 (1985).
29. M. Bini, T.F. Fazzini, G. Poggi, and N. Taccetti, Phys. Rev. Lett. 55, 795 (1985).
30. C.A. Barnes *et al.*, Phys. Rev. Lett. 40, 840 (1978); P.G. Bizetti *et al.*, Lett. Nuovo Cimento 29, 167 (1980); G. Ahrens *et al.*, Nucl. Phys. A390, 486 (1982).
31. B. Desplanques, J.F. Donoghue, and B.R. Holstein, Ann. Phys. (NY) 124, 449 (1980).
32. I.S. Alterev *et al.*, Phys. Lett. 102B, 13 (1981); N.F. Ramsey, Rep. Prog. Phys. 45, 95 (1982). The present Grenoble experiment may improve this limit by an order of magnitude (B. Heckel, private communication.)
33. L.I. Schiff, Phys. Rev. 132, 2194 (1963); G. Feinberg, Trans. N.Y. Acad. Sci. 38, 6 (1977).
34. W.C. Haxton and E.M. Henley, Phys. Rev. Lett. 51, 1937 (1983).
35. V.V. Flambaum, I.B. Khriplovich, and O.P. Sushkov, Zh. Eksp. Teor. Fiz. (JETP) (1984).
36. B.H.J. McKellar and S.R. Choudhury, University of Melbourne preprint UM-P-86/64 (1986).
37. S.P. Mikheyev and A. Yu. Smirnov, Yad. Fiz. 42, 1441 (1985) and Nuovo Cimento 9C, 17 (1986).
38. L. Wolfenstein, Phys. Rev. D 16, 2369 (1978).
39. J.N. Bahcall, B.T. Cleveland, R. Davis, Jr., and J.K. Rowley, Astrophysical J. 292, 279 (1985).
40. S.P. Rosen and J.M. Gelb, Phys. Rev. D 34, 969 (1986); W. Hampel, Max-Planck-Institute preprint (1986); J. Bouchez *et al.*, Saclay preprint DPhPE 86-10 (1986); V. Barger, R.J.N. Phillips, and K. Whisnant, Phys. Rev. D 34, 980 (1986).
41. H.A. Bethe, Phys. Rev. Lett. 56, 1305 (1986).
42. A. Messiah, in Proc. of the Sixth Moriond Workshop on Massive Neutrinos, ed. O. Fackler and J. Trân Thanh Vân (Edition Frontières, 1986), pg. 373.
43. W.C. Haxton, Phys. Rev. Lett. 57, 1271 (1986).

44. S.J. Parke, Phys. Rev. Lett. 57, 1275 (1986).
45. J.N. Bahcall, W.F. Huebner, S.H. Lubow, P.D. Parker, and R.K. Ulrich, Rev. Mod. Phys. 54, 767 (1982); B. Filippone, private communication.
46. L.D. Landau, Physik. Z. Sowjetunion 2, 46 (1932); C. Zener, Proc. Royal Soc. A 137, 696 (1932).
47. J.N. Bahcall, in *Solar Neutrinos and Neutrino Astronomy*, ed. M.L. Cherry, W.A. Fowler, and K. Lande, AIP Conf. Proc. No. 126 (New York, 1985), pg. 60.

Inst. Phys. Conf. Ser. No. 86
Paper presented at Int. Nucl. Phys. Conf., Harrogate, UK, 1986

An upper limit for the electron antineutrino mass from the Zurich experiment on tritium β-decay

W. Kündig, M. Fritschi, E. Holzschuh, J.W. Petersen*, R.E. Pixley, and H. Stüssi

Physics Institute, University of Zurich, 8001 Zurich, Switzerland

Abstract. The endpoint region of the tritium β-spectrum has been measured with 27 eV resolution, using a magnetic spectrometer. The tritium activity was implanted into a thin layer of carbon. The neutrino mass determined is consistent with zero with an upper limit of 18 eV, which includes instrumental and statistical uncertainties as well as uncertainties due to energy loss in the source and the final electronic states.

1. Introduction

In 1930 Wolfgang Pauli postulated the existence of the neutrino, which was finally observed by Reines and Cowan (1953). Until today many aspects of this elusive light particle remain mysterious. In particular, its mass is still in question. Most grand unified theories predict massive neutrinos. Neutrinos heavier than about 1 eV would dominate the mass of our universe and would have dramatic cosmological and astrophysical consequences. The neutrino number density n_ν is directly related to the photon density n_γ, which is obtained by inserting the microwave background temperature T into Planck's radiation law

$$n_\nu = \frac{7}{11} n_\gamma = 13\ T^3\ (\text{cm K})^{-3} \simeq 250\ \text{cm}^{-3} \tag{1}$$

The critical mass density which would just close the Universe, is $\rho_c = 3H^2/8\pi G \simeq 4$ keV cm^{-3}, where G is the gravitational constant and $H \simeq (16 \times 10^9\ y)^{-1}$ the Hubble constant. From ρ_c and n_ν, assuming that there are two neutrino flavours, we get a critical neutrino mass of about 16 eV.

The present approximate limits on the neutrino masses are (Bergkvist 1972, Lubimov et al 1980, Abela et al 1985, ARGUS-Collaboration 1985):

$$m_{\nu_e} < 50\ \text{eV} \qquad m_{\nu_\mu} < 250\ \text{keV} \qquad m_{\nu_\tau} < 70\ \text{MeV}$$

In the following only the mass of the electron neutrino will be considered.

In 1980 a Russian group working at the Institute for Theoretical and Experimental Physics (ITEP) in Moscow reported a finite mass for the electron neutrino in the range 14-46 eV at the 99% confidence level (Lubimov et al 1980). Since their first announcement of a finite neutrino

---

* Present Address: CERN, 1211 Geneva, Switzerland.

mass the ITEP-group has confirmed and considerably improved the experiment. Their present result is 35 ± 2 eV (Boris et al 1985).

The importance of this result has prompted many groups to start measurements aimed at checking the ITEP value.

2. Experimental Methods to Determine the Electron-Neutrino Mass

The standard experiment to determine the neutrino mass is the measurement of the tritium β-spectrum. This method, to be discussed in detail, has the advantage of being independent of any theoretical assumptions. Three additional methods are at present being used to get information about a possible finite mass.

- Neutrino Oscillations of the weak interaction eigenstates (ν_e, ν_μ, ...) may occur if these "physical" neutrinos are superpositions of the mass eigenstates (ν_1, ν_2, ...). Oscillations violate lepton flavour number conservation. For the simple case of two neutrino oscillations the 'physical' states are:

$$\begin{aligned} |\nu_e\rangle &= |\nu_1\rangle\cos\theta + |\nu_2\rangle\sin\theta \\ |\nu_\mu\rangle &= |-\nu_1\rangle\sin\theta + |\nu_2\rangle\cos\theta \end{aligned} \qquad (2)$$

where θ is the mixing angle. The probability for the transformation $\nu_e \rightarrow \nu_\mu$ is (Bethe 1986):

$$\begin{aligned} P &= \sin^2(2\theta)\ [1 - \cos(\alpha\ kx)]/2\alpha \\ k &= [m_2{}^2 - m_1{}^2]\ /\ [2Ehc] = 2.533\ m^{-1}\ \Delta m^2\ (eV^2)/E(MeV) \\ \alpha &= \sqrt{1 + k_R{}^2 - 2k_R\cos(2\theta)} \\ k_R &= 2\sqrt{2}\ G_F N_E\ E/[m_2{}^2 - m_1{}^2] = 2.534 \times 10^{-31}\ E(MeV)N_E(cm^{-3})/\Delta m^2(eV^2) \end{aligned} \qquad (3)$$

E is the neutrino energy, G_F the Fermi coupling constant, N_E the electron density and $\Delta m^2 = m_2{}^2 - m_1{}^2$. For vacuum and normal matter densities $\alpha \simeq 1$. Accelerator experiments looking for the transformation $\nu_\mu \rightarrow \nu_e$ and reactor experiments looking for the disappearance of the electron neutrinos showed (except for unconfirmed results from a French group working at the Bugey reactor) no evidence of neutrino oscillations.

The term α introduced by Wolfenstein (1978) was originally considered to be of no importance until Mikheyev and Smirnov (1985) discovered a resonance type of amplification of the oscillations for neutrinos traversing dense matter. Because the electron neutrino can interact with electrons by the exchange of the $W^\pm$, the charged intermediate vector boson, while the muon neutrino cannot, we may have an explanation for the missing solar electron neutrinos. A sharp resonance appears at a critical electron density $N_{E_c} = [m_2{}^2 - m_1{}^2]\cos(2\theta)/[2\sqrt{2}\ G_F E]$. May be we are missing the 8 MeV ^8B neutrinos due to this resonance effect in the dense solar core (ρ > 100 g cm^{-3}). The solar neutrino counting rate can be explained with assumptions of a small mixing angle θ and a mass difference of about 0.01 eV. This resonance should not affect the low energy neutrinos from the main solar energy production process, the pp reaction, looked at by the planned gallium detectors, so that these experiments should detect the expected full neutrino flux.

- Neutrinoless Double Beta Decay: This decay $[(A,Z) \rightarrow (A,Z+2) + e^- + e^-]$ can occur only if the neutrino has a finite mass and is a Majorana particle, namely it is its own antiparticle. The two-neutrino double β-decay $[(A,Z) \rightarrow (A,Z+2) + e^- + e^- + \bar{\nu} + \bar{\nu}]$ is expected to occur from standard theory; its study is of interest since it might help in estimating the value of the nuclear matrix elements needed to analyze the neutrinoless decay. Double β-decay has been studied in several nuclei both by geochemical techniques (extraction of the (A,Z+2)-daughter from the (A,Z)-parent in an old ore) and with counters. An example, used by several groups, is the decay:

$$^{76}Ge \rightarrow {}^{76}Se + e^- + e^-.$$

In large (a few 100 cm^3) germanium detectors in underground laboratories a spectral line of 2.04 MeV corresponding to the sum of the two electron energies is searched for. The present lower limit for this decay is $\approx 10^{23}$ y, corresponding to an upper limit for a Majorana neutrino mass of a few eV. However, the result depends critically on nuclear model calculations.

- Internal Bremsstrahlung in Electron Capture: Radiative corrections to electron capture allow under certain conditions the decay:

$$(A,Z) \rightarrow (A,Z-1) + \gamma + \nu.$$

The photon spectrum is similar to β-decay given by:

$$N(E_\gamma) = \text{const}\ E_\gamma\ \varepsilon^2 \sqrt{1 - (m_\nu/\varepsilon)^2}\ , \tag{4}$$

where $\varepsilon = E_o - E_\gamma$ is the total neutrino energy and E_o is the endpoint energy. By measuring the bremsstrahlung spectrum it is possible to determine m_ν. A possible candidate is $^{163}Ho \rightarrow {}^{163}Dy$, for which the K- and L-capture is forbidden, and the process gets enhanced by an x-ray resonance process. The upper limit on the electron-neutrino mass obtained with this extremely difficult experiment is about 300 eV.

3. The Tritium β-Decay

Measurement of the endpoint region in the tritium β-spectrum still appears to be the most sensitive method to determine the rest mass m_ν of the electron antineutrino. Tritium decays with a halflife of 12.3 y:

$$^3H \rightarrow {}^3He^+ + e^- + \bar{\nu}\ .$$

The spectrum is given by Fermi's golden rule and is proportional to the electron and neutrino density of states:

$$N(E) = A\ F(E)\ p\ E_T\ \varepsilon^2 \sqrt{1 - (m_\nu/\varepsilon)^2}\ , \tag{5}$$

where p, E, E_T are momentum, kinetic and total energy of the electron. $\varepsilon = E_o - E$ is the total neutrino energy, and E_o the endpoint energy. The recoil energy of about 3 eV is of no importance. The exactly known Fermi function F(E), which takes the Coulomb interaction between the shielded nucleus and the outgoing electron into account, varies only by 1% over the interesting upper 2 keV end of the spectrum. The neutrino mass is determined from a measurement of the spectral shape near the endpoint.

Figure 1 shows the calculated β-spectrum under the assumption $m_\nu = 0$ and $m_\nu = 35$ eV. It illustrates one of the main difficulties of the experiment, the extremely low counting rate near the endpoint. The fraction of events near the endpoint for a zero neutrino mass is $3 \times 10^{-13}\ \varepsilon^3$ (ε in eV). For purely statistical reasons the counting times of the experiment to determine a mass limit are proportional to the sixth power of the mass limit. If a 35 eV limit can be reached with a given spectrometer in 10 days, it takes 50 y to reach 10 eV with the same spectrometer, of course assuming that there are no background and systematic errors.

There are three major sources for possible systematic errors common to most experiments of this type. These are shape and width of the spectrometer resolution function, energy loss when the β-particles leave the source and the electronic final states. A fit to the measured β-spectrum gives little indication of an inadequate representation of any of these distributions and independent knowledge is therefore of prime importance for a reliable determination of m_ν.

4. Spectrometer

The instrument employed consisted of a toroidal field magnetic spectrometer of the Tretyakov type (1975) with 2662 mm average source-detector distance, modified with a radial, electrostatic retarding field around the source. Figure 2 shows a cross section of the approximately cylindrically symmetric arrangement with some electron trajectories. The magnetic field is produced by 36 rectangular current loops (3,4). The electrons leaving the source (1) are decelerated by putting the source at positive high

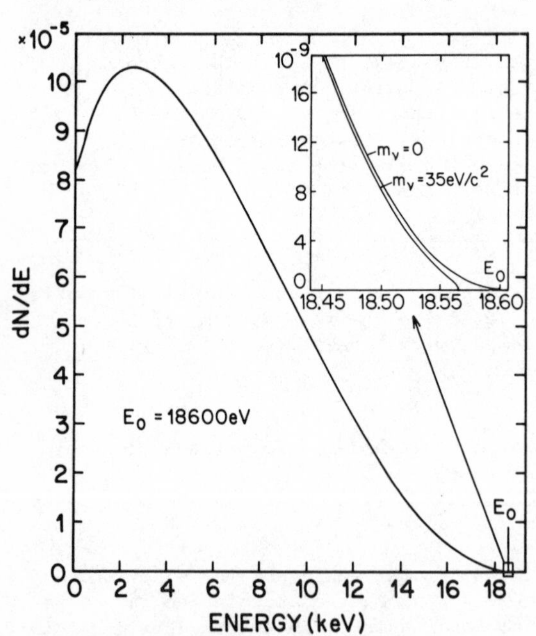

Fig. 1. β-spectrum of tritium. The insert shows the mass sensitive endpoint region magnified by a factor 2000.

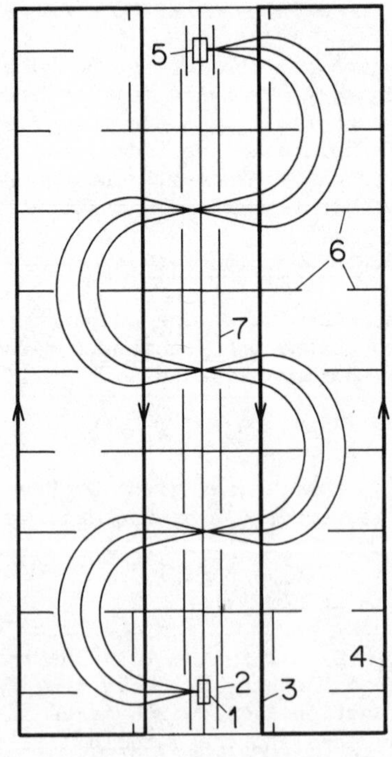

Fig. 2. Scale drawing of the spectrometer.

voltage with respect to a grid (2) near ground potential. The grid consists of 360 fine wires. The accepted solid angle and trajectories are limited by apertures (6,7). The detector (5) is a position sensitive proportional counter, also at high voltage (15 kV) with respect to a second grid of the same type as at the source. Position is determined by charge division, using a resistive anode wire. The diameter of the detector is 6 cm with 10 cm active length. The source assembly consists of 10 rings, each with 5 cm diameter and 1 cm width, i.e. the total active source surface is 157 cm^2. To compensate for the source length a gradient voltage, proportional to the spectrometer dispersion, is applied along the rings. Spectra are recorded by stepping the source high voltage while keeping the analysing energy, E_{mag}, of the magnetic spectrometer constant. For the present investigation we have E_{mag} = 2.2 keV, which gives 27 eV resolution (FWHM) and 1.1 eV/mm dispersion.

In summary, the spectrometer has the following advantages: a) the modest relative resolution (1%) and the small dispersion in the magnetic part require a mechanical precision which can easily be achieved. Also the magnetic and electric fields are known analytically, allowing a reliable Monte-Carlo calculation of the resolution function. This is an important point since a resolution determination from measurements of conversion lines is complicated by shake-off effects (Kündig et al 1984); b) the particles entering the detector have constant energy during a run, implying constant detection efficiency; c) a position sensitive detector has sufficient resolution, i.e. no energy defining slits are necessary and the luminosity is high; d) the resolution is simply determined by the magnet setting and can be chosen for a best compromise between high count rate and high resolution. The obvious disadvantage of the spectrometer is its sensitivity to tritium wall contamination, requiring sources of very high chemical stability. However, background from tritium on the walls is only a problem which reduces mass sensitivity and by itself does not cause systematic errors. In fact the background is a constant as a function of energy, since only the source potential is changed during a measurement which, of course, has no influence on the detector or on tritium on the walls.

5. Sources and Energy Loss Spectrum

The sources were prepared by implantation of T_2^+ ions into carbon, evaporated on to an Al backing. The thickness of the carbon layer was about 2000 Å. The implantation energies were in the range 185 to 400 eV per ion, giving typically an activity of 0.3 mCi/cm^2.

The dominant cause of energy loss of β-particles traversing the carbon substrate is due to excitation of plasmons (Burge and Misell 1968). The probability for exactly n interactions is given by the Poisson distribution

$$P_n(x) = 1/n! \ (x/\lambda)^n \exp(-x/\lambda) \ , \qquad (6)$$

where x is the layer thickness and λ denotes the mean free path. The deflection angle of an inelastically scattered electron is very small and completely negligible in our case. Assuming initially monoenergetic electrons, the energy distribution after the layer is given by

$$f(\Delta,x) = \sum_{n=0}^{\infty} P_n(x) \ \omega^{*n} \ , \qquad (7)$$

where $\omega(\Delta)$ is the normalized energy loss spectrum for a single interaction with energy loss Δ and where

$$\omega^{*n} = \omega * \ldots * \omega \tag{8}$$

denotes the n-fold convolution of $\omega(\Delta)$ with itself. The no-loss term ($n = 0$) is defined by $\omega^{*o} = \delta(\Delta)$. For a distributed source Eq. 7 remains unchanged if P_n is replaced by the right side of Eq. 6 averaged over the source depth. This quantity was measured with 50 Å resolution (FWHM) using a nuclear recoil technique similar to that described by Ross et al (1984). Experimental values for λ and $\omega(\Delta)$ were taken from Burge and Misell (1968) and the energy loss spectra were determined using Eqs. 6 to 8.

6. Measurements and Data Analysis

The data so far available were obtained in four runs (R1 - R4) using three sources (T1 - T3). The measurements were performed with 27 eV resolution (FWHM) in an energy range of about 17.6 keV to 19.3 keV with 5 eV steps. To improve the mass sensitivity, this range was divided into several counting regions with more time spent near the endpoint. A single up-down pass required about one hour. The total measuring time was 27 days. Event data were recorded on magnetic tape and later binned into β-spectra. Figure 3 shows the total dataset used in the evaluation. Figure 4 is the Kurie-plot of the data shown in Fig. 3.

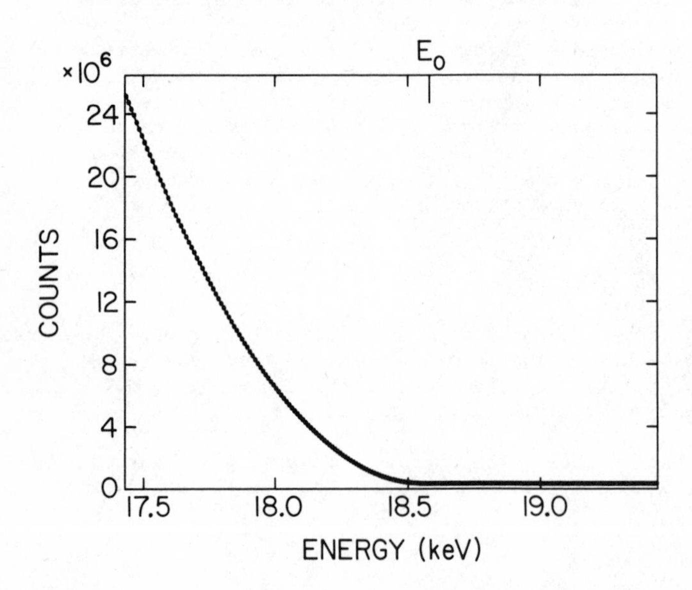

Fig. 3. Measured tritium β-spectrum. The original data of the measured four runs were, after the fitting, binned and added. The vertical scale is correct for the central region of the spectrum. It is ten times less away from the endpoint region, and adjusted at the end for the different energy ranges measured.

The β-spectrum was assumed to have the form

$$N(E) = A\ F(E)\ p\ E_T\ (1 + \alpha\varepsilon_o) \sum_i W_i\ \varepsilon_i^2 \sqrt{1 - m_\nu^2/\varepsilon_i^2}\ , \tag{9}$$

where $\varepsilon_i = E_{oi} - E$. The sum runs over all electronic final states (FS) with branching fractions W_i and endpoints E_{oi}. Each term in the sum is set to zero for E above the point where the term first reaches zero. The E_{oi} are defined by $E_{oi} = E_{oo} - E_{ex,i}$, where E_{oo} corresponds to the endpoint of the electronic groundstate. $E_{ex,i}$ are excitation energies measured from the groundstate. Backscattering in the source substrate is taken into account by the parameter α, which enters in this approximate form by a convolution of a constant backscattering distribution with a β-spectrum proportional to ε_o^2. The fit function is obtained by convoluting Eq. 9 with the spectrometer resolution function and the energy loss spectrum and by adding a constant background term BG. A small correction factor (< 7% variation) for the known changing spectrometer acceptance as a function of retarding voltage has been included.

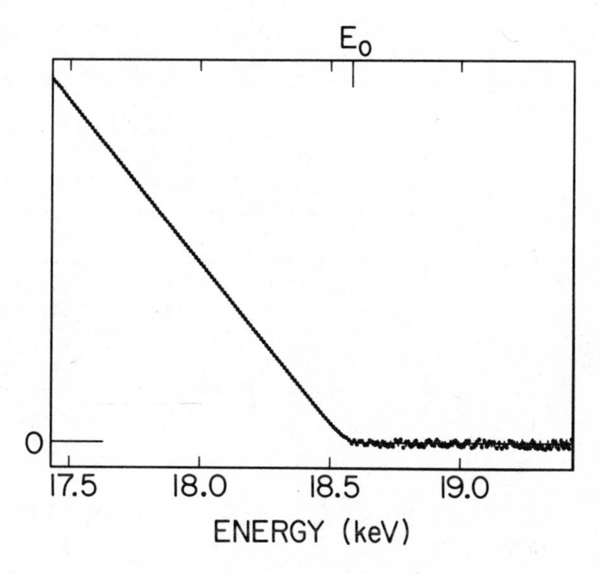

Fig. 4. Kurie-plot of the data of Fig.3. The fitted curve is for $m_\nu = 0$.

The free parameters are α, E_{oo}, BG, an overall normalization A, and the neutrino mass squared m_ν^2. The latter was allowed to take on nonphysical, negative values, in which case the square root in Eq. 9 was replaced by its first order expansion $|1 - m_\nu^2/(2\varepsilon_i|\varepsilon_i|)|$. This gave symmetric, parabola-like curves for χ^2 versus m_ν^2. Our final result is essentially independent of this assumption.

Concerning the final state for our sources we argue as follows. The extremely small diffusion rate of hydrogen isotopes implanted in carbon and the stability of our sources strongly suggest that C-T bonds are formed. Kaplan et al (1982, 1984, 1985) calculated the final state for a variety of molecules with C-T bonds and found very little variation in terms of the groundstate fraction W_o and the mean excitation energy E_{ex}. By fitting various final state distributions we found that just these two quantities are important, whereas details of the final state influence the result very little. We thus take CH_3T, the simplest case, as an adequate representation for the final state of our sources. The corresponding fit results for the four spectra and the combined data set are listed in Table 1. As can be seen, good fits were obtained and m_ν^2 is compatible with zero within one sigma. The nonstatistical variation in the E_{oo} values is caused by uncertainties in the average source-detector distance mainly due to the use of different detectors.

The $m_\nu = 35$ eV hypothesis was tested by fitting all data with fixed values $m_\nu = 0$ and $m_\nu = 35$ eV, giving $\chi^2 = 1370.2$ and $\chi^2 = 1773.2$, respectively, with 1370 degrees of freedom (DOF). As can be seen in Figs. 5 and 6, the latter case can be ruled out without question. In Fig. 7 χ^2 as a function of m_ν^2 of the total dataset is shown.

Possible systematic errors caused by uncertainties in spectrometer resolution, energy loss, and final states, were investigated by varying these distributions within estimated limits and fitting the data again. It was

Table 1. Fit results for four runs with three different sources and the combined set of data using CH_3T final states. Errors indicated are one standard deviation. UL is a statistical upper limit at the 95% confidence level.

Run	Source	χ^2/DOF	E_{oo}-18500 (keV)	α (keV^{-1})	m_ν^2 (eV^2)	UL(95%) (eV^2)
1	T1	328.8/337	82.6±0.3	0.018±0.004	+140±130	356
2	T2	345.6/356	84.5±0.2	0.020±0.002	-9±140	218
3	T2	349.9/356	84.4±0.2	0.022±0.002	-22±140	202
4	T3	343.7/317	77.6±0.2	0.021±0.002	-85± 93	70
all		1370.2/1369			-11± 63	95

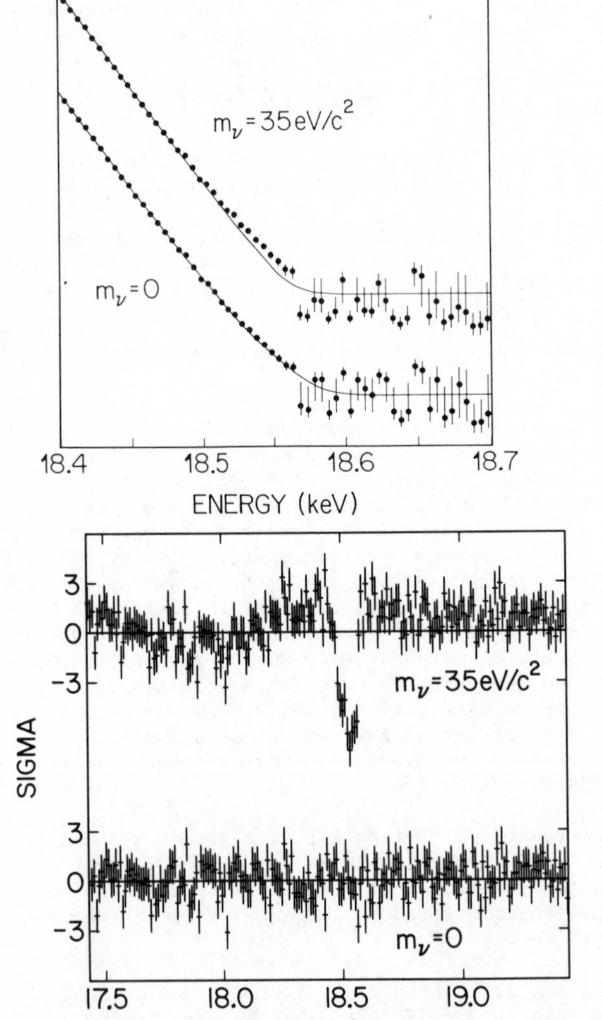

Fig. 5. Curie-plot of the data from run R4 fitted with m_ν=0 and m_ν=35 eV. All other parameters (E_{oo}, BG, A, α) are fitted. Notice the strong correlation of m_ν with the background in the fitted curve.

Fig. 6. Plot of the difference between the fitted function and the data for m_ν=0 and m_ν=35 eV. The four spectra were fitted independently from each other. The figure was obtained by the appropriate binning (after the fit).

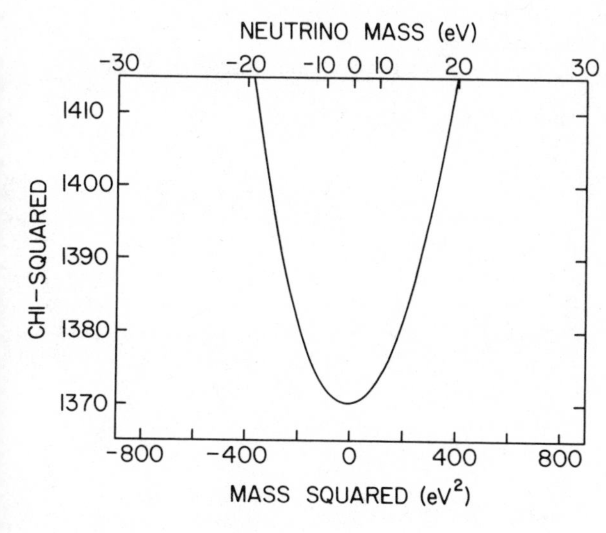

Fig. 7. Plot of χ^2 versus m_ν^2 for the combined set of data, with all other free parameters being fitted. The number of degrees of freedom for these fits is 1370. Assumed are the final states of methane CH_3T.

found that the systematic error for m_ν^2 should be smaller than 204 eV^2. For more details see Fritschi et al (1986). Assuming the best fitted value for m_ν^2 would be exactly zero (and not -11 eV^2) a statistical upper limit of 106 eV^2 for m_ν^2 at the 95% confidence level was calculated. Adding the systematic and the statistical limit linearly we obtain our present result:

$$m_\nu^2 < 310 \text{ eV}^2 \qquad \text{or} \qquad m_\nu < 18 \text{ eV}.$$

7. Conclusion

No indication for a non-zero mass of the electron antineutrino was found, which is in strong contradiction to the ITEP result. We see no possible source of error in our experiment large enough to account for this discrepancy. Presently we are preparing further measurements with new sources, in an attempt to reduce some of our systematic uncertainties.

References

Bergkvist K E 1972 Nucl. Phys. B39 317
Bethe H A 1986 Phys. Rev. Lett. 56 1305
Boris S, Golutvin A, Laptin L, Lubimov V, Nagovizin V, Novikov E, Nozik V, Soloshenko V, Tihomirov I and Tretyakov E 1985 Phys. Lett. 159B 217
Burge R E and Misell D L 1968 Phil. Mag. 18 251
Fritschi M, Holzschuh E, Kündig W, Petersen J W, Pixley R E and Stüssi H 1986 Phys. Lett. 173B 485
Kaplan I G, Smelov G V and Smutnyi V N 1984 DAN USSR 279 1110
Kaplan I G, Smutnyi V N and Smelov G V 1982 Phys. Lett. 112B 417
Kaplan I G, Smelov G V and Smutnyi V N 1985 Phys. Lett. 161B 389
Kündig W, Petersen J W, Pixley R E, Stüssi H and Warden M 1984 Proc. Fourth Moriond Workshop on Massive Neutrinos in Astrophysics and in Physics, ed. J Tran Thanh Van (Edition Frontières, Paris)
Lubimov V A, Novikov E G, Nozik V Z, Tretyakov E F and Kosik V S 1980 Phys. Lett. 94B 266
Lubimov V A, Novikov E G, Nozik V Z, Tretyakov E F, Kozik V S and Myasoedov N F 1981 Sov. Phys. JETP 54 616

Mikheyev S P and Smirnov A Yu 1985 Yad. Fiz. 42 1441
Reines F and Cowan C L 1953 Phys. Rev. 92 830
Ross G G, Terreault B, Gobeil G, Abel G, Boucher C and Veilleux G 1984 J. Nucl. Mat. 128 730
Tretyakov E F 1975 Izv. Akad. Nauk SSSR 39 583
Wolfenstein L 1978 Phys. Rev. D17 2369

Inst. Phys. Conf. Ser. No. 86
Paper presented at Int. Nucl. Phys. Conf., Harrogate, UK, 1986

Summary

D.M. Brink

Department of Theoretical Physics, 1 Keble Road, Oxford OX1 3NP, UK

1. Introduction

Many new developments were reported at the 1986 International Nuclear Physics Conference at Harrogate. In this report I will attempt to summarize the highlights of the Conference. I will discuss the main points raised in the talks of some of the Invited Speakers and comment on some topics which I found to be particularly interesting. The report is organized in a different way to my lecture at the Conference to make it easier to refer to the texts of the talks of the main speakers. The order of the topics in this summary is the same as the order on the Conference programme.

In the opening session of the Conference Professor Bromley spoke about the challenge of new fields of nuclear physics. New experimental facilities present new and exciting possibilities for studying nuclei under extreme conditions, for investigating the quark structure of nuclei and the transition from quark to hadronic matter. Professor Bromley emphasised the importance of the experimental approach and of the role of theory in suggesting new experiments. He spoke about the new results on exotic nuclei and high-spin states which have come out of the experimental effort at Daresbury. He also said that it was characteristic of nuclear physics that it is possible for young people to get to the frontiers of the subject and to make new contributions to it.

Professor Mottelson took a historical approach in his keynote talk. He distinguished three periods in the development of nuclear physics. The important events in the first period (1886-1935) were the discovery of the nucleus and its constituents. The general nature of nuclear dynamics was clarified in the period 1935-1952. In the last period from 1952 to the present, physicists have been discovering the 'feel' of nuclear matter. During the past few years the study of sub-nuclear degrees of freedom has again become one of the central themes of nuclear physics and it will be interesting to see what picture of the nucleus emerges from these new interests. Professor Mottelson reminded us of the coexistence of different theoretical pictures of the nucleus, the shell model and the compound nucleus model, the shell model and the collective model. Various theories picture the nucleus as being composed of nucleons or bosons or clusters. Each approach gives us a different view of its structure.

Professor Vogt looked to the future in his keynote talk. He emphasised that nuclear physics is an experimental science and said that new tools lead to new discoveries. Many doors are open: we can study the

Gamow-Teller response function, pion propagation in the nucleus and collective flow in heavy-ion reactions. Professor Vogt spoke about some of the new experimental facilities which are working now or will be in operation soon. The electron accelerators at Mainz and Bates will deliver CW electron beams up to 1 GeV, while the new CEBAF accelerator will produce electrons with energies up to 5.6 GeV. Relativistic heavy-ion beams are available at the AGS at Brookhaven and at CERN and there is a proposal for a big new accelerator (RHIC) at Brookhaven. There are proposals for kaon factories at TRIUMF, LAMPF and the European Hadron Facility (EHF). Storage rings for cooled heavy-ion beams are being built in a number of laboratories. These accelerators will open up new areas of nuclear physics.

2. Nuclear Probes

The sessions on nuclear probes included talks on proton scattering and on electromagnetic, mesonic and strange probes. In the first part of his talk Dr. Moss spoke about experimental trends in (p,n) and (p,p') reactions and especially about work at Los Alamos for proton energies in the 500 MeV range. He reported on new results on missing Gamow-Teller strength in the (p,n) reaction, and on measurements of polarization parameters which tie in with the relativistic theory of proton scattering reviewed by Professor Wallace in a later session at this Conference. The second part of the lecture described a method to look for signals of the quark structure of nuclei by studying muon pair production with 900 GeV protons. The idea is to study the effects of the nuclear medium by comparing the pair production rates on a deuterium and a calcium target. The third topic was about the measurement of the effective mass of a nucleon in the nucleus by quasielastic proton-nucleus scattering. The relativistic mean field theory discussed by Professor Walecka in a later session predicts that the ratio of the mass of a nucleon inside the nucleus to the free nucleon mass is $M^*/M = 0.6$. The mass ratio affects the spin observable A_y and the experimental data are consistent with $M^*/M = 0.9$.

Professor De Witt Huberts told us about the 100 keV resolution which has been achieved for the (e,e'p) reaction with the new spectrometer set-up at NIKHEF-K in Amsterdam. Now the (e,e'p) reaction has a resolution comparable with that of the $(d,{}^3He)$ reaction. The (e,e'p) experiment can measure a spectral function which is the joint probability, $S(E_m,p_m)$, of finding a proton with energy E_m and momentum p_m in the nucleus. The spectral function is determined in a broad range of momentum values and thus the bound-state wave function can be sampled inside the nucleus as well as at the surface. The NIKHEF group have used their results on the (e,e'p) experiment in conjunction with Groningen data on the $(d,{}^3He)$ reaction to measure the absolute spectroscopic factor for the $f_{7/2}$ level in ${}^{51}V$. They find that the total spectroscopic factor is only 54% of the shell-model sum rule. In another experiment they studied the spectroscopic factor of the $3s_{1/2}$ state in Pb isotopes and found a 20%-30% depletion compared with the shell-model sum rule.

Professor Bowman reported on various experiments using pion beams. Pion elastic and inelastic scattering can give information about isospin mixing. Typical mixing matrix elements in ${}^{12}C$ have a magnitude of about 100 keV. Pion charge-exchange scattering can be used to excite isovector giant resonances. These reactions have the advantage that there is no $T = 0$ background and no spin-flip for forward scattering. It is interesting that isovector monopole and dipole giant resonances are excited quite strongly, but the isovector giant quadrupole state has not been seen.

In his invited talk Professor Dover spoke about reactions initiated by kaon beams as strange probes. Kaon beams can be used to produce hypernuclei either by strangeness exchange or by associated production. The second method uses a pion beam: the pion interacts with a nucleon and produces a positive kaon together with either a lambda or a sigma hyperon which can remain bound in the final nucleus. The main branches for hypernucleus decay are either nucleon or pion emission. Branching ratios for the various decay modes have been measured in some cases. Also gamma rays emitted by excited hypernuclei have been used to look for unnatural parity excited states of hypernuclei. Kaon beams are being used to look for dibaryon states and there is some work on scattering of kaons by nuclei.

3. Nuclear States and Modes of Excitation

On Tuesday we heard talks on the shell model, the IBA, on high-spin states and on giant resonances. Professor Brown brought us up to date on the achievements and goals of recent shell-model calculations for nuclei with A = 16 - 40. Wildenthal and his collaborators obtained an effective interaction by fitting binding energies and excitation energies. Wildenthal's interaction for the (s,d) shell has 63 + 3 parameters and fits the known positive parity levels in the shell up to an excitation energy of 6 MeV to within an rms deviation of 150 keV. The derived wave functions have been used to calculate all measured electromagnetic observables. Calculated E2 matrix elements agree well with experimental values if an effective charge is used. A least square fit gives values for the effective charges of 1.78e for the isoscalar and 0.8e for the isovector operators. Theoretical and experimental magnetic moments agree very well using unrenormalized operators. The experimental GT matrix elements are systematically smaller than the shell model values calculated with the free-nucleon operator by a factor of about 0.77. Drs. Moss and Gaarde both report a quenching of Gamow-Teller strength which is consistent with this value.

Dr. Dieperink discussed the assumptions of the interacting boson model. The simple version of the IBM-2 assumes S-D dominance, meaning only S and D correlated nucleon pairs or s- and d-bosons need be considered. Over the past few years there has been a lot of discussion about the role of G-pairs or g-bosons. A 4^+ excited state in an even nucleus could be a g-boson state or it could consist of two d-bosons coupled together. Dr. Dieperink showed some form factors extracted from electron scattering data which suggest that the two possibilities can be distinguished by electron inelastic scattering experiments. He also discussed the low-energy isovector M1 collective mode in deformed nuclei ('scissors mode') which has been seen in electron scattering experiments. There are new photon scattering experiments on ^{156}Gd which have much better energy resolution than the electron scattering data and which resolve the collective state in a number of components.

The shell model is useful for experimentalists because it can be used to predict properties of new nuclei far from the stable regions of the mass table. At the end of his talk Professor Brown spoke about the future of shell model calculations. To extend the s-d shell calculations to the p-f shell would require a large increase in computing power. An alternative is to find a useful truncation of the shell-model space. In a sense the IBM-2 is a truncation of the shell model with effective interactions and effective operators to a sub-space spanned by S- and D-pairs. Even though it is a very drastic truncation it gives a remarkably good description of the experimental data. It is possible that other less drastic truncations exist which would permit useful large scale shell-model calculations to be made in other parts of the periodic table.

During the past few years there has been a lot of progress in the study of high-spin states. This is due to the development of various detector arrays like the Crystal Ball at Heidelberg and the TESSA arrays at Daresbury. Dr. Nolan spoke about recent results from Daresbury. The most spectacular of these is the observation with the TESSA 3 array of the discrete line spectrum of a superdeformed band in ^{152}Dy. It extends up to an angular momentum of 60 $\hbar$ and accounts for the ridge structure seen previously in continuum gamma-gamma correlations. The Daresbury group has found yet another superdeformed band in ^{132}Ce. The band extends up to about $J = 50\ \hbar$. Dr. Twin continued the discussion in a contributed paper on Tuesday afternoon. He described how the experimental spectrum was extracted from the background by using an isomer subtraction procedure and discussed the assignment of spins to the superdeformed states. States with angular momentum between about 26 $\hbar$ and 60 $\hbar$ are seen in the experiment. Dr. Twin also raised several questions including: how is the band populated? Why does it terminate so suddenly at the low angular momentum end?

In his lecture on giant resonances Dr. Gaarde spoke about the development in the understanding of spin modes in nuclei. He concentrated on spin-isospin excitations and presented new data on the (d,2p) reaction and on heavy-ion charge exchange at intermediate energies. There is a large cross section for the production of the delta resonance in the (d,2p) and the (^{3}He,t) reactions. The experiments show a shift of about 40 MeV in the energy of the delta resonance from the values predicted by simple theories. The shift may be due to a change of properties of the delta in the nuclear medium and is one of the most intriguing aspects of the data. Another intriguing observation concerns the quasi-free peak in the cross section. The peak has the same position and a very similar shape in both the (^{3}He,t) and the (e,e') reactions.

On Tuesday afternoon there was a talk by Dr. Detraz on recent experiments at GANIL on the limits and properties of light nuclei far from stability. The GANIL experiments use the mechanism of projectile fragmentation of a heavy-ion beam incident on a target at high velocity. The Z and N distributions of these fragments, which is governed by statistics, allows observation of exotic nuclear species which could not be obtained by other methods. Many new nuclei have been produced and some of their properties have been measured.

The same afternoon Dr. Lister gave a short invited talk on the gamma spectroscopy of nuclei on the frontiers of stability. He discussed the spectroscopy of ^{80}Zr which is the heaviest known nucleus with N = Z. It is predicted to be a γ-soft deformed nucleus with deformation $\beta = 0.4$ and is produced in very small quantities together with a number of other products in the reaction:

$$^{58}Ni + ^{24}Mg \rightarrow ^{82}Zr^* \rightarrow ^{80}Zr + 2n\ .$$

The gamma spectrum of ^{80}Zr is contaminated by a huge background of lines from other reaction products but Dr. Lister showed that it is possible to subtract the background to reveal two lines from ^{80}Zr.

4. Reaction Mechanisms and Nuclear Structure

This session included talks on clustering, sub-barrier fusion, mean-field physics and intermediate heavy-ion physics.

Dr. Betts gave an invited talk on clustering aspects of nuclear structure and reactions. He began by describing the alpha + ^{16}O cluster model of ^{20}Ne where the alpha particle moves in a deep local potential and certain states of relative motion are excluded by the Pauli principle. This model reproduces the bound-state energies and alpha-scattering phase shifts. He then discussed the relation between this model and the deformed shell model. Recently Marsh and Rae have developed a model of ^{24}Mg based on a rotating alpha-cluster picture. There are a number of equilibrium cluster configurations which seem to correspond to known bands of resonance states seen in ^{12}C + ^{12}C scattering.

A comparative study of fusion excitation functions in the vicinity of the Coulomb barrier shows that they are influenced by nuclear structure effects. In particular the coupling of the fusion process to various direct reaction channels can give a large enhancement in the sub-barrier fusion cross section. Dr. Reisdorf's lecture discussed some recent examples of this phenomenon. In the case of light systems like ^{16}O + ^{13}C conventional quantum tunnelling through a potential barrier can reproduce the experimental data. For heavier systems coupling to collective modes is important. Coupling to vibrational modes enhances sub-barrier fusion and the effect can be calculated starting from a coupled-channels approach. The agreement with the data is remarkably good. Complex new effects show up when systems with $Z_1 Z_2 > 1600$ are studied. For such systems tunnelling is important only for energies well below the Coulomb barrier. When the energy is near the top of the barrier, energy is transferred from relative motion into internal degrees of freedom by frictional processes, and fusion is inhibited. This results in an upward shift of the barrier. Fluctuations spread out the transition region in the fusion excitation function.

Mean-field physics at finite temperature was the second topic to be discussed on Wednesday morning. Dr. Levit gave an overview of phenomena which can be studied in hot nuclei. Then he discussed mean-field theories of temperature-induced shape transitions in rotating nuclei and the upper limit of nuclear binding. He showed how the ideas of Landau's mean-field theory can be used to identify universal features of nuclear shape transitions. The idea is that any excited nucleus has a deformed equilibrium shape. This shape depends on the excitation energy and angular momentum. It can be found by minimizing the free energy with respect to deformation parameters. Dr. Levit writes the mean-field free energy for a nucleus with a given temperature and angular velocity as a polynomial in the shape variables. The coefficients in the expansion depend on the temperature and include effects due to shell structure. The equilibrium shape is determined by minimizing the free energy. When the angular velocity is zero the equilibrium shape is either prolate or oblate, depending on the values of the coefficients in the free energy. Triaxial shapes are possible for non-zero values of the angular velocity, while at high temperature and high angular velocity the shape is always oblate. Alhassid, Levit and Zingman have discussed shape phase transitions and have shown that the equilibrium shape can be read off a universal phase diagram in two dimensionless parameters which are functions of the parameters in the free energy. They have studied the particular example of ^{166}Er and have shown that for angular momenta near 15 $\hbar$ and excitation energies near 60 MeV, there is a very rapid shape transition from almost prolate to oblate.

It is possible that in some cases Dr. Levit's assumptions about the dependence of the free energy on deformation might not be general enough. The free energy surface could have a double minimum. Then an excited nucleus would have two shapes in thermal equilibrium at the same excitation energy and angular momentum. One shape might be triaxial and the other a superdeformed shape in the second well. A situation like this could explain the feeding of the superdeformed band in ^{152}Dy in the experiments described by Drs. Nolan and Twin.

Professor Tamain talked about recent progress in intermediate energy heavy-ion physics. Heavy-ion induced fission has been studied in collisions between heavy ions for projectile energies up to 44 MeV/A. The linear momentum transfer from the projectile to the target can be found by measuring the separation angle between the two fission fragments emitted by the fused system after it has decayed by fission. At lower bombarding energies, below 10 MeV/A, the folding angle distribution indicates that the whole of the projectile fuses with the target nucleus to form a 'compound nucleus' which then decays by fission. At higher energies the folding angle distribution indicates that some kind of incomplete fusion process is occurring. For energies below 44 MeV/A complete fusion dominates. At 44 MeV/A only about 70% of the projectile momentum is transferred to the compound system. For heavy targets and for energies beyond 30 MeV/A only a small fraction of the reaction cross section goes into fusion-like events. These observations indicate that it is difficult to form a compound nucleus with a very high excitation energy and that there might be a maximum excitation energy that the compound nucleus can carry.

5. Relativistic Nuclear Physics

The session on Thursday morning included talks on relativistic nucleon-nucleus scattering, on mean-field theory, on the quark-gluon plasma and on relativistic heavy-ion physics.

Professor Wallace spoke about the relativistic description of proton-nucleus interactions. He began by discussing the framework for a complete characterization of the nucleon-nucleon scattering amplitude. The nucleon-nucleus optical potential can be calculated from the NN amplitudes and the nuclear density by using the impulse approximation. The general expression for the optical potential is quite complicated but simplifies if the incident proton energy is not too high. Then the optical potential can be written as a sum of six terms which include scalar, vector, tensor and spin-orbit components. The resulting potentials have been used to calculate differential cross sections and polarization observables for scattering of protons by ^{40}Ca at several energies between 181 and 800 MeV. The agreement with experimental data is very good.

Professor Walecka discussed recent progress in quantum hadrodynamics. He spoke about relativistic mean-field theory which describes the nucleus as a set of Dirac nucleons moving in self-consistent scalar and vector meson fields. There are six parameters in the model: the masses of one scalar and two vector mesons and three meson-nucleon coupling constants. The vector mesons are assumed to be the rho and omega, and their masses are taken from experiment. The four remaining free parameters are chosen to fit the binding energy, density and symmetry energy of nuclear matter and one property of a finite nucleus (the charge radius of ^{40}Ca). Then many properties of other nuclei, including the spin-orbit coupling strength, can be calculated and compared with experimental values. It is interesting to compare some of the properties of the relativistic theory with non-relativistic Hartree-Fock theory using effective interactions. Both theories give a good description of binding energies and charge distributions of finite nuclei. The mechanism for nuclear saturation is different in the two theories. In the theory described by Professor Walecka it is a relativistic effect and is due to the repulsion between like nucleons and the damping of the scalar meson attraction with increasing density, while in non-relativistic Hartree-Fock theory it is due to a combination of exchange effects and density dependence of the interaction. The ratio of the effective mass to the true nucleon mass in the present theory is $M^*/M = 0.56$ while in the non-relativistic theory it depends on the strength of the exchange part of the effective interaction. (For the Skyrme force SIII it is 0.7.)

Professor Baym began by introducing the concepts of relativistic heavy-ion collisions. He argued that in such collisions it might be possible to study the de-confinement of quarks and gluons. Central collisions at centre-of-mass energies greater than 25 GeV per nucleon may lead to energy deposits of several GeV/fm^3. At these excitation energies the system comes rapidly to local thermal equilibrium and subsequent motion can be described by hydrodynamics. At such energies nuclei are sufficiently transparent to nucleons that nuclei pass through each other, producing two highly excited fragmentation regions, one corresponding to the target and the other to the projectile. These regions are joined by a central rapidity region with a small baryon number but substantial energy density. The evolution of the system can be described by the Landau hydrodynamic model with Lorentz-invariant boundary conditions. The fragmentation regions might also be described by hydrodynamics but the situation is more complex. Immediately after the collision there is not much interaction because of asymptotic freedom. Hydrodynamics starts to operate after a proper time of about 1 fm/c. The initial conditions correspond to a constant energy density and an initial flow velocity $v = z/t$. This initial condition is due to Bjorken and is different from the one used by Landau.

The initial energy density is estimated from pion multiplicities in pp collisions (about 2 GeV/fm^3). This corresponds to a temperature:

$$T = 160\ \varepsilon_o^{1/4} = 200 \text{ MeV}.$$

The initial entropy density is about 14 fm^{-3}

$$s_o = 8.4\ \varepsilon_o^{3/4}$$

For a short time the dominant motion is longitudinal. The longitudinal expansion obeys a simple scaling solution:

$$s(\tau) = s_o \tau_o/\tau\ ,\ v_z = z/t\ .$$

There is also a transverse motion. Both expansions lead to a cooling and there will be a transition to hadronic matter. Entropy may be produced. Then the system becomes a free streaming gas of hadrons.

The proposed programmes for experiments on relativistic heavy-ion collisions were surveyed by Professor Specht. In autumn 1986 several groups at CERN will make a series of experiments using ^{16}O beams with energies up to 225 GeV/A. By 1987 CERN hopes to have beams of ^{32}S and ^{40}Ca. The AGS at Brookhaven will produce 15 GeV/A beams of ^{32}S by the end of 1986. Professor Specht described the various observables which can be studied in a typical reaction and discussed some of the details of the NA34 collaboration which is one of the CERN experiments. This experiment combines 4π calorimeter coverage with measurements of inclusive particle spectra, two-particle correlations, low- and high-mass lepton pairs and photons. A multiwire active target allows maximum interaction rates with a minimum of secondary interactions.

6. Sub-nucleon Physics

On Friday morning there were talks on sub-nucleon degrees of freedom, on quarks in nuclei, on electromagnetic probes and on the EMC effect.

Dr. Isgur spoke about chromodynamics in nuclear physics. He outlined the procedure for making numerical lattice QCD calculations in a Hamiltonian formalism and showed how various quark states on a lattice can be classified by diagrams. In normal nuclear matter, nucleons are grouped into clusters. The repulsion between the clusters is a consequence of the effects of the Pauli principle, of Fermi motion and of the Heisenberg uncertainty principle. Not all six quark systems show clustering into two groups of three clusters.

In his lecture on quarks in nuclei Professor Thomas explained how the quark model has been used to obtain a deeper microscopic understanding of nuclear physics. Several theoreticians have used the resonating group method to calculate N-N scattering completely within the non-relativistic quark model. Others have used bag models or boundary condition models. Some of the results are encouraging but at present they raise more questions than they answer. There is no doubt that the discovery of the EMC effect has stimulated many studies of quark models of nuclei. It now seems that a more careful treatment of nuclear binding and Fermi motion can explain most of the effect. There is no unambiguous experimental evidence that the nucleon gets bigger, or 'swells', inside a nucleus. There is no doubt that the nuclear shell model has been very successful in describing many nuclear properties. Any serious quark model must take this into account.

There are still some doubts regarding the significance of the EMC effect and it is clear that better experimental data are needed. Professor Rith summarized the present evidence and outlined the programme for the future. Professor Sick discussed evidence from electron scattering for meson and quark degrees of freedom in nuclei. During the past 15 years considerable effort has gone into the search for meson degrees of freedom in nuclei. These are visible in electron scattering through the presence of meson exchange currents (MEC). Experience has shown that these are much larger at large momentum transfer. In electron scattering from ^3H they show up for $q^2 > 20\ fm^{-2}$. Professor Sick explained that a combination of careful experiments and good many-body calculations is needed to unravel meson exchange effects. When it comes to quark effects it is at present impossible to make accurate theoretical predictions because no consistent theory exists. In order to learn about the role played by quarks in nuclei we need experiments at higher momentum transfers which look at more exclusive channels and are sensitive to specific wave-function components. This can be done once a CW accelerator like CEBAF is available.

7. Other Topics

A number of other topics were covered in parallel sessions in the afternoons. Professor Kamimura spoke on the three-body coupled-channel theory of scattering and breakup of light and heavy ions. In his theory excitation and breakup degrees of freedom are taken into account explicitly by using the method of coupled discretized continuum channels. Dr. Bohigas spoke about the application of the ideas of quantum chaos to the study of the compound nucleus. It appears that the statistical distributions of level spacings in a compound nucleus have a universal form which is independent of the details of the nuclear forces provided that they are time reversal invariant. Professor Coon reviewed recent progress in our understanding of the effects of three-body forces on nuclear properties. We need to improve our feel for the role of three-body forces in order to extract information about meson exchange currents in nuclei. Dr. Stachel spoke about pion and photon production in medium-energy heavy-ion collisions. These processes are sensitive to coherent effects like clustering and the momentum distribution of nucleons in the nucleus.

Professor Bosch surveyed the planned heavy-ion storage rings. These will be powerful tools for new and precise experiments in nuclear and atomic physics. Such rings are being built in Sweden, Denmark and Germany. They will provide high current beams of highly ionised light and heavy atoms, and will accumulate, store and cool the beams to give a small energy spread and small beam dimensions. It will also be possible to produce beams of radioactive nuclei.

Recently, exciting new results have been found on the emission of positrons from superheavy quasiatoms. Professor Schwalm reviewed the new experiments. The effect occurs in U + U and U + Th collisions for energies near the Coulomb barrier. Narrow positron peaks with energies near 300 keV and widths of about 80 keV were observed several years ago. Now narrow electron peaks have been observed in coincidence with the positrons in the U + Th reaction. The sums of the peak energies take the values 610, 760 and 810 keV. The width of the summed energy of the peaks is less than 40 keV and seems to imply back-to-back, two-body decay of a system at rest in the centre of mass system. There is no agreed explanation about the origin of the effect but it certainly looks interesting.

On the final day Professor Haxton spoke about nuclear tests of fundamental symmetries. He discussed parity non-conservation in the N-N interaction, time-reversal-odd nuclear moments and matter-enhanced neutrino oscillations. This last topic is of astrophysical interest. Mikheyev and Smirnov have discovered a mechanism by which electron neutrinos produced in the solar core could be converted efficiently into neutrinos of a different flavour. It provides a plausible solution to the solar neutrino puzzle. Dr. Kündig presented his new results on the neutrino rest mass. His value for the neutrino mass is consistent with zero with an upper limit of 18 eV which includes instrumental and statistical uncertainties as well as uncertainties due to the energy loss in the source and the final electronic states.

8. Outlook for the Future

The Harrogate Conference has pointed the way to new opportunities in nuclear physics. With existing accelerators it is possible to study nuclei with high excitation energies, high angular momentum and with N and Z values far from stability. The TESSA 3 array at Daresbury has been used for gamma-ray experiments which seemed to be impossible a few years ago, and has revealed new and unexpected phenomena. The heavy-ion cyclotrons in France, Germany and the USA are making discoveries about nuclei under extreme conditions of excitation. It is possible that unusual phase transitions occur in highly excited nuclei. The beautiful experiments at GSI on positron lines produced in U + Th collisions have yielded intriguing and as yet unexplained results. Many new heavy-ion experiments will become possible when the storage rings with cooled heavy-ion beams come into operation. The study of relativistic hadron-nucleus scattering and the relativistic mean-field theory of nuclear structure has made much progress since the last International Conference in Florence. When the new hadron facilities come into operation it should be possible to obtain a clearer understanding of the role of sub-nucleon degrees of freedom in nuclear physics. New electron accelerators which can be used for high momentum transfer scattering experiments will also be able to probe sub-nucleon phenomena. The first of a new series of experiments with relativistic heavy ions will begin later this year. They may be able to excite the quark degrees of freedom in nuclei.

Inst. Phys. Conf. Ser. No. 86
Paper presented at Int. Nucl. Phys. Conf., Harrogate, UK, 1986

Concluding Remarks

H. Feshbach, Chairman, IUPAP Commission on Nuclear Physics and
Massachusetts Institute of Technology

I've been asked to make a few concluding remarks - and I will make them few. To me the most remarkable part of this Conference was the demonstration of the extraordinary range which now comprises the subject called nuclear physics - and the correspondingly wide range in experimental and theoretical techniques of great power and effectiveness. David Brink's summary gave ample demonstration as he ranged from experiments, which I think our local hosts have looked on with justifiable pride, in the uncovering of zirconium-80 and the superdeformed band of dysprosium-152, to those looking for the 'smoking gun' which will tell us how quarks are deconfined in nuclei or <u>if</u> they are deconfined in nuclei. This constitutes an enormous range in phenomena and an equally enormous range in experimental and theoretical ways of tackling such questions. However, there is a danger, namely that the range is so wide that we are no longer able to communicate with each other and that is what makes meetings of this kind so important. They bring together different points of view, different insights, different ways of looking at things, and try to get a consensus of current thinking and developments.

For many years our colleagues who are engaged in particle physics research have justified their support by saying that they were going to discover, or uncover the nuclear forces. This has been a constant theme in their proposals and publications but I think they have failed. They have not done it and have left it behind and gone on to higher energies - as Nathan Isgur said this week, 'they have gone to the holy grail of the golden Lagrangian which will explain the universe in one fell equation!'

To be sure, we have made a lot of progress without knowing what the nuclear forces are in any great detail. It is really amazing how far one can progress with a little empiricism, and the idea that the nuclear forces are essentially short-ranged and strong. One can explain large amounts of data on both a qualitative and a quantitative basis. It involves a very nice interaction between the experimental and theoretical approaches to the phenomena. However, we are now engaged in going beyond that kind of study. We now want to know what the strong interaction is inside the nucleon and what happens to the nucleons when they are inside the nucleus. For example, do they change their shape, becoming 'swollen'. We have many probes and we have many nuclei - these are our strengths. We want to exploit them in order to obtain further information. That is just one of our frontiers and we have many.

David Brink alluded in many ways to the frontiers which are associated with the heavy ions, where we do everything from making new elements to new, and various, dynamic types of reactions. One of the things that intrigued me, is the fixation on the use of equilibrium methods coming from the many-body picture. The words temperature and free energy, and dynamics generally, are used to describe the small systems with which we deal. There are limits on that and I think they need to be explored in order to understand the nature and validity of the results which are obtained using these approximations. There is a complementary type of description which has a similar character: namely the assumption that there are only two types of reactions - those which are called direct and those which lead to compound nuclei, achieving equilibrium. Let me assure you that there is a whole range of reactions that go from direct to compound with all kinds of intermediate states and a whole range of interaction times. If the interaction time is short then you have direct, if the interaction time is long then you have compound but, of course, there is a whole range of interaction times where the situation is in-between. In any given reaction, especially as energy increases, you have to analyse the situation completely rather than fixing on these limiting situations.

These two interests of ours, these two frontiers - the heavy ions, the quarks - 'fuse' when we come to the relativistic heavy ions, when we look to see whether there is a quark-gluon plasma or - let's make it a little less ambitious - whether there is deconfinement. The relativistic heavy ion, of course, is a very daring experiment and the hope that it will produce a quark-gluon plasma is very speculative but of considerable interest. If it does occur, we will be able to explore situations which extend back to the beginning of the universe at the time of the 'Big Bang'. This is a fascinating and a wonderful attempt to get into a new and really unusual area.

One cannot help but be impressed by the vitality of our field, the way it has been growing and increasing the range of its interests. I thought it might be amusing to show you the programme which governed a similar meeting at Gatlinburg in 1966, 20 years ago. This included direct reactions, mostly d,p, intermediate structure in analogue states, electromagnetic interactions - mostly concerning giant dipole resonance. Mesons: - remember this is before Los Alamos, SIN and TRIUMF - mostly pionic x-rays, muonic x-rays, etc. Nuclear models: the shell model and RPA, Bohr-Mottelson theory and so forth. Nucleon-nucleon interaction: in 1966 they had a whole session on nucleon-nucleon interaction - here we sort of sneak it in here and there. Fission, statistical properties of nuclei, properties of individual levels, few-nucleon systems and, in the concluding session, effective forces in nuclei, and nuclear structure with 1 BeV nucleons. If you compare this programme with what we have been discussing this week you get an idea of where we have been going and how fast.

I think this has been a very good Conference and I know you will join me in congratulating all concerned in the smooth and unbelievably efficient way in which the meeting was run. Excellent choices of programme and speakers; the visits to York and Ripley Castle. I must say I couldn't detect one minor glitch during the whole stay, and I have been to many meetings and am very sensitive to glitches. So I would like to take this opportunity to thank, on your behalf and mine, the people who have been responsible for our happy stay here. Remember, of course, you cannot blame them, no matter how hard you try, for the weather! I declare the meeting adjourned.

PARALLEL SESSIONS

Inst. Phys. Conf. Ser. No. 86
Paper presented at Int. Nucl. Phys. Conf., Harrogate, UK, 1986

Antiproton–proton cross-sections at low momenta and the range of the nuclear force

S. Paul

Max Planck Inst. f. Kernphysik, Heidelberg Germany

1. INTRODUCTION

Nucleon-nucleon and pion-nucleon scattering have for a long time been the only source of information about the strong interaction, and in fact, almost everything we know about the nuclear force stems from that source. We have learnt about meson-exchange forces, coupling constants, spin and isospin degrees of freedom, and many physicists believe that our present models describe the low and medium energy phenomena satisfactorily. We might ask the question, what more can we learn from antinucleon-nucleon scattering ?

2. THE POTENTIAL MODEL

For this let me shortly review, how the NN and $\bar{N}N$ interaction are generally described. For the nucleon-nucleon case, the data are very well understood using meson-exchange potentials. The long range part of the interaction is mediated via the exchange of a pion, at shorter distances we can describe the interaction with the exchange of 2π, ρ or an effective 2π exchange like the σ. The existence of stable nuclei tells us, that we need a hard core at very short distances which can be described by a 3π or ω exchange. What happens if we go to the antinucleon-nucleon system ? Since the $\bar{N}N$-system is the G-parity conjugate of the NN system, we only have to transform the NN potential taking into account the correct signs for the different contributions. This leads us from a system with hard core to a very attractive potential. Since the relatively shallow NN potential already can produce a bound np system (namely the deuteron), many speculations have been made about the existence of a whole spectrum of those states in the $\bar{N}N$ system. However, the $\bar{N}N$ system is very much distinguished from the NN system by the existence of an additional reaction channel, namely the annihilation. The question arises, whether any of the bound states can survive ?

Shapiro and collaborators e.g. have argued more than 10 years ago, that the annihilation should be mediated by the exchange of a nucleon or antinucleon and therefore be restricted to short distances <0.2 fm. Solving the Schroedinger equation they predicted numerous states close to threshold [1]. A longer range annihilation however could completely destroy all these states (Myhrer and Gersten [2]). The answer to the question of possible bound states or resonances therefore depends strongly on the extent of the annihilation and its interplay with the boson exchange force and can only be answered experimentally.

Theoretically, the annihilation is taken into account by introducing complex potentials, so called optical potentials. Many authors take the empirical approach using a Woods-Saxon or Yukawa-like potential for the imaginary part and adjusting the annihilation potential parameters to fit the data. The channel of annihilation, however, is a unique chance to connect the $\bar{N}N$ force to the quark degrees of freedom. In this quark picture, annihilation is described with quark rearrangement and quark pair annihilation and creation via gluon exchange. The quark interaction is partially calculated from QCD, but up to now, too many unknown parameters have to be determined empirically by a fit to the data.

Beside the question of resonances, the determination of cross-sections and angular distributions over a wide range of energies is necessary, to test the validity of the present $\bar{N}N$ models. One especially hopes, that going to very low energies, where only S-wave contributes to the scattering, we can probe the central part of the $\bar{N}N$ potentials.

3. EXPERIMENTAL SET-UP

The existence of the low energy antiproton ring (LEAR) at CERN has, for the first time, made possible high statistics and high resolution search for bound states and resonances in the $\bar{p}p$ system. Since the primary energy of the $\bar{p}$ beam can be varied down to very low energies, we are also able to measure the $\bar{p}p$ cross-sections down to center of mass energies of 5 MeV. The Heidelberg group has done an experiment measuring the $\bar{p}p$ differential elastic cross section, the charge-exchange cross section as well as the annihilation into charged and neutral pions in the antiproton momentum range between 600 and 150 MeV/c. Fig. 1 gives a top view of the experimental set up [3]. The $\bar{p}$ beam enters from the left and is defined by two thin scintillation counters. The first one (SD) acts as a start detector for all time-of-flight measurements and limits, by its size, the accepted divergence of the beam. The second detector (TD) is a target defining detector. The liquid hydrogen target itself sits inside a vacuum tank. Two target cells of 2 cm and 7 mm thickness are available, identical dummy targets help to subtract the influence of the target construction itself. Elastically scattered $\bar{p}$ leave the vacuum tank through a mylar window and are defined by means of a semicylindrical MWPC and a ring of forward hodoscopes (FHD). Antineutrons from the charge exchange reaction are detected in the calorimeter-like antineutron counters (ANC). They are distinguished from antiprotons by not giving a signal in the corresponding forward hodoscope slab. Charged pions can be detected with large scintillation counters mounted around the vacuum tank. Together with the FHD they cover about 85% of the solid angle. Neutral pions are identified through their decay gammas in the calorimeters or in one of the 126 leadglass counters mounted hedgehog-like around the detector.

a) top view

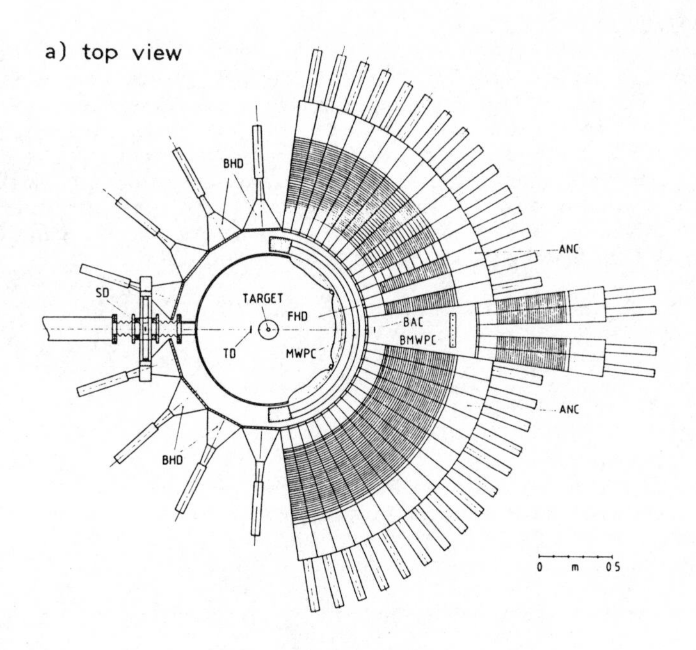

b) side view

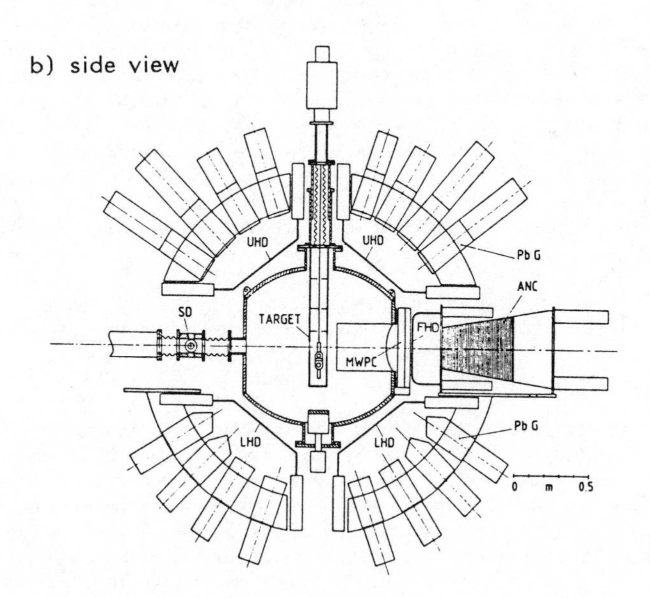

Fig 1. The experimental set-up of PS173.

4. EXPERIMENTAL RESULTS

Fig. 2 shows a spectrum for the differential elastic scattering analyzed at 287 MeV/c. Two independent ways exist to extract physics information from such a spectrum. The first one is the 'Coulomb interference method'. The differential cross-section can be written as the coherent sum of the Coulomb and the nuclear elastic scattering amplitude squared. If we use the standard parametrization for the two amplitudes, we have 3 free parameters which can be adjusted fitting the parametrized cross-sections to the data :

(1) $d\sigma/dt = d\sigma_C/dt + d\sigma_I/dt + d\sigma_N/dt,$

(2) where $d\sigma_C/dt = 4\pi\ (\alpha hc/\beta t)^2\ F(t)^2,$

(3) $d\sigma_I/dt = (\sigma_{tot}/\beta t)\ F(t)\ \exp(-{}^1/_2 bt)(\rho\cos\delta - \sin\delta),$

(4) $d\sigma_N/dt = (\sigma_{tot}/4hc\sqrt{\pi})^2\ (1+\rho^2)\ \exp(-bt),$

(5) $F(t) = (1 + t/0.71\ )^{-4},$

(6) $\delta(t) = -\{\ln(9.5t)+0.5772\}\alpha/\beta.$

α is the fine-structure constant,
β the velocity of the antiproton in the laboratory frame,
$F(t)$ the electromagnetic form factor of the proton,
$\delta(t)$ the phase of the Coulomb amplitude,
and $-t$ the four momentum transfer in units of $(GeV/c)^2$.

These parameters are the total cross-section σ_{tot}, the ratio of the real and imaginary part of the forward elastic scattering amplitude ρ and a slope parameter b. Fig. 3 shows the momentum dependence of the parameter ρ. The open circles stem from our analysis of the first runs and have been published [4]. The filled squares are the preliminary results from the data taken in November '85 and May '86. The fact that the ρ parameter is close to 0 at very low momenta can probably be explained by a cancellation of the S- and P-wave contributions which enter into the ρ parameter with opposite signs. The oscillatory behaviour of the ρ parameter can not easily be explained and could be a hint of a resonance close to threshold. It is striking that none of the theoretical models can describe the data below 300 MeV/c.

The second way of analyzing the elastic scattering spectra is a parametrization of the cross-section with Legendre polynomials.

$$d\sigma/d\Omega_{cm} = \Sigma\ a_i P_i(\cos\theta) \qquad (1)$$

From this we can calculate the angle integrated elastic cross-section and, with the information about the ρ, can extract a spin and isospin averaged partial-wave analysis. The latter one shows that even at momenta of 300 MeV/c we still have about 30-40% P-wave contribution to the elastic scattering [5]. For the total cross-section the P-wave part amounts to even 50%, while in the pp case we are left with almost 85% S-wave scattering. Fig. 4 gives a comparison of the differential elastic spectra with 3 model predictions at 3 different $\bar{p}$ momenta. All models describe the data resonably well.

Fig. 5 shows the result for the differential charge exchange spectra at 4 different momenta, 595, 505, 287 and 183 MeV/c [6]. The curves depict the predictions from 4 different models, namely the Paris model

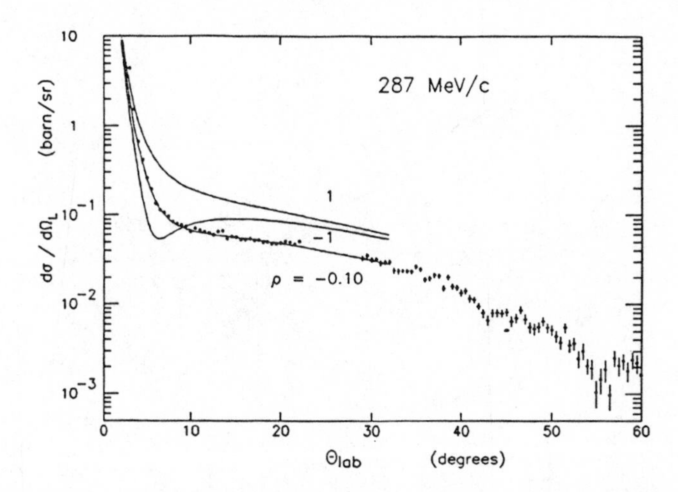

Fig 2. $\bar{p}$ differential elastic scattering at 287 MeV/c. The curves depict the fit using the standard parametrization for the scattering amplitudes. Beside the best fit curve, curves for $\rho=\pm1$ are also drawn in to show the sensitivity of the measurements.

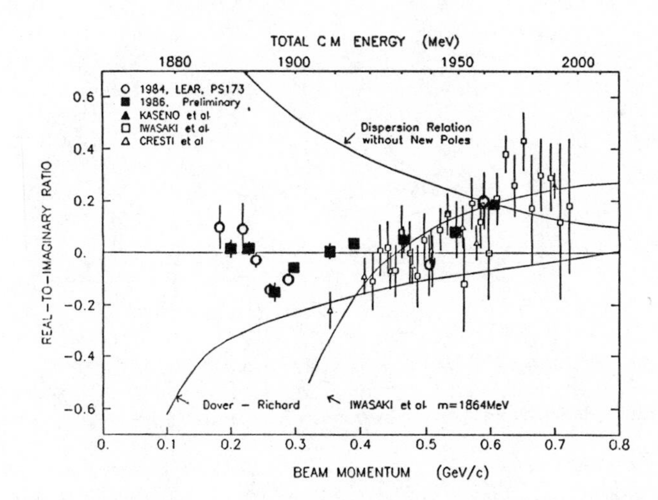

Fig 3. Momentum dependence of the ρ-parameter. The curves show the predictions of various theoretical calculations.

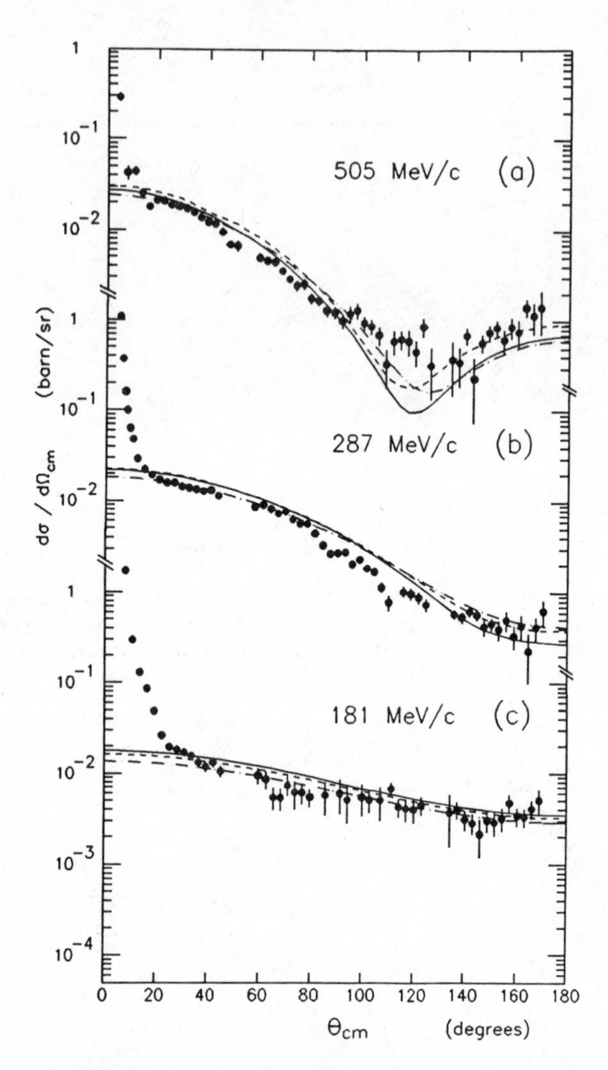

Fig 4. p̄p differential elastic scattering at 505, 287 and 181 MeV/c. The solid, dashed and dash-dotted curves represent the model calculations of Myhrer [10], Lacombe [7] and Timmers [8] respectively.

[7], the Nijmegen model [8], the model by Dover and Richard [9] and the boundary condition model from Myhrer et al [10]. Both, the Nijmegen model and the boundary condition model fit the data very well, while the Paris and the Dover-Richard models fail to describe the charge exchange data. The failure of the two models, which fit the NN data so well, is even more surprising if we assume that the charge exchange reaction is a peripheral process and mainly dominated by one pion exchange with an admixture of 2π exchange. It should be noted in this context, that the boundary condition model leaves only one free parameter.

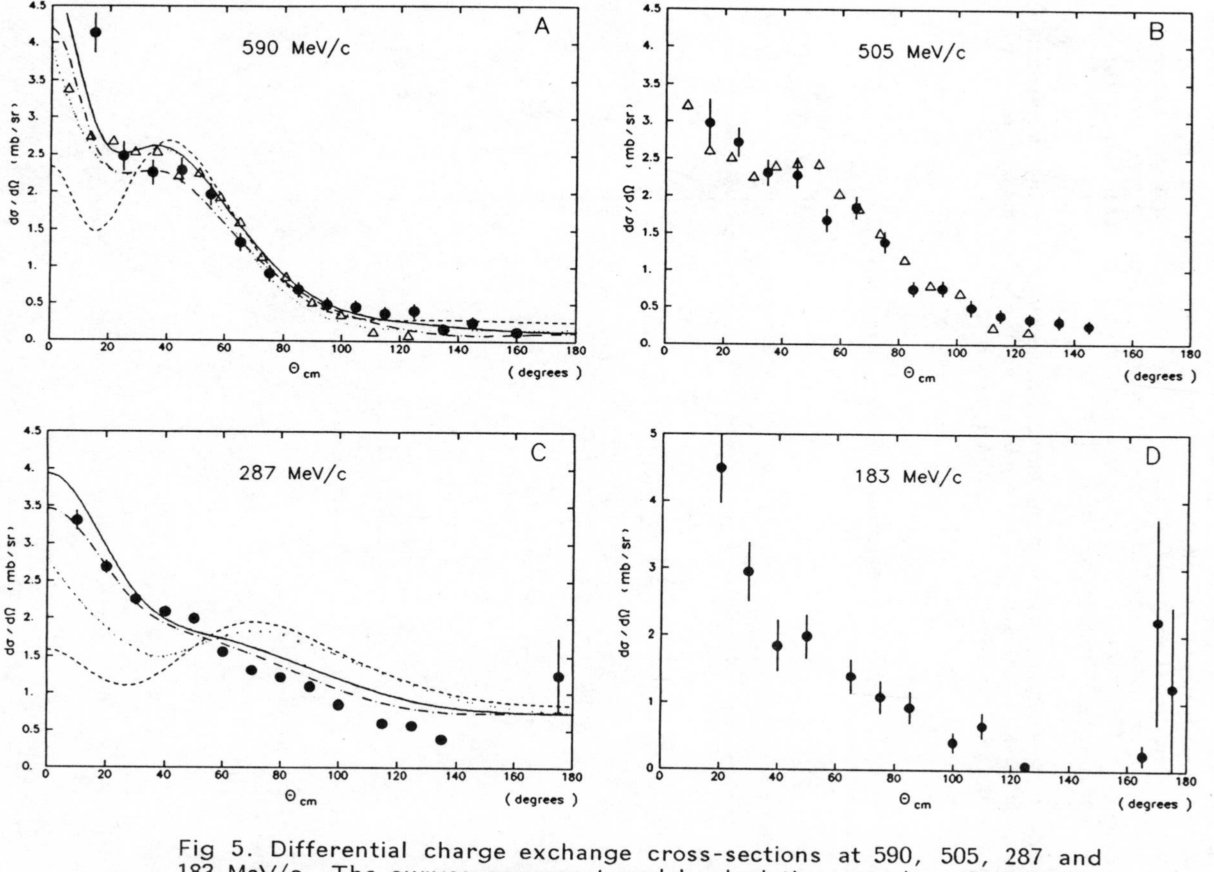

Fig 5. Differential charge exchange cross-sections at 590, 505, 287 and 183 MeV/c. The curves represent model calculations as described in the text. The open triangles give the results from Nakamura et al. [15]

Let us now turn to the annihilation into charged pions (Fig. 6). It was in this channel that an often disputed signal had been seen by many experiments, indicating the existence of a resonance at a $\bar{p}$ momentum of 500 MeV/c corresponding to a mass of 1932 MeV [11]. This signal was often referred to as the S-meson. We have done high statistics measurements at LEAR particularly aiming at the study of all possible systematic effects. The present status of the analysis can not confirm such a resonance with the strength observed by previous experiments. This result is in agreement with a recent total cross-section measurement by Clough et al. [12].

Fig. 7 summarizes the measured cross-sections in the momentum range from 600 to 150 MeV/c, as analyzed so far. All models describe the general trend of the cross-sections correctly.

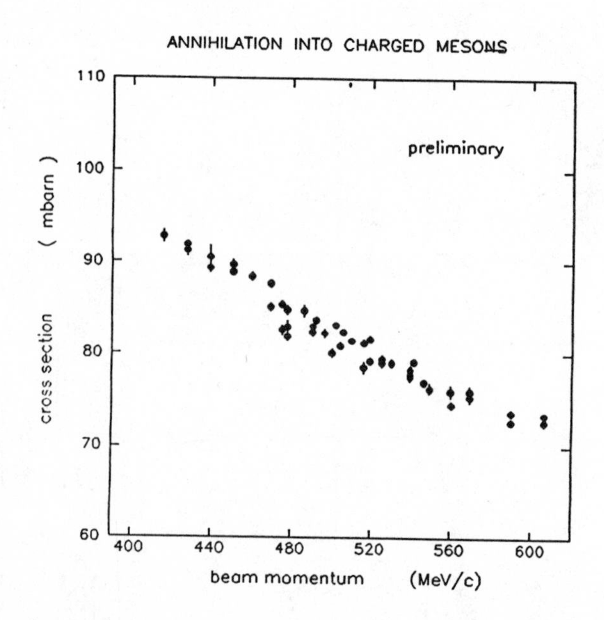

Fig 6. Momentum dependence of the $\bar{p}p$ annihilation into charged mesons.

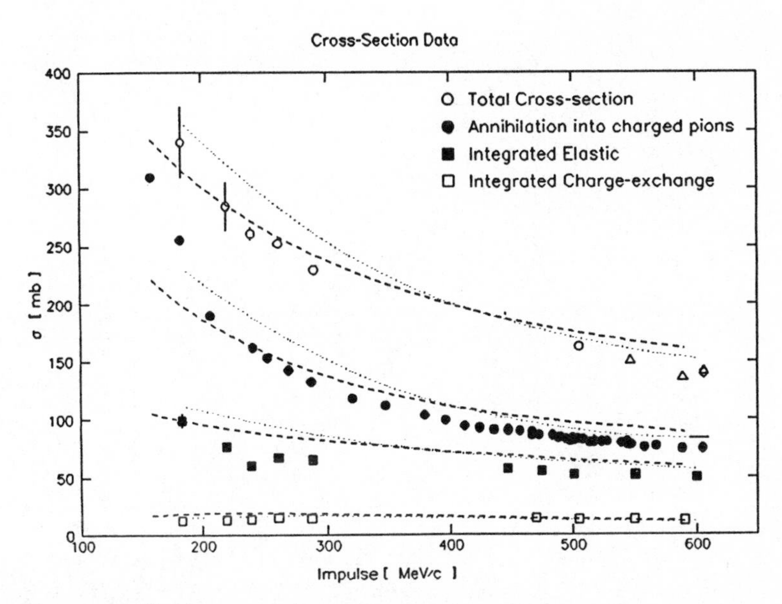

Fig 7. Summary of the measured momentum dependence of the p̄p cross-sections. The dashed and dotted curves show the calculations of Lacombe [7] and Myhrer [10]

5. DISCUSSION

Can we learn from these data something about the annihilation process or its extent and its influence on the boson exchange force?

In the following discussion will try to introduce the concept of the annihilation range. In order to do that, we will show that the range of the hadronic interaction at high energies has the same size as what we will introduce as the annihilation range.

Let us therefore first have a look at the scattering process at very high energies. Fig. 8 gives a comparison of the p̄p and the pp elastic scattering at 3 different ISR energies. The differential cross-sections are characterized by a diffractive peak which is exactly described by scattering on an absorptive disk having the shape of a gaussian profile function. Similar to the low energy case we can parametrize the differential cross-section by $d\sigma/dt = e^{bt}$, b being again the slope parameter. Fig. 9 shows the slope parameter for the p̄p and the pp case. As we can see, b varies only very little with energy and is about the same for p̄p and pp scattering at energies above 50 GeV/c. From the slope parameter we can determine the e-value of the interaction radius, which at high energies has a value of $r = \hbar c\sqrt{2b} = 1.0 \pm .1$ fm. At high energies the slope parameter is determined by the total inelastic cross-section of which the annihilation is only a small fraction. At low energies however the inelasticity is mainly given by the annihilation channel. The constancy of the slope parameter down to these energies indicates that the range of the annihilation is about the

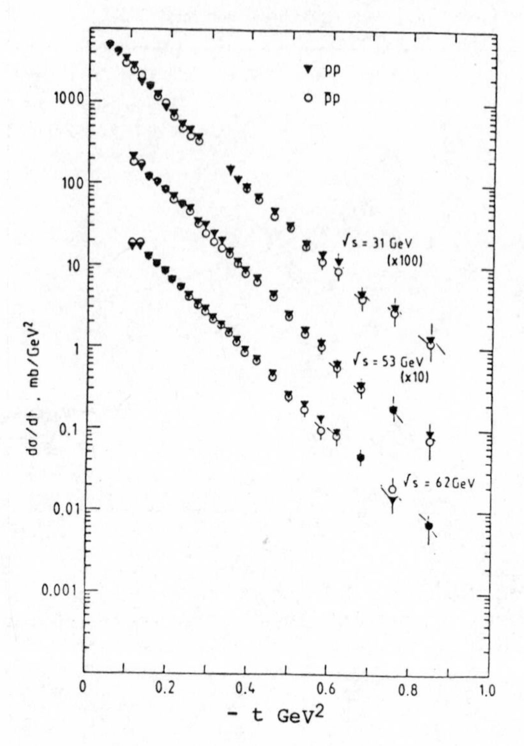

Fig 8. Differential elastic scattering at three different ISR energies for pp and p̄p.

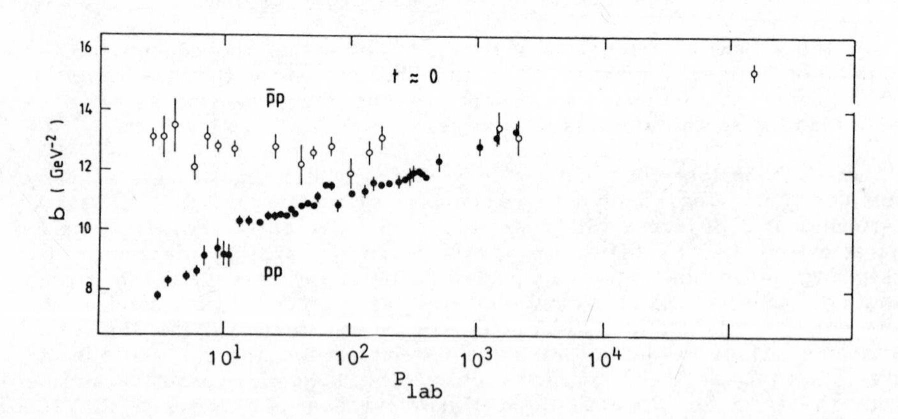

Fig 9. Momentum dependence of the slope parameter b for pp and p̄p scattering.

same as the range of interaction in high energy hadron-hadron collisions. The only problem is that we don't have a good model to relate the profile function to the structure of the nucleon.

Lets however have a closer look to Fig. 10, showing the $\bar{p}p$ differential scattering at 1.6 GeV/c. While the analysis of the slope parameter leads to an intercation radius of 1 fm, a partial wave analysis shows, that the largest contributing angular momentum has a value of l_{max}=13, which can be converted into an interaction radius via $r = \hbar \cdot l_{max}/p_{lab}$ = 1.6 fm. We might therefore assume that the interaction is determined by two different processes and it is tempting to assign the smaller radius to the size of the absorptive disk and the larger one to the size of the pion cloud arising from boson exchange. This interpretation brings us to the low energy data, where the effect of the long range interaction will be enhanced. Here the interaction is usually discussed in terms of potentials [13]. We have therefore fitted our results using an optical potential having the general form $U = V(r) + iW(r)$, where $V(r)$ describes the boson exchange and $W(r)$ is generated by the annihilation. Fig. 11 shows an example of the result of such a fit, comparing the $\bar{p}p$ and $\bar{p}p$ scattering data at 287 MeV/c. Despite the fact that a realistic $\bar{N}N$ potential should be non local we could fit the data over a wide energy range using a simple local potential.

Let us now have a look at the form of our fitted potentials. Two best fit solutions are described in the following section. Solution A uses a one pion exchange potential and a complex potential of Woods-Saxon form. Such a complex annihilation potential also arises from the concept of quark rearrangement. Both real and imaginary parts of the annihilation potential have similar width and depth. Solution B has been calculated using only a complex Woods-Saxon potential. The outcome of this fit is again a shallow but long-range real part to account for the one pion exchange. The purely imaginary annihilation potential however comes out much deeper as in A to provide the necessary annihilation strength. In either case we need a slope of 0.2 fm for the imaginary part. Both solutions show clearly the existence of two interaction regions, a long range part dominated by one pion exchange and a short range part, where annihilation starts to absorb the incoming wave. It also becomes clear that a certain annihilation strength at a critical radius R seems to be necessary to fit the data. Such an annihilation strength can either be produced by a very deep imaginary potential or by a more shallow one, which then has to be combined with a strong attractive real part. The origin of such a short range attraction is not well defined. It could either come out of a 2π exchange or from the annihilation process itself (rearrangement).

Potentials, however, do not provide any direct physical insights. In order to connect the potential to a physical quantity as the inelasticity, we can use the continuity equation and derive the following relation :

$$S = \int S_n dA = -2/h \int \mathrm{Im}V(r) \cdot P(r) d^3r$$

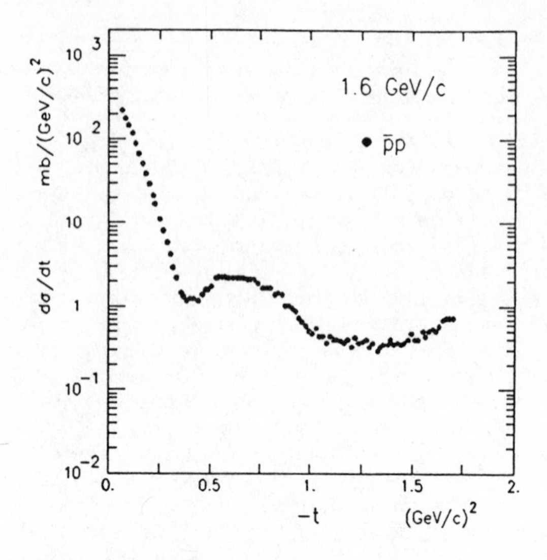

Fig 10. p̄p Differential elastic scattering at 1.6 GeV/c

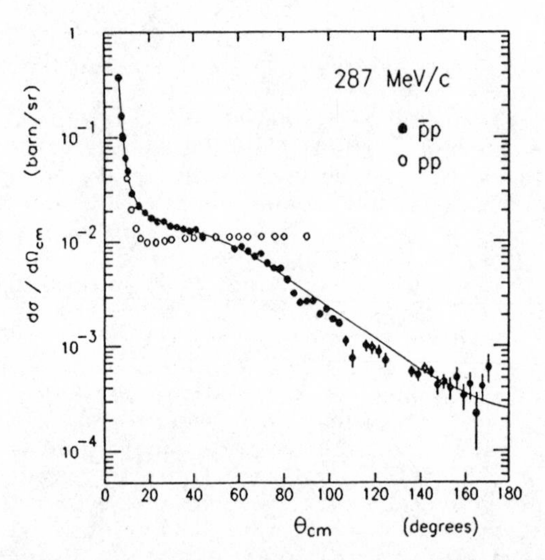

Fig 11. pp and p̄p differential elastic scattering at 287 MeV/c. The solid curve shows the result of a potential fit.

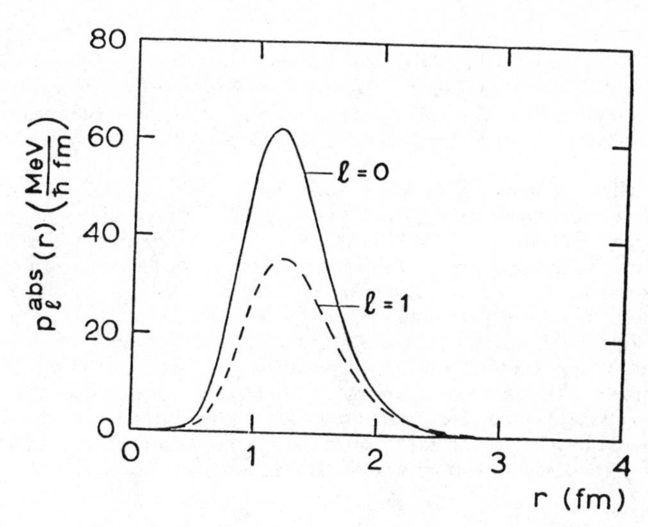

Fig 12. Radial dependence of the annihilation probability as defined in the text.

where $P(r) = \psi^*(r)\cdot\psi(r)$ is the position probability density and S_n is the probability current density. We can now use the quantity

$$P_l^{abs}(r) = 2/h\ \mathrm{Im}V(r)\cdot R_l^*(r)\cdot R_l(r)\cdot r^2$$

which gives the absorption probability as a function of the radius. In order to calculate this quantity $P_l^{abs}(r)$ we have fitted our data using an optical potential. If we now plot the probability for the absorption using the fitted potential parameters we obtain the following picture (Fig 12). It becomes clear that the maximum absorption takes place at distances of about 1.2 fm. At smaller distances of roughly 0.7 fm the incoming flux does fully disappear. Both fit solutions give a similar r-dependence of the absorption probability. This result agrees well with the boundary condition model of Myhrer et al., one of the two models which are able to also give a realistic description of the differential charge exchange cross-sections. This model looks to us as the most appealing one, since it only needs one free parameter to fit the data. This parameter is the distance r_c, where only absorption and no reflection of the incoming wave takes place. The best fit value for r_c is again 0.7-0.8 fm.

The process of annihilation is directly connected to the quark picture and we have seen, that for annihilation to take place we must assume an overlap of the quark wave functions as well as the exchange of gluons. It is tempting to interpret the position of the annihilation maximum at a distance R = 1.2-1.4 fm as the existence of a critical distance R, where quark degrees of freedom start to dominate the $\bar{N}N$ interaction. It might therefore be questioned whether in such conditions, where the incoming wave has almost died out at distances below 1 fm, the 2π exchange can play the same important role as is believed for the $\bar{N}N$ case. This question however can not be answered uniquely with this experiment.

There is however a possibility to bring new physical insights. Since the π and ρ exchange contribute to the polarization with opposite signs, measurements of the spin dependence of $\bar{p}p$ interactions could give a clear answer on the importance of the ρ exchange process.

Together with other institutions the Heidelberg group has proposed a polarization experiment at LEAR [14]. With a high density polarized hydrogen target inside the LEAR ring we want to prepare a polarized $\bar{p}$ beam using the 'filter method'. This method, however, requires a spin dependence of the $\bar{p}p$ absorption cross-sections which have not been measured up to now. A scenario for the experimental set-up is shown in Fig. 13. The antiprotons are polarized by interacting with the polarized target. Once a polarized beam is established we can use the same target to perform the various spin measurements. Since we believe that the charge exchange is a peripheral process, the knowledge of it's spin dependence in particular should give a more definite solution to the importance of the different boson exchange contributions.

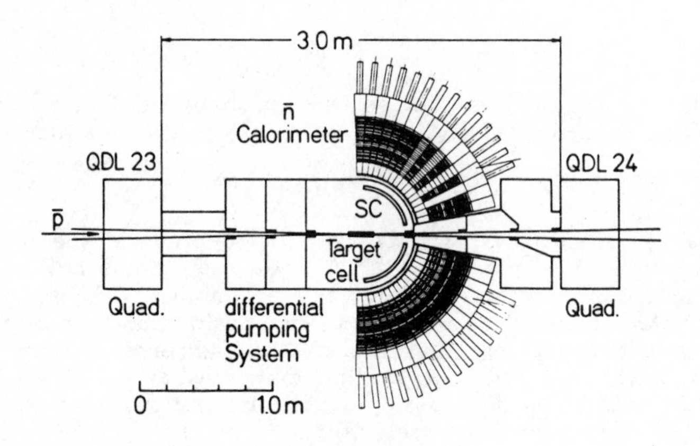

Fig 13. Possible experimental set-up for the proposed $\bar{p}p$ polarization experiment using an internal target in LEAR.

Reference list :
[1] L.N. Bogdanova et al., Ann. of Phys. 84 (1974) 261
[2] F. Myhrer and A. Gersten, Nuovo Cimento A37 21 (1977)
[3] W. Brueckner et al., Proc. of the 7th European Symposium on Antiproton Interactions 1985, ed. M.R. Pennington (Adam Hilger: Bristol) p. 157
[4] W. Brueckner et al., Phys. Lett. 158B (1985) 180
[5] W. Brueckner et al., Phys. Lett. 166B (1986) 113
[6] W. Brueckner et al., Phys. Lett. 169B (1986) 302
[7] J. Cote et al., Phys. Rev. Lett. 48 1319 (1982)
[8] P.H. Timmers et al., Phys. Rev. D29 1928 (1984)
[9] C.B. Dover and J.M. Richard, Phys. Rev. C21 1466 (1980)
[10] O.D. Dalkharov and F. Myhrer, Nuovo Cimento A40 152 (1977)
[11] see for example W. Brueckner et al., Phys. Lett. 67B (1977) 222
[12] A.S. Clough et al., Phys. Lett. 146B (1984) 299
[13] B. Povh and Th. Walcher, Comments Nucl. Part. Phys. 16 (1986) 85
[14] H. Doebbeling et al., Proposal CERN/PSCC/85-80
[15] K. Nakamura et al., Phys. Rev. Lett. 53 (1984) 885

List of Oral Contributions

(In order of presentation: the number in square brackets refers to the contribution in Vol. 1 of the Proceedings)

Production of heavy hypernuclei using stopped antiprotons
J P Bocquet[1], G Ericsson[2], T Johansson[2], J Konijn[3], T Krogulski[4], M Maurel[1], E Monnand[1], J Mougey[5], H Nifenecker[1], P Perrin[1], S Polikanov[6], M Epherre Rey-Campagnolle[7], C Ristori[1] and G Tibell[2]
[1]CEN, Grenoble; [2]University of Uppsala; [3]NIKHEF Amsterdam; [4]University of Warsaw; [5]CEN Saclay; [6]GSI Darmstadt; [7]CSNSM Orsay
[H3]

Electron scattering from transitional nuclei
W Boeglin[1], J M Cavedon[2], P Egelhof[1], J Engel[2], B Frois[2], M Girod[3], D Goutte[2], X-H Phan[2], S K Platchkov[2] and I Sick[1]
[1]University of Basle; [2]CEN Saclay; [3]Centre D'Etudes de Bruyeres-le-Châtel
[C200]

High energy resolution $^{12}C(\gamma,p)^{11}B$
S.V. Springham[1], A C Shotter[1], D Branford[1], J Yorkston[1], B Schoch[2], P Jennewein[2] and J C McGeorge[3]
[1]University of Edinburgh; [2]University of Mainz; [3]University of Glasgow
[C218]

Inst. Phys. Conf. Ser. No. 86
Paper presented at Int. Nucl. Phys. Conf., Harrogate, UK, 1986

Prompt gamma ray spectroscopy on the frontiers of nuclear stability

C J LISTER

Dept. of Physics, Schuster Laboratory, The University, Manchester, M13 9PL.

Introduction

The title of this paper is very general, but it is mostly dedicated to describing one experiment. I should start by being more specific on the subject of the experiment and concentrate on putting it into a broader physics context. The particular experiment concerns the observation of γ-ray transitions between states in $^{80}_{40}Zr_{40}$. This nucleus has 20% fewer neutrons than the nearest stable zirconium isotope ^{90}Zr, is the heaviest N=Z nucleus observed in any kind of experiment, and has been predicted to be one of the most deformed nuclei known. Attempts have been made to predict the behaviour of ^{80}Zr for more than 30 years and searches for evidence of its production for about 15 years, but its final isolation was achieved only a few weeks ago.

Although this experiment is very specific the overall reasons for studying exotic nuclei are much more general and tend to apply to most studies of nuclei with unusual neutron-to-proton ratios. The key concept is the unification of our knowledge of nuclear physics, through various models, to a point where one only needs to know the number of neutrons and protons in a nuclide in order to be able to make a reliable estimate of mass, shape and lowest modes of excitation. The development of these general nuclear models has been particularly rapid in the last five years and has been stimulated by development of experimental techniques which permit spectroscopy of exotic species. The models use various approaches to tackle the nuclear many-body problem but all have parameters which are fitted to the body of the data on nuclei near stability. Thus, it is not surprising that good agreement is found in predicting the behaviour of near stable nuclei, but the crucial tests are in the extrapolation of the models, especially to nuclei very far from stability. Here new experimental data can challenge the models and lead to their refinement to a point where we can hope to predict the properties of nuclei which may never be isolated in the laboratory, but are formed in stars and existed in explosive nucleosynthesis.

Let us return to the specific case of ^{80}Zr and review what has been predicted about it to see how this process works.

Theoretical Estimates of the Shape of $^{80}_{40}Zr_{40}$

Nuclei with both proton and neutron numbers very far from the major shell closures have been expected to be deformed for many years[1]. The quadrupole-quadrupole residual interactions between the valence nucleons were anticipated to lead to polarisation and result in permanent deformation. However, nucleon number 40 has been anticipated to play a special role in stabilising nuclear shapes as it represents the first major shell closure to be destroyed by spin-orbit coupling. The solution of the Schrödinger equation in an axially deformed 3-D harmonic oscillator potential indicates[2] that shell gaps at nucleon number 40 should appear both for spherical and prolate deformed shapes as can be seen from fig. 1. Thus, the total binding energy may be expected to be unusually low for $^{80}_{40}Zr_{40}$ both for spherical and prolate deformed shapes with axis ratio 2:1 (ϵ_2=0.5).

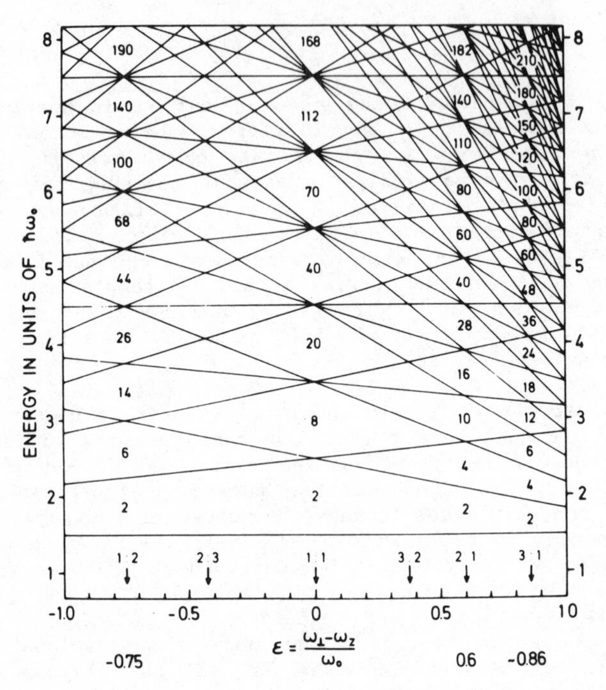

Fig. 1. Single particle levels in a spheroidal harmonic oscillator as a function of axial deformation, ϵ.

More realistic calculations using the Nilsson model reduced the importance of the spherical N=40 gap (though the nucleus $^{90}_{40}Zr_{50}$ retains many properties of a doubly magic nucleus), but a shell gap at N=38 or 40 persisted for prolate deformations of $\epsilon_2 \approx 0.4$ and extra binding for deformed shapes was anticipated[3]. However, the importance of the prolate gap was not initially clear. Calculations of the total binding energy as a function of shape were made by combining the mean liquid drop energy with a shell correction using the Strutinksy formalism. The results of these calculations were contradictory[4,5,6] and a variety of shapes were predicted for nuclei in the A=80 region, from axially deformed prolate to soft oblate. Experimental measurements have provided useful input to development of the models and more recent calculations[7,8,9,10,11] have explored the great sensitivity to the microscopic shell corrections of the nuclear shape which is most tightly bound. Calculations have been made with a variety of Nilsson parameters and with more realistic Woods-Saxon and folded Yukawa potentials. An example of the sophistication of the microscopic-macroscopic technique is shown in fig. 2 which predicts the ground-state quadrupole deformation for more than 4000 nuclides.

Other, more microscopic approaches have also been used. Of particular interest are recent Hartree-Fock calculations[12,13] which make some very detailed predictions of ground-state shapes of Kr, Sr and Zr nuclei. In particular, $^{80}_{40}Zr_{40}$ is predicted[13] to behave similarly to $^{78}_{38}Sr_{40}$ in being a γ-soft but highly prolate deformed rotor, and $^{76}_{38}Sr_{38}$ should be the only good axial rotor in the A=80 region. The similarity of ^{78}Sr and ^{80}Zr is also predicted from IBA calculations[14,15] as they have exactly the same number of bosons, and the calculations predict highly deformed shapes.

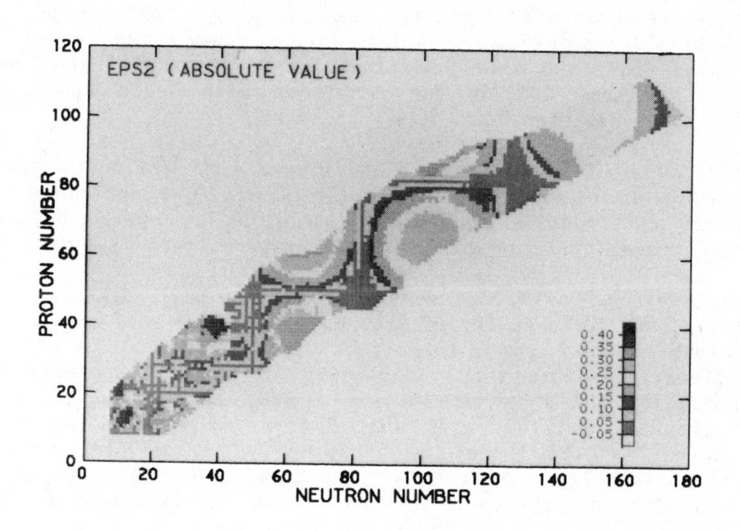

Fig.2. A contour plot of the quadrupole deformation, ϵ_2, as a function of neutron and proton number as predicted by Möller and Nix (ref.6).

Clearly then, an experimental determination of low-lying excited states in ^{80}Zr can put a wide variety of nuclear models to an exacting test and differentiate between those which reliably predict the properties of new exotic species and those which do not. The techniques of gamma-ray spectroscopy are well developed for addressing questions of nuclear shape and can make sensitive and discriminating tests of the models. However, two major experimental difficulties present themselves: firstly the production of states in ^{80}Zr in a suitable nuclear reaction to permit spectroscopy and secondly, the specific identification of their gamma decays.

Production and Indentification of Neutron Deficient Nuclei

Three techniques are commonly used to produce nuclei with unusual neutron-proton ratios; spallation using light or heavy ions, fusion-fission of heavy systems, and cold fusion. All these mechanisms offer the possibility of exotic nuclear production, but all result in the production of many different reaction residues which present enormous experimental difficulties which hamper the study of one particular reaction product. We are going to concentrate on the production of ^{80}Zr through cold fusion and the methods used to isolate this nuclide from a background of other reaction products ≈10^5 times stronger. Many of the experimental problems encountered and the techniques used to solve them are applicable to both neutron poor and neutron rich systems, and to the other reaction mechanisms mentioned above.

Heavy-ion fusion of nuclei near their mutual Coulomb barrier is the optimum method for producing very neutron deficient nuclei. Only a few reactions can lead to the production of ^{80}Zr and calculations using the fusion evaporation codes CASCADE[16] and ALERT[17], together with preliminary experimental measurements, indicated that the ^{24}Mg + ^{58}Ni reaction followed by two neutron evaporation had the largest production cross section. However, even under the best conditions mass A=80 nuclei were expected to be produced in only 20% of the reactions with 3,4,5 and 6 particle evaporation also important.

Two methods of isolating gamma rays associated with ^{80}Zr were contemplated; identification of γ-rays directly associated with two neutron evaporation (through the construction of a large solid angle neutron detector), or through electromagnetic identification of Z=40 residues with mass A=80.

The first method, using a neutron wall or ball detector has been successfully used for a number of studies of isotopes with production cross sections down to the few millibarn (1mb=10^{-27}cm^2) level. The detectors are used to directly measure the multiplicity of light, evaporated particles associated with each γ-ray transition. They permit the study of many reaction channels and thus are ideal for survey work in new regions of exotic nuclei[24]. They have the advantage of being flexible, easy to construct and relatively inexpensive. An example of this type of detector installed at Daresbury is shown in figure 3 and is discussed in ref [18].

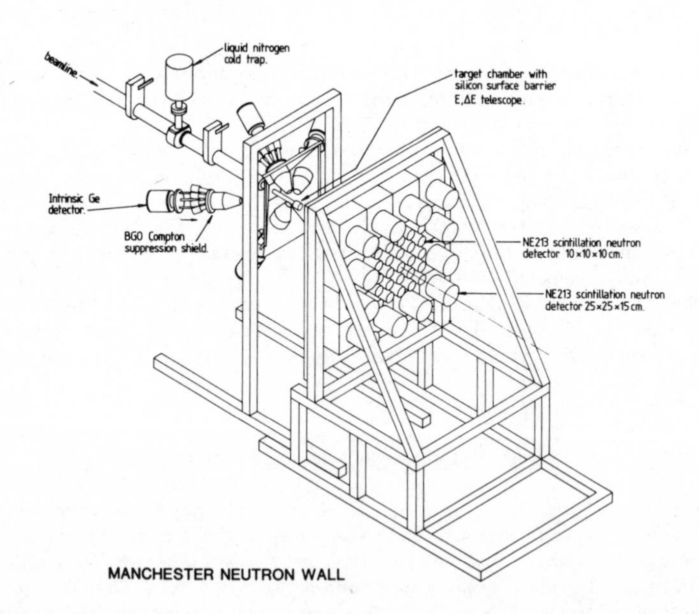

Fig.3. The Manchester Neutron Wall detector used for measuring the multiplicity of light charged particles in coincidence with each gamma ray.

Similar systems are in use at Cologne, Oxford, Stonybrook, Tokyo, Rochester, Berlin and Pennsylvania, while several others are being constructed. Recent experiments have lowered the sensitivity for gamma-ray spectroscopic studies by this method to about 500μb. Groups at Oxford[19] and Pennsylvania[20] have recently studied $^{64}_{32}$Ge using this technique. Below this level of sensitivity many experimental difficulties are encountered, including detector count rate limitations, target impurities and data collection problems, so great advances in sensitivity are not anticipated. However, the evaporation code calculation indicated that ^{80}Zr production may be more than 10 times lower in cross-section than can be observed with multiplicity detectors and so the present studies were made using the newly commissioned Daresbury Recoil Separator[21].

The separator is schematically shown in fig.4 and consists of four main elements. Around the target was situated an array of 14 gamma-ray detectors, which were Compton suppressed[22] to improve the photopeak to background ratio. The target was thin (500μg/cm^2) and reaction products recoiled from the target with a mean velocity of 5.5% of c. The use of an inverse reaction (^{58}Ni beam on ^{24}Mg target) enhanced the focusing of the products into the separator, and thus increased its efficiency. A double Wien filter was used to select a range of recoil velocities ($\Delta v/v = \pm 3\%$) and beam particles were deflected into a tantalum plate at the exit of the first filter. Momentum analysis resulted in dispersion in A/q. A position detector consisting of a carbon foil and position sensitive channel plate detector presented a timing and position signal.

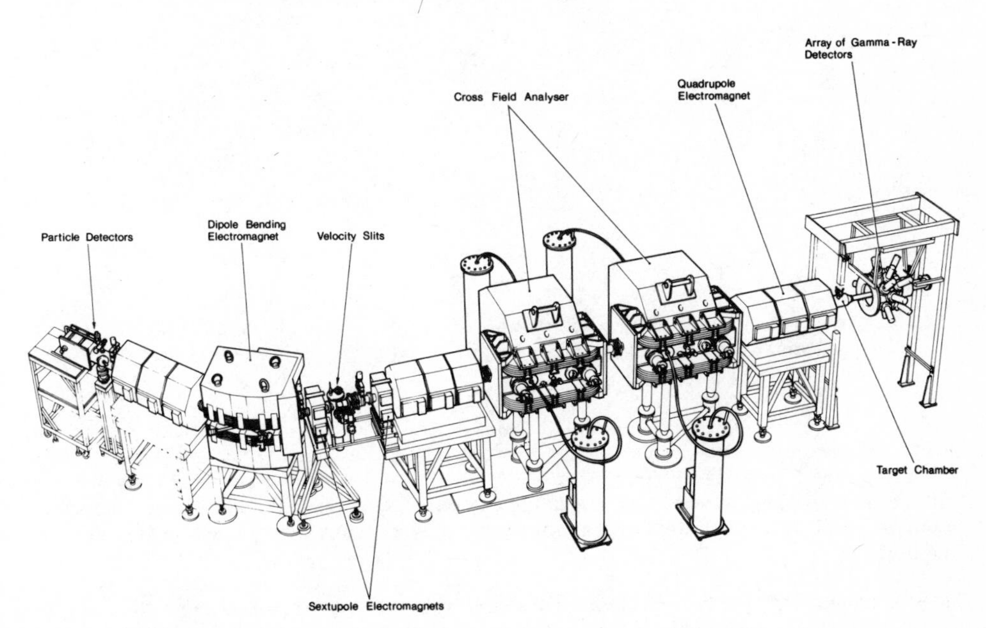

Fig.4. A schematic layout of the main elements of the Daresbury Recoil Separator.

A two dimensional plot of time of flight of residues through the separator against position is shown in fig.5 and clearly reveals the separation of A=79 and A=80 products. The recoils were selected in atomic charge state q=24$^+$.

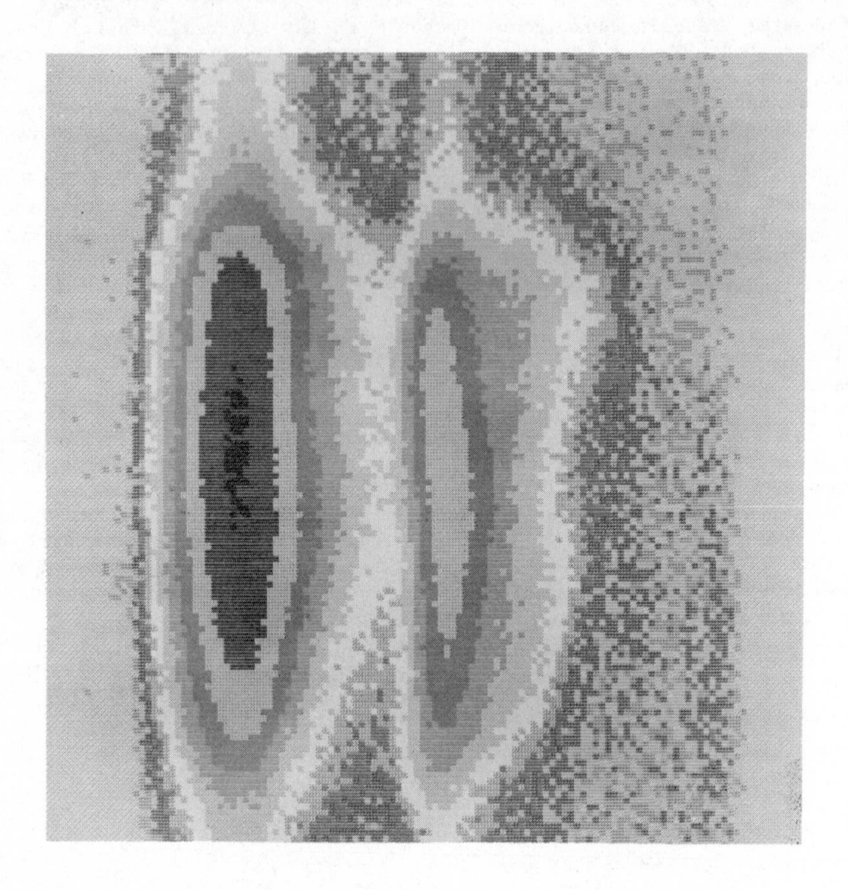

Fig.5. A two dimensional contour plot of recoil position,X, (after dispersion in the dipole magnet) against time of flight through the separator.

The last element of the separator was a two element ion chamber which was used to measure the stopping properties of the residues and hence infer their atomic number, the highest Z ions of a fixed energy having the largest stopping power. A mechanical mask was inserted between position counter and ion chamber to prevent any ions except mass 80 entering the ion chamber.

Fig.6 shows the yield of gamma rays in spectra gated as a function of the size of signal under the first anode. Transitions known to originate from ^{80}Sr and ^{80}Y were resolved, and the anticipated location of ^{80}Zr ions could be found. Unfortunately, the tail due to interactions between the residues and the gas in the ion chamber, together with the inherent resolution of the chamber, meant that an isolated anode signal characteristic <u>only</u> of Zr ions was not possible. However, selection of the optimum size of signal for Zr ions revealed transitions associated with recoils which had the correct stopping characteristics.

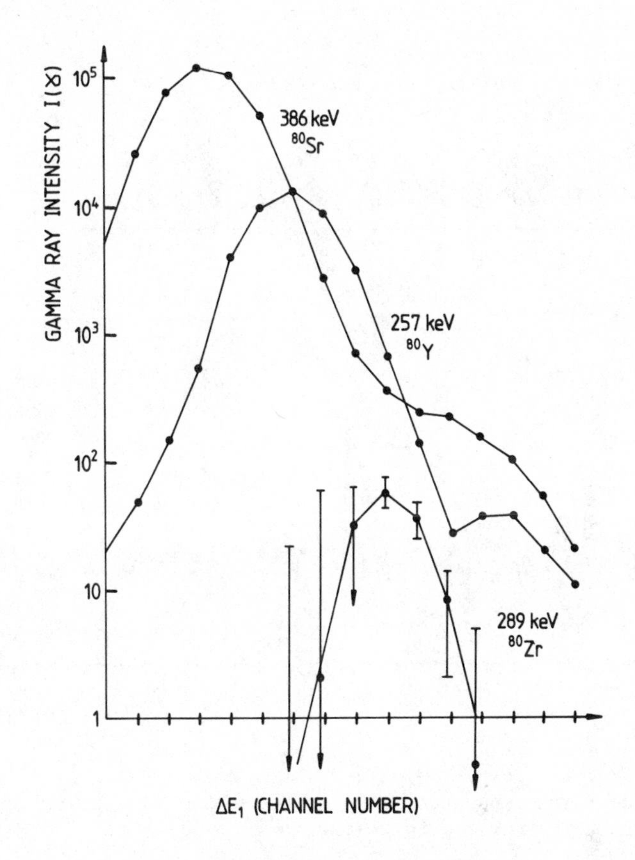

Fig.6. The intensity of γ-ray photopeaks in spectra gated by windows set on the energy loss signal ΔE in the ion chamber. Known ^{80}Sr and ^{80}Y lines are indicated, together with the first excited state in ^{80}Zr.

Subtraction of known Y and Sr lines revealed the first spectrum of $^{80}_{40}Zr_{40}$ γ-rays. (fig.7) These transitions were extremely weak and were observed only after three days of continuous experiment with 10pnA of ^{58}Ni beam. The production cross section of ^{80}Zr was only 10±5μb, five times weaker than the most pessimistic evaporation code calculation and fifty times weaker than the most sensitive previous measurement of prompt γ-rays from neutron deficient isotopes.

The first two states in $^{80}_{40}Zr_{40}$ cast considerable light on the validity of theoretical estimates for this nucleus. As predicted by most of the recent calculations, ^{80}Zr is indeed extremely deformed and its low first excited state at 289keV is consistent[23] with a deformation of ϵ_2=0.4. The low $J^{\pi}=2^+$ state and the $E(4^+)/E(2^+)$ ratio of 2.86 are both remarkably similar to the ^{80}Zr isotone ^{78}Sr, as predicted by HF and IBA calculations[13,15].

Neither of these nuclei appear to be good axial rotors, and it will be intriguing to test the HF calculations that $^{76}_{38}Sr_{38}$ alone in this region is a good axial rotor.

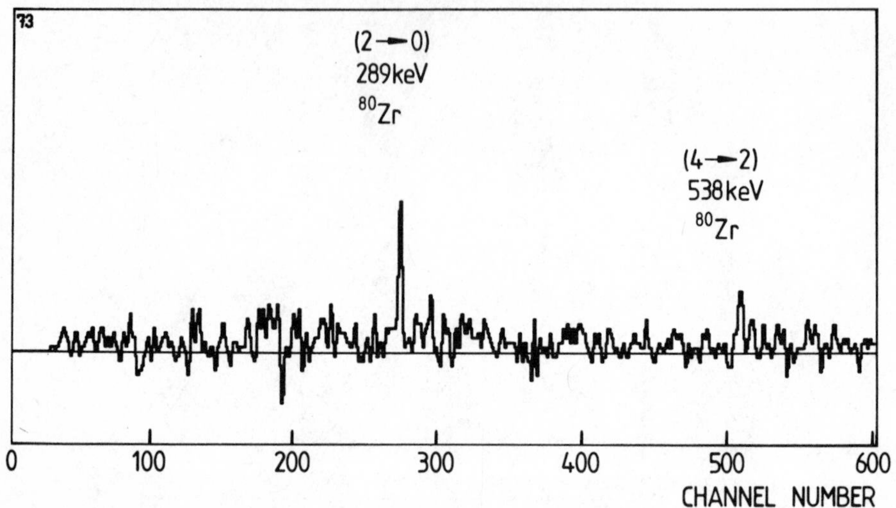

Fig.7. The γ-ray spectrum associated with ^{80}Zr after subtraction of transitions known to belong to ^{80}Y and ^{80}Sr.

Conclusions

In all, considerable progress has been made. The increased sophistication of detector systems has meant that nuclei which could only just be studied five years ago can now be examined in great detail. The frontiers of exotic nuclei have been pushed to the N=Z line which is predicted to lie only 4 nucleons away from the proton drip line where no binding remains. A little has been learned about ^{80}Zr, but perhaps more importantly a technique has been developed which allows us to study nuclei far from stability with two orders of magnitude greater sensitivity than was possible three years ago. Indeed the prospects for exotic studies look bright!

Acknowledgements

The research described in this paper has involved many people from Manchester, Liverpool and Chalmers Universities and from the NSF at Daresbury. I would like to thank Mr A.A. Chisti, M. Campbell, R. Moscrop and Drs. A.N. James, T. Morrison, W. Gelletly, L. Goettig, H.G. Price, B.J. Varley, O. Skeppstedt and J. Simpson. The work was supported by grants from the UK Science and Engineering Research Council.

References

1. See for example O. Nathan and S.G. Nilsson in Vol. 1 of α- β- γ-ray spectroscopy (Amsterdam, North Holland, 1965).
2. E. Marshalek, L. Person and R.K. Sheline, Rev. Mod. Phys. 35 (1963) 108.
3. I. Ragnarsson, S.G. Nilsson and R.K. Sheline, Phys. Reports 45 (1978) 1.
4. D. Bucurescu, G. Constantinescu and M. Ivascu, Rev. Romanian Physics 24 (1979) 971.
5. S. Aberg, Phys. Scr. 25 (1982) 23.
6. P. Möller and J.R. Nix, At. Data and Nucl. Data Tables 26 (1981) 165.
7. W. Nazarewicz et. al. Nucl. Phys. A435 (1985) 397.
8. K. Heyde, J. Moreau and M. Waroquier, Phys. Rev. C9 (1984) 859.
9. I. Ragnarsson and R.K. Sheline, Phys. Scr. 29 (1984) 385.
10. R. Bengtsson, P. Möller, J.R. Nix and J. Zhang, Phys. Scr. 29 (1984) 402.
11. D. Galeriu, D Bucurescu and M Ivascu. J. Phys. G12 (1986) 326.
12. D.P. Ahalpara, K.H. Bhatt, R. Sahu, J. Phys. G11 (1985) 735 and D.P. Ahalpara, A. Abzauzi and K.H. Bhatt, Nucl. Phys. A445 (1985) 1.
13. P. Bonche, H. Flocard, P.H. Heenen, S.J. Krieger and M.S. Weiss, Nucl. Phys. A443 (1985) 39.
14. D. Bucurescu et. al. Nucl. Phys. A401 (1983) 22.
15. S.L. Tabor, Phys. Rev. C34 (1986) 311.
16. F. Pühlhofer, Nucl. Phys. A280 (1977) 267.
17. M. Beckermann and M. Blann, University of Rochester Report UR-NSRL-135.
18. W.F. Piel et. al. Phys. Rev. C28 (1983) 209 and L. Goettig et. al. Submitted to Nucl. Phys. (1986).
19. S.S.L. Ooi et. al. accepted for publication in Phys. Rev. C (1986).
20. T. Chaprun et. al. Private Communication (1986).
21. Daresbury Technical Memorandum DL/NSF/TM38 (1978) and A.N. James et. al. to be published (1986).
22. P.J. Nolan, D.W. Gifford and P.J. Twin, Nucl. Inst. Meth. A236 (1985) 95.
23. Deformation estimated from Grodzins formula: $\epsilon^2=1176/[E(2)A^{7/3}]$ and extrapolated from the known collectivity of the isotone ${}^{78}_{38}Sr_{40}$ (C.J. Lister et. al. Phys. Rev. Lett. 49 (1982) 308).
24. C.J. Lister et. al. Phys. Rev. Lett. 55 (1985) 810.

List of Oral Contributions

(In order of presentation: the number in square brackets refers to the contribution in Vol. 1 of the Proceedings)

Observation of a discrete line superdeformed band up to 60 h in ^{152}Dy
P J Twin[1], B M Nyako[2], A H Nelson[1], J Simpson[1], M A Bentley[3], H Cranmer-Gordon[3], P D Forsyth[3], D Howe[3], A R Mokhtar[3], J D Morrison[3], J F Sharpey-Schafer[3] and G Sletten[4]
[1]SERC Daresbury Laboratory; [2]Inst. Nucl. Res., Debrecen; [3]University of Liverpool; [4]Niels Bohr Institute, Copenhagen
[B148]

Pairing fluctuations in rapidly rotating nuclei
Y R Shimizu[1], R A Broglia[1,2], M Gallardo[3] and J D Garrett[1]
[1]Niels Bohr Institute, Copenhagen; [2]University of Milan and INFN Milan; [3]University of Seville
[C66]

The $h_{11/2}$ proton and neutron excitations in $^{142}_{64}Gd_{78}$
A Facco[1], S Lunardi[1], M Morando[1], F Soramel[1], G De Angelis[2], W Gast[2], R Julin[2], P Kleinheinz[2], R Lieder[2], B Rubio[2], W Starzecki[2] and W Urban[2]
[1]University of Padua and INFN Padua; [2]KFA Jülich
[B36]

γ-ray spectroscopy with 1% of the fusion cross section using a 4π charged particle detector
T Chapuran, D P Balamuth, J Görres and J Arrison
University of Pennsylvania
[B44]

Ternary processes from 12 to 19 MeV/u
A Olmi[1], A A Stefanini[1], P R Maurenzig[1], J Albinski[2], A Gobbi[2], S Gralla[2], N Herrmann[2], K D Hildenbrand[2], H Stelzer[2], W J F Müller[2], J Kuzminski[2] and M Petrovici[2]
[1]University of Florence and INFN Florence; [2]GSI Darmstadt
[C153]

Investigation of the fission decay of giant multipole resonances in uranium isotopes by an (e,e'f)-reaction
T Weber[1], H J Emrich[2], R D Heil[1], T Kihm[3], U Kneissl[1], K T Knoepfle[3], M Pecho[1] and W Wilke[1]
[1]University of Giessen; [2]University of Mainz; [3]MPI Heidelberg
[B65]

Cluster radioactivity and clustering formation in nuclei
M Iriondo, D Jerrestam and R J Liotta
Res. Inst. Physics, Stockholm
[B82]

Inst. Phys. Conf. Ser. No. 86
Paper presented at Int. Nucl. Phys. Conf., Harrogate, UK, 1986

Three-body coupled-channel theory of scattering and breakup of light and heavy ions

M. Kamimura, Y. Sakuragi[*], Y. Iseri[**], M. Yahiro[+], H. Kameyama, M. Kawai and M. Tanifuji[++]

[*]Department of Physics, Kyushu University, Fukuoka, Japan
[**]Institute for Nuclear Study, University of Tokyo, Tanashi, Tokyo, Japan
[+]Chiba College of Economics, Chiba, Japan
[++]Shimonoseki University of Fisheries, Shimonoseki, Japan
Department of Physics and Reseach Center of Ion Beam Technology, Hosei University, Tokyo, Japan

Abstract. It is shown that the method of coupled discretized continuum channels (CDCC) based on the three-body model for direct reactions is very successful in explaining the following, recently developed experiments using deuteron, ^{6}Li and ^{7}Li projectiles whose breakup threshold energies are very low: (i) Precise measurement of all the possible analyzing powers in elastic scattering of polarized deuteron at 56 MeV, (ii) scattering of polarized deuteron at intermediate energies, (iii) deuteron projectile breakup at 56 MeV, (iv) scattering of polarized ^{7}Li at 20 and 44 MeV and (v) projectile breakup of ^{6}Li at 178 MeV and ^{7}Li at 70 MeV. The CDCC analyses of those data are made transparently with no adjustable parameters.

1. Introduction

Breakup processes of deuteron, ^{3}He and light heavy-ion projectiles are of current interest from the viewpoint of the three-body model for nuclear direct reactions. Among the various theoretical approaches based on the three-body model, the method of coupled discretized continuum channels (CDCC) (Johnson and Soper 1970, Rawitscher 1974, Farrel et al 1976, Yahiro et al 1982) has been recognized as a useful and practical tool for non-Faddeev theoretical studies of scattering and breakup of the d, ^{3}He, ^{6}Li, ^{7}Li and ^{12}C projectles (c.f., for example, review talks of Kamimura et al 1983, 1984, 1985; references therein).

In this paper we further apply the method of CDCC to the scattering of $\vec{d}$+^{58}Ni, ^{208}Pb at 56 MeV, $\vec{d}$+^{58}Ni at 400 MeV and the scattering of $^7\vec{Li}$+^{12}C at 21 MeV and $^7\vec{Li}$+^{120}Sn at 44 MeV. We shall discuss the effects of the virtual excitation and breakup of the projectile on the polarization observables as well as the cross sections with emphasis placed on the effective spin-dependent interactions induced dynamically by those effects. The method of CDCC is known to be useful in describing the elastic-breakup of light ions such as (d,pn) and (^{3}He,dp) reactions (Yahiro et al 1984, Kamimura et al 1984, Yahiro 1985, Iseri 1985). It is further tested here for the elastic breakup of light heavy-ions in the reactions of ^{12}C,^{208}Pb(^{6}Li,αd) at 178 MeV and ^{12}C,^{120}Sn(^{7}Li,αt) at 70 MeV.

Presented by M. Kamimura

2. Method of CDCC

We briefly sketch the method of CDCC taking the case of deuteron projectile as an example. The hamiltonian of the p+n+A system may be given by

$$H = h_d + T_R + U_{pA} + U_{nA} , \tag{1}$$

Here h_d (= $T_r + V_{pn}$) is the internal hamiltonian of the projectile nucleus, and U_{pA} and U_{nA} are assumed to be the p-A and n-A optical potentials at $E_p=E_n=E_d/2$. The total wave function is expanded in terms of the complete set of eigen-functions of the internal Hamiltonian h_d. In order to avoid the difficulty of solving the continuum-state coupled-channel equations, we truncate and discretize the p-n continuum states as follows. The relative angular and linear momenta, ℓ and k respectively, are restricted to ℓ=0 and 2 (hence, $I^{\pi}=1^+, 2^+, 3^+$) and $0 \le k \le 1.0\ fm^{-1}$. The k-continuum of each (ℓ, I) is discretized into a finite number of bins; four bins are quite satisfactory for the CDCC calculation of elastic scattering, while eight bins for breakup reactions (Yahiro et al 1982, 1984). The exact p-n wave function is averaged within each bin. The averaged wave function is assumed to be the wave function of the discretized breakup states corresponding to that bin.

The form factors are calculated by folding U_{pA} and U_{nA} into the discretized states of the p-n system. It is to be stressed that there is <u>no freely adjustable parameter</u>. We then solve the coupled-channel equations which are similar to those for the elastic and the inelastic scattering to projectile discrete states, and obtain discrete S-matrix elements. Convergence of the elastic and breakup S-matrix elements was examined successfully with respect to increasing ℓ_{max} and k_{max} and narrowing Δk (Yahiro et al 1982).

Continuous S-matrix elements for the breakup into the continuum states may be given, as a function of k, by smoothly interpolating the discrete S-matrix elements with respect to k. The continuous S-matrix elements immediately give the elastic breakup cross sections for (d,pn) reactions. Observed breakup cross sections of ^{12}C, ^{51}V, ^{118}Sn(d,pn) reactions were well explained by the CDCC calculation (Yahiro et al 1984). The same treatment has been applied to the projectile breakup of ^3He (Iseri 1985), ^6Li and ^7Li (Sakuragi, 1985, 1986, Sakuragi et al 1986c); it is assumed that the projectile nuclei are composed of the d+p, α+d and α+t clusters, respectively.

Since the coupling to the rearrangement channels are known to be enough weak for $E_d \gtrsim 40$ MeV (Kawai 1982, Kawai et al 1986), the rearrangement cross sections may be calculated by Born approximation using the CDCC wave functions (Iseri et al 1983, Iseri 1985 for (d,p) and (p,d) cross sections).

3. Scattering of polarized deuteron at 56 MeV and 400 MeV

Matsuoka et al (1986) precisely measured the cross sections and the vector and tensor analyzing powers of deuteron scattering at 56 MeV from several targets. Their data for ^{58}Ni and ^{208}Pb targets are shown in Figs.1 and 2. In the hamiltonian (1), the optical potentials U_{pA} and U_{nA} are taken from those at $E_p=E_n$=28 MeV (Becchetti and Greenlees 1969). The proton potential at 30.3 MeV reproduces very well the observed cross sections and vector analyzing powers of $\vec{p}+^{58}Ni$ and $\vec{p}+^{208}Pb$ at the energy. As mentioned above, we consider the deuteron breakup states of S wave (S^*) and D wave (D^*).

The deuteron ground state is assumed to be composed of S wave (S_0) and D wave (D_0) which are determined by using the Reid soft-core potential.

As for the ^{58}Ni target case, our CDCC calculation (Iseri et al 1986) gives the solid curves in Fig. 1 which well reproduce the cross section and A_{yy}, A_{xx}, A_{xz} and X_2 (except $\theta_d^{cm} \gtrsim 100°$ of A_{xz}) and explain the gross structure of A_y. The dashed curves are given by a single-channel (folding model) calculation in which all the deuteron breakup states are ignored. In the dotted curves, the D-wave component of the deuteron ground state is further omitted from the calculation of the dashed curves. The difference between the solid curve and dashed one may be identified as the effect of the

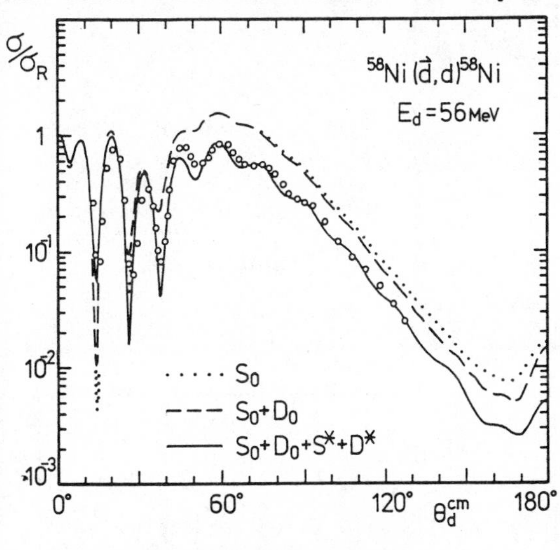

Fig. 1. Observed (Matsuoka et al 1986) and calculated (Iseri et al 1986) cross section and vector and tensor analyzing powers for $\vec{d}+^{40}$Ca at 56 MeV.

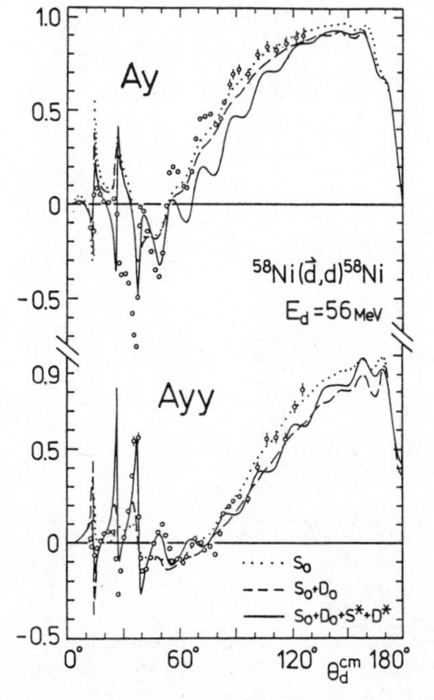

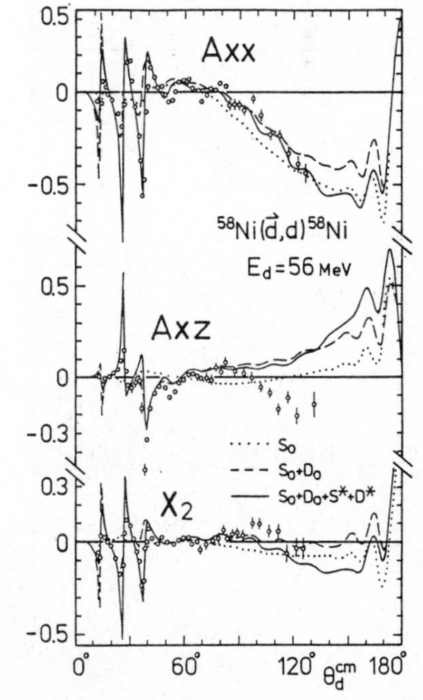

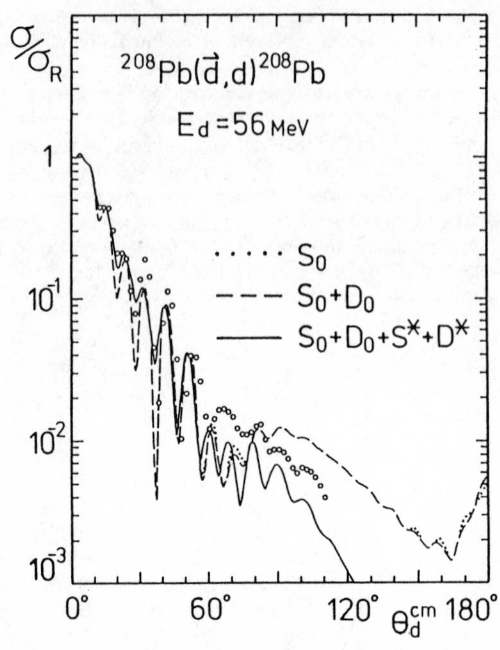

Fig. 2. Observed (Matsuoka et al 1986) and calculated (Iseri et al 1986) cross sections and vector and tensor analyzing powers for $\vec{d}+^{208}Pb$ at 56 MeV.

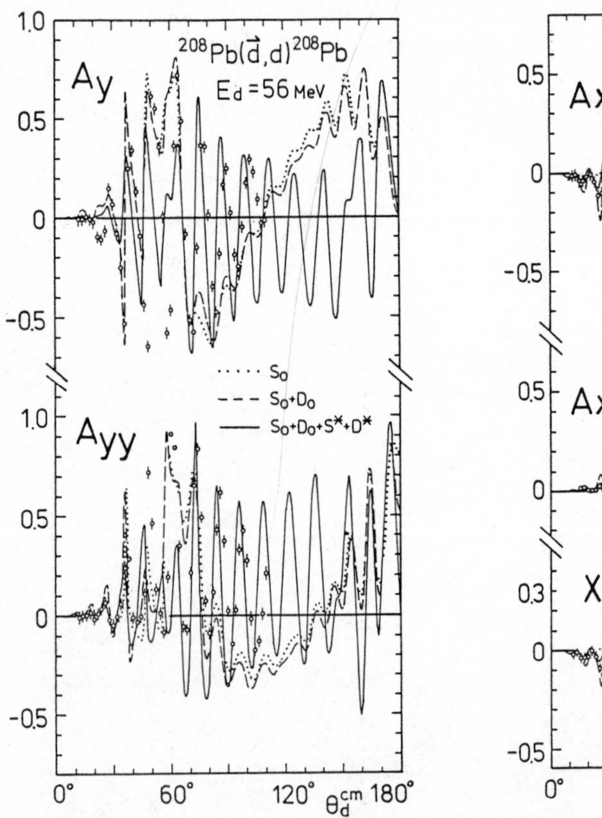

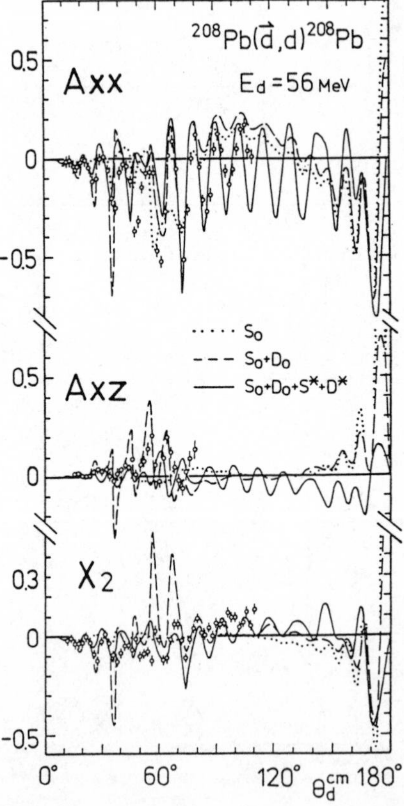

breakup channels, while the difference between the dashed curve and the dotted one as the effect of the deuteron D-wave component. The breakup effect is very large in the cross section, significant in A_y and rather small in the tensor analyzing powers. The D-wave component in the deuteron ground state is important in all the tensor analyzing powers at forward angles ($\theta_d^{cm} \gtrsim 60°$).

An absolutely important role of the breakup channels on the analyzing powers is seen in the case of $\vec{d}+^{208}Pb$ at 56 Mev in Fig. 2. A characteristic feature of this case is the very rapid oscillation in the angular distribution of the analyzing powers. The CDCC calculation (solid curves) reproduces satisfactorily well the experimental data of the cross section and vector and tensor analyzing powers. The breakup effect is absolutely necessary to reproduce the rapid oscillation in the angular dependence of the analyzing powers. This oscillation is found (Iseri et al 1986) to come from a strong interference between the near-side and far-side scattering; the deuteron breakup effect gives rise to an additional repulsive potential in the surface region, which makes the near-side and far-side contributions come closer much in this scattering. The breakup effect is also seen evidently in attenuating A_{xz} and X_2 toward the observed strength at $\theta_d^{cm} \lesssim 70°$.

The deuteron breakup effect is expected to become less important as the incident energy increases. In the elastic scattering of $\vec{d}+^{58}Ni$ at 400 MeV we examine the effect as well as the validity of the three-body model at intermediate energies. The proton potential U_{pA} is determined (Yahiro et al 1985) so as to fit the 200-MeV proton scattering (σ and A_y), and $U_{nA}=U_{pA}$ is assumed except the Coulomb part. The U_{pA} employed is a wine-bottle type, which is derived from the Dirac optical model in its effective Schrödinger equation form.

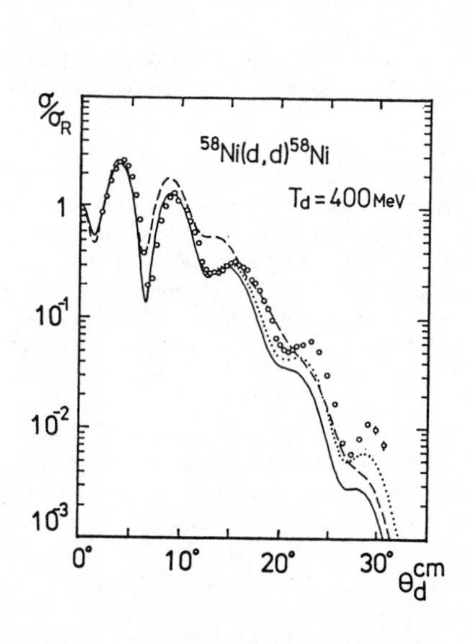

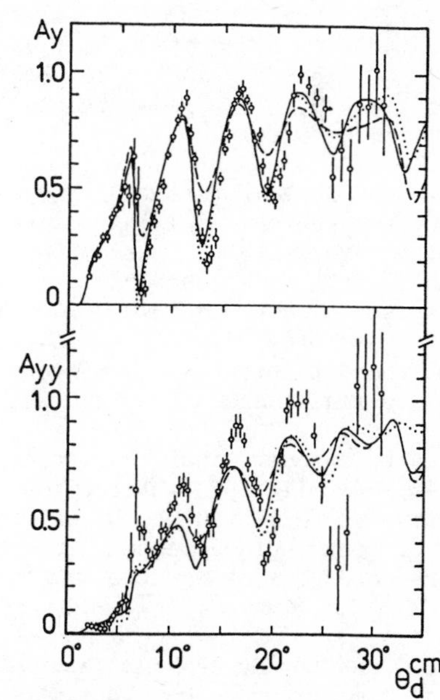

Fig. 3. Observed (van Sen et al 1985) and calculated (Yahiro et al 1985, 1986) cross sections and analyzing powers for $\vec{d}+^{58}Ni$ at 400 MeV.

Calculated (Yahiro et al 1985, 1986 with Schrödinger equation with the relativistic kinematics correction) and observed (van Sen et al 1985) cross sections and analyzing powers for the $\vec{d}+^{58}Ni$ at 400 MeV are shown in Fig. 3. Even in the single-channel calculation without the breakup channel (solid curve), the agreement with the data is satisfactory. The difference between the solid curve and the dotted curve (CDCC calculation) is identified as the effect of the deuteron virtual breakup. For $\theta_d^{cm} \gtrsim 20°$, the breakup effect enhances the cross section almost to account for the observed one. In contrast, the effect is very small in the cross section at forward angles up to ~15° and in A_y and A_{yy} up to ~25°.

It is found (Yahiro et al 1985) that, in the deuteron scattering at 400 MeV, U_{pA} of the wine-bottle type is superior to that of the best-fit Woods-Saxon type (dashed curve in Fig. 3), though both potentials reproduce equally well the proton scattering at 200 MeV.

4. Scattering of polarized ^7Li much above the Coulomb barrier

A lot of investigations both experimental and theoretical have recently been concentrated on scattering of polarized ^6Li and ^7Li by nuclei (Johnson 1985, Moroz 1985, Kamimura et al 1985, further references therein). Among various theoretical approaches, the three-body coupled-channel (CC) studies with folding models of projectile-target interactions have provided a successful understanding of the experimental data of cross sections and vector and second-rank tensor analyzing powers for both of elastic and projectile-inelastic scattering of $^6\vec{Li}$ and $^7\vec{Li}$ from ^{58}Ni at $E_{lab} \simeq 14$ and 20 MeV (Ohnishi et al 1984, Nishioka et al 1984).

In the CC studies, the projectile ^7Li is assumed to consist of an α-particle and a triton, and α-target and t-target optical potentials at $E_\alpha \simeq 4/7 \cdot E_{lab}$ and $E_t \simeq 3/7 \cdot E_{lab}$ respectively are folded into the $3/2^-$ ground state, the $1/2^-$ bound state at $E_x = 0.478$ MeV and the $7/2^-$ and $5/2^-$ resonance states at $E_x = 4.63$ and 6.68 MeV (without non-resonant breakup continuum states in Fig. 7). Four-channel CC calculations were performed with no adjustable parameter.

At 14 and 20 MeV, less than and comparable to the Coulomb-barrier height, most of the vector analyzing powers of elastic scattering were explained by effective spin-orbit interactions from both of the projectile virtual excitations and higher orders of the ground-state tensor potential, while the tensor analyzing powers were mainly explained by the tensor potential itself, and the virtual-excitation effects are very small. In this case the spin-orbit potential obtained by the folding has little effects in both the analyzing powers.

It may be expected that, at incident energies much higher than the Coulomb barrier, details of interactions and reaction mechanisms concerned are effectively reflected in the calculated observables, and as a result the role of each spin-dependent interactions is generally different from that at the energies comparable to or below the barrier. This is the case for $^7\vec{Li}+^{12}C$ at 21 MeV and $^7\vec{Li}+^{120}Sn$ at 44 MeV.

In the scattering of $^7\vec{Li}+^{12}C$ at 21 MeV, the folding spin-orbit potential (FSOP) almost produces the magnitude of the observed vector analyzing powers as long as the projectile excitations are switched off. However, when they are switched on once, FSOP hardly contributes to the analyzing power and the observed analyzing power is attributed to the effective spin-

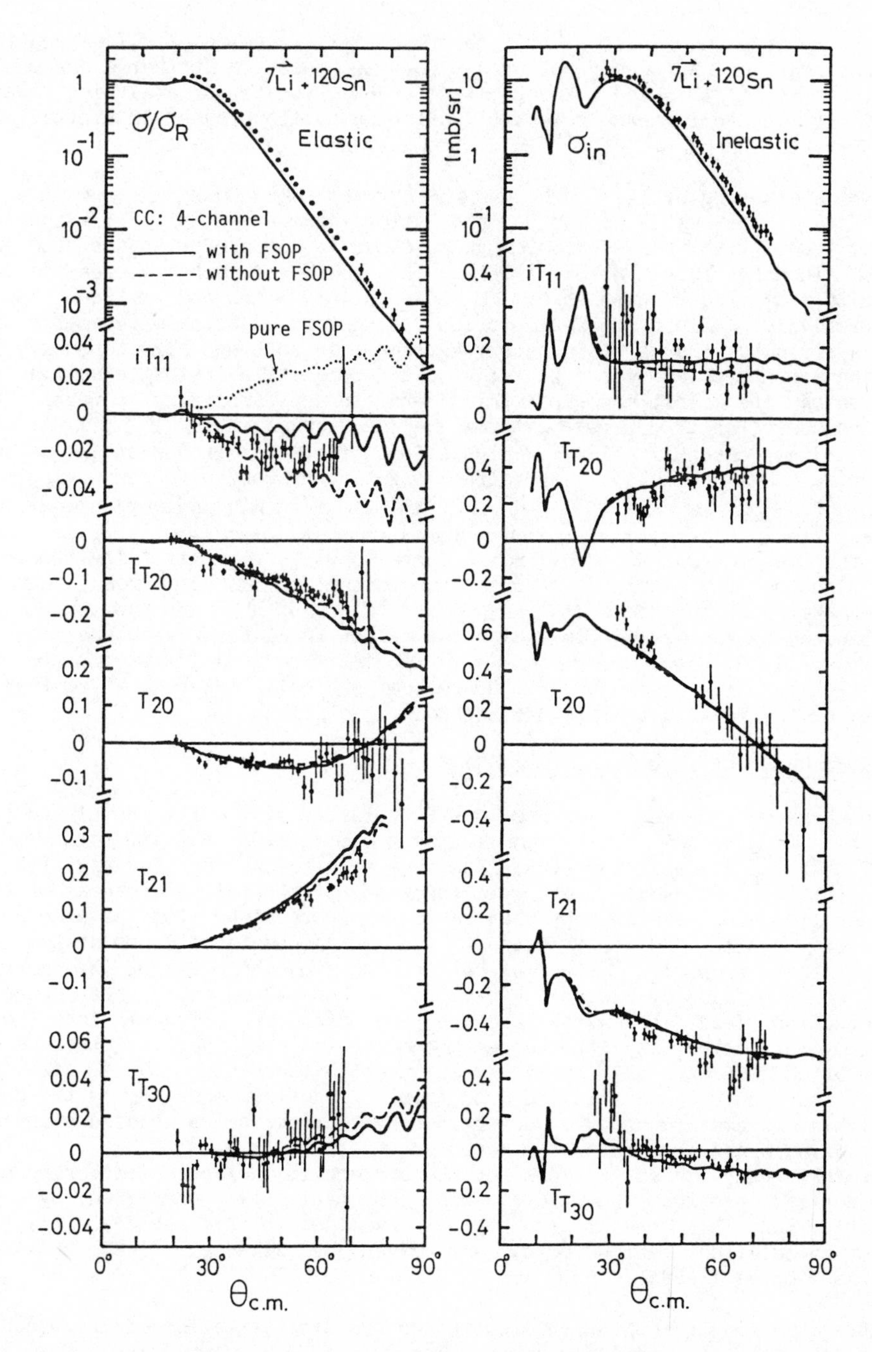

Fig. 4. Observed (Tungate 1985, Moroz et al 1985) and calculated (Kamimura et al 1985, Sakuragi et al 1986b) cross sections and vector, second- and third-rank tensor analyzing powers for $^{7}\vec{Li}+^{120}Sn$ at 44 MeV; elastic scattering (left) and the projectile-inelastic ($1/2^{-}$) scattering (right).

orbit interaction induced by the projectile virtual excitations and the higher order of the tensor potential; this mechanism is further dicussed by Kamimura et al 1985 and Sakuragi et al 1986a. The tensor analyzing powers in this scattering are remarkably affected by the projectile excitations contrary to the case of the ^{58}Ni target (Sakuragi et al 1985).

In the scattering of $^{7}\vec{Li}+^{120}$Sn at 44 MeV, the role of FSOP becomes evident. In Fig.4, the result of the CC calculation (Kamimura et al 1985, Sakuragi et al 1986b) with the same framework as above is compared with the observed data (Tungate 1984, Moroz 1985). The solid and dashed curves are the results of the four-channel CC calculation with and without FSOP, respectively. Since the calculation includes no adjustable parameter, it is surprising that a satisfactory agreement is obtained simultaneously in so many observables for the elastic and projectile-inelastic scattering including the first-observed third-rank tensor analyzing powers. The dotted curve for iT_{11} (elastic) shows the pure contribution of FSOP given by the single-channel ($3/2^-$) calculation with the ground-state tensor potential switched off; it is rather large in contrast with the negligible role of FSOP in $^{7}\vec{Li}+^{58}$Ni at 20 MeV. The contribution of the tensor interactions to iT_{11} in the four-channel calculation without FSOP is shown by the dashed curve. The solid curve for iT_{11} shows that the coherent contribution is nearly the sum of the dotted and dashed curves. We thus note that in $^{7}\vec{Li}+^{120}$Sn at 44 MeV the role of FSOP is evident and is not attenuated by the projectile-excitation effect in contrast with the case of $^{7}\vec{Li}+^{12}$C at 21 MeV. A deep insight of the effects of the spin-dependent interactions is given by a theoretical analysis (Sakuragi et al 1986b) based on the invariant amplitude method.

5. Breakup of ^{6}Li and ^{7}Li projectiles

A number of observed cross sections for elastic scattering of ^{6}Li and ^{7}Li at $E_{lab}\simeq 50\sim 170$ MeV have beautifully been reproduced with the microscopic CDCC method (Sakuragi et al 1983, Kamimura et al 1983,1984, Sakuragi 1985). The totally antisymmetrized wave functions of the relevant states of the projectile are described by the resonating group method for the α+d (α+t) cluster system, and the resonant and non-resonant continuum states are discretized properly. The real part of the projectile-target interaction is constructed by doubly folding the so-called M3Y inter-nucleon interaction (modified with knock-on-type exchange effects) into those projectile states and the target ground state. In order to introduce an absorption effect, the imaginary part with the factor iN_I and the same shape as the real part is added to this interaction, where N_I is the only flexible parameter and is determined by fitting the calculated elastic-scattering cross section to the measured.
Therefore, the cross sections for the projectile-inelastic scattering and projectile breakup reactions are automatically predicted by the calculation. This type of prediction is examined in this section. All the CDCC calculations shown below are given by Sakuragi (1985,1986) and Sakuragi et al (1986c).

Figure 5 illustrates the truncation and the discretization of the breakup continuum of ^{6}Li employed in the microscopic CDCC calculation. Observed cross sections of elastic scattering of ^{6}Li at 156 MeV from ^{12}C and ^{208}Pb are well reproduced by the CDCC calculations; the only adjustable parameter N_I is chosen to be 0.47 and 0.60 for ^{12}C and ^{208}Pb, respectively. The value of N_I at 156 MeV is also used at 178 MeV since the elastic cross sections at 178 MeV from ^{12}C and ^{208}Pb are not measured and the energy dependence of N_I for the ^{6}Li scattering is known to be very weak at the energies. The breakup cross sections are then absolutely predicted by the CDCC calculation.

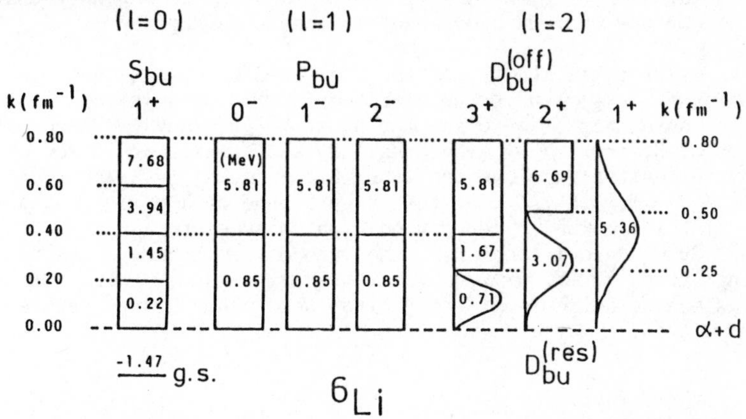

Fig. 5. Truncation and discretization of the α-d breakup continuum of ^{6}Li

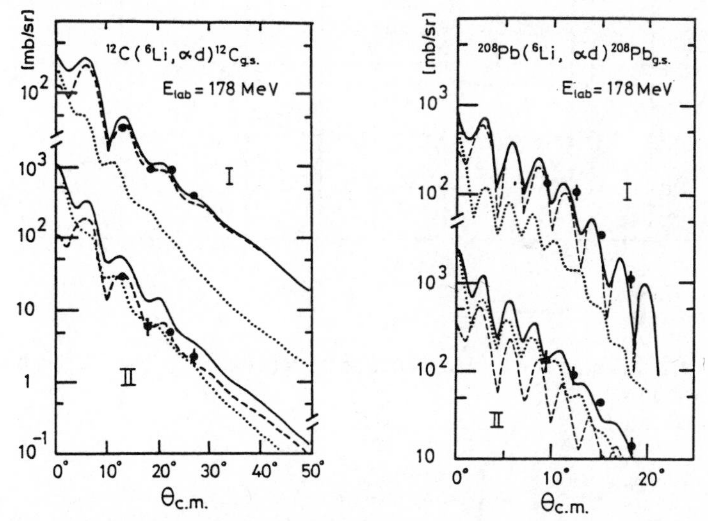

Fig. 6. Observed (Shimoda et al 1986) and calculated (Sakuragi 1986, Sakuragi et al 1986) cross sections for the elastic breakup of ^{6}Li projectile from ^{12}C and ^{208}Pb at 178 MeV.

The elastic breakup cross sections for ^{12}C(^{6}Li,αd) and ^{208}Pb(^{6}Li,αd) at 178 MeV were measured (Shimoda et al 1986) for breakup from the α-d continuum at $0<\varepsilon<1.0$ MeV (region I) and at $1.0<\varepsilon<4.0$ MeV (region II) with the integration over ε in each region, ε being the energy of the α-d relative motion. In Fig. 6, the agreement between the calculated (solid curves) and observed cross sections for the regions I and II is satisfactory. In the region I the contribution of the breakup from the 3^+ resonance (dashed curves) is much larger (by about one order) than that from the non-resonant states (dotted curves), while in the region II the non-resonant contribution (dashed curves) is comparable with the 2^+ resonance contribution (dotted curves) for ^{12}C target and even exceeds the latter for

^{208}Pb. It is found that the coupling between the resonant and non-resonant states is strong and works to reduce the cross section for breakup from the resonant states by a factor of two to three; this suggests that a DWBA calculation of the breakup cross sections seems dangerous.

Figure 7 illustrates the truncation and the discretization of the breakup continuum of ^{7}Li employed in the microscopic CDCC calculation. Since there exists no data for elastic scattering at E_{lab}=70 MeV, we have determined the value of N_I (=0.45) by averaging the values which are fixed by fitting the elastic scattering cross sections (Fig. 8) at E_{lab}=63 MeV (N_I=0.40) and at 78.7 MeV (N_I=0.50). In the figure the projectile-inelastic cross section ($1/2^-$) is well reproduced, which suggests a reliability of the following CDCC calculation for the projectile breakup. As for the ^{7}Li scattering from ^{120}Sn, no data has been reported at ~70 MeV, but we employ N_I=0.65 since elastic scattering from some other heavy targets are well explained with N_I~0.6.

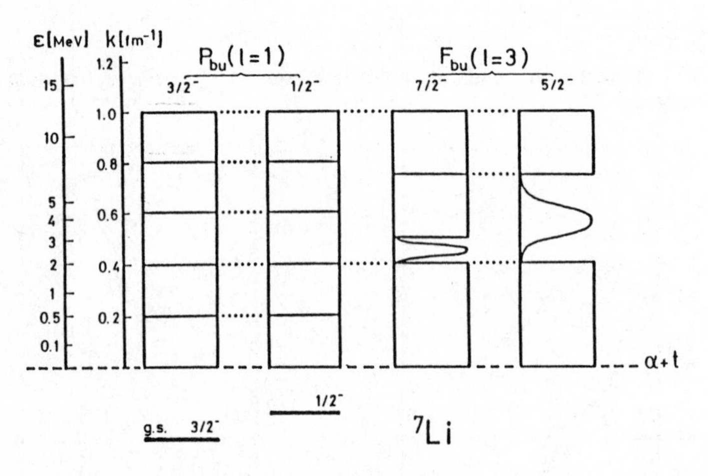

Fig. 7. Truncation and discretization of the α-t breakup continuum of ^{7}Li

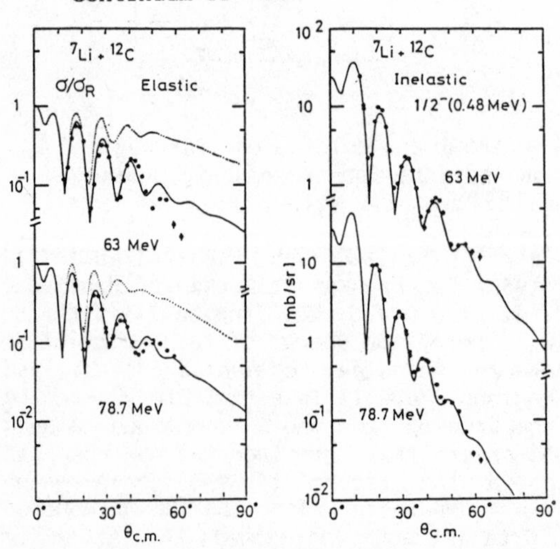

Fig. 8. Observed (Zeller et al 1980) and calculated (Sakuragi et al 1986c) cross sections for elastic and projectile-inelastic ($1/2^-$) scattering of ^{7}Li from ^{12}C at 63 and 79 MeV. Solid and dotted curves are the results of the CDCC calculation and the single-channel calculation.

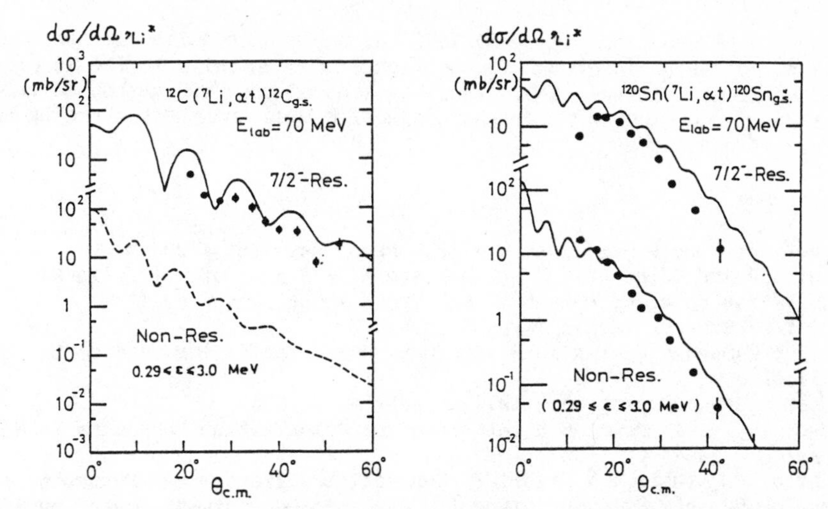

Fig. 9. Observed (Shotter et al 1981, 1984) and calculated (Sakuragi 1986 and Sakuragi et al 1986c) cross sections for the elastic breakup of ^{7}Li projectile from ^{12}C and ^{120}Sn at 70 MeV.

In Fig. 9 the calculated result for the elastic breakup of ^{7}Li is compared with the measured cross sections (Shotter et al 1981, 1984) for breakup from the $7/2^{-}$ resonance at E_x=4.63 MeV (ε=2.16 with respect to the α-t breakup threshold) and from the non-resonant continuum at $0.29<\varepsilon<3.0$ MeV. In the case of ^{120}Sn target, the observed cross sections for breakup from the $7/2^{-}$ resonance and from the non-resonant states are comparable in magnitude, and they are reproduced satisfactorily well by the CDCC calculation.

For ^{12}C target, the observed breakup from the $7/2^{-}$ resonance was fairly well reproduced by the calculation, whereas the breakup from the non-resonant continuum is negligible in the experiment but, in the calculation, it is comparable to the breakup from the $7/2^{-}$ resonance; this discrepancy is to be investigated in future. In the calculation of this section, breakup form factors due to the Coulomb interaction are not included, and therefore in the case of low bombarding energies for heavy targets the breakup due to the Coulomb interaction is a future subject of the CDCC calculation (a preliminary calculation has been done by Sakuragi et al 1986c).

6. Concluding remarks

We have shown that the CDCC approach based on the three-body model for direct reactions works well in understanding the scattering and breakup of deuteron, ^{6}Li and ^{7}Li projectiles which have very low threshold energies for breakup into the constituent clusters. Since the approach treats the projectile breakup channels explicitly, it has been precisely investigated that the projectile virtual breakup affects significantly not only on cross sections but also on spin observables of elastic scattering; the virtual breakup has been found to induce effectively not only a central potential but also a spin-dependent one with substantial strengths.

Further application of CDCC to new data for ^{7}Li breakup by Edinburgh group (contributions to this conference, C215-C217, and private communications) will be interesting.

In a (d,p) reaction at E_d=80 MeV, the neutron transfer via the deuteron breakup channels is known to contribute substantially to the (d,p) cross section (Iseri et al 1983). Applications of the CDCC method to transfer reactions in a wide range of the projectile mass and the bombarding energy will be of particular interest.

References

Becchetti F D Jr., Greenlees G W 1969 Phys. Rev. **182** 1190

Farrel J P Jr., Vincent C M and Austern N 1976 Ann. of Phys. **114** 93

Iseri Y, Yahiro M and Nakano M 1983 Prog. Theor. Phys. **69** 13

Iseri Y 1985 Ph.D. Thesis (Kyushu Univ.)

Iseri Y, Kameyama H, Yahiro M, Kamimura M and Tanifuji M 1986 to be published

Johnson R C and Soper P J R 1970 Phys. Rev. **C1** 976

Johnson R C 1985 Proc. 6th Int. Symp. on Polarization Phenomena in Nuclear Physics, Osaka, pp 7-30

Kamimura M, Yahiro M, Iseri Y, Sakuragi Y, Nakano M and Fukushima Y 1983 Proc. RCNP Int. Symp. on Light Ion Reaction Mechanisms, Osaka, pp 558-580

Kamimura M, Yahiro M, Iseri Y, Nakano M and Sakuragi Y 1984 Proc. 4th Int. Conf. on Clustering Aspects of Nuclear Structure, Chester, pp 181-197

Kamimura M, Sakuragi Y, Yahiro M and Tanifuji M 1985 Proc. 6th Int. Symp. on Polarization Phenomena in Nuclear Physics, Osaka, pp 205-220

Kawai M 1982 Proc. 3rd Int. Conf. on Nuclear Reaction Mechanisms, Varenna

Kawai M, Kamimura M and K. Takesako 1986 Prog. Theor. Phys. Supplement **87** to be published

Matsuoka N et al 1986 Nucl. Phys. **A455** 413

Moroz Z 1985 Proc. 6th Int. Symp. on Polarization Phenomena in Nuclear Physics, Osaka, pp 221-233

Nishioka H, Tostevin J A Johnson R C and Kubo K -I 1984 Nucl. Phys. **A415** 230

Ohnishi H, Tanifuji M, Kamimura M, Sakuragi Y and Yahiro M 1984 Nucl. Phys. **A415** 271

Rawitscher G H 1974 Phys. Rev. **C9** 2210

Sakuragi Y, Yahiro M Kamimura M 1983 Prog. Theor. Phys. **70** 1047

Sakuragi Y 1985 Ph.D. Thesis (Kyushu Univ.)

Sakuragi Y, Kamimura M, Tanifuji M and Yahiro M 1985 Phys. Lett. **153B** 372

Sakuragi Y 1986 Contributions to this conference

Sakuragi Y, Kamimura M, Tanifuji M and Yahiro M 1986a Phys. Lett. in press

Sakuragi Y, Yahiro Y, Kamimura M and Tanifuji M 1986b Nucl. Phys. to be published

Sakuragi Y, Yahiro M and Kamimura M 1986c Prog. Theor. Phys. Supplement **87** to be published

Shimoda T et al 1986 Contribution to this conference C159 and private communication

Shotter A C et al 1981 Phys. Rev. Lett. **46** 12

Shotter A C et al 1984 Phys. Rev. Lett. **53** 1539

Tungate G 1984 private communication

van Sen N et al 1985 Phys. Lett. **156B** 185 and private communications

Yahiro M, Nakano M, Iseri Y and Kamimura M 1982 Prog. Theor. Phys. **67** 1467

Yahiro M, Iseri Y, Kamimura M and Nakano M 1984 Phys. Lett. **141B** 19

Yahiro M 1985 Ph.D. Thesis (Kyushu Univ.)

Yahiro M, Kameyama H, Iseri Y, Kamimura M and Kawai M 1985 Proc. 6th Int. Symp. on Polarization Phenomena in Nuclear Physics, Osaka, pp 198-202

Yahiro M, Kameyama H, Iseri Y, Kamimura M and Kawai M 1986 Physics Letters to be published

Zeller A F et al 1980 Phys. Rev. **C22** 1534

Inst. Phys. Conf. Ser. No. 86
Paper presented at Int. Nucl. Phys. Conf., Harrogate, UK, 1986

Light nuclei far from stability: limits and properties

Claude Détraz

GANIL, Caen, France

1. Why and how to study light exotic nuclei.

The production and the study of light nuclei either neutron- or proton-rich has benefited from several breakthroughs. Although the ingenuity of physicists often provided ad-hoc ways to produce a rare, sought-after nuclear species, a few mechanisms were particularly prolific. Fission was, for a long time, the most efficient source of isotopes far from the valley of β-stability (Lysekil 1966). In the early 70's, deep inelastic collisions provided a new way to enrich the neutron number of light nuclei. Tens of new neutron-rich isotopes were observed (Artukh 1971, Volkov 1973). Two other mechanisms continously contributed to the steady increase in the number of known isotopes. Compound nucleus formation followed by the evaporation of some nucleons, mostly neutrons, is one of them. Well suited to the use of advanced spectroscopic methods, it keeps playing a major role in the study of exotic isotopes (Armbruster 1982, Roeckl 1983). Also target fragmentation, as induced by energetic projectiles, usually light (p, α, π, ...), associated with mass spectroscopy techniques, has been a rich source of results (Thibault 1981).

A few years ago, a new method appeared, potentially as general as those mentioned above. It uses the mechanism of projectile fragmentation. A heavy-ion projectile impinging on a target at high velocity, typically a few hundreds of MeV per nucleon, is likely to experience fragmentation. The Z and N distributions of these fragments, which are governed by statistics, allow observation of exotic nuclear species which could not be reached otherwise, as demonstrated in the pioneering work done at the Bevalac (Symons et al 1979, Westfall et al 1979).

The availability of new heavy-ion accelerators, with energies reaching 100 MeV per nucleon and intensities much higher than those obtainable at the Bevalac, opens a possibility to further extend that method. Fragmentation at, say, 50 MeV per nucleon is certainly not the clear process that makes up for the total reaction cross section at much higher energies. The momentum distribution of the fragments, which in first order results from the inner momenta of participant nucleons within the projectile, is relatively broader at GANIL than at Bevalac energies, hence the collection of fragments is less efficient. This distribution is further broadened by dissipation effects, reminiscent of low-energy mechanisms, which persist at a few tens of MeV per nucleon, as it is now well documented (Guerreau 1985). Yet the fact that the intensities available at GANIL for the projectiles of interest reach 5×10^{11}pps, and are expected to increase over the years, offers an unmatched opportunity to produce new exotic isotopes.

This raises the difficult question of determining which new isotopes, and which of their properties, are the most relevant to study. There are indeed some three or four thousand bound isotopes yet unobserved and it is worthwhile establishing which of them can shed light on the fundamental properties of nuclei and can contribute to the better understanding of existing nuclear models. The analysis of the evolution of nuclear properties along a chain of isotopes of given Z might help clarify this question. Figure 1 presents the variation of log $T_{1/2}$ with the neutron number of isotopes. Several regions can be identified. For instance, it seems that the detailed study of one more nucleus on the long slope of region e, towards the neutron drip line, cannot be expected to bring any but local information.

One would expect more from two other types of research:

First, the exploration of new types of radioactivity should be pursued. In the latter years new exotic processes have been discovered. Among the most spectacular results is certainly proton radioactivity (Hoffmann 1984). Several new β-delayed emissions were also observed : two neutrons (Azuma et al. 1979), two protons (Cable et al. 1983), three neutrons (Azuma et al. 1980, Langevin et al. 1981), triton (Langevin et al. 1984) and also β-delayed α emission from a neutron-rich isotope (Détraz et al. 1983). New radioactivities are now within experimental reach, as discussed below (§3). The wealth of information further gathered on the already known β-delayed proton -and even more, neutron-emission has led to realize that these are widespread processes, which rapidly dominate all decay processes far from β-stability (Klapdor et al. 1984) and hence play a major role in explosive stellar nucleosynthesis.

Second, it appears that a most fruitful contribution from the study of exotic nuclei would come from the knowledge of basic nuclear parameters on long series of isotopes. These parameters are, for instance, the values of the binding energies as obtained from mass measurements ; $T_{1/2}$ values ; β- delayed neutron emission probability (P_n) values ; excitation energies of the lowest 2^+ levels of even-even isotopes. Static properties such as magnetic moments or quadrupole deformations (Jacquinot and Klapisch 1979, Otten 1981) although difficult to measure systematically far from stability, are also of major importance. The point is that the reliable knowledge of such properties, even with a limited accuracy, provides a severe test of current models as long as it bears on large numbers of neighbouring isotopes.

2. Towards the proton and neutron drip lines

Knowing that an isotope is bound or not for proton or neutron emission puts a limit on its binding energy, and allows a comparison with the large number of theoretical mass predictions (Maripuu 1975). The fragmentation-like process of projectiles from GANIL has been used to push the limits of experimentally observed nuclei further toward the drip lines.

The prime requirement of such investigations is to detect and identify fast projectile fragments with as high an efficiency and as low a background as possible. A double magnetic system called LISE (Langevin and Anne 1985) was built to best meet this requirement. Fragments

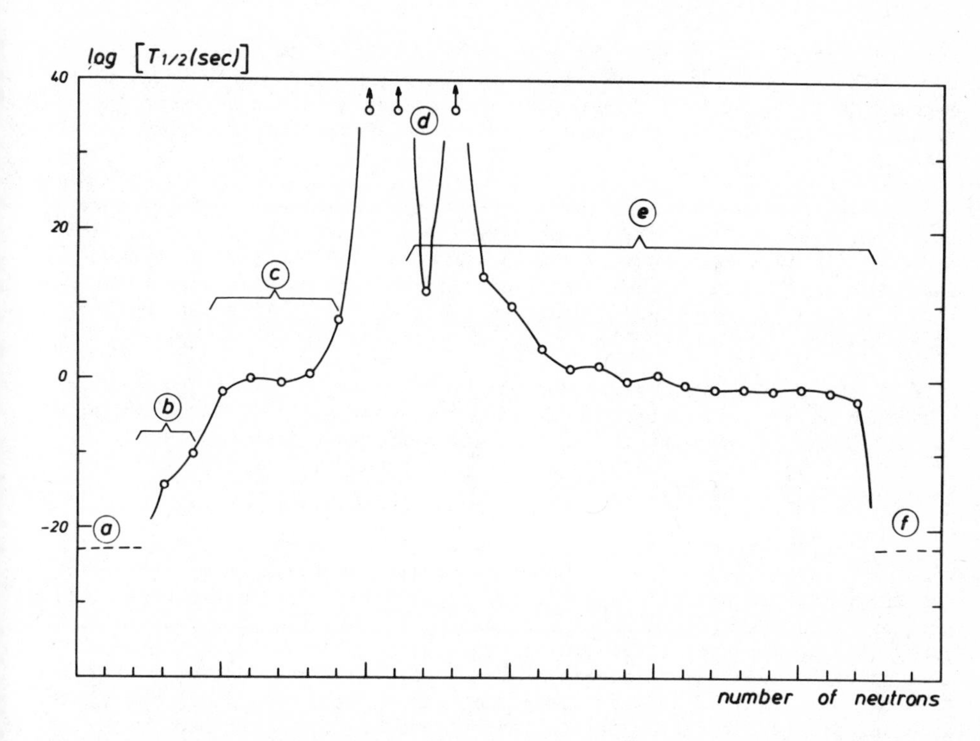

Fig. 1 Typical variation with N, the neutron number, of the radioactive half lives of isotopes of given Z. So-called stable nuclei (region d) have $T_{1/2}$ values of the order of magnitude of the proton half life itself for which only a lower limit is known. Furthermore the half lives of nuclei much heavier than ^{56}Fe are also limited by the fact that fission-like disintegration processes are open, even if it is with exceedingly long half lives (Sandulescu et al. 1985). If the nucleus is bound for hadron emission but can β-decay, $T_{1/2}$ ranges from 1 ms up, the shortest half-life yet observed being 1.5 ms for ^{35}Na (Langevin et al. 1983 a). This corresponds to regions c and e, and these isotopes are usually said to "exist" or "to be bound". It is in these regions that a growing number of β-delayed emission processes have been observed (see text). At last if the nucleus is unbound for neutron emission, $T_{1/2}$ falls brutally down to some 10^{-23} s. A staggering odd-even effect can occur at the border of the neutron drip line. On the proton-rich side the fall from about 10^{-2}s to about 10^{-23}s is not as drastic due to the Coulomb barrier which inhibits the emission of low-energy protons. Thus, for available proton energies smaller than about 1 MeV (region b) proton radioactivity can take place and was actually observed (Hofmann 1984). That is also where two-proton radioactivity is expected. The limit between this region b and region a where unbound protons leave the nucleus before it is actually observed obviously depends upon detection techniques. Hence the limit between proton-active and proton-unbound nuclei cannot be but somewhat arbitrary. It should be emphasized that, for most Z values, except the very lowest ones, only a small fraction of this curve is known.

selected according to their A/Z values are collected in a low-background room by a triple-focusing system within the 5 % momentum acceptance of the magnetic system. Further fragment selection, accomplished by using an energy degrader between the two dipoles, has been successfully realized (Dufour et al. 1986 a).

In successive runs over the last two years, many new isotopes were observed. From an ^{40}Ar projectile, ^{23}N, ^{29}Ne, ^{30}Ne (Langevin et al. 1985) and even ^{22}C (Pougheon et al. 1986 a) were observed, while the unbound character of ^{21}C and ^{25}O was established. The fragmentation of neutron-rich ^{86}Kr projectiles yielded fourteen new isotopes (Guillemaud-Mueller et al. 1985).

On the proton- rich side, the use of ^{40}Ca projectiles allowed the observation of $T_z = -5/2$ isotopes, ^{23}Si, ^{27}S, ^{31}Ar and ^{35}Ca (Langevin et al 1986). Lastly a run with a ^{58}Ni beam was performed. Although the data analysis is not final, twelve new isotopes can already be reported (Pougheon et al. 1986 b)(fig.2).

As for the production yields, several points came out of this work. The complexity of the reaction mechanism, at variance with the simple fragmentation process, was clearly established. For instance the influence of the neutron richness of the target was observed (Guerreau et al. 1983). Furthermore the large yields observed from transfer mechanisms, even at relatively high incident energy, definitely open new possibilities (see fig. 2). It also becomes clear that the maximum efficiency for yielding new isotopes far from the valley of β-stability is reached for Z-values not too much smaller than the Z-value of the projectile. Figure 2 shows a good example of that general trend which might be explained by the increasing surface excitation energy associated with an increasing size of the bite operated on the projectile, in the framework of the abrasion-ablation model. This increased excitation energy for lower Z values would indeed result in the evaporation of more nucleons when the quasi-projectile deexcites, hence to a return towards the valley of stability for the final cold nuclei.

There are at least indications that, with heavier projectiles, the fall of yields with increasing N for a given Z might be more moderate than the factor close to 10 which is observed for each step further away from stability in the case of lighter projectiles. This is possibly related to the onset of new reaction mechanisms, like fission for instance. An analysis of the yields observed, which is essential for determining the most efficient choice of the projectile, of its energy, and of the magnetic setting of LISE, is under way (Guerreau 1986).

The experimental results obtained so far yield useful informations on the proton and neutron drip lines. On the proton-rich side, the observation of the four $T_z = -5/2$ nuclei allows to reach the drip line for most Z values up to Z = 20 (see fig.4). As shown in figure 1, the drip line is not as sharp on that side as on the neutron-rich one. Hence the so-called existence of a few light proton-rich nuclei is still open. A most interesting case concerns ^{22}Si since Garvey-Kelson calculations (Jänecke 1976) lead to a near-zero binding energy.

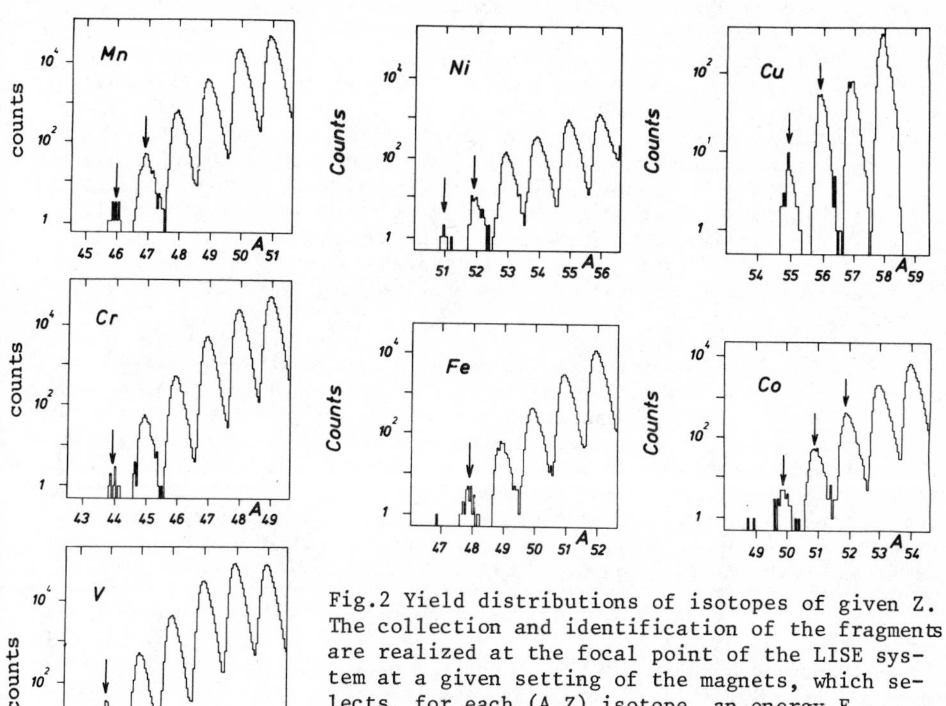

Fig.2 Yield distributions of isotopes of given Z. The collection and identification of the fragments are realized at the focal point of the LISE system at a given setting of the magnets, which selects, for each (A,Z) isotope, an energy E within the acceptance of the system. Thus these yield distributions do not reflect the cross section distributions. The yet unknown isotopes are marked by arrows. These are preliminary results from an on-going analysis of the data collected with a 55A MeV ^{58}Ni beam from GANIL. Note that most new isotopes are obtained for elements with Z values only slightly smaller than the projectile one, and that Cu isotopes result from the addition of one proton to the projectile.

On the neutron-rich side, the neutron drip line appears to be reached all the way up to Z = 7 (Pougheon et al. 1986 a). The relevance of the type of results reported here is well illustrated by the case of ^{29}Ne that all mass predictions but one (Uno and Yamada 1982) found unbound and which was definitely observed (Langevin et al. 1985). It might be relevant to note that with Z = 10 and N = 19, this isotope, for which a binding energy larger than predicted by systematics or extrapolations is observed, lies very close to the Z = 11-12, N = 20 nuclei. This is an area of very strong deformation (Détraz 1981) at the expected location of a closed shell. That case indeed certainly provided a clear example of the unique and fundamental information on nuclear models that the study of exotic nuclei can contribute.

3. The search for new radioactivities

The concept of radioactivity has greatly benefited in the last few years from the discovery of new radioactive processes. Those have been mentioned above (§1). Some new ones might become attainable in the near future.

A very fascinating one would be two-proton radioactivity. Pairing energy effects make the binding energy of many even-Z proton-rich nuclei stronger against one-proton than two-protons emission. At the proton drip line it might happen that a nucleus is thus bound for one proton emission but is able to decay by the emission of two protons. This process was discussed long ago by Goldanski (1961, 1966) and Jänecke (1964). The probability of this emission is governed by barrier penetration which itself strongly depends upon the kinematics of the two protons. It is generally found that the configuration most likely to speed up the crossing of the barrier corresponds to two correlated protons equally sharing the available energy.

Some light nuclei which can be produced from GANIL projectiles are good candidates to exhibit 2p radioactivity. The negative binding energy (E_{2p}) should not lie below about -0.8 or -1 MeV so that the $T_{1/2}$ (2p) value is not so short as to make the nucleus decay before it is detected. And it should not lie above about -0.3 or -0.4 MeV so that the $T_{1/2}$ (2p) value is not that much longer than the competing $T_{1/2}$ (β) value as to make the 2p branching ratio vanishingly small. Thus a narrow energy window, say $-0.4 > E_{2p} > -0.8$ MeV, exists for potential candidates for 2p radioactivity with Z values around 15 or 20.

The nucleus ^{31}Ar, recently produced with good statistics (Langevin et al. 1986) appears not to be unbound enough since Garvey Kelson calculations (Jänecke 1976) predict E_{2p} of the order of - 200 keV. One should yet remember that large deviations sometimes occur for new masses measured away from stability (Haustein 1984), so that 2p emission from ^{31}Ar cannot be ruled out. Another candidate is ^{39}Ti with a predicted - 700 keV 2p value,but which is still to be produced (see fig.2).

Candidates for neutron radioactivity were identified long ago by Pecker et al (1971) as long-lived isomers with very high spin, among neutron-rich nuclei of the fp shell. The neutron radioactivity process depends drastically upon the difference between the energy locations of such a high-spin configuration and the neutron emission threshold. Thus precise predictions are particularly difficult. Yet such situations as found by Pecker et al are probably bound to occur. This makes the search for neutron radioactivity legitimate, even if difficult, especially since, among the candidates identified, some of them, such as ^{67}Fe, can be obtained at GANIL. It should be remembered that proton radioactivity was first observed from an isomeric state (Jackson 1970, Cerny 1970).

At last it seems that at least one more β-delayed particle emission process might be possible close to the proton drip line. It concerns ^{3}He emission and requires that the binding energy of ^{3}He in the daughter nucleus is significantly smaller than the Q_{β}-value of the parent nucleus. Energy windows of 3 MeV or more can be found for ^{22}Si, ^{27}S, ^{31}Ar, ^{35}Ca, ^{39}Ti or ^{43}Cr for example.

4. Mass measurements

The knowledge of the binding energy of a nucleus provides a stringent test of nuclear forces, especially in the case of exotic nuclei which have an unusual imbalance of protons and neutrons. Indeed predictions widely differ between themselves far from stability and, even with limited accuracy, experimental results are quite discriminatory and enlightening. The variation of the binding energy along a series of isotopes also reveals fundamental nuclear properties such as shell closure effects or onsets of deformation (Audi et al. 1979, Epherre et al. 1979).

Thus the fact that new isotopes are produced at GANIL with large yields opens the possibility of a fruitful and broad investigation of binding energies of light nuclei far from stability. The absolute measurement of the nuclear mass must be accomplished with an uncertainty much smaller than 1 MeV to provide useful information. For a nucleus of interest, all the collected events must then be free of spurious background counts, and the relative FWHM of the two parameters necessary to identify the isotope must be at most a few 10^{-4}. The accuracy on the peak centroid, a portion of the FWHM which decreases with increasing statistics, can then reach the required level of a few 10^{-5}.

A very powerful method to reach that goal has been developed at GANIL. The nature and energy of a fragment produced are extracted from two parameters known with a remarkably high accuracy. Its magnetic rigidity is measured by the SPEG spectrometer with a FWHM of 10^{-4} (Birien and Valero 1981). Its time-of-flight is determined along a flight path some 100 meters long since the target is exceptionally located near the exit of the second and last sector-separated cyclotron of GANIL, while the fragment is detected in the focal plane of SPEG. New mass values with accuracy better than 500 keV (fig.3) have already been reported (Gillibert et al. 1986) for $^{20-21}N$, ^{23}O, $^{24-25-26}F$ and new data are being analyzed. They are expected to give improved results, with uncertainties around 200 keV, for some 12 new isotopes (Gillibert and Mittig 1986). They will bring new information on the behaviour of neutron-rich nuclei near N = 20, where the very strong deformation observed at Z = 11 (as mentionned in §2) seems to quickly disappear for increasing values of Z (Fifield 1985, Fifield 1986).

There is clearly a concrete possibility of obtaining systematically the binding energy surface of light nuclei, which should stimulate theoretical efforts in this field.

A special attention should be directed to the case of neutron-rich isotopes with very low Z, especially since a calculation of their masses from shell-model or Hartree-Fock techniques is rather unreliable. Although the non-occurrence of light neutral nuclei (Turkevich et al. 1977, De Boer et al. 1980) and the unbound character of ^{10}He (Volkov 1973) are supported by strong experimental evidence, the theoretical understanding of the nuclei close to the N-axis of the isotope chart is still limited. In this respect, the difficult mass measurements recently performed at Dubna (Belozyorov et al. 1985) for the unbound neutron-rich isotopes of hydrogen are very useful.

5. Other quantitative information on exotic nuclei.

While a production yield of one nucleus per day might suffice to determine that an isotope exists, i. e. is bound, much higher yields are needed to derive quantitative information about them. For an absolute measurement of its mass, as accomplished by the method described in the preceeding

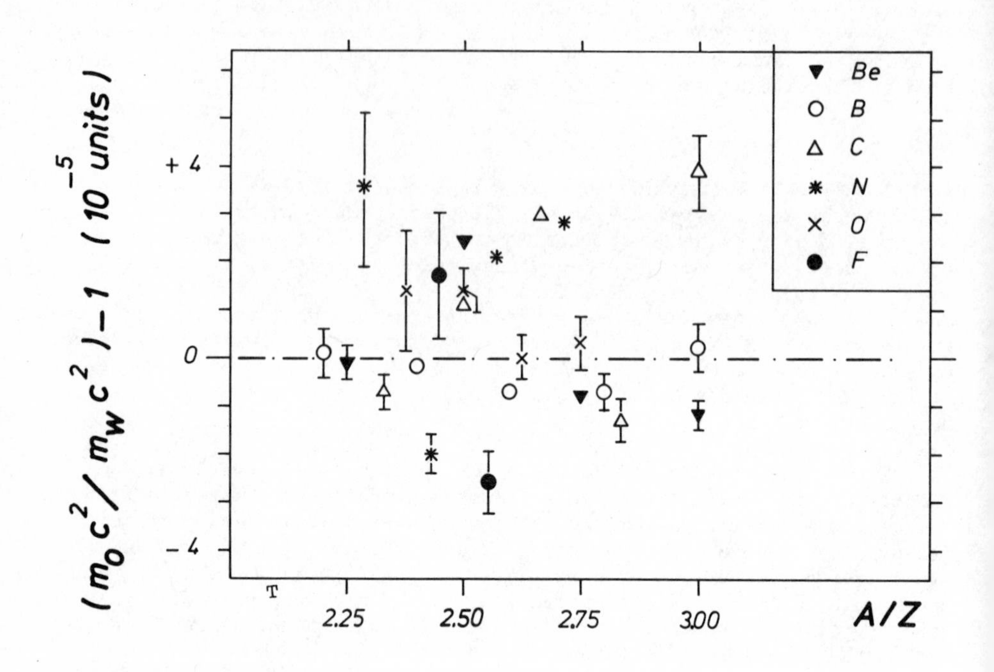

Fig. 3 An analysis of the deviation of the absolute masses of known nuclei as measured at GANIL where m_o refers to the measured value and m_W to the compilation by Wapstra and Audi (1985). This figure is extracted from Gillibert et al. (1986). The dispersion of the values around zero has a FWHM of 4×10^{-5}.

section, it is difficult to obtain accurate values if the yield is much lower than one nucleus per second. Some kind of simple γ - spectroscopy still requires a yield two orders of magnitude higher. It also requires that the fragments selected by LISE according to their A/Z value be further selected to reduce the number of simultaneously collected nuclides in the same run. An energy degrader between the two dipoles provides such a second independent selection. It has been put to use by the Bordeaux group with excellent results (Dufour et al. 1986 a). Indeed, for the nuclei of interest, the energy-loss selection operates according to a $A^{2.5}/Z^{1.5}$ law which, combined with the magnetic selection which follows the A/Z law, drastically restricts the number of nuclear species collected at the focal plane of LISE. Although simple in its principles, this method requires considerable care to conserve the achromaticity properties of the instrument. Furthermore, changing charge states in the energy degrader might reduce its efficiency and considerably complicate its use for high-Z fragments. Yet, for nuclei up to about Z = 20, it has brought excellent results. The half lives and γ-energies for the β-decay of thirteen new neutron-rich isotopes have been reported (Dufour et al. 1986), as far from stability as, e. g. ^{24}F, ^{26}Ne, ^{36}Si or ^{40}S (see fig. 4). A similar production mechanism, i.e. fragmentation, together with another type of isotope separation technique, has allowed to measure the half lives of the ^{14}Be and ^{17}C fragments from 30A MeV ^{18}O projectiles at MSU(Curtin et al.1986).

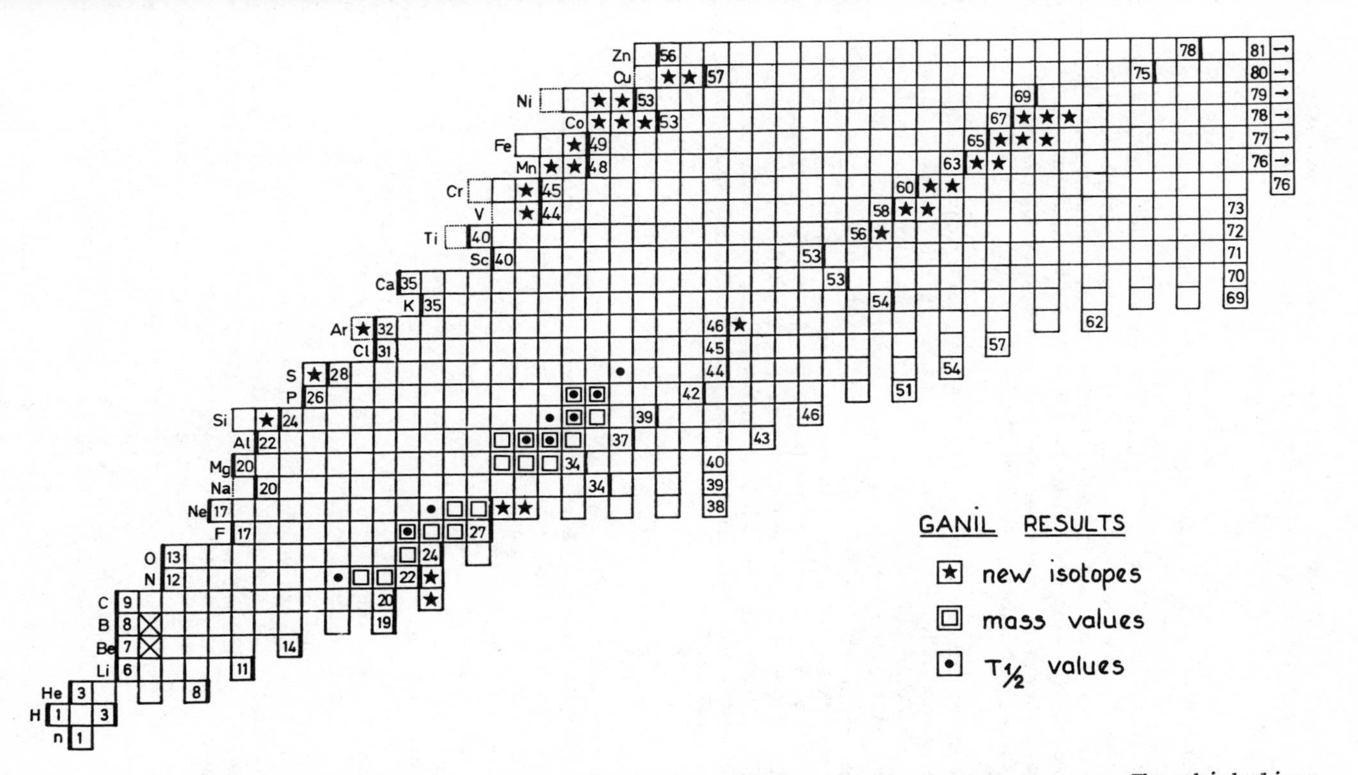

Fig.4 Chart of the light nuclei showing the results obtained at GANIL over the last two years. The thick lines indicate the limits of previously known isotopes. The neutron drip-line is predicted by Uno and Yamada (1982), and the proton-drip line is calculated by the Garvey-Kelson procedure (Jänecke 1976). The proton-rich isotopes represented by squares with dotted lines are those which are calculated to be only slightly unbound for p and 2p emission ($0 > S_p > -0.4$ MeV and $0 > S_{2p} > -0.8$ MeV) so that they are candidates for p and 2p radioactivities.

Some indications that isomeric states are formed can be and will be examined by this method. It is clear that in this region the occurrence of shape or high-spin isomers would constitute very valuable information.

The fruitfulness of such a systematic measurement of $T_{1/2}$ values should not be overlooked, since the comparison of the experimental trends with model calculations can bring new information. For instance, $T_{1/2}$ measurements recently performed at GSI with the on-line mass spectrometer indicate a systematic over estimation of the half lives by the calculations, whether or not they include shell effects (Bosch et al 1985), in the region of fp-shell neutron-rich nuclei. On the contrary, the $T_{1/2}$ values measured by the same group for neutron-rich nuclei in the rare-earth and radium regions agree well with the predictions.

Although they are only starting, two experimental programs launched at GANIL should be mentioned. Very proton- (or neutron-)rich isotopes have high probabilities of β-delayed proton (or neutron)emission. The study of these two processes is indeed of the highest importance : β-delayed proton emission brings direct information on the location of high-isospin isobaric states and on the quenching of the Gamow-Teller resonance (Bjornstad et al. 1985) ; β-delayed neutron emission provides overall information on the location of shell-model configurations in the daughter nuclei (Langevin et al. 1983 b). Such studies can be performed with yields in excess of about 10 nuclei per second, which makes them practicable at GANIL for a large number of isotopes.

6. Summary

During the last two years, the fragmentation-like process of projectiles from the high intensity beams of GANIL, with energies in the range of 40 to 80 MeV per nucleon, has produced scores of yet unknown isotopes. Their spectroscopic study has been undertaken (fig.4). This new powerful method should certainly allow to further extend the field of known nuclear properties away from the valley of β-stability. It can directly benefit from the use of new beams, such as ^{48}Ca, and from the improvement of accelerated intensities. Since fragmentation-like processes occur at energies well in excess of 30 or 35 MeV per nucleon, the higher energies planned for the heavier ions at MSU, GANIL and GSI will allow an extension of the region of known nuclei all the way up to uranium.

This certainly represents another breakthrough in the study of exotic nuclei. Yet, one should not overlook that well-established methods can still bring results of the highest importance. In particular, the on-line isotope separators have the unique advantage of providing nuclei at rest, which opens the possibility of implementing the most effective spectroscopic techniques. And the deep inelastic collisions might still be the best suited tool in specific cases. The structure effects observed in ^{238}U induced dissipative collisions (Mayer et al. 1985) might favour the production of magic exotic nuclei such as the much sought-after ^{100}Sn.

It has been emphasized above that fruitful information can be derived from the study of exotic nuclei if systematic measurements of the most basic properties are made for whole areas of the nuclidic chart. Examples were given of such result bearing on the mass, the $T_{1/2}$ value, the probability for β-delayed neutron emission, the γ-energies. One more parameter might have been unduly neglected so far, the total reaction cross section

(Bruandet 1986) which yet brings unique information on the relative radii of isotopes and can be determined by simple techniques well suited to the scanty yield of exotic nuclei. The recent results obtained at Berkeley (Tanihata et al. 1985) for the He and Li isotopes will obviously be extended to others.

References

Armbruster P 1982 Report GSI 82-30
Artukh A G et al 1971 Nucl. Phys. A176 284
Audi G et al 1979 AMCO 6 Conf.ed. Nolen and Benenson (NY: Plenum Pr.) p 281
Azuma, R E et al 1979 Phys.Rev. Lett. 43 1652
Azuma, R E et al 1980 Phys. Lett. 96B 31
Belozyorov A V et al 1985 Dubna report E7-85-966 and Nucl. Phys. to be published
Birien P and Valero 1981 rapport CEA-N-2215
Bjornstad T et al report CERN-EP/85-23 and submitted to Nucl. Phys. A
Bosch U et al 1985 Phys. Lett. 164B, 22
Bruandet J P 1986 invited talk at the HICOFED Conference (Caen), J. de Phys. C4 125
Cable M D et al 1983 Phys. Rev. Lett. 50 404
Cerny J et al 1970 Phys. Lett. 33B 284
Curtin M S et al 1986 Phys. Rev. Lett. 56 34
De Boer F W N et al 1980 Nucl. Phys. A350 149
Détraz C 1981 4th Int. Conf. on Nuclei far from stability (Helsingør) CERN report 81.09 p 361
Détraz C et al 1983, Nucl. Phys. A402 301
Dufour J P et al 1986a Nucl. Instr. Methods A248 267
Dufour J P et al, 1986b CENBG report 86.12 (Bordeaux) and submitted to Z. Phys. A
Epherre M et al 1979 AMCO 6 Conf. ed. Nolen and Benenson (NY: Plenum Pr.) p 299
Fifield L K et al 1985 Nucl. Phys. A440 531
Fifield L K et al 1986 Nucl. Phys. A453 497
Gillibert A et al 1986 GANIL report 86.16 and Phys. Lett.B to be publisched
Gillibert A and Mittig W 1986, private communication
Goldanski V I 1961 Nucl. Phys. 27 648
Goldanski V I 1966 Sov. Phys. Usp.8 770
Guerreau D et al 1983 Phys. Lett. 131B 293
Guerreau D 1985 in Nucleus Nucleus Collisions II (Visby) North Holland ed. p 37c
Guerreau D 1986 GANIL report P86.07, invited talk at HICOFED (Caen) J. de Phys. C4 207
Guillemaud-Mueller D et al 1985 Z. Phys. A322 415
Haustein P E 1984 AMCO 7 Conference, Tech. Hoch. Darmstadt ed. p 413
Hofmann S et al 1984 AMCO 7 Conference, Tech. Hoch. Darmstadt ed. p 184
Jackson K P et al 1970 Phys. Lett. 33B 281
Jacquinot P and Klapisch R 1979 Rep. Progr. Phys. 42 773
Jänecke J 1964 Nucl. Phys. 61 326
Jänecke J 1976 At.Data and Nucl.Data Tables 17 455 (see especially equation 9)
Klapdor H V et al 1984 At. Data and Nucl. Data Tables 31 81
Langevin M et al 1981 Nucl. Phys. A366 449
Langevin M et al 1983a Phys. Lett.125B 116
Langevin M et al 1983b Phys. Lett.130B 251

Langevin M et al 1984 phys. Lett. 146B 176
Langevin M and Anne R 1985 Instrumentation for heavy ion research vol.7 (Harwood Acad. Publ.) p 191
Langevin M et al 1985 Phys. Lett. 150B 71
Langevin M et al 1986 Nucl. Phys. A455 149
Lysekil 1966 : Conference on Nuclides far off the stability line Almqvist and Wiksell ed. Stockholm, especially p 267-323
Maripuu S 1975 ed. At. Data and Nucl. Data Tables 17 411
Mayer W et al 1985 Phys. Lett. 152B 162
Otten E W 1981 Proc. of the Int. Conf. on Nucl. Physics (Berkeley) North Holland ed. 471c
Pecker L K et al 1971, Phys. Lett. 36B 547
Pougheon F et al 1986a Report IPNO-DRE 86.05 (IPN Orsay) and Europhysics Letters to be published
Pougheon F et al 1986b to be published
Roeckl E 1983 Nucl. Phys. A400 131c
Sandulescu A et al 1985 Phys. Rev. Lett. 54 490
Symons P J et al 1979 Phys. Rev. Lett. 42 40
Tanihata I et al 1985 Phys. Lett. 160B 380 and Phys. Rev. Lett. 55 2676
Thibault C 1981 4th Int. Conf. On Nuclei far from stability (Helsingør) CERN report 81.09 p 47 and ref.
Turkevich A et al 1977 Phys. Lett. 72B 163
Uno M and Yamada M 1982 INS report NUMA 40 (Tokyo)
Volkov V V 1973 Int. Conf. on Nuclear Physics (Munich) vol.1 279
Wapstra A H and Audi G 1985 Nucl. Phys. A432 1
Westfall G D et al 1979 Phys. Rev. Lett. 43 1859

List of Oral Contributions

(In order of presentation: the number in square brackets refers to the contribution in Vol. 1 of the Proceedings)

Giant resonance excitation by inelastic scattering of medium energy ^{12}C ions
J Arvieux[1], B Bonin[2], G Bruge[2], M Buenerd[3], C Cerruti[1], J Chauvin[3], J Y Hostachy[3], D Lebrun[3], J C Lugol[2], L Papineau[2] and P Roussel[2]
[1]CEN Saclay, LNS; [2]CEN Saclay, DPh.N; [3]ISN Grenoble
[B56]

Polarisation phenomena in one nucleon transfer reactions induced by 44 MeV ^{7}Li ions on ^{120}Sn
G Tungate[1], K Becker[1], K Blatt[1], G Grawert[1], H Leuker[1], D Fick[1], I M Turkiewicz[2], K Rusek[2], Z Moroz[2], R Butsch[3], B Heck[3], H Jänsch[3], D Krämer[3], K-H Möbius[3], W Ott[3], E Steffens[3] and A Weller[3]
[1]Phillips-University, Marburg; [2]INS Warsaw; [3]MPI Heidelberg
[C15]

Polarization potentials in oxygen-lead collisions
I J Thompson[1], J S Lilley[2], M A Nagarajan[2], D W Banes[2] and B R Fulton[3]
[1]University of Bristol; [2]SERC Daresbury Laboratory; [3]University of Birmingham
[C227]

Isotopic separation with LISE (GANIL). Application to studies of exotic nuclei
D Jean[1], J P Dufour[1], R Del Moral[1], H Emmerman[1], F Hubert[1], C Poinot[1], M S Pravikoff[1], A Fleury[1], H Delagrange[2], K H Schmidt[3] and H Geissel[3]
[1]CEN Bordeaux-Gradignan; [2]GANIL; [3]GSI Darmstadt
[C179]

High energy gamma-ray emission in heavy ion collisions
J Stevenson, K B Beard, W Benenson, J Clayton, E Kashy, A Lampis, D J Morrissey, M Samuel, R J Smith, A Tam and J S Winfield
Nat. Superconducting Cyclotron Laboratory and Michigan State University
[C161]

Fragment angular distributions in fission and fission-like reactions induced by medium energy light and heavy ions
V S Ramamurthy and S S Kapoor
Bhaba Atomic Research Centre
[C105]

Cold fragmentation of ^{40}Ar nuclei at 380 MeV
A Budzanowski, M Bürgel, J Galin, D Hilscher, H Homeyer, G Ingold, U Jahnke and J Uckert
Hahn-Meitner Inst., Berlin
[C6]

Heavy-ion charge exchange at intermediate energies
N Anantaraman[1], J S Winfield[1], S M Austin[1], Ziping Chen[1], A Galonsky[1], L H Harwood[1], J van der Plicht[1], H-L Wu[1], A F Zeller[1], C C Chang[2], C Ciangaru[2] and S Gales[3]
[1]Nat. Superconducting Cyclotron Laboratory and Michigan State University; [2]University of Maryland; [3]IPN Orsay
[C83]

Reaction Mechanisms and Nuclear Structure

Symmetry axes of angular momentum transfer in heavy-ion reactions
W Dünnweber[1], R Ritzka[1], A Glaesner[1], W Hering[1], W Trombik[1],
A Budzanowski[2], M Bürgel[2] and H Homeyer[2]
[1]University of Munich; [2]Hahn-Meitner Inst., Berlin
[C25]

Peripheral collisions in heavy-ion reactions with ^{40}Ar at 44 MeV/u
R Hingmann[1], W Kühn[1], V Metag[1], R Mühlhans[1], R Novotny[1], A Ruckelshausen[1],
H Emling[2], H J Wollersheim[2], R Kulessa[3], B Haas[4], J P Vivien[4],
H Delagrange[5], C Grégoire[5], Y Schütz[5]
[1]University of Giessen; [2]GSI Darmstadt; [3]University of Cracow;
[4]CRN Strasbourg; [5]GANIL
[C143]

Dubna experiments relevant to the synthesis of exotic nuclei and to the search for them in nature
G N Flerov
JINR Dubna
[Post-deadline contribution]

Symmetric fission of ^{24}Mg
B R Fulton[1], S J Bennett[1], C A Ogilvie[1], J S Lilley[2], D W Banes[2],
W D M Rae[3], S C Allcock[3], R R Betts[3] and A E Smith[3]
[1]University of Birmingham; [2]SERC Daresbury Laboratory; [3]University of Oxford
[C48]

Quasi-elastic transfer and sub-barrier fusion in ^{32,36}S + ^{58,64}Ni
A M Stefanini[1], G Montagnoli[1], G Fortuna[1], R Menegazzo[1], S Beghini[2],
C Signorini[1,2], A DeRosa[3], G Inglima[3], M Sandoli[3], F Rizzo[4] and
G Pappalardo[4]
[1]INFN Legnaro; [2]University of Padua; [3]University of Naples; [4]University of Catania
[C31]

Investigation of the reactions of ^{84}Kr and ^{129}Xe with heavy targets around 20 MeV/u
R Lucas[1], M Berlanger[1], J Albinski[2], C Cerruti[1], S Chiodelli[3], D Dalili[1],
A Demeyer[3], A Gobbi[2], O Granier[1], D Guinet[3], N Hermann[2], G Imme[4],
S M Lee[1], P L'Hénoret[1], S Leray[1], C Mazur[1], J Natowitz[5], C Ngô[1], A Olmi[2],
G Raciti[4], M Ribrag[1], C Spinella[4], T Suomijarvi[1], E Tomasi[1] and
L Vagneron[3]
[1]CEN Saclay; [2]GSI Darmstadt; [3]IPN Lyon; [4]INFN Catania;
[5]A & M Texas University
[C102]

Nucleon transfer reactions to rotational states induced by ^{206,208}Pb projectiles
H J Wollersheim[1], F W N de Boer[1], H Emling[1], H Grein[1], E Grosse[1],
W Spreng[1], G Eckert[2], Th. W.Elze[2], K Stelzer[2] and Ch. Lauterbach[3]
[1]GSI Darmstadt; [2]University of Frankfurt; [3]University of Munich
[C257]

Reaction Mechanisms and Nuclear Structure

High-spin alpha-cluster resonances in ^{16}O and ^{20}Ne

S C Allcock[1], W D M Rae[1], P R Keeling[1], S Marsh[1], A E Smith[1], B R Fulton[2] and D.W. Banes[3]
[1]University of Oxford; [2]University of Birmingham; [3]SERC Daresbury Laboratory
[B20]

Inst. Phys. Conf. Ser. No. 86
Paper presented at Int. Nucl. Phys. Conf., Harrogate, UK, 1986

The compound nucleus: a classically chaotic quantum system

O. Bohigas

Division de Physique Théorique*, Institut de Physique Nucléaire,
F-91406 Orsay Cédex

1. Introduction

The purpose of the present contribution is to explain its title, or, more precisely, connections between the different ingredients appearing in it (we more or less understand separately the meaning of classical system, classically chaotic motion, quantum system, compound nucleus). The emphasis will be put on spectral properties of quantum systems, and properties related to wave functions and strength functions will not be discussed.

Which experimental quantities are interesting for the present purpose ? The positions of resonances of the compound nucleus. What we want are many *complete* and *pure* sequences of resonance energies corresponding to states having the same quantum numbers J^{π} in a given energy range. Indeed very characteristic spectral features are rapidly lost if the sequence of levels is incomplete (missing levels) and/or polluted by spurious levels due to erroneous spin-parity assignments. The relevant information comes from neutron resonance spectroscopy and from high resolution proton scattering experiments, resulting from an impressive long term collective effort performed mainly at Columbia University (Liou et al 1975) and Duke University (Wilson et al 1975).

What is the situation from a theoretical point of view ? At the region, for instance, of neutron threshold, the nuclear density is so high that one must give up a description of microscopic detail, a description of individual states. The aim of nuclear models at this and higher excitation energies is rather to describe average behaviours like level densities and special states like giant resonances, analogue states, etc., which have a peculiar structure. For the rest of the states, for the overwhelming and silent majority of anonymous states, what is needed is a statistical theory of energy levels which will not predict the detailed sequence of levels of each nucleus but which will describe the degree of irregularity of the level sequence that is expected to occur in any nucleus. In other words a statistical theory of level fluctuations (departures of the spectrum from its average behaviour). The random matrix theory initiated by Wigner and fully developed by Dyson, Mehta and others serves this purpose (see Porter (1965), Mehta (1967), Brody et al (1981)). One considers the Hamiltonian H as an NxN stochastic matrix (the matrix elements are random variables); the random matrix ensemble is specified by the probability density P(H)dH and one is interested in the eigenvalue fluctuations in the limit of large N. The underlying space-time symmetries obeyed by the system restrict the admissible matrix ensembles. If the

*Laboratoire Associé au CNRS

Hamiltonian is time-reversal invariant and has integer angular momentum (or is rotation invariant with 1/2-odd-integer angular momentum), the Hamiltonian matrix can be chosen real and symmetric. When no other information except these general space-time symmetries is taken into account, one is led to the Gaussian Orthogonal Ensemble (GOE) of real symmetric matrices.

2. Fluctuation measures. Comparison of predictions with experiment

To discuss fluctuation measures, we shall refer to

1) The *spacing distribution* p(x) between adjacent levels
2) Quantities directly related to the number statistic n(L) : given an interval of length L, it counts the number of levels contained in the interval. If one scales the spectra in such a way that the mean spacing is unity, the average value of n(L) is L. We shall consider higher moments or cumulants of n(L) : *variance* $\Sigma^2(L)$, *skewness* $\gamma_1(L)$ and *excess* $\gamma_2(L)$.
3) The Δ_3 *statistic of Dyson and Mehta*. Let N(E) be the staircase function giving the number of levels up to energy E. Δ_3 measures, given an interval of length L, the least-square deviation of the staircase function from the best straight line fitting it. Its average value $\overline{\Delta}_3$ is related to the variance $\Sigma^2(L)$ of n(L).

In general $\Sigma^2(L)$ and $\overline{\Delta}_3(L)$ can be expressed in terms of the 2-level correlation function, whereas $\gamma_1(L)$ ($\gamma_2(L)$) depends also on the 3-level (3- and 4- level) correlation function(s).

For the sake of comparison, let us first consider two limiting cases

1) Take a random variable s whose probability density p(x) is exp(-x). Construct a sequence $\{x_i\}$ as follows : $x_1=1/2$, $x_{i+1}=x_i+s_i$ (i=1,2,..), where s_i are outcomes of independent trials of the variable s. The resulting spectrum is what is called a *Poisson spectrum* (case of *maximum randomness*). One has $p(x)=e^{-x}$, $\Sigma^2(L)=L$, $\overline{\Delta}_3(L)=L/15$.
2) A *picket fence spectrum* or spectrum of the harmonic oscillator in one dimension $x_{i+1}=x_i+1$. It is the most ordered spectrum one can imagine, with *no randomness*. One has $p(x)=\delta(x-1)$, $\Sigma^2(L)=0$, $\overline{\Delta}_3(L)=1/12$.

The predictions of 'non trivial' models of level fluctuations, namely of GOE, are as follows. The spacing distribution is very well approximated by

$$p(x) = (\pi/2)\, x \exp(-(\pi/4)x^2)$$

It shows *level repulsion*, i.e. tendency to avoid level clustering or small probability of small spacings (p(x) vanishes at the origin). The number variance for $L \gtrsim 1$ is given by

$$\Sigma^2(L) = (2/\pi^2)\ln L + 0.44$$

The logarithmic increase with L is to be compared to the linear increase for a Poisson spectrum. One can speak of *spectral rigidity* or of semi-crystalline nature of the spectrum, showing long range order. For large L, the average of $\Delta_3(L)$ is given by

$$\overline{\Delta}_3(L) \simeq (1/\pi^2)\ln L - 0.007$$

How well do GOE predictions compare with experimental data ? Results are reproduced in Fig.1. As can be seen, all the fluctuation measures considered, which include a thorough study of 2-point measures and to some extent more than 2-point measures as well, are fully consistent with GOE predictions.

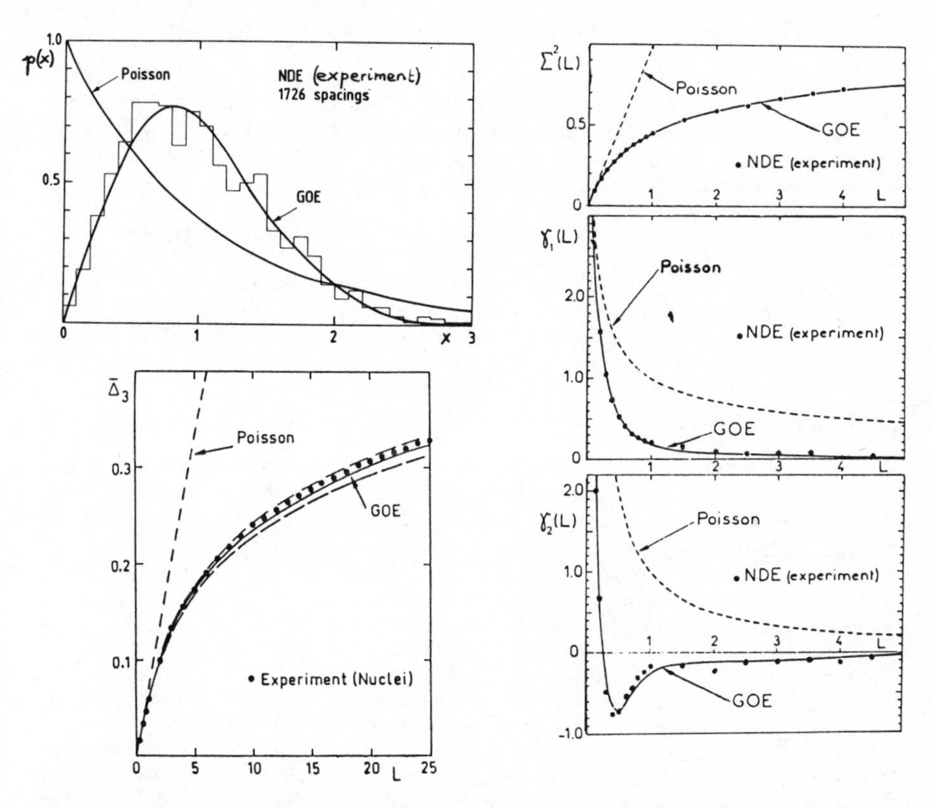

Fig. 1 Results of fluctuation measures of nuclear resonances. Poisson and GOE predictions are given for comparison (taken from Haq et al (1982), Bohigas et al (1983), Bohigas et al (1985)).

Now, from the theoretical point of view, are GOE-fluctuations specific to GOE or, on the contrary, are there other models that give GOE fluctuation patterns ? And, from the experimental one, are the fluctuations of the nuclear resonances specific to nuclei or, on the contrary, are they also observed in other systems ? Concerning the first question, one knows that there is a whole variety of models of random matrix ensembles that give GOE-fluctuations (some results are analytical, the others are obtained from Monte Carlo calculations). Concerning the second one, some atomic spectra have also been analyzed (Camarda et al 1983). And the outcome is that the fluctuations seem to be consistent with GOE predictions, although the statistical significance is much lower than for the nuclear case. We therefore see that, on the one hand, the spectra of very different systems (nuclei, atoms), when properly scaled, seem to have identical fluctuation patterns, even though they are governed by very different force laws (short range interactions and Coulomb long range forces). On the other hand, these characteristic fluctuation patterns, although not specific of, are well reproduced by GOE, a *parameter-free theory*. Thus a simple picture emerges : the level fluctuation laws seem to be *universal*, both from the experimental and from the theoretical point of view.

3. Spectral fluctuations of classically chaotic quantum systems

To obtain some clues on the origin of the universality of level fluctuation laws and also to investigate how 'complicated' a system must be in order to show GOE-fluctuations (Bohigas and Giannoni 1984a) let us very briefly remind the general scheme characterizing the route towards chaos for classical conservative Hamiltonian systems.

All conservative Hamiltonian systems with n degrees of freedom have in common three essential properties :
i) for a given set of initial conditions, the dimensionality of the accessible surface in phase space is less or equal to 2n-1 ; since the system is conservative, the energy is constant along this energy surface S_E.
ii) the volume element in phase space is conserved (Liouville's theorem) ; in other words, the Hamiltonian flow is incompressible.
iii) trajectories in phase space cannot cross.

Apart from these features the motion in phase space can exhibit a great variety of behaviours. For instance, one may ask how does a given volume element evolve with time. Does it tend to cover the whole energy surface S_E as time goes to infinity or does it remain in a restricted part of S_E ? Does it conserve approximately its initial shape or does it display more or less dramatic deformations with time ? According to the answers to such questions, one can define a hierarchy of regularity going from

$$\text{integrable} \rightarrow \text{ergodic} \rightarrow \text{mixing} \rightarrow \text{K-system}$$

in the sense of regularity towards chaoticity. The integrable systems, which are the more regular ones and can be used as clocks, possess as many integrals of motion as number n of degrees of freedom. The motion in phase space of an integrable system is restricted to an n-dimensional torus, instead of a (2n-1)-dimensional energy surface for a generic system. For two-dimensional systems (n=2) one therefore has that a generic system will move on a 3-dimensional energy surface embedded in the 4-dimensional phase space, whereas an integrable system will move on a 2-dimensional torus.

In contrast with integrable systems, almost every trajectory of an ergodic system passes through almost every point of the energy surface, spending equal times in equal areas : the energy surface is asymptotically uniformly covered by a typical orbit.

Does ergodicity imply chaotic motion ? For the motion to be irregular, erratic, a given volume element has to deform with time allowing for instability with respect to a perturbation. One therefore has to ask for a stronger property than simple ergodicity. Mixing systems are such that any volume element tends to 'dilute' uniformly in S_E as time goes to infinity, in the same way as a solute dilutes in a solvent if two liquids are miscible. Consequently, the distance between two points initially close to each other may become arbitrarily large as time is running.

The mixing property, however, tells nothing of the rate of separation of orbits. It only contains the concept of asymptotic equilibration. Ergodic systems which possess the strongest degree of irregularity (K-systems) have a further property besides mixing : their orbits separate exponentially with time. As a consequence of such a dramatic instability, long time predictions on the system are impossible, the memory of the initial state vanishing with time. Notice that such systems are deterministic in the sense that they are governed by causal equations : in principle they are predictable. However, due to the finite precision available for any practical purpose, one cannot follow their time evolution beyond some critical time. In the language of communication theory, K-systems are sources which continuously produce information (the so-called Kolmogorov entropy is positive), in contrast with integrable systems, whose motion is periodic or quasi-periodic, and for which knowing the history of any orbit during some given time interval is sufficient to determine with probability one its future evolution. Integrable systems considered as sources can be compared with records, which infinitely repeat the same message, whereas K-systems can be compared with a broadcast station, which is supposed to produce indefinitely new information.

There are very few explicit examples of dynamical systems for which it has been mathematically proved that they are K-systems. Among them, let us pay special attention to two-dimensional billiards. A billiard consists in the motion of a free point particle of mass m in a domain Γ of the plane of arbitrary shape. The particle is elastically reflected when it hits the boundary of Γ, according to the laws of specular reflection. There is at least one constant of the motion, namely the energy $E = (1/2)mv^2$. For particular shapes of the boundary, the system may be, at one extreme, integrable, or at the other extreme a K-system. For instance (see Fig.2) the circular billiard is integrable (the angular momentum is a second constant of the motion) whereas Sinai's and Bunimovich's stadium billiards are K-systems.

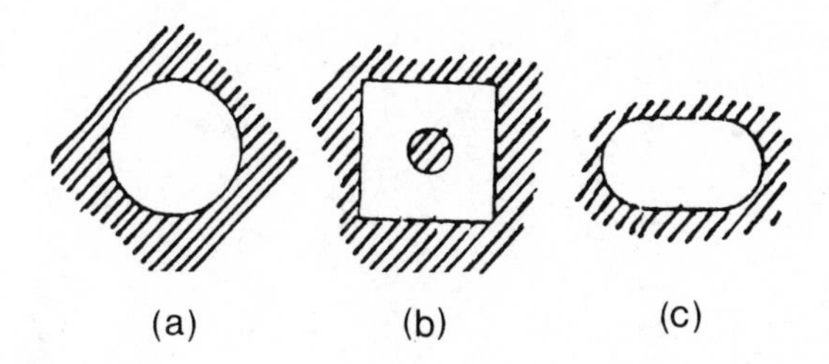

Fig. 2 Different boundaries in which a free particle is enclosed : (a) circular billiard ; (b) Sinai's billiard ; (c) Bunimovich's stadium billiard.

What are the spectral fluctuations of the corresponding quantum system? We have to solve the Schrödinger equation

$$(-\hbar^2/2m)\,\Delta\,\psi_n(\underset{\sim}{r}) = E_n\,\psi_n(\underset{\sim}{r})$$

with Dirichlet boundary conditions, namely vanishing of the wave function $\psi_n(\underset{\sim}{r})$ at the boundary of Γ. Results of the spectral fluctuations are reproduced in Fig. 3. The message seems fairly clear : The system whose classical analogue is integrable shows Poisson fluctuations (Berry and Tabor 1977) whereas the systems whose classical analogues are fully chaotic show GOE fluctuation patterns. By now, a considerable amount of numerical experience has accumulated and it supports the conclusions drawn from the

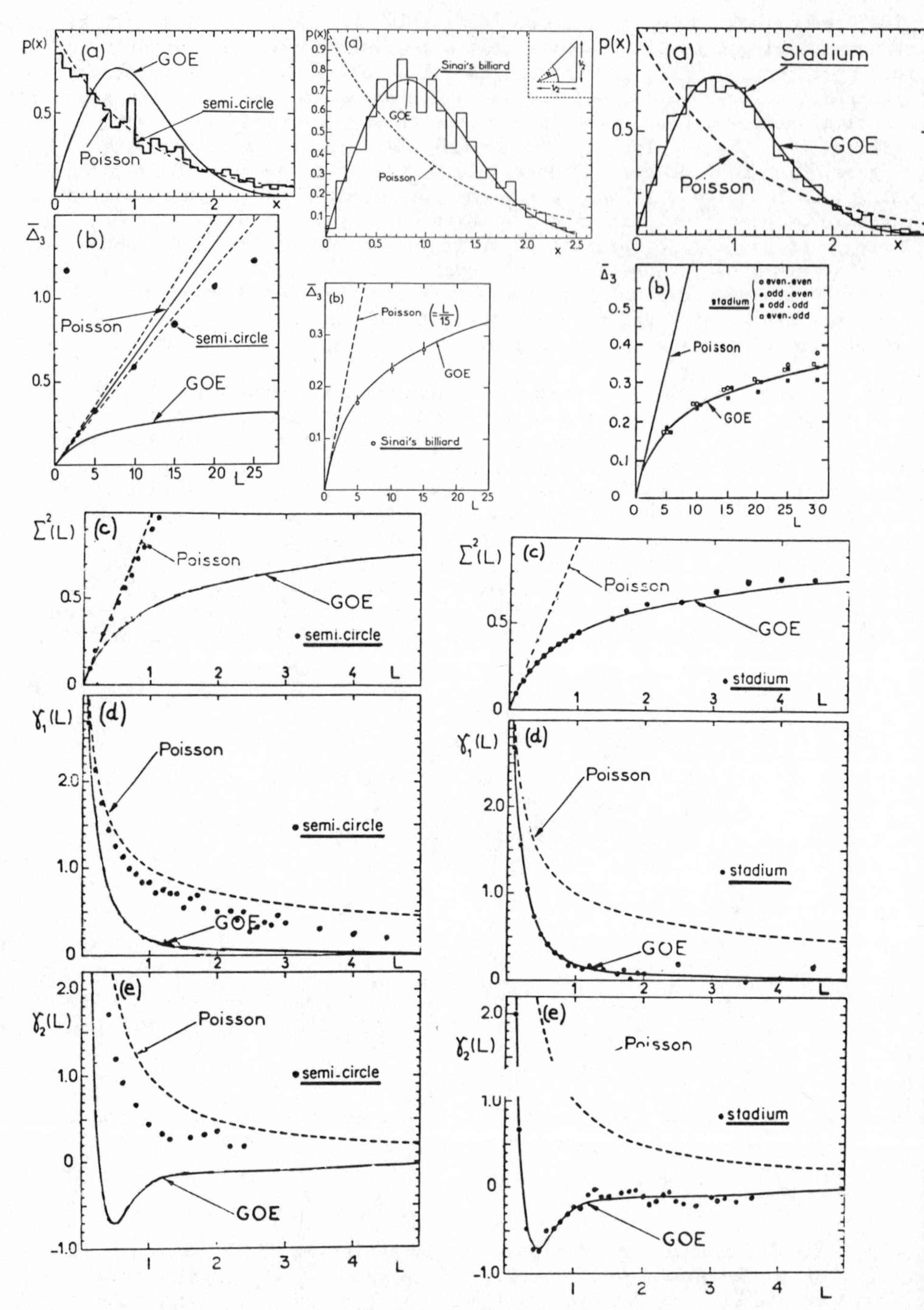

Fig. 3 Spectral fluctuations of the systems shown on Fig. 2 : (a) spacing distributions p(x) ; (b) $\overline{\Delta}_3(L)$ statistic ; (c) number variance $\Sigma^2(L)$; (d) number skewness $\gamma_1(L)$; (e) number excess $\gamma_2(L)$ (taken from Bohigas et al 1984b, Bohigas et al 1984c).

preceding examples (see Seligman et al 1986). To summarize, these results establish the universality of the laws of spectral fluctuations and explain the success of GOE in describing the fluctuations of the compound nucleus resonances.

What is important now is to understand theoretically these findings. Recently Berry (1985) has given a semiclassical derivation of the spectral rigidity. He employs the semiclassical techniques introduced by Gutzwiller, Bloch and Balian, which consists in an asymptotic representation of the spectral density as a sum over all the periodic orbits of the classical system. He predicts universality in the regime $1 \ll L < L_{max}$, with L_{max} determined by the shortest classical closed orbit ; for $L \gg L_{max}$ there is a non-universal saturation value of $\overline{\Delta}_3(L)$. In the universal regime, the long range predictions coincide with Poisson , (GOE) predictions for classically integrable , (chaotic and time-reversal) systems, in agreement with what is found numerically. It is important to notice that, in deriving the results, random matrix theory is not used and no statistical assumptions are made.

4. Transition regimes. Hydrogen atom in a strong magnetic field.

The classification of possible motions of classical systems mentioned above is not complete in the following sense : the phase space of a generic system will show in general islands of regularity surrounded by regions of chaoticity, that is, the structure of the phase space will be neither purely regular nor purely chaotic. How is this reflected in the spectral fluctuations of the corresponding quantum system (Seligman et al 1984) ? Based on semiclassical arguments Berry and Robnik (1984) suggest that the spectral fluctuations should result from independently superposing a Poisson spectrum with relative weight μ, given by the fraction of phase space of classical regular motion, and a GOE spectrum with relative weight $\overline{\mu}$ given by the fraction of phase space of classical chaotic motion ($\mu+\overline{\mu}=1$). The resulting spacing distribution, for instance, is

$$p(x)=\mu^2 e^{-\mu x}\ \mathrm{erfc}\left(\frac{\sqrt{\pi}}{2}\overline{\mu}x\right)+\left(2\mu\overline{\mu}+\frac{\pi}{2}\overline{\mu}^3 x\right)\ \exp\left(-\mu x-\frac{\pi}{4}\overline{\mu}^2 x^2\right)$$

which interpolates between the Poisson result ($\overline{\mu}=0$) and the GOE result ($\overline{\mu}=1$).

The study of the hydrogen atom in a strong magnetic field, an extremely interesting system for its fundamental 'simplicity' and for its astrophysical relevance, is also especially well suited in the present context (see Gay (1985) and references therein). The conceptual importance of the problem follows from the fact that the two limiting cases of zero and infinite field strength, namely the Coulomb and the Landau (oscillator like) problems respectively, which have very different symmetries, are the only three-dimensional problems which are exactly soluble, both classically and quantum mechanically. The motion of the electron submitted to the action of the Coulomb and Lorentz forces is governed, in the symmetrical gauge with vector potential $\vec{A}=(1/2)(\vec{r}\times\vec{B})$ (the magnetic field is in the z-direction), by the Hamiltonian (in a.u.)

$$H = \frac{p^2}{2} - \frac{1}{r} + \frac{\gamma}{2}L_z\ \frac{\gamma^2}{8}\ (x^2+y^2)$$

γ is the reduced magnetic field strength B/B_c, $B_c = 2.35 \times 10^5$ T. The two last terms are the paramagnetic interaction associated with the normal Zeeman effect which is trivial in the present context and the diamagnetic interaction. L_z (and parity) are the only constants of the motion besides the energy. The importance of the diamagnetic effects are characterized by the ratio of the diamagnetic to Coulomb energy, which is proportional to $\gamma n^3 \propto \gamma |E|^{-3/2}$, where n is the principal quantum number and E is the energy. To magnify diamagnetic effects one can : i) increase the magnetic field (values of γ up to 10^{-4} can be achieved under laboratory conditions) ii) work with Rydberg atoms (one can achieve typically values of $n \simeq 50$).

The phase space structure of this system (diamagnetic Kepler problem) has been recently studied in detail by Delande and Gay (1986). By increasing the value of $\gamma / |E|^{3/2}$, which is the significant parameter, one evolves from a motion which is fully regular, then a connected chaotic region appears and keeps increasing until it occupies the whole surface of energy, for a critical value of the parameter. How is this behaviour reflected in the spectral fluctuations of the corresponding quantum system? One expects a transition from the Poisson to the GOE regime (due to the particular spatial symmetry of the problem, although the system is not T-invariant one expects GOE-fluctuations, Robnik and Berry 1986). And this is indeed what is found when analyzing the computed spectrum. The results are illustrated in Fig. 4, where the fluctuation measures of states having $L_Z = 0$ and even parity are reproduced (similar results have been obtained by Wintgen et al 1986 and Wunner et al 1986).

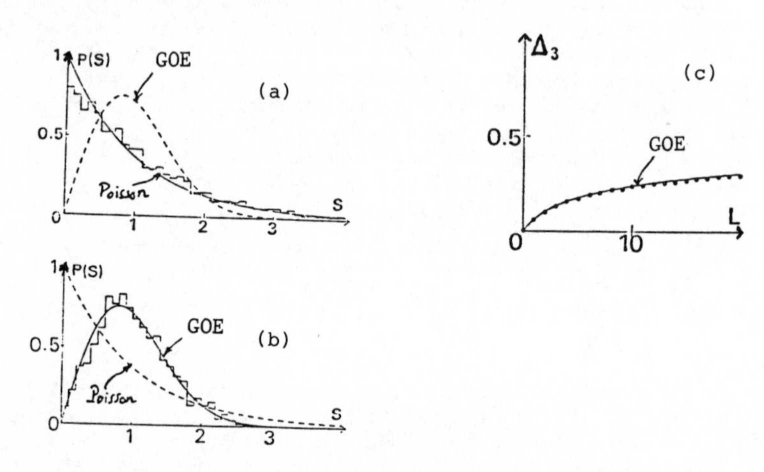

Fig. 4 Hydrogen atom in a magnetic field : (a) spacing distribution in the regular region ; (b) id. in the fully chaotic region ; (c) spectral rigidity in the fully chaotic region (taken from Delande and Gay 1986).

In summary, the Kepler diamagnetic problem offers a unique opportunity for studying chaotic motion : i) it is an interesting problem from the point of view of symmetries, ii) the classical as well as the quantum properties can be computed accurately, iii) there is a transition from regular to chaotic motion when the field strength is increased, iv) it can be studied in the laboratory. For instance, the Bielefeld group (Holle et al 1986) performs measurements with magnetic field strengths $B \simeq 6T$ and Rydberg atoms $N \simeq 40$, which correspond already to the chaotic regime.

5. Remarks and conclusions

In the last few years, two areas - random matrix physics and the study of chaotic motion - that remained disconnected for a long time, have interpenetrated each other. We know now that the fluctuations exhibited by the nuclear resonances, which gave rise to a large extent to the development of random matrix theories, are not specific to nuclei. On the contrary, they are generic of fully chaotic (time-reversal invariant) systems as demonstrated by the study of many low-dimensional systems. It has also been possible to study in low-dimensional systems transitions from Poisson to GOE fluctuations corresponding to the regular to chaotic regime transition. The view that 'disorder' is a macroscopic concept, a property of systems made up of many particles, is incorrect both classically and quantum-mechanically.

The hydrogen atom in a strong magnetic field deserves special mention : for its richness concerning symmetries, for its experimental feasibility, and because it can be theoretically treated in detail, it will probably become the chaotic system 'par excellence'.

At the risk of repetition, we stress that the belief that fluctuation patterns of the compound nucleus resonances are characteristic of the compound nucleus concept is incorrect. One can hardly imagine a system as far from the compound nucleus as the hydrogen atom (in a magnetic field). And still, the level fluctuations are the same.

The discussion presented in this contribution results from a collaborative effort with M.J. Giannoni and C. Schmit.

References

Berry M V and Tabor M 1977 Proc. Roy. Soc. Lond. A356, 375
Berry M V and Robnik M 1984 J. Phys. A17, 2413
Berry M V 1985 Proc. Roy. Soc. Lond. A400, 229
Bohigas O, Haq R U, Pandey A 1983 in Nuclear Data for Science and Technology, Böckhoff K H (ed.), Reidel, Dordrecht
Bohigas O and Giannoni M J 1984a in Mathematical and Computational Methods in Nuclear Physics, Dehesa J S et al (eds.), Lecture Notes in Physics 209, Springer Verlag
Bohigas O, Giannoni M J and Schmit C 1984b Phys. Rev. Lett. 52, 1
Bohigas O. Giannoni M J and Schmit C 1984c J. Physique Lett. 45, L-1015
Bohigas O, Haq R U and Pandey A 1985 Phys. Rev. Lett. 54, 1645
Brody T A, Flores J, French J B, Mello P A, Pandey A and Wong S S M 1981 Rev. Mod. Phys. 53, 385
Camarda H S and Georgopulos P D 1983 Phys. Rev. Lett. 50, 492
Delande D and Gay J C 1986 Phys. Rev. Lett. 57, 2006
Gay J C 1985 in Photophysics and Photochemistry in the Vacuum Ultraviolet, Mc Glynn et al (eds.), Reidel, Dordrecht
Haq R U, Pandey A and Bohigas O 1982 Phys. Rev. Lett. 48, 1086
Holle A, Wiebusch G, Main J, Hager B, Rottke H and Welge K H 1986 Phys. Rev. Lett. 56, 2594
Liou H I, Hacken G, Rainwater J and Singh U N 1975 Phys. Rev. C11, 462 and references therein

Mehta M L 1967 Random Matrices and the Statistical Theory of Energy Levels, Academic Press
Porter C E 1965 (ed.) Statistical Theories of Spectra : Fluctuations, Academic Press
Robnik M and Berry M V 1986 J. Phys. A19, 669
Seligman T H, Verbaarschot J J M and Zirnbauer M R 1984 Phys. Rev. Lett. 53, 215
Seligman T H and Nishioka H (eds.) 1986 Quantum Chaos and Statistical Nuclear Physics, Lecture Notes in Physics 263, Springer Verlag
Wilson W M, Bilpuch E G and Mitchell G E 1975 Nucl. Phys. A245, 285 and references therein
Wintgen D and Friedrich H 1986 preprint
Wunner G, Woelk U, Zech I, Zeller G, Ertl T, Geyer F, Schweitzer W and Ruder H 1986, preprint.

Inst. Phys. Conf. Ser. No. 86
Paper presented at Int. Nucl. Phys. Conf., Harrogate, UK, 1986

List of Oral Contributions

(In order of presentation: the number in square brackets refers to the contribution in Vol. 1 of the Proceedings)

Shell-model calculations with modern effective interactions
F Brut[1], E Osnes[2] and D Strottman[3]
[1]ISN Grenoble; [2]University of Oslo; [3]Los Alamos National Laboratory
[B51]

Parity doublets in nuclear rotational bands
N Rowley[1], M A Nagarajan[1], D M Brink[2] and B Buck[2]
[1]SERC Daresbury Laboratory; [2]University of Oxford
[B131]

The choice of wave equation for kaonic atoms
J Law[1], M J Turner[2] and R C Barrett[2]
[1]University of Guelph; [2]University of Surrey
[H8]

Renormalisation theory on an s-d-g-boson hamiltonian
K Amos, I Morrison and X J Chen
University of Melbourne
[B8]

Critical remarks on the IBM and the structure of nonrotational states in deformed nuclei
V G Soloviev
JINR Dubna
[B156]

Shell-model evidence for the exchange term in the interacting boson-fermion hamiltonian
T Otsuka[1], N Yoshida[2], P Van Isacker[2], A Arima[2] and O Scholten[3]
[1]JAERI Tokai; [2]University of Tokyo; [3]Michigan State University
[B78]

Low-lying $K^{\pi} = 1^{+}$ states in deformed rare-earth nuclei
R Nojarov, A Faessler and O Civitarese
University of Tübingen
[B14]

A self-consistent study of triaxial deformations in heavy nuclei
N Redon[1], J Meyer[1], M Meyer[1], P Quentin[2], M S Weiss[3], P Bonche[4], H Flocard[5] and P-H Heenen[6]
[1]IPN Lyon; [2]LPTh. Bordeaux; [3]Lawrence Livermore Laboratory; [4]DPThe. Saclay; [5]DPNTh Brussels
[B49]

Inst. Phys. Conf. Ser. No. 86
Paper presented at Int. Nucl. Phys. Conf., Harrogate, UK, 1986

Three-body forces and light nuclei

Sidney A. Coon

Department of Physics, University of Arizona, Tucson, Arizona 85721, USA

Abstract. The concept and features of three-body forces are discussed by illustrative examples from other many-body systems. Long-range model-independent triple-dipole molecular forces play an important role in rare gas systems. The construction of two-meson-exchange three-nucleon forces is constrained by chiral-symmetry breaking and by hadronic phenomenology. The trinucleon bound state can be solved exactly with these rather complex forces. The binding energies are improved, the charge form factors are not. Experiment in the continuum is not yet matched by calculation.

1. Introduction

Why should such a seemingly esoteric subject as three-body forces be on the agenda of a general nuclear physics conference? Everyone knows that the traditional model in which nonrelativistic nucleons interact via two-body or pair-wise forces has resulted in a semiphenomenological description of nuclear physics which enjoys considerable success. However, one can no longer ignore the continuing discrepancies between calculated and experimental nuclear properties, especially for the three-nucleon bound state where the calculated ^{3}H and ^{3}He nuclei have less binding energy and larger radii than experiment. The calculation can be done exactly, given a two-body interaction such as the dispersion theoretic NN potentials, which successfully describe the expanding data base of NN scattering parameters. The discrepancies may be due to relativistic corrections, many-body forces, or suppression of mesonic or quark degrees of freedom. As we shall see, these topics are theoretically intertwined and experimentally hard to pin down.

Before plunging into the evidence for and models of three-body forces in nuclei, I first discuss such forces in other systems. The demand for a nonrelativistic potential theory of a many-body system leads immediately to forces between two bodies, between three bodies, between four bodies, etc. Already in 1939, Primakoff and Holstein[1] demonstrated the many-electron force that arises when one wishes to reexpress the Hamiltonian of a system of nonrelativistic electrons in a form in which the radiation variables are absent. It corresponds to the time-ordered Z graph, or particle-antiparticle "pair" graph of a relativistic formalism. The Primakoff-Holstein three-body force depends in an irreducible way on the simultaneous coordinates of three electrons when only electron degrees of freedom are taken into account. This is the archtypal example of a three-body force arising from frozen degrees of freedom (in this case, radiation variables) in a system of elementary particles with no discernable substructure.

Liquid helium, and the rare gas systems in general, provide an example of a many-body system in which the constituents have substructure. An atom-atom potential can be understood microscopically in terms of the electrons, nuclei, and photons comprising the atoms. The potential model thus produces the physics of the finite size of the constituents in a definite and correct way. In addition to two-body forces, the potential model incorporates the effect of suppressed degrees of freedom through the presence of three-body and higher many-body forces. Such models accurately predict the bulk properties of gaseous, liquid, and solid forms of rare gases.

When applied to nuclei, potential theory offers the same possibility of including the physics of substructure. At long range, the substructure is presumably nucleonic excitations and the force is attributed to exchange of mesons. The correct description of the strong interaction is undoubtedly the QCD picture of quarks and gluons. At short distances, the freezing out of quark degrees of freedom may lead to many-body forces different in character from the traditional long-range three-body forces of nuclear physics. To set the stage for an account of including the physics of substructure in nucleons, where there is no lack of alternate approaches and controversy, I will first review the rather stable situation in the potential description of rare gas systems. In Section 3, I will proceed to a description of contemporary models of nuclear three-body forces (TBF's). The consequences of the most studied TBF's in the three- and four-body bound states and the three-body continuum will comprise Sections 4 and 5. Some technicalities of current theoretical interest will be discussed in Section 6, and a summary section closes this report.

2. A Three-Body Force with Observable Effects

The potentials used in molecular physics are models whose connection with fundamental theory is becoming closer with the advent of large scale computing. The long-range potential is taken from second-order perturbation theory without exchange; the physics is a Coulomb-induced dipole-dipole interaction leading to a $1/r^6$ attractive force between the atoms. The short-range repulsive part of the pair potential is borrowed from first-order perturbation theory with exchange. It is due to the orbital rearrangement of the electrons forced by the Pauli principle. The resulting model is then fit to scattering data, two-body bound-state data, and the second virial coefficient of the gaseous phase. These pair potentials are not adequate to describe all the properties of noble gas systems, especially those of neon, argon, krypton, and xenon.

Long-range three-body forces were calculated from third-order perturbation theory without exchange by Axilrod and Teller and independently by Muto.[2] Physically, they correspond to a triple-dipole interaction induced by the exchange of photons between neutral particles with substructure (Fig. 1). The strength of the Axilrod-Teller-Muto (ATM) force is given by a sum over dipole oscillator strengths for all possible electronic transitions from the ground state. The oscillator strength can be determined from the absorption spectrum supplemented by refractive indices and polarizabilities, so it is very accurately known and model-independent. The best two-body potentials give good agreement with experiment for dense fluid and solid argon, krypton, and xenon, if they are used with the ATM three-body interaction and not otherwise.[3] For example, the ATM correction to the binding energy of neon and argon crystals is 4% and 8%, respectively. Similar agreement with experiment for the third virial coefficient of the gas phase (which depends only on clusters of three atoms) is achieved with a good pair potential and the ATM force.[3]

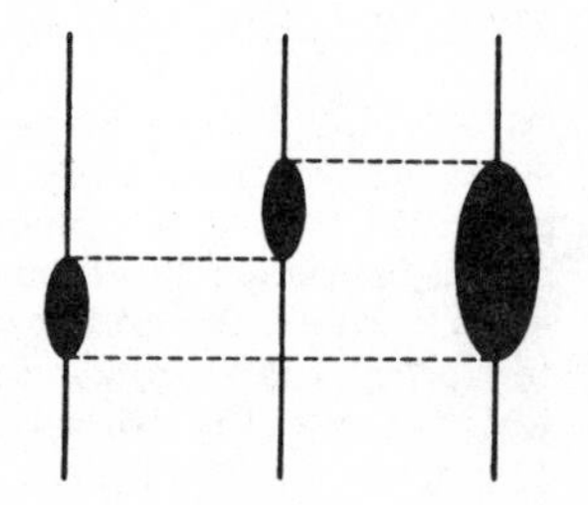

Figure 1

In addition to the ATM force, one would expect short-range three-body forces from third-order perturbation theory with exchange (overlap of electron orbitals). There is a controversy in the current literature about the role of these exchange forces.[3,4] These attractive forces and the ATM force cancel for nearest neighbor equilateral triangles.[4] However, overlap-dependent many-body interactions are vastly more difficult to calculate accurately than the long-range interactions, which are directly related to oscillator strengths. A new experimental probe of many-body forces utilizes the quantum mechanical scattering of He atoms from a krypton overlayer on graphite.[5] An experimental many-body correction thus obtained can be compared to the sum of overlap-dependent and ATM three-body forces. The effect of the ATM force was an order of magnitude larger than the former. On the other hand, scattering of neutrons from dense-gaseous krypton to determine the structure function indicates a need for an additional repulsive TBF over and above the ATM force.[6] The theoretical short-range TBF's are attractive.[4] At this stage, experiment seems to tell us that overlap-dependent many-body interactions must add up to something close to zero for each of the rare gases; neon, argon, krypton, and xenon.

When scaled in length by r_0 (the radius of a sphere containing one particle at equilibrium density) and in energy by $\hbar^2/mr_0$, the nuclear $^1S_0$ potential and the rare gas atomic potentials are qualitatively similar. In particular, the pair potentials both have strong repulsive cores that induce strong short-range correlations (Fig. 2). Quantum fluids can be characterized by (eight times) the fraction of the total volume occupied by the hard cores. This ratio is ~1.2 for ^{4}He and 1.1 for Kr, but only 0.1 for nuclear matter. Liquid krypton is very much denser than nuclear matter. Yet, the ATM force, which is rigorously correct only asymptotically appears to be the overwhelmingly dominant many-body correction.

In the foregoing discussion, substitute pion exchange for the dipole-dipole or triple-dipole force due to the exchange of photons, and substitute "quark overlap" for electron overlap potentials. Just as the long-range atomic forces can be determined by experiment, we shall see that the long-range many-nucleon forces are nearly completely determined by experimental knowledge (on the pion-nucleon interaction) and chiral constraints. Just as the short-range exchange two- and three-body forces are difficult to calculate reliably in atomic systems, nuclear physicists are still a long way from making accurate ab initio calculations of the short-range two-body force, let alone three-body forces. There will be soon a controversy over the relative importance of exchange TBF's in nuclear physics, akin to the current one in molecular physics. I venture to predict that, because nuclei are so dilute, the long-range TBF's will prove to be the important ones.

3. Three-Nucleon Force Models

In contrast with the molecular case, where it is the bulk properties that confirm the need for three-body forces, it is in the small clusters of nucleons that the discrepancies with experiment are the sharpest. We seem to have a satisfactory theory of the long- and intermediate-range parts of the nucleon pair interaction.[7] These dispersion theoretic pair potentials are based on a type of perturbation theory of the strong

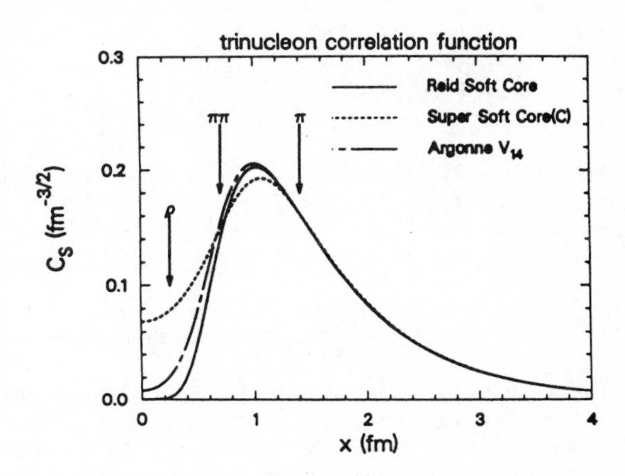

Figure 2

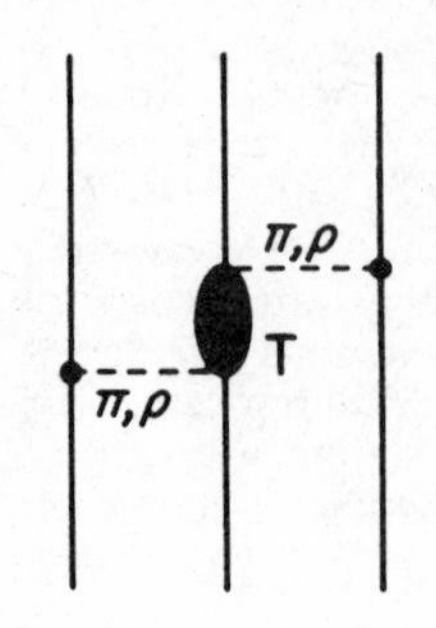

Figure 3

interaction at the hadronic level which is constrained by crossing, unitarity and analyticity along with experimental data from low-energy πN and ππ scattering. Their short-range correlations in the triton are similar to those due to the model potentials[8] depicted in Fig. 2. Experience with meson exchange currents and the short-distance suppression of the pair correlations shown in Fig. 2 lead one to hope that the 2π-exchange three-nucleon force (2π TBF), which is of π range, will be the dominant correction. The 2π potential suggested by the Feynman diagram of Fig. 3 is of longest range and, hence, the earliest and most extensively studied.[9] The shaded oval T represents anything except a forward-propagating nucleon state. The exchanged pions are space-like and "off-mass-shell" ($q^2 \neq \mu^2$).

The 2π TBF closest in spirit to the dispersion theoretic pair potentials is based on analytic πN scattering amplitudes which satisfy the low-energy theorems of the current algebra-PCAC description of axial-vector nucleon amplitudes analogous to those of Compton scattering.[10] To lowest order, the off-mass-shell πN amplitudes depend only on experimentally determined on-mass-shell quantities.[11] The resulting model independent TBF[12] differs from those based on a separation of the πN amplitude into partial waves and separate models for s- and p-wave interactions.[13,14] The Δ-isobar does not dominate the potential; instead the current algebra terms and the Δ-isobar terms play an equally important role.

This TBF, dubbed the Tucson-Melbourne (TM) three-body force, follows the dictum of Brown, Green, and Gerace[15] that, no matter how one chooses to model the πN scattering amplitude imbedded in Fig. 3, the model should be consistent with (broken) chiral symmetry. That is, the model should possess well-defined chiral ($q^2 \to 0$) limits. The additional imposition of appropriate hadronic phenomenology (which includes chiral-symmetry-breaking information) leads to the representation of the amplitude as $T = T_{pole} + \Delta T + q \cdot C \cdot q'$, which is shown in Fig. 4.

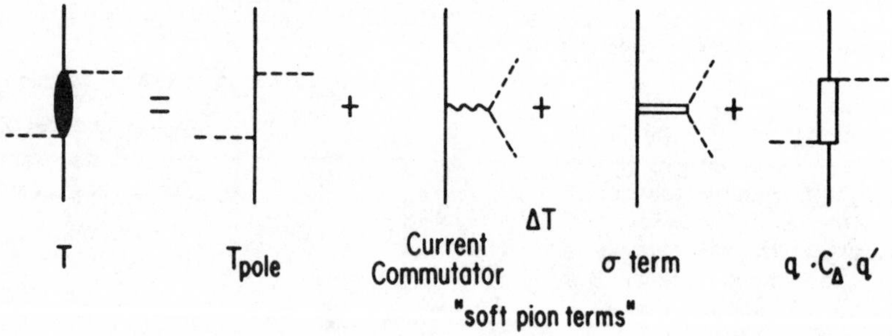

T = T_{pole} + Current Commutator + σ term + $q \cdot C_\Delta \cdot q'$

ΔT

"soft pion terms"

Figure 4

The short-range cutoff in the trinucleon pair correlation corresponds to a maximum momentum transfer of $q^2 \lesssim -9\,\mu^2$. This kinematic region ($-9\,\mu^2 < q^2 < 0$) is below the threshold for physical πN scattering. An expansion of the πN amplitude $(q/m)^2$ about the nucleon pole ($q^2 = q'^2 = 0$) appears appropriate because $\mu^2/m^2 \sim 1/50$. The chiral constraints, or soft pion ($q^2=0$) theorems, are satisfied by the sum of T_B, the nucleon pole term and ΔT. The higher-order terms are given by the model dependent background $q \cdot C \cdot q'$, which for πN scattering is dominated by the Δ-isobar.[10] The background amplitude can be obtained by writing a dispersion relation for C and saturating it with the on-shell Δ resonance. Dispersing on C rather than $q \cdot C \cdot q'$ avoids disturbing the soft pion results. Also shown in Fig. 4 are the covariant pole model Feynman diagrams which would arise from a chiral Lagrangian in a model-building approach.[16,17]

Current algebra instead deals with the chiral-symmetry breaking πN σ term,[10,18] which manifestly combines the σ-meson pole with the nucleon pole to enforce the delicate cancellation referred to as pair suppression. A second cancellation of the σ term against the Δ-dominated $q \cdot C \cdot q'$ makes the s-wave πN scattering length extremely small. This cancellation does not occur in the sub-threshold kinematic region. Hence, an extrapolation from threshold which does not properly account for the σ term and the background amplitude should be considered obsolete.[13,14] A third delicate cancellation in the relevant kinematical region is between the "momentum dependence" of the σ term and the Δ-dominated background $q \cdot C \cdot q'$. This cancellation is not an accident, but is demanded by the Adler zero and PCAC. To first order in the expansion, these cancellations are automatic if the entire blob on the left-hand side of Fig. 4 is parametrized by fitting the chiral constraints and the on-shell data constraints. This parametrization is unique up through order q^2/m^2. To this extent, the TM TBF is then independent of any model of the πN interaction.

A second approach, already mentioned above, seeks to construct a dynamical model for the πN amplitude from the pole diagrams of a chiral Lagrangian.[16,17] The resulting amplitude, when the chiral constraints are respected,[18] is nearly the same as the model-independent TM amplitude, although the interpretation of the TBF (labeled Brazil 2π TBF) is different. If the πN amplitude is contracted to include just the nucleon and delta poles, this suggests that three-nucleon forces can be incorporated into the calculations of nuclear physics if the Hilbert space is extended to include states in which some particles are deltas.[19] The πN amplitude implied by the "deltas-in-the-wavefunction" model is a very poor representation of the on-shell amplitude in the subthreshold region.[11] This "deltas-in-the-wavefunction" approach ignores the chiral constraints, and the difficulties with the propagator and the $\pi N \Delta$ vertex which, as is well known,[20] are ambigious in field theory away from the resonance. The "delta" one sees in diagrams of this approach is on-mass shell and propagating forward in time; the extent to which it approximates the off-shell delta of Fig. 4 depends precisely upon the size of the delta Z-graph and upon the unknown parameter in the $\pi N \Delta$ vertex.[21] (Obviously the field theory ambiguity remains in any model of the πN amplitude. The chiral Lagrangian practitioners simply choose some parameter value. The dispersion theorists found that the ambiguity can be isolated in the isospin-even, forward amplitude and absorbed into the phenomenological "slope" of the σ term.)

The coupling of the off-mass-shell pions to the outer nucleons of Fig. 3 is softened by a πNN form factor, $g(q^2)=gF_{\pi NN}(q^2)$. In most calculations reported here, the Goldberger-Treiman slope between $q^2=\mu^2$ and $q^2=0$ is parametrized by the monopole form factor $F_{\pi NN}(q^2) = (\Lambda^2 - \mu^2)/(\Lambda^2 - q^2)$, where $\Lambda=5.8\ \mu$. This value of Λ is determined by our present understanding of the Goldberger-Treiman discrepancy, a second measure of chiral-symmetry breaking.[22] As we shall see in Section 4, TBF effects are strongly dependent on the form factor.

Two-meson exchange TBF's of shorter range can be envisaged by replacing one or both of the pions in Fig. 3 by a ρ meson. These have been constructed according to the delta-dominance picture,[23] and by extending the current algebra-PCAC approach of the Tucson-Melbourne group.[24] The latter $\rho\pi$ TBF is not so dissimiliar to the sum of the nucleon-antinucleon pair term emphasized by the chiral Lagrangian approach advocated by the Brazilian group,[25] and the earlier[23] delta contribution. This is because the chiral-symmetry-breaking term in the $\rho N \to \pi N$ amplitude, in contrast to the πN σ term, is negligible. The off-shell $\rho N \to \pi N$ amplitude was obtained by adapting, via the current-field identity, a current algebra analysis of pion photoproduction.[26] The off-shell $\rho N \to \rho N$ amplitude is, of course, analogous to the Compton amplitude $\gamma N \to \gamma N$. Both the $\rho\pi$ and the $\rho\rho$ TBF's of the Tucson-Melbourne derivation have important pair terms and, unfortunately, important model-dependent $q \cdot C \cdot q'$ background terms. These forces have not been rigorously applied to nuclear systems as the 2π TBF has.

The nuclear analogue of the ATM molecular force (Fig. 1) corresponds to the exchange of three virtual pions and is thus of $\pi\pi$ range on Fig. 2. Such forces have been constructed before with pair theory alone,[27] with πN model amplitudes based on σ exchange,[28] and with πN model amplitudes based on the Δ resonance.[29] The individual contributions to the binding energy of nuclear matter were of opposite sign and larger than that of the 2π TBF. This is simply another manifestation of the cancellations previously discussed in the πN amplitude. It is imperative that these terms be considered together before a 3π-exchange three-body force can be believed.

Recently, a nuclear analogue of the Primakoff-Holstein TBF was constructed from the exchange of a σ meson and ω meson (of the Walecka quantum hadrodynamic Lagrangian).[30] The effects of this TBF in the calculable A=3,4 systems would then give an indication of the consequence of treating the nucleon as a four-component Dirac particle. The σ meson is the chiral partner of the pion, so one must be very careful to consider all appropriate diagrams at this level. Doing so leads one to 3σ forces with attendant cancellations which, however, are not controlled by data on the $\sigma N \to \sigma N$ amplitude as they are in the πN case.[31]

A final TBF within the picture of nucleons and mesons is the Lisbon force,[32] which is viewed as a distortion of the pion cloud between two nucleons due to the presence of a third one. In a molecular dynamics approach, both a pair potential and a TBF can be extracted from the system of nucleons and a real pion which interacts with each nucleon via a phenomenological πN potential. This approach is still being developed, but shows promise of treating the pair and triple forces on the same footing.

QCD-motivated three-body forces have proved just as hard to evaluate reliably as their electron overlap analogues in the rare gas atom case. For example, quark exchange forces within potential models have been worked out carefully by two groups.[33,34] Yet a change confined to the structure of the short-range piece of the confinement potential changed the total quark-exchange three-body contribution to ^{3}He from +0.25 MeV[34] to -0.25 MeV.[33] Bag model TBF's have been constructed.[35] The Skyrme Lagrangian yields up three-Skyrmion forces.[36] Much more will be done along these lines in the near future. In evaluating this activity, we must remain aware that only the ATM triple dipole molecular force and the two-meson-exchange nuclear forces (in particular the TM forces) have been normalized to the empirical evidence of asymptotic scattering measurements.

4. Effects of Three-Nucleon Forces on A=3 and A=4 Bound States

A great advance in few-body theory in the last three years is the demonstrated ability to solve the Faddeev equations with a given pair potential and a three-body force to obtain exact results. Calculations with the Faddeev reformulation of the Schroedinger equation are traditionally organized by expanding the pair potentials in a series (of infinite extent), each term of which acts only in a single partial wave (e.g., $^1S_0$), truncating the series, and then solving the truncated problem "exactly" in a numerical sense. The binding energy has converged in a 34-channel solution (all $j \leq 4$). Such 34-channel calculations have been made for the TM 2π TBF with various pair potentials,[37,38] and a perturbative treatment of the TBF includes 18 channels.[39] The RSC pair potential binds ^{3}H at 7.35 MeV compared to the experimental value of 8.48 MeV. If the TM 2π TBF (Λ=5.8 μ) is included, the total binding energies differ by only 0.5 MeV: E_T~8.86 from the Iowa-Los Alamos group,[37] E_T~8.98 (estimated) from the Bochum group[39] and E_T~9.23 (extrapolated) from the Sendai group.[42] It is quite an achievement to reach such agreement on so small an effect (the ratio $\langle V_{TBF}\rangle/\langle V_{pair}\rangle$ ~ 1.5 MeV/60 MeV~0.025). Sorting among the RSC and newer pair potentials such as the dispersion theoretic Paris[7] and phenomenological[8] Argonne V14 potential, one finds a consensus among Refs. 37-39 that the TM 2π TBF overbinds the triton by about 0.4 to 0.9 MeV for Λ=5.8 μ. The dependence on the πNN form factor is shown in Table 1 from Ref. 37 and will be discussed in Section 6.

TABLE 1. Triton binding energies (in MeV) for the RSC/TM model using the W_1 Faddeev equation decomposition with various values of the πN factor cutoff Λ

No. of channels	Λ = 4.1μ	Λ = 5.8μ	Λ = 7.1μ
5	6.93	7.55	8.75
9	7.27	8.33	10.30
18	7.44	8.93	11.40
34	7.46	8.86	11.16

I have plotted the effect of the TM 2π TBF on binding energy and charge radii of the nuclei ^{3}H, ^{3}He, and ^{4}He in Fig. 5. Binding energy per particle is plotted to get the results on one plot. The ends of the arrows show the effect of adding the TBF to a given pair potential. The points from the A=3 isodoublet are taken from the thesis of Chia-Rong Chen,[40] which contains a valuable summary of the Iowa-Los Alamos formalism and numerical results up to 1985. In contrast to the trinucleon, there is no consensus on the energy of ^{4}He even for the same pair potential; results from the exp(S) expansion[41] and the ATMS variational method[42] are shown. Some

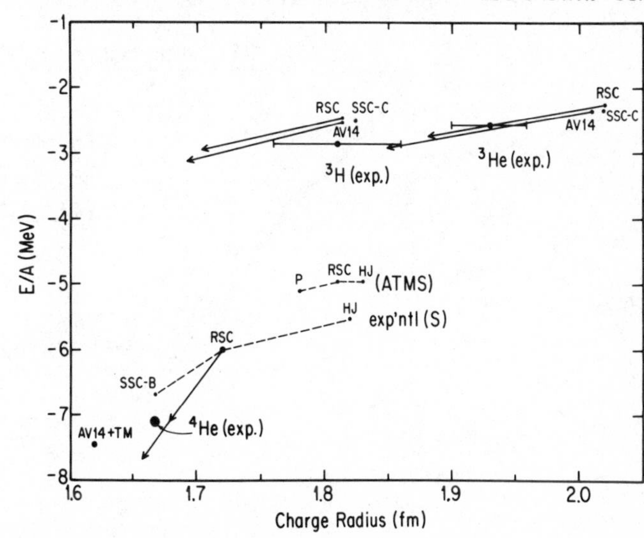

Figure 5

years ago, an exp(S) self-consistent calculation of ^{4}He was made with the RSC pair potential and an effective two-body force obtained from the TM 2π TBF by a nuclear matter average.[43] The displayed effect of the TBF is consistent with later variational calculations[14,44] with the AV14 pair potential which kept the three-body character of the TM force. Detailed comparison reveals that the ratio of TBF contributions to the energy of ^{4}He versus ^{3}H is about 5 rather than the expected combinatorial factor of 4. The Hokkaido group has suggested that this is primarily due to the smaller size of the ^{4}He nucleus.[45] Whatever the ratio, it is the same for different spin-isospin parts of the TBF,[46] suggesting that improved A=4 ground-state calculations would not tell us much more about the TBF than we have learned from the trinucleon calculations.

Returning to the trinucleon, we note that quantities such as charge radii are primarily sensitive to the outer parts of wave functions, and therefore sensitive to the binding energy. Empirical scaling relations have been demonstrated for radii, Coulomb energies, probabilities of various wave-function components, and asymptotic normalization constants.[47,48] Up to subtleties discussed in these papers, it would appear that the TBF's studied so far affect the binding energy, and all the other interesting features of the trinucleon simply follow.

The dramatic exception to this trend is the high-momentum-transfer parts of the charge form factors of ^{3}He and ^{3}H. It has long been hoped that a particular angular dependence[49] of pion exchange TBF's would simultaneously solve the binding energy problem and the charge-form-factor problem. Figure 6 from Ref. 50 shows that, quantitatively, the inclusion of TBF's does help increase the size of the secondary maximum, but the calculated position of the minimum and the secondary maximum move outward, worsening the disagreement with experiment. This is true for both ^{3}He and ^{3}H. What is needed is the ad hoc addition to either form factor of a component which vanishes at $q^2 = 0$ and is negative in the region of the minimum and secondary maximum. Some meson exchange currents (Fig. 6 was calculated in impulse approximation) have this property, but no calculation has yet treated exchange currents and relativistic corrections consistently. We note in passing that the photon-quark interaction coupled with a possible presence of six-quark clusters in the trinucleon produces exactly the modification required.[51] The latter picture of charge form factors is in a very early stage of development and cannot yet be considered settled.

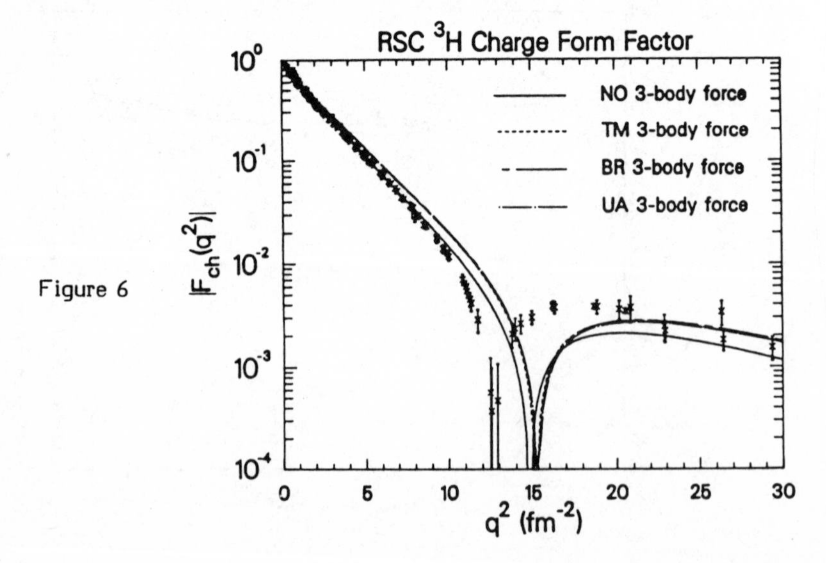

Figure 6

After years of arduous labor by strongly committed groups, it appears that the net effect of including the two-pion exchange three-body force in the nuclear Hamiltonian is to uncover an adjustable parameter in the theory. That is, Table 1 and Fig. 5 show that one can match the binding energy and radii by altering the πNN form factor parameter Λ away from the accepted value. This cannot be a satisfying conclusion. What of the exchange of heavier mesons? A perturbative study with a 3-channel wave function indicated that the Tucson-Melbourne $\rho\pi$ TBF had a repulsive effect about 20% that of the 2π TBF, and the $\rho\rho$ TBF was negligible.[52] The lack of a tensor force (and D state in the triton) and the severely truncated wavefunction make this result somewhat problematic. It does agree, however, with the trend of 50% repulsion of the $\rho\pi$ TBF in a nuclear matter estimate which did include tensor correlations.[24] Exact calculations are needed.

Effects of the TM 2π TBF on spin-orbit splittings, magnetization density of ^{17}O, the shell-model effective interaction, and nuclear matter are reviewed in Ref. 53. Some of the effects are larger and more interesting than those reviewed here, but the calculations were crude and the many-body theories more approximate.

5. Three-Body Forces in the Continuum

The close relationship that exists between the n-d and p-d doublet s-wave scattering lengths and the trinucleon binding energies has long been known as the Phillips curve.[54] This curve goes through the experimental datum for n-d threshold scattering. Inclusion of the 2π TBF to a RSC model trinucleon does not change the shape or position of either Phillips curve; the TBF increases the binding, and both scattering lengths decrease accordingly.[55]

Breakup reactions such as n+d→n+n+p possess a greater variety of final-state configurations which might be able to furnish more information on the action of a TBF. The calculations of the Bochum group[56] have already indicated that the TM 2π TBF could have effects in star, collinear, and quasi-free scattering regions in the pnn final-state phase space. The counter-intuitive results of these Faddeev calculations cast doubts on previous suggestions of the effect of a TBF on the nn scattering length extracted from three-nucleon final-state reactions.[57] The Bochum calculations have guided experimental searches for such effects at Koln[58] and Bochum.[59] So far, little evidence for a three-body force effect has been found, but the two-body forces used were separable approximations to simple pair potentials. The present confrontation between calculation and experiment shows that only the best available pair potential should be used, including exact Coulomb effects and higher partial waves to allow for polarization predictions.[60] Faddeev breakup codes with local realistic pair potentials are being developed and tested;[61] they are at least an order of magnitude harder than the bound-state codes.

6. Technical aspects of Meson-Exchange Three-Body Forces

Much of the applied work of Sections 4 and 5 utilized the TM 2π TBF, which was written down in its present form already 10 years ago.[12] Since then there have been refinements and discussions of this TBF. Some of this work has been prompted by the extreme sensitivity of the TBF effects to the πNN form factor (Table 1). This sensitivity is largely due to terms in the force which would be Dirac delta functions or derivatives of delta functions if the form factor was absent. The form factor spreads out these singularities so that the terms in the expansion demanded by chiral symmetry breaking are very sensitive to the parameter Λ. It has been urged that such singularities be dropped from a two-meson exchange TBF.[17] This somewhat ad hoc prescription has large numerical consequences in perturbation estimates in the triton[62] and negligible consequences in an exact calculation.[37] In any case, one lesson is clear: we must obtain independent information on the πNN form factor. There may be other ways of taming the singularities than simply dropping them. One approach currently being worked on is to deal directly with s, t, and u channel poles in the "blob" of Fig. 3.[63]

The consistency of the various corrections to a pair-potential Hamiltonian has often been questioned[64,65] For example, the forward-propagating nucleon term corresponds to the iteration of the one pion exchange potential (OPEP) and should be subtracted from the blob of Figs. 3 and 4. The result is a correction of relativistic order. At the same time, one should consider time-ordered graphs of overlapping pion exchanges.[66] Meson exchange currents do not appear on Fig. 6 because they should be linked to the pion exchange part of the nuclear charge and isoscalar current operators and cannot be because of the nature of the phenomenological pair potentials underlying the calculation. These exchange currents are also relativistic corrections. These three theoretical problems show that a consistent treatment of relativistic corrections thus require a simultaneous analysis of the nonstatic OPEP, the charge or current operator, and those parts of the TBF which can be constructed only from pions and nucleons. This has recently been done.[67] The TBF from pair terms and overlapping, retarded pion-exchanges differs from the original TM force only in the presence of several new nonlocal terms, and in the specification of the choice of relativistic ambiguity parameters.[64] The numerical impact of the new nonlocal terms has not been determined.

7. Summary

The two-pion exchange force of nuclear physics is analogous to the triple dipole force of molecular physics. Each is the longest range three-body force in its own system. The structure of each is given by perturbation theory. The strength of each force can be normalized to experimental data. Overlap-dependent three-body forces of shorter range are difficult to calculate accurately in each system. The triple dipole force appears to be exactly what is needed to match a variety of experimental facts in the rare gas systems, including the very dense quantum liquids. A similar statement cannot yet be made about the 2π TBF or, indeed, any nuclear three-body force. However, correct calculations can be made in the trinucleon bound state with a realistic pair potential and a two-meson exchange TBF. Most aspects of the A=3 bound state simply follow the binding energy, which is increased a little too much by the 2π TBF. The 2π TBF does not solve the charge form factor problem. There are some indications that one will learn little more about TBF's from more accurate calculations of the ^{4}He ground state. Two-meson exchange TBF's of shorter range and an even richer structure are waiting in the wings.

The future of three-body physics and three-body forces lies in the continuum. Calculations are still in their infancy. Somehow, amplitude probing experiments must be thought of that can take advantage of the rich spin-isospin structure of the meson-exchange three-body forces. This has proved difficult. We may take encouragement from recalling that real progress in the theory of three-body forces came only when attention turned away from πN total cross sections (dominated by the (3,3) resonance) toward the subthreshold πN amplitudes relevant to the problem.

References

1. H. Primakoff and T. Holstein, Phys. Rev. 55 (1939) 1218; C. Chanmugam and S. S. Schweber, Phys. Rev. A 5 (1970) 1369.
2. B. M. Axilrod and E. Teller, J. Chem. Phys. 11 (1943) 299; Y. Muto, Proc. Phys. Math. Soc. (Japan) 17 (1943) 629.
3. J. A. Barker, Phys. Rev. Lett. 57 (1986) 230; J. A. Barker, Mol. Phys. 57 (1986) 755.
4. W. J. Meath and R. A. Aziz, Mol. Phys. 52 (1984) 225; H. Jonsson and J. H. Weare, Faraday Discuss. Chem. Soc. (1985) 80.
5. H. Jonsson and J. H. Weare, Phys. Rev. Lett. 57 (1986) 412.
6. A. Teitsma and P. A. Egelstaff, Phys. Rev. A 21 (1980) 367; P.A. Egelstaff et al., Phys. Rev. A 30 (1984) 374.
7. P. Signell, in The (p,n) Reaction and the Nucleon-Nucleon Force, ed. C. D. Goodman et al. (Plenum Press, New York, 1979), p. 1; M. Lacombe et al., Phys. Rev. C 21 (1980) 861.
8. R. V. Reid, Ann. Phys. (N.Y.) 50 (1968) 411; R. de Tourreil and D. W. L. Sprung, Nucl. Phys. A242 (1975) 445; R. B. Wirenga, R. A. Smith, and T. L. Ainsworth, Phys. Rev. C 29 (1984) 1207.
9. Early TBF models are reviewed by B. H. K. Mckellar, in Few Body Dynamics, ed. A. N. Mitra et al. (North-Holland, Amsterdam, 1976), p. 508; and B. H. J. Mckellar and R. Rajaraman, in "Mesons and Nuclei," ed. M. Rho and D. Wilkinson (North-Holland, Amsterdam, 1979), p. 357.
10. M. Scadron, in Few-Body Dynamics, ibid., p. 325, S. A. Coon, M. D. Scadron, and B. R. Barrett, Nucl. Phys. A242 (1975) 467.
11. G. Höhler, H. P. Jakob, and R. Strauss, Nucl. Phys. B39 (1972) 237; G. Höhler, Landolt-Bornstein, Vol. I/9B, Pion-Nucleon Scattering, ed. H. Schoppor (Springer-Verlag, New York, 1983).
12. S. A. Coon, B. R. Barrett, M. D. Scadron, D. W. E. Blatt, and B. H. J. McKellar, in Few-Body Dynamics, ed. A. N. Mitra et al. (North-Holland, Amsterdam, 1976), p. 739; S. A. Coon et al., Nucl. Phys. A317 (1979) 242; S. A. Coon and W. Glöckle, Phys. Rev. C 23 (1981) 1790.
13. J. Fujita and H. Miyazawa, Prog. Theor. Phys. 17 (1957) 360.
14. J. Carlson, V. R. Pandharipande, and R. B. Wirenga, Nucl. Phys. A401 (1983) 59.
15. G. E. Brown, A. M. Green and W. J. Gerace, Nucl. Phys. A115 (1968) 435.
16. S. N. Yang, Phys. Rev. C 10 (1974) 2067.
17. M. R. Robilotta et al., Phys. Rev. C 31 (1985) 646; M. R. Robilotta and M. P Isidro Filho, Nucl. Phys. A451 (1986) 581.
18. K. Goring et al., Proc. of Lake Louise Conf., Canada, May 24-31, 1986.
19. A. M. Green, Rep. Prog. Phys. 39 (1976) 1109; M. Müther, Prog. Part. Nucl. Phys. 14 (1985) 123; P. U. Sauer, invited talk at the Int. Symp. on the Three-Body Force in the Three-Nucleon System, George Washington University, Washington, D.C., April 24-26, to appear in Lec. Notes in Physics (GWU TBF Symp., 1986).
20. R. D. Peccei, Phys. Rev. 176 (1968) 1812.
21. B. H. J. McKellar, invited talk at Int. Workshop on Three-Body Forces, Bochum, West Germany (1983), unpublished.
22. S. A. Coon and M. D. Scadron, Phys. Rev. C 23 (1981) 1150.
23. M. Martzolff, B. Loiseau, and P. Grange, Phys. Lett. 92B (1980) 46; M. Martzolff, Ph.D. dissertation, University of Strasbourg (1980).
24. R. G. Ellis, S. A. Coon, and B. H. J. McKellar, Nucl. Phys. A438 (1985) 631.
25. M. R. Robilotta and M. P. Isidro Filho, Nucl. Phys. A414 (1984) 394.
26. J. T. MacMullen and M. D. Scadron, Phys. Rev. D 20 (1979) 1069.
27. S. Drell and K. Huang, Phys. Rev. 91 (1953) 1527.

28. T. Ueda, T. Sawada, and S. Takagi, in Few Body Dynamics, ed. A. N. Mitra et al. (North-Holland, Amsterdam, 1976), p. 67.
29. J. Fujita, M. Kawai, and M. Taifuji, Nucl. Phys. 29 (1962) 252.
30. B. D. Keister and R. B. Wiringa, Argonne preprint, PHY-4723-TH-86.
31. B. H. J. McKellar, invited talk at GWU TBF Symp, 1986.
32. A. C. Fonseca and M. T. Peña, in Few Body Problems in Physics, Vol. II, ed. B. Zeitnitz (North-Holland, Amsterdam, 1984), p. 489; GWU TBF Symp., 1986.
33. Y. Suzuki and K. T. Hecht, Phys. Rev. C 29 (1984) 1586.
34. K. Maltman, Nucl. Phys. A439 (1985) 648.
35. V. E. Kuzmichev, Nucl. Phys. A430 (1984) 636.
36. U.-G. Meissner and U. B. Kaulfuss, Phys. Rev. C 30 (1984) 2058.
37. C. R. Chen, G. L. Payne, J. L. Friar, and B. F. Gibson, Phys. Rev. C 33 (1986) 1740.
38. T. Sasakawa and S. Ishikawa, in Few Body Systems 1 (1986) 3.
39. A. Bömelburg, Phys. Rev. C 34 (1986) 14.
40. C. R. Chen, Ph.D. dissertation, University of Iowa (1985) .
41. H. Kummel, K. H. Luhrmann, and J. G. Zabolitzky, Phys. Rep. 36C (1978) 1; J. G. Zabolitzky, Phys. Lett. 100B (1981) 5.
42. T. Katayama and M. Sakai, in Few Body Problems in Physics, ed. A. N. Mitra et al. (North-Holland, Amsterdam, 1976), pp. 435, 437.
43. S. A. Coon, J. G. Zabolitzky, and D. W. E. Blatt, Z. Phys. A281 (1977) 137.
44. R. B. Wirenga, Nucl. Phys. A401 (1983) 86.
45. M. Sato, Y. Akaishi, and H. Tanaka, Prog. of Theor. Phys. Supp., No. 56 (1974) 930.
46. M. Sato, private communication.
47. J. L. Friar et al., Phys. Lett. 161B (1985) 241.
48. S. Ishikawa and T. Sasakawa, Phy. Rev. Lett. 56 (1986) 317.
49. M. Fabre de la Ripelle, C. R. Acad. Sci. (Paris) 288 (1979) 325.
50. J. L. Friar, B. F. Gibson, G. L. Payne, and C. R. Chen, submitted to Phys. Rev. C.
51. J. P. Vary, S. A. Coon and H. J. Pirner, in Hadronic Probes and Nuclear Interactions, ed. J. R. Comfort et al. (AIP, New York, 1985), p. 83.
52. S. A. Coon, M. T. Peña, and R. G. Ellis, Phys. Rev. C 30 (1984) 1366.
53. S. A. Coon, invited talk at the GWU TBF Symp., 1986.
54. A. C. Phillips, Rep. Prog. Phys. 40 (1977) 905.
55. C. R. Chen, G. L. Payne, J. L. Friar, and B. F. Gibson, Phys. Rev. C 33 (1986) 401.
56. W. Meier and W. Glöckle, Phys. Lett. 138B (1984) 329.
57. I. Slaus, Y. Akaishi, and H. Tanaka, Phys. Lett. 48 (1982) 993.
58. M. Karus et al., Phys. Rev. C 31 (1985) 1112.
59. B. Bannach et al., GWU TBF Symp., 1986; K. Bodek et al., ibid.
60. H. Paetz gen. Schieck et al., ibid.
61. R. A. Brandenburg, ibid.
62. M. Sato and S. A. Coon, Bull. Am. Phys. Soc. 30 (1985) 765.
63. J. L. Friar and S. A. Coon, work in progress.
64. J. L. Friar, Ann. Phys. (N.Y.) 104 (1977) 380.
65. J. L. Friar, B. F. Gibson, and G. L. Payne, Ann. Rev. Nucl. Part. Sci. 34 (1984) 403.
66. W. Glöckle and S. N. Yang, Phys. Rev. C 33 (1986) 1774.
67. S. A. Coon and J. L. Friar, Phys. Rev. C (Sept. 1986).

List of Oral Contributions

(In order of presentation: the number in square brackets refers to the contribution in Vol. 1 of the Proceedings)

Momentum distribution and correlations in ^{3}He (e,e'p) experiment
C Marchand[1], M Bernheim[1], M Brussel[2], G P Capitani[3], E de Sanctis[3], P Dunn[4], S Frullani[3], G Garibaldi[3], A Gerard[1], A Magnon[1], Z E Meziani[5], J Morgenstern[1], J Mougey[6], J Picard[1], D Reffay[1], S Turck-Chieze[1] and P Vernin[1]
[1]CEN Saclay; [2]University of Illinois; [3]INFN Frascati and University of Rome; [4]Massachusetts Inst. Technology; [5]Stanford University; [6]CEBAF
[D18]

Deuteron electrodisintegration with polarized electrons d($\vec{e}$,e'N)N
W Leidemann and H Arenhövel
Johannes Gutenberg University, Mainz
[E17]

Quasi-free absorption of pions on ^{3}He
G Backenstoss[1], M Izycki[1], P Salvisberg[1], M Steinacher[1], P Weber[1], H J Weyer[1], S Cierjacks[2], S Ljungfelt[2], H Ullrich[2], M Furić[3] and T Petković[3]
[1]University of Basle; [2]KFA Karlsruhe; [3]University of Zagreb
[A19]

Inst. Phys. Conf. Ser. No. 86
Paper presented at Int. Nucl. Phys. Conf., Harrogate, UK, 1986

The planned heavy-ion storage-cooler rings: powerful tools for new experiments in nuclear and atomic physics

Fritz Bosch

GSI D-6100 Darmstadt, F.R. Germany

1. Introduction

At present, there is a boom in plans and proposals for heavy-ion storage-cooler rings. Although none of these rings are operational, now is certainly a good time for discussing the aims and perspectives of such facilities within the framework of nuclear and atomic physics. In analyzing these plans, however, we should critically compare the cost with the expected progress in physical science.

I am going to present this survey in four parts: First, I will try to give some arguments for starting immediately on cooler rings, and then proceed to a short discussion of the most important aims and hopes connected with them. The design of the experimental storage ring ESR at GSI, Darmstadt, will be shown later in more detail. This Facility is yet to be built - but is already approved and funded - and is tailored for storing and cooling all of the few-electron heavy ions up to bare uranium (U^{92+}) over an energy range from a few MeV/u up to about 1 GeV/u.

Success or failure of heavy-ion cooler rings depends on achieving a significant enhancement of phase space density of the stored ions by momentum exchange with a collinear, 'cold' electron beam of the same mean velocity. For this reason the basic ideas of this ingenious concept of electron cooling will be discussed in the third section. The problems which may occur when extending this well-proven method of electron cooling of protons to the cooling of heavy ions will also be outlined.

In the last, and main part, selected experiments already proposed for the ESR will be presented, emphasizing their encouraging perspectives for doing quite new and exciting experiments in the fields of nuclear, atomic, and fundamental physics. This draft of future experiments also covers - to some extent - the proposals made for the other planned heavy-ion cooler rings. Most of them are designed - in contrast to the ESR - for storing only light or at best medium-heavy ions and with a research programme focused mainly on atomic physics.

2. An Incomplete Survey of Planned Heavy-Ion Storage-Cooler Rings

Spectacular success has been achieved in both the storage and the phase-space enhancement of the lightest baryons ($p,\bar{p}$). The overwhelming success of stochastic cooling of antiprotons at CERN, which was the indispensable condition for observing the long-searched for W- and Z-bosons for the first time, is well known.

Probably not as widely known is the steady progress attained in electron cooling of protons at Novosibirsk, where this revolutionary method was invented by Budker (1967), and also at other facilities (e.g. ICE at CERN, Fermilab (Poth, 1984)).

The clear demonstration of a hitherto unknown quality of proton beams, with a relative momentum spread $\Delta p/p$ of the order of 10^{-5}, coincided in time with the availability of new, powerful ion-sources (e.g. EBIS, CRYEBIS), delivering highly charged heavy ions at current densities reaching the space-charge limits. Furthermore, at the same time, many low- or medium-energy heavy-ion accelerators were available all over the world, and so it became almost self-evident to join everything together - sources, accelerators and cooling - and to dream, at least, of experiments with objectives hitherto inaccessible and with unprecedented precision.

Some years ago GSI planned to open up the exciting possibility of re-stripping highly charged heavy ions to completely bare nuclei, and to inject, accumulate, and cool them to the highest phase-space density in a storage-cooler ring. Thus the idea of an additional experimental storage-cooler ring for high currents of the heaviest, highly stripped ions, on their SIS18 synchrotron, was born. From this time researchers at GSI dreamed of the possibility of 10^{10} stored bare uranium particles at 500 MeV/u energy and with a relative momentum spread smaller than 10^{-5}. Moreover, the comparatively large cross sections for projectile fragmentation at high energies suggested the possibility of producing exotic nuclei far from stability, the separation of specific isotopes by a fragment separator, and the injection of these mono-isotopic secondary beams into the ESR for accumulation and cooling purposes. Then utilizing the luminosity gain due to an internal target intersecting this circulating beam, precision spectroscopy and reaction studies of those astrophysically significant particles seem to become feasible.

Altogether, the boom in heavy-ion cooler projects became more or less inevitable. Plans and experiment, dream and reality, however, are still separated by a deep sea on which no one has yet sailed: the sea of electron cooling of heavy ions. Indeed, the crucial point is whether electron cooling of heavy ions will work or not. In this sense, everything I will describe later on must, necessarily, be open-ended.

This survey is summarized in Table 1 which shows the main features of most of the planned heavy-ion storage-cooler rings. Neither highly advanced proton or light-ion cooler rings (e.g. Bloomington, Indiana) nor the old-fashioned antiproton cooler ring LEAR at CERN are incorporated. Also, projects which are either not yet funded or still under modification (e.g. Oak Ridge, Berkeley) are not shown.

Common features of all planned heavy-ion cooler rings are:

a) The storage of a high particle number up to the space-charge limit.

b) The enhancement of phase-space density by cooling devices.

c) The luminosity gain with respect to single pass experiments by using internal targets.

The maximum charge number of the ions to be stored, the degree of ionization and the maximum energy are quite different, however, from ring to

Table 1
Planned heavy-ion storage-cooler rings (Lynen, 1984)

Name and Locality	Domain of ions to be stored	Cooling method(s)	Merging beams	Secondary beams
ASTRID Aarhus	light (Z < 18) modestly ionized	electrons laser	probably yes	no
CELSIUS Uppsala	very light (Z < 10) up to Ne^{7+}	electrons	no	no
COSY Jülich	very light (Z < 7) up to N^{7+}	electrons	no	no
CRYRING Stockholm	light/medium up to Xe^{44+}	electrons laser	yes H^-, H^0...	no
ESR Darmstadt	all ions up to U^{92+}	electrons stochastic (laser ?)	yes same isotope	yes nuclei far from stability
TARN II Tokyo	light to heavy	electrons stochastic	no (?)	probably yes
TSR Heidelberg	light/medium up to I^{47+}	electrons	no	no

ring. Also the use of a 'merging-beam' technique - either by crossing two beams of the same isotope at slightly different energies (ESR) or by crossing different species of particles (CRYRING) - is restricted to only a few cases. Similarly, there are few plans for utilizing secondary beams due to the rather high energy needed for their efficient production. On the other hand, in the case of only modestly ionized atoms (ASTRID, CRYRING), it looks quite promising that the proposed 'laser-cooling' (Channel 1981, Javanainen 1985, Herlander 1984) will reach momentum spreads $\Delta p/p$ smaller than 10^{-6} or even 10^{-7}. Later on, I will come back to this point. Examples for the maximum number of particles which can circulate in one storage-cooler cycle are shown in Fig. 2.

3. The Design of the Storage-Cooler Ring ESR at GSI

An outline of the new facilities to be built at GSI is shown in Fig. 1. The main parts are the heavy-ion synchrotron SIS 18 with a circumference of 206.4 m and a maximum magnetic bending power $B\rho$ of 18 T m; the cooler ring ESR with a maximum $B\rho$ of 10 T m, and exactly half the circumference of the SIS; and, between them, the fragment-separator FS for separation of secondary beams to be injected into the ESR.

The ions are injected at 11.4 MeV/u into SIS 18 from the existing UNILAC accelerator and further accelerated to maximum energies depending on the charge state of the ions as depicted in Fig. 2.

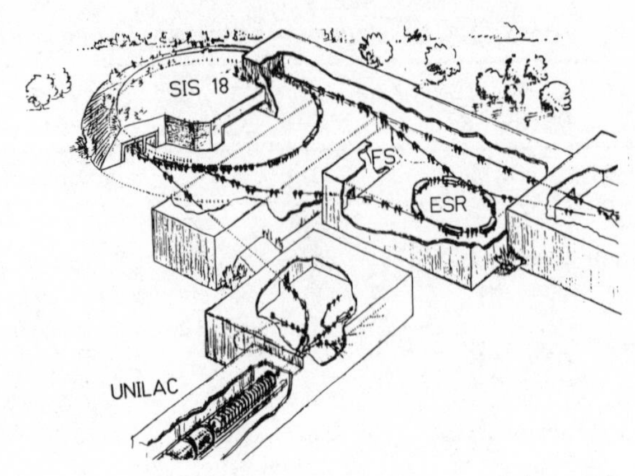

Fig. 1 Layout of the synchrotron SIS 18, the experimental storage-cooler ring ESR, and between them the fragment separator FS.

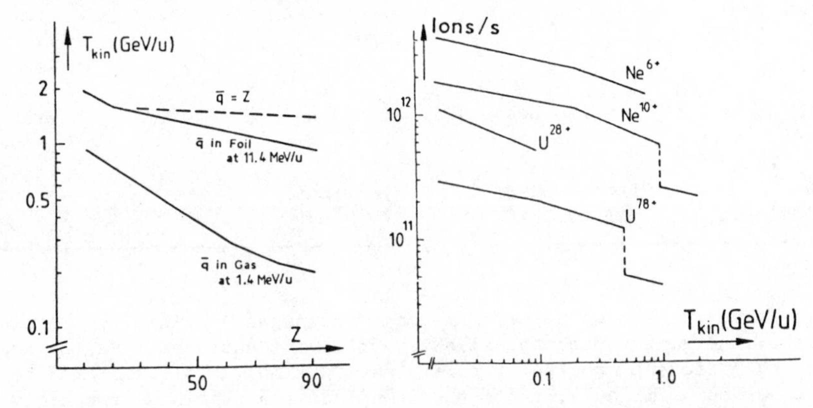

Fig. 2 Left: maximum particle energies in the SIS obtained when gas stripping or foil stripping is applied (the dashed curve denotes the case of reinjection of fully stripped ions into SIS). Right: maximum particle numbers obtained for neon and uranium as a function of energy (the drop denotes a change in the repetition rate from 3 Hz to 1 Hz).

For uranium ions of charge state q = 78 the maximum achievable energy is 1 GeV/u. The injection into the synchrotron is accomplished during 10 to 30 turns. Space-charge limits are reached at injected particle currents of 2×10^{14} s^{-1} (U^{78+}) and 2×10^{15} s^{-1} (Ne^{10+}). These high currents may be provided by a newly developed (Keller, 1983) high intensity ion-source for low charge states (U^{2+}) in connection with a new RFQ-structure at the first domain of acceleration.

As well as slow extraction over a period of 100 ms a fast extraction from SIS 18 is foreseen. The latter allows a convenient transfer of one half of the SIS turn into one ESR turn at the second harmonic rf frequency of the SIS, thereby preserving the bunch structure of the beam. Between SIS and ESR the beam may be stripped once again in a foil to the highest desired charge state. From Bevalac data (Gould, 1984) a 60% efficiency for bare uranium (at 556 MeV/u) and even a 100% efficiency for bare xenon (at 609 MeV/u) in a copper target of several g/cm^2 thickness can be extrapolated.

Alternatively, a production target for 'cold' projectile fragmentation, induced by the large transverse Coulomb field, may be installed. When using light target atoms (e.g. C) the kinematics focuses most of the fragments - some of which may have a very unusual N/Z ratio - into a very small 'cone' close to the beam direction and the beam velocity, thus allowing an effective separation of both nuclear masses and nuclear charges in the special fragment separator FS between SIS 18 and ESR. These separated fragments can be accepted by the ESR in a quite large momentum band ($\Delta p/p = \pm 2\%$) and with a radial acceptance of 140π mm mrad. Therefore it is hoped to accept, to accumulate, and to cool quite high currents of exotic radioactive nuclei for the purpose of in-beam mass determination, precision spectroscopy and reaction studies.

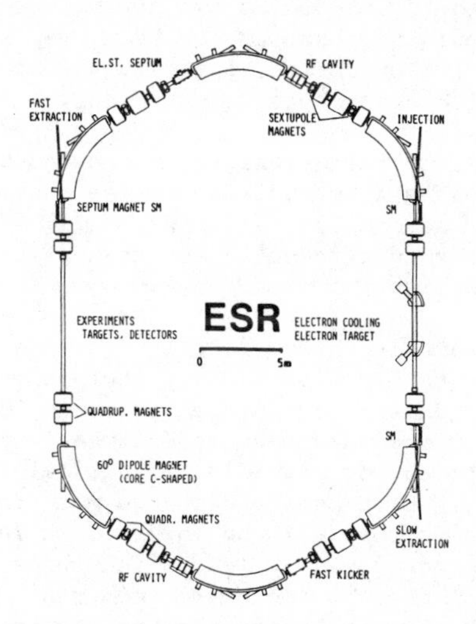

Fig. 3 The layout of the storage-cooler ring ESR with a circumference of 103.2 m and 10 T m bending power. Two straight sections of 9.5 m length are foreseen for installing the electron cooler and the experiments (internal target). The rf cavities are for acceleration, deceleration, and bunching.

In connection with SIS 18 and the FS, the ESR (cf. Fig. 3) offers a variety of attractive features: as a storage ring it can store completely stripped heavy ions at high energies (834 MeV/u for Ne^{10+}, 556 MeV/u for U^{92+}); via a fast extraction, these bare nuclei may be injected back into the SIS for further acceleration. The flexible ion optics of the ring essentially allows three modes of operation:

a) By applying a constant dispersion on the straight sections (9.5 m long), the 'hot' secondary beams provided by the FS with large momentum spread can be accepted.

b) At zero dispersion on the straight section multicharge operation (e.g. U^{89+} up to U^{92+}) becomes possible.

c) With large dispersion two beams of slightly different energies can be stored simultaneously and can be brought to intersect at a small angle of about 100 mrad. This 'merging beam' mode may be applied to investigations of interactions of ions with a well-defined number of electrons and at energies corresponding to a fixed-target equivalent energy near the Coulomb barrier (~ 7 MeV/u).

As a stretcher ring the ESR may store coasting beams with a duty factor of up to about 90%. On the other hand it should also be possible to confine the beam into bunches of 1 ns duration by using the rf cavities installed. The latter may also serve for deceleration purposes, down to UNILAC energies. Bunching as well as deceleration, however, requires phase-space cooling for reducing the emittance growth.

Internal targets (normal or polarized atomic and ionic, or electron beams) which will be installed in one of the two straight sections of the ESR, together with the circulating, cooled ion beam, may become very important for the study of reaction processes, owing to the tremendous gain in luminosity compared with single pass experiments.

Without doubt, the most exciting perspectives open up with the ESR as a cooler ring. Two complementary cooling devices can be operated simultaneously. For the 'hot' secondary beams ($\Delta p/p \sim 2\%$) stochastic pre-cooling may be used - which is much more efficient than electron cooling for those beams of low phase density.

As the 'normal' ion beams injected directly from the SIS 18 already have a small momentum spread of $\Delta p/p \sim 5 \times 10^{-3}$, electron cooling (cf. section 4) is foreseen for those beams in an interaction zone of 2.5 m length. Electrons travelling at the same mean velocity as the ions, and collinearly with them, are confined, by a guiding solenoidal field, within an area of about 5 cm diameter. By momentum exchange with the ions, the phase-space density of the electrons will be lowered, while that of the ions will be enhanced correspondingly. Electron energies from 2 keV to 350 keV at current densities of about 1 A/cm^2 are required. As a figure of merit one expects, for heavy ions, by extrapolating the results attained in electron cooling of protons (cf. section 4), relative momentum spreads smaller than 10^{-5} and emittances of about 0.1π mm mrad. As most of the proposed experiments are based on such small momentum spreads, it is not an exaggeration to state that successful electron cooling denotes the cornerstone of the new facility at GSI and of all other planned facilities of this kind.

4. Phase-space Cooling of Ion Beams

4.1. Electron Cooling

The basic idea of electron cooling of ions, first proposed by Budker (1967), is simple and ingenious. Electron cooling denotes a repeated interaction of an electron beam of very small spread in the longitudinal as well as in the transverse momentum, with a collinear ion beam of the same mean velocity, but with a much higher original spread in momentum. By momentum exchange between ions and electrons caused by the long-range Coulomb forces, ions slower than electrons are accelerated, faster ones are decelerated. Ion momenta transverse to the direction of the electron beam are reduced. This leads to a progressive assimilation of the properties of the two beams until thermal equilibrium is reached and the ion temperature T_i becomes equal to the electron temperature T_{el}. Since the scattered 'hot' electrons are steadily substituted by 'cold' ones from the electron gun, the ion beam finally reaches the small temperatures of the original electron beam, i.e. a 'transverse' temperature $T_\perp$ in the order of 0.1 eV (given by the cathode temperature of about 1200 K) and a 'longitudinal' temperature $T_\parallel$ - greatly reduced compared with $T_\perp$ - caused by the acceleration of the electrons, which is given by

$$T_\parallel = T_\perp \cdot \frac{T_\perp}{2\gamma^2\beta^2 m_e c^2} \sim 10^{-6} \times T_\perp \quad (\beta = 0.8) \tag{1}$$

where β and γ are the usual relativistic factors. Moreover, the guiding solenoidal field may tend to freeze the transverse degrees of freedom of the electrons - which could lead to a very fast 'supercooling' (Debrenev 1978, Parkhomchuk 1984).

Starting from equal electron and ion temperatures at equilibrium, it can be easily derived that the relative momentum spread of the ion beam becomes smaller than that of the electron beam, in proportion to the square root of their mass ratio:

$$\frac{(\Delta p)\text{ion}}{(p)\text{ion}} = \sqrt{\frac{m_e}{m_{ion}}} \times \frac{(\Delta p)e}{(p)e} \tag{2}$$

From this equation and from a typical value measured (Poth, 1984) for the momentum spread of cooled protons of $\Delta p/p \sim 6 \times 10^{-5}$, a final momentum spread smaller than 10^{-5} may be expected for heavy ions such as uranium.

A simple estimate of the cooling time τ_c can be derived from the Bethe-Bloch equation for the stopping power of an ion in an electron gas, yielding (Poth, 1984)

$$\tau_c \sim A/Z^2 \cdot 1/n_e \cdot T_{el}^{3/2} \tag{3}$$

where A and Z are the ion mass and nuclear charge, respectively, and n_e is the electron density.

Equation (3) shows the very important fact that the cooling time τ_c gets shorter for heavy ions. For the ESR typical cooling times range from 0.15 to about 600 ms for 30 MeV/u U^{92+} and 500 MeV/u Ne^{10+}, respectively.

In a storage ring various heating processes are inevitable which tend to reduce the cooling efficiency. Among them are scattering in the residual gas ($\sim 10^{-11}$ mbar) and in internal gas targets ($N < 10^{14}$ cm^{-2}) and intra-beam scattering (Piwinski, 1974).

Most restricting, however, are electron capture processes in the cooler region itself by radiative electron capture (REC), the time-mirrored photoeffect. An electron, travelling with a kinetic energy T_e in the ion rest frame, is transferred into a vacant bound state thereby emitting a photon, the energy of which is the sum of the binding energy and the kinetic energy T_e. The cross section σ_R for this REC process scales (Stobbe, 1930) as $\sigma_R \propto Z^2/T_e$. In Fig. 4 the expected ratio of the cooling cross section σ_c and σ_R is shown (in per cent) as a function of T_e. Since for 'normal' beams, injected from SIS 18, the kinetic energy T_e of the cooling electrons in the ion rest frame is of the order of a few eV and

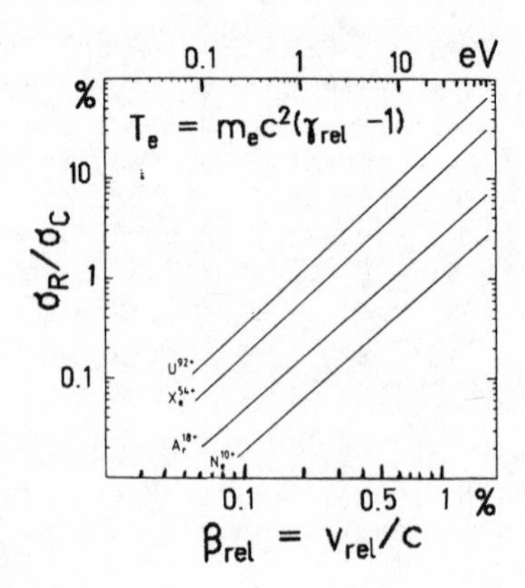

Fig. 4 Ratio (in per cent) of the cross sections σ_R for radiative electron capture and σ_C for cooling, as a function of the relative velocity of electrons in the ion rest frame (in per cent of the velocity of light).

even for the 'hot' secondary beams is about 10 to 20 eV, it may be safely concluded that recombination times are one to three orders of magnitude larger than the corresponding cooling times. We can therefore state, finally, that all heavy ions may be cooled down within some ten or a few 100 ms and may then be stored for at least a few seconds or even some 10 s without significant losses. But, I have to insist, there is at present no experience in electron cooling of ions heavier than protons.

4.2 Laser Cooling

Laser cooling of circulating beams adopts the principle that was very successfully applied in cooling thermal atomic beams or ions in a rf trap (Phillips, 1983). Absorption of highly directional photons on the one hand, from a laser tuned to within the absorption energy of an atomic resonance, and isotropic re-emission of the light on the other hand, lead to a net momentum transfer. This may accomplish a velocity narrowing which proceeds even faster than electron cooling and to even smaller momentum spreads ($\Delta p/p < 10^{-7}$).

For efficient laser cooling, however, a device is needed that guarantees simultaneously longitudinal as well as transverse cooling, and atomic resonance-levels that decay promptly. For this reason, laser cooling of circulating ion beams has been seriously proposed only for 'ASTRID' and 'CRYRING', where modestly ionized ions are foreseen which may offer several ionic levels suitable for a fast, resonant excitation and a prompt decay. Whether laser cooling can also be applied to highly stripped heavy ions (ESR), where in principle only the 1s hyperfine splitting is available for resonant excitation (with rather large lifetimes of the order of milliseconds) is a question still under discussion.

4.3 'Phase-transition' of a Cooled Ion-beam

An observation made in Novosibirsk (Didansky, 1984) that the longitudinal momentum spread of cooled protons becomes - for values of $\Delta p_\parallel/p_\parallel$ of the order of 10^{-6} - independent of the number of stored particles, opened a highly productive discussion as to whether a phase transition of a cooled circulating beam into an ordered travelling 'crystal' might occur, whereby the positions of the ions in their rest frame become frozen. Extending the interpretation of the Russian authors who claimed to have seen a one-dimensional phase transition, J P Schiffer and P Kienle (Schiffer, 1985) have discussed the possibility of a phase transition of a heavy-ion beam into a three-dimensional lattice, a travelling 'Coulomb-solid'.

In a beam of particles, bent by electromagnetic fields into their orbits, there is a steady competition of a long-range regulating force, the Coulomb repulsion, on the one hand, and a disorder due to the chaotic temperature motion on the other. From simple thermodynamic considerations the decisive parameter for a phase transition is derived (Hansen, 1975), namely the ratio Γ of the Coulomb energy of an ion-pair to the temperature, with

$$\Gamma = \frac{Z^2e^2/a}{kT} \quad (a \propto N^{-1/3}) \tag{4}$$

where a is the mean distance between two ions, and N the particle density. A phase transition is predicted to occur at $\Gamma > 150$ (Hansen, 1975).

Inserting the more or less realistic assumption of 10^{11} U particles bunched to a 1 ns pulse (= 20 cm length) of 1 mm diameter, and assuming an ionic temperature of 0.1 eV, we get a mean distance of $a \sim 10^{-4}$ cm, and a value for Γ of ~ 140, near the critical value. Hence, this discussion may be more than academic. As clear as it is that this phase transition must occur under idealized conditions, it is unclear whether this ordering may be maintained during the deflection of the ions in the dipole magnets. An ordered ionic state might be detected, e.g. by observing coherently scattered light when using a suitable wavelength.

5. Sketch of the Experimental Programme at the ESR

The ESR is designed mainly for the physics of ions at very high phase-space density with both nuclear and ionic charge being large. Experiments on the most simple nuclear and atomic systems as well as on rather exotic species will become feasible at well-defined conditions and with the possibility of a unique precision. The following is necessarily a crude and incomplete outline of a few experiments currently under consideration and a minute exposition is not intended. They rather serve as representative examples showing the quite new and far-reaching perspectives of a heavy-ion cooler ring, and this presentation should, by no means, restrict your own power of imagination.

5.1 Nuclear Physics with Radioactive Beams

Nuclei far from stability may be produced very efficiently by fragmentation of high-energy ions. Utilizing the fragment separator which will combine an achromatic dipole system and a wedge-like degrader in the first focal plane, similarly to LISE at GANIL (Dufour, 1984), it should become possible to separate out fragments of single masses and to inject them into the ESR. There, they may be accumulated and cooled in order to finally obtain a highly brilliant radioactive beam. For those particles, circulating two million times a second, a mass determination can be performed by a steadily repeated Fourier analysis of their phase-space distribution ('Schottky-scan'). A mass resolution of the order of $\Delta M/M \sim 10^{-6}$ is anticipated.

Furthermore, the accumulated exotic nuclei of high phase-space density allows detailed study of nuclear structure far from stability, when colliding with internal atomic beams like H, D, ^3He, ^4He. For large mass ratios of projectile and target nucleus, and for not too large Q-values, the energy E_R of the recoiling nucleus is given by

$$E_R = A_p/A_R \cdot E \cos^2 \theta_R + 2Q \tag{5}$$

For the fixed recoil angle θ_R, the recoil energy is determined solely by the Q-value. Measuring E_R and, first of all, θ_R precisely (at best around $\theta_R = 90°$), a Q-value resolution of the order of 70 keV could be attained. If the radioactive beams can be polarized - and several schemes are under discussion - magnetic and quadrupole moments of both groundstate and (short-lived) excited states can be determined with standard techniques such as NMR or perturbed angular correlation.

Based on Bevalac data of Westfall (1979), first estimates of the production yield of exotic nuclei are now available (Sümmerer, 1986). Some of them are shown in Table 2. In particular, the rather high source strengths for proton-rich (e.g. ^{26}P, ^{47}Mn, ^{30}Cl) as well as neutron-rich (e.g. ^{72}Ni, ^{76}Ni, ^{44}S, ^{130}Cd) nuclei should be noticed.

Table 2
Estimated production rates of selected isotopes by projectile fragmentation (1 GeV/u, 1 g/cm^2 ^{12}C target)

Beam	Intensity (s^{-1})	Nuclide	Cross-section (μb)	Production Rate (s^{-1})
^{40}Ca	2×10^{12}	^{30}Cl	10	1×10^6
		^{26}P	3	3×10^5
		^{20}Mg	1	1×10^5
^{48}Ca	2×10^{12}	^{44}Ar	5000	5×10^8
		^{44}S	5	5×10^5
^{64}Zn	1×10^{12}	^{56}Cu	20	1×10^6
		^{47}Mn	1	5×10^4
^{76}Ge	1×10^{12}	^{68}Ni	1000	5×10^7
		^{72}Ni	8	4×10^5
^{86}Kr	5×10^{11}	^{76}Ni	$3 \cdot 10^{-3}$	1×10^2
^{136}Xe	5×10^{11}	^{130}Cd	0.1	3×10^3

5.2 Bound β^--decay: The Weak Interaction and Astrophysics

There is a still undiscovered mode of β^--decay, which must have played a key role during nucleosynthesis: the β^--decay into bound final electron states. This decay is the time-mirrored electron-capture (EC) process. For EC with a small Q-value (e.g. if only electron capture from L or higher shells occurs) a bound β-decay into the K-shell is allowed in principle, but in a neutral atom there is no vacant state. If, however, the atom is highly ionized - as during nucleosynthesis at temperatures from ~ 5 keV to 80 keV - there is a high probability that this bound

β-decay may happen. Hence, stable neutral atoms can become unstable if their electrons are removed.

Furthermore, an allowed but slow β-decay can be tremendously enhanced in a highly ionized atom. This is due to the fact that a bound β-decay may allow the transition to an excited state of the daughter nucleus with a ΔI^{π} selection rule which is more favourable for this transition than for the 'normal' transition of a neutral atom.

Most striking, perhaps, is the nucleus ^{187}Re, where the lifetime is supposed to change from 4×10^{10} y to a few years (Takahasi, 1984) if a bound β-decay to the first excited state of ^{187}Os (E = 10 keV) is allowed in a completely ionized atom. Now, the present ratio of ^{187}Re to ^{187}Os is one out of only two or three 'clocks' available for estimating the age of our galaxy (Luck, 1984). And, certainly, this present Re/Os ratio must have been significantly influenced by those enhanced bound β-decays during nucleosynthesis.

Heavy-ion storage rings, and especially the ESR, will offer the very first opportunity to recover those aspects of nucleosynthesis on the earth by searching for bound β-decay of stored and stripped heavy ions. Three nuclei seem to be most suitable for this: ^{163}Dy, ^{205}Tl (important as a neutrino 'detector') and ^{187}Re. Lifetime estimates for them are between 0.1 y (Dy) and 6 y (Re) (Takahasi, 1984), yielding some 500 to 30000 decays/s in the ESR at 10^{11} particles stored.

Measuring the number of daughter nuclei as a function of the storage time immediately gives the sought-after lifetime of bound β-decay. However, since the daughter nuclei have nearly exactly the same q/A ratio as the mother nuclei, a direct separation of the former in the ring itself seems to be impossible. Therefore, other methods for their detection have to be applied - e.g. resonant hyperfine excitation, or detection of characteristic K x-rays, or implantation followed by resonant multi-step photoionization. This is still a matter of discussion.

5.3 Hyperfine Interaction

The ESR is uniquely suitable for studies of the 1s hyperfine splitting in hydrogen-like heavy atoms for several reasons. Firstly, hydrogen-like atoms can be delivered in abundance as circulating beams without any other 'noisy' species of atom. Secondly, 1s hyperfine transitions might be induced - despite the rather small splitting and the long lifetime - by resonant collinear laser-excitation, due to the huge shift of a wavelength λ_0(lab) in a fast moving frame (λ):

$$\lambda = \lambda_0 \cdot (1 + \beta \cos\alpha)/(1 - \beta^2)^{1/2} \tag{6}$$

For atoms of 620 MeV/u ($\beta = 0.80$), for example, a wavelength λ_0 of 600 nm will be shifted in the moving frame to 1800 nm and 200 nm, for photons travelling parallel ($\alpha = 0°$) and antiparallel ($\alpha = 180°$), respectively. For this reason, within a broad range of nuclei ($50 < Z < 92$) hyperfine transitions can be induced resonantly by powerful laser devices. Furthermore, if the ions are cooled to $\Delta p/p \sim 10^{-5}$, the resolution will be limited only by the typical resolution of the fluorescence-light detector (Fabry-Perot) of $\Delta\lambda/\lambda \sim 2 \times 10^{-5}$. Expected rates for this fluorescence light are of the order of 1 s^{-1} (for a solid angle of $\Omega = 10^{-6}$).

There is even speculation that, due to a large cross-section σ_{ex} for electron spin exchange in the cooler section ($\sigma_{ex} \sim 10^{-15}$ cm^2), the upper 1s hyperfine level might always be saturated. In this case, laser excitation would not be needed at all for observing a sufficient number of hyperfine transitions. However, the presumed spin-exchange cross section σ_{ex} is extrapolated from light atoms and has never been proven for heavy atoms.

Circulating cooled beams with a groundstate spin I = 0 also offer a long sought-after possibility to make precise measurements of the g-factor of a strongly bound 1s electron as a function of Z. Applying a transverse magnetic field one observes resonant Zeeman transitions induced by microwaves. In this case the special properties of wavelength transformations in a fast moving frame could be helpful in improving the precision significantly. An accurate determination of the bound electron g-factor would be fundamental within the framework of quantum electrodynamics.

5.4 Atomic Physics

Probably the main attraction of the ESR lies in the field of atomic physics, because the abilities of the ESR fit extremely well to those needed for several 'ideal' experiments in atomic physics. All hydrogen-like atoms, i.e. the most simple and, at the same time, the most fundamental systems will be provided. They are, furthermore, at well-controlled and precisely defined conditions, e.g. their velocity spread is nearly equal to that of ions trapped in a rf cavity. Finally, very specific tools for their special preparation (polarized internal targets) or for their resonant excitation and de-excitation (collinear lasers, cooling electrons) are available. Thus, the experimental programme in the field of atomic physics may have an extremely wide scope which I can present here, at best, in very broad outline.

5.4.1 Precision spectroscopy

Radiative electron capture (REC) may serve as a sensitive diagnostic of the cooling process, because the width of a specific REC transition is determined mainly by the temperature of the ion beam. In the case of a well-cooled, fully stripped ion beam, on the other hand, an observation of the REC transition into the K-shell delivers the 1s binding energy of a one-electron atom with precision, because all the usual line-broadening effects caused by a finite lifetime are absent. It is anticipated that the 1s binding energy of a heavy hydrogen-like atom could be measured up to a relative accuracy of 10^{-6} (for ion-beam temperatures of ~ 0.1 eV) in this way.

Similarly, the binding energy of any atomic state in helium-like or lithium-like atoms might be set accurately using resonant di-electronic recombination (DR). This process (the time-mirrored Auger-effect) happens whenever a free electron is captured and simultaneously a bound electron gets excited to a higher-lying bound state by balancing exactly the transition energy of the former. For this purpose a monoenergetic electron beam has to be tuned to these resonances. In this way, each bound state of any few-electron atom may be populated resonantly. The spectroscopic precision finally achieved, however, depends crucially on how well the heavy-ion cooling works.

'Classical' x-ray spectroscopy of one- or few-electron atoms may also be brought to a precision hitherto unknown by combining the properties of a cooled ESR ion-beam and a (polarized) internal target, the former delivering monoenergetic bare nuclei, the latter bound (polarized) electrons. This spectroscopy can address the importance of higher order terms $\sim (\alpha Z)^n$ in the Lamb shift which are only at the 10^{-6} level in hydrogen but dominate in hydrogen-like uranium.

The efficient production of high-charge low-energy recoil ions (E < 10 eV) in an internal gas target offers a quite different approach to precision spectroscopy. Estimated cross sections for producing bare recoil ions are of the order of 10^{-19} cm^2 to 10^{-22} cm^2 for 200 MeV/u U^{92+} impinging on neutral Ar and Xe, respectively. With 10^{10} U ions stored in the ring and a target thickness of 10^{12} cm^{-2}, 10^9 Ar^{18+} and 10^6 Xe^{54+} s^{-1} may be produced.

5.4.2 Resonance processes - fundamental questions

The combination of a high-velocity, brilliant ion beam and of powerful devices for the selective preparation of (polarized) atomic states can bring back to the stage some old-fashioned fundamental problems for their much more accurate re-examination: the velocity of light emitted from a fast moving source and the transverse Doppler-effect. For example, as discussed by Poulsen (1981), a two-photon excitation can be tuned to a real intermediate level by absorbing one (laser) photon travelling parallel (red shifted) and another one antiparallel (blue shifted). According to the huge shift of the wavelength in a fast-moving frame (cf. eq. 6), a wide range of relative spacings in a three-level system can be accessed depending on the size of β. If, vice versa, these energies are precisely known and if the velocity spread of the moving frame (ion beam) is small ($\Delta v/v \sim 10^{-5}$), then the wavelength transformation predicted by eq. 6 (in other words, the Lorentz transformation) may be checked with an accuracy at least one order of magnitude better than the present state of the art. If two photons are re-emitted, on the other hand, the velocity of light, emitted parallel and antiparallel with respect to the fast moving source, can also be compared accurately using a suitable detection arrangement.

There is speculation (Winnacker, 1984) about the possibility of enhancing radiative capture processes from a collinear electron beam with laser light and thus populating highly selectively Rydberg states of highly ionized atoms. As well as its spectroscopic impact, this method could provide a powerful source for monochromatic radiation in the UV or even x-ray regime. Since, furthermore, high-lying states of a few-electron ion denote, by definition, a population inversion, even amplification of this radiation might be achieved in the case of densely bunched ion beams.

Summary

Heavy-ion storage-cooler rings, and amongst them not least the ESR, are without doubt, powerful tools for doing quite new, exciting and hitherto inaccessible experiments which address fundamental questions in many fields of physics. Their feasibility in most of the experiments discussed above depends decisively, however, on the success of enhancing significantly the phase-space density of those stored beams by electron cooling.

Acknowledgement

It is a pleasure for me to thank all my colleagues for many fruitful discussions. I am particularly indebted to P Kienle, who first inspired the concept of the ESR as described in this contribution.

References

Budker G I 1967 Atomic Energy 22 346
Channel P J 1981 J. Appl. Phys. 52 3791
Debrenev Y S et al 1978 Part. Accelerators 8 235
Didansky N S et al 1984 in: ECOOL (KFK Karlsruhe) 275 ed. H Poth
Dufour J P et al 1984 in: 7th High Energy Heavy Ion Study (GSI Darmstadt)
Gould H et al 1984 Phys. Rev. Lett. 52 180
Hansen J P 1975 Phys. Rev. A11 1025
Herlander C J 1984 in: Workshop on the physics with heavy ion cooler rings, B4, Heidelberg
Javanainen J et al 1985 J. Opt. Soc. Am. B2
Keller R et al 1983 in: Proc. Int. Ion Engineering Congress, Kyoto (Japan)
Luck J M et al 1980 Nature 283 256
Lynen U 1984 in: Workshop on the physics with heavy ion cooler rings, D13, Heidelberg
Parkhomchuk V V 1984 in: ECOOL (KFK Karlsruhe) 71 ed. H Poth
Phillips W D et al 1983 Atomic Physics 9, 338 (eds. Van Dyck, Fortson)
Piwinski A 1974 Conf. on High-Energy Accelerators, Stanford 405
Poth H 1984 in: Workshop on the physics with heavy ion cooler rings, A4, Heidelberg
Poulsen O et al 1981 Phys. Rev. Lett. 47 1522
Schiffer J P and Kienle P 1985 Z. Phys. A321 181
Stobbe M 1930 Ann. Phys. 7 661
Sümmerer K 1986 private communication
Takahasi K and Yokoi K 1983 Nucl. Phys. A404 578
Westfall G D et al 1976 Phys. Rev. Lett. 37 1202
Winnacker A 1984 in: Workshop on the physics with heavy ion cooler rings, C9, Heidelberg

Inst. Phys. Conf. Ser. No. 86
Paper presented at Int. Nucl. Phys. Conf., Harrogate, UK, 1986

List of Oral Contributions

New Experimental Techniques and Nuclear Sizes: Oral Contributions
(In order of presentation: the number in square brackets refers to the contribution in Vol. 1 of the Proceedings)

Ultrasensitive laser spectroscopy and nuclear charge radii of radioactive strontium isotopes
D A Eastham[1], P M Walker[1], J R H Smith[1], J A R Griffith[2], D E Evans[2], S A Wells[2], M J Fawcett[3] and I S Grant[3]
[1]SERC Daresbury Laboratory; [2]University of Birmingham; [3]University of Manchester
[A18]

Nuclear charge radii of neutron-deficient short-lived gold-isotopes
K Wallmeroth, G Bollen, M J G Borge, J Campos, A Dohn, P Egelhof, J Grüner, H-J Kluge, U Krönert, R B Moore, A Rodriguez, A Venugopalan, J Wood and the ISOLDE Collaboration
Mainz, CERN, Madrid, Montreal, Atlanta, Bombay Co-operation
[C145]

Recent advances in the nuclear orientation of nuclei far from stability
L Vanneste, N Severijns, D Vandeplassche, E van Walle, J Wouters
Leuven University
[B146]

List of Oral Contributions

(In order of presentation: the number in square brackets refers to the contribution in Vol. 1 of the Proceedings)

Axial charge transitions in relativistic nuclear models and nonrelativistic meson exchange currents
I S Towner
Chalk River Nuclear Laboratories
[D6]

The e.m. polarizability and the swelling of nucleons in the nucleus
M Ericson[1] and M Rosa-Clot[2]
[1]IPN Lyon and CERN; [2]University of Pisa
[E5]

Measuring the antiquark distribution function in nuclei
G T Garvey
Los Alamos National Laboratory
[Post-deadline contribution]

Isobar and quark degrees of freedom in the deuteron
E L Lomon, P Blunden and P Sitarski
Massachusetts Inst. Technology
[E9]

Nuclear binding effects in inelastic electron scattering
P J Mulders
NIKHEF-K, Amsterdam
[E14]

Inst. Phys. Conf. Ser. No. 86
Paper presented at Int. Nucl. Phys. Conf., Harrogate, UK, 1986

Pion and photon production in medium energy heavy-ion reactions

Johanna Stachel

Physics Department, SUNY Stony Brook, NY 11794, USA

1. Introduction

While in nucleon-nucleon collisions pion production is limited to projectile energies larger than 290 MeV, in nucleus-nucleus collisions this process has been observed at substantially lower energies per nucleon. The threshold there is defined by the requirement that the centre-of-mass energy still exceed the pion rest mass of e.g. 135 MeV for a π°. For $^{12}C + ^{12}C$ collisions this would correspond to a beam energy in the laboratory of 22.5 MeV/nucleon. We will report and summarize here results on production close to and actually reaching this phase-space limit.

Since the pion production cross sections in this energy regime are very small - of the order of nb - a detection method with a large angular and kinetic energy acceptance is required. We therefore concentrated on the detection of neutral pions where, to some extent at the expense of energy resolution, solid angles of $\geqslant 1$ sr can be achieved. On the other hand ultimate pion kinetic energy resolution is not necessary for the heavy systems studied where final-state level densities are high and the pion spectra structureless. The emphasis in the experiments presented here is in studying reactions where a major fraction or all of the projectile's kinetic energy is transformed in the creation of one new, energetic particle. The expectation is that collective or coherent effects in these medium energy heavy-ion collisions, if they exist, should be most visible in such processes.

As compared to the observation of nucleons or nuclear fragments the advantage is that complications in the interpretation due to knock-out or spectator contributions etc. are avoided. On the other hand it would be desirable to study a probe that is not, or only weakly, reinteracting in the surrounding nuclear medium. As will be outlined below, in the energy range of interest here, the pion scattering cross section is small (Nakai 1980), while the pion absorption levels off at cross sections corresponding to a mean free path of 2-3 fm and is, consequently, not negligible for heavy target nuclei (and projectiles). We, therefore, extended our investigations to the observation of inclusive photon production in the same reactions and at gamma energies $\geqslant 35$ MeV.

2. Experiments

2.1 Pion production

The detection method used for neutral pions is to detect the two decay gamma rays in coincidence in a Pb-glass spectrometer. Because of the high background suppression requirements arrangements are chosen which consist of short (1-2 radiation lengths) converter blocks backed by deep (~ 15 radiation lengths) absorber blocks. Two such set-ups have been used in medium energy heavy-ion reactions, details can be found in Stachel (1986), Noll (1984).

We have studied inclusive pion production in reactions of 35 MeV/nucleon ^{14}N + Al,Ni,W and 25 MeV/nucleon ^{16}O + Al,Ni at the MSU K = 500 superconducting cyclotron and the ORNL HHIR facility, respectively. Data on the reactions 22 MeV/nucleon ^{32}S + Al,Ni are presently being analyzed.

The response of the spectrometer, consisting of 20 Pb-glass telescopes, has been studied with tagged photons in the energy range of interest and both the acceptance of the spectrometer and distortions of the various differential cross sections due to cuts necessary in the analysis to suppress background have been modelled in detailed Monte-Carlo simulations and all the data shown are corrected for such effects (for details see again Stachel (1986)). Pions are identified by their invariant mass (see Fig. 1). The beam correlated and cosmic-ray background that has to be subtracted is of the order of 10 and 1 nb at 35 and 25 MeV/nucleon, respectively. It is evident that, with this detection technique, inclusive studies that presently reach the 1 nb level for pion production cannot be extended to much smaller limits.

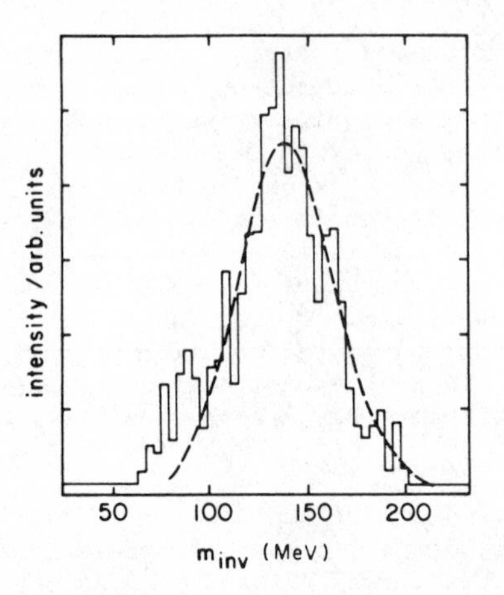

Fig. 1 Experimental pion invariant mass spectrum for 35 MeV/nucleon ^{14}N + Ni. Also shown by the dashed line is the result of a Monte-Carlo simulation.

2.2 Photon production

To detect high-energy gamma rays produced in the same nuclear reactions an array consisting of the large Pb-glass absorber blocks (see above) has been used (Alamanos 1986). Since the photons, unlike pions, cannot be identified by an invariant mass the precise knowledge of the response of the detection system is more important. The energy response of our system was measured to be linear between 20 and 120 MeV and, more importantly, the resolution for a given gamma energy was found to be determined by the photoelectron statistics and it can be described by a Poisson distribution over 2-3 orders of magnitude (the resolution at E_γ = 100 MeV is 30 MeV (FWHM)). Nevertheless, it should be emphasized that the strength of this detection system is at high gamma energies (E ⩾ 50-100 MeV) where shower statistics is improving and other techniques fail increasingly. The low energy part of the γ-spectrum contains contributions from statistical γ-emission (e.g. from a compound nucleus). These production cross sections are known and their presence requires thresholds in the 'high-energy' photon spectra that increase with decreasing beam energy (typically, 35 MeV at 35 MeV/nucleon and 45-50 MeV at 25 MeV/nucleon). Both the effects of finite energy resolution and sideways leakage of the electromagnetic showers have been modelled with detailed Monte-Carlo simulations and can be corrected for. Another advantage of this detection method lies in the good time resolution (FWHM ~ 1.4 ns) that proves to be essential for background suppression.

3. Experimental Results

3.1 Pion production

Only the main features of the data will be summarized here. For details see Stachel (1986).

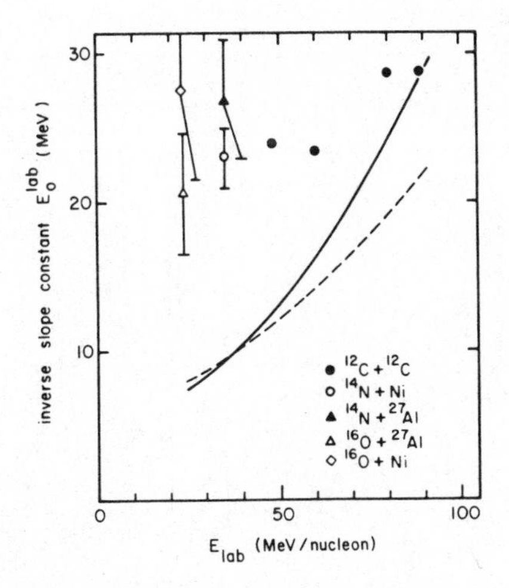

Fig. 2 Experimentally determined slope constants E_0 of pion kinetic energy spectra plotted as a function of the beam energy per nucleon (for the C + C data see Noll (1984)). The solid line corresponds to a thermal model calculation (Prakash 1986).

(i) Pion kinetic energy spectra: The spectra are peaked at low kinetic energies (T_π = 10-15 MeV) and decay exponentially towards higher T_π. It is remarkable that the decreasing trend in the inverse slope constants E_0 that has been observed (Noll 1984) for beam energies decreasing from 84-60 MeV does not continue at lower beam energies. Instead, the slope constants level off at, within the resolution, a constant value of E_0 = 23 MeV (see Fig. 2). Probably, this indicates a change in pion production mechanism at low projectile energies. In the reaction 25 MeV/nucleon ^{16}O + Al the creation of a pion at rest requires 54% of the total cm energy. Pions have been observed in the kinetic energy bin 70-110 MeV which, within the energy resolution, reaches the phase-space limit where all cm energy is transferred to one pion.

(ii) We find angular distributions $d\sigma/d\Omega_{lab}$ (Fig. 3) that are forward peaked for the Al target and forward-backward symmetric with a minimum at 90° for the Ni target. At higher beam energies (60-84 MeV/nucleon) angular distributions forward-backward symmetric in the cm frame with minima at 90° have been observed (Noll 1984) with no notable dependence on the target mass.

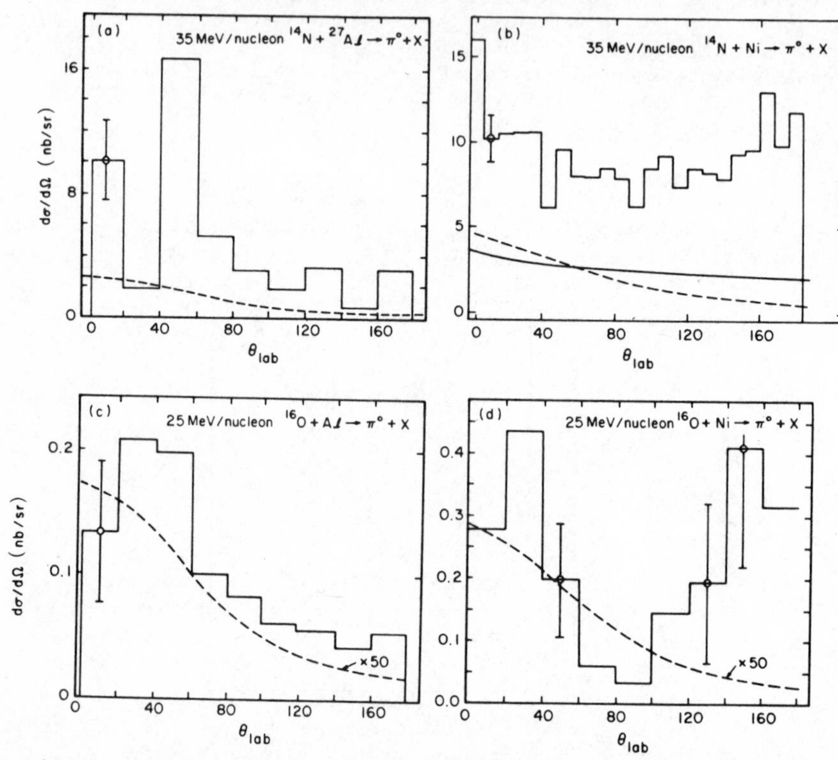

Fig. 3 Experimental pion angular distributions (solid line, see Fig. 2).

(iii) The rapidity distributions $d\sigma/dy$ have centroids corresponding to velocities clearly slower than the nucleon-nucleon cm system. Within the uncertainties, we find $\langle y \rangle \sim 0$ for all systems. In Noll (1984) values intermediate between the expected numbers for the nucleon-nucleon cm and the nucleus-nucleus cm have been observed.

(iv) The distributions of pion momentum perpendicular to the beam direction $d\sigma/dp_\perp$ are compatible in shape with thermal distributions as e.g. prescribed in the Hagedorn model (Hagedorn 1965). The same is found at higher beam energies (Noll 1984). The corresponding temperatures are in general lower by a few MeV than what would be consistent with the kinetic energy spectra (about 4.5 MeV at 35 MeV/nucleon ^{14}N + Ni).

(v) The target mass dependence has been found (Noll 1984) to be consistent with $\sigma \propto A_T^{2/3}$. We observe a pronounced flattening for heavier targets at 35 MeV/nucleon. This trend is already indicated at higher beam energies and increases with decreasing projectile energy.

As outlined in detail in Stachel (1986) the beam energy and A_T dependence of the integrated cross sections and the shape of the angular distributions as well as the centroids and shapes of the rapidity distributions can be understood to arise from a combination of stopping of the projectile and pion re-absorption. From measured pion absorption cross sections (Nakai 1980) a mean free path of the order of 2-3 fm is extracted. Hence, for heavy projectiles and targets re-absorption can certainly not be neglected and the primary pion production cross sections are estimated to lie factors of 2 to 5 above the measured secondary values for Ni and U targets. The effects of pion rescattering are expected to be more complicated. From the energy dependence of measured pion scattering cross sections (Nakai 1980) we estimate these effects to be small below T_π = 100 MeV.

3.2 Photon production

The first data on high-energy photon production in medium-energy heavy-ion collisions have been published by E. Grosse and collaborators (Grosse 1984), who used the same detection system as for neutral pions. Cross sections at E_γ = 100-150 MeV are found to be comparable to those for π° production in the same reactions and at low T_π. The slope constants of the exponential energy spectra are 5 MeV larger than in the corresponding π° spectra.

We find (Alamanos 1986) for the reaction 35 MeV/nucleon $^{14}N + Ni \rightarrow \gamma + X$ a photon production cross section of 300 ± 75 µb for $E_\gamma \geqslant$ 35 MeV and associated photon spectra with slope constants of 14.5 ± 1.0 MeV. The angular distribution is shown in Fig. 4. It has a pronounced maximum around 90° and is not consistent with an isotropic angular distribution in the cm frame, but rather it requires a dipole-like contribution. We are presently analyzing data from reactions of 25 MeV/nucleon ^{16}O, 22 MeV/nucleon ^{32}S and 16 MeV/nucleon ^{58}Ni on targets of Al, Ni, Cd and Au, each with about 10 different angles for gamma detection between 17 and 168°. The finding of a strong dipolar component in the angular distributions persists and is found also to be invariant to the beam intensity and the target-detector distance.

An increasing dipole component with decreasing beam energy is found in Grosse (1984). A comparable dipole component of 30% is also found in the recent studies of Bertholet (1986), although in the latter case reactions involving a much heavier projectile (Kr + C, Ag, Au) have been studied and it is not clear to what extent the results are comparable. Another study of a heavy system (Hingmann 1986), 44 MeV/nucleon $^{40}Ar + ^{158}Gd$, using BaF_2 for gamma detection finds a photon production cross section of 2.0 mb for E $\geqslant$ 30 MeV and a slope constant E_0 = 12.6 MeV. For that system so far no information about the angular distribution is available. Stevenson et al (1986) use a CsI/Lucite combination for photon detection. While their data should be comparable to our systems, the corresponding angular distributions show a strong forward peaking in the laboratory system and are claimed (Stevenson et al 1986) to be consistent with isotropic emission in the source's rest frame. The origin of these discrepancies is not settled at the time being.

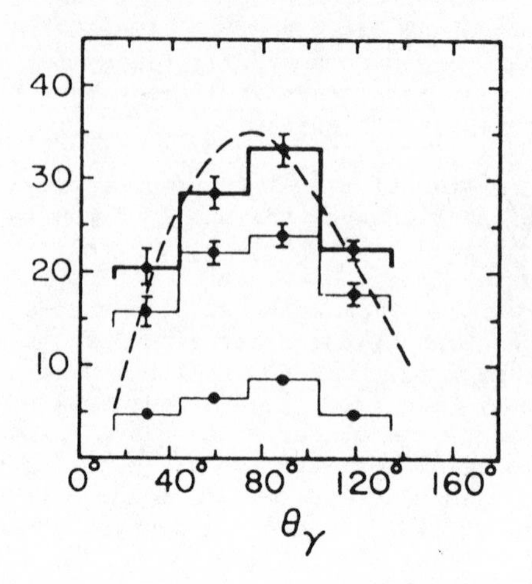

Fig. 4 Single photon angular distribution in the lab. frame for 35 MeV/nucleon ^{14}N + Ni at E_γ = 35-155 MeV (top) and E_γ = 35-55 and 55-155 MeV (two lower histograms). The errors shown here are statistical. The systematic uncertainties are ≈ 25%.

4. Theoretical Interpretation

Various approaches have been used to understand inclusive pion production in the reactions described above. They range from hard scattering of individual target and projectile nucleons to coherent nucleus-nucleus bremsstrahlung. Some understanding has been achieved, but not at a basic level. The reader is referred to the discussion in Stachel (1986) and the references given there. Here only the results of a thermal model (Prakash 1986) will be discussed briefly in order to show where such an approach works and where the limits are, beyond which a treatment in a coherent or collective model is needed.

The thermal model (Prakash 1986) that we have chosen is an extension of an ansatz proposed by Aichelin and Bertsch (1984). The geometric overlap region between target and projectile forms a 'hot' system of impact parameter dependent size. The thermal characteristics are calculated within the Fermi gas model and the decay is treated as in the compound nucleus model. In addition to statistical factors and level densities, the inverse cross sections, i.e. the cross sections for pion absorption on the respective nuclei, enter. These are taken from experiment for $T_\pi \geq 37$ MeV and extended towards $T_\pi = 0$ using an optical potential. For details see Prakash (1986). This model is based on the assumption of local equilibrium in the overlap region but, other than that, has no free parameters. In particular, the assumption of chemical equilibrium is not needed. The resulting cross sections for the various systems and $E_{lab}/A \leq 100$ MeV are in agreement with the experimental data within a factor of two. As can be seen from Fig. 2, the slope constants from such a calculation are, for lower beam energies, systematically too low. In fact, the data are described quantitatively for pion production consuming less than 50% of the total cm energy while these thermal model calculations fail increasingly in the rare reactions where the majority of the projectile energy is transferred to one pion. We expect that this is where a collective approach is needed.

Figure 5 visualizes how, as the end of phase space is approached, all different data merge and scale with the energy available in the cm system (as they should). On the other hand, one can see that at higher energies different scaling applies, the energy of one or a few nucleons there being the relevant quantity. This is where one also expects an $A_T^{2/3}$ scaling to apply. It would be desirable to achieve a microscopic and yet quantitative understanding of the collective features giving rise to pion production close to pionic fusion and in heavy systems. Possibly an approach based on coherent Δ-hole excitations proposed for peripheral collisions at higher beam energies (Prakash 1985) could be extended to describe such cases.

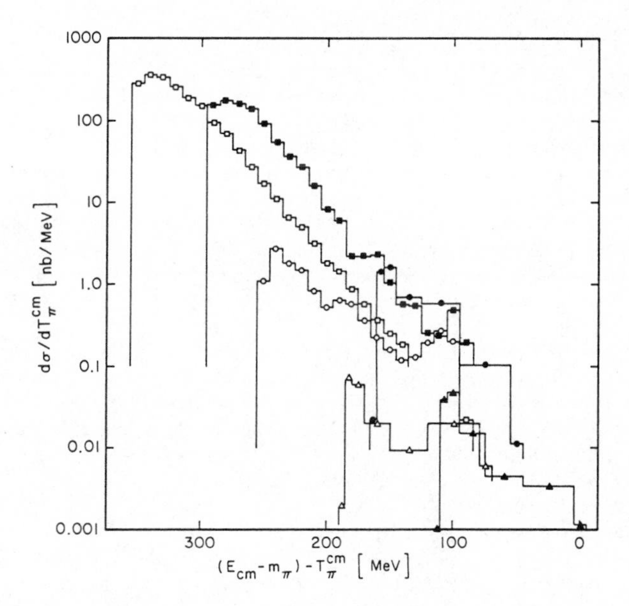

Fig. 5 Pion kinetic energy spectra as a function of the energy available in the cm system after pion emission. Experimental values: 25 MeV/nucleon O + Al, Ni (closed and open triangles, (Young 1986)), 35 MeV/nucleon N + Al, Ni (closed and open dots, (Stachel 1986)), 74 and 84 MeV/nucleon C + C (closed and open squares, (Noll 1984)).

The observed cross sections and energy spectra for inclusive photon production are reproduced quite well with the same thermal model (see Alamanos (1986)), the only difference being that experimental photon absorption cross sections have to be used there. A similar approach is used in Neuhauser (1986). On the other hand, models invoking incoherent nucleon-nucleon bremsstrahlung (Nifenecker 1985) seem to give comparable results. It should be noted that both types of calculation so far cannot account for a strong dipolar shape of the angular distributions. Although relatively crude, an approach of coherent nucleus-nucleus bremsstrahlung (Vasak 1986) indicates how such dipolar angular distributions might be obtained. It is, however, not clear whether such a conclusion persists under a more elaborate treatment of the slowing down process.

The results presented here have been obtained in collaborations with N Alamanos, T C Awes, P Braun-Munzinger, R L Ferguson, D Fox, R F Freifelder, F E Obershain, P Paul, F Plasil, M Prakash, R Ronningen, S Sen, P DeYoung, G R Young, L Waters and P H Zhang. Financial support of the NSF and the US DOE (contract DEAC05-840R21400 with Martin Marietta Energy Systems) is gratefully acknowledged.

References

Aichelin J and Bertsch G F 1984 Phys. Lett. B138 350; Aichelin J 1984 Phys. Rev. Lett. 52 2340.
Alamanos N et al 1986 Phys. Lett. B173 392.
Bertholet R et al 1986 Nouvelles de GANIL 16 juin.
Grosse E 1984 in Fundamental Problems in Heavy Ion Collisions, ed. N Cindro, W Greiner and R Caplar World Scientific 347; Grosse E et al 1986 Europhysics Lett. 1 9.
Hagedorn R 1965 Nuovo Cim. Suppl. III 147.
Hingmann R et al 1986 contribution to this conference and Nouvelles de GANIL in print.
Nakai K et al 1980 Phys. Rev. Lett. 44 1446; Ashery D et al 1980 Nucl. Phys. A335 385; Navon I et al 1982 Phys. Rev. C28 2548.
Neuhauser D and Koonin S E 1986 Nucl. Phys. in print.
Nifenecker H and Bondorf J P 1985 Nucl. Phys. A442 478; Bauer W et al Phys. Rev. C, in print.
Noll H et al 1984 Phys. Rev. Lett. 52 1284.
Prakash M, Braun-Munzinger P and Stachel J 1986 Phys. Rev. C33 937.
Prakash M, Guet C and Brown G E 1985 Nucl. Phys. A447 625c.
Stachel J et al 1986 Phys. Rev. C33 1420; Braun-Munzinger P et al 1984 Phys. Rev. Lett. 52 255
Stevenson J et al 1986 Phys. Rev. Lett. 57 555.
Vasak D 1986 Phys. Lett. B176 276.
Young G R et al 1986 Phys. Rev. C33 742.

Inst. Phys. Conf. Ser. No. 86
Paper presented at Int. Nucl. Phys. Conf., Harrogate, UK, 1986

List of Oral Contributions

(In order of presentation: the number in square brackets refers to the contribution in Vol. 1 of the Proceedings)

π^o production near threshold with monochromatic photons
R Beck[1], G Breitbach[2], G Koch[2], C McGeorge[3], V Metag[2], R Owens[3], G Schäfer[1], B Schoch[1], H Ströher[2], T Tschesche[2] and P Wallace[3]
[1]University of Mainz; [2]University of Giessen; [3]University of Glasgow
[E29]

Photoexcitation of low-lying collective dipole states in ^{160,162,164}Dy
C Wesselborg[1], U E P Berg[2], P Von Brentano[1], B Fischer[3], R D Heil[2], H Hollick[3], U Kneissl[2], D Kollewe[3], S Lindenstruth[2], H H Pitz[2], U Seemann[2], R Stock[2] and K O Zell[1]
[1]University of Cologne; [2]University of Giessen; [3]University of Stuttgart
[B62]

Tagged photon measurements of correlated neutron-proton pairs from the ^{12}C(γ,np) reaction
I Anthony[1], J R M Annand[1], G I Crawford[1], S N Dancer[1], S J Hall[1], J D Kellie[1], J C McGeorge[1], I J D MacGregor[1], R O Owens[1], P Wallace[1], D Branford[2], A C Shotter[2], S Springham[2], B Schoch[3], J Vogt[3], R Beck[3] and G Liesenfeld[3]
[1]University of Glasgow; [2]University of Edinburgh; [3]Johannes-Gutenberg University, Mainz
[C148]

Inst. Phys. Conf. Ser. No. 86
Paper presented at Int. Nucl. Phys. Conf., Harrogate, UK, 1986

On the narrow positron lines observed in heavy ion collisions

D.Schwalm

Physikalisches Institut der Universität Heidelberg, Philosophenweg 12, D6900 Heidelberg

Abstract: A summary of the status of the experiments concerning the observation of peak structures in the positron energy spectra from heavy-ion collisions is given.

1. Introduction

The experiments on positron production in collisions between very heavy ions (Backe et al. 1978, Kozhuharov et al. 1979, Schweppe et al. 1983, Clemente et al. 1984, Cowan et al. 1985, Tsertos et al. 1985, Cowan et al. 1986a) have been motivated by the expectation that spontaneous positrons (for a list of references relevant to the phenomena see Greiner 1983) should be observable under favorable conditions (Rafelski et al. 1978, Reinhardt et al. 1981) as peak-like structures in the positron energy spectra in addition to the continuous contributions from dynamic atomic processes and from internal pair conversion decays of excited nuclear states. In fact, peak structures have been observed in the e^+energy spectra of the heaviest collision systems (Schweppe et al. 1983,Clemente et al. 1984, Tsertos et al. 1985) but followup systematic studies (Cowan et al. 1985, Bokemeyer et al. 1985, Bosch et al. 1986, König et al. 1986) have shown that a connection of these lines with the process of spontaneous positron production is highly unlikely; moreover, nuclear conversion processes seem to be an unlikely cause of these lines. To search for further clues regarding the origin of the structures, electron-positron coincidence measurements have been performed recently (Cowan et al. 1986a). The exciting observation that the e^+ peak structures seem to be correlated with electrons of similar energy has prompted considerable experimental and theoretical activities, but so far no completely consistent picture has been found to account for all of the observed properties. In the following summary I will therefore present mainly the available experimental data concerning this puzzle. A more comprehensive summary of the earlier experiments and theory can be found e.g.in the proceedings of the Lahnstein meeting (Greiner et al. 1983) and an up-to-date collection of the experimental and theoretical work will be contained in the proceedings of the 1986 Summer School at Maratea(Greiner 1986).

2. The data

The main contributions to the positron spectra observed in heavy-ion collisions at bombarding energies around the Coulomb barrier seem to be well understood: Positrons from the internal pair conversion decay of excited nuclei give rise to so-called nuclear positrons, which dominate in lighter systems ($Z_u=Z_1+Z_2\leq160$); they can be calculated in a semi-empirical way from the corresponding γ spectra (Meyerhof et al. 1977, Backe et al. 1978). In heavier systems, positrons from pairs produced by the time dependent Coulomb field of the two nuclear charges (so-called dynamic atomic positrons) are getting increasingly important (see U. Müller et al. 1983 and references therein), and in the heaviest collision system measured so far ≈ 75% of the observed positron yield is due to this contribution. Both processes yield smooth, bell-shaped energy spectra with maxima around 500 keV; the gross properties of the measured shapes, yields, and impact-parameter dependences are well accounted for by these two contributions (see e.g.Schwalm 1984).

Comprehensive experimental studies of differential properties of the positron energy spectra, however, have revealed additional structure. These investigations have been mainly pursued by two groups working at the UNILAC accelerator of the Gesellschaft für Schwerionenforschung, GSI, at Darmstadt, namely the EPOS collaboration (GSI - Yale -Frankfurt - Heidelberg - Mainz - Brookhaven; present members: H. Backe, K. Bethge, H.Bokemeyer, T. Cowan, H. Folger, J.S. Greenberg, K. Sakaguchi, P. Salabura, D. Schwalm, J. Schweppe, K. Stiebing, P. Vincent) and the ORANGE collaboration (GSI - Munich;present members: E. Berdermann, F. Bosch, M. Franosch, S. Huchler, P. Kienle, W. König, C. Kozhuharov, H. Tsertos, W. Wagner). The set-up used by the EPOS group to collect the data shown in figs. 1 and 2 is schematically displayed in the upper panel of fig. 1 (see also Bokemeyer et al. 1983a, Greenberg and Vincent 1985); in all collision systems seen so far in greater detail prominent line-like structures have been observed (The data shown in figs. 1,2 are from Schweppe et al. 1983, Cowan et al. 1985 and Bokemeyer et al. 1985). The set-up used by the ORANGE group (see also Kienle 1986) in their investigation of the U+Th and U+U systems (Clemente et al. 1984, Tsertos et al. 1985) is shown schematically in the upper panel of fig. 3 ;as displayed in the lower panel at fig. 3, prominent peaks are observed in these experiments as well.

Attempts to explain these structures by internal pair conversion (ICP) processes in one of the collision products have been unsuccessful mainly because of the following experimental findings: (i) Identifying the e^+ peak with IPC processes requires the presence of corresponding structure in the γ-spectra and electron spectra which is not observed. This is exemplified in fig. 4 for the EPOS data on the U+Cm system (Bokemeyer et al. 1983b) using calculated branching ratios (see Schlüter et al. 1983 and references therein). Similar analyses - leading to the same conclusions - have been performed by both the EPOS and the ORANGE group for all other systems where peak structures have been observed. (ii) The width of the lines are too narrow when compared to the triangular shape of the e^+ spectrum following normal IPC of a single prominent nuclear transition (see e.g.Schweppe et al. 1983). Monoenergetic IPC seems to be excluded as the e^+ line widths do not show the expected dependence on the heavy ion scattering angle . The relevant EPOS data are displayed in fig. 5. Monoenergetic IPC in superheavy giant nuclear complexes has also been shown to lead to serious discrepancies with the experimental data (Schlüter et al. 1986).

Attempts to explain the positron lines within the standard scenario of spontaneous positron decay predicted to occur in the strong field of long-living, supercritical nuclear complexes have also been unsuccessful as the predicted peak-energy dependence (Reinhardt et al. 1981, Müller et al. 1986) on the combined nuclear charge Z_u is not observed. This is shown in fig. 6, where the average peak energies in the c.m. system relevant for this discussion are plotted as a function of Z_u (EPOS data from Cowan et al. 1985, 1986a; Bokemeyer et al. 1985; ORANGE data from Tsertos et al. 1985; Bosch et al. 1986; König 1986) together with theoretical expectations (Reinhardt et al. 1981). In particular the observation of peaks in subcritical systems ($Z_u \leq 173$), where spontaneous positron production is no longer possible, by the EPOS group (Bokemeyer et al. 1985), and by the ORANGE group (Bosch et al. 1986, König 1986) strongly argues against an interpretation in terms of spontaneous positron decay.

One of the remaining possibilities the data presented so far seems to be consistent with is the production of a common source(s) in these collision systems which spills out monoenergetic positrons at one or two or even more different energies. Thus the EPOS group began to consider the alternative that the e^+ lines may be the product of a two-body

EPOS SPECTROMETER

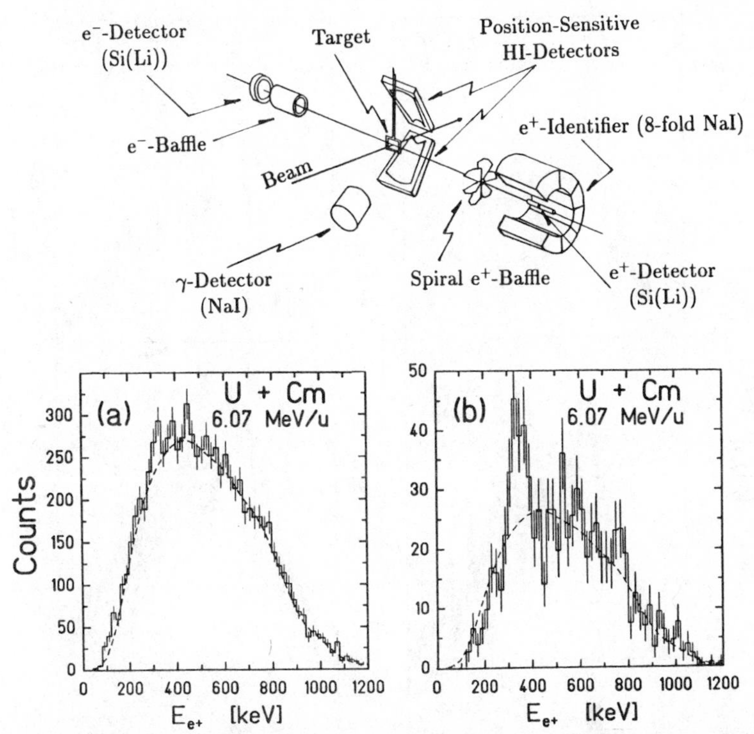

Fig.1 **Upper panel:** Schematic view of the EPOS set-up, a broad band, high efficiency and high resolution positron detection system. The scattered as well as the recoiling ions are detected in coincidence by two parallel plate avalanche detectors each spanning an angular range of $20^{o} \leq \vartheta \leq 80^{o}$ and $\Delta\varphi = 60^{o}$ and which allow to determine the ion scattering angle ϑ with respect to the beam axis to $\leq 1^{o}$. By the angle-angle correlation (quasi-) elastic collisions can be separated from reactions involving larger kinetic energy losses and/or mass transfers ($\Delta Q \geq 30$ MeV, $\Delta A \geq 5$ amu for symmetric collision system). A solenoidal magnetic field, arranged perpendicular to the scattering plane transports positrons to a cylindrical Si (Li) detector with an energy resolution of ≈ 10 keV at 500 keV. The strength of the magnetic field is adjusted as to collect all positrons emitted with angles $\Theta \leq 135^{o}$ with respect to the target e^{+}- detector direction. Simultaneous transport of electrons is strongly suppressed by a spiral baffle (exploiting the opposite spiral directions of e^{+} and e^{-}) while the positrons are further identified by their annihilation radiation. The resulting e^{+} efficiency is a smooth function of energy with $\varepsilon \geq 0.75\ \varepsilon_{max}$ for energies between 150 and 850 keV and $\varepsilon_{max} = 20\%$ of 4π. The γ as well as the e^{-} detector are used for auxiliary measurements described in fig. 4 and in the text.
Lower panel: (a) Typical total e^{+} energy spectrum observed in coincidence with elastic events ($-45^{o} \leq \vartheta_1 - \vartheta_2 \leq 45^{o}$, $90^{o} \leq \vartheta_1 + \vartheta_2 \leq 94^{o}$) when bombarding a 0.5 mg /cm^{2} ^{248}Cm target with ^{238}U ions of 6.07 MeV/ u. The dashed curve represents the calculated shape of the e^{+} spectrum expected from dynamic atomic and nuclear e^{+} production processes. Additional intensity seems to be present between 300 and 400 keV; its ratio to the smooth spectra can be improved by selecting smaller kinematical regions. This is demonstrated by the spectrum displayed in (b), which is obtained from the same data as shown in (a) but requiring $-10^{o} \leq \vartheta_1 - \vartheta_2 \leq +18^{o}$ and $90^{o} \leq \vartheta_1 + \vartheta_2 \leq 91.5^{o}$, the cuts being optimized for the best peak-to-total ratio; a narrow, ≈ 70 keV broad peak is obtained with a centroid energy of $\langle E_{e^+} \rangle = 335 \pm 15$ keV.

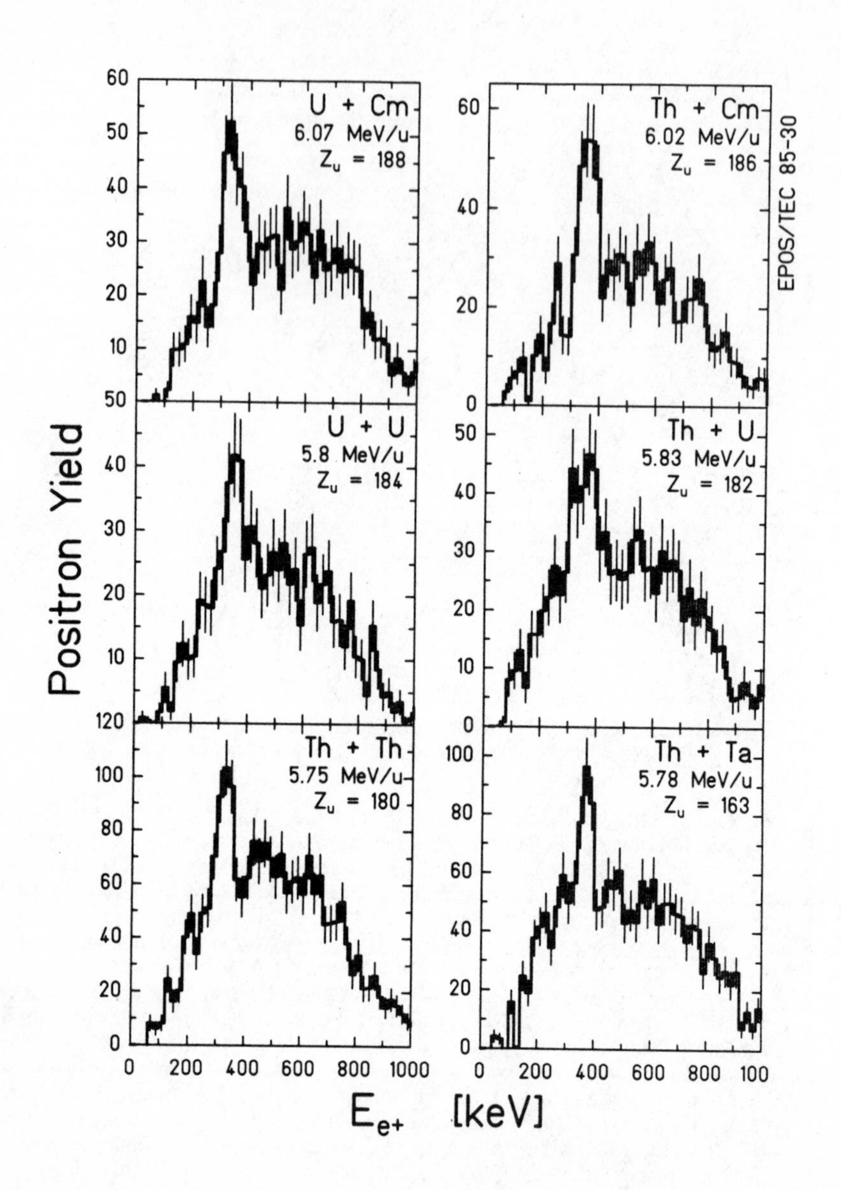

Fig.2 Positron energy spectra observed with the EPOS set-up for six collision systems with a combined charge $Z_u = Z_{proj.} + Z_{target}$ ranging from $Z_u = 188$ to $Z_u = 163$ and for bombarding energies corresponding to a marginal nuclear contact in head-on collisions. A pronounced peak appears in each spectrum in addition to the continuous atomic and nuclear background; as discussed in fig. 1, the peaks could be enhanced relative to this background by choosing constraints on the scattering kinematic. Average peak energies are between 315 and 375 keV, their widths are about 70-80 keV,and their c.m. differential cross-sections are similar and of the order of 5-10 μb/sr assuming the cross-sections to be constant over the target thickness (≈ 0.15 Mev/u).

ORANGE SPECTROMETER

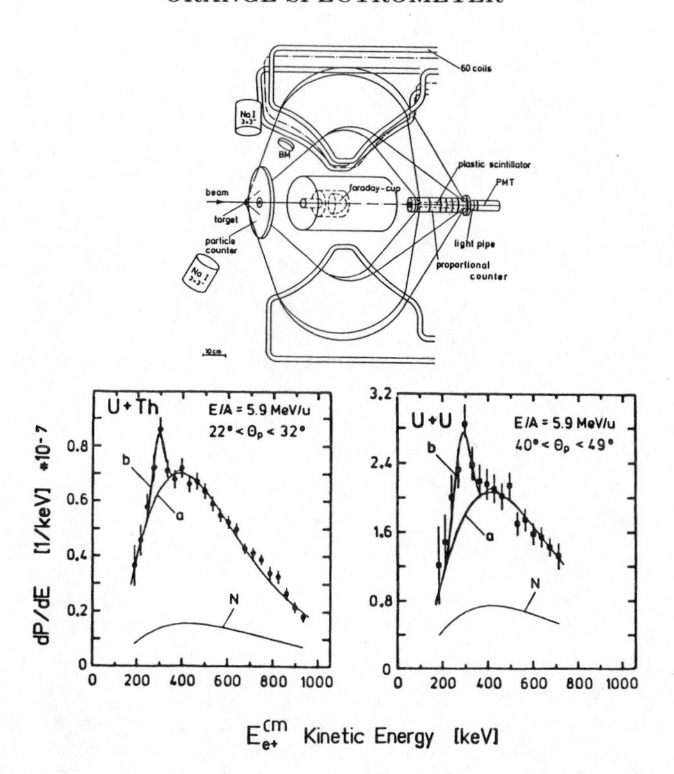

Fig.3 Upper panel: Schematic view of the ORANGE set-up which makes use of an iron-free "orange-type" ß-spectrometer. Positrons emitted from the target between $30^o \leq \vartheta_{e^+} \leq 70^o$ relative to the beam direction are focussed by a toroidal magnetic field produced by 60 copper coils on a detector system, which consists of a plastic scintillator surrounded by a proportional detector. The anode foil of this detector is subdivided such that the total accepted momentum bite of 14% can be divided into smaller bins for better momentum resolution ($\Delta p/p \geq 0.03$) corresponding to an energy resolution of $\geq$ 15 keV for 300 keV positrons. The total efficiency is up to 15% of 4π in the accepted momentum range; a total e^+ spectrum is measured by varying the magnetic field. Positrons are recorded in coincidence with scattered projectiles or recoils from the target by means of an annular parallel plate avalanche detector covering a scattering angular region of $12^o \leq \vartheta \leq 51^o$ and $\Delta\varphi = 2\pi$, the ϑ resolution being of the order of 2° - 3°.

Lower panel: Positron energy spectra from the bombardment of a ^{232}Th target (left) and a ^{238}U target (right) with 5.9 MeV/u ^{238}U-ions, selecting ion scattering regions as indicated. The observed e^+- energies have been transformed to the c.m. system which amounts to a shift of about -18 keV at E_{e^+} = 300 keV. The nuclear background (N) has already been subtracted in these cases. Curve (a) represent a calculation for dynamic atomic positrons, normalized by a factor of 0.8, while curve (b) is a superimposed fit of a line function. Again narrow ($\approx$80keV) lines are observed with mean energies of $\langle E_{e^+} \rangle_{cm} = 277 \pm 6$ keV (U+Th) and $\langle E_{e^+} \rangle_{cm} = 280 \pm 6$ keV (U + U) and mean c.m. differential cross sections of $\approx$7 and $\approx$15 µb/sr, respectively, assuming the cross-sections to be constant over the target thickness ($\approx$ 0.1-0.2 MeV/u).

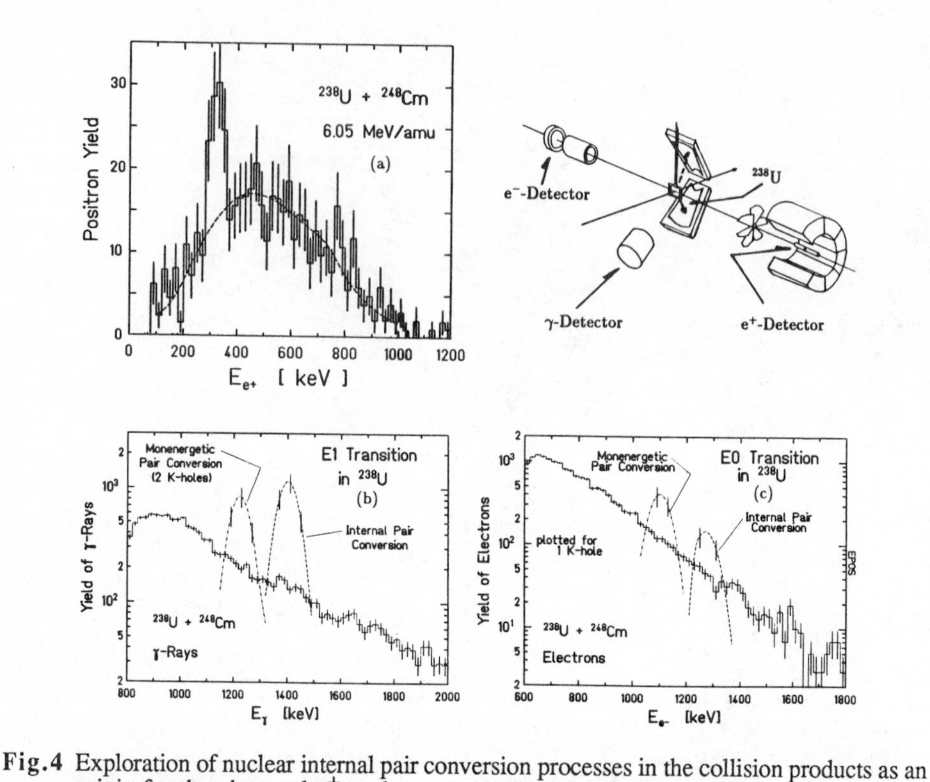

Fig.4 Exploration of nuclear internal pair conversion processes in the collision products as an origin for the observed e^+ peak structures can be carried out using γ spectra and e^- conversion-electron spectra measured under the same kinematical constraints as the positron spectra. Internal pair conversion (IPC), internal conversion (IC) and γ -ray emission are all related by branching ratios which are in first order independent of nuclear structure effects . Thus by using calculated branching ratios the e^+ peak intensity can be related to peaks in the e^- and γ spectra which should have been observed in case the assumed nuclear source is indeed the origin of the e^+ line. The method is exemplified using the data obtained by the EPOS group on the ^{238}U + ^{248}Cm system. Electrons were detected using a Si(Li) detector positioned on the solenoid axis in the second arm of the EPOS set-up; to minimize kinematic broadening effects the acceptance of the e^- detector was limited by a cylindrical baffle in front of the e^- detector to small angles with respect to the solenoid axis. Gamma-rays were recorded using NaI detectors located in the scattering plane.

In (b) the γ - ray spectrum observed under the same conditions as the e^+ data shown in (a) is displayed. Also shown are the expected γ peaks assuming the e^+ line to be due to normal IPC or to monoenergetic IPC (where the e^- is captured in a K-shell vacancy and the accompanying e^+ thus has a well defined energy) in the U-like reaction product, and assuming the most favorable multipolarity (E1) and an empty (!) K-shell. The intensities of the expected γ - peaks are obviously at variance with the measured γ - ray spectrum and the same conclusion is true when considering Cm-like reaction products. To investigate also the role of E0 transitions, in (c) an e^- spectrum is shown measured simultaneously with the e^+ data displayed in fig. 1b. For the intensities of the expected IC lines to be consistent with the measured e^- spectrum more than 1.85 vacancies have to be present in the K-shell at the moment of the nuclear decay. Since neutral U-atoms have K-vacancy lifetimes of $\approx 10^{-17}$ sec as compared to nuclear lifetimes of $\approx 10^{-13}$sec or longer, it is difficult to imagine even for a highly ionized U atom to have the K-shell nearly empty at the time the nuclear transition occurs. Thus both processes are a highly unlikely cause for the observed e^+ lines.

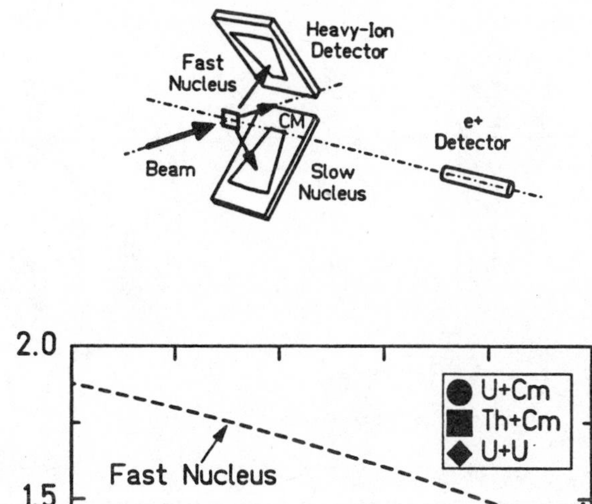

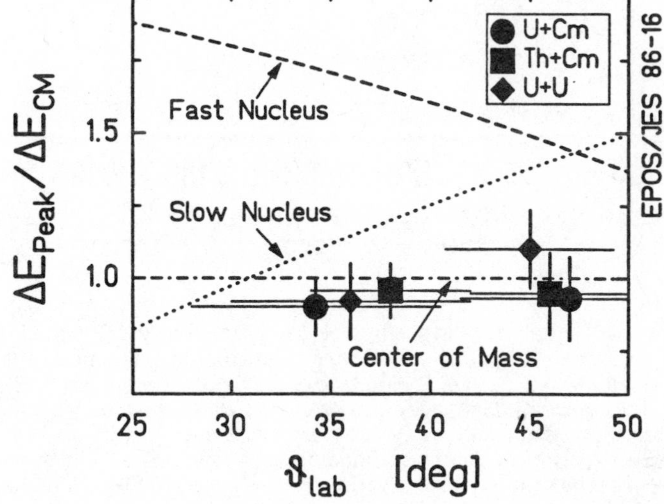

Fig.5 Information about the velocity of the system emitting the e^+ lines can be extracted from their widths ΔE_{peak} observed in the EPOS experiments. This is illustrated in the lower panel for the widths measured in the U+Cm, Th+Cm and U+U systems, where the lines could be observed for two sufficiently different scattering - angle regions. The experimental widths are compared to those expected from kinematical considerations in case the line positrons are emitted from the fast or slow product of the binary scattering event or from the c.m. system, respectively, assuming isotropic emission of the positrons in the respective rest systems and an intrinsic line width of $\leq$40 keV. The calculated widths are only plotted for the fast and slow ejectiles of the U + Cm system at 6.05 MeV/u as very similar curves are obtained for the other systems considered. Thus the narrow peak widths are not only at variance with normal internal pair conversion in one of the reaction products, the widths being mainly determined in this case by the triangular shape of the corresponding intrinsic e^+ distribution, but also at variance with monoenergetic internal pair conversion in either one of the two reaction products as the expected ϑ dependence is not observed. On the other hand, the data are consistent with the assumption of an isotropic, monoenergetic e^+ emission from a source moving with the c.m. velocity. It bears emphasis, however, that this is not a proof for c.m. emission as the statement depends on the assumptions that the emitter moves perpendicular to the solenoid axis and the e^+ emission is isotropic.

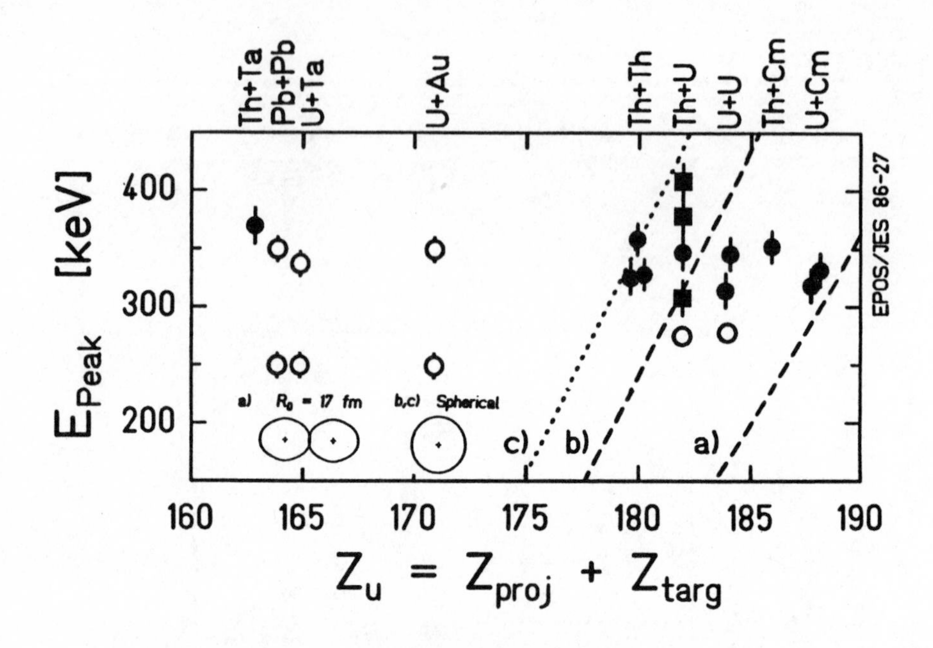

Fig.6 Average c.m. energies for the positron peaks observed by the EPOS (closed symbols) and by the ORANGE collaboration (open symbols) as a function of the combined atomic charge $Z_u = Z_{proj} + Z_{target}$. In case of the EPOS experiments the points actually represent the observed laboratory energies as the transformation from the lab. system into the c.m. system results only in a very small energy shift of -1.3 keV. In the ORANGE experiments the corresponding shift amounts to -18 keV for the two systems Th+U and U + U (see also fig. 3). For the other ORANGE data shown the sign of the shift is reversed as these experiments were performed in a modified set-up where the ORANGE has been turned around by 180^o(König et al. 1986); the positrons are now detected in a detector system through which the beam can pass and which is built up of Si-detector chips mounted like the roof of a pagoda.

The most striking feature of the data is the seeming independence of the energies of the peak(s) over the full range of Z_u, which strongly argues against an interpretation of these peaks as being due to the spontaneous positron-production mechanism. Curves (a) to (c) display the theoretical predictions for the standard scenario of spontaneous positron production from a nuclear complex built up by the two colliding nuclei for times longer than 10^{-19} sec. Such long nuclear delay times are needed in order to account for the narrow intrinsic line width of $\leq$40 keV (see fig. 5). The three curves reflect the strong dependence of the spontaneous positron peak energy on Z_u as well as its dependence on the nuclear charge distribution and the screening effect of the atomic electrons. Curves (a) and (b) show limiting cases of nuclear configurations for an expected ionization of about 50^+: (a) is for two deformed nuclei sticking together end-to-end, their centers being separated by $\approx$17fm, (b) is for a nuclear configuration with the two nuclei being merged into a spherical form. Curve (c) displays the additional effect of screening by assuming that all electrons are stripped away from the spherical nuclear complex. While some accidental balance between ionization and nuclear shape could still be employed to explain the peak positions for the very heavy systems by spontaneous e^+ production, the observation of peaks in systems below the limiting curve (c) clearly shows the inadequacy of such an interpretation.

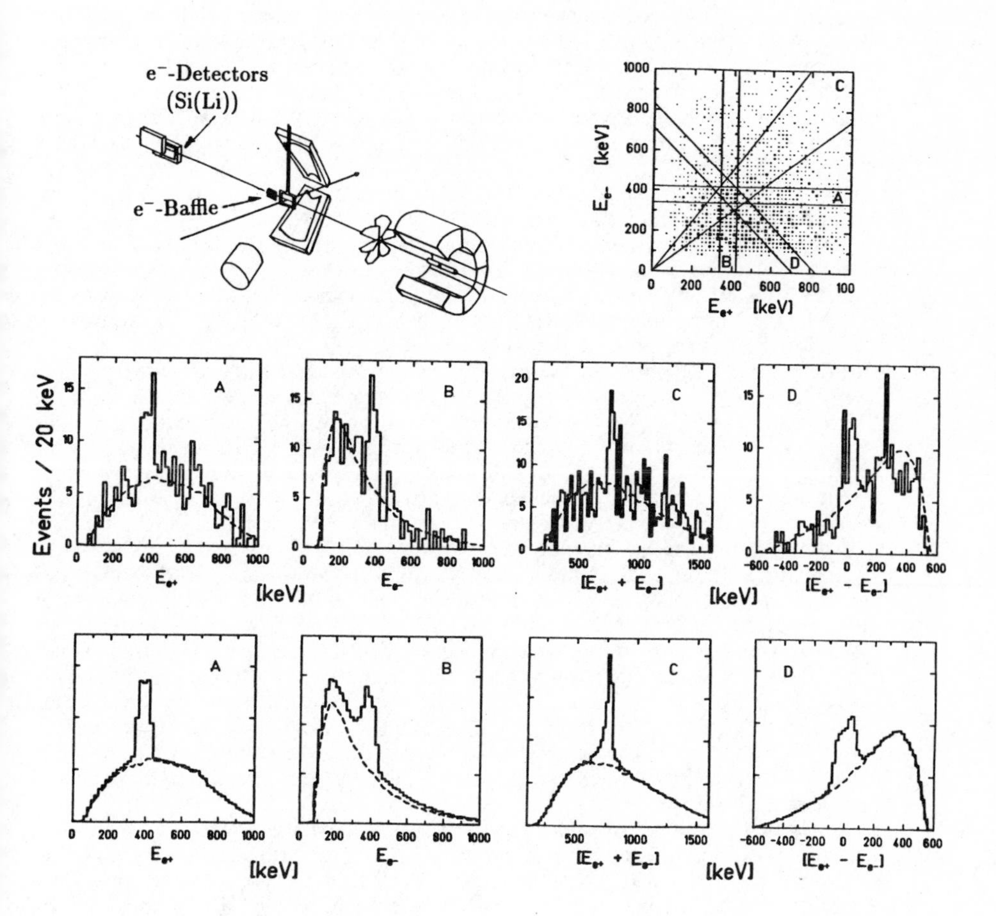

Fig.7 Upper panel (left): Modified EPOS set-up to detect positrons in coincidence with electrons and both scattered ions. While the heavy-ion detectors and the e^+ detection system was taken over from the e^+ singles measurements (see fig. 1), the second arm of the solenoid was rearranged to allow for an efficient transportation and detection of coincident electrons: The magnetic field was reshaped such that positrons emitted at angles $\leq 95^o$ with respect to the target -e^+-detector system are still transported to the e^+ detector while electrons emitted in the other half-sphere are transported towards the two e^- detectors consisting of 65 x 32 x 3 mm^2 planar Si(Li) crystals. The two detectors are positioned symmetrically off the solenoid axis to allow the large flux of low - energy

(continued next page)

Fig. 7 (continuation)
δ-electrons to spiral past between the two detectors; they are oriented with their sensitive faces parallel to the beam axis and are slightly staggered in order to suppress the detection of oppositely spiraling positrons. The suppression of low-energy electrons was further improved by a small sheet baffle mounted close to the target. The resulting e^+ - e^- coincidence peak efficiency , $\varepsilon_{e^+e^-}$, depends on the spatial correlation between e^+ and e^- . For the back-to-back decay of a hypothetical particle of mass 1.8 MeV - being at rest in the c.m. system - into an e^+- e^- pair, the coincidence efficiency relative to the singles peak efficiency, ε_{e^+}, amounts to $\varepsilon_{e^+e^-}/ \varepsilon_{e^+} \approx 15$ % as compared to ≈6 % for spatially uncorrelated pair - decays.

Upper panel (right): Intensity distribution of coincidence events as a function of the kinetic energies of the positrons, E_{e^+}, and electrons, E_{e^-}, for collisions between Th and U ions of 5.83 MeV/u ; note that the matrix includes the (quasi-) elastic events for all ion-scattering angles between 20° and 70°. Gates A through D are used to project out the spectra shown in the middle panel.

Middle panel: Projections of the measured concidence events on (A) the positron energy E_{e^+}, (B) the electron energy E_{e^-}, (C) the sum-energy $E_{e^+} + E_{e^-}$, and (D) the difference-energy $E_{e^+} - E_{e^-}$ axis using gates A through D. The dashed lines are the corresponding projections of a matrix obtained from a Monte Carlo calculation using the shapes of the total electron - positron coincidence spectra to represent the dynamic atomic and nuclear background and normalizing to the total event rate. In each spectrum a peak is clearly visible above the smooth background. In the positron spectrum (A) the peak has an energy of $\langle E_{e^+} \rangle = 380 \pm 15$ keV and a width of 80 ± 15 keV similar to the peak found in this system in the previous EPOS studies (see figs. 2,6) . No peak is observed if the gate A is shifted to higher or lower electron energies. In the electron spectrum (B) the peak has an energy of $\langle E_{e^-} \rangle = 375 \pm 15$ keV and a width of 75 ± 15 keV; again no peak is found if the gate B is shifted to other e^+ energies. The correlation between the electron and positron peak is even more clearly displayed in the sum - energy (C) and difference - energy spectra (D), showing a sum - energy line at $\langle E_{e^+} + E_{e^-} \rangle = 760 \pm 20$ keV and a difference-energy line at $E_{e^+} - E_{e^-} \approx 0$. The intensities of the correlated peaks deduced from the 4 spectra agree well within their statistical errors; the correlated peaks have a statistical significance of about 6σ.

Lower panel: In order to understand the response of the apparatus to different electron - positron production processes, detailed Monte Carlo calculations have been performed for a series of possible scenarios. While all internal pair conversion scenarios seem to be ruled out on intensity grounds and because they do not lead to a peak in the difference-energy spectrum (see also fig. 8) , the two-body decay of an object having a mass of 1.8 MeV , being at rest in the c.m. system, and decaying isotropically into an e^+ - e^- pair results in spectra displayed in the lower panel which very much resemble the experimental ones. In particular, only 3 % of the total positron yield has to be assumed to be due to this decay to explain the observed peak intensities. Moreover, comparing corresponding spectra obtained by additional constraints on the heavy-ion scattering angles as to pull out the e^+ peak also in the e^+-singles spectrum, the ratio of the coincident peak intensity to the peak intensity in the e^+-single spectrum is consistent with the coincidence efficiency expected for back-to-back emission of the e^--e^+ pair.

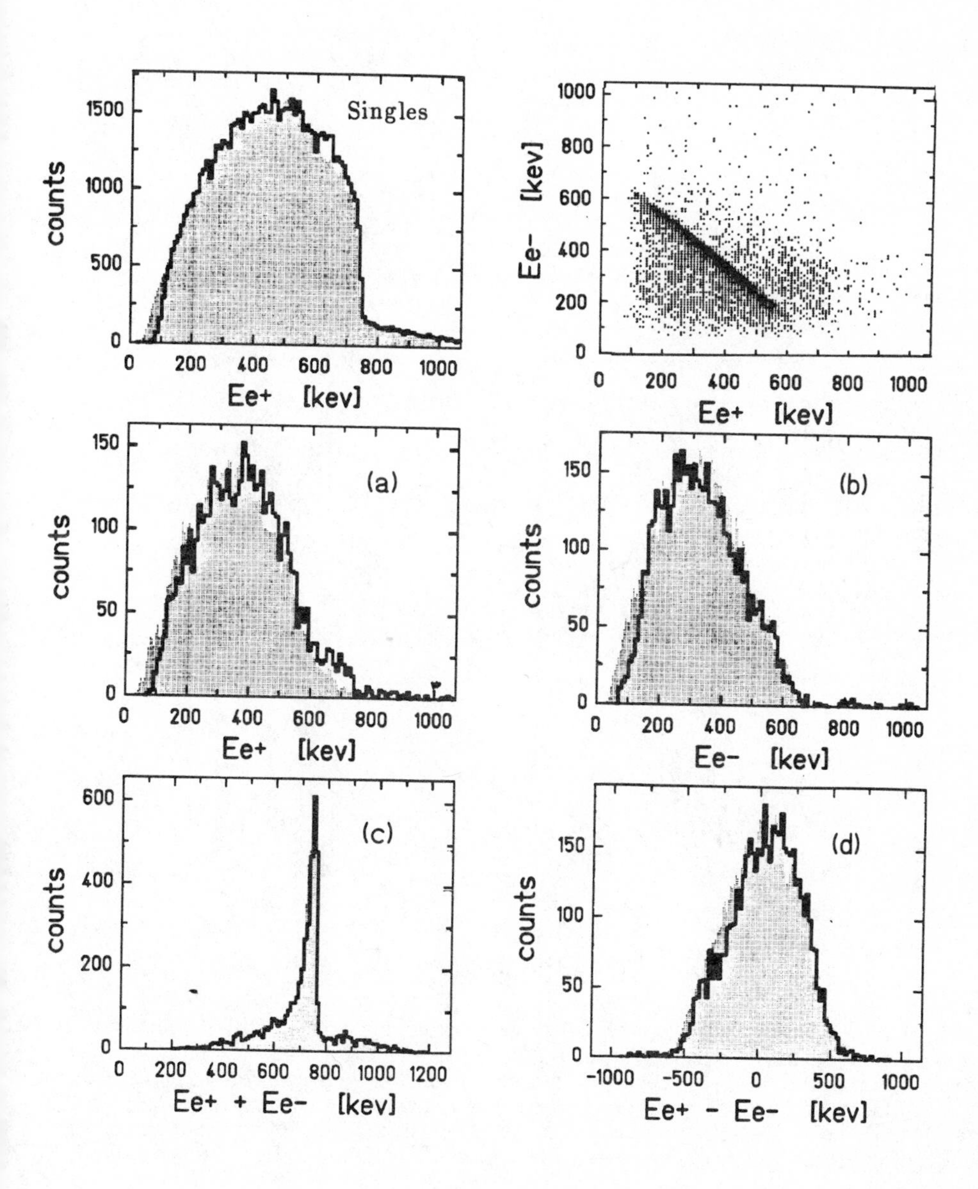

Fig.8 Electron-positron energy correlation measurement using the EPOS set-up shown in fig. 7 for the internal pair conversion decay of the 0^+, 1.76 MeV state in ^{90}Zr, populated in the β^- decay of a ^{90}Sr (^{90}Y) source. Figs (a) through (d) display the projections of the total coincidence matrix shown in the upper RH corner on the E_e+, E_e-, $E_e+ + E_e-$, and $E_e+ - E_e-$ axes, respectively. Also shown are the results of a Monte Carlo calculation (shaded spectra) the only parameter adjusted being the total number of counts in the positron singles spectrum shown in the upper left corner of the figure. Note, that the difference spectrum is structureless around $E_e+ - E_e- \approx 0$.

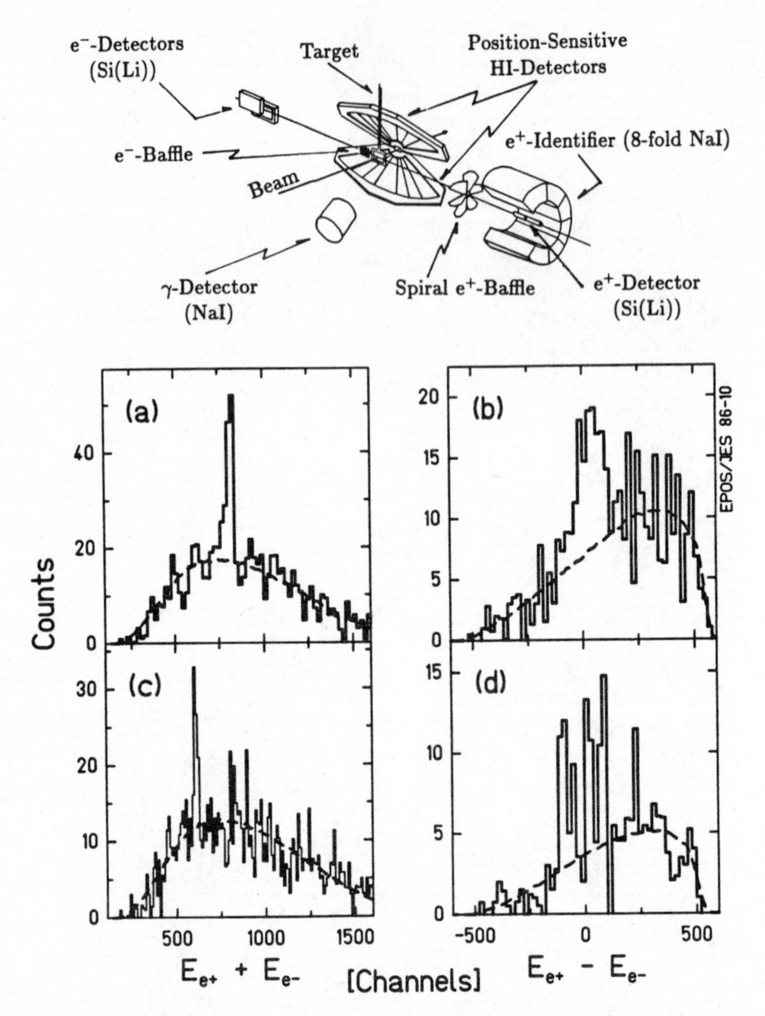

Fig.9 Results of a preliminary analysis of a new measurement on the Th + U collision system at bombarding energies around 5.87 MeV/u by the EPOS collaboration. In this experiment the φ - acceptance of the heavy - ion detector was increased by more than a factor of 2 and the resolution of the electron detectors was improved to ≈ 10 keV at 500 keV. The upper panels and lower panels display the energy-sum and energy - difference spectra (using cuts similar to C and D explained in fig. 7) for two subsets of the data gated on beam energy, heavy-ion scattering angle and positron /electron time-of-flight to enhance the two sum lines at ≈620 keV and ≈810 keV, respectively, visible already in the ungated energy - sum spectrum; weaker sum - lines may be present in this measurement including the line at ≈760 keV observed before (fig. 7). The data confirm the evidence from the earlier coincidence measurements (Cowan et al. 1986a) that there are more than one correlated peak structures with $E_{e^+} \approx E_{e^-}$ and absolute e^+ peak energies lying within the band of values shown in fig. 6. Moreover, the new data being recorded with an improved intrinsic e^- resolution substantiate the hint from the earlier measurement that the energy - sum lines are significantly narrower than the corresponding positron and electron lines. This implies a correlated cancellation of kinematical energy shifts as expected e.g. in cases of back-to-back e^+-e^- decays of objects being at rest in the c.m. system.

decay of e.g. a neutral system(s), being at rest in the c.m. system, into an electron-positron pair.Obviously the signature of such a decay would be a back-to-back emission of a monoenergetic electron and positron with identical energies. Fig. 7 displays schematically the modified EPOS set-up to look for coincident positron-electron emission together with first experimental results (Cowan et al. 1986a) obtained for the U+Th collision system. The data show evidence that the peak positrons are indeed correlated with electrons of similar energy. Similar, although statistically less significant observations were made for the Th+Th and Th+Cm systems (Cowan et al. 1986a). Detailed Monte-Carlo calculations have been performed to study the performance of the apparatus and to test different scenarios (Cowan 1986b). The only consistent picture found so far which seems to describe the experimental results is the two-body decay of an object with a mass of 1.8 MeV into an $e^+ e^-$ pair, which moves with a lab. velocity of the order of the c.m.velocity of $v/c \approx 0.05$ (see lower panel in fig. 7). Note, that the correlated observation of e^+e^-pairs rules out monoenergetic IPC as well as spontaneous positron decay; normal IPC is ruled out as it does not produce a peak in the difference-energy spectrum; this is clearly established by the source measurement and the corresponding Monte-Carlo calculation shown in fig. 8 . New data obtained by the EPOS collaboration on the U+Th collision system at slightly higher bombarding energies is displayed in fig. 9; they imply the presence of more than one sum-line with $E_{e^+} \approx E_{e^-}$, hinted already in the first measurements (Cowan et al. 1986a). Moreover, the sum-energy lines are found to be significantly smaller than the individual electron and positron peaks, again consistent with a two-body decay of slowly moving objects into e^+-e^-pairs.

3. Summary and Conclusion

The present experimental situation may be summarized as follows (*italics* are used to indicate that nothing else has been looked at so far): Narrow peaks are observed in the positron energy spectra from *collision systems with* $163 \leq Z_u \leq 188$. Observed centroid energies are between 250 and 400 keV, experimental line widths are ~ 70 to 80 keV independent of the heavy-ion scattering angles, differential c.m. line cross-sections in the heaviest collision systems are 5-10μb/sr, being approximately the same in lighter systems (EPOS) / decreasing for lower Z_u systems (ORANGE, see Kienle 1986); it should be emphasized, however, that the quoted cross-sections depend on the assumption that they are constant over the beam-energy range scanned in the targets. The heavy-ion scattering is *near the Coulomb barrier* and close to elastic. Positron peaks have been found to be correlated with electron peaks in the *U+Th (Th+Th and Th+Cm) collision systems* with $\langle E_{e^+} \rangle \approx \langle E_{e^-} \rangle$ and $\Delta E_{e^+} \approx \Delta E_{e^-}$; several distinct sum-energy lines with different energies are observed, which are significantly narrower than the contributing e^+ and e^-peaks. The peaks are allegedly not due to internal pair conversion or spontaneous positron decay. The experimental findings are apparently consistent with a two-body decay of objects of different masses moving slowly in the lab.system, however, despite considerable theoretical activities, attempts to understand the experimental findings in terms of a new, previously undetected particle(s) (see references listed under Schäfer et al. 1985) with a mass of ~3 m_{e^-}, in particular of an axion (see references listed under Mukhopadhyay and Zehnder 1986), of new bound states of electron-positron pairs (Müller et al. 1986,Wong 1986b) or of soliton in the QED vacuum (Celenza 1986) have not come up with a convincing solution which could account for all of the observed properties - in particular the multiplicity of lines, the observed intensity, or the low velocity of the decaying state. In view of this it might also be worthwhile (see also Scharf 1986) to reconsider the problem of pair creation in the strong electric and magnetic fields (up to 1 MV/ fm and 10^{12} Tesla, respectively) of the two colliding nuclei, which has not yet been solved in a fully dynamical way.

It is obvious that both the theoreticians but in particular also we experimentalists have to go back to our working benches. Questions which should and could be answered in the next round of experiments are: What are the precise production conditions? Are collision systems with high Z_u and/or is nuclear contact needed? What is the dependence of the line cross section on the heavy-ion kinematic? Are resonance or threshold effects present as hinted by several observations such as the occurrence of different sum-lines at different beam energies? Moreover, a new experimental set-up should be prepared which allows to measure not only the energies but also the emission angles of the electrons and positrons in order to be able to determine e.g. the invariant mass of the e^+e^- pairs or to look for angular correlation effects.

Acknowledgements: I would like to thank all my colleagues from the EPOS collaboration for the interesting and exciting times we are having together; in particular I would like to take the opportunity to thank the former and present Ph.D. students, Dr. A. Gruppe, Dr. J. Schweppe, T. Cowan, K. Sakaguchi and P. Salabura who as usual have been carrying the main load of our experiments.Moreover, I thank the ORANGE group for providing me with their newest results. Part of this work has been supported by the Bundesministerium für Forschung und Technologie (BRD), by the U.S. Department of Energy,and by the National Science Foundation.

References

Backe H, Handschug L, Hessberger F, Kankeleit E, Richter L, Weik R, Willwater R,Bokemeyer H, Vincent P, Nakayama Y, Greenberg J S 1978 Phys. Rev. Lett. **40** 1443

Bokemeyer H, Bethge K, Folger H, Greenberg J S , Grein H, Gruppe A, Ito S,Schulé R, Schwalm D, Schweppe J, Trautmann N, Vincent P, Waldschmidt M 1983a in Greiner W 1983 p.273

Bokemeyer H, Folger H, Grein H, Cowan T, Greenberg J S, Schweppe J, Bethge K, Gruppe A, Stiebing K E, Schwalm D, Vincent P, Begemann M, Kluever M,Trautmann N 1983 in Proc. Intern. Conf. on Nucl. Physics, Florence

Bokemeyer H, Folger H, Grein H, Cowan T, Greenberg J S, Schweppe J. Balanda A, Bethge K, Gruppe A, Sakaguchi K, Stiebing K.E, Schwalm D, Vincent P, Backe H, Begemann M,Klüver M, Trautmann N 1985 in GSI Annual Report 1984 p.177

Bosch F, Kienle P, König W, Koshuharov C, Berdermann E, Franosch M, Huchler S, Kemmer J, Tsertos H, Wagner W 1986, GSI Annual Report 1985 p.179

Celenza LS et al, 1986 Phys. Rev. Lett **57** 55

Clemente M, Berdermann E, Kienle P, Tsertos H, Wagner W, Kozhuharov C,Bosch F, König W 1984 Phys. Lett **B137** 41

Cowan T, Backe H, Begemann M, Bethge K, Bokemeyer H, Folger H, Greenberg J S, Grein H, Gruppe A, Kido Y, Klüver M, Schwalm D, Schweppe J, Stiebing K E , Trautmann N, Vincent P 1985 Phys. Rev. Lett **54** 1761

Cowan T, Backe H, Bethge K, Bokemeyer H, Folger H, Greenberg J S , Sakaguchi K, Schwalm D, Schweppe J, Stiebing K E , Vincent P 1986a Phys. Rev. Lett. **56** 444

Cowan T 1986b in Greiner W 1986, to be published

Greenberg J S, Vincent P 1985 in Treatise on Heavy Ion Science, ed. Bromley (New York: Plenum Press) p.139

Greiner W 1983 ed. Quantum Electrodynamics of Strong Fields (New York:Plenum)

Greiner W 1986 ed. Proc.of the Summer School on Physics of Strong Fields, Maratea, to be published

Kienle P 1986 in Greiner W 1986, to be published

König W 1986 in Greiner W. 1986,to be published

Kozhuharov C, Kienle P, Berdermann E, Bokemeyer H, Greenberg J S ,Vincent P, Backe H, Handschug L, Kankeleit E 1979 Phys. Rev. Lett. **42** 376

Mukhopadhyay N C, Zehnder A 1986 Phys. Rev. Lett **56** 206

Peccei RD et al 1986 Phys. Lett **B172** 435

Kraus L M, Wilczek F 1986 Phys. Lett **B 173** 189

Suzuki M 1986 Phys. Lett **B 175** 364

Kraus L M, Wise M B 1986 Phys. Lett **B 176** 483

Müller B et al. 1986 J. Phys. **G 12** L109
Müller U. Soff G, de Reus T, Reinhardt J, Müller B, Greiner W 1983 Z.Phys. **A 313** 263,and priv. communication
Müller U, de Reus T, Reinhardt J. Müller B,Greiner W 1986 Z.Phys. **A 323** 261
Rafelski J, Müller B, Greiner W 1978 Z. Phys. **A 285** 49
Reinhardt J, Müller U, Müller B, Greiner W 1981 Z. Phys. **A 303** 17
Schäfer A et al. 1985 J. Phys. **G 11** L 69
Balentekin A B et al. 1985 Phys. Rev. Lett. **55** 461
Schäfer A et al. 1985 J. Phys. **G 11** 169
Reinhard J 1986 Phys. Rev. **C33** 194
Chodos A, Wijewardhana L C R 1986 Phys. Rev. Lett **56** 302
Lane K 1986 Phys. Lett. **B169** 97
Müller B, Reinhardt J 1986 Phys. Rev. Lett. **56** 2108
Ma E 1986 Phys. Rev. **D 34** 293
Carrier D et al 1986 Phys. Rev. **D 34** 1332
Mageras G et al. 1986 Phys. Rev. Lett. **56** 2672
Bowcock T et al 1986 Phys. Rev. Lett. **56** 2676
Zee A 1986 Phys. Lett **B172** 377
Schäfer A et al. 1986 Mod. Phys. Lett **A1**, 1
Schäfer A et al. 1986 Z. Phys. **A 324** 243
Schröder UE 1986 Mod. Phys. Lett. **A1** 157
Scharf G 1986 Phys. Lett. **B 177** 429
Schlüter P, de Reus T, Reinhardt J, Müller B, Soft G 1983 Zeitsch. f. Phys. **A 314** 297
Schlüter P, Müller U, Soff G, de Reus T, Reinhardt J. Greiner W 1986 Z.Phys. **A 323** 139
Schwalm D 1984 in Electronic and Atomic Collisions,Eichler J, Hertels I.V., Stolterfoht edts. (Amsterdam: Elsevier Science Publ. B.V.)p. 295
Schweppe J, Gruppe A, Bethge K, Bokemeyer H, Cowan T, Folger H,Greenberg J S , Grein H, Ito S, Schulé R, Schwalm D, Stiebing K E,Trautmann N, Vincent P, Waldschmidt M, 1983 Phys. Rev. Lett. **51** 2261
Tsertos H, Berdermann E, Bosch F, Clemente M, Kienle P, König W, Kozhuharov C, Wagner W 1985 Phys. Lett **B162** 273
Wong C M 1986a Phys. Rev. Lett **56** 1047
Chu M C, Ponisch V 1986 Phys. Rev. **C 33** 2222
Wong C M et al. 1986b, to be published

List of Oral Contributions

(The number in square brackets refers to the contribution in Vol. 1 of the Proceedings)

The renormalisation of the axial-vector strength in nuclei: experiments on superallowed beta decay
M J G Borge[1], P G Hansen[2,5], B Jonson[2], S Mattson[2], O B Neilsen[3], G Nyman[2], H L Ravn[1], A Richter[3] and K Riisager[5]
[1]CERN; [2]Chalmers Univ. Technology; [3]Niels Bohr Institute, Copenhagen; [4]Inst. Kernphysik, Darmstadt; [5]University of Aarhus
[B105]

$0\nu\beta\beta$ decay and a limit on the right-handed leptonic current
T Tomoda[1], A Faessler[1], K W Schmid[1] and F Grümmer[2]
[1]University of Tübingen; [2]KFA Jülich
[F3]

Mikheyev-Smirnov-Wolfenstein enhancement of oscillations as a possible solution to the solar neutrino problem
S P Rosen and J M Gelb
Los Alamos National Laboratory
[F2]

Charge symmetry breaking in neutron-proton elastic scattering at 477 MeV
R Abegg[1,2], D Bandyopadhyay[3], J Birchall[3], E B Cairns[2], H Coombes[2], C A Davis[3], N E Davison[3], P P J Delheij[1], P W Green[1,2], L G Greeniaus[1,2], H P Gubler[3], D C Healey[1], C Lapointe[2], W P Lee[3], W J McDonald[2], C A Miller[1,2], G A Moss[2], G R Plattner[4], P R Poffenberger[3], W D Ramsay[3], G Roy[2], J Soukup[2], J P Svenne[3], R Tkachuk[2], W T H van Oers[3], G D Wait[1] and Y P Zhang[3]
[1]TRIUMF, Vancouver; [2]University of Alberta; [3]University of Manitoba; [4]University of Basle
[A1]

LIST OF POSTERS
SELECTED FOR PRESENTATION

Inst. Phys. Conf. Ser. No. 86
Paper presented at Int. Nucl. Phys. Conf., Harrogate, UK, 1986

Poster Session 1

Poster contributions are listed in the order they were presented at the conference. This is largely based on the PACS subject classification. The letters A-H refer to the session topics used in Volume 1 of the Proceedings as follows:

- A Nuclear Probes
- B Nuclear States and Modes of Excitation
- C Reaction Mechanisms and Nuclear Structure
- D Relativistic Nuclear Physics
- E Sub-Nucleon Aspects of Nuclear Physics
- F Fundamental Interactions in Nuclei
- G Nucleon-Nucleon Interactions
- H Hypernuclei and Instrumentation

POSTER SESSION 1: 26 August 1986

1E The colour-dielectric model: numerical solutions
A.G. WILLIAMS, L.R. Dodd, A.W. Thomas

2E Scaling and scale changes in DIS on nuclei
R.P. BICKERSTAFF, A.W. Thomas

3G Ratio of spin transfer parameters D_t/R_t in $d(\vec{p},\vec{n})pp$ quasi-elastic scattering
C.A. DAVIS, J. Birchall, D. Bandyopadhyay, N.E. Davison, S.A. Page, W.D. Ramsay
W.T.H. Van Oers, R. Abegg, P.W. Green, L.G. Greeniaus, C.A. Miller, C. Lapointe, G.A. Moss
R. Tkachuk

4G Contributions to the deuteron binding energy
H.F.K. ZINGL, L. Crepinsek, K. Schwarz

5E Improved limit for the electric polarizability of the neutron
F.W.K. FIRK, J.W. Kruk

6F The Oxford neutrino mass experiment
N.A. JELLEY, P.A. Holmes, S.S. Williams

7C Near-complete occupancy of the 3s proton shell in ^{208}Pb
G.J. WAGNER, H. Clement, P. Grabmayr, S. Klein, H. Röhm

8B On-line nuclear orientation and relaxation measurements on Cs isotopes
T.L. SHAW, V.R. Green, C.J. Ashworth, J. Rikovska, N.J. Stone, P.M. Walker, I.S. Grant
L. Goettig

9B Nuclear spin and magnetic moment of ^{11}Li
G. ULM, E. Arnold, D. Bauer, J. Bonn, R. Gegenwart, R. Neugart, E.W. Otten, K. Wendt

10B Nuclear moments and charge radii in the range around Z=50
G. ULM, U. Dinger, J. Eberz, G. Huber, H. Lochmann, R. Menges, R. Neugart, K. Wendt
D. Kirchner, O. Klepper, T. Kühl, D. Marx

11D Jensen-Mayer term in nuclear Dirac phenomenology
M. BAWIN, G.L. Strobel

12B Systematics of the particle-rotor model around mass number A=153
A.M.J. SPITS

13C α-backbending, nucleon-nucleon correlation and shape in nuclei
GU JINNAN

14B The wall formula as a limit for the damping of surface oscillations of a Fermi liquid
C. FIOLHAIS

15C Temperature dependent effective interaction
A. LEJEUNE, P. Grange, J. Cugnon

16B Quantum effects in the Vlasov theory
A. DELLAFIORE, F. Matera

17B $K^{\pi}=0^{+}$ and $K^{\pi}=1^{+}$ collective excitation in deformed nuclei
P.J. BRUSSAARD, L.C. De Winter, N.R. Walet, K. Allaart, A.E.L. Dieperink

18B g-factors of ^{158}Er at high spins
A. ANSARI, S.C.K. Nair

19B Calculation for odd-odd gold nuclei in IBFFM/OTQM
V. LOPAC, S. Brant, V. Paar

20B Nuclear spins, moments, and differences in charge radii in a series of radon isotopes
R. NEUGART, G. Ulm, H.T. Duong, K. Wendt, W. Borchers, E.W. Otten

21C The quenching of pair correlations in Yb nuclei
J.D. GARRETT, G.B. Hagemann, B. Herskind, M.A. Riley, R. Chapman, J.C. Lisle, J.N. Mo
J.F. Sharpey-Schafer, J. Simpson, Zhang Jingye

22C A two-level model with competing pairing interactions
J.A. EVANS, G.G. Dussel, E.E. Maqueda, R.P.J. Perazzo

23B A survey of nuclear mass regions to identify additional states of a newly discovered alpha-cluster-like type
H.J. DALEY

24C Electron scattering from ^{7}Li
R.A. BALDOCK, P.J.B. Hopkins, B. Buck

25B Description of the new 1^{+} mode by the PHF method
E. MOYA DE GUERRA, P. Sarriguren, J.A. Caballero, A.E.L. Dieperink

26B Spreading width of giant resonances at finite temperature
P.F. BORTIGNON, R.A. Broglia, M. Gallardo

27B Description of nuclear excitations in BPA and RPA
E. BOEKER, K. Allaart, W. Hengeveld, W.H. Dickhoff, M.G.E. Brand

28B Spherical and deformed states in high-spin yrast region of ^{110}Sn and ^{112}Sn
J. KASAGI, H. Harada, T. Murakami, K. Yoshida, T. Inamura, T. Kubo

29B Collectivity of low lying states in ^{134}Sm and ^{136}Sm
P.J. BISHOP, M.J. Godfrey, A. Kirwan, P.J. Nolan, D.J. Thornley, J.M. O'Donnell
R. Wadsworth, D.L. Watson, D.J.G. Love, A.H. Nelson

30B Excited states of ^{109}Rh
D. DE FRENNE, N. Kaffrell, P. Hill, J. Rogowski, H. Tetzlaff, N. Trautmann, E. Jacobs
P. De Gelder, K. Heyde, G. Skarnemark, S. Hoglund, J. Alstad, N. Blasi, M.N. Harakeh
W.A. Sterrenburg, K. Wolfsberg

31C Level structure of ^{131}Cs populated in the decay of ^{131}Ba
LIU YUNZUO, Sun Huibin, Hu Dailing, Zhou Jiewen

32B High spin states in the odd proton transitional nucleus ^{159}Tm
J. SIMPSON, B.M. Nyakó, A.R. Mokhtar, M. Bentley, H.W. Cranmer-Gordon, P.D. Forsyth
J.D. Morrison, J.F. Sharpey-Schafer, M.A. Riley, J.D. Garrett, C.H. Yu, A. Johnson
J. Nyberg, R. Wyss

33B Study of ^{53g}Co based on separation by beta recoil effect
K. ESKOLA, A. Hautojärvi, K. Vierinen, J. Aystö, J. Honkanen, P. Taskinen, S. Messelt

34F Investigation of neutrino properties in a radioactive decay
K.YA. GROMOV

35C Equivalent local potentials for the short-range NN-interaction from quark exchange kernels
G. PANTIS

36C Measurement of the annihilation cross section for $\bar{n}$ in Fe at 540 MeV/C
F. IAZZI, T. Bressani, E. Chivassa, S. Costa, G. Dellacasa, M. Gallio, P. Macciotta
S. Marcello, A. Masoni, B. Minetti, M. Morandin, A. Musso, G. Puddu, S. Serci, C. Voci

37C A modified CCBA for the study of direct reactions in deformed nuclei
H. MASSMANN, G. Ramirez, M. Bernath, O. Dragún

38B Recoil distance lifetime measurements in ^{134}Nd and ^{136}Nd
J. BILLOWES, K.P. Lieb, J.W. Noe, W.F. Piel Jr., S. Rolston, G.D. Sprouse, O.C. Kistner
F. Christancho

39C Deep inelastic and quasi-free processes in ^{51}V(α,α'p) reactions
H. MACHNER, M. Nolte, J. Jacoby, P. Jahn, M. Rogge, G. Seniwongse, H. Stockhorst, P. Turek

40C Temperature dependence of the Skyrme forces
U. LOMBARDO, M. Baldo, G. Giansiracusa

41C α - transfer between ^{16}O nuclei near the barrier
J. CARTER, B.O. Carragher, R.G. Clarkson, V. Hnizdo, J.P.F. Sellschop

42C Simple analytic formulas for light-nucleus (A<19) folded and Coulomb potentials
P.-P. PROVENCHER, R. Roy

43C A RPA approach to (γ,N) reactions
M. WAROQUIER, J. Ryckebusch

44C The deformed spin-orbit potential and inelastic scattering of polarized protons
at intermediate energy from light nuclei
R. DE SWINIARSKI

45F Unconservation of space parity in nuclear reactions with polarized neutrons
P.A. KRUPCHITSKY

46B Investigation of isoscalar giant resonances in ^{6}Li-scattering
W. EYRICH, A. Hofmann, A. Lehmann, B. Muhldorfer, H. Schlösser, H. Wirth, H.J. Gils
H. Rebel, S. Zagromski

47B Study of giant resonances in hot Dy nuclei using particle-γ coincidences
H.P. MORSCH, B. Bochev, T. Kutsarova, R. Lieder, W. Gast, G. Hebbinghaus, A. Kramer-Flecken
J.P. Didelez

48C Level spacing of doorway resonances
C.C. HSU, S.C. Yeh

49C Isospin mixing in the unified exciton model
D. RYCKBOSCH, R. Van de Vyver, E. Kerkhove, P. Berkvens, H. Ferdinande, P. Van Otten
A. De Graeve, L. Van Hoorebeke

50C Eikonal approximation for elastic scattering of polarized deuterons
NGUYEN VAN SEN, Ye Yanlin

51D Nuclear collision mechanisms at high energies and their manifestation in
secondary particle spectra
S. BELYAEV

52C Correlated fluctuations in the ^{89}Y(^{19}F,x)y reaction excitation
functions
R. LUCAS, B. Berthier, M.C. Mermaz, T. Suomijärvi, J.-P. Coffin, G. Guillaume, B. Heusch
F. Jundt, F. Rami

53C Large dimension expansion for potential scattering
R. DUTT

54C A search for dispersive effects in elastic electron scattering off ^{12}C
C.W. DE JAGER, E.A.J.M. Offermann, L.S. Cardman, H.J. Emrich, G. Fricke, H. Miska
D. Rychel, H. De Vries

55B Determination of the nuclear convection current in ^{20}Ne by inelastic
electron scattering
A.P. STAMP

74C α-emission from the incomplete fusion of ^{20}Ne and ^{27}Al
K.A. GRIFFIOEN, E.A. Bakkum, C.P.M. Van Engelen, R.J. Meijer, R. Kamermans

75C Binary reactions near the Coulomb barrier in systems with dynamically hindered fusion
R. BELLWIED, H. Keller, J.V. Kratz, K. Lutzenkirchen, W. Bruchle, K.J. Moody, M. Schadel K. Sümmerer, G. Wirth, W. Reisdorf

76C The side-splash in light systems
H.R. JAQAMAN

77C Sequential and incomplete fusion reaction mechanisms in ^{7}Li induced break-up reactions at 10 MeV/A
T. DAVINSON, A.C. Shotter, V. Rapp, D. Branford

78E A comparison of $\pi\Delta$ interaction mechanism with the double charge exchange experimental data on self-conjugate nuclei
T.E.O. ERICSON, C.R. Ching, T.H. Ho, W.Q. Zhao

79B Isospin mixing in ^{4}He from inelastic pion scattering
D. DEHNHARD, C.L. Blilie, D.B. Holtkamp, S.J. Seestrom-Morris, S.K. Nanda W.B. Cottingame, D. Halderson, C.L. Morris, C.F. Moore, P.A. Seidl, H. Ohnuma, K. Maeda

80D Observation of $\pi^{o}\pi^{o}$ interference and size of pion emission volume in π^{-}-Xe interactions at 3.5 GeV/c
J. PLUTA

81C Wide angle fragment correlations for the reactions $^{12}C(^{7}Li,\alpha+t)^{12}C$, $^{120}Sn(^{7}Li,\alpha+t)^{120}Sn$, $^{208}Pb(^{7}Li,\alpha+t)^{208}Pb$ at 70 MeV
J. YORKSTON, A.C. Shotter, T. Davinson, D. Branford, E.W. Macdonald

82C Skyrme-Hartree-Fock calculation of the mass-asymmetric fission barrier of ^{240}Pu
F. TONDEUR

83C Application of the quasi-molecular theory of fission
N. QUMRI

84C Proton induced fission of uranium isotopes
W. SCOBEL, M. Strecker, R. Wien

85C Study of the energy, mass and charge distribution of the $^{241}Pu(n_{th},f)$ fragments
P. SCHILLEBEECKX, P. Geltenbort, C. Wagemans, A. Oed, F. Gönnenwein

86B The photofission and (γ,n) reactions in ^{238}U from 5-10 MeV
N.P. HAWKES, D.J.S. Findlay, M.R. Sené

87C Low energy antiproton annihilation on nuclei
M.P. BUSSA, F. Balestra, Yu.A. Batusov, G. Bendiscioli, S. Bossolasco, F.O. Breivik L. Busso, I.V. Falomkin, L. Ferrero, C. Guaraldo, A. Haatuft, A. Halseinsild, T. Jacobsen E. Lodi Rizzini, A. Maggiora, K. Miklebost, J. Olsen, D. Panzieri, G. Piragino G.B. Pontecorvo, A. Rotondi, P. Salvini, M.G. Sapozhnikov, S.O. Sorensen, F. Tosello A. Zenoni

88B Gamma and particle decays of ^{24}Mg high spin states above 14 MeV
I. WRIGHT, D.M. Pringle, W.J. Vermeer, W.N. Catford

89C Studies of low-lying states of neutron-rich phosphorus isotopes via multi-nucleon transfer reactions
C.L. WOODS, W.N. Catford, L.K. Fifield, T.R. Ophel, N.A. Orr, D.C. Weisser

90B Studies of odd-odd silver nuclei using the neutron capture reaction
T.D. MAC MAHON

91B Neutron dependence of the Z=64 shell closure - identification of neutron deficient ^{142}Gd and $^{142,144}Dy$ nuclei
L. GOETTIG, W. Gelletly, C.J. Lister, R. Moscrop, B.J. Varley, R. Wadsworth

92B High-spin shell model and octupole excitations in $^{148}_{84}Gd_{64}$
P. KLEINHEINZ, S. Lunardi, M. Piiparinen, M. Ogawa, G. De Angelis, J. Blomqvist

93B Gamma-ray decays in ^{124}Ce studied using the Daresbury recoil separator
A.N. JAMES, T.P. Morrison, P.J. Nolan, K.L. Ying, K.A. Connell, J. Simpson, H.G. Price

94H Response function for ^{3}He neutron spectrometers
O. TENGBLAD, K-H. Beimer, G. Nyman

95C Anomalous strong interaction effects in pionic ^{181}Ta,Re and ^{209}Bi
J. KONIJN, C.T.A.M. De Laat, P. David, H. Janszen, T. Krogulski, L.A. Schaller
L. Schellenberg, A. Taal

96E The hidden colour channel of baryon-baryon interaction from the viewpoint of lattice QCD
W. FEILMAIR, M. Faber, H. Markum, M. Meinhart

97F O(5)xU(1) electro weak gauge theory and the exact relation between the Cabibbo and the Weinberg angles
M. SAMIULLAH

Inst. Phys. Conf. Ser. No. 86
Paper presented at Int. Nucl. Phys. Conf., Harrogate, UK, 1986

Poster Session 2

1E Structure functions in the chiral bag model
V. VENTO, V. Sanjosé, S. Noguera

2E Centre of mass separation for Dirac equation for harmonic potential
S.B. KHADKIKAR, S.K. Gupta

3E The octet baryon magnetic moments
S.B. KHADKIKAR, S.K. Gupta

4E Radial excitations in nucleon-nucleon scattering
O. BOHIGAS, J. Carbonell, B. Silvestre-Brac, C. Gignoux

5E Determining the properties of six-quark resonances
E.L. LOMON, P. González, P. La France

6G The violation of charge symmetry in n-p scattering
A.W. THOMAS, A.G. Williams, G.A. Miller

7G Cross section and analysing power in free $\vec{n}$p forward elastic scattering at energies from 400 to 1100 MeV
Y. TERRIEN, G.A. Korolev, A.V. Khanzadeev, J.C. Lugol, G.E. Petrov, J. Saudinos B.H. Silverman, E.M. Spiridenkov, A.A. Vorobyov, F. Wellers

8B Gamma-decaying states at high excitation in ^{19}F
W.J. VERMEER, D.M. Pringle

9B The mass of ^{33}Al
P.J. WOODS, R. Chapman, J.L. Durell, J.N. Mo, R. Smith, B.R. Fulton, R.A. Cunningham P.V. Drumm, L.K. Fifield

10B Mass and excited states of ^{51}Ca
W.N. CATFORD, L.K. Fifield, N.A. Orr, T.R. Ophel, D.C. Weisser, C.L. Woods

11C A modification of the Skyrme interaction and the RMS radii of nuclei
S. MASSEN

12B Nuclear and electronic g-factors of ^{211}Fr and the nuclear single-particle structure in the range $^{207-228}$Fr
C. EKSTRÖM, L. Robertsson, A. Rosen

13B Quadrupole moments of the first excited states of ^{134,136,138}Ba
A.M. BAXTER, S.M. Burnett, P.J. Rothschild, M.P. Fewell, G.J. Gyapong, R.H. Spear

14B Doubly decoupled bands in ^{176,178}Re
A.J. KREINER, J. Davidson, M. Davidson, M. Debray, G. Falcone, D. Hojman, I. Mayans, C. Pomar D. Santos

15B Gross structures in triton induced reactions
L. ZYBERT, J.B.A. England, O. Karban, R. Zybert, D. Barker, M. Mannion, J.M. Nelson C. Ogilvie, L. Potvin, N. Clarke, K. Pearce, D.L. Watson

16B Effective interactions for the sd and pf shells
W.A. RICHTER, B.A. Brown, M. Van der Merwe, R.E. Julies

17C Vector analyzing powers of the reactions ^{2}H($\vec{d}$,n)^{3}He and ^{2}H($\vec{d}$,p)^{3}H
N. KNIEST, E. Pfaff, G. Reiter, G. Clausnitzer

18C Deuteron wavefunctions
S.G. COOPER, M.W. Kermode, S. Klarsfeld

19G Quark-quark potential in the nucleon-nucleon interaction
A. OSMAN

20C Core breaking in nuclei around mass 40
B.J. COLE, W.L. Gadinabokao

41C The effect of memory in deeply inelastic collisions
D. PAL, S. Chattopadhyay, K. Kar

42B Excitation of giant resonances in ^{90}Zr by inelastic scattering of 270 MeV ^{3}He particles
V.S. RAMAMURTHY, A. Saxena, S. Kailas, P.P. Singh, P. Schwandt, E.J. Stephenson, Q. Chen D.L. Friesel

43C Study of $^{6}Li(\alpha,\alpha d)^{4}He$ at E_{α} = 45 MeV
A. ROY, R.K. Bhowmik, C.V.K. Baba, M.G. Betigeri, R. Mythili, A.K. Jain

44C Role of direct and pre-equilibrium processes in description of threshold reaction excitation functions
V.N. MANOKHIN

45C Non-equilibrium many-body theory for thermal equilibration of hot nuclei
J. BAR-TOUV, Yu L. Khait

46C Resonant states in α-^{40}Ca elastic scattering
J.P.F. SELLSCHOP, A. Zucchiatti, L. Mirman, M. Gering, S. Di Salvo

47C Emission of LRA particles in the spontaneous fission of ^{244}Pu
M. VERBOVEN, B. Thierens, P. D'hondt, E. Jacobs, D. De Frenne

48A Energy and target dependence of the breakup contribution to the composite particle optical potential
A.A. IOANNIDES, R.S. Mackintosh

49C ^{4}He D-state effects in the $^{2}H(d,\gamma)^{4}He$ reaction at low energies
A.M. EIRÓ, F.D. Santos, A. Arriaga

50G Precise polarization measurements in neutron-proton scattering at 25 MeV
J. SROMICKI, W. Haeberli, D. Holslin, P.A. Quin

51C The 0° photodisintegration of the deuteron at low energy
A. DE GRAEVE, A. Zieger, R. Van de Vyver, H. Ferdinande, P. Berkvens, B. Ziegler D. Ryckbosch, L. Van Hoorebeke, P. Van Otten

52C Nuclear structure effects on the E1 strength located in the giant dipole resonance of heavy nuclei
R. NOLTE, A. Baumann, K.W. Rose, D. Nowotny, F. Schröder, P. Rullhusen, F. Smend M. Schumacher

53B The measurements for total widths of several levels in ^{24}Mg and ^{40}Ca
GUO ZHENDI, Gu Xiliang, Gu Mu, Hu Yude, Yang Fujia

54A The reaction $D(\gamma,p)n$ at intermediate energies
P. WALLACE, J.R.M. Annand, I. Anthony, G.I. Crawford, S.N. Dancer, S.J. Hall, J.D. Kellie J.C. McGeorge, A. McPherson, I.J.D. MacGregor, R.O. Owens, B. Schoch, J. Vogt, R. Beck D. Branford, S.V. Springham

55E Measurements of the deuteron magnetic form factor at high Q^2
P. BOSTED, R.G. Arnold, L. Clougher, G. DeChambrier, A.T. Katramatou, A. Lung, G. Petratos A. Rahbar, S.E. Rock, Z.M. Szalata, R.A. Gearhart, B. Debebe, R.S. Hicks, A. Hotta G.A. Peterson, J. Alster, J. Lichtenstadt, J. Lambert

56B Effects of core polarization on the Coulomb form factors in Ni isotopes
A. YOKOYAMA, K. Ogawa

57D Nuclear medium effects in $^{40}Ca(e,e'p)^{39}K$ experiment
A. GERARD, M. Bernheim, A. Magnon, C. Marchand, J. Morgenstern, J. Mougey, J. Picard D. Reffay, S. Turck-Chieze, P. Vernin, H.E. Jackson, G.P. Capitani, E. De Sanctis S. Frullani, G. Garibaldi

58B Inclusive charge longitudinal response in finite nuclei
A. MOLINARI, W. Alberico, P. Czerski, M. Ericson

59B Stretched state excitations in the $^{26}Mg(p,n)^{26}Al$ reaction at 134 MeV
B.D. ANDERSON, C. Lebo, R. Madey, J.W. Watson, C.C. Foster

77B The polarization of high-energy gamma rays from hot $^{125,127}Cs$ compound nuclei
CHR. BARGHOLTZ, Th. Arctaedius, R.E. Ekström, K. Fransson, B. Ritzen, P.-E. Tegnér

78C Angular momentum distribution of the $^{58}Ni+^{124}Sn$ fusion reaction
B.A. ROBSON, Y. Kondo, J.J.M. Bokhorst, D.J. Hinde, J.R. Leigh

79C Recoil studies of incomplete fusion
D.J. PARKER, J. Asher, J.J. Hogan

80F Effects of the vacuum polarization on sub-coulomb ^{12}C-^{12}C-scattering
P. EGELHOF, D. Vetterli, G. Baur, W. Boeglin, R. Henneck, A. Klein, H. Mühry, G.R. Plattner F. Rösel, I. Sick, D. Trautmann, A. Weller, M. Jaskóla

81C Cross sections of fast neutron induced reactions on aluminium and silicon
J. CSIKAI, Zs. Lantos, S. Sudár, T. Chimoye, T. Vilaithong, N. Chirapatpimol

82C A study of breakup for neutron rich projectiles
E.W. MACDONALD, A.C. Shotter, D. Branford, J. Rahighi, J. Yorkston

83C Impact parameter dependence of the specific entropy and the cluster formation in high-energy heavy-ion collisions
V.D. TONEEV, K.K. Gudima, G. Röpke, H. Schulz

84C Neutron and proton contributions to excited states in ^{39}K by π^+ and π^- scattering
M.J. SMITHSON, S. Mordechai, D.S. Oakley, C.F. Moore, R. Gilman, J.D. Zumbro, H.T. Fortune P.H. Kutt, C.L. Morris, K.S. Dhuga, S.J. Seestrom-Morris, D.L. Watson

85A Total reaction cross sections of 50 and 65 MeV pions on nuclei
E. FRIEDMAN, O. Meirav, A. Altman, D.R. Gill, R.R. Johnson

86C Re-evaluation of fission yields and charge distribution in $^{235}U(n_{th},f)$
CHUNG CHIEN

87C Fission fragment spins from prompt γ-ray multiplicities
D.E. HOOK, J.L. Durell, Y.A. Abdelrahman, G.G. Colvin, W. Gelletly, J.O. Newton W.R. Phillips

88C Photofission cross-sections of nuclei at intermediate energies
J.B. MARTINS, O.A.P. Tavares, J.D. Pinheiro Filho, I.O. De Souza

89C Hard photon production in nucleus-nucleus collisions at 30 MeV/u and 44 MeV/u
J.A. PINSTON, R. Bertholet, M. Kwato Njock, M. Maurel, E. Monnand, H. Nifenecker, P. Perrin F. Schussler, D. Barnéoud, C. Guet, Y. Schütz

90C Momentum distribution of protons in ^{6}Li
A.M. VAN DEN BERG, J.B.J.M. Lanen, J.F.J. Van den Brand, H. Hendriks, J.W.A. Den Herder E. Jans, P.H.M. Keizer, G.J. Kramer, L. Lapikás, E.N.M. Quint, P.K.A. De Witt Huberts H.P. Blok, R. Ent, G. Van der Steenhoven

91B Negative parity bands in ^{98}Tc and ^{100}Rh
A.M. BIZZETI-SONA, P. Blasi, A.A. Stefanini, G. Galeazzi

92B Octupole multiplets in the one-proton-hole nucleus ^{145}Eu
B. RUBIO, P. Kleinheinz, A. Ercan, R. Julin, M. Lach, L.G. Mann, W. Stöffl, E.A. Henry, V. Dave J. Blomqvist

93H The Karlsruhe polarized neutron facility POLKA
P. DOLL, G. Fink, R.W. Finlay, T.D. Ford, W. Heeringa, H.O. Klages, G. Schmalz, H. Skacel H. Krupp, Chr. Maier

94H High resolution thermal detectors
G. NYMAN, H.H. Andersen, P. Delsing, P.G. Hansen, B. Jonson, H.L. Ravn, K. Riisager

95C The (t,p) L=0 strength to excited states in gadolinium isotopes
G. LØVHØIDEN, T.F. Thorsteinsen, E. Andersen, M. Kiziltan, D.G. Burke

96A Measurements of tensor observables in πd elastic scattering
N.R. STEVENSON, Y.M. Shin

97C Measurement of transfer reactions below the Coulomb barrier using a recoil mass separator
A.E. SMITH, R.R. Betts, P.M. Evans, C. Pass, N. Poffé, L. Stuttge, J.S. Lilley, D.W. Banes
K.A. Connell, J. Simpson, J.R.H. Smith, B.R. Fulton, S.J. Bennett, P.J. Woods, A.N. James

98B Energy and angular distribution of ^{252}Cf fission neutrons in the low energy range
D. SEELIGER, A. Lajtai, H. Marten, D. Richter, W. Neubert

99A 3-Nucleon-absorption of pions in ^{3}He
G. BACKENSTOSS, M. Izycki, P. Salvisberg, M. Steinacher, P. Weber, H.J. Weyer, S. Cierjacks
S. Ljungfelt, H. Ullrich, M. Furić, T. Petković

LIST OF DELEGATES

Abdelrahman,Y.A.	Univ. Manchester	Dept. Phys.	Manchester M13 9PL	GBR
Abriola,D.	CNEA	Dept. Fisica	1429 Buenos Aires	ARG
Abu-Kamar,S.	Univ. Surrey	Dept. Phys.	Guildford GU2 5XH	GBR
Agodi,C.	INFN, Lab. Naz. del Sud		95129 Catania, Sicily	ITA
Ait-Tahar,S.	Univ. Oxford	Dept. Nucl. Phys.	Oxford OX1 3RH	GBR
Ajzenberg-Selove, F.	Univ. Pennsylvania	Dept. Phys.	Philadelphia,PA 19104	USA
Akiyama,Y.	Nihon Univ.	Dept. Phys.	Setagaya-ku,Tokyo 156	JPN
Alder,K.	Univ. Basel	Inst. f. Theor.Phys	4056 Basel	SWI
Allcock,S.C.	Univ. Oxford	Dept. Nucl. Phys.	Oxford OX1 3RH	GBR
Amos,K.	Univ. Melbourne	School Phys.	Parkville 3052	AUS
Anagnostatos,G.S.	NRC Demokritos	Tandem Acc. Lab	Aghia-Paraskevi-Attiki	GRC
Anantaraman,N.	Michigan State Univ.	Cyclotron Lab.	East Lansing MI 48824	USA
Anderson,B.D.	Kent State Univ.	Dept. Phys.	Ohio 44242	USA
Andres,V.	Univ. Sevilla	Dept. Fis. Nucl.	41080 Sevilla	ESP
Annand,J.R.M.	Univ. Glasgow	Dept. Nat. Phil.	Glasgow G12 8QQ	GBR
Ansari,A.	Inst. Phys.		Bhubaneswar 751005	IND
Arctaedius,Th.	Univ. Stockholm	Dept. Phys.	113 46 Stockholm	SWE
Arena,N.	INFN, Lab. Naz. del Sud		95129 Catania, Sicily	ITA
Arve,P.O.	SERC Daresbury Lab.	TCS Div.	Warrington WA4 4AD	GBR
Asher,J.	AERE Harwell	Nucl. Phys. Div.	Didcot, Oxon OX11 0RA	GBR
Austern,N.	Univ. Pittsburgh	Dept. Phys.	Pittsburgh, PA 15260	USA
Austin,S.M.	Michigan State Univ.	Cyclotron Lab.	East Lansing, MI 48824	USA
Baldock,R.A.	Univ. Oxford	Dept. Theor. Phys.	Oxford OX1 3NP	GBR
Ball,G.C.	AECL Chalk River Lab.		Ontario, K0J 1J0	CAN
Banes,D.W.	SERC Daresbury Lab.	NSF Div.	Warrington WA4 4AD	GBR
Bargholtz,Chr.	Univ. Stockholm	Dept. Phys.	113 46 Stockholm	SWE
Barker,F.C.	Aust. Nat. Univ.	Dept. Theor. Phys.	Canberra, ACT 2601	AUS
Barnett,A.R.	Univ. Manchester	Dept. Phys.	Manchester M13 9PL	GBR
Barrett,B.R.	National Sci. Foundation	Phys. Div.	Washington DC 20550	USA
Barrett,R.C.	Univ. Surrey	Dept. Phys.	Guildford GU2 5XH	GBR
Bartholomew,G.A.	AECL Chalk River Lab.	Nucl. Phys. Div.	Ontario K0J 1J0	CAN
Bar-Touv,J.	Ben-Gurion Univ.	Dept. Phys.	Beer-Sheva	ISR
Bauhoff,W.	Univ. Hamburg	Theor. Kernphysik	D 2000 Hamburg 50	FRG
Bawin,M.	Univ. Liège	Inst. Phys. B5	4000 Liège	BEL
Baxter,A.M.	Aust. Nat. Univ.	Dept. Phys.	Canberra, ACT 2601	AUS
Baym,G.A.	Univ. Illinois	Dept. Phys.	Urbana, IL 61801	USA
Becha,M.B.	Univ. Birmingham	Dept. Phys.	Birmingham B15 2TT	GBR
Beghini,S.	Univ. Padova e INFN	Dipt. Fisica	35131 Padova	ITA
Bellwied,R.	Univ. Mainz	Inst. f. Kernchemie	6500 Mainz	FRG
Bergqvist,I.	Univ. Lund	Dept. Phys.	223 62 Lund	SWE
Berrier-Ronsin,G.	IPN Orsay		91405 Orsay Cedex	FRA
Bertulani,C.A.	KFA Jülich	Inst. f. Kernphys.	5170 Jülich	FRG
Betts,R.R.	Argonne Nat. Lab.	Phys. Div.	Argonne IL 60439	USA
Beyer,M.	Johannes Gutenberg Univ	Inst. f. Kernphys.	6500 Mainz	FRG
Bharuth-Ram,K.	Univ. Durban-Westville	Dept. Phys.	Durban 4000	RSA
Bickerstaff,R.P.	Univ. Adelaide	Dept. Phys.	Adelaide SA 5001	AUS
Birse,M.C.	Univ. Manchester	Dept. Phys.	Manchester M13 9PL	GBR
Bishop,P.J.	Univ. Liverpool	Dept. Phys.	Liverpool L69 3BX	GBR
Bishop,R.F.	UMIST	Dept. Maths.	Manchester M60 1QD	GBR
Bizzeti,P.G.	Univ. Firenze e INFN	Dipt. Fisica	50125 Firenze	ITA
Bizzeti-Sona,A.M.	Univ. Firenze e INFN	Dipt. Fisica	50125 Firenze	ITA
Blachot,J.	CEN Grenoble	DRF- Lab.Phys.Nucl.	38041 Grenoble	FRA
Blasi,P.	Univ. Firenze	Dipt. Fisica	50125 Firenze	ITA
Blok,H.P.	Vrije Univ.	Natuurkundig Lab.	1007 MC Amsterdam	NLD
Blyth,C.O.	Univ. Birmingham	Dept. Phys.	Birmingham B15 2TT	GBR
Boeker,E.	Vrije Univ.	Natuurkundig Lab.	1081 HV Amsterdam	NLD
Bohigas,O.	IPN Orsay	Div. de Phys. Nucl.	91406 Orsay Cedex	FRA
Bortignon,P.F.	Inst. G. Galilei	Dipt. Fisica	35131 Padova	ITA
Bosch,F.	GSI Darmstadt		6100 Darmstadt 1	FRG
Bosted,P.	SLAC		Stanford, CA 94305	USA
Bowman,J.D.	Los Alamos National Lab.	Dept. Nucl. Phys.	Los Alamos NM 87545	USA
Brandenburg,R.	Univ. Basel	Inst. f.Theor. Phys	4056 Basel	SWI
Brandolini,F.	Univ. Padova	Dipt. Fisica	35131 Padova	ITA
Branford,D.	Univ. Edinburgh	Dept. Phys.	Edinburgh EH9 3JZ	GBR
Brink,D.M.	Univ. Oxford	Dept. Theor. Phys.	Oxford OX1 3PH	GBR
Bromley,D.Allan	Yale Univ.	Dept. Phys.	New Haven CT 06511	USA
Broude,C.	Weizmann Insitute	Dept. Phys.	Rehovot 76100	ISR
Brown,B.A.	Michigan State Univ.	Cyclotron Lab.	East Lansing MI 48824	USA
Bruge,G.	CEN Saclay	DPhN/ME	91191 Gif-sur-Yvette	FRA
Brussaard,P.J.	Rijksuniv. Utrecht	Fysisch Lab.	3508 TA Utrecht	NLD
Budzanowski,A.	Inst. of Nuclear Physics		31-342 Krakow	POL
Burcham,W.E.	Univ. Birmingham	Dept. Phys.	Birmingham B15 2TT	GBR

Burde,J.	Hebrew Univ.	Racah Inst. Physics	Jerusalem	ISR
Bussa,M.P.	INFN Torino		10125 Torino	ITA
Butler,P.A.	Univ. Liverpool	Dept. Phys.	Liverpool L69 3BX	GBR
Cardella,G.	INFN, Lab. Naz. del Sud		95129 Catania, Sicily	ITA
Carter,J.	Univ. Witwatersrand	Dept. Phys.	Johannesburg 2001	RSA
Catford,W.N.	Aust. Nat. Univ.	Dept. Nucl. Phys.	Canberra, ACT 2601	AUS
Cavallaro,Sl.	Univ. Catania	Dipt. Fisica	95129 Catania, Sicily	ITA
Cerny,J.	Univ. California	Dean's Office	Berkeley, CA 94720	USA
Ceuleneer,R.	Univ de l'Etat à Mons	Fac. des Sciences	7000 Mons	BEL
Chant,N.S.	Univ. Maryland	Dept.Phys. & Astron	College Park, MD 20742	USA
Chapman,R.	Univ. Manchester	Dept. Phys.	Manchester M13 9PL	GBR
Chapuran,T.	Univ. Pennsylvania	Dept. Phys.	Philadelphia PA 19104	USA
Chen, B.Q.	Univ. Hannover	Inst. Theor. Phys.	3000 Hannover 1	FRG
Cheng,X.Wu	Academia Sinica	Inst. Nucl. Res.	Shanghai	CHN
Chung Chien	National Tsing Hua Univ.	Inst. of Nucl. Sci.	Hsinchu, Taiwan 30043	CHN
Chung,K.C.	Cen. Bra. Pesq. Fis.		Rio de Janeiro 22290	BRA
Cizewski,J.A.	Rutgers Univ.	Serin Phys. Lab.	Piscataway, NJ 08854	USA
Clarke,N.M.	King's College London	Dept. Phys.	London WC2R 2LS	GBR
Close,F.E.	Rutherford Appleton Lab.	HEP Div.	Chilton, Oxon OX11 0RA	GBR
Cohen,L.	Institute of Physics		London SW1X 8QX	GBR
Cohler,M.D.	Univ. York	Dept. Phys.	York, YO1 5DD	GBR
Cole,B.J.	Univ. Witwatersrand	Dept. Phys.	Johannesburg 2001	RSA
Coley,D.A.	Univ. Surrey	Dept. Phys.	Guildford GU2 5XH	GBR
Conlon,T.W.	AERE Harwell	Nucl. Phys. Div.	Didcot, Oxon OX11 0RA	GBR
Coon,S.A.	Univ. Arizona	Dept. Phys.	Tucson, Arizona 85721	USA
Cooper,S.G.	Univ. Liverpool	DAMTP	Liverpool L69 3BX	GBR
Copnell,J.	Univ. Sussex	School MAPS	Brighton BN1 9QH	GBR
Couvert,P.	CEN Saclay	DPhN/ME	91191 Gif-sur-Yvette	FRA
Covello,A.	Univ. Napoli	Dipt. Fisica Nucl.	80125 Napoli	ITA
Cranmer-Gordon,H.W.	Univ. Liverpool	Dept. Phys.	Liverpool L69 3BX	GBR
Csikai,J.	Kossuth Univ.	Inst. Exp. Physics	Debrecen 4001	HUN
Daley,H.J.	SERC Daresbury Lab.	TCS Div.	Warrington WA4 4AD	GBR
Dalmas,J.	CEN Bordeaux		33170 Gradignan	FRA
Davidson,W.F.	NRC Canada		Ottawa K1S 0P5	CAN
Davinson,T.	Univ. Edinburgh	Dept. Phys.	Edinburgh EH9 3JZ	GBR
Davis,C.A.	TRIUMF		Vancouver, BC V6T 2A3	CAN
Davis,E.D.	Univ. Witwatersrand	Dept. Phys.	Johannesburg 2001	RSA
Davis,N.J.	Univ. Edinburgh	Dept. Phys.	Edinburgh EH9 3JZ	GBR
Dawson,J.F.	Univ. New Hampshire	Dept. Phys.	Durham, NH 03842	USA
De Frenne,D.	Rijksuniv. Gent	Nucl. Phys. Lab.	9000 Gent	BEL
De Graeve,A.	Rijksuniv. Gent	Nucl. Phys. Lab.	9000 Gent	BEL
De Jager,C.W.	NIKHEF-K		1009 AJ Amsterdam	NLD
De Sanctis,E.	INFN Frascati		00044 Frascati	ITA
De Swiniarski,R.	Univ. Grenoble	Inst. Sciences Nucl	38026 Grenoble-Cedex	FRA
De Witt Huberts,P.K.A.	NIKHEF-K		1009 AJ Amsterdam	NLD
Dehnhard,D.	Univ. Minnesota	Dept. Phys.& Astron	Minneapolis, MN 55455	USA
Dellafiore,A.	Univ. Firenze	Dipt. Fisica	50125 Firenze	ITA
Détraz,C.	GANIL		14021 Caen -Cedex	FRA
Diaz,J.	Univ. Valencia	Faculty of Physics	Valencia	ESP
Dieperink,A.E.L.	KVI Groningen		9747 AA Groningen	NLD
Dillon,G.	INFN Genova	Dipt. Fisica	16146 Genova	ITA
Diószegi,I.	Hungarian Acad. Sciences	Inst. of Isotopes	1525 Budapest	HUN
Dodd,L.R.	Univ. Adelaide	Dept. Math. Phys.	Adelaide SA 5001	AUS
Dojo,M.	Kyushu Univ.	Cen. Adv. Ins. Anal	Kasuga-shi, 816	JPN
Doll,P.	KFA Karlsruhe	Inst.f.Kernphysik I	7500 Karlsruhe	FRG
Doubre,H.	GANIL		14021 Caen-Cedex	FRA
Dover,C.B.	Brookhaven National Lab.	Dept. Phys.	Upton, New York 11973	USA
Dreiss,G.J.	Physical Review		Ridge, New York 11961	USA
Dünnweber,W.	Univ. München	Sektion Physik	8046 Garching	FRG
Durell,J.L.	Univ. Manchester	Dept. Phys.	Manchester M13 9PL	GBR
Eastham,D.A.	SERC Daresbury Lab.	NSF Div.	Warrington WA4 4AD	GBR
Eckle,F.J.	Univ. München	Sektion Physik	8046 Garching	FRG
Eckle,G.	Univ. München	Sektion Physik	8046 Garching	FRG
Egelhof,P.	Univ. Mainz	Inst. f. Physik	6500 Mainz	FRG
Eiró,A.M.	Univ. Lisboa	Centro Fisica Nucl.	1699 Lisboa Codex	PRT
Ekström,A.C.	NFL - Studsvik	Nucl. Tech. Div.	61182 Nyköping	SWE
Ekström,L.P.	Univ. Lund	Dept. Phys.	22362 Lund	SWE
Elliott,J.P.	Univ. Sussex	School MAPS	Brighton BN1 9QH	GBR
Ericson,T.E.O.	CERN	Theor. Phys. Div.	1211 Geneva 23	SWI
Eskola,K.	Univ. Helsinki	Dept. Phys.	00170 Helsinki	FIN
Evans,J.A.	Univ. Sussex	School MAPS	Brighton BN1 9QH	GBR
Evans,P.M.	Univ. Oxford	Nucl. Phys. Lab.	Oxford OX1 3PH	GBR

Eyrich,W.	Univ. Erlangen-Nürnberg	Phys. Inst.	8520 Erlangen	FRG
Fatemian,M.	Univ. Oxford	Dept. Theor. Phys.	Oxford OX1 3PH	GBR
Fawcett,M.J.	Univ. Manchester	Dept. Phys.	Manchester M13 9PL	GBR
Feilmair,W.	Tech. Univ. Wien	Inst. f. Kernphys.	1020 Wien	AUT
Ferrero,J.L.	Inst. Fis. Corpusc.		Burjasot, Valencia	ESP
Feshbach,H.	Massachusetts Inst. Tech	Dept. Phys.	Cambridge, MA 02139	USA
Fewell,M.P.	Aust. Nat. Univ.	Dept. Nucl. Phys.	Canberra, ACT 2601	AUS
Figuera,A.S.	Univ. Catania e INFN	Dipt. Fisica	95129 Catania, Sicily	ITA
Findlay,D.J.S.	UKAEA-Harwell	Nucl. Phys. Div.	Didcot, Oxon OX11 ORA	GBR
Fiolhais,C.	Univ. Coimbra	Dept. Fisica	3000 Coimbra	PRT
Firk,F.W.K.	Yale Univ.	Dept. Phys.	New Haven, CT 06520	USA
Flerov,G.N.	JINR, Dubna	Lab. Nucl.Reactions	Moscow	USR
Fornal,B.	INFN Legnaro		35020 Legnaro, Padova	ITA
Forsyth,P.D.	Univ. Liverpool	Dept. Phys.	Liverpool L69 3BX	GBR
Franco,V.	Brooklyn College	Dept. Phys.	Brooklyn, NY 11210	USA
Frank,I.M.	JINR, Dubna		Moscow	USR
Friedman,E.	Hebrew Univ.	Racah Inst. Phys.	Jerusalem 91904	ISR
Fukuda,K.	Osaka Prefecture	Radiation Centre	Sakai, Osaka	JPN
Fulton,B.R.	Univ. Birmingham	Dept. Phys.	Birmingham B15 2TT	GBR
Gaarde,C.	Niels Bohr Institute	Tandem Acc. Lab.	2100 Copenhagen	DNK
Gadinabokao,W.L.	Univ. Bophuthatswana		Mafeking 8670	RSA
Gadioli,E.	Univ Milano e INFN	Dipt. Fisica	20133 Milano	ITA
Garrett,J.D.	Niels Bohr Institute		2100 Copenhagen	DNK
Garvey,G.T.	Los Alamos Nat. Lab.	Dept. Nucl. Phys.	Los Alamos, NM 87545	USA
Gelletly,W.	Univ. Manchester	Dept. Phys.	Manchester M13 9PL	GBR
Gérard,A.	CEN Saclay	DPhN/HE	91191 Gif-sur-Yvette	FRA
Gerl,J.	Univ. Heidelberg	Phys. Institut	6900 Heidelberg	FRG
Gersten,A.	Ben-Gurion Univ.	Dept. Phys.	Beer-Sheva 84105	ISR
Ghazarian,A.	Univ. York	Dept. Phys.	York, YO1 5DD	GBR
Giannatiempo,A.	Univ. Firenze	Dipt. Fisica	50125 Firenze	ITA
Giansiracusa,G.	Univ. Catania	Ist. Fisica	95129 Catania, Sicily	ITA
Gidefeldt,L.	Swedish Nat.Sci.Res.Coun.		113 85 Stockholm	SWE
Gils,H.J.	KFA Karlsruhe	Inst.Kernphysik III	7500 Karlsruhe	FRG
Glaudemans,P.W.M.	Univ. Utrecht	Fysisch Lab.	3508 TA Utrecht	NLD
Goettig,L.	Univ. Manchester	Dept. Phys.	Manchester M13 9PL	GBR
Goldfarb,L.J.B.	Univ. Manchester	Dept. Phys.	Manchester M13 9PL	GBR
Goldring,G.	Weizmann Inst.of Science	Dept. Phys.	Rehovot 76100	ISR
Gomes,P.R.S.	Univ. Fed. Fluminense	Dept. Fisica	Niteroi, RJ 24000	BRA
Gomez-Camacho,J.	Univ. Surrey	Dept. Phys.	Guildford, GU2 5XH	GBR
Greaves,D.	Nuclear Physics	c/o Nordita	2100 Copenhagen	DNK
Green, L.L.	SERC Daresbury Lab.	Director's Office	Warrington WA4 4AD	GBR
Greenlees,G.W.	Univ. Minnesota	Dept. Phys.	Minneapolis,MN 55455	USA
Griffioen,K.A.	Univ. Utrecht	Van de Graaff Lab.	3508 TA Utrecht	NLD
Griffith,J.A.R.	Univ. Birmingham	Dept. Phys.	Birmingham B15 2TT	GBR
Gromov,K.Ya	JINR, Dubna		Moscow	USR
Guardiola,R.	Univ. Granada	Dept. Fis. Nucl.	18001 Granada	ESP
Guo Zhendi	Fudan Univ.	Dept. Nucl. Science	Shanghai	CHN
Gyarmati,B.	Inst. Nuclear Research		4001 Debrecen	HUN
Gysin,C.	Univ. Basel	Inst. Phys.	4056 Basel	SWI
Hadjimichael,E.	Fairfield Univ.	Dept. Phys.	Fairfield, CT 06430	USA
Hagemann,G.B.	Niels Bohr Institute	Tandem Acc. Lab.	2100 Copenhagen	DNK
Hamburger,E.W.	Univ. São Paulo	Inst. Fisica	São Paulo	BRA
Hammans,M.	Univ. Basel	Inst. Phys.	4056 Basel	SWI
Hartwig,D.	GSI Darmstadt		6100 Darmstadt	FRG
Hasegawa,M.	Fukuoka Dental College	Lab. of Physics	Fukuoka 81401	JPN
Hawkes,N.P.	AERE Harwell	Nucl. Phys. Div.	Didcot, Oxon OX11 ORA	GBR
Haxton,W.	Univ. Washington	Dept. Phys.	Seattle, WA 98195	USA
Hefter,E.F.	Springer-Verlag		6900 Heidelberg	FRG
Heide,P.	Tech. Univ. Berlin	Inst. f. Kernphys.	1000 Berlin 12	FRG
Helppi,H.	Lappeenranta Univ. Tech.	Dept. Phys.	SF-53851 Lappeenranta	FIN
Hessey,N.P.	Univ. Birmingham	Dept. Phys.	Birmingham B15 2TT	GBR
Hintz,N.M.	Univ. Minnesota	School Phys.	Minneapolis, MN 55455	USA
Hodgson,P.E.	Univ. Oxford	Nucl. Phys. Lab.	Oxford OX1 3PH	GBR
Hoernle,R.F.A.	Univ. Witwatersrand	Schonland Res. Ctre	Johannesburg 2001	RSA
Holbrow,C.H.	Colgate Univ.	Dept. Phys.	Hamilton, NY 13346	USA
Holmes,P.A.	Univ. Oxford	Dept. Nucl. Phys.	Oxford OX1 3PH	GBR
Holmgren,H.D.	S.E. Univ. Res. Assoc.		Washington DC 20036	USA
Hook,D.E.	Univ. Manchester	Dept. Phys.	Manchester M13 9PL	GBR
Horiuchi,H.	Kyoto Univ.	Dept. Phys.	Kyoto 606	JPN
Huby,R.	Univ. Liverpool	Dept. Phys.	Liverpool L69 3BX	GBR
Hunn,C.	Univ. Bergen	Dept. Phys.	5000 Bergen	NOR
Iazzi,F.	INFN Torino	Dip.Fisica-Politec.	10129 Torino	ITA

Igarashi,M.	Tokyo Medical College	Dept. Phys.	Tokyo 160	JPN
Inamura,T.	RIKEN	Cyclotron Lab.	Saitama 351-01	JPN
Ioannides,A.A.	The Open University	Dept. Phys.	Milton Keynes MK7 6AA	GBR
Iori,I.	Univ. Milano	Dept. Phys.	20133 Milano	ITA
Irvine,J.M.	Univ. Manchester	Dept. Phys.	Manchester M13 9PL	GBR
Isgur,N.	Univ. Toronto	Dept. Phys.	Ontario M5S 1A7	CAN
Izycki,M.	Univ. Basel	Dept. Phys.	4056 Basel	SWI
Jacobs,E.	Rijksuniv. Gent	Nucl. Phys. Lab.	9000 Gent	BEL
Jain,B.K.	Bhabha Atom. Res. Centre	Nucl. Phys. Div.	Bombay 400 085	IND
James,A.N.	Univ. Liverpool	Dept. Phys.	Liverpool L69 3BX	GBR
Jean,D.	CEN- Bordeaux		33170 Gradignan	FRA
Jelley,N.A.	Univ. Oxford	Dept. Nucl. Phys.	Oxford OX1 3PH	GBR
Johansen,E.F.	Univ. Oslo	Inst. Phys.	0316 Oslo 3	NOR
Johnson,R.C.	Univ. Surrey	Dept. Phys.	Guildford GU2 5XH	GBR
Jones,G.D.	Univ. Liverpool	Dept. Phys.	Liverpool L69 3BX	GBR
Jonson,B.N.J.	Chalmers Univ. of Tech.	Dept. Phys.	412 96 Göteborg	SWE
Kalfas,C.	NRC Demokritos	Tandem Acc. Lab.	Aghia Paraskevi 15310	GRC
Kamimura,M.	Kyushu Univ.	Dept. Phys.	Fukuoka 812	JPN
Karban,O.	Univ. Birmingham	Dept. Phys.	Birmingham B15 2TT	GBR
Kasagi,J.	Tokyo Inst. Technology	Dept. Phys.	Meguro, Tokyo	JPN
Kato,N.	Kyushu Univ.	Dept. Phys.	Fukuoka 812	JPN
Keeling,P.R.	Univ. Oxford	Nucl. Phys. Lab.	Oxford OX1 3PH	GBR
Keinonen,J.	Univ. Helsinki	Accelerator Lab.	00550 Helsinki	FIN
Kerek,A.	Research Inst. Physics		104 05 Stockholm 50	SWE
Kermode,M.W.	Univ. Liverpool	DAMTP	Liverpool L69 3BX	GBR
Khadkikar,S.B.	Physical Res. Lab.		Ahmedabad 380 009	IND
Khazaie,F.	Univ. Manchester	Dept. Phys.	Manchester M13 9PL	GBR
Kim,C.O.	Korea Univ.	Dept. Phys.	Seoul 132	KOR
Kirson,M.Z.	Weizmann Inst.of Science	Dept. Nucl. Phys.	Rehovot 76100	ISR
Kirwan,A.	Univ. Liverpool	Dept. Phys.	Liverpool L69 3BX	GBR
Klein,A.	Univ. Pennsylvania	Dept. Phys.	Philadelphia,PA 19104	USA
Kleinheinz,P.	KFA Jülich	Inst. f. Kernphys.	5170 Jülich 1	FRG
Kniest,N.	Justus Liebig Univ.	Inst. f. Kernphys.	6300 Giessen	FRG
König,V.	Inst.Medium Energy Phys		8023 Zürich	SWI
Konijn,J.	NIKHEF-K		1009 AJ Amsterdam	NLD
Koo,W.K.	Univ. Sci. Malaysia	School Phys.	11800 USM, Penang	MYS
Kopecky,J.	Neths. Energy Res.Found.		1755 ZG Petten	NLD
Korner,H.J.	Tech. Univ. Munchen	Dept. Phys. E12	8046 Garching	FRG
Kreiner,A.J.	CNEA	Dipt. Fisica	1429 Buenos Aires	ARG
Krupchitsky,P.A.	ITEP		117259 Moscow	USR
Kundig,W.	Univ. Zürich	Phys. Inst.	8001 Zürich	SWI
Lancman,H.	Brooklyn College, CUNY	Dept. Phys.	Brooklyn, NY 11210	USA
Landowne,S.	Argonne National Lab.	Phys. Div.	Argonne, IL 60439	USA
Lanza,E.G.	INFN, Lab. Naz. del Sud		95129 Catania, Sicily	ITA
Lawrie,J.J.	National Accel. Centre		Faure 7131	RSA
Lecolley,J.F.	Univ. Caen	Lab. Phys. Nucl.	14000 Caen	FRA
Leidemann,W.	Johannes Gutenberg Univ.	Inst. f. Kernphys.	6500 Mainz	FRG
Leitner,W.	Univ. Tübingen	Phys. Inst.	7400 Tübingen	FRG
Lejeune,A.	Univ. Liège	Inst. Physique	4000 Liège 1	BEL
Lenske,H.	Univ. München	Sektion Physik	8046 Garching	FRG
Leonardi,R.	Univ. Povo	Dept. Phys.	38050 Povo (Trento)	ITA
Leslie,J.R.	Queen's Univ.	Dept. Phys.	Kingston, ON K7L 3N6	CAN
Levit,S.	Weizmann Inst.of Science	Dept. Nucl. Phys.	Rehovot 76100	ISR
Li Shounan	Inst. of Atomic Energy		Beijing	CHN
Lieb,K.P.	Univ. Göttingen	Dept. Phys.	3400 Göttingen	FRG
Lilley,J.S.	SERC Daresbury Lab.	NSF Div.	Warrington WA4 4AD	GBR
Liotta,R.J.	Research Inst. Phys.-AFI		10405 Stockholm	SWE
Li-Scholz,A.	State Univ.of New York	Dept. Phys.	Albany, NY 12222	USA
Lisle,J.C.	Univ. Manchester	Dept. Phys.	Manchester M13 9PL	GBR
Lister,C.J.	Univ. Manchester	Dept. Phys.	Manchester M13 9PL	GBR
Liuti,S.	Ist. Superiore di Sanita	Sezione INFN	00161 Rome	ITA
Logan,B.A.	Univ. Ottawa	Dept. Phys.	Ottawa K1K 2A6	CAN
Lombardo,U.	Univ. Catania	Dipt. Fisica	95129 Catania, Sicily	ITA
Lomon,E.L.	Massachusetts Inst. Tech		Cambridge, MA 02139	USA
Longo,G.	Univ. Bologna e INFN	Dipt. Fisica	40126 Bologna	ITA
Lopac,V.	Univ. Zagreb	Technoloski Fakulte	41000 Zagreb	YUG
Love,D.J.G.	SERC Daresbury lab.	NSF Div.	Warrington WA4 4AD	GBR
Lovhoiden,G.	Univ. Bergen	Inst. Phys.	5000 Bergen	NOR
Lowe,J.	Univ. Birmingham	Dept. Phys.	Birmingham B15 2TT	GBR
Lucas,R.	CEN Saclay	DPhN/MF	91191 Gif-sur-Yvette	FRA
Lunardi,S.	Univ. Padova	Dipt. Fisica	35131 Padova	ITA
Luukko,A.	Lappeenranta Univ. Tech.	Dept. Phys.	53851 Lappeenranta	FIN

Mac Mahon,T.D.	Imperial College	Reactor Centre	Ascot, Berks. SL5 7PY	GBR
Macdonald,E.W.	Univ. Edinburgh	Dept. Phys.	Edinburgh EH9 3JZ	GBR
MacGregor,I.J.D.	Univ. Glasgow	Dept. Nat. Phil.	Glasgow G12 8QQ	GBR
Machner,H.	KFA Jülich	Inst. f. Kernphys.	5170 Jülich	FRG
Mackintosh,R.S.	The Open University	Dept. Phys.	Milton Keynes MK7 6AA	GBR
Manfredi,V.	Univ. Padova	Dipt. Fisica	35131 Padova	ITA
Manokhin,V.N.	Fiziko-Energet. Inst.		Kaluga Region,Obninsk	USR
Maqueda,E.E.	CNEA	Dept. Phys.	1429 Buenos Aires	ARG
Marinov,A.	Hebrew Univ.	Racah Inst. Physics	Jerusalem 91904	ISR
Marsh,S.	Univ. Oxford	Nucl. Phys. Lab.	Oxford OX1 3PH	GBR
Mason,G.R.	Univ. Victoria	Dept. Phys.	Victoria, BC V8W 2Y2	CAN
Mason,J.P.	AERE Harwell	Nucl. Phys. Div.	Didcot, Oxon OX11 ORA	GBR
Massen,S.	Univ. Thessaloniki	Dept. Theor. Phys.	Thessaloniki 54006	GRC
Massmann,H.	Univ. Chile	Dept. Phys.	Casilla 653, Santiago	CHL
Matsuki,S.	Kyoto Univ.	Inst. Chemical Res.	Sakyo, Kyoto 606	JPN
Matthes,W.K.	JCR Euratom		21027 Ispra (Varese)	ITA
McGeorge,J.C.	Univ. Glasgow	Dept. Nat. Phil.	Glasgow G12 8QQ	GBR
McGovern,J.A.	Univ. Manchester	Dept. Phys.	Manchester M13 9PL	GBR
McKee,J.S.C.	Univ. Manitoba	Dept. Phys.	Winnipeg, Manitoba	CAN
McMurray,W.R.	National Accel. Centre		Faure 7131	RSA
McNeil,J.	Univ. Manchester	Dept. Phys.	Manchester M13 9PL	GBR
Merz,F.	Univ. München	Sektion Physik	8046 Garching	FRG
Metag,V.	Justus Liebig Univ.	II Phys. Institut	6300 Giessen	FRG
Mines,J.R.	Univ. Liverpool	DAMTP	Liverpool L69 3BX	GBR
Mitchell,E.W.J.	SERC	Chairman's Office	Swindon, SN2 1ET	GBR
Modarres,M.	Univ. Manchester	Dept. Phys.	Manchester M13 9PL	GBR
Molchadzki,A.	Hebrew Univ.	Dept. Phys.	Jerusalem	ISR
Molinari,A.	Univ. Torino	Dip. Fis. Teor.	10125 Torino	ITA
Mondragon,A.	Univ.Nac.Autonoma Mexico	Inst. Fisica	01000 Mexico DF	MEX
Morando,M.	Univ. Padova	Dipt. Fisica	35131 Padova	ITA
Moro,R.	Univ. Napoli	Dipt. Fisica Nucl.	80125 Napoli	ITA
Morrison,G.C.	Univ. Birmingham	Dept. Phys.	Birmingham B15 2TT	GBR
Morrison,J.D.	Univ. Liverpool	Dept. Phys.	Liverpool L69 3BX	GBR
Morsad,A.	Centre Recherches Nucl.	Groupe PNIN	67037 Strasbourg	FRA
Morsch,H.P.	KFA Jülich	Inst. f. Kernphys.	5170 Jülich	FRG
Mosconi,B.	Univ. Firenze	Dipt. Fisica	50125 Firenze	ITA
Moscrop,R.	Univ. Manchester	Dept. Phys.	Manchester M13 9PL	GBR
Moss,J.M.	Los Alamos National Lab.	Group P2, MS D 456	Los Alamos NM 87545	USA
Mottelson,B.R.	Niels Bohr Institute		2100 Copenhagen	DNK
Moya de Guerra,E.	CSIC	Inst. Estructura	28006 Madrid	ESP
Mühldorfer,B.	Univ. Erlangen-Nürnberg	III Phys. Institut	8520 Erlangen	FRG
Mulders,P.J.	NIKHEF-K		1009 AJ Amsterdam	NLD
Mullins,S.M.	Univ. York	Dept. Phys.	York, YO1 5DD	GBR
Nagarajan,N.A.	SERC Daresbury Lab.	TCS Div.	Warrington WA4 4AD	GBR
Naudé,W.J.	Univ. Stellenbosch	Dept. Phys.	Stellenbosch 7600	RSA
Nedjadi,Y.	Univ. Oxford	Dept. Phys.	Oxford OX1 3RH	GBR
Nelson,J.M.	Univ. Birmingham	Dept. Phys.	Birmingham B15 2TT	GBR
Neugart,J.	Univ. Mainz	Dept. Phys.	6500 Mainz	FRG
Nguyen Van Sen	ISN Grenoble		38026 Grenoble Cedex	FRA
Nojarov,R.	Univ. Tübingen	Inst. Theor. Physik	7400 Tübingen 1	FRG
Nolan,P.J.	Univ. Liverpool	Dept. Phys.	Liverpool L69 3BX	GBR
Nolte,R.	Univ. Göttingen	II Phys. Institut	3400 Göttingen	FRG
Novotny,R.	Univ. Giessen	II Phys. Institut	6300 Giessen	FRG
Nybo,K.	Univ. Bergen	Dept. Phys.	5000 Bergen	NOR
Nyman,G.	Chalmers Univ. Tech.	Dept. Phys.	412 96 Göteborg	SWE
O'Donnell,J.M.	Univ. York	Dept. Phys.	York, YO1 5DD	GBR
Offermann,E.A.J.M.	NIKHEF-K		1009 AJ Amsterdam	NLD
Ogilvie,C.	Univ. Birmingham	Dept. Phys.	Birmingham B15 2TT	GBR
Okolowicz,J.	Inst. of Nucl. Phys.		152 Krakow	POL
Olmi,A.	INFN Firenze		50122 Firenze	ITA
Olsson,N.	NFL - Studsvik	Nucl. Tech. Div.	61182 Nykoping	SWE
Orlandini,G.	Univ. Trento	Dipt. Fisica	38050 Povo(Trento)	ITA
Osman,A.	Univ. Cairo	Dept. Phys.	Giza, Cairo	EGY
Osnes,E.	Univ. Oslo	Dept. Phys.	0316 Oslo 3	NOR
Ouichaoui,S.	USTHB	Istitut de Physique	Babezzouar, Alger	ALG
Owens,R.O.	Univ. Glasgow	Dept. Nat. Phil.	Glasgow G12 8QQ	GBR
Paetz Gen Schieck,H.	Univ. Notre Dame	Dept. Phys.	Notre Dame, IN 46556	USA
Pal, D.	Saha Inst. Nucl. Phys.	TNP Division	700009 Calcutta	IND
Pal, K.F.	SERC Daresbury Lab.	TCS Div.	Warrington WA4 4AD	GBR
Pal, M.K.	Saha Inst. Nucl. Phys.	TNP Division	700009 Calcutta	IND
Pálla,G.	Central Res. Institute		1525 Budapest	HUN
Parker,D.J.	AERE Harwell	Nucl. Phys. Div.	Didcot, Oxon OX11 ORA	GBR

Pascholati,P.R.	Univ. de São Paulo	Inst. de Fisica	01498 São Paulo	BRA
Pass,C.	Univ. Oxford	Dept. Nucl. Phys.	Oxford OX1 3RH	GBR
Paul.P.	State Univ. New York	Dept. Phys.	Stony Brook, NY 11794	USA
Paul,S.	Max Planck Institut	Inst. f. Kernphys.	6900 Heidelberg	FRG
Pearce,K.	King's College London	Dept. Phys.	London WC2R 2LS	GBR
Peker,L.K.	Brookhaven National Lab.	Phys. Div.	Upton, NY 11973	USA
Pinheiro Filho,J.D.	Univ. Fed. Fluminense	Inst. de Fisica	24210 Niteroi, RJ	BRA
Pinston,J.A.	Centre d'Etudes Nucl.	Lab. Phys. Nucl.	38041 Grenoble Cedex	FRA
Pluhar,Z.	Max Planck Institut	Inst. f. Kernphys.	6900 Heidelberg	FRG
Pluta,J.	JINR, Dubna	Lab. High Energies	Moscow	USR
Poletti,A.R.	Univ. Auckland	Dept. Phys.	Auckland	NZL
Preston,M.A.	McMaster Univ.	Dept. Phys.	Hamilton, L8S 4M1	CAN
Price,H.G.	SERC Daresbury Lab.	NSF Div.	Warrington WA4 4AD	GBR
Pringle,D.M.	Univ. Oxford	Nucl. Phys. Lab.	Oxford OX1 3RH	GBR
Provencher,P.-P.	Univ. Laval	Lab. Van de Graaff	Quebec G1K 7P4	CAN
Quentin,P.	Univ. Bordeaux I	Lab. Phys. Theor.	33170 Gradignan	FRA
Qumri,N.	Univ. Jordan	Dept. Phys.	Amman	JOR
Rabey,A.	Univ. Surrey	Dept. Phys.	Guildford,GU2 5XH	GBR
Rae,W.D.M.	Univ. Oxford	Dept. Nucl. Phys.	Oxford OX1 3RH	GBR
Ramamurthy,V.S.	Bhabha Atom. Res. Centre	Nucl. Phys. Div.	Bombay 400 085	IND
Ranyuk,Yu.N.	Ukrainian Acad. Sciences	Inst. Phys. & Tech.	Kharkov 310 108	USR
Reisdorf,W.	GSI Darmstadt		6100 Darmstadt 1	FRG
Richter,W.A.	Univ. Stellenbosch	Dept. Phys.	Stellenbosch 7600	RSA
Rith,K.	Max Planck Institut	Inst. f. Kernphys.	6900 Heidelberg	FRG
Rizzo,F.	Univ. Catania	Dipt. Fisica	95129 Catania, Sicily	ITA
Robbins,A.	Rutgers Univ.	Serin Phys. Lab.	Piscataway, NJ 08854	USA
Robinson,R.L.	Oak Ridge National Lab.	Building 6000	Oak Ridge, TN 37831	USA
Robinson,S.J.	Institut Laue-Langevin		38042 Grenoble	FRA
Robson,B.A.	Aust. Nat. Univ.	Dept. Theor. Phys.	Canberra, ACT 2601	AUS
Robson,D.	Florida State Univ.	Dept. Phys.	Tallahassee, FL 32306	USA
Rolph,P.M.	Univ. Birmingham	Dept. Phys.	Birmingham B15 2TT	GBR
Rose,K.W.	Univ. Göttingen	II Phys. Institut	3400 Göttingen	FRG
Rosen,S.P.	Los Alamos National Lab.	T-Division, MS B288	Los Alamos, NM 87545	USA
Rowley,N.	SERC Daresbury Lab.	TCS Div.	Warrington WA4 4AD	GBR
Roy,A.	Tata Institute	NR Group	Colaba, Bombay 400 005	IND
Rubio,B.	KFA Jülich	Inst. f. Kernphys.	5170 Jülich 1	FRG
Ryckbosch,D.	Rijksuniv. Gent	Nucl. Phys. Lab.	9000 Gent	BEL
Ryde,H.	Univ. Lund	Dept. Phys.	223 62 Lund	SWE
Samiullah,M.	Yarmouk Univ.	Dept. Phys.	Irbid	JOR
Santos,F.D.	Univ. Lisboa	Centro Fisica Nucl.	1699 Lisboa	PRT
Satchler,G.R.	Oak Ridge National Lab.	Phys. Div.	Oak Ridge, TN 37831	USA
Satpathy,L.	Institute of Physics		Bhubaneswar 751005	IND
Schillebeeckx,P.	Rijksuniv. Gent	Nucl. Phys. Lab.	9000 Gent	BEL
Schmidt,F.	Univ. Washington	Dept. Phys.	Seattle, WA 98195	USA
Schramm,D.N.	Univ. Chicago	Enrico Fermi Inst.	Chicago, IL 60637	USA
Schutte,D.	Univ. Bonn	Inst.Theor.Kernphys	53 Bonn	FRG
Schwalm,D.	Univ. Heidelberg	Phys. Institut	6900 Heidelberg	FRG
Scobel,W.	Univ. Hamburg	Inst. Exper. Physik	2000 Hamburg 50	FRG
Seeliger,D.	Tech. Univ. Dresden	Physics Section	8027 Dresden	DDR
Segato,G.	Univ. Padova	Dipt. Fisica	35131 Padova	ITA
Sellschop,J.P.F.	Univ. Witwatersrand	Schonland Res.Ctre.	Johannesburg 2001	RSA
Sené,M.R.	AERE Harwell	Nucl. Phys. Div.	Didcot, Oxon OX11 ORA	GBR
Sergiwa,S.	Univ. Manchester	Dept. Phys.	Manchester M13 9PL	GBR
Sharpey-Schafer,J.F.	Univ. Liverpool	Dept. Phys.	Liverpool L69 3BX	GBR
Shaw,T.L.	Univ. Oxford	Clarendon Lab.	Oxford OX1 3PU	GBR
Sheldon,E.	Univ. Lowell (N. Campus)	Dept. Phys.	Lowell, MA 01854	USA
Shimizu,Y.R.	Kyushu Univ.	Dept. Phys.	Fukuoka 812	JPN
Shotter,A.C.	Univ. Edinburgh	Dept. Phys.	Edinburgh EH9 3JZ	GBR
Sick,I.	Univ. Basel	Inst. f. Physik	4056 Basel	SWI
Siemssen,R.H.	KVI Gröningen		9747 AA Gröningen	NLD
Simmonds,P.J.	King's College London	Dept. Phys.	London WC2R 2LS	GBR
Simpson,J.	SERC Daresbury Lab.	NSF Div.	Warrington WA4 4AD	GBR
Smith,A.E.	Univ. Oxford	Nucl. Phys. Lab.	Oxford OX1 3RH	GBR
Smith,J.R.H.	SERC Daresbury Lab.	NSF Div.	Warrington WA4 4AD	GBR
Smithson,M.J.	Los Alamos National Lab.		Los Alamos, NM 87545	USA
Soloviev,V.G.	JINR, Dubna		Moscow	USR
Spear,R.H.	Aust. Nat. Univ.	Dept. Nucl. Phys.	Canberra, ACT 2601	AUS
Specht,H.-J.	Max Planck Institut	Inst. f. Kernphys.	6900 Heidelberg	FRG
Spits,A.M.J.	SCK/CEN		2400 Mol	BEL
Sprouse,G.D.	State Univ. New York	Dept. Phys.	Stony Brook, NY 11794	USA
Stachel,J.	State Univ. New York	Dept. Phys.	Stony Brook, NY 11794	USA
Stamp,A.P.	Univ. Auckland	Dept. Phys.	Auckland	NZL

Stefanini,A.M.	INFN Legnaro		35020 Legnaro(Padova)	ITA
Steinmayer,M.	Univ. München	LMU	8000 München 2	FRG
Stevenson,J.	Michigan State Univ.	Cyclotron Lab.	East Lansing,MI 48824	USA
Stevenson,N.R.	TRIUMF		Vancouver BC V6T 2A3	CAN
Street,A.M.	Univ. Oxford	Nucl. Phys. Lab.	Oxford OX1 3RH	GBR
Stuttge,L.	Univ. Oxford	Dept. Nucl. Phys.	Oxford OX1 3RH	GBR
Sun Zuxun	Inst. Atomic Energy		PO Box 275, Beijing	CHN
Talmi,I.	Weizmann Inst. Science	Dept. Nucl. Phys.	76100 Rehovot	ISR
Tamain,B.	Univ. Caen	Lab. Phys. Corpusc.	14032 Caen Cedex	FRA
Taras,P.	Univ. Montreal	Lab. Phys. Nucl.	Montreal H3C 3J7	CAN
Temmer,G.M.	Rutgers Univ.	Dept. Phys.	New Brunswick NJ 08903	USA
Tengblad,O.	Chalmers Univ. Tech.	Dept. Phys.	41296 Göteborg	SWE
Testoni,J.E.	CNEA - Tandar	Dept. Fisica	1429 Buenos Aires	ARG
Thomas,A.W.	Univ. Adelaide	Dept. Phys.	Adelaide SA 5001	AUS
Thompson,I.J.	Univ. Bristol	Dept. Eng. Maths.	Bristol BS8 1TR	GBR
Thorsteinsen,T.F.	Univ. Bergen	Dept. Phys.	5000 Bergen	NOR
Tibell,G.	Uppsala Univ.	Gustaf Werner Inst.	75121 Uppsala	SWE
Tkachuk,R.	Univ. Alberta	Dept. Phys.	Edmonton, T6G 2N5	CAN
Tolsma,L.D.	Eindhoven Univ. Tech.	Dept. Phys.	5600 MB Eindhoven	NLD
Tomoda,T.	Univ. Tübingen	Inst. Theor. Phys.	7400 Tübingen	FRG
Tondeur,F.	Univ. Libre de Bruxelles		1050 Bruxelles	BEL
Toneev,V.D.	JINR, Dubna		Moscow	USR
Towner,I.S.	AECL Chalk River Lab.	Phys. Div.	Ontario K0J 1J0	CAN
Trautmann,D.	Univ. Basel	Inst. Theor. Phys.	4056 Basel	SWI
Tripathi,R.K.	Univ. Liège	Inst. Physique B5	4000 Liège 1	BEL
Tulip,J.	Univ. Alberta	Dept. Phys.	Edmonton T6G 2G7	CAN
Tungate,G.	Univ. Birmingham	Dept. Phys.	Birmingham B15 2TT	GBR
Turkiewicz,J.	Inst. Nuclear Studies		Warsaw	POL
Twin,P.J.	SERC Daresbury Lab.	NSF Div.	Warrington WA4 4AD	GBR
Ulm,G.	CERN	EP Division	1211 Geneva 23	SWI
Van Assche,P.H.M.	SCK/CEN		2400 Mol	BEL
Van de Vyver,R.	Rijksuniv. Gent	Nucl. Phys. Lab.	9000 Gent	BEL
Van den Berg,A.M.	Rijksuniv. Utrecht	Van de Graaff Lab.	3508 TA Utrecht	NLD
Van Hees,A.G.M.	Rijksuniv. Utrecht	Fysisch Lab.	3508 TA Utrecht	NLD
Van Hoorebeke,L.	Rijksuniv. Gent	Nucl. Phys. Lab.	9000 Gent	BEL
Van Isacker,P.	Univ. Sussex	School MAPS	Brighton BN1 9QH	GBR
Van Middelkoop,G.	NIKHEF		1009 DB Amsterdam	NLD
Vanneste,L.	Univ. Leuven	I.K.S.	3030 Leuven	BEL
Vannucci,L.	INFN Legnaro		35020 Legnaro(Padova)	ITA
Varley,B.J.	Univ. Manchester	Dept. Phys.	Manchester M13 9PL	GBR
Vento,V.	Univ. Valencia, CSIC	Fac. Fisiques IFIC	Burjassot, Valencia	ESP
Verboven,M.L.K.	Rijksuniv. Gent	Nucl. Phys. Lab.	9000 Gent	BEL
Vermeer,W.J.	Univ. Oxford	Nucl. Phys. Lab.	Oxford OX1 3RH	GBR
Vernin,P.	CEN Saclay	DPhN/HE	91191 Gif-sur-Yvette	FRA
Vetterli,D.	Univ. Basel	Inst. Phys.	4056 Basel	SWI
Viennot,M.	Lab. de Phys. Nucl.	Faculty of Science	Rabat Agdal	MOR
Viesti,G.	Univ. Padova	Dipt. Fisica	35131 Padova	ITA
Viollier,R.D.	Univ. Cape Town	Dept. Phys.	Rondebosch 7700	RSA
Vitturi,A.	Univ. Padova	Dipt. Fisica	35131 Padova	ITA
Vogt,E.	TRIUMF		Vancouver BC V6T 2A3	CAN
Vuaridel,B.	ETH Zürich	Inst. Med. En. Phys	8093 Zürich	SWI
Wadsworth,R.	Univ. York	Dept. Phys.	York, YO1 5DD	GBR
Wagner,G.J.	Univ. Tübingen	Phys. Institut	7400 Tübingen 1	FRG
Walecka,J.D.	Stanford Univ.	Dept. Phys.	Stanford, CA 94305	USA
Walker,P.M.	SERC Daresbury Lab.	NSF Div.	Warrington WA4 4AD	GBR
Wallace,P.A.	Univ. Glasgow	Dept. Nat. Phil.	Glasgow G12 8QQ	GBR
Wallace,S.J.	Univ. Maryland	Dept. Phys.	College Park,MD 20742	USA
Wapstra,A.H.	NIKHEF-K		1009 AJ Amsterdam	NLD
Waroquier,M.	Rijksuniv. Gent	Inst. Nucl. Phys.	9000 Gent	BEL
Watson,D.L.	Univ. York	Dept. Phys.	York, YO1 5DD	GBR
Weber,T.	Strahlenzentrum	Inst. f. Kernphys.	6300 Giessen	FRG
Wesselborg,C.	Univ. Cologne	Inst. f. Kernphys.	5 Cologne 41	FRG
Weyer,H.J.	Univ. Basel	Phys. Inst.	4056 Basel	SWI
Wilets,L.	Univ. Washington	Dept. Phys.	Seattle, WA 98195	USA
Wilkinson,Sir D.H.	Univ. Sussex	Vice Chancellor	Brighton, BN1 9RH	GBR
Williams,A.G.	Univ. Adelaide	Dept. Phys.	Adelaide, SA 5001	AUS
Williams,S.S.	Univ. Oxford	Dept. Nucl. Phys.	Oxford OX1 3RH	GBR
Winfield,J.S.	Michigan State Univ.	Dept. Phys.	East Lansing,MI 48824	USA
Wollersheim,H.J.	GSI Darmstadt		6100 Darmstadt	FRG
Woods,C.L.	Aust. Nat. Univ.	Dept. Nucl. Phys.	Canberra, ACT 2601	AUS
Woods,P.J.	Univ. Birmingham	Dept. Phys.	Birmingham B15 2TT	GBR
Woods,S.A.	National Physical Lab.		Teddington TW11 0LW	GBR

Wormald,M.R.	AERE Harwell	Nucl. Applic. Group	Didcot, Oxon OX11 ORA	GBR
Wright,I.	Univ. Manchester	Dept. Phys.	Manchester M13 9PL	GBR
Yamada,S.	Kyoto Univ.	Res. Reactor Inst.	Osaka 590-04	JPN
Ying,K.L.	Univ. Liverpool	Dept. Phys.	Liverpool L69 3BX	GBR
Yokoyama,A.	Teikyo Univ.	Phys. Lab.	Hachioji, Tokyo 192	JPN
Yorkston,J.	Univ. Edinburgh	Dept. Phys.	Edinburgh EH9 3JZ	GBR
Zeidman,B.	Argonne National Lab.	Phys. Div.	Argonne, IL 60439	USA
Zeldes,N.	Hebrew Univ.	Racah Institute	Jerusalem	ISR
Zingl,H.F.K.	Univ. Graz	Inst. Theor. Phys.	8010 Graz	AUT
Zurmuhle,R.W.	Univ. Pennsylvania	Dept. Phys.	Philadelphia,PA 19104	USA
Zvara,I.	JINR, Dubna		Moscow	USR
Zybert,L.	Univ. Birmingham	Dept. Phys.	Birmingham B15 2TT	GBR

Author Index

*Invited speakers